Juran's Quality Handbook

Juran's Quality Handbook

The Complete Guide to Performance Excellence

Joseph A. De Feo

Seventh Edition

McGraw Hill Education

New York Chicago San Francisco
Athens London Madrid
Mexico City Milan New Delhi
Singapore Sydney Toronto

Cataloging-in-Publication Data is on file with the Library of Congress.

Juran's Quality Handbook, Seventh Edition

1 2 3 4 5 6 7 8 9 DOC 21 20 19 18 17 16

ISBN 978-1-265-82165-4
MHID 1-26-582165-8

Sponsoring Editor
Robert Argentieri

Editorial Supervisor
Donna M. Martone

Acquisitions Coordinator
Lauren Rogers

Project Manager
Jyotsna Ojha,
Cenveo® Publisher Services

Copy Editor
Cenveo Publisher Services

Proofreader
Cenveo Publisher Services

Indexer
Cenveo Publisher Services

Production Supervisor
Pamela A. Pelton

Composition
Cenveo Publisher Services

Art Director, Cover
Jeff Weeks

About the Editor

Joseph A. De Feo is a leading business management consultant, quality management practitioner, and successor to Dr. Juran as president of Juran Institute, Inc. He is recognized worldwide for his expertise in enabling organizations to achieve business excellence. Mr. De Feo's varied areas of expertise include coaching executives to understand key factors in driving organization-wide change programs and deploying breakthrough management principles like Lean, Six Sigma, strategic planning, and business process improvement.

Contents

Contributors

Marcos E.J. Bertin *Director, Bertin Quality Consulting* (CHAPTER 10, THE BOARD OF DIRECTORS AND ORGANIZATIONAL EXCELLENCE)

R. Kevin Caldwell *Executive Vice President, Juran Institute, Inc.* (CHAPTER 14, LEAN TECHNIQUES AND THE SHINGO PRIZE; CHAPTER 16, ROOT CAUSE ANALYSIS METHODS)

Joseph A. De Feo *Chairman and CEO, Juran Institute, Inc.* (INTRODUCTION, QUALITY AND PERFORMANCE EXCELLENCE; CHAPTER 1, UNIVERSAL PRINCIPLES OF QUALITY; CHAPTER 2, DEVELOPING AN EXCELLENCE CULTURE; CHAPTER 3, DEVELOPING THE QUALITY AND EXCELLENCE OFFICE; CHAPTER 4, QUALITY PLANNING AND DESIGN OF NEW GOODS AND SERVICES; CHAPTER 5, QUALITY IMPROVEMENT AND BREAKTHROUGH PERFORMANCE; CHAPTER 6, QUALITY CONTROL TO ASSURE COMPLIANCE TO CUSTOMER REQUIREMENTS; CHAPTER 7, STRATEGIC PLANNING AND PERFORMANCE EXCELLENCE; CHAPTER 8, ORGANIZATION ROLES TO SUPPORT A QUALITY AND EXCELLENCE CULTURE; CHAPTER 9, QUALITY ASSURANCE AND AUDITS; CHAPTER 11, ISO 9000 QUALITY MANAGEMENT SYSTEM; CHAPTER 15, SIX SIGMA: BREAKTHROUGH TO IN-PROCESS EFFECTIVENESS; CHAPTER 16, ROOT CAUSE ANALYSIS METHODS; CHAPTER 17, CONTINUOUS INNOVATION USING DESIGN FOR SIX SIGMA; CHAPTER 21, MANAGING QUALITY IN OPERATIONS: SERVICE; CHAPTER 22, MANAGING QUALITY IN OPERATIONS: MANUFACTURING, CHAPTER 23, MANAGING THE SUPPLY CHAIN; CHAPTER 24, INSPECTION, TEST, AND MEASUREMENT; CHAPTER 25, ORGANIZATION-WIDE ASSESSMENT OF QUALITY)

John F. Early *Senior Associate, Juran Institute, Inc.* (CHAPTER 4, QUALITY PLANNING AND DESIGN OF NEW GOODS AND SERVICES; CHAPTER 12, ROLE OF STATISTICS; CHAPTER 15, SIX SIGMA: BREAKTHROUGH TO IN-PROCESS EFFECTIVENESS; CHAPTER 17, CONTINUOUS INNOVATION USING DESIGN FOR SIX SIGMA; CHAPTER 20, ACCURATE AND RELIABLE DATA AND MEASUREMENT SYSTEMS)

Rick Edgeman, *Ph.D., Research Director, Shingo Institute* (CHAPTER 7, STRATEGIC AND HOSHIN PLANNING AND PERFORMANCE EXCELLENCE; CHAPTER 14, LEAN TECHNIQUES AND THE SHINGO PRIZE)

Leonard Heflich *Vice President of Global Food Safety, Quality and Crisis Management, Grupo Bimbo, Mexico City* (CHAPTER 13, MANAGING RISK)

Joseph M. Juran *Chairman Emeritus, Juran Institute, Inc.* (CHAPTER 1, UNIVERSAL PRINCIPLES OF QUALITY; CHAPTER 6, QUALITY CONTROL TO ASSURE COMPLIANCE TO CUSTOMER REQUIREMENTS)

Er Ralston *Senior Associate, Juran Institute, Inc.* (CHAPTER 25, ORGANIZATION-WIDE ASSESSMENT OF QUALITY)

Peter Robustelli *Executive Vice President, Juran Institute, Inc.* (CHAPTER 8, ORGANIZATION ROLES TO SUPPORT A QUALITY AND EXCELLENCE CULTURE)

Kenneth Snyder *Executive Director, Shingo Institute* (CHAPTER 7, STRATEGIC AND HOSHIN PLANNING AND PERFORMANCE EXCELLENCE; CHAPTER 14, LEAN TECHNIQUES AND THE SHINGO PRIZE)

Brian A. Stockhoff, *Ph.D., Senior Associate, Juran Institute, Inc.* (CHAPTER 19, GRAPHICAL TOOLS TO IMPROVE PROCESS PERFORMANCE)

Brad Wood, *Ph.D., Executive Vice President, Juran Europe Ltd.* (CHAPTER 18, BENCHMARKING BEST PRACTICES FOR MARKET LEADERSHIP)

Additional Contributions and Editing

Therese Lederer, Christopher De Feo, Ph.D., Zubair Anwar, Michael Stamp, Gerhard Plenert

Juran Global Staff

Tina Frigeri, Brittany Ernst, Michelle Millar, Jeremy Hopfer, Kaitlin Tyer, Audra D'Agostino, Joseph M. De Feo, Dr. David Fearon, Ralph Ingriselli

Preface

Preface by Joseph A. De Feo, Editor-in-Chief

Joseph A. De Feo on the new Edition of and Uses of the Handbook

There are many stakeholders and situations, which create challenges to practitioners to establish a linkage between their own situations and those from which the expertise in this handbook was derived. This linkage is established by discovering the commonality which makes them both members of one species and makes it your own. My main message is that one must adopt and adapt the right method at the right time to get the right outcome. This handbook is a set of lessons learned and adaptations from many years of experience.

We set out to make this 7th edition handbook a reference for all who are leading, creating, producing, and delivering high quality products, services, and processes to attain superior customer experiences and business results. Experience with the first six editions has shown that "those who are involved" include:

1. All *levels in the organizational hierarchy*—from the board members, chief executives, and operational managers to the workforce. It is a mistake to assume that the sole purpose of the book is to serve the needs of just the quality managers and engineers. The purpose of the book is to serve the entire organization including the workforce at all levels of the organization.
2. The various *functions* engaged in producing products (goods, services, or information): such as research and development, market research, finance, operations, marketing and sales, human resources, supplier relations, customer service, administration, and support activities.
3. The various *specialists* associated with the processes carrying out the strategic and tactical tasks to ensure all products, services, and business processes are properly designed, controlled, and continually improved to meet customer and society needs.
4. The various *industries* that make up the global economy: manufacturing, construction, services of all kinds—hi-tech, transportation, communication, oil and gas, energy, utilities, financial, health care, hospitality, government, and so on.

The handbook is also an aid to certain "gatekeepers and stakeholders" who, though not directly involved in leading, producing, or marketing products and services, nevertheless have "a need to know" about the qualities produced and the associated positive and negative side effects. These stakeholders include the executive leadership, quality officers, and engineers that are given the responsibility to improve manage for quality day to day; customers looking to better understand how to improve their suppliers, the supply chain, the users, the public, the owners, the media, and even government regulators.

We have conducted many focus groups and learned that our readers, the practitioners, make a wide variety of uses of *Juran's Quality Handbook*. Experience has shown that usage is dominated by the following principal motives:

1. To use as a "one-stop shopping" reference guide for the methods, tools, and roadmaps to create a sustainable business operation driven by high quality products and services.
2. To utilize as a guide to educate their specialists such as: quality management and assurance departments, systems engineering, organizational and operational effectiveness departments, finance and the like.
3. To find special tools or methods and examples of their use on topics such as leadership's role in leading quality, incorporating the voice of the customer in the design of products and services, reliability engineering, design of experiments, or statistical tools.
4. To study the narrative material as an aid to solving their own organizational and business problems.
5. To review subject matter for specific self-training or to secure material for teaching or training others.

Use of the handbook happens more frequently during times of change such as when developing new business initiatives, working on new processes and projects, organizing departments and functions, or just trying out new ideas.

Organization of the Handbook

Irrespective of intended use, the information provided in this handbook was designed to make finding things easy—but it is not always easy to do that. The problem for the user becomes one of (1) knowing where to find it and (2) adapting the information to his or her specific needs. While there is a great deal of know-how in this book it is important to understand how it is organized. There a total of 25 chapters; each chapter is a topic on its own and collectively makes up a robust performance excellence system. Based on frequent reader feedback we decided to make changes in this edition.

First, there is no section specific to Healthcare, Insurance, Defense, Government Agencies, etc. An industry-focus would be somewhat redundant in today's business as the methods and tools are similar. For instance, there is no healthcare section because we felt that this industry requires a Handbook in itself to meet the many challenges to provide high quality health outcomes, population management, and many regulatory requirements. Dedicating one chapter to this important industry would not do it justice. We are working on a new publication to meet these needs.

Secondly, based on many requests, we put back many of the important statistical tables for practitioners to use. Because we are listening to our customers and we want this handbook to be a "one-stop shopping" resource, we have attempted to improve our marketplace.

A Note on Abbreviations

Abbreviations of names or organizations are usually used only after the full name has previously been spelled out, for example, American Society for Quality (ASQ). In any case, all such abbreviations are listed and defined in the index.

The text of the handbook emphasizes the "main road" of quality management know-how, that is, the comparatively limited number of usual situations which nevertheless occupy the bulk of the time and attention of practitioners. Beyond the main road are numerous "side roads," that is, less usual situations which are quite diverse and which require special solutions. (The term "side road" is not used in any derogatory sense. The practitioner who faces an unusual problem must nevertheless find a solution for it.) As to these side roads, the handbook text, while not complete, nevertheless points the reader to available solutions. This is done in several ways:

1. **Citations.** The handbook cites numerous papers, books, and other bibliographic references. In most cases these citations also indicate the nature of the special contribution made by the work cited in order to help the reader to decide whether to go to the original source for elaboration.
2. **Special Bibliographies.** Some sections provide supplemental lists of bibliographical material for further reference. The editors have attempted to restrict the contents of these lists to items which: (1) bear directly on the subject matter discussed in the text or (2) are of uncommon interest to the practitioner. A special bibliography in Appendix III lists quality standards and specifications.
3. **Literature Search.** Papers, books, and other references cited in the handbook contain further references which can be found for further study. Use can be made of available abstracting and indexing services. Various other specialized abstracting services are available on such subjects as reliability, statistical methods, research and development, and so on.
4. **The Internet.** It is now possible to find almost any book or article in print and many that are out of print in just a few minutes using a web search. Using search engines, one can find thousands of articles on numerous topics or by selected authors. Many special sites focus on performance excellence and quality management in its broadest sense. A simple e-mail contact with a web site author may bring forth even more unpublished works or research in progress. Sites developed by university departments doing research in quality are especially useful for searching for specific examples and new methods and tools.
5. **Author Contact.** The written book or paper is usually a condensation of the author's knowledge; that is, what he or she wrote is derived from materials, which are one or two orders of magnitude more voluminous than the published work. In some cases it is worthwhile to contact the author for further elaboration. Most authors have no objection to being contacted, and some of these contacts lead not only to more information but also to visits and enduring collaboration.
6. **Societies and Other Sources:** Resourceful people are able to find still other sources of information relating to the problem at hand. They contact the editors of journals to discover which organizations have faced similar problems, so that they may contact these organizations. They contact suppliers and customers to learn if competitors have found solutions. They attend meetings—such as courses, seminars, and conferences of professional societies—at which there is discussion of the problem. There is hardly a problem faced by any practitioner which has not already been actively studied by others.

Adapting the Handbook to Your Special Needs

In many cases a practitioner is faced with adapting to a specific situation, knowledge derived from a totally different technology, that is, a different industry, product, or process. Making this transition requires that he or she identify the commonality, that is, the common principle to which both the specific situation and the derived knowledge correspond.

Often the commonality is managerial in nature and is comparatively easy to grasp. For example, the concept of self-control is a universal management concept and is applicable to any person in any organization.

Commonality of a statistical nature is even easier to grasp, since so much information is reduced to formulas which are indifferent to the nature of the technology involved. Even in technological matters, it is possible to identify commonalities despite great outward differences. For example, concepts such as process capability apply not only to manufacturing processes, but to health care, services and administrative, and support processes as well. In like manner, the approaches used to make improvements by discovering the causes of failures have been classified into specific categories that exhibit a great deal of commonality despite wide differences in technology.

Dr. Juran on the Creation of the *Juran's Quality Handbook*

The idea of a handbook on quality control originated late in 1944 and was part of my decision to become a freelance consultant after World War II. I had in mind a whole series of books: the *Quality Control Handbook,* which was to be a comprehensive reference book; and separate from all other publications. The handbook was created as a compendium of knowledge in the field of managing for quality which later became managing business and operational excellence. The emphasis was to document only on the universals, "the principles that are valid no matter what the product, the service, the process or function." I was only dimly aware that I would be contributing to the evolution of a new science-managing for quality; in the 1st edition I ended up with fifteen chapters. I wrote six of those; other authors wrote the remaining nine, which I often edited. From the outset, the handbook became the "bible" of managing for quality and has increasingly served as international reference book for professionals and managers in the field... With the publication of the 5th edition the name was changed to the *Juran's Quality Handbook*; the joint editors-in-chief were Dr. A. Blanton Godfrey and me. For the 6th edition and beyond, Mr. Joseph A. De Feo will be Editor-in-Chief. As a fine consultant and practitioner, he has earned the recognition that comes with the publication of such a book as this.

A Word of Thanks

There are many people that contributed to this handbook in many ways. I may miss a few but my intent was to include everyone. A Handbook that is published every 7 to 10 years requires success in the marketplace. Therefore, my first thank you is to all of the great customer and supplier relationships that were fostered to provide incredible examples and success for our use. This list includes organizations such as Molex, Duracell, GE, ABB, Doosan, JR Simplot, Aramark, Alta Resources, Lowe's Home Improvement, Almarai, MedStar Health, Christiana Care Health Services, Kaiser Permanente, John Deere Co., Astellas, Aramark, and Sekisui Diagnostics.

A word of thanks to my advisers and mentors: the late Dr. Juran, Dr. Gryna, Dr. Barnard, Dr. Greg Watson, Dr. Blanton Godfrey, David Bluestein, Paul Borawski, (past ASQ President), and Harry Hertz.

Thank you to my team that pulled this book together in the usual excellence fashion:

Tina Frigeri, Brittany Ernst, Michelle Millar, Jeremy Hopfer, Kaitlin Tyer, Audra D'Agostino, Dr. David Fearon, Ralph Ingriselli, Therese Lederer, Christopher De Feo, Ph.D., Mark A. De Feo, Joseph M. De Feo, Zubair Anwar, Michael Stamp, Gerhard Plenert.

If you have suggestions please feel free to visit the McGraw Hill website or contact us directly at info@juran.com.

Juran's Quality Handbook

INTRODUCTION

Quality and Performance Excellence

Joseph A. De Feo

High Points of This Chapter

1. Organizations that attain superior results, by designing and continuously improving the quality of their goods and services, are often called world class, best practices, and most recently performance excellence.
2. The quality of goods and services that meet customer needs must be a top leadership priority. Customer satisfaction must be the chief operating goal embedded in the company vision and strategic plans.
3. Leaders must make quality a strategic priority. The business plan was opened up to include quality goals and balanced scorecards, year after year.
4. Leaders must create a culture of performance excellence. All organizations that have attained superior quality and results did so with a systematic approach not a haphazard one.

5. Developing a quality management system to enable continuous innovation, compliance and control and continual process improvement is good business practice. The business plan must include goals for improvement.
6. Educate the workforce in the methods to manage for quality and excellence. Training was extended beyond the quality department to all functions and levels, including upper managers.
7. Engage employees. This includes training and empowering the workforce to participate in planning and improvement of the "useful many" opportunities. Motivation was supplied through extending the use of recognition and rewards for responding to the changes demanded by the quality revolution. Measurements were developed to enable upper managers to follow progress toward providing customer satisfaction, meeting competition, improving quality, and so on. Upper managers took charge of managing for quality by recognizing that certain responsibilities were not delegable—they were to be carried out by the upper managers, personally.
8. Conduct organization-wide assurance audits. This focus is on improving and ensuring that all goods, services, processes, and functions in an organization are of high quality.
9. Continually benchmark best practices to set new targets. This approach was adopted to set goals based on superior results already achieved by others.

Why Quality Management? A Brief History Lesson

Quality is a word with multiple meanings. It is often used to describe the term a customer uses when they like something they tried or purchased. It is a word that is used synonymously with an organization's mission—to provide superior quality to customers and stakeholders. It is also a word used to describe many methods to improve business performance, such as this text's title, *Juran's Quality Handbook*. A brief look at history will help better understand where quality management is today and where it may be headed tomorrow.

Practitioners are urged to learn about the different approaches to quality, choose the best methods to fit your business needs, and then customize the framework for your own organizations. In this book we will present the body of knowledge that works.

We will refer to examples of the most popular frameworks such as the U.S. Malcolm Baldrige National Award for Excellence (and all similar frameworks around the globe) and the ISO Quality Management System to name a few.

Two dramatic examples illustrate how quality has an impact on both sales revenue and costs. First, two examples of poor quality. An international organization that claimed to have strong customer service refused to accept sales orders for delivery in less than 48 hours—even though competitors honored delivery requests of 24 hours. Imagine the millions of dollars of sales revenue lost each year because this company did not recognize the need of its customers. Second, listen to the president of a specialty manufacturing company: "Our scrap and rework costs this year were five times our profit. Because of those costs, we have had to increase our selling price and we subsequently lost market share. Quality is no longer a technical issue; it is a business issue." Samsung Electronics used quality as a mantra, as a method to attain $1.6 billion in savings by focusing on improving the quality of flat-screen TV sets, making them the best practice globally. Everyone knows about Toyota quality. Why? Because its products have a reputation of being the best.

There are many stories like this. The conclusion is the same. If an organization focuses on the relentless pursuit of quality, it will attain a sustainable business performance and positive global reputation. Our forefathers knew—as we know—that quality is important. Metrology, specifications, inspection—all go back many centuries.

Then came the twentieth century. The pace quickened with a lengthy procession of "new" activities and ideas launched under a bewildering array of names: quality control, continuous quality improvement, defect prevention, statistical process control, reliability engineering, quality cost analysis, zero defects, total quality management, supplier certification, quality circles, quality audit, quality assurance, quality function deployment, Taguchi methods, competitive benchmarking, Lean and Six Sigma. This book discusses all these concepts and places them into a context that is needed to sustain business performance.

Following World War II, two major forces emerged that have had a profound impact on quality.

The first force was the Japanese revolution in quality. Prior to World War II, many Japanese products were perceived, throughout the world, as of poor quality. To help sell their products in international markets, the Japanese took some revolutionary steps to improve quality:

1. Upper level managers personally took charge of leading the revolution.
2. All levels and functions received training in the quality disciplines.
3. Quality improvement projects were undertaken on a continuing basis—at a revolutionary pace.

The Japanese success has been almost legendary.

The second major force to affect quality was the prominence of product quality in the public mind. Several trends converged to highlight this prominence: product liability cases; concern about the environment; some major disasters and near disasters; pressure by consumer organizations; and the awareness of the role of quality in trade, weapons, and other areas of international competition. This emphasis on quality has been further accented by the emergence of national awards such as the Baldrige and European Quality Awards.

Quality is not limited to the manufacturing sector. Quality concepts are applied to other sectors such as health care, education, not-for-profit organizations, and governments. Product quality is not the only focus. Service quality, process quality, and data quality are now being measured, controlled, and improved.

During the twentieth century, a significant body of knowledge emerged on achieving superior quality. Many individuals contributed to this knowledge, and five names deserve particular mention: Juran, Deming, Feigenbaum, Crosby, and Ishikawa.

J. M. Juran emphasizes the importance of a balanced approach using managerial, statistical, and technological concepts of quality. He recommends an operational framework of three quality processes, quality planning, quality control, and quality improvement. The foundation for this book is the Juran approach. This book makes frequent references to *Juran's Quality Handbook*, sixth edition, which is denoted JQH6, as well as the fifth edition, JQH5.

W. Edwards Deming also had a broad view of quality, which he initially summarized in 14 points aimed at the management of an organization. These 14 points rest on a system of "profound knowledge" that has four parts: the systems approach, understanding of statistical variation, the nature and scope of knowledge, and psychology to understand human behavior.

A. V. Feigenbaum emphasizes the concept of total quality control throughout all functions of an organization. Total quality control really means both planning and control. He urges creating a quality system to provide technical and managerial procedures that ensure customer satisfaction and an economical cost of quality.

Philip Crosby defines quality strictly as "conformance to requirements" and stresses that the only performance standard is zero defects. His activities demonstrated that all levels of employees can be motivated to pursue improvement but that motivation will not succeed unless tools are provided to show people how to improve.

Kaoru Ishikawa showed the Japanese how to integrate the many tools of quality improvement, particularly the simpler tools of analysis and problem solving.

The approaches of these gurus have similarities as well as differences—particularly in the relative emphasis on managerial, statistical, technological, and behavioral elements. This book draws upon the contributions of these and other experts.

Juran (1995) provides a comprehensive history of managing for quality for different time spans (ancient, medieval, modern), geographical areas, products, and political systems.

The major forces affecting managing for quality led to a changing set of business conditions.

Watson (2016) states that the true meaning of quality is much deeper and broader than these specific definitions imply. What is most salient about quality is the purpose for which it is pursued. What does this mean? What is the "objective function" of quality efforts? Quality is the persistent pursuit of goodness coupled tightly with a simultaneous relentless avoidance of badness.

All of the above definitions are useful to understanding how to manage quality. However, the most important definition of quality is what your customers mean by it. This is the essence of this handbook. Quality is defined by your customers. This handbook points out what has worked for others that may work for you.

Quality and Business Performance

The impact that the *quality* of products, services, and processes has on business performance and performance excellence has been a dominant force in the last decade. "A business that has goods and services that are superior to its competitors' in features and value will outperform those that do not."

Every organization can achieve performance excellence and sustainable business results through by understanding Juran's universal principles and methods to manage for quality. These methods:

- The incorporation of customer focus in the business strategic plans and policies.
- Developing a quality management system that includes a system to continuously create innovative new goods and services; controls to ensure what was designed is produced and a system to continually improve the processes that produce goods, services, and information.

Leaders have learned that to be competitive today the organization must move from thinking that managing quality is a "fad" or that "we already did that" to "quality is a necessity and a given." The Juran universal principles of the management of quality is as important as managing for finance. In both case the body of knowledge has evolved and advanced. In this newest edition of *Juan's Quality Handbook*, we will highlight this evolution and provide clarity on some of the older methods.

Superior *quality* goods and services will result in sustainable financial results because goods and services that are superior to the competitors' are salable. Goods and services that are salable because of quality continually drive revenue and maintain lower costs, leading to greater profitability. The pursuit of superior quality by understanding the underlying universal methods will transform the business and create a favorable quality and performance excellence culture.

Culture change does not happen haphazardly. Superiority in quality from a customer perspective does not just happen. The business leaders must make quality happen. Quality happens when the organization sets the strategic direction with a relentless pursuit to be the best in quality.

Organizations that attain superior results by designing, controlling, and continuously improving the quality of their goods and services are often called *world-class*. They have achieved a state of performance excellence. Organizations that have attained a state of superior quality are well respected by customers because their products and services exceed customers' expectations, which leads to sustainable business results. This pursuit of excellence through quality management methods creates greater customer, stakeholder, and employee satisfaction, which enables the organization to sustain performance over a longer term.

Managing Quality Should Not Be Optional

A common argument occurs among managers when asked, Does high quality cost more, or does high quality cost less? Seemingly there is usually disagreement. One-half agree that it costs more, and the other half feel it costs less. Can they are both right? Yes! The culprit is the word "quality." It is spelled the same way and pronounced the same way, but has multiple meanings. To manage for superior quality and results, leaders must define the word "quality" from the perspective of customers—those people who buy the goods, services, and even the reputation of your organization.

At one financial services company, the leaders would not support a proposal to reduce wasteful business processes because the staff had labeled them *quality improvement*. Some of the leaders felt improving quality would cost more money. In their view, higher quality meant higher cost. Others felt it would cost less. The subordinates were forced to rename the proposal *productivity improvement* to secure approval and avoid confusion. Such confusion can be reduced if each organization makes clear the distinction between the multiple meanings of the word "quality." However, some confusion is inevitable as long as we use a single word to convey very different ideas.

Leaders must have a common understanding of quality so they can manage it. First, agree on the meaning of the word "quality" as it applies to your business and its customers. Once defined, then it can be managed. If it can be managed, then it can be provided to the satisfaction of customers and stakeholders. Without a common understanding of the word "quality," the organization will continue to make many short-term initiatives to improve quality and it will lead to "initiative overload."

There have been efforts to clarify matters by adding supplemental words. There also have been efforts to coin a short phrase that would clearly and simultaneously define both major meanings of the word "quality." A popular definition was first presented by Joseph M. Juran. *Quality* was defined as meaning "fitness for use." Dr. Deming used "conformance to requirements." Robert Galvin, Chairman Emeritus of Motorola, used *Six Sigma* to distinguish the high level of quality as it related to defects. Others stated that quality means world-class excellence, best-in-class and performance excellence.

To better reflect the service economy we have settled on quality to mean all goods and services must be "fitness for purpose." The purpose is always defined by the customer using the good or service. Customers judge the quality of goods and services by how well they meet their needs. These needs drive the purchase of goods and services. If an organization understands the needs of its many customers, it should be able to design goods and services that are fit for purpose. No matter what the organization produces—a good or a service—it must be fit for its purpose. To be fit for purpose, every good, service, and interaction with customers must have the right features (characteristics of the good or service that satisfies customer needs) and be free of failure (no rework, no waste, no complaints).

It is unlikely that any short phrase can provide the depth of meaning needed by leaders and managers who are faced with choosing a course of action to improve business

Features That Meet Customer Needs	Freedom from Failures
Higher quality enables organizations to • Increase customer satisfaction • Make products salable • Meet competition • Increase market share • Provide sales income • Secure premium prices • Reduce risk	Higher quality enables organizations to • Reduce error rates • Reduce rework, waste • Reduce field failures, warranty charges • Reduce customer dissatisfaction • Reduce inspection, test • Shorten time to put new products on the market • Increase yields, capacity • Improve delivery performance
Major effect is on revenue.	Major effect is on costs.
Higher quality costs more.	Higher quality costs less.

Source: Juran, J. M., De Feo, Joseph A., *Juran's Quality Handbook*, 6th ed., McGraw-Hill, New York, 2010, p. 6.

Table 0.1 The Meaning of Quality

performance and quality. The best you can do is to understand the distinctions set out in Table 0.1 and define quality based on these distinctions.

Table 0.1 presents two of the many meanings of the word "quality." These two are of critical importance to managing for quality.

Quality Impacts Costs and Revenue and Costs

First, quality has a big effect on *costs*. In this case, *quality* has come to mean freedom from troubles traceable to office errors, factory defects, field failures, and so on. *Higher quality* means fewer errors, fewer defects, and fewer field failures. When customers perceive a service or good as low quality, they usually refer to the failures, the defects, the poor response times, etc. To increase this type of quality, an organization must master the universal of quality improvement. This is often called *breakthrough* or *Six Sigma*. It is a systematic method to reduce the number of such deficiencies or the "costs of poor quality" to create a greater level of quality and fewer costs related to it.

Second, quality has an effect on *revenue*. In this case, *higher quality* means delivery of those features of the good or service that respond better to customer needs. Such features make the product or service salable. Since the customers value the higher quality, they buy it and you get revenue from it. It is well documented that being the quality leader can also generate premium prices and greater revenue.

The effects on costs and on revenue interact with one another. Not only do goods or services with deficiencies add to suppliers' and customers' costs, but also they discourage repeat sales. Customers who are affected by field failures are, of course, less willing to buy again from the guilty supplier. In addition, such customers do not keep this information to themselves—they publicize it so that it affects the decisions of other potential buyers, with negative effects on the sales revenue of the supplier.

The effect of poor quality on organizational finances has been studied broadly. In contrast, study of the effect of quality on revenue has lagged. This imbalance is even more surprising, since most upper managers give higher priority to increasing revenues than to reducing costs. This same imbalance presents an opportunity for improving organization economics through better understanding of the effect of quality on revenue.

Quality, Earnings, and the Stock Market

At the most senior levels of management and among board members, there is keen interest in financial metrics such as net income and share price. It is clear that different levels of quality can greatly affect these metrics, but so do other variables. Variables such as market choices, pricing, and financial policy can influence these metrics. Separating out the market benefits of managing for quality has just become feasible.

During the early 1990s, some of the financial presses published articles questioning the merits of the Malcolm Baldrige National Quality Award, Six Sigma, and other similar initiatives to improve performance. These articles were challenged with an analysis of the stock price performance of organizations known to practice these methods. The Baldrige winners were compared to the performance of the S&P 500 as a whole. The results were striking. The Malcolm Baldrige National Quality Award winners outperformed the S&P 500. The Baldrige winners had advanced 89 percent, as compared to only 33 percent for the broad Standard & Poor's Index of 500 stocks ("Betting to Win on the Baldie Winners," 1993, p. 8). This set of winners became known as the "Baldie Fund."

The impact of the quality universals is also clear for organizations that are not measured by the performance of their asset values. Michael Levinson, City Manager of 2007 Award Recipient for the City of Coral Springs, stated it this way: "People ask, 'Why Baldrige?' My answer is very simple: Triple A bond rating on Wall Street from all three ratings agencies, bringing capital projects in on time and within budget, a 96 percent business satisfaction rating, a 94 percent resident satisfaction rating, an overall quality rating of 95 percent, and an employee satisfaction rating of 97 percent ... that's why we're involved with Baldrige."

Building Market Leadership

Market leadership is often the result of entering a new market first and gaining superiority. However, once superiority is gained, it must be maintained through continuing product or service improvements, or it could be lost if another organization decides to redefine that market by improving its quality. That supplier will gain superiority over the market leader and will become the *quality leader*. Organizations that have attained this leadership have usually done so on the basis of a strategic choice. They adopted a positive strategy to establish leadership as a formal business goal and then set out the means to reach that goal. Once attained, quality leadership endures until there is clear cumulative evidence that some competitor has overtaken the leadership. Lacking such evidence, the leadership can endure for decades and even centuries. However, superior quality can also be lost through some catastrophic change.

The growth of competition in quality has stimulated the expansion of strategic business planning to include planning for quality and quality leadership.

Quality and Share of Market

Growth in market share is often among the highest goals of upper managers. Greater market share means higher sales volume. In turn, higher sales volume accelerates return on investment disproportionally due to the workings of the break-even chart.

In Fig. 0.1, to the right of the break-even line, an increase of 20 percent in sales creates an increase of 50 percent in profit, since the fixed costs do not increase. (Actually, constant costs do vary with volume, but not at all in proportion.) The risks involved in increasing market share are modest, since the technology, product or service, facilities, market, and so on are already in existence and of proved effectiveness.

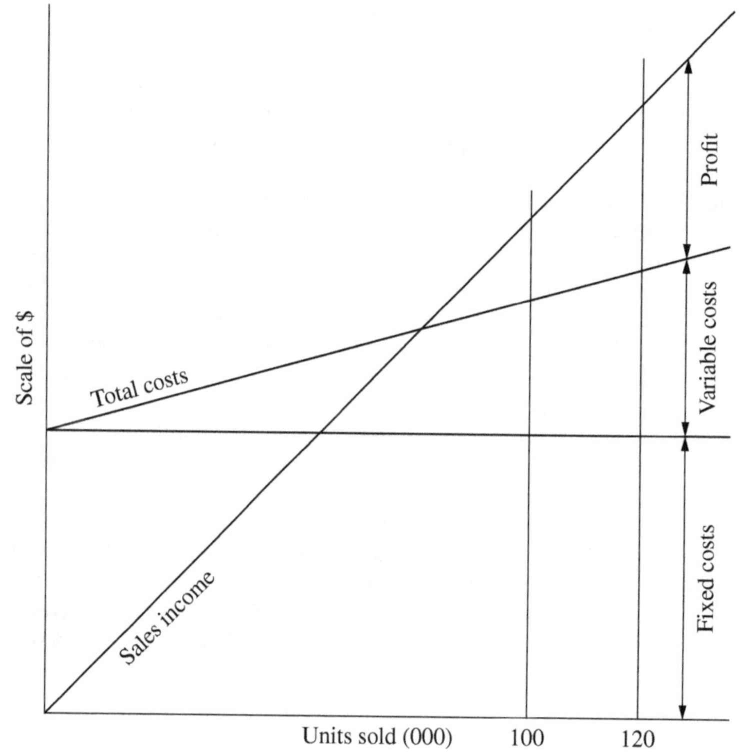

Figure 0.1 Break-even chart. (Juran, J. M., *Juran's Quality Handbook*, 5th ed., McGraw-Hill, New York, 1999, p. 7.13.)

Effect of Quality Superiority

Quality superiority can often be translated into higher share of market, but it always requires a special effort to do so. Superior quality must be clearly based on the voice of the customer and the benefits the customer is seeking. If quality superiority is defined only in terms of the company's internal standards, the customer may not perceive the value. For example, patients may be willing to pay the extra cost to travel long distances to a top health care system such as the Mayo Clinic in the United States rather than visit a local hospital because they perceive the superior clinical outcomes available at the Mayo Clinic.

Superiority Obvious to the Buyer

In such cases, the obvious superiority can be translated into higher share of market. This concept is fully understood by marketers, and from time to time they have urged product or service developers to come up with product or service features, which can then be promoted to secure higher share of market. Examples of such cases are legion.

Superiority Translatable into Users' Economics

Some products or services are outwardly alike but have dissimilar performances. An obvious example is the difference in the electric power consumption of appliances with otherwise identical features. In this and similar examples, it is feasible to translate the technological difference into the language of money. Such translation makes it easier for amateurs in technology to understand the significance of the quality superiority.

The superior reliability of a power tool can be translated into the language of money to secure a price premium. The superior reliability could also have been used to secure higher share of market. The superior quality of a truck tire can be translated into cost per unit of distance traveled.

The initiative to translate may also be taken by the buyer. Some users of grinding wheels keep records on wheel life. This is then translated into money—grinding wheel costs per 1000 pieces processed. Such a unit of measure makes it unnecessary for the buyer to become expert in the technology of abrasives.

Collectively, cases such as the above can be generalized as follows:

- There is in fact a quality difference among competing product or services.
- This difference is technological so that its significance is not understood by many users.
- It is often possible to translate the difference into the language of money or into other forms within the users' systems of values.

Superiority Minor but Demonstrable

In some cases, quality superiority can secure added share of market even though the competitive "inferior" product is nevertheless fit for purpose.

A manufacturer of antifriction bearings refined its processes to such an extent that its product or services were clearly more precise than those of the competitors. However, the competitors' product or services were fit for purpose, so no price differential was feasible. Nevertheless, the fact of greater precision impressed the client's engineers and secured increased share of market.

In consumer goods or services, even a seemingly small difference may be translated into increased market share if the consumers are adequately sensitized to the differentials and value them.

An executive of a manufacturer of candy-coated chocolates seized on the fact that his product did not create chocolate smudge marks on consumers' hands. He dramatized this in television advertisements by contrasting the appearance of the children's hands after eating his and the competitors' (uncoated) chocolate. His share of market rose dramatically.

Superiority Accepted on Faith

Consumers can be persuaded to accept, on faith, assertions of good or service superiority, which they themselves are unable to verify. An example was an ingenious market research on electric razors. The sponsoring organization (Schick) employed an independent laboratory to conduct the tests. During the research, panelists shaved themselves twice, using two electric razors, one after the other. On one day, the Schick razor was used first and a competing razor immediately after. On the next day, the sequence was reversed. In all tests, the contents of the second razor were weighed precisely. The data clearly showed that when the Schick was the second razor, its contents weighed more than those of the competitors. The implication was that Schick razors gave a cleaner shave. Within a few months, the Schick share of market rose as follows:

- September, 8.3 percent
- December, 16.4 percent

In this case, the consumers had no way to verify the accuracy of the asserted superiority. They had the choice of accepting it on faith or not at all. Many accepted it on faith.

No Quality Superiority

If there is no demonstrable quality superiority, then share of market is determined by marketing skills. These take such forms as persuasive value propositions, attractive packaging, and so on. Price reductions in various forms can provide increases in share of market, but this is usually temporary. Competitors move promptly to take similar action. Such price reduction can have a permanent effect if the underlying cost of production has also been reduced as the result of process improvements that give the company a competitive cost edge over its competitors.

Carryover of Failure-Prone Features

Market leadership can be lost by perpetuating failure-prone features of predecessor models. The guilty features are well known, since the resulting field failures keep the field service force busy restoring service. Nevertheless, there has been much carryover of failure-prone features into new models. At the least, such carryover perpetuates a sales detriment and a cost burden. At its worst, it is a cancer that can destroy seemingly healthy product or service lines.

A notorious example was the original xerographic copier. In that case, the "top 10" list of field failure modes remained essentially identical, model after model. A similar phenomenon existed for years in the automobile industry.

The reasons behind this carryover have much in common with the chronic internal wastes that abound in so many organizations:

- The alarm signals are disconnected. When wastes continue, year after year, the accountants incorporate them into the budgets. That disconnects the alarm signals—no alarms ring as long as actual waste does not exceed budgeted waste.
- There is no clear responsibility to get rid of the wastes. There are other reasons as well. The technologists have the capability to eliminate much of the carryover. However, those technologists are usually under intense pressure from the marketers to develop new product or service and process features in order to increase sales. In addition, they share distaste for spending their time cleaning up old problems. In their culture, the greatest prestige comes from developing the new.

The surprising result can be that each department is carrying out its assigned responsibilities, and yet the product or service line is dying. Seemingly nothing short of upper management intervention—setting goals for getting rid of the carryover—can break up the impasse.

Changes in Customer Habits

Customer habits can be notoriously fickle. Obvious examples are fashions in clothing or concerns over health. Consumerism is now driving lifestyles. Many people have reduced their consumption of beef and increased that of poultry and fish. Such shifts are not limited to consumers. Industrial organizations often launch "drives," most of which briefly take center stage and then fade away. The associated buzzwords similarly come and go.

The Twentieth Century and Quality

The twentieth century witnessed the emergence of some massive new forces that required responsive action. These forces included an explosive growth in science and technology;

threats to human safety, health, and the environment; the rise of the consumerism movement; and intensified international competition in quality.

Explosive Growth in Technology

This growth made possible an outpouring of numerous benefits to human societies: longer life spans, superior communication and transport, reduced household drudgery, new forms of education and entertainment, and so on. Huge new industries emerged to translate the new technology into these benefits. Nations that accepted industrialization found it possible to improve their economies and the well-being of their citizenry.

The new technologies required complex designs and precise execution. The empirical methods of earlier centuries were unable to provide appropriate product and process designs, so process yields were low and field failures high. Organizations tried to deal with low yields by adding inspections to separate the good from the bad. They tried to deal with field failures through warranties and customer service. These solutions were costly, and they did not reduce customer dissatisfaction. The need was to prevent defects and field failures from happening in the first place.

Threats to Human Safety, Health, and the Environment

With benefits from technology came uninvited guests. To accept the benefits required changes in lifestyle, which, in turn, made quality of life dependent on continuity of service. However, many products were failure-prone, resulting in many service interruptions. Most of these were minor, but some were serious and even frightening—threats to human safety and health as well as to the environment.

Thus the critical need became quality. Continuity of the benefits of technology depended on the quality of the goods and services that provided those benefits. The frequency and severity of the interruptions also depended on quality—on the continuing performance and good behavior of the products of technology. This dependence came to be known as "life behind the quality dikes."

Consumer Movement

Consumers welcomed the features offered by the new products but not the associated new quality problems. The new products were unfamiliar—most consumers lacked expertise in technology. Their senses were unable to judge which of the competing products to buy, and the claims of competing organizations often were contradictory.

When products failed in service, consumers were frustrated by vague warranties and poor service. "The system" seemed unable to provide recourse when things failed. Individual consumers were unable to fight the system, but collectively they were numerous and hence potentially powerful, both economically and politically. During the twentieth century, a "consumerism" movement emerged to make this potential a reality and to help consumers deal more effectively with these problems. This same movement also was successful in stimulating new government legislation for consumer protection.

Intensified International Competition in Quality

Cities and countries have competed for centuries. The oldest form of such competition was probably in military weaponry. This competition then intensified during the twentieth century under the pressures of two world wars. It led to the development of new and terrible weapons of mass destruction.

A further stimulus to competition came from the rise of multinational organizations. Large organizations had found that foreign trade barriers were obstacles to export of their products. To get around these barriers, many set up foreign subsidiaries that then became their bases for competing in foreign markets, including competition in quality.

The most spectacular twentieth-century demonstration of the power of competition in quality came from the Japanese. Following World War II, Japanese organizations discovered that the West was unwilling to buy their products—Japan had acquired a reputation for making and exporting shoddy goods. The inability to sell became an alarm signal and a stimulus for launching the Japanese quality revolution during the 1950s. Within a few decades, that revolution propelled Japan into a position of world leadership in quality. This quality leadership in turn enabled Japan to become an economic superpower. It was a phenomenon without precedent in industrial history.

The Twenty-First Century and Quality

The cumulative effect of these massive forces has been to "move quality to center stage." Such a massive move logically should have stimulated a corresponding response—a revolution in managing for quality. However, it was difficult for organizations to recognize the need for such a revolution—they lacked the necessary alarm signals. Technological measures of quality did exist on the shop floors, but managerial measures of quality did not exist in the boardrooms. Thus, except for Japan, the needed quality revolution did not start until very late in the twentieth century. To make this revolution effective throughout the world, economies will require many decades—the entire twenty-first century. Thus, while the twentieth century has been the "century of productivity," the twenty-first century will be known as the "century of quality."

The failure of the West to respond promptly to the need for a revolution in quality led to a widespread crisis. The 1980s then witnessed quality initiatives being taken by large numbers of organizations. Most of these initiatives fell far short of their goals. However, a few were stunningly successful and produced the lessons learned and role models that will serve as guides for the West in the decades ahead.

Today all countries can attain superiority in quality. The methods, tools, and know-how exist. A country that is an emerging country today may provide higher quality than one that has been producing it for centuries. Today, and into the foreseeable future, all organizations in all industries must continue to strive for perfection.

An organization that creates high-quality goods and services will be positively impacted in two ways. First, quality can affect financial results because superior products and services are more salable; thereby, increasing sales and lowering costs and thus leading to greater profitability. Second, the pursuit of high quality transforms a culture. This happens after repeated success in eliminating poor quality, process waste, and customer dissatisfaction. The transformational changes required of an organization do not happen haphazardly. They are a result of an organization's unyielding pursuit to be the best in quality and implementing a systematic method to get there. This distinction has had multiple names over the decades. Organizations that attain superior results, by designing and continuously improving the quality of their goods and services, are often called *world class, best practices, vanguard companies*, and most recently *performance excellence*. We define this as an organization that has *attained a state of performance excellence because its products and services exceed customers' expectations; they are regarded by their peers and have superior, sustainable results.*

This quest of performance excellence through quality creates high stakeholder and employee satisfaction, which enables the organization to sustain the pursuit for the long term. These organizations have reached a state of *performance excellence.*

Market Quality Leadership and Business Strategy

Market leadership is often the result of entering a new market first and gaining superiority that marketers call a *franchise*. Once gained, this franchise can be maintained through continuing product or service improvement and effective promotion. However, another organization may decide to redefine that market by improving the performance of the good or service—improving its quality—and gaining superiority over the market leader. Then it becomes the "quality leader" in the eyes of the customers. Organizations that have attained this leadership have usually done so on the basis of two principal strategies:

1. Let nature take its course. In this approach, organizations apply their best efforts, hoping that in time these efforts will be recognized as the leader creates a failure or gives up its position.
2. Help nature out by adopting a positive strategy—establish leadership as a formal business goal and then set out to reach that goal. That goal, once attained, can lead to superior results and sustain that position for long periods.

Those who decided to take action to make superior quality a formal goal soon found that they also had to answer the question, "Leadership in what?" Leadership in quality can exist in any of the multiple aspects of fitness for purpose, but the focus of the organization will differ depending on which aspects are chosen. If quality leadership is to consist of

- Superior quality of design
- Superior quality of conformance
- Availability
- Guarantees
- Speed of field repairs

Then the organization must focus on

- Product development of its goods and services
- Strong quality control and systematic quality improvement
- Operational controls
- Reliability and maintainability programs
- Creation of a field service capability that is rapid and free of defects

Once attained, quality leadership endures until there is clear cumulative evidence that some competitor has overtaken the leadership. Lacking such evidence, the leadership can endure for decades and even centuries. However, superior quality can also be lost through some catastrophic change.

A brewery reportedly changed its formulation in an effort to reduce costs. Within several years, its share of market declined sharply. The original formula was then restored but market share did not recover. (See "The Perils of Cutting Quality," 1982.)

In some cases, the quality reputation is built not around a specific organization but around an association of organizations. In that event, this association adopts and publicizes some mark or symbol. The quality reputation becomes identified with this mark, and the association goes to great lengths to protect its quality reputation.

The medieval guilds imposed strict specifications and quality controls on their members. Many medieval cities imposed "export controls" on selected finished goods in order to protect the quality reputation of the city (Juran 1995).

The growth of competition in quality has stimulated the expansion of strategic business planning to include planning for quality and quality leadership. One approach to superior quality is through product development in collaboration with the leading user of the goods or services—a user who is influential in the market and hence is likely to be followed. For example, in the medical field, an individual is "internationally renowned; a chairman of several scientific societies; is invited to congresses as speaker or chairman; writes numerous scientific papers" (Ollson 1986). Determining the identity of the leading user requires some analysis. (In some respects, the situation is similar to the sales problem of discovering who within the client organization is the most influential in the decision to buy.) Ollson lists 10 leader types, each playing a different role.

Consumer Preference and Share of Market

Consumers rely heavily on their own senses to aid them in judging quality. This fact has stimulated research to design means for measuring quality by using human senses as measuring instruments. This research has led to development of objective methods for measuring consumer preference and other forms of consumer response. A large body of literature is now available, setting out the types of sensory tests and the methods for conducting them.

At first, these methods were applied to making process control and product or service acceptance decisions. But the applications were soon extended into areas such as consumer preference testing, new-product or new-service development, advertising, and marketing.

For some products or services, it is easy to secure a measure of consumer preference through "forced-choice" testing. For example, a table is set up in a department store and passersby are invited to taste two cups of coffee, A and B, and to express their preference. Pairs of swatches of carpet may be shown to panels of potential buyers with the request that they indicate their preferences. For comparatively simple consumer goods or services, such tests can secure good data on consumer preference. More complex products such as insurance or financial instruments may require more sophisticated analysis such as conjoint analysis or discrete choice methods.

The value of consumer preference data is greatly multiplied through correlation with data on share of market. Figure 0.2 shows such a correlation for 41 different packaged consumer food products. This was an uncommonly useful analysis and deserves careful study.

Each dot on Fig. 0.2 represents a food product sold on supermarket shelves. Each product has competitors for the available shelf space. The competing products sell for identical prices and are packaged in identically sized boxes containing identical amounts of product or service. What may influence the consumer?

- The contents of the package, as judged by senses and usage, which may cause the consumer to prefer product A over product B
- The marketing features such as attractiveness of the package, appeal of prior advertising, and reputation of the manufacturer

In Fig. 0.2 the horizontal scale shows consumer preference over the leading competitor as determined by statistically sound preference testing. The vertical scale shows the share of market versus the leading competitor, considering the two as constituting 100 percent.

In Fig. 0.2 no product showed a consumer preference below 25 percent or above 75 percent. The 75/25 preference levels mean that the product is so superior (or inferior) that three users out of four can detect the difference. Since all other factors are essentially equal,

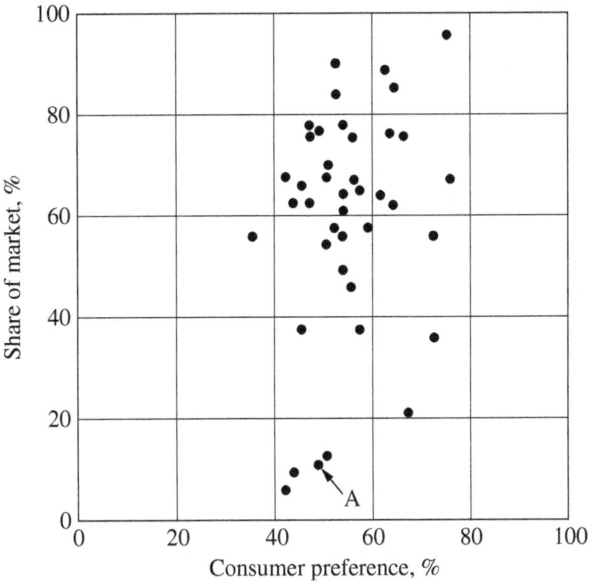

FIGURE 0.2 Consumer preference versus share of market. (Juran, J. M., *Juran's Quality Handbook*, 5th ed., McGraw-Hill, New York, 1999, p. 7.17.)

this result implies that a product that is preferred by more than 75 percent of consumers eventually takes over the entire market, and its competition disappears.

In contrast to the vacant areas on the horizontal scale of consumer preference, the vertical scale of share of market has data along the entire spectrum. One product (marked A in Fig. 0.2) lies squarely on the 50 percent consumer preference line, which probably means (under forced-choice testing) that the users are guessing as to whether they prefer that product or service or that of its competitor. Yet product or service A has only 10 percent share of market and its competitor has 90 percent. In addition, this inequality in share of market has persisted for years. The reason is that the 90 percent organization was the first to bring that product to market. As a result, it acquired a "prior franchise" and has retained its position through good promotion.

The conclusion is that when competing products or services are quite similar in consumer preference, any effect of such small quality differentials is obscured by the effect of the marketing skills. In consequence, it is logical to conclude that when quality preferences are evident to the user, such quality differences are decisive in share of market, all other things being equal. When quality differences are slight, the decisive factor in share of market is the marketing skills.

As a corollary, it appears that organizations are well advised to undertake quality improvements, which will result in either (1) bringing them from a clearly weak to an acceptable preference or (2) bringing them from an acceptable preference to a clearly dominant preference. However, organizations are not well advised to undertake quality improvements that will merely make minor improvements that are not largely perceived and valued by their customers, since marketing skill is usually the dominant factor in determining the share of market when the differences in quality are small.

It is easy for technologists to conclude that what they regard as important in the good or service is also of prime concern to the user. In the carpet industry, the engineers devote much effort to improving wear qualities and other technological aspects of fitness for purpose. However, after a market research study was conducted, it was determined that consumers' reasons for selecting carpets were primarily sensory, and not durability:

- Color 56 percent
- Pattern 20 percent
- Other sensory qualities 6 percent
- Non-sensory qualities 18 percent

For more complex consumer goods or services it is feasible, in theory, to study the relation of quality to market share by securing quantitative data on (1) actual changes in buying patterns of consumers and (2) actions of suppliers that may have created these changes. In practice, such information is difficult to acquire. It is also difficult to conclude, in any one instance, why the purchase was of model A rather than B. What does emerge are demographic patterns, that is, age of buyers, size of family, and so on, that favor model A rather than B. For goods or services sold through merchants, broad consumer dissatisfaction with quality can translate into *merchant preference*, with extensive damage to share of market.

A maker of household appliances was competitive with respect to product or service features, price, and promptness of delivery. However, it was not competitive with respect to field failure, and this became a major source of complaints from consumers to the merchants. Within several years the maker (B) lost all its leadership in share of market, as shown in the table below. Table 0.2 stimulated the upper managers of organization B to take action to improve its product reliability.

Industrial Products and Share of Market

Industrial goods or services are sold more on technological performance than on sensory qualities. However, the principle of customer preference applies, as does the need to relate quality differences to customer preference and to share of market.

Quality and Competitive Bidding

Many industrial products are bought through competitive bidding. Most government agencies are required by law to secure competitive bids before awarding large contracts. Industrial organizations require their purchasing managers to do the same. The invitations to bid usually include the parameter of quality, which may be specified in detail or through performance specifications.

	Leaders in Market Share during:			
Model Price	**Base Year**	**Base Year Plus 1**	**Base Year Plus 2**	**Base Year Plus 3**
High	A	C	C	C
Medium	B	B	C	C
Low	C	C	C	C
Special	B	B	B	C

Table 0.2 Consumer Preference

To prospective suppliers the ratio of awards to bids is of great significance. The volume of sales and profit depends greatly on this ratio. In addition, the cost of preparing bids is substantial. Finally, the ratio affects the morale of the people involved. (Members of a winning team fight with their competitors; members of a losing team fight with one another.) It is feasible to analyze the record of prior bids in order to improve the percentage of successful bids. Table 0.3 shows such an analysis involving 20 unsuccessful bids.

To create Table 0.3, a multifunctional team analyzed 20 unsuccessful bids. It identified the main and contributing reasons for failure to win the contract. The team's conclusions show that the installation price was the most influential factor—it was a contributing cause in 10 of the 14 cases. This finding resulted in a revision of the process for estimating the installation price and an improvement in the bidding/award ratio.

	Bid Not Accepted Due to				
Contract Proposal	**Quality of Design**	**Product Price**	**Installation Price**	**Reciprocal Buying**	**Other**
A1		X	X		X
A2			XX		
A3	XX	X			
A4	XX		X		
A5	XX				
A6	XX				
A7		XX			
A8		XX			
A9			XX		
A10			XX		
B1	X		X		
B2				XX	
B3				XX	
B4				XX	
B5		X	X		
B6		X	XX		
B7	XX				
B8		X	X		
B9				X	
B10	X	X	X		
Totals	7	8	10 (out of 14)	4	1

X = contributing reasons; XX = main reason.
Only 14 bids were made for installation.
Source: Juran, J. M., *Juran's Quality Handbook*, 5th ed., McGraw-Hill, New York, 1999.

TABLE 0.3 Analysis of Unsuccessful Bids

Macroeconomic Influences on Results

The ability of an organization to secure revenue is strongly influenced by the economic climate and by the cultural habits that have evolved in various economies. These overriding influences affect product or service quality as well as other elements of commerce.

National Affluence and Organization

The form of a nation's economy and its degree of affluence strongly influence the approach to its problems.

Subsistence Economies

In subsistence economies, the numerous impoverished users have little choice but to devote their revenue to basic human needs. Their protection against poor quality is derived more from their collective political power than from their collective economic power. Much of the world's population remains in a state of subsistence economy.

Shortages and Surpluses

In all economies, a shortage of goods (a "sellers' market") results in a relaxing of quality standards. The demand for goods exceeds the supply, so users must take what they can get (and bid up the price to boot). In contrast, a buyers' market results in a tightening of quality standards.

Life with the Risk of Failure

As societies industrialize, they revise their lifestyle in order to secure the benefits of technology. Collectively, these benefits have greatly improved the quality of life, but they have also created a new dependence. In the industrial societies, great masses of human beings place their safety, health, and even their daily well-being behind numerous "quality dikes." For instance, many pharmaceuticals often enable a person to receive quick short-term health benefits, but in the long term the illness may get worse.

Voluntary Obsolescence

As customers acquire affluence, economic organizations increasingly bring out new goods or services (and new models of old products) that they urge prospective users to buy. Many of the users who buy these new models do so while possessing older models that are still in working order. This practice is regarded by some reformers as a reprehensible economic waste.

In their efforts to put an end to this asserted waste, the reformers have attacked the organizations who bring out these new models and who promote their sale. Using the term "planned obsolescence," the reformers imply (and state outright) that the large organizations, by their clever new models and their powerful sales promotions, break down the resistance of the users. Under this theory, the responsibility for the waste lies with the organizations that create the new models.

In the experience and judgment of the author, this theory of planned obsolescence is mostly nonsense. The simple fact, obvious to both producers and consumers, is that the consumer makes the decision (of whether to discard the old product or service and buy the new). Periodically, this fact is dramatized by some massive marketing failure.

- The early models of home refrigerators lacked many features of modern models: freezer compartments, ice cube makers, shelves in the door, and so on. As these features were added to new models, homeowners who had bought the original models became increasingly unhappy until they bought a new model despite the fact that the old model was still running. Note that the decision to buy the new model was made by the customer, not by the manufacturer.

- The latter half of the 1970s saw the introduction of recorded entertainment into the home of the consumer with the creation of the video cassette recorder (VCR). For two decades this invention was a staple in millions of people's domestic lives across the globe. The introduction of the digital video disc (DVD) player, a machine whose ultimate utility was the same as a VCR, replaced it within years, not decades. Offered improved quality and additional features but with the same basic function as a VCR, consumers chose the DVD even when they already had an operating appliance with equivocal functionality in their homes. New forms of downloadable video through the Internet are beginning to revolutionize this market again.

Involuntary Obsolescence

A very different category of obsolescence consists of cases in which long-life products contain failure-prone components that will not last for the life of the product or service. The life of these components is determined by the manufacturer's design. As a result, even though the user decides to have the failed component replaced (to keep the product or service active), the manufacturer has made the real decision because the design determined the life of the component.

This situation is at its worst when the original manufacturer has designed the product or service in such a way that the supplies, spare parts, and so on are nonstandard, so that the sole source is the original manufacturer. In such a situation, the user is locked into a single source of supply. Collectively, such cases have lent themselves to a good deal of abuse and have contributed to the consumerism movement.

Contrast in Views: Customers' and Producers'

Industrial organizations derive their revenue from the sale of their goods or services. These sales are made to "customers," but customers vary in their functions. Customers may be wholesalers, processors, ultimate users, and so on, with resulting variations in customer needs. Response to customer needs in order to sell more goods or services requires a clear understanding of just what those needs are and how the organization can meet them.

Human needs are complex and extend beyond technology into social, artistic, status, and other seemingly intangible areas. Suppliers are nevertheless obliged to understand these intangibles in order to be able to provide products or services that respond to such needs.

The Spectrum of Affluence

In all economies the affluence of the population varies across a wide spectrum. Suppliers respond to this spectrum through variations in product or service features. These variations are often called *grades*.

For example, all hotels provide overnight sleeping accommodations. Beyond this basic service, hotels vary remarkably in their offerings, and the grades (deluxe suites, four-star, and so on) reflect this variation. In like manner, any model of automobile provides the basic service of point-to-point transportation. However, there are multiple grades of automobiles. There are luxury brands such as Porsche, BMW, Mercedes, Cadillac, and Lexus; and there are more affordable ones such as GM, Hyundai, Ford, and Toyota. The higher grades supply services beyond pure transportation. They may provide more features that result in higher levels of safety, comfort, appearance, and status.

Fitness for Purpose and Conformance to Specification

Customers and suppliers sometimes differ in their definition of quality. Such differences are an invitation to trouble. To most customers, quality means those features of the product or service that respond to customer needs. In addition, quality includes freedom from failures, plus good customer service if failures do occur. One comprehensive definition for the above is "fitness for purpose."

In contrast, for years many suppliers had defined quality as conformance to specification at the time of final test. This definition fails to consider numerous factors that influence quality as defined by customers: packaging, storage, transport, installation, reliability, maintainability, customer service, and so on.

Table 0.4 tabulates some of the differences in viewpoint as applied to long-life goods.

The ongoing revolution in quality has consisted in part of revising the suppliers' definition of quality to conform more nearly to the customers' definition.

Cost of Use

For consumable goods or many services, the purchase price paid by the customer is quite close to the cost of using (consuming) the good or service. However, for long-lived product or services, the cost of use can diverge considerably from the purchase price because of added factors such as operating costs, maintenance costs, downtime, depreciation, license fees, new releases, and transaction service charges.

The centuries-old emphasis on purchase price has tended to obscure the subsequent costs of use. One result has been suboptimization; that is, suppliers optimize their costs rather than the combined costs of suppliers and customers.

	Principal Views	
Aspects	**Of Customers**	**Of Producers**
What is purchased?	A product needed by the customer	Goods made by the producer
Definition of quality	Fitness for purpose during the life of the product or service	Conformance to specification on final test
Cost	Cost of use, including • Purchase price • Operating costs • Maintenance • Downtime • Depreciation • Loss on resale	Cost of producers
Responsibility for keeping in service	Over the entire useful life	During the warranty period
Spare parts	A necessary evil	A profitable business

Table 0.4 Contrasting Views: Customers' and Producers'

The concept of life-cycle costing offers a solution to this problem, and progress is being made in adopting this concept.

Degrees of User Knowledge

In a competitive market, customers have multiple sources of supply. Quality is an obvious consideration when making a purchasing choice or decision. However, customers vary greatly in their ability to evaluate quality, especially prior to purchase. Table 0.5 summarizes the extent of customer knowledge and strength in the marketplace as related to quality matters.

The broad conclusions that can be drawn from Table 0.5 are as follows:

- Original equipment manufacturers (OEMs) can protect themselves through their technological and/or economic power as much as through contract provisions. Merchants and repair shops must rely mainly on contract provisions supplemented by some economic power.
- Small users have very limited knowledge and protection. The situation of the small user requires some elaboration.

Aspects of the Problem	Original Equipment Manufacturers (OEMs)	Dealers and Repair Shops	Consumers
Makeup of the market	A few, very large customers	Some large customers plus many smaller ones	Very many, very small customers
Economic strength of any one customer	Very large, cannot be ignored	Modest or low	Negligible
Technological strength of customer	Very high; has engineers and laboratories	Low or nil	Nil (requires tech assistance)
Political strength of customer	Modest or low	Low or nil	Variable, but can be very great collectively
Fitness for purpose is judged mainly by:	Qualification testing	Absence of consumer complaints	Successful usage
Quality specifications dominated by:	Customers	Manufacturer	Manufacturer
Use of incoming inspection	Extensive test for conformance to specification	Low or nil for dealers; in-use tests by repair shops	In-use test
Collection and analysis of failure data	Good to fair	Poor to nil	Poor to nil

Source: Juran, J. M., *Juran's Quality Handbook*, 5th ed., McGraw-Hill, New York, 1999, p. 7.5.

Table 0.5 Customer Influences on Quality

Perception of the User, the Product or Service Is	The Resulting Income to the Supplier Is
Not fit for purpose	None, or in immediate jeopardy
Fit for purpose, but noticeably inferior to competitive products	Low due to loss of market share or need to lower prices
Fit for purpose and competitive	At market prices
Noticeably superior to competitive products	High due to premium prices or greater share of market

TABLE 0.6 Consumer Preference Effect on Income

With some exceptions, small users do not fully understand the technological nature of the product or service. The user does have sensory recognition of some aspects of fitness for use: the bread smells fresh-baked, the radio set has clear reception, the shoes are good-looking. Beyond such sensory judgments, and especially concerning the long-life performance of the product or service, the small user must rely mainly on prior personal experience with the supplier or merchant. Lacking such prior experience, the small user must choose from the propaganda of competing suppliers plus other available inputs (neighbors, merchants, independent laboratories, and so on).

To the extent that the user does understand fitness for use, the effect on the supplier's revenue is influenced as follows:

In Table 0.6, the terms "fitness for purpose," "inferior," "competitive," and "superior" all relate to the situation as seen by the user. (The foregoing table is valid as applied to both large customers and small users.)

Stated Needs and Real Needs

Customers state their needs as they see them, and in their language. Suppliers are faced with understanding the real needs behind the stated needs and translating those needs into suppliers' language.

It is quite common for customers to state their needs in the form of goods, when their real needs are for the services provided by those goods (see Table 0.7). For example:

Preoccupation with selling goods can divert attention from the real needs of customers.

In the classic, widely read paper "Marketing Myopia," Levitt (1960) stressed service orientation as distinguished from product orientation. In his view, the railroads missed an opportunity for expansion due to their focus on railroading rather than on transportation. In like manner, the motion picture industry missed an opportunity to participate in the growing television industry due to its focus on movies rather than on entertainment (Levitt 1960).

Stated Needs of Customer	Real Needs of Customer
Food	A pleasant taste and nourishment
An automobile	Transportation, safety, comfort
A flat screen TV	Entertainment, news, movies in the home
Toothpaste	Clean teeth, sweet breath, etc.
7/24 Banking	Ability to deposit or get money anytime as needed

TABLE 0.7 Stated Needs versus Real Needs

Understanding the real needs of customers requires answers to questions such as these: Why are you buying this product or service? What service do you expect from it?

Psychological Needs

For many products or services, customer needs extend beyond the technological features of the good or service; customer needs also include matters of a psychological nature. Such needs apply to both goods and services. A person in need of a haircut has the option of going to (1) a "shop" inhabited by "barbers" or (2) a "salon" inhabited by "hair stylists." Either way, she/he is cut by a skilled artisan. Either way, her/his resulting outward appearance is essentially the same. What differs is her/his remaining assets and her/his sense of well-being (Juran 1984).

What applies to services also applies to physical goods. There are factories in which chocolate-coated candies are conveyed by a belt to the packaging department. At the end of the belt are two teams of packers. One team packs the chocolates into modest cardboard boxes destined for budget-priced merchant shops. The other team packs the chocolates into satin-lined wooden boxes destined to be sold in deluxe shops. The resulting price for a like amount of chocolate can differ by several fold. The respective purchasers encounter other differences as well: the shop decor, level of courtesy, promptness of service, sense of importance, and so on. However, the goods are identical. Any given chocolate on that conveyer belt has not the faintest idea of whether it will end up in a budget shop or in a deluxe shop.

Technologists may wonder why consumers are willing to pay such price premiums when the goods are identical. However, for many consumers, the psychological needs are perceived as real needs, and the consumers act on their perceptions. Most suppliers design their marketing strategies to respond to customers' perceived needs.

"User-Friendly" Needs

The "amateur" status of many users has given rise to the term "user-friendly" to describe a condition that enables amateurs to use technological and other complex product or services with confidence. Consider the following example.

The language of published information should be simple, unambiguous, and readily understood. Notorious offenders have included legal documents, owners' operating manuals, forms to be filled out, and so on. Widely used forms (such as federal tax returns) should be field-tested on a sample of the very people who will later be faced with filling out the forms.

Goods or services should be broadly compatible. Much of this has been done through standardization committees or through natural monopolies. An example of the lack of such compatibility during the 1980s was the personal computer—many personal computers were able to "talk" to computers made by the same manufacturer but not to computers made by other manufacturers.

The Need to Be Kept Informed

Customers sometimes find themselves in a state of uncertainty: Their train is late, and they don't know when to expect it; there is a power outage, and they don't know when power will be restored. In many such cases, the supplier organization has not established the policies and processes needed to keep customers informed. In actuality, customers, even if kept informed, usually have no choice but to wait it out. Nevertheless, being kept informed reduces the anxiety—it provides a degree of assurance that human beings are aware of the problem and that it is in the process of being solved.

The New York City subway system rules require conductors to explain all delays lasting 2 minutes or more. One survey reported that this rule was followed only about 40 percent of the time. A City Hall report concluded that "shortage of information is a significant source of public antagonism toward the Transit Authority" (Levine 1987).

In contrast, some airlines go to pains to keep their customers informed of the reasons for a delay and of the progress being made in providing a remedy.

A different category of cases involves organizations secretly taking actions adverse to quality but without informing the customer. The most frequent are those in which goods or services not conforming to specification are shipped to unwary customers. In the great majority of such cases, the products or services are fit for use despite the nonconformance. In other cases, the matter may be debatable. In still other cases, the act of shipment is at least unethical and at worst illegal.

The partnership between Firestone Tires and Ford Explorer SUVs in the late 1990s created one of the most defective and ultimately deadly relationships in modern automotive history. Firestone tires were continually failing under the frame of the Ford Explorer, often causing the SUV to flip and roll. More than 250 people lost their lives because of this defect while 3000 other incidents were reported due to this imperfection. What made this situation sordid was the fact that neither Ford nor Firestone took responsibility for this obvious problem. Ford Explorers with Firestone tires were still being sold to the general public even when, in the early stages, unusually high rates of crashes were taking place. Instead of recalling the models in question, the manufacturer let this problem persist for years until the incident rate and death count became so high that the problem could not be ignored further.

Once discovered, any secretive actions tend to arouse suspicions, even if the product or service is fit for customer use. The customers wonder, "What else has been done secretly without our being informed?"

The usual reason for not informing the customer is a failure to raise the question, What shall we tell the customers? It would help if every nonconformance document included a blank space titled "What is to be communicated to the customers?" The decision may be to communicate nothing, but at least the question has been asked.

Cultural Needs

The needs of customers, especially internal customers, include cultural needs—preservation of status, continuity of habit patterns, and still other elements of what is broadly called the *cultural pattern*. Some of the inability to discover customer needs is traceable to failure to understand the nature and even the existence of the cultural pattern.

Cultural needs are seldom stated openly—mostly they are stated in disguised form. A proposed change that may reduce the status of some employee will be resisted by that employee. The stated reasons for the resistance will be on plausible grounds, such as the effect on costs. The real reason will not emerge. No one will say, "I am against this because it will reduce my status." Discovery of the real needs behind the stated needs is an important step toward a meeting of the minds. (For elaboration on the nature of cultural patterns and the "rules of the road."

Needs Traceable to Unintended Use

Many quality failures arise because the customer uses the product or service in a manner different from that intended by the supplier. This practice takes many forms:

- Untrained patient care workers are assigned to processes requiring trained workers.
- Equipment is overloaded or is allowed to run without adherence to maintenance schedules.
- The product or service is used in ways never intended by the supplier. For instance, a screwdriver is used as a hammer. It was not designed to hit things! It can break and cause harm to the user.

All this influences the relationship between quality and revenue. The critical question is whether the product or service or service development should be based on intended use or actual use. The latter often requires adding a factor of safety during the development. For example:

- Fuses and circuit breakers are designed into electrical circuits for protection against overloads.
- Spell-check software is designed to detect grammar and spelling errors.
- Public utility invoicing may include a check of customers' prior usage to guard against errors in reading the meters.

Such factors of safety may add to the cost. Yet they may well result in an optimal overall cost by helping to avoid the higher cost arising from actual use or misuse.

Needs Related to Dissatisfaction

When products or services fail to meet the needs of their customers, a new set of customer needs arises—how to restore service and get compensated for the associated losses and inconvenience. These new needs are communicated through customer complaints, which then are acted on by special departments such as customer service or call centers. Inadequate organizational response to consumer complaints and to the terms of warranties has contributed importantly to the rise of the "consumerism" movement.

Studies of how to respond to customer complaints have identified the key features of a response system that meets customer needs.

Complaints also affect product or service salability. This has been researched in studies commissioned by the U.S. Office of Consumer Affairs. The findings may be summarized as follows:

- Of customers who were dissatisfied with products or services, nearly 70 percent did not complain. The proportions varied with the type of product or service involved. The reasons for not complaining included these: the effort to complain was not worth it; the customers believed that complaining would do no good; customers lacked knowledge of how to complain.
- Over 40 percent of the complaining customers were unhappy with the responsive action taken by the suppliers. Here again the percentage varied according to the type of product or service involved.

Future salability is strongly influenced by the action taken on complaints. Figure 0.3 shows broadly the nature of consumer behavior following dissatisfaction with a purchase. This strong influence extends to brand loyalty. Figure 0.4 shows the extent of this influence as applied to large-ticket durable goods, financial services, and automobile services, respectively.

That same research concluded that an organized approach to complaint handling provides a high return on investment. The elements of such an organized approach may include

- A response center staffed to provide 24/7 access by consumers
- A toll-free telephone number
- A computerized database
- Special training for the personnel who answer the telephones
- Active solicitation of complaints to minimize loss of customers in the future

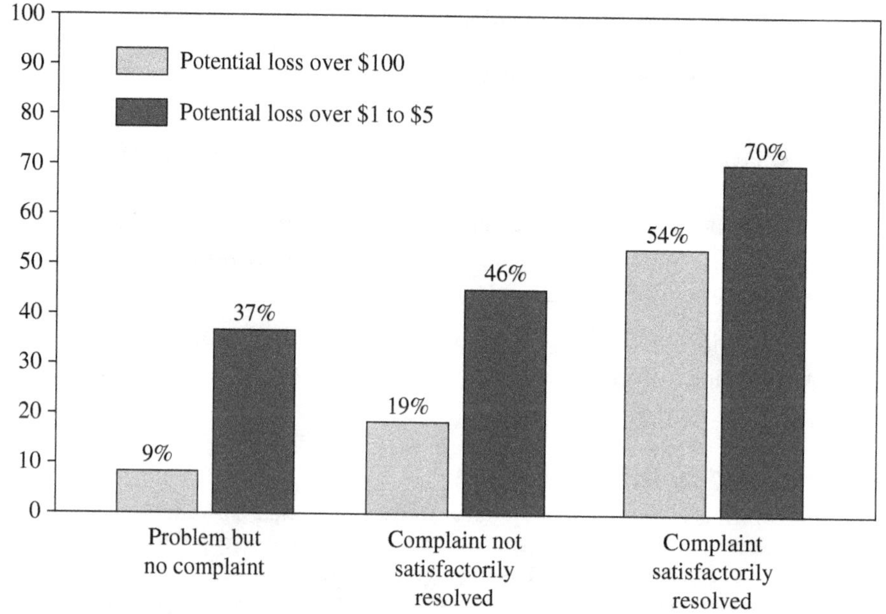

Figure 0.3 Behaviors of consumers after they experience product dissatisfaction. (*Planning for Quality*, 2d ed., Juran Institute, 1990, pp. 4–12.)

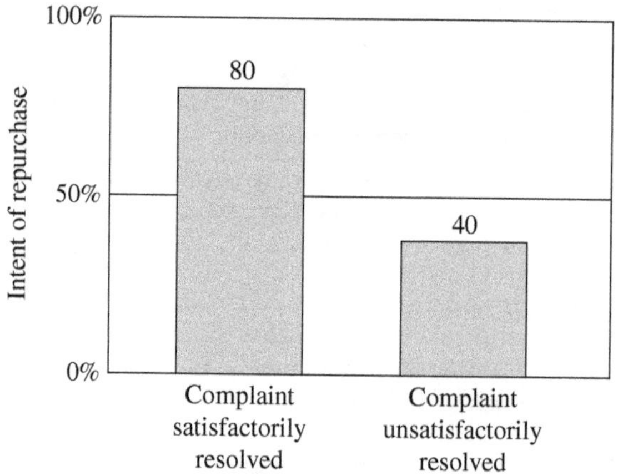

Figure 0.4 Consumer loyalty versus complaint resolution: large ticket durable goods, financial services, automotive services. (*Planning for Quality*, 2d ed., Juran Institute, 1990, pp. 4–14.)

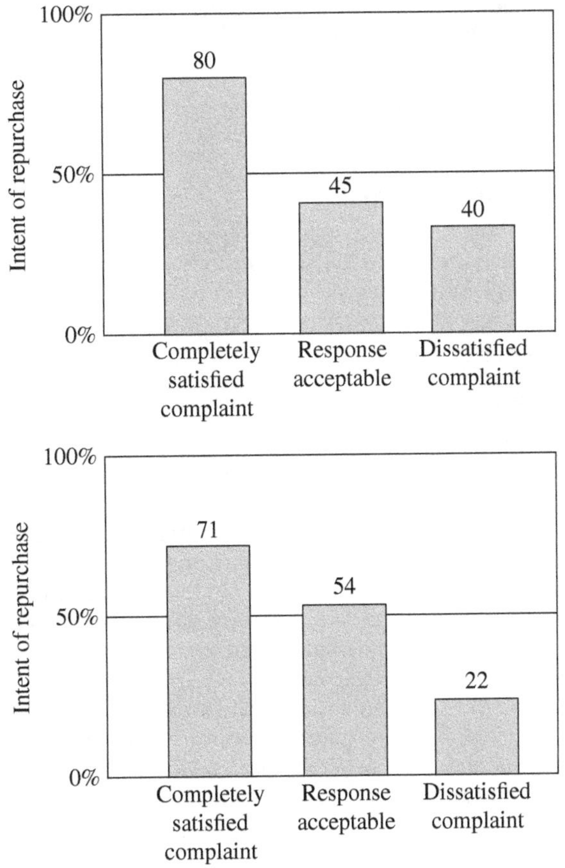

FIGURE 0.4 *(Continued)*

Discovering Hidden Customer Needs

The most simplistic assumption is that customers are completely knowledgeable as to their needs and that market research can be used to extract this information from them. In practice, customer knowledge can be quite incomplete. In some cases, the customer may be the last person to find out. It is unlikely that any customer ever expressed the need for a Walkman (a miniature, portable audiotape player) before such devices came on the market. However, once they became available, many customers discovered that they needed one.

These gaps in customer knowledge are filled in mainly by the forces of the competitive market and by the actions of entrepreneurs.

Inferior Available Product or Services

When available products are perceived as inadequate, a vacuum waiting to be filled emerges. Human ingenuity then finds ways to fill that vacuum:

- The number of licensed New York taxicabs has remained frozen for years while the population has increased. The resulting shortage of cabs has been filled by unlicensed cabs, limousines, buses, and even bicycles.

- Government instructions for filling out tax forms have been confusing to many taxpayers. One result has been the publication of some best-selling books and software on how to prepare tax returns.
- The service provided by tradesmen has been widely regarded as expensive and untimely. One result has been the growth of a large do-it-yourself industry.

Reduction of Time for Service

Some cultures exhibit an urge to "get it over with." In such cultures, those who can serve customers in the shortest time are rewarded by a higher share of market. A spectacular example of this urge is the growth of the fast-food industry. This same need for fast food and quick service is an essential element in the urge of a producer to employ "just-in-time" manufacturing techniques.

Perfectionism

The human being exhibits an instinctive drive for precision, beauty, and perfection. When unrestrained by economics, this drive has created the art treasures of the ages.

In the industrial society, there are many situations in which this urge for perfection coincides with human needs. In food and drug preparation, certain organisms must be completely eliminated, or they will multiply and create health hazards. Nuclear reactors, underground mines, aircraft, and other structures susceptible to catastrophic destruction of life require a determined pursuit of perfection to minimize dangers to human safety. So does the mass production of hazardous materials.

However, there are numerous other situations in which the pursuit of perfection is antagonistic to society, since it consumes materials and energy without adding to fitness for purpose, either technologically or aesthetically. This wasteful activity is termed "perfectionism" because it adds cost without adding value.

Acknowledgments

This chapter has drawn extensively from the following:

De Feo, J. A., and Barnard, W. W. (2004). *Juran Institute's Six Sigma Breakthrough and Beyond*. McGraw-Hill, New York.

Gryna, F. M., Chua, R. C. H., and De Feo, J. A. (2007). *Quality Planning and Analysis*, 5th ed. McGraw-Hill, New York.

Juran, J. M. (ed.) (1995). *A History of Managing for Quality*. Sponsored by Juran Foundation, Inc. Quality Press, Milwaukee, WI.

Juran, J. M. (2004). *Architect of Quality*. McGraw-Hill, New York.

Juran, J. M., and Godfrey, A. B. (1999). *Juran's Quality Handbook*, 5th ed. McGraw-Hill, New York.

The author is grateful to the copyright holders for permission to quote from these works.

References

Betting to Win on the Baldie Winners. (1993). *Business Week*, October 18.

Juran, J. M. (1964, 1995). *Managerial Breakthrough*. McGraw-Hill, New York.

Juran, J. M. (ed.) (1995). *A History of Managing for Quality*. Quality Press, Milwaukee, WI.

Levine, R. (1987). Breaking Routine: Voice of the Subway. *The New York Times*, January 15.
Levitt, T. (1960). Marketing Myopia. *Harvard Business Review*, July–August, pp. 26–28ff.
Ollson, J. R. (1986). The Market-Leader Method; User-Oriented Development. *Proceedings 30th EOQC Annual Conference*, pp. 59–68. European Organization for Quality Control.
The Perils of Cutting Quality. (1982). *The New York Times*, August 22.
Watson, G. H. (2016). Understanding the Role of Quality. Presentation at *Croation Society for Quality*. Porec, Croatia. May 20.

CHAPTER 1

Universal Principles of Quality Management

Joseph M. Juran & Joseph A. De Feo

High Points of This Chapter

1. "Managing for quality" is a set of universal methods that an enterprise, a business, an agency, a university, a hospital, or any organization can use to attain superior results by ensuring that all goods, services, and processes meet stakeholder needs.
2. "Quality" has two meanings that must be clearly understood and communicated. The first relates to how well the features of the services or products you produce meet customer needs and thereby provide them satisfaction. "Quality" also means freedom from failure.
3. The Juran Trilogy embodies the universal principles needed to create high-quality goods, services, and processes.
4. Implementing new processes to create innovative products and services by discovering the voice of the customer will enable every organization to understand the customers' needs better and then create or design products that meet these needs.

5. Implement processes to ensure that products conform to the design criteria when they are produced. We must control quality and predict how it will perform in the marketplace.
6. Implement a systematic approach to improving quality or creating breakthroughs to eliminate those failures that are chronic in our processes or products.

The Concept of Universals from the Voice of Dr. Joseph M. Juran

"During my studies of algebra and geometry, I stumbled across two broad ideas that I would put to extensive use in later years. One was the concept of "universals"; the other was the distinction between theory and fact.

My study of algebra exposed me for the first time to the use of symbols to create generalized models. I knew that 3 children plus 4 children added up to 7 children, and 3 beans plus 4 beans added up to 7 beans. Now by using a symbol such as x, I could generalize the problem of adding 3 + 4 and state it as a universal:

$$3x + 4x = 7x$$

This universal said that 3 + 4 always equals 7 no matter what x stands for—children, beans, or anything else. To me the concept of universals was a blinding flash of illumination. I soon found out that universals abounded, but they had to be discovered. They had various names—rules, formulas, laws, models, algorithms, patterns. Once discovered, they could be applied to solve many problems.

By 1954, in my text *Managerial Breakthrough*, I outlined the beginnings of the many universals that led to superior result. The first was the universal of control—the process for preventing adverse change. The second was the universal sequence for breakthrough improvement. The latter went on to become known as Six Sigma today. By 1986, I discovered that there was another universal. This was the planning for quality, at the strategic level and product and service design levels. I also came to realize that those three managerial processes (planning, control, and improvement) were interrelated, so I developed the Juran Trilogy diagram, to depict this interrelationship. The Juran Trilogy embodies the core processes by which we manage for quality. As a corollary, these same core processes constitute an important sector of science in managing for quality. To my knowledge, there is growing awareness in our economy that mastery of those universal processes is critical to attaining leadership in quality and superior results."

What Does Managing for Quality Mean?

For many decades, the phrase used to define quality was simply "fitness for use." It has been generally accepted that if an organization produced goods that were "fit for use" as viewed by the customer, then those goods were considered of high quality. Throughout most of the twentieth century this definition made sense because it was easy to grasp. Simply put, if customers purchased a good and it worked, they were pleased with the quality of it. To the producers of that product it was easy to produce as long as the producer had a clear understanding of the customer requirements.

Managing for quality therefore meant to "ensure product conformance to requirements." The majority of tasks largely fell on the operations and quality departments. These functions were responsible to produce, inspect, detect, and ensure the product met requirements.

Two developments have led us to modify this time-honored definition. The first was the realization that the quality of a physical good, its fitness for use, was broader than just its conformance to specifications. Quality was also determined by the design, packaging, order fulfillment, delivery, field service, and all the service that surrounded the physical good. The operations and quality departments could not manage quality alone.

The second development was a shift in the economy from production dominated by goods to production heavily concentrated in services and information. As stated in the Introduction, Quality and Performance Excellence and to reflect these changes, the authors of this handbook (seventh edition) have chosen to use the phrase "fit for purpose" instead of "fit for use" to define the quality of a product. We will use the term "product" to refer to goods, services, and information. Regardless of whether a product is a good, a service, or information, it must be "fit for its purpose" by the customers of that product. The customer is not just the end user but all those whom the product impacts, including the buyer, the user, the supplier, the regulatory agencies, and almost anyone who is affected by the product from concept to disposal. With such an expanding set of customers and their needs, the methods and tools to manage for quality must grow as well.

For the twenty-first century "managing for quality" has been defined as "a set of universal methods that any organization, whether a business, an agency, a university, or a hospital, can use to attain superior results by designing, continuously improving, and ensuring that all products, services, and processes meet customer and stakeholder needs."

Management of quality is not the only set of universal methods to manage an organization. It is one set of managerial methods that successful organizations have used and others should use if they want to assure their products: goods, services, and information meet customer requirements. This evolution will continue as more industries adopt the methods and tools used to manage the quality of goods and services. Emerging organizations and countries will create new means to adopt management methods to their unique needs. Today, a full range of industries, including hospitals, insurance organizations, medical laboratories, and financial service organizations, are managing for business superior performance excellence.

The accelerated adoption of techniques to manage for quality began in the late 1970s when U.S. businesses were badly affected by many Japanese competitors. Japanese manufactured goods were generally viewed by the purchasers of those goods as having higher quality. This led to the definition of "Japanese or Toyota quality." These terms have become synonymous with higher quality that is required to meet the needs of the customers. As consumers or customers had a better choice, it forced some U.S. organizations into bankruptcy and others to compete at a new level of performance. Eventually many American and then later European organizations regained lost markets with higher quality.

One of the first to accomplish that was Motorola. Motorola was affected by Japanese organizations such as NEC, Sony, and others. The improved quality resulted in Motorola becoming the first winner of the U.S. Malcolm Baldrige National Quality Award. Motorola itself evolved the universal quality improvement model and created the Six Sigma model for quality improvement. Since then American quality improved, and the quality revolution continued into a global revolution. From 1986 to today this model of quality improvement has become the most valued model for many industries around the globe. Today organizations such as Samsung, Quest Diagnostics, Oracle, and Telefonica have become more competitive and are among quality leaders in their industries.

Each of these organizations and others have all contributed to the methods to manage quality. They all used the basic tools of Six Sigma and quality management to expand business processes and all parts of the supply chain. Now quality is not the quality department's responsibility. It is the responsibility of the entire hierarchy. Managing quality has become the way to manage an entire organization. It has become the driving force of many strategies. To be the best in my industry, to have the highest quality, and to provide the highest level of customer delight are all business strategies. If achieved, these strategies will enable these organizations to attain financial success, cultural change, and satisfied customers.

In this seventh edition of the *Juran's Quality Handbook*, we aim to provide a concise, simpler, and hopefully clear set of methods and tools to manage "quality." This will include not only the quality of goods or services but also the quality of process and function, which lead to overall organizational quality.

As the needs of customers and society have changed, the means for meeting their needs also changed. The methods of managing quality in 1980 may not work for your organization today. What works today may not work tomorrow. Even the universals that continue to deliver superior results may one day need to be modified. This handbook will present the best of what works and the lessons learned for those that did not. One lesson learned was that many organizations that were once quality leaders failed to sustain their successful performance over time. Why did this happen? Did they fail to sustain results because of weak leadership? Was it a poor execution of their strategies? Were those external forces? These questions have haunted many professionals who have had to defend their "quality programs." We will try to provide answers to these questions and more in this version of the handbook.

In Table 1.1, which addresses the meaning of quality, we have presented two of the many meanings of the word "quality" as they relate to goods and services. These two are of critical importance to managing for quality:

1. Quality as it relates to how well the features of a service or good meet customer needs and thereby provide them with satisfaction. In this meaning of the word, higher quality usually costs more.
2. Quality as it relates to freedom from failures. In this sense, the meaning of the word is oriented to costs, and "higher quality usually costs less."

By adopting these simple definitions of quality as it relates to goods and services one can create a systematic approach to manage quality by

- Creating processes to design goods and services to meet needs of its stakeholders (external and internal). Every organization must understand what the customers' needs are and then create or design services and goods that meet those needs.
- Creating processes to control quality. Once goods and services are designed we must ensure compliance to the design criteria.

Features Which Meet Customer Needs	Freedom from Failures
Higher quality enables organizations to	*Higher quality enables organizations to*
Increase customer satisfaction	Reduce error rates
Meet societal needs	Reduce rework, waste
Make products and services salable	Reduce failures, warranty charges
Exceed competition	Reduce customer dissatisfaction
Increase market share	Reduce inspection, test, and audits
Provide salesrevenue	Shorten time to develop new products
Secure premium prices	Increase yields, capacity
	Improve delivery performance
The major effect is on revenue	*The major effect is on costs*
Usually higher quality costs more.	Usually higher quality costs less.

Source: Juran Institute, Inc., 2009.

Table 1.1 The Meaning of Quality

- Creating a systematic approach for improving continuously or creating breakthroughs. Services, goods, and the processes that produce them suffer from chronic failures that must be discovered and remedied.
- Creating a functions to ensure you continue to do the three things listed above.

By designing quality, controlling it during operations, and then continuously improving on it, any organization can be on its way to becoming a "quality organization." The global quality leaders as described above are relentless in their pursuit of ensuring that all their goods and services meet or exceed their customer requirements—but not at all costs. Attaining quality that satisfies customers but not the business stakeholders is not a good business to be in. To be truly a quality organization, the products and services must be produced at costs that are affordable to the producer and its stakeholders. The quality-cost-revenue relationship, however, must be properly understood in making these judgments. Increased feature quality must generate enough revenue to cover the added costs of additional features. But higher quality from lower failures will usually reduce cost and thereby improve financial performance. For organizations that do not generate revenue, feature quality must not cost more than your budget allows, but quality improvement against failures will almost always improve financial health.

By using these two definitions of quality and by understanding the impact of good or poor quality on an organization's performance, one can create long-term plans to maintain high quality of goods, services, processes, and financial performance. Managing over the long term also requires that the organization set up systems to ensure that the changing needs of its customers are well understood to avoid the failure to sustain performance that plagues even the most successful organizations.

Performance and Organizational Excellence Programs

Performance Excellence, Operational Excellence, Lean Six Sigma, Toyota Production System, and Total Quality Management (TQM) are names for "quality" methods, and some may find them synonymous with the universals to managing for quality. As Juran's universals of managing for quality become embedded and used in many new industries, a new brand may be formed. Most of the time, these new brands are useful because they help advance the needs to improve performance. Just as the early guilds led to quality standards, society and changing customer needs also require the universals to be adapted. One common problem with the methods to manage quality was found in the service sector. Service organizations always felt that the word "quality" meant product quality. Many services do not see their products as goods. They are services. Therefore, they substitute the words "service quality" with "service excellence." Over time, this phrase catches on and we have a new brand. Most of the time, this new brand builds positively on the previous brand. Other times, the alterations to the methods result in less positive outcomes and shunning of the brand. This happened to TQM. Total Quality Management was the brand in the 1990s. It was replaced with Six Sigma. Why? The methods of managing for quality were evolving as many organizations were trying to regain competitiveness. The problem with TQM was that it was not measurable or as business-focused as needed. Over time, it lost its luster. However, there were many organizations that improved their performance immensely, and they continue with TQM today. Others move on to the new brand. At the time of this writing, Lean Six Sigma and Performance Excellence are in vogue. They too will change over time. In the end, it does not matter what you call your processes to manage for quality as long as you do what is needed to attain superior results. The universals live on. No matter the industry, country, or century, meeting and exceeding the needs of your customers will drive your results.

Our Glossary of Key Terms

In the world of managing for quality, there is still a notable lack of standardization of the meanings of keywords. However, any organization can do much to minimize internal confusion by standardizing the definitions of keywords and phrases. The basic tool for this purpose is a glossary. The glossary then becomes a reference source for communication of all sorts: reports, manuals, training texts, and so on.

The definitions of "quality" include certain keywords that themselves require definition. A few are important before we continue:

- *Organizations*. In this handbook, we use this word to mean any entity, business, company, institution, agency, business unit, hospital, bank, Internet provider, casino, etc., that provides an output—a product, service, or information—to a customer, whether for profit or not for profit.
- *Universal management methods and tools*. A universal management method, tool, or process means it can be used in any industry, any function, any organization, and in any culture. It is truly universal. To most employees of an organization the word "manage" means to assign resources, set goals, establish controls, and review results with respect to products, processes, and people. To organizations that are "world-class," it means a full sequence of activities that produce the intended results to meet customer and societal needs. Managerial processes not limited to just finance, human resources, technology, and operations. They also include managerial processes to understand customer needs; to design new products and services to meet those needs; to have systems and controls in place to ensure the needs are met over time; to have systems and initiatives in place to continually improve all of them, to ensure society's needs are not negatively impacted.
- *Product: goods, services, or information*. These are the outputs of any process that meet the needs the customers. To economists, products include both goods and services and can also include information. However, under popular usage, "product" sometimes means goods only. The authors will generally use "product" to refer to both goods and services.
    - A product can also be a physical good such as a toy, a computer, or a document containing information such as a proposal, an architectural drawing, or a website on the World Wide Web.
    - A product can also be a service, which is work that is performed for someone else. A carpenter builds a home for a homeowner, the user; an automotive technician repairs cars for their owners; a nurse cares for patients; and a web browser provides fast information to meet the needs of its users.
- *Feature*. A feature is a property or characteristic possessed by a good or service that responds to customer needs. A feature of an automobile can be the fidelity with which its stereo system meets the listening needs of its driver. This may have little or no impact on how the automobile drives and performs, but it does meet the needs of the customer in other ways. A feature can be the emergency-room fast track for critical patients in need of immediate medical attention. Features are what a company, an organization, a system, or an agency must include in the design of a good or service to meet customer needs. Features must be created through understanding exactly how to meet the most important needs.

- *Cost of poor quality (COPQ).* These are the costs that would disappear in the organization if all failures were removed from a product, service, or process. They are measured as a percentage of sales or total costs.
- *Customer.* A customer is anyone external to your company, organization, system, or agency who is affected by the use of the product or service. A customer receives value for the product of the organization. A customer may be the ultimate user of the product or an intermediate customer external to your organization but not the user, such as a parent who purchases a game for a child or a surgeon who implants a device in the patient. A good or service can have many customers. A common practice is to differentiate between "external" customers and "internal" customers. External customers are defined as above, and internal customers are users within the organization. The term "customer" alone will generally refer to the external customer. Customers are sometimes called *stakeholders*. This term is typically intended to encompass external customers and internal customers, shareholders, management, and employees. Since this wide range of roles will have divergent and even conflicting needs for the organization, we will generally discuss each group separately rather than place them all in one category.
- *Processor.* Processors are employees, departments, functions, business units, and agencies that produce or carry out a process within the organization. To achieve superior results, an organization must clearly focus on the external customers but ensure that all processors are able to complete their work as designed, on time, every time.
- *Customer satisfaction.* Customer satisfaction is the positive state of customers when their needs have been met by the good or service they purchase or use. Satisfaction is mainly driven by the features of the good or service produced.
- *Customer dissatisfaction.* This is the negative state of a customer when a good or service has a failure that results in an unmet need and, consequently, customer annoyance, a complaint, a claim, or returned goods.
- *Failure.* Failure is any fault, defect, failure, or error that impairs a service or product from meeting the customer needs. These can be stated as too many defects or failures (in goods or services). Failures take such forms as waiting too long for phone responses from a call center, errors on invoices, warranty claims, power outages, failures to meet delivery dates, and inoperable goods.
- *Customer loyalty.* This is the delighted state of a customer when the features of the good and service meet their needs and are delivered free from failure. Loyalty is also relative to the offerings of the competition. A loyal customer continues to purchase or use your organization's goods and services. Loyalty is a strategic financial measure of customer satisfaction. Creating loyal customers is the goal of a superior performer.
- *Customer disloyalty.* This is the very negative state of customers who no longer want your products or services. They find better-performing products and services and then become disloyal to the producer to whom they had once been loyal.
- *Superior performers, world class, or best in class.* These various labels are used in the marketplace for organizations with products that are generally accepted as having the highest quality. These organizations become the de facto comparative benchmark for others to attain. Examples include Toyota Motor Company, Samsung Electronics, the Mayo Clinic, and Google, to name a few.

Management of Quality: The Financial and Cultural Benefits

Features Effect on Revenue

Revenue can include several types of transactions: (1) money collected from selling a good or service, (2) taxes collected by a government, or (3) donations received by a charity. Whatever the source, the amount of the revenue relates in varying degrees to the ability of the good or service features produced to be valued by the recipient—the customer. In many markets, goods and services with superior features are able to attract more revenue through some combination of higher market share and premium pricing. Services and products that are not competitive with features often are sold at lower prices.

Failures Effect on Income

The customer who encounters a deficiency may take action that creates additional cost for the producer, such as file a complaint, return the product, make a claim, or file a lawsuit. The customer also may elect instead (or in addition) to stop buying from the guilty producer as well as to publicize the deficiency and its source. Such actions by multiple customers can seriously damage to a producer's revenue.

Failures Effect on Cost

Deficient quality creates excess costs associated with poor quality. "Cost of poor quality" (COPQ) is a term that encompasses all the costs that would disappear if there were no failures—no errors, no rework, no field failures, and so on. Juran Institute's research on the cost of poor quality demonstrates that for organizations that are not managing quality aggressively, the level of COPQ is shockingly high.

Calculating the costs of poor quality can be highly valuable for an organization. COPQ shows enterprise leaders just how much poor quality has inflated their costs and consequently reduced their profits. Detailed COPQ calculations provide a road map for rooting out those costs by systematically removing the poor quality that created them.

In the early 1980s, it was common for many business leaders to make a statement that their COPQ was about 20 to 25 percent of sales revenue. This astonishing number was backed up by many independent organizations calculating their own costs. For this handbook, we conducted additional research to determine a more precise and current estimate for COPQ in the economy.

The task was not as easy as it may sound. Many sources were in disagreement as to what costs should be included in the total. In addition, the actual form of the statistic was presented in myriad ways: percentage of sales, percentage of operating expenses, percentage of value added, an absolute dollar value, a dollar value per employee, and even a number of deaths in the health care industry. While many sources provided hard primary data, others would cite vague "experts" or "studies." When specific sources were cited, they sometimes referred to one another in circular fashion.

Based on findings and extrapolations from published literature, as well as a report conducted by the Midwest Business Group on Health and Juran Institute, and the reasoned judgment of knowledgeable health care practitioners, it was estimated that 30 percent of all direct costs of health care are the result of poor-quality care, consisting primarily of overuse, misuse, and waste in the system. The impact of underuse on costs is not clear. With national health expenditures of roughly \$1.4 trillion in 2001, the 30 percent figure translates into \$420 billion spent each year as a direct result of poor quality. In addition, the indirect costs of poor quality (e.g., reduced productivity due to absenteeism) add an estimated 25 to

50 percent, or $105 to $210 billion, to the national bill. Private purchasers absorb about one-third of these costs. In fact, we estimate that poor-quality health care costs the typical employer between $1900 and $2250 per covered employee each year. Even if these figures are off by 50 percent, poor-quality health care exacts a several hundred-billion-dollar toll on our nation each year (Midwest Business Group on Health et al. 2003).

Our best synthesis suggests that by the year 2003, the COPQ was in the range of 15 to 20 percent for manufacturing organizations, with many achieving even lower levels as the result of systematic programs to reduce it. For service organizations COPQ as a percentage of sales was still a staggering 30 to 35 percent of sales. These numbers included the costs of redoing what had already been done, the excess costs to control poor processes, and the costs to correctly satisfy customers. Failures that occur prior to sale obviously add to a producer's costs. Failures that occur after sale add to customer's costs as well as to producer's costs. In addition, post sale failures reduce producers' future sales because customers may be less apt to purchase a poor quality service.

How to Manage for Quality: A Financial Analogy

To manage quality, it is good to begin by establishing a *vision* for the organization, along with policies, goals, and plans to attain that vision. This means that quality goals and policies must be built into the organization's strategic plan. (These matters are treated elsewhere in this handbook, especially in Chap. 7, Strategic Planning and Performance Excellence.) Conversion of these goals into results (making quality happen) is then achieved through established managerial processes—sequences of activities that produce the intended results. Managing for quality makes extensive use of three such managerial processes:

- Designing or planning for quality
- Compliance, controlling or assuring quality
- Improving or creating breakthroughs in quality

These three processes are interrelated and are known as the *Juran Trilogy*. They parallel the processes long used to manage for finance. These financial processes consist of the following:

Financial planning. This process prepares the annual financial and operational budgets. It defines the deeds to be done in the year ahead. It translates those deeds into money—revenue, costs, and profits. It determines the financial benefits doing all those deeds. The final result establishes the financial goals for the organization and its various divisions and units.

Financial control. This process consists of evaluating actual financial performance, comparing this with the financial goals and taking action on the difference—the accountant's "variance." There are numerous subprocesses for financial control: cost control, expense control, risk management, inventory control, and so on.

Financial improvement. This process aims to improve financial results. It takes many forms: cost reduction projects, new facilities, and new-product development to increase sales, mergers, and acquisitions, joint ventures, and so on.

These processes are universal—they provide the basis for financial management, no matter what type of organization it is.

The financial analogy can help leaders realize that they can manage for quality by using the same processes of planning, control, and improvement. Since the concept of the trilogy

Quality Planning	Quality Control	Quality Improvement
Establish goals	Determine the control subjects	Prove the need with a business case
Identify who are the customers	Measure actual performance	Establish a project infrastructure
Determine the needs of the customers	Compare actual performance to the targets and goals	Identify the improvement projects
Develop features which respond to customers' needs		Establish project teams
Develop processes able to produce the products	Take action on the difference	Provide the teams with resources, training, and motivation to: Diagnose the causes Stimulate remedies
Establish process controls transfer the plans to the operating forces	Continue to measure and maintain performance	Establish controls to hold the gains

TABLE 1.2 Managing for Quality

is identical to that used in managing for finance, leaders are not required to change their conceptual approach.

Much of their previous training and experience in managing for finance is applicable to managing for quality.

While the conceptual approach does not change, the procedural steps differ. Table 1.2 shows that each of these three managerial processes has its own unique sequence of activities. Each of the three processes is also a universal—it follows an unvarying sequence of steps. Each sequence is applicable in its respective area, no matter what the industry, function, culture, etc.

Implementing the Juran Trilogy

The Juran Trilogy Diagram

The three processes of the Juran Trilogy are interrelated. Figure 1.1 shows this interrelationship. The Juran Trilogy diagram is a graph with time on the horizontal axis and cost of poor quality on the vertical axis. The initial activity is quality planning. The market research function determines who the customers are and what their needs are. The planners or product realization team then develops product features and process designs to respond to those needs. Finally, the planners turn the plans they created over to operations: "Run the process, produce the features, deliver the product to meet the customers' needs."

Chronic and Sporadic

As operations proceed, soon it is evident that the processes that were designed to deliver the good or service are unable to produce 100 percent quality. Why? Because there are hidden failures or periodic failures that require rework and redoing. Figure 1.1 shows an example where more than 20 percent of the work processes must be redone owing to failures. This waste is considered chronic—it goes on and on until the organization decides to find its root

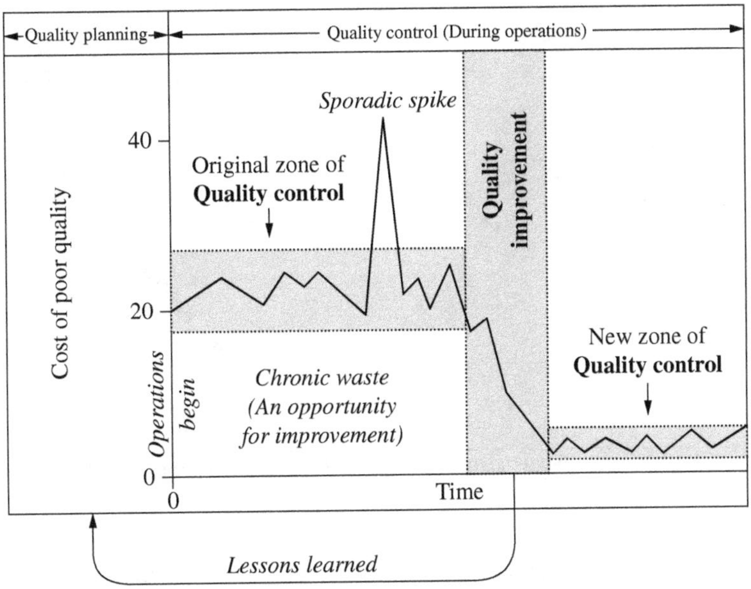

FIGURE 1.1 Juran Trilogy.

causes. Why do we have this chronic waste? Because it was planned that way. The planners could not account for all unforeseen obstacles in the design process.

Under conventional responsibility patterns, the operating forces are unable to get rid of this planned chronic waste. What they can do is to carry out control—to prevent things from getting worse, as shown in Fig. 1.1. It shows a sudden sporadic spike that has raised the failure level to more than 40 percent. This spike resulted from some unplanned event such as a power failure, process breakdown, or human error. As a part of the control process, the operating forces converge on the scene and take action to restore the status quo. This is often called *corrective action, troubleshooting, fire-fighting*, and so on. The end result is to restore the error level back to the planned chronic level of about 20 percent.

The chart also shows that in due course the chronic waste was driven down to a level far below the original level. This gain came from the third process in the trilogy—improvement. In effect, it was seen that the chronic waste was an opportunity for improvement, and steps were taken to make that improvement.

The Trilogy Diagram and Failures

The trilogy diagram (Fig. 1.1) relates to product and process failures. The vertical scale therefore exhibits units of measure such as cost of poor quality, error rate, percent defective, service call rate, waste, and so on. On this same scale, perfection is at zero, and any increase is bad. The results of reducing failures equals a reduction in the cost of poor quality, meeting more delivery promises, reduction of the waste, decrease in customer dissatisfaction, and so on.

Allocation of Time within the Trilogy

An interesting question for managers is, "How do we design our functions and allocate their time relative to the processes of the trilogy?" Figure 1.2 is a model designed to show this interrelationship in a Japanese company (Itoh 1978).

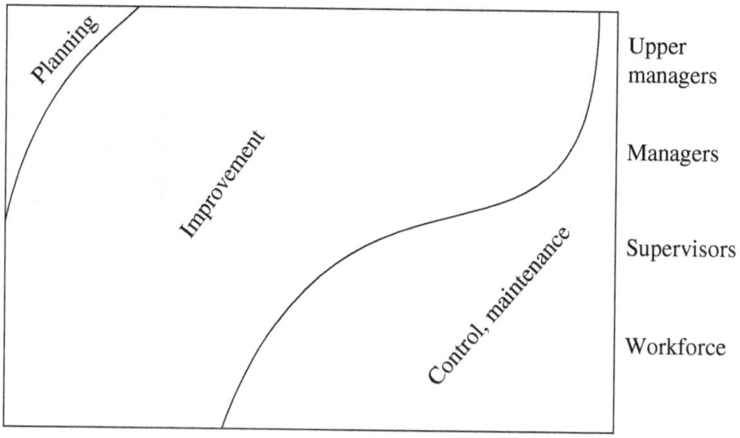

FIGURE 1.2 Itoh model. (Adapted from *Management for Quality*, 4th ed., Juran Institute, Inc., 1987, p. 18.)

In Fig. 1.2, the horizontal scale represents the percentage allocation of any person's time and runs from 0 to 100 percent. The vertical scale represents levels in the organizational hierarchy. The diagram shows that the upper managers spend the great majority of their time on planning and improvement. They spend a substantial amount of time on strategic planning. The time they spend on control is small and is focused on major control subjects.

At progressively lower levels of the hierarchy, the time spent on strategic planning declines, whereas the time spent on control and maintenance grows rapidly. At the lowest levels, the time is dominated by control and maintenance, but some time is still spent on planning and improvement.

Figure 1.2 shows these unvarying sequences in abbreviated form. Extensive detail is provided in other chapters of this handbook: Chap. 4, Quality Planning and Design of New Goods and Services; Chap. 5, Quality Improvement and Breakthrough Performance; and Chap. 6, Quality Control to Assure Compliance to Customer Requirements.

References

Itoh, Y. (1978). "Upbringing of Component Suppliers Surrounding Toyota." International Conference on Quality Control, Tokyo.

Midwest Business Group on Health, Juran Institute, Inc., and The Severyn Group, Inc. (2003). "Reducing the Costs of Poor-Quality Health Care through Responsible Purchasing Leadership."

CHAPTER 2

Developing an Excellence Culture

Joseph A. De Feo

High Points of This Chapter

1. Creating a Culture of Excellence requires that the organization attain five breakthroughs to assure sustained results.
2. The required breakthroughs are leadership and management, organization and structure, current performance, culture, and agile sustainability.
3. Creating excellence in organizations will enable our global society to avoid technological failures from harming the environment and, ultimately, its people.
4. Organizational change attempted in isolation from the whole organization and without systems thinking can easily create more problems than existed previously.

Culture Defined

Your organization is a society. A society is "an enduring and cooperating social group whose members have developed organized patterns of relationships through interaction with each other ... a group of people engaged in a common purpose," according to Webster's. A society consists of habits and beliefs ingrained over long periods of time. Your workplace is a society, and, as such, it is held together by the shared *beliefs* and *values* that are deeply embedded in the personalities of the society's members. (A workplace whose workforce is segmented into individuals or groups who embody conflicting beliefs and values does not hold together. Various social explosions will eventually occur, including resistances, revolts, mutinies, strikes, resignations, transfers, firings, divestitures, and bankruptcies.)

Society members are rewarded for conforming to their society's beliefs and values—its norms—and they are punished for departing from them. Not only do norms encompass values and beliefs, they also include enduring systems of relationships, status, customs, rituals, and practices.

Societal norms are so strong and deeply embedded that they lead to customary patterns of social behavior sometimes called "cultural patterns." In the workplace, one can identify performance-determining cultural patterns such as participative versus authoritarian management styles, casual versus formal dress, conversational styles ("Mr./Ms." and "Sir/Madam" versus first names), and a high trust level that makes it safe to say what you really think versus low trust level/suspiciousness that restricts honest or complete communication and breeds game playing, deceit, and confusion.

What Does Culture Have to Do with Performance Excellence?

To achieve a performance breakthrough, it is desirable—if not necessary—that the organization's norms and cultural patterns support the organization's performance goals. Without this support, performance goals may well be diluted, resisted, indifferently pursued, or simply ignored. For these reasons, the characteristics of your organization's culture are a vital matter that your management needs to understand and be prepared to influence. As we shall see, this is easier said than done; but it *can* be done.

J. M. Juran provides a timely example of the influence of culture on an organization's performance. Here are excerpts from his description of a management challenge currently facing managers as it has for many years:

> There has been great difficulty in getting production operators and supervisors to accept control charts as a shop tool. I believe this to be a statement of fact, based on extensive firsthand observation of the shockingly high mortality rate of control charts when actually introduced on the shop floor. This difficulty is not merely a current phenomenon. We encountered it back in the late 1920s in the pioneering effort to use control charts on the production floor of the Hawthorne Works of the Western Electric Organization. Neither is it merely an American phenomenon, since I have witnessed the same difficulty in Western Europe and in Japan as well … It is my belief that the failure of the control chart to secure wide acceptance on the factory floor is due mainly to lack of adaptation into the culture of the factory, rather than to technical weaknesses in the control chart … There are a number of problems created by the control chart, as viewed by the shop supervisor:
>
> - The control chart lacks "legitimacy" (i.e., it is issued by a department not recognized as having industrial legislative powers).
> - The control chart conflicts with the specification, leaving the operator to resolve the conflict.
> - The control chart conflicts with the specification, leaving the operator to resolve the conflict.
> - The control chart calls for a pattern of operator action that differs from past practice, but without solving the new problems created as a result of disturbing this past practice.

Legitimacy of the Metrics and the Control Chart

The human passion for "law and order" does not stop at the organization's gate. For example, within a factory there is the same human need for a predictable life, free from unpleasant surprises. Applied to the workforce, this concept of law and order resolves into various principles:

- There must be one and only one personal supervisor (boss) to whom an employee is responsible.
- There is no limit to the number of impersonal bosses (manuals, drawings, routines), but each boss must be legitimate; that is, it must have clear official status.
- When there is a conflict between the orders of the personal boss and an impersonal boss, the former prevails.
- When there is a conflict between something "legitimate" and something not established as legitimate, the former prevails.

Dr. Juran stated, "There can be no quarrel with these principles, since they are vital to law and order on the factory floor ..." Introduction of control charts to the factory floor results in a series of changes in the cultural pattern of the shop:

- A new source of industrial law is opened up, without clear evidence of its legitimacy.
- This new industrial law conflicts with long-standing laws for which there have been no clear repeal through recognized channels of law.
- New sources of factual information are introduced without clear disposition of old sources.
- New duties are created without clear knowledge of their effect on employees who are to perform those duties.

Conclusions

The introduction of modern techniques has an impact on the organization in two aspects:

1. The technical aspect, involving changes in processes, instrument records, and other technical features of the operation
2. The social aspect, involving changes in humans, status, habits, relationships, scale of values, language, and other features of the cultural pattern of the shop

> *The main resistance to change is due to the disturbance of the cultural pattern of the workplace.*
>
> —J. M. Juran

Transforming a Culture

Changing a culture is difficult and usually unsuccessful unless a comprehensive approach exists to achieve and sustain it. The Juran Transformation Model and Roadmap describes five separate and unique types of breakthroughs that must occur in an organization before sustainability is attained. Without achieving these breakthroughs, an organization usually attains good short term results, but the results may not be sustainable for long periods of time. If performance excellence is the state in which an organization attains superior results through the application of the universal quality management universals, then an organization must ensure that these methods are used successfully. The journey from where your organization is to where it wants to go may require a transformational change. This change will result in the ability of the organization to sustain its performance, attain world-class status, and market leadership.

The five breakthroughs are listed as follows:

1. Leadership and management
2. Organization and structure
3. Current performance
4. Culture
5. Agile sustainability

The Juran Transformation Model

The Juran Transformation Model (Fig. 2.1) is based on research from Joseph M. Juran and the Juran Institute has demonstrated that every organization must complete five organizational breakthroughs, when complete, help produce a state of performance excellence. Each breakthrough addresses a specific organizational subsystem that must change. Each is essential for supporting organizational life; none by itself is sufficient. In effect, the breakthroughs all empower the operational subsystem whose mission is to achieve technological proficiency in producing the goods, services and information for which customers will pay for or use. There is some overlap and duplication of activities and tasks among the different breakthrough types. This is to be expected because each subsystem is interrelated with all the others, and each is affected by activities in the others. The authors acknowledge that some issues in each type of breakthrough may have already been

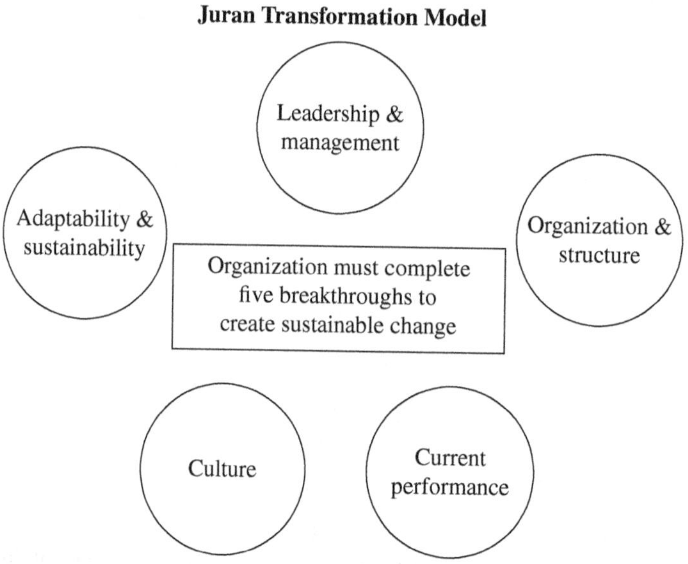

FIGURE 2.1 Juran transformation model. (*Juran Institute, Inc. 2009.*)

addressed by the reader's organization—so much the better. If this is the case, if you did not start your organization's performance excellence journey from the beginning, pick up the journey from where your organization presently finds itself. Closing the gaps will likely be part of your organization's next strategic business planning cycle. To close the gaps, design strategic and operational goals and projects to reach those goals and deploy them to all functions and levels.

Breakthrough and Transformational Change

Breakthroughs can occur in an organization at any time, usually as the result of a specific initiative, such as a specific improvement project (e.g., a Lean or Six Sigma Improvement project; a design of a new service, or the invention of a new technology). These changes can produce sudden explosive bursts of beneficial change for your organization and society. But they may not be enough to cause the culture to change or sustain itself in relation to the changes that occurred. This is because it may not have happened for the right reason. It was not purposeful. It came about through chance. Change by "chance" is not predictable or sustainable. What an organization needs is predictable change.

Today's organizations operate in a state of perpetual, unpredictable change that requires the people in them to produce continuous adaptive improvements as pressure mounts for new improvements to be made from the outside. These improvements may take months or even years to accomplish because it is the cumulative effect of many coordinated and interrelated organizational plans, policies, and breakthrough projects. Taken together, these diligent efforts gradually transform the organization.

Organizations that do not intend to change usually will when a crisis—or a fear of impending crisis—triggers a need for change within an organization. Consider the following scenario.

Two of the largest competitors have introduced new products that are better than ours. Consequently, sales of products X and Y are heading steadily down, and taking our market share along. Our new product introduction time is much slower than the competition, making the situation even worse. The new plant cannot seem to do anything right. Some equipment is often down, and, even when in operation, produces too many costly defective items.

Too many of our invoices are returned because of errors, with the resulting postponement of revenue and a growing number of unsatisfied customers, not to mention the hassle and costs of rework. Accounts receivable have been much too high and are gradually increasing. We are becoming afraid that the future may offer additional threats we need to ward off or, more importantly, plan for so they can be prevented altogether. Leadership must take action, or the organization is going to experience a loss of market share, customer base, and revenue.

Breakthroughs Are Essential to Organizational Vitality

There are four important reasons why an organization cannot survive very long without the medicinal renewing effects of continual breakthrough:

1. *Excess cost due to poor customer satisfaction continues to increase.* The cost associated with poor quality is often too high. One reason is that organizations are plagued by a continuous onslaught of crises precipitated by mysterious sources of chronic high costs of poorly performing processes. As previously stated the total chronic levels of COPQ have been reported to be as high as 20 percent or more of the costs of goods sold. This number varies by type of industry and organization. It is not unusual for these costs at times to exceed profit or be a major contributor to losses. In any case,

the average overall level is appalling (because it is substantial and *avoidable*), and the toll it takes on the organization can be devastating. COPQ is a major driver of many cost-cutting initiatives, not only because it can be so destructive if left unaddressed but also because savings realized by reducing COPQ directly affect the bottom line. Furthermore, the savings continue, year after year, as long as remedial improvements are irreversible, or controls are placed on reversible improvements.

2. It makes good business sense that *mysterious and chronic causes of waste must be discovered, removed, and prevented from returning*. Breakthrough improvement becomes the preferred initial method of attack because of its ability to uncover and remove specific root causes and to hold the gains—it is designed to do just that. One could describe breakthrough improvement methodology as applying the scientific method to solving performance problems. Breakthrough improvement methodology closely resembles the medical model of diagnosis and treatment.
3. *Chronic and continuous change.* Breakthroughs are required for organizational survival because of the state of chronic. Unrelenting change has become so powerful and so pervasive that no constituent part of an organization finds itself immune from its effects for long. Because any or all components of an organization can be threatened by changes in the environment, if an organization wishes to survive, it is most likely to be forced into creating basic changes that are powerful enough to bring about accommodation with new conditions. Performance breakthroughs, consisting as it does of several specific types of breakthrough in various organization functions, is a powerful approach that is capable of determining countermeasures sufficiently effective to prevail against the inexorable forces of change. An organization may have to reinvent itself. It may even be driven to reexamine, and perhaps modify, its core products, business, service, or even its customers.
4. *Without continuous improvement, organizations die.* Another reason why breakthroughs are essential for organizational survival is found in knowledge derived from scientific research into the behavior of organizations. Leaders can learn valuable lessons about how organizations function and how to manage them by examining open systems theory. Among the more important lessons taught by this theory is the notion of *negative entropy*. Negative entropy refers to characteristics that human organizations share with biological systems such as the living cell, or the living organism (which is a collection of cells). *Entropy* is the tendency of all living things—and all organizations—to head toward their own extinction. Negative entropy consists of countermeasures that living systems and social systems take to stave off their own extinction. Organisms replace aging cells, heal wounds, and fight disease. Organizations build up reserves of energy (backlogs and supplies) and constantly replace expended energy by acquiring more energy (sales and raw materials) from their environment. Eventually, living organisms lose the race. So do organizations if they do not continually adapt, heal "wounds" (make performance breakthrough improvements), and build up reserves of cash and goodwill. The Juran Transformation Model is a means by which organizations can stave off their own extinction.

Systems Thinking and Transformational Change

Organizations are like living organisms. They consist of a number of subsystems, each of which performs a vital specialized function that makes specific, unique, and essential

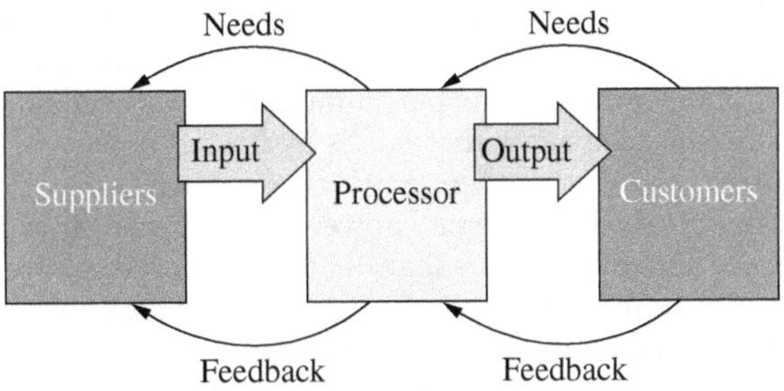

Figure 2.2 The triple role. (*Juran Institute, Inc. 2009, p. 8.*)

contributions to the life of the whole. A given individual subsystem is devoted to its own specific function such as design, production, management, maintenance, sales, procurement, and adaptability. One cannot carry the biological analogy very far because living organisms separate subsystems with physical boundaries and structures (e.g., cell walls, the nervous system, the digestive system, the circulatory system, etc.). Boundaries and structure of subsystems in human organizations, on the other hand, are not physical; they are repetitive events, activities, and transactions. The repetitive patterns of activities are, in effect, the work tasks, procedures, and processes carried out by organizational functions. Open systems theorists call these patterns of activities roles. A role consists of one or more recurrent activities out of a total pattern of activities, which, in combination, produce the organizational output.

Roles are maintained and carried out in a repetitive, relatively stable manner by means of mutually understood sets of expectations and feedback loops, shown in Fig. 2.2. The Triple Role Open Systems theory and Juran's model focuses particularly on the technical methods, human relationships, organization structures, and interdependence of functional roles associated with these activities and transactions. Detailed knowledge of the repetitive transactions between the organization and its environment, and also within the organization itself, is essential in accomplishing breakthroughs because these transactions determine the effectiveness and efficiency of performance.

Figure 2.2 shows a model that applies equally to an organization as a whole, to individual subsystems and organizational functions (e.g., departments and workstations within the organization, and to individual organizational members performing tasks in any function or level). All these entities perform three more or less simultaneous roles, acting as supplier, processor, and customer. Acting as a processor, charged with the duty of transforming imported energy, organizations receive raw materials—goods, information, and/or services—from their suppliers, who may be located inside or outside the organization. The processor's job consists of transforming the received things into a new product of some kind—goods, information, or service. In turn, the processor supplies the product to his or her customers who may be located within or outside the organization.

Each of these roles requires more than merely the exchange of things. Each role is linked by mutually understood expectations (i.e., specifications, work orders, and procedures) and feedback as to how well the expectations are being met (i.e., complaints, quality reports, praise, and rewards). Note that in the diagram, the processor must communicate (shown by arrows) to the supplier a detailed description of his or her needs and requirements. In addition, the processor provides the supplier with feedback on the extent to which the expectations are being met. This feedback is part of the control loop and helps to ensure consistent adequate performance by the supplier. The customer bears the same responsibilities to his or her processors who, in effect, are also suppliers (not of the raw materials but of the product).

When defects, delays, errors, or excessive costs occur, causes can be found somewhere in the activities performed by suppliers, processors, and customers, in the set of transactions between them, or perhaps in gaps in the communication of needs and feedback. Breakthrough efforts must uncover the precise root causes by deep probing and exploration. If the causes are really elusive, discovering them may require placing the offending repetitive process under a microscope of unprecedented power and precision, as is done in Six Sigma. Performance excellence initiatives require that all functions and levels be involved, at least to some extent, because each function's performance is interrelated and dependent to some degree on all other functions. Moreover, a change in the behavior of any one function will have some effect on all the others, even though it may not be apparent at the time. This interrelatedness of all functions has practical day-to-day implications for a leader at any level, that is, the imperative of using "systems thinking" when making decisions, particularly decisions to make changes.

Because an organization is an open system, its life depends on (1) successful transactions with the organization's external environment and (2) proper coordination of the organization's various specialized internal functions and their outputs.

The proper coordination and performance of the various internal functions is dependent on the management processes of planning, controlling, and improving and on human factors such as leadership, organizational structure, and culture. To manage in an open system (such as an organization), management at all levels must think and act in systems terms. Managers must consider the impact of any proposed change not only upon the whole organization but also the impact on the interrelationships of all the parts. Failure to do so, even when changing seemingly little things, can make some pretty big messes. Leaders need to reason as follows: "If there is to be a change in x, what is required (inputs) from all functions to create this change, and how will x affect each of the other functions, and the total organization as well (ultimate output/results)?" Organizations will not change until the people in them change, regardless of the breakthrough approach.

There are three important lessons learned from the experience of the authors:

1. *All organizations need a systematic approach to ensure that change happens.* The problems that appear in one function or step in a process often have their origin upstream from that function or step in the process. People in a given work area cannot necessarily solve the problem in their own work area by themselves—they need to involve others in the problem-solving process. Without systematic involvement of the other functions, suboptimization will occur. Suboptimization results in excess costs and internal customer dissatisfaction—the exact opposite of what is intended.
2. *Change can only be created with active participation of all employees from the top on down and over time.* This includes not only individuals who are the source of a problem but also those affected by the problem and those who will initiate changes to remedy the problem (usually those who are the source of the problem, and perhaps others).
3. *Functional change alone is not sufficient to transform an organization.* Breakthroughs attempted in isolation or within a structure from the whole organization and without systems thinking can easily create more problems than existed at the start of the breakthrough attempt.

Attempts to bring about substantial organizational change such as performance excellence requires not only changing the behavior of individuals (as might be attempted by training) but also of redefining their roles in the social system. This requires, among other things, changing the expectations that customers have for their processors and changing expectations that processors have for their suppliers. In other words, performance breakthroughs require a capability of organizational design to produce consistent, coordinated behavior to support specific organizational goals. Modifications will likely also be made to other elements that define roles such as job descriptions, job fit, work procedures, control plans, other elements of the quality system, and training. To achieve a breakthrough, it is not sufficient simply to train a few Black Belts in the martial arts as experts and complete a few projects. Although this will probably result in some improvement, it is unlikely to produce long-term culture change and sustainability. The authors believe that too many organizations are settling for simple improvements when they should be striving for breakthroughs.

As we have seen, attaining a performance excellence state consists of achieving and sustaining beneficial changes. It is noteworthy that having a bright idea for a change does not, by itself, make change actually happen. People must understand why the change is needed and see the impact it will have on them before they can change what they do and perhaps how they do it. The very persons who could benefit most from it, especially if they have been successful in doing things the usual way often resist beneficial changes, sometimes. Leading change can be a perplexing and challenging undertaking. Accordingly, individuals trying to implement change should acquire know-how in how to do it.

Breakthroughs in Leadership

Breakthroughs in leadership occur when managers answer two basic questions:

1. How does management set performance goals for the organization and motivate the people in the organization to reach them and be held accountable?
2. How do managers best use the power of the workforce and other resources in the organization and how should they best manage them?

Issues with leadership are found at *all* levels, not just at the top of an organization. A breakthrough in leadership and management results in an organization characterized by unity of purpose and shared values as well as a system that enables engagement of the workforce.

Each work group knows what its goals are and, specifically, what performance is expected from the team and the individuals. Each individual knows specifically what he or she is to contribute to the overall organizational mission and how his or her performance will be measured. Few erratic or counterproductive behaviors occur. Should such behaviors occur, or should conflict arise, guidelines to behavior and decision making are in place to enable relatively quick and smooth resolution of the problem. There are two major elements to leadership: (1) leaders must decide and clearly communicate where they want their employees to go; and (2) leaders must entice them to follow the path by providing an understanding of why this is a better way. In this handbook, the words "leader" and "manager" do not necessarily refer to different persons. Indeed, most leaders are managers, and managers should be leaders. The distinctions are matters of intent and activities, not players. Leadership can and should be exercised by managers; leaders also need to manage. If leadership consists of influencing others in a positive manner that attracts others, it follows that those at the top of the managerial pyramid (CEOs and C-suite) can be the most effective leaders because they possess more formal authority than anyone else in an organization. In fact, top managers are usually the most influential leaders. If dynamics change, such as introducing Lean Six Sigma into an organization, the most effective approach by far is for the CEO to lead the charge. Launching Lean Six Sigma is helped immensely if other leaders, such as union presidents also lead the charge. The same can be said if senior and middle managers, first-line supervisors, and leaders of nonmanagement work crews "follow the leader" and support a performance excellence program by word and actions. Leadership is not dictatorship because dictators make people afraid of behaving in "incorrect" ways, and perhaps they occasionally provide public treats (free gasoline, for example, as has happened in Turkmenistan); freeing prisoners; or staging public spectacles that, together with propaganda, are designed to make people follow the leader. Dictators do not necessarily inspire desired behavior from their followers but rather motivate through fear.

The Roles of Managers to Attain a Breakthrough in Leadership

Strategic Planning and Deployment: Moving from Good to Great

The first step in strategic planning is to determine the organization's mission. (What business are we in? What services do we provide?) Next, a vision for the desired future state of the organization is formulated and published (e.g., "We will become the supplier of choice, worldwide, of product X or service Y"). After proclaiming the basic reason for the organization's existence, and the overall general goal the organization seeks to achieve in the future, senior management generates a few key strategies the organization is to implement to fulfill the mission and realize the vision (e.g., ensure a reliable source of high-quality raw materials, ensure a stable well-qualified workforce at all levels for the foreseeable future, and/or reduce our overall costs of poor quality by 50 percent of last year's annual cost by the end of the year). Now the process becomes more precise. For each key strategy, a small number of quantified strategic goals (targets) are listed; the resources and people available can accomplish a few that. These quantified strategic goals are further divided into goals for this year, goals for the next 2 years, and so on. Finally, for each quantified strategic goal, a practical number of operational goals are established that describe exactly who is to do what to reach each specific strategic goal. Normally, operational goals are specific projects to be accomplished (such as Six Sigma projects), specific performance targets to be reached by each function or work group, for example.

Strategic deployment is the process of converting goals into specific precise actions, each action designed to realize a specific goal. Deployment occurs in two phases: one phase is *during* the strategic planning process; the other phase is *after* the strategic plan is completed. During the strategic planning process, after the management team determines its key strategies, the team circulates these strategies to others in the organization: department heads, functional heads, process owners, and the like. They, in turn, may circulate the strategies further out to supervisors, team leaders, and so on. These individuals, in turn, may circulate the strategies to everyone they supervise. Each party is asked to contribute ideas and suggestions concerning what activities could be undertaken to carry out the strategies, what the specific quantified strategic goals should be, and what resources would be required. These responses are conveyed to the senior management team, who use the responses to promulgate more specific strategic and operational goals. This exchange of proposed activities to reach goals may take place several times. Some individuals call these iterations and reiterations "catch ball." With each cycle, various goals are refined, becoming more specific, more practical, and quantified. Finally, a set of precise strategic and operational goals emerges, each with owners. In addition, which to measure performance toward goals and to provide managers at all levels with a scorecard of progress devises metrics. Most significantly, these goals have been established with the participation of leaders who will be responsible for carrying them out and are accountable for the results.

Emerging from this event is an organization united in its commitment to reaching the same goals. All functions and levels have been included. This is highly significant because leadership is not considered to be something exercised by one person at the top of an organization. It is ideally performed at any level and in any function, by anyone who influences others. With a well-deployed strategic plan, specific acts of leadership (attempts to influence others) should be relatively consistent from leader to leader, from function to function, and from time to time. Decisions made at different levels or in different functions should not conflict with one another very often. That, at least, is the ideal.

Providing Employee Engagement and Empowerment

When managers do everything they can to provide the means for everyone to be engaged in improving the organization it leads to empowered employees. This empowered state is often called the state of self-control, which will greatly enhance their credibility and the level of trust followers will feel toward them. This will happen because when an individual is empowered to be in self-control, that individual has at his or her disposal all of the elements necessary to be successful on his or her job. When a leader does this, followers will feel gratitude and respect toward that leader and will be inclined to follow that leader because "My leader comes through for me. My leader doesn't just talk; my leader delivers!"

A brief review of these elements follows because they can be so instrumental in demonstrating leadership. A person is in a state of self-control if that person

- *Knows exactly what is expected.* The standard of performance for the process; who does what and who decides what and to know how he or she is doing compared to the standards
- Receives timely feedback to have the ability to *regulate the process*
- Has a capable process that includes the necessary tools, equipment, materials, maintenance, time, and the authority to *adjust the process when it is nonconforming*

A person in a state of self-control has, at his or her disposal, all the means necessary to perform work tasks successfully. Management must provide the means because only management controls the required resources needed to put someone in self-control. Persons who have long been suffering from lack of self-control and its associated inability, through no fault of their own, to perform as well as they would like, are especially grateful to a leader/manager who relieves them from the suffering by making self-control possible. These persons come to respect and trust such a manager, and they tend to become an enthusiastic follower of that manager, mindful of the good things—including enhanced self-confidence and self-esteem—that have flowed from that manager.

Performing Periodic Audits

Conducting periodic audits performed by leaders and managers is a superb method of demonstrating commitment to and support for an effort to change. Leaders and managers, especially senior executive managers, enhance their credibility and power to lead by personally walking around the organization and talking to the people about what they do, and how they do it. A management audit has both formal and informal aspects. The formal aspect consists of asking each person being audited to answer certain specific written questions and to produce data and other evidence of performance that conforms to the formal controls. The informal aspect is simply talking with the people who are being audited about what is on their mind and sharing with them what is on the manager's mind. The management audit is roughly the equivalent of senior generals visiting the troops in the field. It is a chance for managers to demonstrate their interest in how things are going: what is going well and/or what needs corrective action. It is a splendid opportunity to listen to what people have to say and to show respect for them. If managers follow up on the suggestions and complaints they hear, that is yet another way to demonstrate that they care enough about "the troops" to provide them with needed support and assistance. It grants to anyone in the organization a direct line of communication with the top, something that makes many people feel important, and motivates them to keep performing at their best. Importantly, the managers' ability to lead is reinforced.

Conferring Public Rewards and Recognition

Leaders can assist their followers to take desired new norms and patterns of behavior upon themselves as their own, if doing so is rewarding to the followers consistently and over time. The effect of rewards and recognition can be magnified when

- Rewards and recognition are awarded in public, with fanfare and ceremony
- Leaders are in the presence of individuals whose behavior the leader is seeking to influence
- The award is accompanied by an explanation of its connection to a specific desired new behavior toward which the leader is attempting to extract from the followers. For example, after launching a Six Sigma initiative, your organization decides to have an all-organization special assembly to recognize the seven original Six Sigma project teams. Each team makes a presentation of its just-completed project, complete with slides, handouts, and exhibits

Carrying Out the Nondelegable Managerial Practices

- Creating and serving on an executive council to lead and coordinate the performance breakthrough activities
- Forming policy to allow time to participate in breakthrough teams
- Establishing organizational infrastructure
- Providing resources (especially time)
- Reviewing the progress of performance toward goals, including the progress of projects
- Removing obstacles, dissolving resistance, and providing support and other corrective action if progress is too slow

Provide Resources to Continually Innovate and Improve

Teams, project by project, produce breakthroughs. These teams are assigned goals to attain by using project charters. Each project is formally chartered, in writing, by the Executive Council. The Executive Council also provides the project teams with the people and other resources the teams need to carry out their missions.

The manager's role is one of managing the organization so that high standards are met, proper behavior is rewarded—or individuals are held accountable—facilities and processes are maintained, and employees are motivated and supported. Performance toward goals is measured and tracked for all functions and levels (i.e., overall organization, function, division, department, work group, and individual). Performance metrics are regularly summarized, reported, and reviewed to compare actual performance with goals. Management routinely initiates corrective action to address poor performance or excessively slow progress toward goals. Actions may include establishing performance breakthrough improvement projects, providing additional training or support, clearing away resistance, providing needed resources, and performing disciplinary action. Leaders and management must do the following:

- Create and maintain systems and procedures that ensure the best, most efficient, and effective performance of an organization in all functions and levels
- Reward (and hold people accountable, if necessary) appropriate behavior
- Consistently uphold and demonstrate high standards
- Focus on stability

Breakthroughs in Organizational Structure

Creating a breakthrough in organizational structure does the following:

- Designs and puts into place the organization's operational systems (i.e., quality management system, clear vision and strategy, leadership participation, orientation of new employees, training, communication processes, and supply chains)
- Designs and puts into practice a systematic structure that integrates each function with all the others and sets forth relative authority levels and reporting lines (e.g., organization charts and the means to manage across it)
- Aligns and coordinates the respective interdependent individual functions into a smooth functioning, integrated organization

Creating a breakthrough in organization structure is a response to the basic question: "How do I set up organizational structures and processes to reap the most effective and efficient performance toward our goals?"

Trends in this area are clear. Project teams perform more and more work. Job tasks may be described by *team project* descriptions rather than, or in addition to, *individual* job descriptions. Performance evaluation is often related to the accomplishments of one's team instead of or in addition to, one's individual accomplishments.

Management structure consists of cross-functional *processes* that are managed by process owners, as well as vertical *functions* that are managed by functional managers. Where both vertical and horizontal responsibility exists, potential conflicts are resolved by matrix mechanisms that require negotiated agreements by the function manager and the cross-functional (horizontal) process owner.

Unity and consistency in the operation of *both* cross-functional processes and vertical functions is essential to creating performance breakthroughs and is essential to continued organizational survival. All members of leadership teams at all levels simply must be in basic agreement as to goals, methods, priorities, and styles. This is especially vital when attempting performance breakthrough improvement projects because the causes of so many performance problems are cross functional, and the remedies to these problems must be designed and carried out cross functionally. Consequently, one sees in a Lean or Six Sigma implementation, for example, quality or executive councils, steering committees, champions (who periodically meet as a group), cross-functional project teams, project team leaders, Black Belts, and Master Black Belts. These roles all involve dealing with change and teamwork issues. There is also a steady trend toward fewer authority or administrative levels and shorter reporting lines.

The rate of change in the business world is not going to slow down anytime soon. If anything, competition in most industries will probably speed up over the next few decades. Enterprises everywhere will be presented with even more terrible hazards and wonderful opportunities, driven by the globalization of the economy along with related technological and social trends (John P. Kotter 1996).

There are three accepted basic types of organization for managing any function work and one newer, emerging approach. The most traditional and accepted organization types are functional, process, and matrix. They are important design baselines because these organizational structures have been tested and studied extensively and their advantages and disadvantages are well known. The newer, emerging organizational designs are network organizations.

Function-Based Organization

In a function-based organization, departments are established based on specialized expertise. Responsibility and accountability for process and results are usually distributed piecemeal among departments. Many firms are organized around functional departments that have a well-defined management hierarchy. This applies both to the major functions (e.g., human resources, finance, operations, marketing, and product development) and also to sections within a functional department. Organizing by function has certain advantages—clear responsibilities and efficiency of activities within a function. A function-based organization typically develops and nurtures talent and fosters expertise and excellence within the functions.

Therefore, a function-based organization offers several long-terms benefits. However, this organizational form also creates "walls" between the departments. These walls—sometimes visible, sometimes invisible—often cause serious communications barriers. However, function-based organizations can result in a slow, bureaucratic decision-making apparatus as well as the creation of functional business plans and objectives that may be inconsistent with overall strategic business unit plans and objectives. The outcome can be efficient operations *within* each department but with less-than-optimal results delivered to external (and internal) customers.

Process-Managed Organizations

Many organizations are beginning to experiment with an alternative to the function-based organization in response to today's "make it happen fast" world. Businesses are constantly redrawing their lines, work groups, departments, and divisions, even entire companies, trying to increase productivity, reduce cycle time, enhance revenue, or increase customer satisfaction. Increasingly, organizations are being rotated 90 degrees into process-based organizations.

In a process organization, reporting responsibilities are associated with a process, and accountability is assigned to a process owner. In a process-based organization, each process is provided with the functionally specialized resources necessary.

This eliminates barriers associated with the traditional function-based organization, making it easier to create cross-functional teams to manage the process on an ongoing basis.

Process-based organizations are usually accountable to the business unit or units that receive the benefits of the process under consideration. Therefore, process-based organizations are usually associated with responsiveness, efficiency, and customer focus.

However, over time, pure process-based organizations run the risk of diluting and diminishing the skill level within the various functions. Furthermore, a lack of process standardization can evolve, which can result in inefficiencies and organizational redundancies. Additionally, such organizations frequently require a matrix-reporting structure, which can result in confusion if the various business units have conflicting objectives. The matrix structure is a hybrid combination of functional and divisional archetypes.

Merging Functional Excellence with Process Management

What is required, however, is an organization that identifies and captures the benefits of supply chain optimization in a responsive, customer-focused manner while promoting and nurturing the expertise required managing and continuously improving the processes on an ongoing basis.

This organization will likely be a hybrid of functional and process-based organizations, with the business unit accountable for objectives, priorities, and results, and the functional department accountable for process management and improvement and resource development.

According to the late Dr. Frank Gryna, the Center for Quality at the University of Tampa, Florida, the organization of the future will be influenced by the interaction of two systems that are present in all organizations: the technical system (equipment, procedures) and the social system (people, roles)—thus the name "sociotechnical systems" (STSs).

Much of the research on sociotechnical systems has concentrated on designing new ways of organizing work, particularly at the workforce level. For example, supervisors are emerging as "coaches"; they teach and empower rather than assign and direct. Operators are becoming "technicians"; they perform a multi-skilled job with broad decision making, rather than a narrow job with limited decision making. Team concepts play an important role in these new approaches. Some organizations now report that, within a given year, 40 percent of their people participate on a team; some organizations have a goal of 80 percent. Permanent teams (e.g., process team, self-managing team) are responsible for all output parameters, including quality; ad hoc teams (e.g., a quality project team) are typically responsible for improving quality. The literature on organizational forms in operations and other functions is extensive and increases continuously. For a discussion of research conducted on teams, see Katzenbach and Smith (1993). Mann (1994) explains how managers in process-oriented operations need to develop skills as coaches, developers, and "boundary managers." The attributes associated with division managers, functional managers, process managers, and customer service network managers are summarized in Table 2.1. There is emerging evidence

Attributes of Roles	Division Manager	Function Manager	Process Manager	Network Leader
Strategic orientation	Entrepreneurial	Professional	Cross-functional	Dynamic
Focus objectives	Customer adaptability	Internal efficiency	Customer effectiveness	Variable adaptability, speed
Operational responsibility	Cross-functional	Narrow, parochial	Broad, pan-organizational	Flexible
Authority	Less than responsibility	Equal to responsibility	Equal to responsibility	Ad hoc, based on leadership
Interdependence	May be high	Usually high	High	Very high
Personal style	Initiator	Reactor	Active	Proactive
Ambiguity of task	Moderate	Low	Variable	Can be high

(*Sources:* The first two columns are adapted from the work of Financial Executive Research Foundation, Morristown, NJ. The last two columns represent the work of Edward Fuchs.)

TABLE 2.1 Attributes of Various Roles

that divisional and functional organizations may not have the flexibility to adapt to a rapidly changing marketplace or to technological changes.

Design a system that promotes employee empowerment and involvement. Traditional management was based on Frederick Taylor's teachings of specialization. At the turn of the twentieth century, Taylor recommended that the best way to manage manufacturing organizations was to standardize the activity of general workers into simple, repetitive tasks and then closely supervise them (Taylor 1947). Workers were "doers"; managers were "planners." In the first half of the twentieth century, this specialized system resulted in large productivity increases and a very productive economy. As the century wore on, workers became more educated, and machinery and instruments more numerous and complicated. Many organizations realized the need for more interaction among employees. The training and experience of the workforce was not being used. Experience in team systems, where employees worked together, began in the latter half of the twentieth century, although team systems did not seriously catch on until the mid-1970s as pressure mounted on many organizations to improve performance. Self-directed teams began to emerge in the mid-1980s. For maximum effectiveness, the work design should require a high level of employee involvement.

Engagement and Empowerment

Workers who have been working under a directive command management system where the boss gives orders and the worker carries them out cannot be expected to adapt instantly to a highly participative, high-performance work system. There are too many new skills to learn and too many old habits to overcome. According to reports from numerous organizations that have used high-performance work systems, such systems must evolve. This evolution is carefully managed, step by step, to prepare team members for the many new skills and behaviors required of them.

The first stage of involvement is the consultative environment, in which the manager consults the people involved, asks their opinions, discusses their opinions, and then takes unilateral action. A more advanced state of involvement to appoint a special team or project team to work on a specific problem, such as improving the cleaning cycle on a reactor. This involvement often produces in team members' pride, commitment, and sense of ownership.

An example of special quality teams is the "blitz team" from St. Joseph's Hospital in Paterson, NJ. Teams had been working for about a year as a part of the total quality management (TQM) effort there. Teams were all making substantial progress, but senior management was impatient because the TQM was moving too slowly. Recognizing the need for the organization to produce quick results in the fast-paced marketplace, the team developed the blitz team method (from the German word for lightning). The blitz team approach accelerated the standard team problem-solving approach by adding the services of a dedicated facilitator. The facilitator reduced elapsed time in three areas: problem-solving focus, data processing, and group dynamics.

Because the facilitator was very experienced in the problem-solving process, the team asked the facilitator to use that experience to provide more guidance and direction than is normally the style on such teams. The result was that the team was more focused on results and took fewer detours than usual. In the interest of speed, the facilitator took responsibility for the processing of data between meetings, thus reducing the time that elapsed between team meetings. Furthermore, the facilitator managed the team dynamics more skillfully than might be expected of an amateur in training within the organization. The team went from first meeting to documenting root causes in 1 week. Some remedies were designed and implemented within the next few weeks.

The team achieved the hospital's project objectives by reducing throughput delays for emergency room (ER) patients. ER patients are treated more quickly, and worker frustrations have been reduced (Niedz 1995). Special teams can focus sharply on specific problems. The team's success depends on assigning team people who are capable of implementing solutions quickly.

Breakthrough Project Teams and Daily Value Engagement Team

Employees need time to improve process. Unfortunately unless they are given time nothing really happens. To this end management must create the time by establishing "team." Time is necessary to organize and make sure the team members know what they are doing, why they are doing it, how to organize the work, and who will be involved. However, schedules are so short; there is never time to organize the work team.

Many managers create false teams which start working, believing they know what to do, and taking direct action. They do not have time to get support from others, or to determine the right goals for the team, or to create and implement a plan that allows them to achieve the proposed goals or to plan how to work together. Nevertheless, what these teams all have in common is that no one understands in the same way what they are doing, why, how, and with whom. Linking the effort of the team to five critical success factors can solve this problem.

Management must have employees participate on cross-functional breakthrough team such as Lean and Six Sigma teams or allow them to participate in daily functional teams or as we like to call them *value engagement teams*. VET are small teams of people that continually focus on assuring their processes are effective and efficient.

Leadership Style

Empowered team members share leadership responsibilities, sometimes willingly and sometimes reluctantly. Decision making is more collaborative, with consensus as the objective. Teams work toward win-win agreements. Teamwork is encouraged. Emphasis is more on problem solution and prevention, rather than on blame. During a visit to Procter & Gamble's plant in Foley, FL, the host employee commented that in the past he would not have believed he would ever be capable of conducting this tour. His new leadership roles had given him confidence to relate to customers and other outsiders.

Positivity and Citizenship

Honesty, fairness, trust, and respect for others are more readily evident. In mature teams, members are concerned about each other's growth in the job (i.e., members reaching their full potential). Members share their experiences more willingly and coach each other, as their goal is focused on the team success, rather than on their personal success. Members recognize and encourage each other's (and the team's) successes more readily.

Reasons for High Commitment

As previously stated, empowered team members have the authority, capability, desire, and understanding of the organization's goals. In many organizations, they believe that this makes members feel and behave as if they were owners and makes them more willing to accept greater responsibility. Empowered team members also have greater knowledge, which further enhances their motivation and willingness to accept responsibility.

Means of Achieving High Performance

It has been observed that as employees accept more responsibility and have more motivations, and greater knowledge, they freely participate more toward the interests of the business. They begin to truly act like owners, displaying greater discretionary effort and initiative. Empowered team members have the authority, the capability, and the desire to understand the organization's direction. Consequently, members feel and behave as if they were owners and are willing to accept greater responsibility. They also have greater knowledge, which further enhances their motivation and willingness to accept responsibility.

Enough progress has been made with various empowered organizations that we can now observe some key features of successful efforts. These have come from experiences of various consultants, visits by the authors to other companies, and published books and articles. These key features can help us learn how to design new organizations or redesign old ones to be more effective. The emphasis is on key features, rather than a prescription of how each organization is to operate in detail. This list is not exhaustive, but it is a helpful checklist, useful for a variety of organizations.

Focus on External Customers

The focus is on the external customers, their needs, and the products or services that satisfy those needs when

- The organization has the structure and job designs in place to reduce variation in process and product
- There are few organizational layers
- There is a focus on the business and customers
- Boundaries are set to reduce variances at the source
- Networks are strong
- Communications are free flowing and unobstructed
- Supplier and customer input are used to manage the business
- Employees understand who the critical customers are, what their needs are, and how to meet customer needs with their own actions. Thus, all actions are based on satisfying the customer. The employees (e.g., operator, technicians, and plant manager) understand that they work for the customer rather than for the plant manager

In empowered organizations, managers create an environment to make people great, rather than control them. Successful managers are said to "champion" employees and make them feel good about their jobs, their organization, and themselves. When he was head of the Nissan plant in Smyrna, TN, Marvin Runyon stressed, "management's job is to provide an environment in which people can do their work" (Bernstein 1988).

Organization and Knowledge Management

Broken down into its simplest form, the learning process consists of observation–assessment–design–implementation, which can vary along two main dimensions:

1. *Conceptual learning*. The process of acquiring a better understanding of cause and effect relationship, leading to "know-why."
2. *Operational learning*. The process of obtaining validation of action outcome links, leading to "know-how."

Professor M. Lapré, Assistant Professor of Operations Management at Owen Graduate School of Management at Vanderbilt University, Nashville, TN, and L. Van Wassenhove, the Henry Ford Chaired Professor of Manufacturing at INSEAD (Institut Européen d'Administration des Affaires), a multicampus international graduate business school and research institution, show that it is possible to accelerate factories' learning curves through focused quality and productivity improvement efforts.

Breakthroughs in Current Performance

Breakthroughs in current performance (or improvement) do the following:

- Significantly improve current levels of results that an organization is attaining. This happens when a systematic project-by-project improvement system of discovering root causes of current chronic problems and implements solutions to eliminate them.
- Devise changes to the "guilty" processes and reduce the costs of poorly performing processes.
- Install new systems and controls to prevent the return of these root causes.

A system to attain breakthroughs in current performance addresses the question "How do we reduce or eliminate things that are wrong with our products or processes, and the associated customer dissatisfaction and high costs (waste) that consumes the bottom line?" A breakthrough improvement program addresses *quality* problems—failures to meet specific important needs of specific customers, internal and external. (Other types of breakthroughs address other types of problems.) Lean, Six Sigma, Lean Six Sigma, Root Cause Corrective Action, and other programs need to be part of a systematic approach to improve current performance. These methods address a few specific types of things that always go wrong:

- Excessive number of defects
- Undue number of delays
- Unnecessary long cycle times
- Unwarranted costs of the resulting rework, scrap, late deliveries, dissatisfied customers, replacement of returned goods, loss of customers, and loss of goodwill

Lean and Six Sigma teams are all methods to improve performance. They are all project based and require multifunctional teams to improve current levels of performance. Each requires a systematic approach to complete the projects.

A systematic approach to improving performance of processes is to

- Define the problem (performed by the champions and executive council)
- Measure (performed by the project team)
- Analyze (performed by the project team)
- Improve (performed by the project team, often with help of others)
- Control (performed by the project team and the operating forces is)

Breakthroughs in current levels of performance problems are attained using these methods. The Lean and Six Sigma method will place your ailing processes under a microscope of unprecedented precision and clarity and make it possible to understand and control the relationships between input variables and desired output variables.

Your organization does have a choice as to what "system" to bring to bear on your problems: a "conventional" weapon system (quality improvement) or a "nuclear" system (Six Sigma). The conventional system is perfectly effective with many problems and much cheaper than the more elaborate and demanding nuclear system. The return on investment is considerable from both approaches, but especially so from Six Sigma if your customers are demanding maximum quality levels.

Breakthroughs in current performance solve problems such as excessive number of defects, excessive delays, excessively long time cycles, and excessive costs.

Breakthroughs in Culture

The result of completing many improvements creates a habit of improvement in the organization. Each improvement starts to create a quality culture because collectively it does the following:

- Creates a set of new behavior standards and social norms that best supports organizational goals and climate
- Instills in all functions and levels the values and beliefs that guide organizational behavior and decision making
- Determines organizational cultural patterns such as style (e.g., informal versus formal, flexible versus rigid, congenial versus hostile, entrepreneurial/risk-taking versus passive/risk adverse, rewarding positive feedback versus punishing negative feedback), extent of internal versus external collaboration, and high energy/morale versus low energy/morale

Performance breakthrough in culture is a response to the basic question: "How do I create a social climate that encourages organization members to align together eagerly toward the organization's performance goals?"

As employees continue to see their leadership "sticking to it" culture change happens. An organization is not yet at a sustainable level for transformational change. There are still issues that must be addressed, including

- Reviewing the organization's vision, mission, and values
- Orienting new employees and training practices
- Rewarding and recognizing policies and practices
- Human resource policies and administration
- Quality and customer satisfaction policies
- Fanatic commitment to customers and their satisfaction
- Commitment to continuous improvement
- Standards and conduct codes, including ethics
- No "sacred cows" regarding people, practices, and core business content
- Community benefit and public relations

An organization's culture exerts an extraordinarily powerful impact on organizational performance. The culture determines what is right or wrong, what is legitimate or illegitimate, and what is acceptable or unacceptable. Consequently, a breakthrough in the culture is profoundly influential in achieving a performance breakthrough. It is also probably the most difficult and time-consuming breakthrough to make happen. It is also so widely misunderstood that attempts to pull it off often fail.

A breakthrough in culture (1) creates a set of behavior standards, and a social climate that supports organizational goals, (2) instills in all functions and levels the values and beliefs that guide organizational behavior and decision making, and (3) determines organizational cultural patterns such as *style* (informal versus formal, flexible versus rigid, authoritarian top-down versus participative collaboration, management driven versus leadership driven, and the like), the organization's *caste system* (the relative status of each function), and the *reward structure* (who is rewarded for doing what).

How Are Norms Acquired?

New members of a society—a baby born into a family or a new employee hired into the workplace—are carefully taught who is who and what is what. In short, these new members are taught the norms and the cultural patterns of that particular society. In time, they discover that complying with the norms and cultural patterns can be satisfying and rewarding. Resistance or violation of the norms and cultural patterns can be very dissatisfying because it brings on disapproval, condemnation, and possibly punishment. If an individual receives a relatively consistent pattern of rewards and punishments over time, the beliefs and the behaviors being rewarded gradually become a part of that individual's personal set of norms, values, and beliefs. Behaviors that are consistently disapproved or punished will gradually be discarded and not repeated. The individual will have become socialized.

How Are Norms Changed?

Note that socialization can take several years to take hold. This is an important prerequisite for successfully changing an organization's culture that must be understood and anticipated by agents of change, such as senior management. The old patterns must be extinguished and replaced by new ones. This takes time and consistent, persistent effort. These are the realities. Consider what the anthropologist Margaret Mead has to say about learning new behaviors and beliefs:

> An effective way to encourage the learning of new behaviors and attitudes is by consistent prompt attachment of some form of satisfaction to them. This may take the form of consistent praise, approval, privilege, improved social status, and strengthened integration with one's group, or material reward. It is particularly important when the desired change is such that the advantages are slow to materialize—for example, it takes months or even years to appreciate a change in nutrition, or to register the effect of a new way of planting seedlings in the increased yield of an orchard. Here the gap between the new behavior and results, which will not reinforce the behavior until they are fully appreciated, has to be filled in other ways.

She continues:

> The learning of new behaviors and attitudes can be achieved by the learner's living through a long series of situations in which the new behavior is made highly satisfying—without exception if possible—and the old not satisfying.
>
> New information psychologically available to an individual, but contrary to his customary behavior, beliefs, and attitudes, may not even be perceived. Even if he is actually forced to recognize its existence, it may be rationalized away, or almost immediately forgotten.

As an individual's behavior, beliefs, and attitudes are shared with members of his cultural group, it may be necessary to effect a change in the goals or systems of behavior of the whole group before any given individual's behavior will change in some particular respect.

This is particularly likely to be so if the need of the individual for group acceptance is very great—either because of his own psychological makeup or because of his position in society.

To achieve effective breakthroughs in culture, the entire management team at all levels must share, exhibit, and reinforce desired new cultural norms and patterns of behavior—and the norms must be consistent, uninterrupted, and persistent.

Do not expect cultural norms or behavior to change simply because you publish the organization's stated values in official printed material or describe them in speeches or exhortations. Actual cultural norms and patterns may bear no resemblance at all to the values described to the public or proclaimed in exhortations. The same is true of the actual flow of influence compared to the flow shown on the organization chart. (New employees rapidly learn who is really who and what is really what, in contrast to and in spite of the official publicity.)

A forceful leader-manager can, by virtue of his or her personality and commitment, influence the behavior of individual followers in the *short term* with rewards, recognition, and selective exclusion from rewards. The authors know of organizations that, in introducing a Six Sigma or similar effort, have presented messages to their employees along the following lines:

> The organization cannot tell you what to believe, and we are not asking you to believe in our new Six Sigma initiative, although we hope you do. We can, however, expect you to behave in certain ways with respect to it. Therefore, let it be known that you are expected to support it, or at least get out of its way, and not resist. Henceforth, rewards and promotions will go to those who energetically support and participate in the Six Sigma activities. Those who do not support it and participate in it will not be eligible for raises or promotions. They will be left behind, and perhaps even replaced with others who do support it.

This is fairly strong language. Such companies often achieve some results in the short term. However, should a forceful leader depart without causing the new initiative to become embedded in the organization's cultural norms and patterns (to the extent that individual members have taken on these new values and practices as their own), it is not unusual for the new thrust to die out for lack of consistent and persistent reinforcement.

Resistance to Change

Curiously, even with such reinforcement, change—even beneficial change—will often be resisted. The would-be agent of change needs to understand the nature of this resistance and how to prevent or overcome it.

The example of the control chart case drew the conclusion that the main resistance to change is due to the disturbance of the cultural pattern of the shop when a change is proposed or attempted. People who are successful—and therefore comfortable—functioning in the current social or technical system do not want to have their comfortable existence disrupted, especially by an "illegitimate" change.

When a technical or social change is introduced into a group, group members immediately worry that their secure status and comfort level under the new system may be very different (worse) than under the current system. Threatened with the frightening possibility of losing the ability to perform well or losing status, the natural impulse is to resist the change. Group members have too much at stake in the current system. The new system will require them not only to let go of the current system willingly but also to embrace the uncertain, unpredictable new way of performing. This is a tall order. It is remarkable how profoundly even a tiny departure from cultural norms will upset society members.

What Does Resistance to Change Look Like?

Some resistance is intense, dramatic, and even violent. Dr. Juran reminds us of some examples: When fourteenth-century European astronomers postulated a sun-centered universe, this idea flew in the face of the prevailing cultural beliefs in an earth-centered universe. Their ancestors, religious leaders, grandparents, and parents had passed down this belief for many generations. (Furthermore, on clear days, one could see with one's own eyes the sun moving around the earth.) Reaction to the new "preposterous" unacceptable idea was swift and violent. If the sun-centered believers are correct, then the earth-centered believers are incorrect—an unacceptable, illegitimate, wrong-headed notion. To believe in the new idea required rejecting and tossing out the old. But the old was deeply embedded in the culture. So the "blasphemous" astronomers were burned at the stake.

Another example from Dr. Juran: When railroads converted from steam-powered to diesel-powered locomotives in the 1940s, railroad workers in the United States objected. It is unsafe, even immoral, they protested, to trust an entire trainload of people or valuable goods to the lone operator required to drive a diesel. Two people, an engineer who drove, and a fireman who stoked the fire had "always" operated locomotives. If one *were* incapacitated, the other could take over. But what if the diesel engineer had a heart attack and died? So intense were the resulting strikes that an agreement was finally hammered out to keep the fireman on the job in the diesels! Of course, the railroad workers were really protesting the likely loss of their status and jobs.

Norms Helpful in Achieving a Cultural Transformation

Transforming a culture requires a highly supportive workforce. Certain cultural norms appear to be instrumental in providing the support needed. If these norms are not now part of your culture, some breakthroughs in culture may be required to implant them. Some of the more enabling norms are as follows:

- *A belief that the quality of a product or process is at least of equal importance, and probably of greater importance than the mere quantity produced.* This belief results in decisions favoring quality: defective items do not get passed on down the line or out the door; chronic errors and delays are corrected.
- *A fanatical commitment to meeting customer needs.* Everyone knows who his or her customers are (those who receive the results of their work), and how well he or she is doing at meeting those needs (they *ask*). Organization members, if necessary, drop everything and go out of their way to assist customers in need.
- *A fanatical commitment to stretch goals and continuous improvement.* There is always an economic opportunity for improving products or processes. Organizations who practice continuous improvement keep up with, or become better than, competitors.

Organizations that do not practice continuous improvement fall behind and become irrelevant or worse—go out of business. Six Sigma product design and process improvement is capable, if executed properly, of producing superb economical designs and nearly defect-free processes to produce them, resulting in very satisfied customers and sharply reduced costs. The sales and the savings that follow show up directly on the organization's bottom line.

- *A customer-oriented code of conduct and code of ethics.* This code is published, taught in new employee orientations, and taken into consideration in performance ratings and in distributing rewards. Everyone is expected at all times to behave and make decisions in accordance with the code. The code is enforced, if needed, by managers at all levels. The code applies to everyone, even board members—perhaps especially to them considering their power to influence everyone else.

- *A belief that continuous adaptive change is not only good but necessary.* To remain alive, organizations must develop a system for discovering social, governmental, international, or technological trends that could impact the organization. In addition, organizations will need to create and to maintain structures and processes that enable a quick, effective response to these newly discovered trends.

Given the difficulty of predicting trends in the fast-moving contemporary world, it becomes vital for organizations to have such processes and structures in place and operating. If you fail to learn and appropriately adapt to what you learn, your organization can be left behind very suddenly and unexpectedly and end up in the scrap heap. The many rusting, abandoned factories the world over testifies to the consequences of not keeping up and consequently being left behind.

Policies and Cultural Norms

Policies are guides for managerial action and decision making. Organization manuals typically begin with a statement of the organization's quality policy. This statement rates the relative worth that organization members should place on producing high-quality products, as distinguished from the mere quantity of products produced. ("High-quality products" are goods, services, or information that meets important customer needs at the lowest optimum cost with few, if any, defects, delays, or errors.) High-quality products produce customer satisfaction, sales revenue, repeat demand or sales, and low costs of poor quality (unnecessary waste). Here, in that one sentence, are reasons for attempting quality improvement. Including a value statement in your organization's quality manual reinforces some of the instrumental cultural norms and patterns essential for achieving a "quality culture" and, ultimately, performance breakthroughs.

Keep in mind that if the value statement, designed to be a guide for decision making, is ignored and not enforced, it becomes worthless, except perhaps as a means of deceiving customers and employees in the short term. You can be sure, however, that customers and employees will soon catch on to the truth and dismiss the quality policy, waving it away as a sham that diminishes the whole organization and degrades management credibility.

Human Resources and Cultural Patterns

Human resources play a significant role in reinforcing cultural norms. It does so by several means that include

- *Recruiting.* Advertisements contain descriptions of desirable traits (e.g., dependable, energetic, self-starter, creative, analytic), as well as characterizations of the organization (e.g., service oriented, customer oriented, committed to being a world leader in quality, progressive, world class, and equal opportunity). Organizational values are often featured in these messages.
- *Orientation and training.* It is customary when providing new employees with an introduction to an organization to review with them expected modes of dress, behavior, attitudes, and traditional styles of working together.
- *Publishing employee handbooks.* The handbooks distributed to new employees, and to everyone annually, are replete with descriptions of organizational history, traditional

policies and practices, and expectations for organization members. All of these topics express directly or indirectly detailed elements of the official culture.

- *Reward and recognition practices.* In our rapidly changing world, management teams find themselves agonizing over what kind of employee behavior should be rewarded. Whatever the behavior is, and when it is rewarded, the reward reinforces the cultural norms embodied in that behavior, and it should induce more of the same behavior from the ones who are rewarded, as well as attract others to do the same.
- *Career path and promotion practices.* If you track the record of those promoted in an organization, you are likely to find either (1) behavior that conforms to the traditional cultural norms in their background or (2) behavior that resembles desired new cultural norms required for a given organizational change, such as launching a Six Sigma effort. In the former case, management wants to preserve the current culture; in the latter case, management wants to create breakthroughs in culture and bring about a new culture that is at least somewhat altered. In both cases, the issue of the relationship of the person being promoted to the organizational culture is a significant factor in granting the promotion.

Breakthroughs in Agile Sustainability

Agile sustainability is a phrase that described when an organization can respond to the question "How do I prepare my organization to respond quickly and effectively to unexpected change?" and sustaining it over time.

Creating a breakthrough in agile sustainability requires

- creating structures and processes that uncover and predict changes or trends in the environment that are potentially promising or threatening to the organization,
- creating processes that evaluate information from the environment and refer it to the appropriate organizational person or function,
- participation in creating an organizational structure that facilitates rapid adaptive action to exploit the promising trends or avoid the threatening disasters, and
- a response to the question "How do I prepare my organization to respond quickly and effectively to unexpected change?"

The survival of an organization, like all open systems, depends on its ability to detect and react to threats and opportunities that present themselves from within and from outside. To detect potential threats and opportunities, an organization must not only gather data and information about what is happening but also discover the (often) elusive meaning and significance the data hold for the organization. Finally, an organization must take appropriate action to minimize the threats and exploit the opportunities gleaned from the data and information.

To do all this will require appropriate organizational structures, some of which may already exist (an intelligence function, using an adaptive cycle, a "Big Data or Information Quality Council") and a data quality system. The Information Quality Council acts, among other things, as a "voice of the market." Dates are defined as "facts" (such as name, address, and age) or "measurements of some physical reality, expressed in numbers and units of measure that enable our organization to make effective decisions by." These measurements are the raw material of information, which is defined as "answers to questions" or the "meaning revealed by the data, when analyzed." The typical contemporary organization appears to the authors to be awash in data but bereft of useful information. Even when an organization possesses multiple databases, much doubt exists regarding the quality of the data and, therefore, the organization's ability to tell the truth about the question it is supposed to answer.

Managers dispute the reliability of reports, especially if the messages contained in the data are unfavorable. Department heads question the accuracy of financial statements and sales figures, especially when they bring bad tidings.

Often, multiple databases will convey incongruent or contradictory answers to the same question. This is because each individual database has been designed to answer questions couched in a unique dialect or based on the unique definitions of terms used by one particular department or function, but not all functions. Data often are stored (hoarded?) in isolated unpublicized pockets, out of sight of the very people in other functions who could benefit from them if they knew they existed. Anyone who relies on data for making strategic or operational decisions is rendered almost helpless if the data are not available or are untrustworthy. How can a physician decide on a treatment if X-rays and test results are not available? How can the sales team plan promotions when it does not know how its products are selling compared to the competition? What if these same sales people knew that the very database that could answer their particular questions already exists but is used for the exclusive benefit of another part of the organization? It is clear that making breakthroughs in adaptability is difficult if one cannot get necessary data and information or if one cannot trust the truthfulness of the information one does get. Some organizations for which up-to-date and trustworthy data are absolutely critical go to great lengths to get useful information. However, in spite of their considerable efforts, many organizations nevertheless remain plagued by chronic data quality problems.

The Route to Agile Sustainability and Its Prerequisites

Creating a breakthrough to create agile sustainability requires structures and processes that do the following

- detect changes or trends in the internal or external environment that are potentially threatening or promising to the organization,
- interpret and evaluate the information, and
- refer the distilled information to empowered functions or persons within the organization who take action to ward off threats and exploit opportunities. This is a continuous perpetual cycle.

The cycle might more precisely be conceptualized as a *spiral*, as it goes round and round, never stopping (see Fig 2.3). Several prerequisite actions are needed to set the cycle in motion and create breakthroughs in adaptability. Although each prerequisite is essential, and all are sufficient, perhaps the most crucial is the Information Quality Council and the data quality system. Everything else flows from timely trustworthy data—data that purport to describe truthfully the aspects of reality that is vital to your organization.

Prerequisites for the Agile Sustainable Breakthroughs

- Leadership and management
- Organization and structure
- Current performance
- Culture

A Journey around the Agile Sustainability Cycle

An intelligence function gathers data and information from the internal and external environment. At minimum, we need to know some of the following basic things.

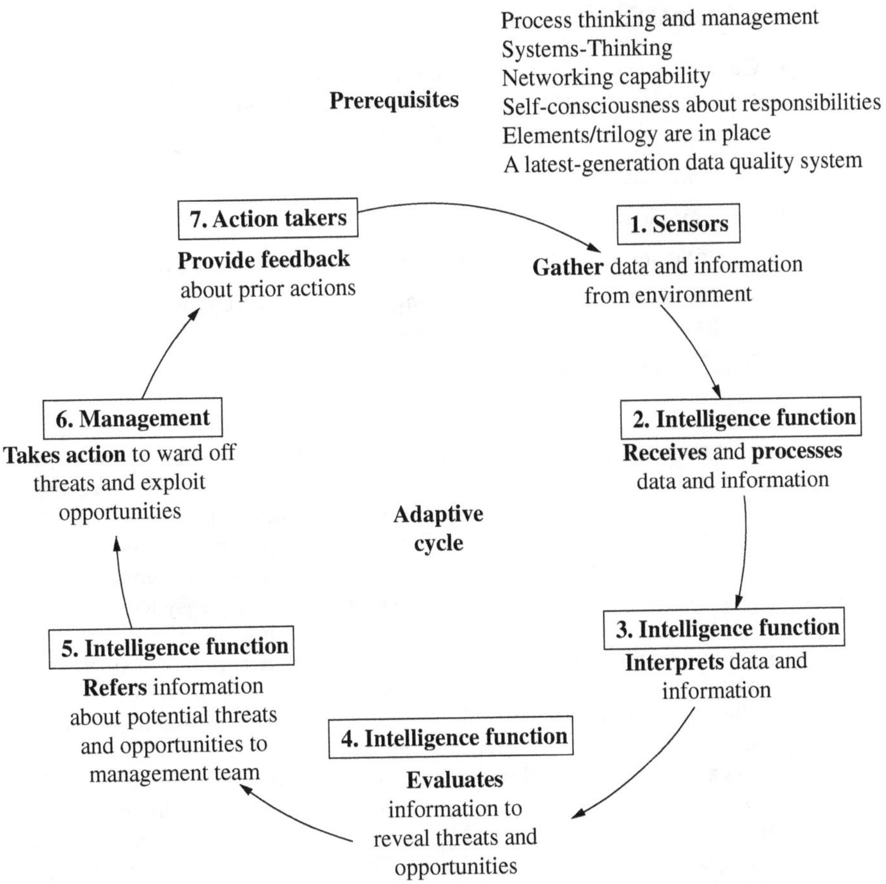

FIGURE 2.3 Adaptive cycle—to detect and to react to organizational threats and opportunities. (*De Feo and Barnard 2004, p. 291.*)

From the Internal Environment

- Process capability of our measurement and data systems
- Process capability of our key repetitive processes
- Performance of our key repetitive processes (human resources, sales, design, engineering, procurement, logistics, production, storage, transportation, finance, training, etc.; yields, defect types and levels, and time cycles)
- Causes of our most important performance problems
- Management instrument panel information: score cards (performance toward goals)
- Internal costs and costs of poor quality (COPQ)
- Characteristics of our organizational culture (how much does it support or subvert our goals)
- Employee needs
- Employee loyalty

From the External Environment

- Customer needs, now and in the future (what our customers or clients and potential customers or clients want from us or our products)
- Ideal designs of our products (goods, services, and information)
- Customer satisfaction levels
- Customer loyalty levels
- Scientific, technological, social, and governmental trends that can affect us
- Market research and benchmarking findings (us compared to our competition; us compared to best practices)
- Field intelligence findings (how well our products or services perform in use)

You may add to this list other information of vital interest to your particular organization. This list may seem long. It may seem expensive to get all this information. (It can be.) You may be tempted to wave it away as excessive or unnecessary. Nevertheless, if your organization is to survive, there appears to be no alternative but to gather this kind of information, and on a regular, periodic basis. Fortunately, as part of routine control and tracking procedures already in place, your organization probably gathers much of this data and information. Gathering the rest of the information is relatively easy to justify, given the consequences of being unaware of, or deaf or blind to, vital information.

Information about internal affairs is gathered from routine production and quality reports, sales figures, accounts receivable and payable reports, monthly financial reports, shipment figures, inventories, and other standard control and tracking practices. In addition, specially designed surveys—written and interviews—can be used to gain insights into such matters as the state of employee attitudes and needs. A number of these survey instruments are available off the shelf in the marketplace. Formal studies to determine the capability of your measurement systems and your repetitive processes are routinely conducted if you are using Six Sigma in your organization. Even if you do not use Six Sigma, such studies are an integral part of any contemporary quality system. Scorecards are very widely utilized in organizations that carry out annual strategic planning and deployment. The scores provide management with a dashboard, or instrument panel, which indicates warnings of trouble in specific organizational areas. Final reports of operational projects from quality improvement teams, Six Sigma project teams, and other projects undertaken as part of executing the annual strategic business plan, are excellent sources of "lessons learned" and ideas for future projects. The tools and techniques for conducting COPQ studies on a continuing basis are widely available. The results of COPQ studies become powerful drivers of new breakthrough projects because they identify specific areas in need of improvement. In sum, materials and tools for gathering information about your organization's internal functioning are widely available and easy to use.

Gathering information about conditions in the external environment is somewhat more complex. Some approaches require considerable know-how and great care. Determining customer needs is an example of an activity that sounds simple but actually requires some know-how to accomplish properly. First, it is proactive. Potential and actual customers are personally approached and asked to describe their needs in terms of benefits they want from a product, services, or information. Many interviewees will describe their needs in terms of a problem to be solved or a product feature. Responses like these must be translated to describe the benefits the interviewee wants, not the problem to be solved or the product feature they would like. Tools and techniques for determining ideal designs of current and future products or services are also available. They require considerable training to acquire the skills, but the payoffs are enormous. The list of such approaches includes Quality Planning, Design for Six Sigma (DFSS), TRIZ, a technique developed in Russia for projecting future customer needs and product features. Surveys are typically used to get a feel for customer satisfaction. A "feel" may be as close as you can get to knowledge of customer feelings and perceptions. These glimpses can be useful if they reveal distinct patterns of perceptions whereby large proportions of a sample population respond very favorably or very unfavorably to a given issue. Even so, survey results can hardly be considered "data," although they have their uses if suitable cautions are kept in mind. The limitations of survey research methodology cloud the clarity of results from surveys. (What really is the precise difference between a rating of "2" and a rating of "3"? A respondent could answer the same question different ways at 8:00 A.M. and at 3:00 P.M. for example.) (A satisfaction score increase from one month to another could be meaningless if the group of individuals polled in the second month is not the exact same group that was polled the first month. Even if they were the same individuals, the first objection raised above would still apply to confound the results.)

A more useful approach for gauging customer "satisfaction," or more precisely, their detailed responses to the products or services they get from you, is the customer loyalty study, which is conducted in person with trained interviewers every six months or so on the same people. The results of this study go way beyond the results from a survey. Results are quantified and visualized. Customers and former customers are asked carefully crafted standard questions about your organization's products and performance. Interviewers probe the responses with follow-up questions and clarifying questions. From the responses, a number of revealing pieces of information are obtained and published graphically. Not only do you learn the features of your products or services that cause the respondents happiness and unhappiness but also such things as how much improvement of defect X (late deliveries, for example) it would take for former customers to resume doing business with you. Another example:

You can graphically depict the amount of sales (volume and revenue) that would result from given amounts of specific types of improvements. You can also learn what specific "bad" things you'd better improve, and the financial consequences of doing so or not doing so. Results from customer loyalty studies are powerful drivers of strategic and tactical planning, and breakthrough improvement activity.

Discovering scientific, technological, social, and governmental trends that could affect your organization simply requires plowing through numerous trade publications, journals, news media, websites, and the like, and networking as much as possible. Regular searches can be subcontracted so you receive, say, published weekly summaries of information concerning very specific types of issues of vital concern to you. Although there are numerous choices of sources of information concerning trends, there appears to be little choice of whether to acquire such information. The trick is to sort out the useful from the useless information.

A basic product of any intelligence function is to discover how the sales and performance of our organizations' products, services, and sales compare with our competitors and potential competitors. Market research and field intelligence techniques are standard features in most commercial businesses, and books on those topics proliferate.

Many organizations undertake benchmarking studies to gather information on world-class best practices. They study the inner workings of repetitive processes such as design, warehousing, operating oil wells, and mail order sales—almost anything. The processes studied are not necessarily those of your competitors; they need only be the very best (efficient, effective, and most economical). Benchmarking studies are classic intelligence detective work, and are often conducted on a subcontract basis with organizations that specialize in benchmarking. The results are typically published and shared with all participants. When you have discovered best practices, you can compare your performance with them and describe gaps between theirs and yours, thus identifying breakthrough opportunities.

Completing the adaptive cycle will enable the organization to attain a breakthrough in adaptability and lead to sustainability. Skipping a breakthrough may not indicate a problem in the short term, only in the long term. Consider the economic crisis that hit the global economy in 2008. There were many global organizations that we considered leaders in their markets—when business was good. During the crisis, so many top performers of the past went out of business, were merged with others, or went into bankruptcy only to emerge a different organization. Why did so many organizations have trouble? Our theory was that although these organizations were good at responding to their customer needs they were not watching society's needs. This led to a lack of information that, if it was available, would have provided enough time to "batten down the hatches," to ride the crisis out. To avoid this from happening, creating a high performing, adaptable organization may lead to better performance when things are not so good.

Sustainability

The second part of this breakthrough is sustainability. Sustainability has two important meanings. The first is to sustain the benefits of the transformational changes that took place. The second is to assure the organization is sustainable from an environmental point of view. At the time of publication, we felt we would only focus on long-term results.

Sustainability is the return to evaluate performance annually based on the findings of the Information Council. With this information, leaders can adjust the organization to ensure that it stays ahead of its customers and can sustain itself for the long term.

The Juran Transformation Roadmap

There are five phases in the Juran Transformation Roadmap, each one corresponding to the breakthroughs that are described in this chapter. Each phase is independent, but the beginning and end of each phase are not clearly delineated. Each organization reacts differently to changes. This means that one business unit in an organization may remain in one phase longer than another unit. These phases once again are a managerial guide to change, not a prescription.

The five phases of the Transformation Model and Roadmap are shown in the Fig. 2.4. The road starts at the *Decide Phase.* This phase begins when someone on the executive team decides that something must be done or else the organization will not meet shareholder expectations or will not meet its plan and ends with a clear plan for change.

In the Decide Phase, the organization will need to create new information or better information than it may have had about itself. This information can come from a number of reviews or assessments. Our experience shows that the more *new* information an organization has, the better its planning for change. Some of the important areas that should be reviewed are as follows:

- Conduct a Customer Loyalty Assessment to determine what they like or dislike about your products and services.
- Identify the areas of strength and uncover possible problems in the organization's performance.
- Understand employee attitudes toward the proposed changes.
- Understand the key business processes and how the changes will affect them.

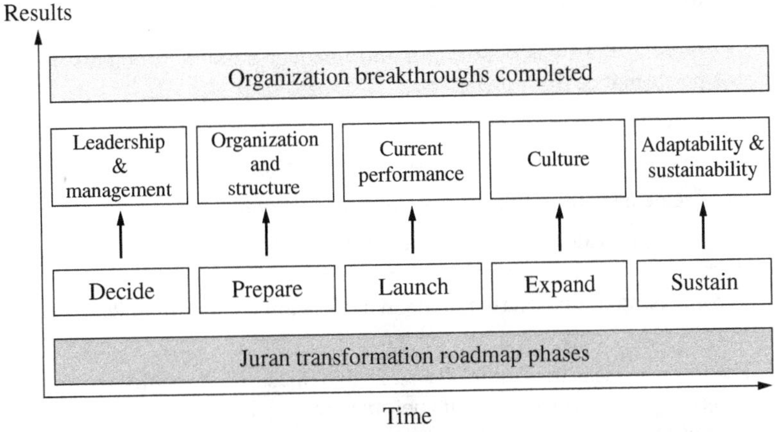

FIGURE 2.4 Juran roadmap and breakthroughs.

- Conduct a cost analysis of poorly performing processes to determine the financial impact of these costs on the bottom line.
- Conduct a world-class quality review of all business units to understand the level of improvement needed in each unit.

A comprehensive review of the organization prior to launch is essential for success. We show a typical review that we recommend to all organizations embarking upon a Six Sigma transformational initiative. From these assessments and reviews, the executive team now has qualitative and quantitative information to define the implementation plan for its organization.

The deployment plan must include the following items:

- Infrastructure that is needed to steer the changes
- Methodology and tools that will be used throughout the implementation
- Goals and objectives of the effort
- Detailed milestones for achieving results

The conclusion of this phase results in the breakthrough in leadership and management.

The second phase is the *Prepare Phase*. In this phase, the executive team begins to prepare for the changes that will take place. It focuses on developing a pilot effort to try the change in a few business units before carrying it out in the organization as a whole.

This phase begins by deploying the plan created in phase one and it ends after a successful launch of pilot projects in phase three. From here, the organization begins to identify the improvement projects that must be carried out to meet the desired goals established in the Decide Phase. In this phase, the organization launches the pilot projects, reviews the projects' progress, and enables the projects' success. Upon completion of the pilot projects, executives evaluate what has worked and what has not. Then executives either abandon their efforts or change the plan and expand it throughout the organization.

The following actions can be taken for your organization:

- Identify the areas of strength and uncover possible problems in the organization's performance from phase one.
- Identify value streams and key business processes that need improvement.
- Select multifunctional pilot or demonstration projects and create project charters.
- Create a training plan and set of learning events to train the teams.
- Communicate the steps taken in this phase to the workforce.

The conclusion of this phase results in a breakthrough in organizational structure.

The third phase is the *Launch Phase*. In this phase, the executive team begins demonstration projects in a few business units before carrying them out in the total organization. Each project will require a project charter, a team and an effective launch, reviewing the progress and maintaining the gains before results are attained. The length of this phase depends on the number of projects and results expected. For most organizations, this phase completion takes less than 1 year. As each project is completed, and results are attained, leaders can then evaluate the lessons learned and expand by launching more projects.

The conclusion of this phase results in a breakthrough in current performance.

Expansion can take months or years, depending on the size of the organization. An organization of 500 employees will require less time to deploy a plan across the organization than an organization of 50,000. The *Expand Phase* may take 3 to 5 years. Note that positive financial results will occur long before cultural changes take place. Staying in the Expand Phase is not a bad thing. An organization must continue to implement its plan, business unit by business unit, until the organization has had enough time to implement the desired changes. The final phase is the *Sustain Phase* when the organization has a fully integrated operation. All improvement and Six Sigma goals are aligned with the strategy of the organization. Key business processes are defined and well managed, and process owners are assigned to manage them. Employee performance reviews and compensation are in line with the changes required. Those who comply with the change are rewarded. The executives and business unit heads conduct regular reviews and audits of the change process. This may result in a discussion or even a change in the strategy of the organization.

The organization may have learned more about its capabilities and more about its customers that may lead to a change in strategy.

The conclusion of this phase results in a breakthrough in culture.

The *Sustain Phase* also lasts as long as the organization is meeting its strategic and financial goals. Deviations from expected results, possibly due to macroeconomic events outside the organization, require a review of the scorecard to determine what has changed. When this is determined, the organization makes the changes, continues, and sustains itself at the current level.

The conclusion of this phase results in a breakthrough in agile sustainability.

Lessons Learned in Deploying the Transformation Roadmap

As you begin your journey down this road, note the many lessons learned from organizations that have led a change process and failed initially. These failures can be avoided by suitable planning, listed as follows:

- All organizations and their units are at different levels of maturity regarding performance.
- Champions and internal experts (such as Six Sigma Black Belts) become drivers who propel their organization to superior performers or best in class.
- Extensive training in tools and techniques for all employees ensures that learning has taken place and that they can use the tools to improve performance.
- Systematic application and deployment through proven methodologies such as Six Sigma Improvement (DMAIC) and Design (DFSS) are necessary to create a common language and create results in current performance.
- Focusing improvements on the customer first will enable cost reduction, and delighted customers will enable breakthrough bottom-line results.
- Significant increase in customer satisfaction happens only when you improve the processes and services that impact them.
- No organization has ever successfully implemented a plan without the leadership and commitment of the executive team—they are the ones who control the resources and provide the communication that will change the culture.

With this roadmap and the lessons learned, all organizations should be able to achieve sustainable results well into the future. If more organizations get on board with positive, customer-focused change initiatives, we will be able to create a global society that reduces our dependence on the quality dikes we have built over the years.

As your organization continues to renew itself annually through the strategic planning process, this cycle of improvement should continue. Barring any leadership changes or crisis, your organization should be on its way to attaining superior and sustainable results.

References

Bernstein, P. (1988). "The Trust Culture," *SAM Advanced Management Journal*, Vol. 53:3 pp. 4–8.

De Feo, J. A., and Barnard, W. W. (2004). *Juran Institute's Six Sigma Breakthrough and Beyond: Quality Performance Breakthrough Methods*. McGraw-Hill, New York.

Juran Institute, Inc., (2009). "Quality 101: Basic Concepts and Methods for Attaining and Sustaining High Levels of Performance and Quality, version 4," Juran Institute, Inc., Southbury, CT.

Katzenbach, J. R., and Smith, D. K. (1993). *Wisdom of Teams: Creating the High Performance Organization*, Harvard Business School Press, Boston, MA.

Kotter, J. P. (1996). *Leading Change*. Harvard Business School Press, Cambridge, MA.

Mann, D. W. (1994). "Reengineering the Manager's Role," ASQC Quality Congress Transactions 1994, American Society for Quality, Milwaukee, WI, pp. 155–159.

Mead, M. (1955). "Cultural Patterns and Technical Change." *Population Studies*, Vol. 8, No. 2.

Niedz, B. A. (1995). "The Blitz Team" IMPRO95 Conference Proceedings, Juran Institute, Inc. Southbury, CT.

Taylor, F. W. (1947). *The Principles of Scientific Management*, Harper and Row, New York.

CHAPTER 3

Developing the Quality and Excellence Office

Joseph A. De Feo

High Points of This Chapter

1. The Quality and Excellence Office plays an important role at both the strategic and tactical levels in enabling the organization to strive for quality and performance excellence.
2. A distinction is often made between managing *quality of product* and *creating transformational change*, which has led to the recognition of parallel departments or divisions within the organization that often leads to lackluster results. A department that combines resources and speaks one language is the best practice for the future.
3. The role of the Quality and Excellence Office, its responsibilities, and how it is organized is driven by its mission. This mission is to ensure that the organization designs, controls, and continuously improves performance of the product, the process, and its people to meet the changing needs of its customers. The Office has direct and indirect functions to carry out that mission.
4. The Office must be an enabler—enabling the organization to drive quality and performance excellence at the business and organization levels.

5. Capable experts are needed and developed to have the right competencies and capabilities, based on what customers throughout the organization need, in order for them to achieve quality and performance excellence.
6. Skills assessment, education, and certification will close the gaps and enable today's Offices to transform themselves into the office the organization needs to sustain itself.

Introduction to the Office of the Future

Computer Sciences Corporation, one of the largest technology solutions and outsourcing organizations in the world, understands the Office of the future. Darryl W. Bonadio, Master Black Belt and Director of MSS Quality Management and Improvement, describes quality and the Office this way:

> Quality is important to everyone at CSC. Quality must become a cultural attitude that results in personalization of our five standards: Positive Talk, Confident Perspective, Outcomes Oriented, Be Accountable, and Respect Our Client. CSC's Managed Services Sector manages quality through its Global Process Innovation and Quality Excellence organization. This is our Office. The Enterprise Excellence Program we have implemented meets this challenge by addressing three important tenets. The first is Delivery Excellence, which is the integration of quality principles, tools, and approaches that enable achievement of business objectives and promote client advocacy. Second, Passionate Delivery, where engaged employees are the most effective quality assurance factor in our service-oriented environment. And finally, Enterprise Performance Management that provides an end-to-end analysis and a feedback loop for continuous improvement.

You may be tasked to define and develop the Quality or Performance Excellence office, as some call it, in your organization. You may even be the Director or Vice President of Quality. You are faced with the challenge of organizing and driving quality and performance excellence in your business unit. Key questions you may be pondering and that must be addressed include

- What is the role of this office?
- What are its mission and responsibilities?
- What types of personnel are needed?
- What competencies and capabilities are needed?
- How do you develop capable experts to drive quality and performance excellence?
- How should the office be organized to develop and deploy the functions needed?

These questions are the motivation for this chapter. An underlying purpose of this chapter is to assist the reader in asking the right questions and to provide guidance in defining and developing the quality and performance excellence capabilities in any organization.

The chapter is organized as follows:

- We begin with key operational definitions and terms.
- We then address the approach for coordinating quality and performance excellence. We discuss the role and responsibilities of the Office.
- We provide guidance on how to organize the Office.
- We showcase an example on how the Office in a global organization is organized, including organization charts, job profiles, and descriptions of various quality functions.

- We conclude this chapter with the means to address the challenge of developing capable experts to drive quality and performance excellence using a customer-focused development approach by describing skills assessment, reeducation, and certification requirements.

Proof of the Need to Improve Office Performance

The Juran Institute develops and regularly conducts a competency and skills assessment of Office personnel. The results of these assessments convinced us and our clients that a change is needed. Here is one review of its assessment and findings.

There are eight steps used to carry out the Quality and Performance Excellence Skills Assessment from Juran Institute. Since 1994, we have been conducting this assessment and keeping track of the results. Our findings led us to raise awareness of the need to improve the skill set of the quality professionals. It also led us to help close the gaps. The eight steps are

1. Reviewed skills questions and an organization's competency models to identify the organization's needs and its internal customers
2. Translated and edited competency questions for organization terminology
3. Developed competency questions around seven major topic areas:
   - Assessment and auditing
   - Critical thinking
   - Process control
   - Quality theory
   - Root cause analysis
   - Statistics and data analysis
   - Teamwork and coaching
4. Conducted written and oral reviews using the developed competency questions covering the major topic areas
5. Summarized the results by question (effectiveness of answering the questions) and by individual (each individual's score for the oral and written reviews)
6. Compared responses of staff skills to the industry best-practice skills, as defined by multiple scores of assessments conducted by Juran Institute
7. Analyzed, interpreted, summarized the findings and presented to management
8. Developed a training and development plan to close the gaps

An anchored scale of 1 to 5 was used for scoring the participant responses:

1. no response or very poor response
2. a weak or poor response
3. an acceptable response, but with little detail
4. a strong response, with detail
5. a very strong, outstanding detailed and understanding response

Typical Findings from Our Experience

Strengths

1. Most respondents are well versed in their organization's products, technology, work area, and programs.
2. Most understand the importance of Quality and Performance Excellence as the means to satisfy customers and attain superior results.
3. Competency and skill capability (scores) improve with successive job grades, from lowest levels to the directors.
4. Almost 100 percent of the respondents felt this assessment was a welcome activity to help them obtain the awareness that the Office needs continued education.

Opportunities for Improvement

1. Of the seven categories, three areas are always scoring low:
   a. Root cause analysis graphical tools
   b. Data analysis and basic statistics
   c. Quality management methods and theory
2. Over 75 percent of quality professionals were not aware of quality management history, methods, and early practitioners such as Deming, Shewhart, and Juran.
3. Many lower-level job-grade quality professionals lack the knowledge of quality methods and tools—even some basic ones—that are required in the Organization Quality and Performance Excellence (Q&PE) Competency Models.
4. Over 70 percent of individuals answered oral and written questions—in total—unsatisfactorily.
5. Our experience indicates that some of the staff may not be able to drive Quality and Performance Excellence and/or improvements in their work areas with limited knowledge of quality management tools and techniques (see summary charts).
6. Many are not able to articulate responses to oral questions. They had difficulty expressing themselves with clarity and precision. Some answered questions not asked, giving the appearance of not listening or not understanding what was asked.

The authors feel a change is needed and a program to improve performance of the Office staff is in order.

Operational Definitions and Key Terminology

The Quality and Excellence Office

Traditionally, the Office is given the responsibility of ensuring that the functions required for establishing and producing quality products and services meet the customers' specified requirements. This office is called by various names; however, it is most commonly known as the Quality or Quality Assurance Department. In regulated industries, such as the pharmaceutical or the medical device industry, it is often called QA/RA, short for Quality Assurance/Regulatory Assurance. Within defense industries, it may be known as Mission Assurance or Quality and Mission Assurance. In other industries, it might be called Quality and Safety Compliance. In this chapter, the traditionally limited role of the Office, as mentioned, is challenged. The modern definition of the Office is expanded to a business level and is called the Performance Excellence Office, discussed later in this chapter.

The Quality Functions

Quality functions are the actions or activities that are carried out on a daily basis according to the three universal processes of the Juran Trilogy: quality planning, improvement, and control.

Quality planning activities include joint supplier planning, designing or redesigning processes, new development of products or services, design reviews, toll-gate reviews, and quality plans.

Quality improvement activities include problem solving, root cause analysis, and projects to remove waste or improve process capability.

Quality control activities include implementing quality standards, carrying out source inspection, testing, in-process inspection, final inspection, and audits.

These functions may or may not be performed by quality personnel alone, and usually require the participation and input of employees (the community of practitioners) throughout an organization.

Quality Management Principles

ISO 9000 identifies eight quality management principles that sum up what can be used by the Office in order to lead the organization toward performance excellence.

1. *Customer focus*. Organizations depend on their customers and, therefore, should understand current and future customer needs, should meet their customer requirements, and should strive to exceed customer expectations.
2. *Leadership*. Leaders establish unity of purpose and direction of the organization. They should create and maintain the internal environment in which people can become fully involved in achieving the organization's objectives.
3. *Involvement of people*. People at all levels are the essence of an organization, and their full involvement enables their abilities to be used for the organization's benefit.
4. *Process approach*. A desired result is achieved more efficiently when activities and related resources are managed as a process.
5. *System approach to management*. Identifying, understanding, and managing interrelated processes as a system contributes to the organization's effectiveness and efficiency in achieving its objectives.
6. *Continual improvement*. Continual improvement of the organization's overall performance should be a permanent objective of the organization.
7. *Factual approach to decision making*. Effective decisions are based on the analysis of data and information.
8. *Mutually beneficial supplier relationships*. An organization and its suppliers are interdependent, and a mutually beneficial relationship enhances the ability of both to create value.

Quality Management System

A formal definition of a Quality System can be found in the ISO 9000 series of standards. ISO 9000 defines a Quality Management System as a "management system to direct and control an organization with regard to quality." Since quality is an organization wide function, the Quality System is therefore organization wide. While the Quality and Performance Excellence Office plays a major role, the Quality System is much larger in scope and, therefore, may be "directed and controlled" from multiple offices.

Approach to Coordinate Quality and Performance Excellence

Control versus Creating Beneficial Change

Gryna et al. (2007) state that the approach used to coordinate quality and performance excellence activities take two major forms:

1. Coordination for *control* is achieved by the regular line and staff departments, primarily through the use of formal procedures and feedback loops. Feedback loops take such forms as audits of execution versus plans, sampling to evaluate process and product quality, control charts, and reports on quality.
2. Coordination for *creating change* is achieved primarily through the use of quality project teams, Six Sigma or Lean project teams, rapid improvement (or kaizen) events, and other organizational forms for creating change.

Coordination for control is often the focus of the Office. However, more and more often, such a focus is so preoccupying that the Office is unable to make major strides in creating change. As a result, some "parallel organizations" for creating change have evolved. This evolution, in our opinion, has taken place because many of the quality experts lacked the skills required to speak the language of management. As Dr. Juran often stated, "They did not speak the language of money, they spoke the language of things" (Juran 1954, *Managerial Breakthrough*).

Parallel Organizations for Creating Change

All organizations are engaged in creating beneficial change as well as in preventing adverse change ("control"). Much of the work of creating change consists of processing small, similar changes. An example is the continual introduction of new products consisting of new colors, sizes, shapes, and so on. Coordination for this level of change can often be handled by carefully planned procedures.

Nonroutine and unusual programs of change generally require new organizational forms. These new forms are called "parallel organizations." *Parallel* means that these organizational forms exist in addition to and simultaneously with the regular "line" organizations.

Examples of parallel organizational forms for achieving change in quality and performance excellence are process teams, project teams, and performance excellence steering committees or councils. Parallel organizations may be permanent or ad hoc [a business process team or value stream management (VSM) team is permanent; in contrast, a Six Sigma project team or a Lean rapid improvement team is ad hoc and disbands when its mission is accomplished].

The IT, finance, and human resource functions often are asked to create programs to reduce costs, improve processes, and improve the skills of staff to drive quality. In some cases, they succeed. In other cases, they may have lacked the knowledge of how to manage for quality, and management soon is tired by the lack of results.

It is the responsibility of executive management to drive performance excellence throughout the organization. Performance excellence should not be viewed as the responsibility of the Office, even though it is integral to enabling performance excellence to occur. However, this office should be seen as the leaders of change to create a sustainable future. It is time to take it back.

Role of the Office

Many organizations have traditionally centralized the quality functions to an Office (or Quality Department). Over the decades, quality and performance excellence tasks have been assigned to other functional groups. For example, process capability studies were transferred from the Quality Department to a Process Engineering Department. Also, as the definition of quality broadened from operations only (little q) to all activities (Big Q), most organizations now have personnel in various functional departments trained and responsible for implementing quality and performance excellence. Authority to make decisions is now delegated to lower levels. Partnering with key suppliers and customers is becoming increasingly common. Also, organizations have become flatter, and cross-functional work teams and project teams are used to solve performance-related problems.

So, what is or what should be the role of the Office? Is the Office limited to a tactical role in the organization? Is there or should there be a strategic role for the Office? What are the responsibilities of the Office? What authority should it have? Many researchers and authors, such as Gryna (2002), Crosby (2000), Imler (2006), and Watkins (2005) have discussed the changing role of the Office. Discussion over the changing role of quality professionals has also occurred by the likes of Spichiger (2002) and Westcott (2004). The role of the Office is best discussed by considering it at two levels: tactical and strategic.

> Successful strategy execution depends on both satisfying the customer today and achieving excellence in the future. We continue to use the Juran Trilogy® to establish a solid foundation for achieving and sustaining breakthroughs throughout the entire organization of the Builder Cabinet Group at Masco Corporation. This has enabled us to continually meet or exceed the expectations of customers today, develop innovative breakthrough products and processes for tomorrow, and most importantly, provided the means for developing team members throughout my company to execute the strategy to action. As the executive of the Office, we will continue to drive our performance towards sustainability. (Steve Wittig, VP of Six Sigma, Masco, BCG 2009)

Tactical Level

At the tactical level, the traditional role was and still is to provide an independent evaluation of product quality or service quality. Inspections, testing, and product or service audits are examples. In this role, the Office is often viewed as being limited to supporting operations, providing an independent evaluation to ensure that productivity targets are not pursued at the expense of not meeting quality specifications.

Examples include

- Inspection and checking
- Product testing
- Supplier quality
- Root cause corrective action
- Quality system audits

As technology and the needs of the organization have changed, the role of the Office must expand beyond this traditional view. At the tactical level, the Office should play a role in *enabling* others in the organization to carry out the management of quality and performance excellence, either independently of or in collaboration with other functions. Examples of expanded tasks include

- Calculate the costs due to poor quality and use this to marshal the right resources to reduce these costs.
- Conduct process capability studies in nonmanufacturing processes.
- Participate in design reviews of new products as early as possible in the development cycle.
- Work with the supplier chain and provide evaluations of supplier selection and ongoing performance.
- Monitor customer satisfaction and create and manage the corrective action process.
- Be involved with the engineering change process.
- Reduce the inspection by creating in-process checking and self-inspection by the workforce.
- Work with functions to implement error-proofing efforts.
- Participate in kaizen or rapid improvement events.
- Use data-driven methods, such as statistical process control, to monitor performance.
- Identify improvement projects, such as Lean and Six Sigma projects.
- Integrate with the environment, health, and safety programs.
- Develop a skilled and competent function of experts that can consult with all functions.

Strategic Level

In many organizations, the Office is often viewed as having only a tactical role. That the Office can perform a useful strategic role is not well recognized by many organizations. However, enlightened organizations have recognized the strategic value of the Quality Office. They view the Office as a strategic asset with a key role in shaping, planning, and enabling the deployment of strategies, strategic goals, and business plans of the organization. A strategic approach for the Office of the future is to think of their role as providing "Enterprise Assurance" (Juran Institute White Paper, 2008).

The following is a list of strategic activities in which the Office needs to play an important role:

- Vision, mission, and policy development
- Assisting upper management with strategic planning and goal-setting
- Recommending to upper managers how to reduce the costs of poor quality
- Providing the organization with the most effective methods and/or tools to reduce these costs
- Being involved in annual business planning to incorporate improvement goals
- Demonstrating how quality can affect social responsibility and the environment
- Organizationwide assessment and planning to close the gaps

- Organizationwide transformation and improvement initiatives such as Lean Six Sigma
- Innovation of major business processes (such as demand creation, product development, order fulfillment, supply chain, and HR management processes)
- Developing and deploying balanced scorecards and data systems to support all key business processes, policy deployment, and improvement efforts

These roles are consistent with the views presented by Gryna et al. (2007), where the Office should have both tactical and strategic roles.

Moving from Office to the Quality and Excellence Office of the Future

To better understand responsibilities, we need to make an important distinction. There are *direct* responsibilities and *indirect* responsibilities of a separate function called the Quality and Excellence Office. We will refer to this new function as "the Office." Direct responsibilities are those activities and results over which the Office has control because they are executed by full-time personnel who report to the Quality Executive. Indirect responsibilities are those over which the Office has an influence, but little or no direct control. Indirect responsibilities are those that the Office enables others (nonquality personnel) to do. The Office provides the infrastructure and the means for these indirect responsibilities to be carried out by personnel in other functions. For example, the execution of process control plans to ensure quality compliance is one where the Office enables others (in operations) to carry out this important task. Another important example is the development and deployment of a change management program to attain performance excellence. This is one where the Office has both direct and indirect responsibility—*direct* because quality personnel are performance excellence specialists; *indirect* because selected nonquality personnel from other functions are also performance excellence specialists. In both cases, the Office plays an integral part in developing and enabling all performance excellence specialists (such as Six Sigma and Lean Six Sigma Green Belts, Black Belts, and Lean experts) to contribute to the improvement of the organization.

Consider Its Mission

Who should lead the Office? At what level of management? To whom is the Office accountable, and to whom should the Office report? These are important questions to ask when setting up the Office and are best answered by first considering the mission of the Office. Ideally, we want to ensure that the mission and associated responsibilities are supported by the corresponding management level and the appropriate level of authority is given to the Office. Authority must be consistent with responsibility; otherwise, we have a paralyzed office not able to carry out its duties.

For example, at an organization that provides customer care and call center services outsourced by major corporations, the **mission** of the Office is to contribute to the organization's financial growth by providing services that

- Enable the efficient delivery of exceptional customer experiences
- Foster a culture of fact-based leadership and continuous improvement

Not surprisingly, in this organization, the Office is led by the Quality Director, who reports to the Chief Financial Officer. To support that mission, responsibilities of this office reflect both the tactical and strategic role of the Office discussed earlier. Functions that are the responsibility of this Office are

- Process improvement (Six Sigma and Lean)
- Workforce management (contact center volume forecasting, staffing, and intraday staff management)
- Customer contact quality (call and e-mail quality)

Note: The Office should always be part of the executive management team or a direct report to it to be effective at leading performance excellence. Otherwise, the office will have little credibility.

Size and Scope Must Be Consistent with Its Mission

How large should the Office be? To determine size and scope, we have to consider what quality functions should be done versus what functions should be enabled by the Office. To the maximum extent possible, the Office should enable others in the organization to perform as many quality and performance excellence functions as possible. Process ownership of those functions increases when the Office enables others to carry them out. The experience of recent decades has shown that the best way to implement quality and performance excellence is through line organizations rather than through a staff department. Having an elitist department of experts and adopting a "corporate seagull" approach should be avoided. (The late Quality guru Phillip Crosby used this analogy of seagulls. It is used here to illustrate the effect. Seagulls fly into your area and make a lot of noise. And when they are done, they fly off ... and leave a mess behind!)

The Office should plan to enable others as much as possible to drive toward achieving what we have called to be engaged and in self-control, where (1) what is expected is clearly known, (2) actual performance is known through short or immediate feedback loops, and (3) the means and ability to regulate is available so that actual performance meets expectations.

The groups or areas that comprise the Office must be consistent with its mission and responsibilities as well. Referring back to the call center example, the organization chart shows the areas that report to the Quality Director (Fig. 3.1), and the responsibilities are listed in Table 3.1.

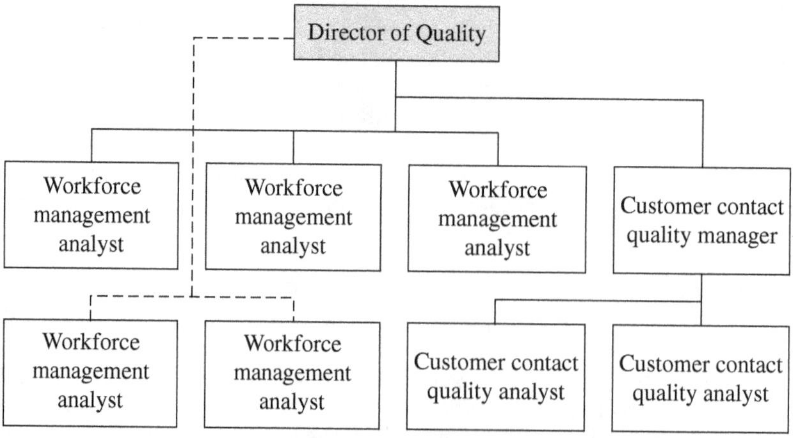

FIGURE 3.1 Quality Department organization chart of a call center company.

Title	Responsibility
Quality Director	• Research, develop, and execute the organizations corporate wide Lean, Six Sigma, quality management, and process improvement efforts. • Understand the business environment and ensure deployment initiatives are relevant and timely. • Assist in identifying continuous improvement opportunities within existing processes. • Work with multiple stakeholders to understand the key problem areas and work with this team to develop solutions. • Train and mentor the teams to drive improvement initiatives across the organization. • Use DMAIC/ Lean as a methodology to drive improvements. • Drive Six Sigma/Lean cultural change among the middle and lower management. • Understand lean concepts and identify and facilitate elimination of non-value-added activities. • Implement sustained solutions—set up all necessary control mechanisms such as dashboard, review procedures, and responsibility matrix. • Report project status and results to Manager and Operations. • Review program plans to establish appropriate quality assessment (metrics planning) and improvement methodology. • Demonstrate team leadership and manage team dynamics in all aspects of the DMAIC model (define, measure, analyze, improve, control) and maintain Six Sigma principles. • Perform other initiatives beyond process improvement projects which involve setting up right metrics to measure the process performance, automation initiatives, failure modes elimination and early warning systems, etc.
Workforce Management Analyst	• Has extensive experience in the call center environment, with a strong understanding of the Aspect eWorkforce Management application. • Responsible for insuring optimal forecasting, staffing and schedule compliance levels for existing programs and forecast future staffing needs to support Customer Contact Quality Manager recruitment efforts. • Works closely with Program Directors to assess impact of daily absences, changes in call volumes, etc.
Customer Contact Quality Manager	• Responsible for managing the entire organization's call quality process. • Works with teams at all locations to completely implement the call quality process. • Works with teams and clients on call calibrations. • Is responsible for all reporting and analysis of the call quality process. • Works with the training group to ensure training dollars are focused on specific needs identified in the call quality process. • Works directly with clients to ensure their needs are being met. • Improves the overall quality of the product that the organization delivers on behalf of clients each and every day.
Customer Contact Quality Specialist	• Responsible for supporting multiple clients in the area of interaction quality. • Responsible for conducting call auditing, sales and customer service training, and developing and motivating both front line agents as well as management. • Leads by example the value and necessity of interaction quality.

Table 3.1 Responsibilities of the Office of the Call Center Company

Quality Office of Today	Quality Office of Tomorrow
Vice President of Quality	Vice President of Enterprise Assurance
Director of Quality Assurance	Director of Operational Excellence
Quality Managers	Management as Champions
Quality Engineers	Quality Engineers and Master Black Belts
Quality Auditors	Auditors: Quality, Finance, Safety, Environment
Technical Analysts	Technical: Analysts, Green Belts, Lean Experts

TABLE 3.2 Comparisons of Quality Officers

Table 3.2 is a list of positions recommended for the Office of the future. Titles vary by industry, organization, and culture. We have identified the most common needs.

Example: Organizing the Office in a Multinational Global Organization

The scope, size, and structure become more challenging when the organization is global, with multiple divisions and regions operating around the world. To what extent should the Office be centralized (versus decentralized)? How much of the structure should be in a matrix? To what extent should there be direct reporting relationships versus indirect dotted-line reporting relationships to the head of the Office? Should a plant quality manager report directly or indirectly to the local plant's general manager?

To illustrate how the Office can be organized, we will look at an example of a multinational global organization. Figure 3.2 shows the organizational chart of a global organization with multiple locations and divisions around the globe. Given the nature of the responsibilities, the Chief Officer (who is the Vice President of Quality) reports to the President and COO (who in turn reports to the CEO) at a high enough level of authority that is consistent with the global responsibilities of the Office (Fig. 3.3). The organizational chart for the Office is shown in Fig. 3.4. The plant-level Quality organization can be seen in Fig. 3.5.

The Office is structured to drive global standards in areas with significant opportunity (such as reducing the number of suppliers and increasing global coordination for key customer quality improvements), to leverage best practices (such as by maintaining a library of Lean Six Sigma projects and FMEA libraries to capture engineering experience), and to revitalize use of quality audits to highlight continued improvements.

To support the global business divisions of this company, the Office is structured vertically in the divisions (Global Division Quality) from which selected Officers will then coordinate and drive best practices across the divisions (Global Functional Quality), as seen in Fig. 3.6.

Elaborating on the structure found in Fig. 3.6, the global quality head is a member of the division management team and the global quality leadership team. Plant quality managers are direct reports. The five functional heads coordinate best practices globally in the following areas: Customer Quality, Supplier Quality, Lean Six Sigma, Audit and Certification, Project Management, and Annual Quality Improvement Plans (AQIPs). Specific objectives, action plans, and metrics are established for each global functional area.

Global Division Quality has a flat structure, where the division owns the resources and quality functional heads coordinate globally. Global Division Quality is responsible and accountable for execution of global processes and AQIP content and achievement.

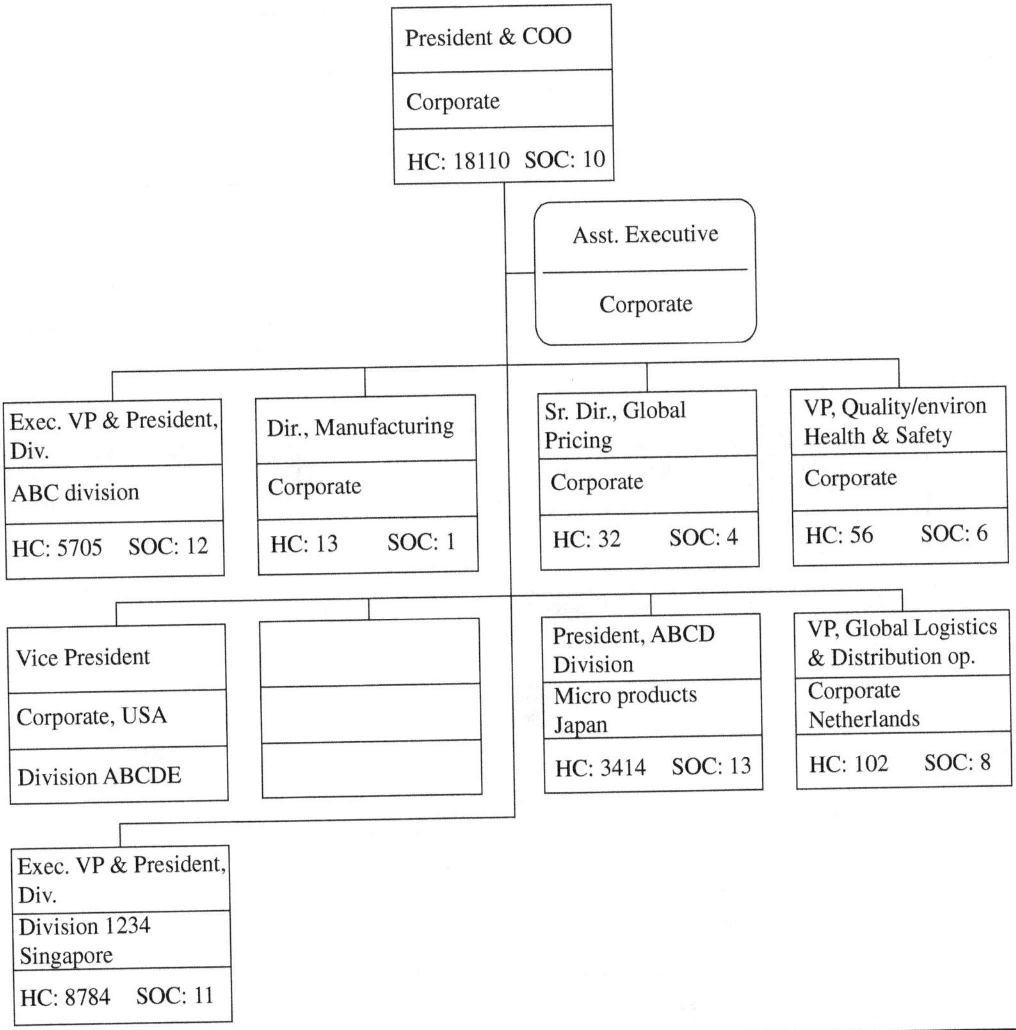

Figure 3.2 Organization structure of the COO office of a multinational global company.

Functional heads are part of each division's team. Some regional responsibilities will remain for division quality heads. Responsibilities of Global Functional Quality are

- Customer quality
  - To coordinate consistent proactive and reactive customer support
- Supplier quality
  - To enforce one global standard for suppliers and contract manufacturers
- Lean Six Sigma
  - To eliminate waste
  - To improve product and process quality

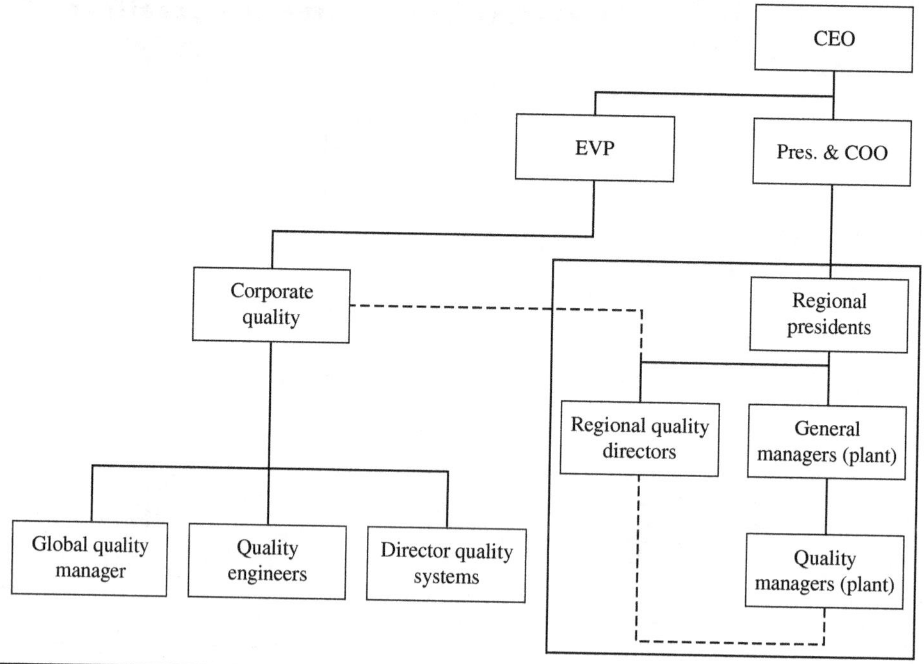

Figure 3.3 Global quality organization of a multinational global company.

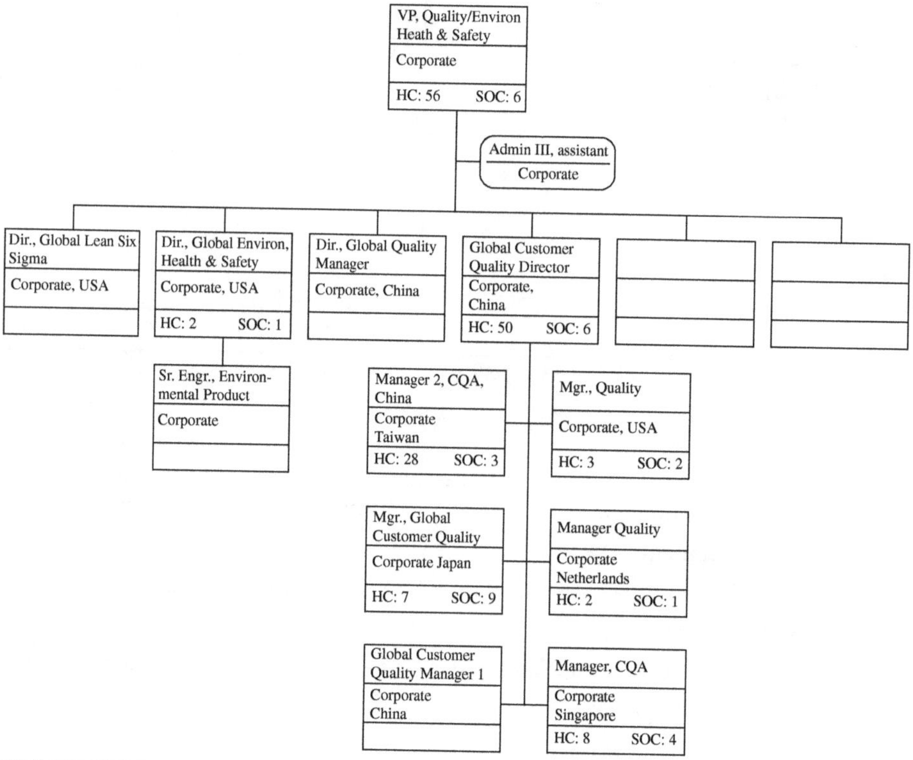

Figure 3.4 Global QEHS organization of a multinational global organization.

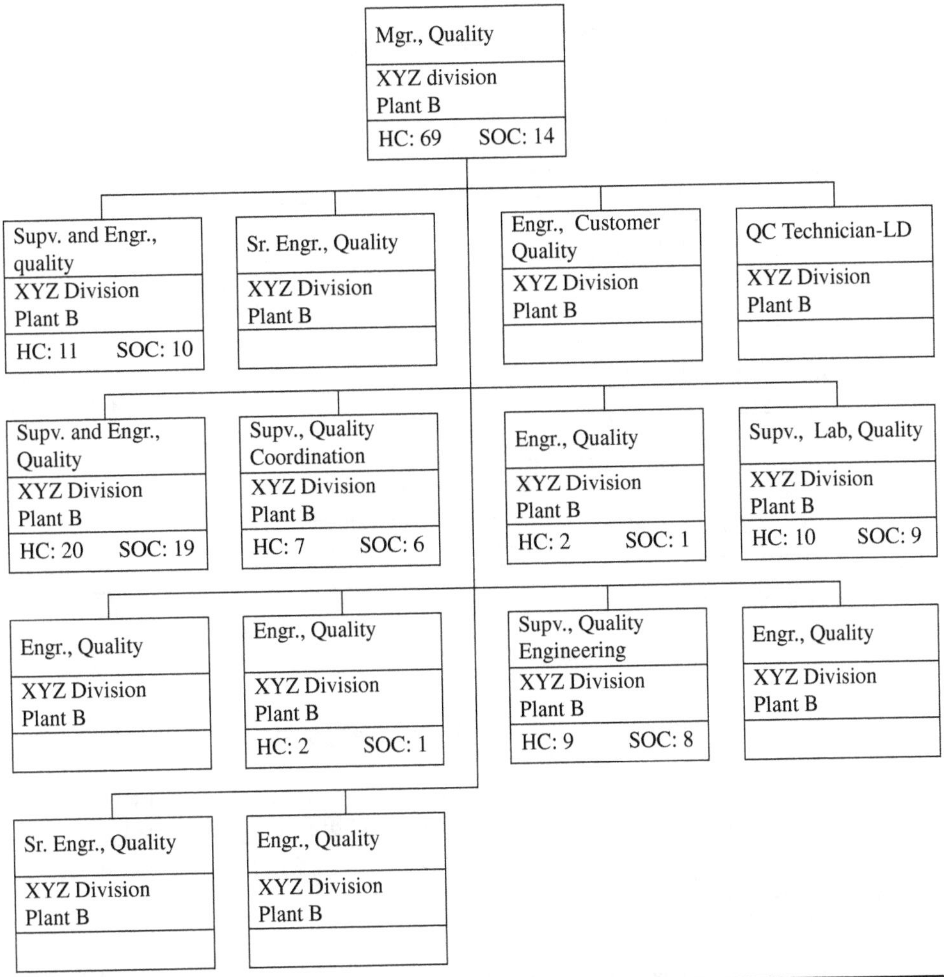

FIGURE 3.5 Plant-level quality organization of a multinational global company.

- Audit
  - Continuous evolvement of quality systems
  - Global utilization of IQRS audit and related improvement plans
- Project management
  - To coordinate global development and implementation of systems, tools, and quality process improvements
- Rules of engagement
  - The manufacturing plant has responsibility for manufacturing related ERA (Emergency Response Action), containment, customer complaints, and 8D actions and responses.
  - The selling entity is responsible for communicating the ERA from seller to customer and to enter the complaint in SAP and inform the manufacturing plant.

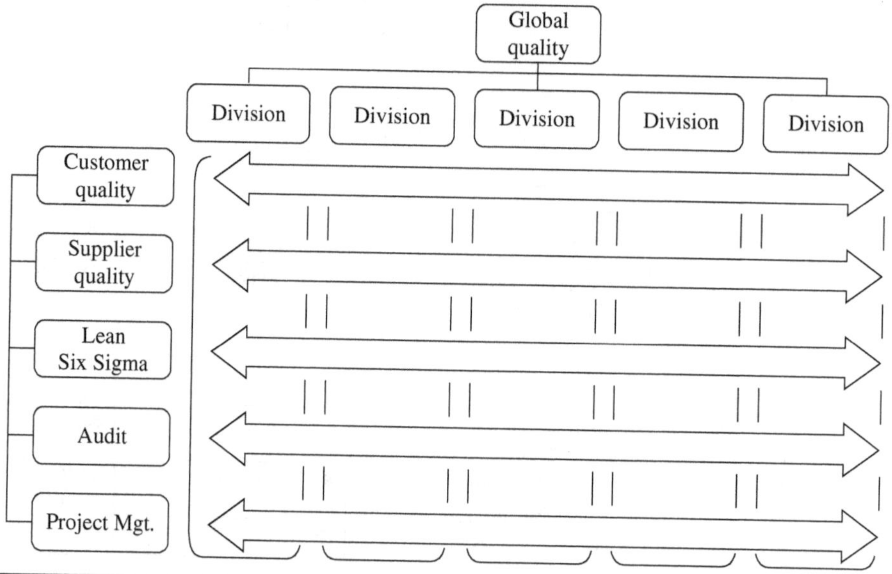

FIGURE 3.6 Global quality structure.

- Escalation above the plant quality manager goes first to the division quality head for the manufacturing division.
- When a customer requires an on-site visit to discuss a quality problem, it is the responsibility of the closest (geographic) division quality head.

Developing Capable Experts

The Quality Director (The Leadership)

The competencies of the Quality Engineer and the Quality Director are described in this section. Bloom's Taxonomy, first proposed by Bloom (1956), and later revised by Anderson and Krathwohl (2001), is utilized in the competency matrices to describe the levels of knowledge and expertise required for each role. Each matrix is composed of two main sections. The first provides a list of competencies, which are further defined by tools and concepts. In the second section, the tools and concepts are paired with anywhere from zero to all of the six levels of Bloom's taxonomy. These levels describe the cognitive demands required to understand and utilize the competencies. For a complete display of the matrices, refer to "Competencies Matrices" in the appendix.

The six levels of Bloom's revised taxonomy (Anderson and Krathwohl 2001, beginning with the lowest order to the highest order, is as follows:

1. *Remembering*. Retrieving, recognizing, and recalling relevant knowledge from long-term memory
2. *Understanding*. Constructing meaning from oral, written, and graphic messages through interpreting, exemplifying, classifying, summarizing, inferring, comparing, and explaining

3. *Applying*. Carrying out or using a procedure through executing or implementing
4. *Analyzing*. Breaking material into constituent parts and determining how the parts relate to one another and to an overall structure or purpose through differentiating, organizing, and attributing
5. *Evaluating*. Making judgments based on criteria and standards through checking and critiquing
6. *Creating*. Putting elements together to form a coherent or functional whole; reorganizing elements into a new pattern or structure through generating, planning, or producing

The Quality Director (or Quality Vice President or Quality Manager, depending on organization size and titles) is someone who leads the Office and works with senior management to ensure that quality is planned, ensured, controlled, and improved. The competencies of a Quality Director are

All of the competencies of a Quality Engineer, including the following:

- *Leadership and management*. Motivating and influencing senior management and others, managing departments, and leading the Office
- *Strategic planning and deployment*. Developing and aligning strategies and goals, and incorporating quality and performance excellence into strategic planning and deployment
- *Customer relationship management*. Identifying customers and needs, utilizing customer feedback, and improving customer satisfaction and loyalty
- *Supply chain management*. Evaluation and selection of suppliers, management and improvement of suppliers, and supplier certification and partnering
- *Quality information system*. Establish metrics, monitoring, and evaluation protocols
- *Training and development*. Skills assessment, training needs analysis, and development of personnel

As mentioned, the Quality Director should also possess all six of the taxonomy levels.

The Quality Engineer or Qualitist: The Technical Experts

The Quality Engineer, or if you like, a "qualitist," as referred to in *Architect of Quality* (2003) by Dr. Juran, is a person who studies the management of quality and is an expert at the deployment of it. The Quality Engineers should have the requisite knowledge, skills, and experience to be a critical enabler, capable of carrying out the three universal processes for managing quality (quality planning, improvement, and control) within the context of a quality system. The competencies of the Quality Engineer are listed under seven categories in the Competency Matrix in the appendix. The seven categories are

1. Basic concepts of quality
2. Quality systems
3. Organization wide assessment
4. Quality planning
5. Quality control
6. Quality improvement
7. Change management

The Competencies for the Officer Engineer require, at a bare minimum, the first three taxonomy levels: remembering, understanding, and applying knowledge. Most competencies, however, require the addition of the next two: analyzing and evaluating. (For more details on both the Quality Engineer and Quality Director, refer to the Competency Matrices in Appendix II.)

For specific examples of duties, desired attributes, skills, experience, and educational backgrounds required for different Officer positions, the reader is referred to the example job profiles in the appendix. These examples showcase the descriptions and requirements for a Quality Vice President, Director, Manager, and Engineer in a multinational global company.

The Performance Excellence Practitioners: The Belts and Change Agents

In addition to the Quality Director and Engineers, the Office of the future will include full-time practitioners: Master and Black Belts, Lean Experts, and Change Agents. These are not new to many organizations. What is new is that their role should be tied to the Office to ensure that these positions continue to keep up with the needs of the organization. For more information on these roles see Chapter 12, The Role of Statistics and Probability.

The following is a list of prerequisite attributes and capabilities that can be used to identify candidates for development into future Office and performance excellence experts (such as Black Belts, Master Black Belts, Lean Experts, Lean Masters, and various other titles).

The candidate

- Has a managerial or technical specialist position
- Has a deep knowledge of business practices and understands your organization's business plan
- Has high-level or university-level math and reading skills; has basic education in data analysis and statistics
- Has been trained in the management of quality methods and tools
- Has a track record of superior performance, including leadership of new initiatives or change initiatives, or has demonstrated capability of it
- Welcomes accountability and challenges, and is willing to take prudent risks
- Has solid technical skills and knowledge in the target environment
- Makes decisions based on facts and data, and searches for "best practices"
- Has perseverance, stability, and is creative yet pragmatic
- Is a good coach, mentor, and teacher; ideally has experience leading teams
- Has credibility at all levels of the organization
- Has experience using common technology and software; not averse to learning new tools

In addition to the skills and experience prerequisites, there are a number of characteristics and traits that will naturally enhance the candidate's ability to drive change. It is beneficial if the candidate

- Is a clear communicator
- Manages stress effectively
- Learns new concepts quickly

- Has a "can-do" attitude and the ability to manage multiple assignments
- Is goal driven and plans ahead
- Is solution oriented
- Is able to work effectively with all levels of the organization
- Stays on top of key aspects and is timely and efficient

Example profiles of key performance excellence roles can be found in Appendix II.

Professional Certification for Key Officers

The introduction of Six Sigma in the past decade led to an insurgence in certification of Belts. This was largely due to a lesson learned from the Total Quality Management (TQM) era. During TQM, many so-called experts were trained in the "methods of TQM." Unfortunately, few were trained in the tools to collect and analyze data. As a result, numerous organizations did not benefit from the TQM programs. Motorola introduced a core curriculum that all Six Sigma practitioners needed to learn. That evolved into a certification program that went beyond the borders of Motorola. As a result, there are many "certifiers" that will provide a certification as a Master Black Belt, Black Belt, Green Belt, and so on. Most certifications state that the person certified is an "expert" in the skills of Six Sigma or Lean or both. Certification did lead to improved performance, but also to some weak experts due to no oversight of the certifiers, many of which were consulting companies or universities not well versed in the methods or tools of Six Sigma and Lean.

The American Society for Quality (ASQ) has, for many years, offered certification for quality technicians, quality auditors, quality engineers, and quality managers. As the Six Sigma movement grew, the ASQ and its affiliates around the world began to certify Black Belts. Although not perfect, the ASQ is in a better position to monitor the certifications than self-serving firms. Certification must be based on legitimacy to be effective. Having too many firms certifying Belts will only lead to a weaker certification process.

ASQ's Certified Quality Engineer (CQE) program is for people who want to understand the principles of product and service quality evaluation and control. For a detailed list of the CQE body of knowledge, the reader is referred to the certification requirements for Certified Quality Engineer at www.asq.org.

ASQ also offers a certification for Officers at the quality management level, called Certified Manager of Quality/Organizational Excellence. ASQ views the Certified Manager of Quality/Organizational Excellence as "a professional who leads and champions process-improvement initiatives—everywhere from small businesses to multinational corporations—that can have regional or global focus in a variety of service and industrial settings. A Certified Manager of Quality/Organizational Excellence facilitates and leads team efforts to establish and monitor customer/supplier relations, supports strategic planning and deployment initiatives, and helps develop measurement systems to determine organizational improvement. The Certified Manager of Quality/Organizational Excellence should be able to motivate and evaluate staff, manage projects and human resources, analyze financial situations, determine and evaluate risk, and employ knowledge management tools and techniques in resolving organizational challenges."

No matter what organization you use to certify your experts, here are some lessons learned about certification:

- One project is not enough to make someone an expert.
- Passing a written test that is not proctored is no guarantee that the person who is supposed to be taking the test is actually taking it.

- Getting your organization to sign off on the success of the Belt project is no guarantee, unless someone in the organization is knowledgeable about the methods of Six Sigma.
- Select a reputable certifying body.

Responsibilities for the Development of Capable Experts

The training and development of quality and performance excellence experts can succeed only if there is accountability and responsibility for its implementation and effectiveness. This accountability and responsibility lies with the same group that it does in any other key competitive or developmental strategy—with the leadership team. It is their responsibility to agree on the strategy and ensure that it will support the other operational, cultural, and financial corporate strategies. They are not responsible for the planning, design, and execution of the development strategy; this responsibility generally lies with a component of the human resource function, with technical support provided by the Office. The responsible parties are executive leadership, human resources, and the Office.

Executive leadership. The executive team bears the responsibility for creating a quality and performance excellence culture in the organization. A quality and performance excellence culture is a product of behaviors, skills, tools, and methods as they are applied to the work. These changes do not come about without showing people "how" to implement and sustain this culture. Therefore, the executive team must become educated in quality and performance excellence, and stimulate their professional development team to offer options for training and development for quality and performance excellence. On the basis of these options, the executive team will then develop and approve a strategy and strategic goals for the training and development effort. This effort may be organization wide and long term (3 to 5 years), or narrowly focused on a particular segment of the organization or product/service line and planned for a relatively short duration.

Human resources. The human resources (HR) function (or subfunction) bears the responsibility for implementing the quality and performance excellence training and development strategy. The implementation activities include the selection of subject matter, training design and delivery, and establishing an evaluation process. This is integrated with other corporate training and development activities, and follows the same implementation process. The subject matter may be internally sourced or may be outsourced to external quality and performance excellence training providers. The major difference between how this is approached now compared to the past is that there is a strong trend to seamlessly integrate the quality and performance excellence training into the professional development curriculum and to include a high degree of customization to reflect the organization's culture. This is especially true for organizations that have a mature quality and performance excellence system in place.

The Office. The Office is responsible for collaborating with the HR professionals to share their technical expertise on quality and performance excellence, much the same as key sales professionals would share their expertise in identifying and developing the curriculum for sales training. This is a departure from the past, when organizations had elaborate (and sometimes very large) quality departments that identified, developed, and delivered quality and performance excellence training, separate from the training department. This created barriers in the implementation of performance excellence as an integral part of all activities (big Q) and contributed to the "quality versus real work" dilemma of the late 1980s and early 1990s.

An underlying principle of quality and performance excellence is to have an unwavering focus on the customer. Training and development for quality and performance excellence demands the same. A clear understanding of who the customers are, what their needs are, and what the features should be of a training and development strategy, the subsequent subject matter that responds to those needs are critical components.

A clear understanding of the customer means that all of those who will participate or benefit from the training must be considered in the design and delivery. Responsive organizations carefully approach this identification of customers and their objectives, and communication of how the training can help achieve those objectives. Many times, because of the lack of such clear definition, organizations waste huge amounts of time and money providing training on tools and techniques that they will never use. For example, providing training on advanced statistical tools to Champions or team members of Lean Six Sigma projects is wasteful. It was not uncommon in the past for organizations to measure success in quality and performance excellence in terms of the number of individuals they trained and the number of subjects in which they were trained!

Customer-Focused Development Approach

By focusing on the customer, the following approach is recommended, based on the design methodology. For more details, the reader is referred to Chap. 14, Lean Techniques and The Shingo Prize. The approach has three major steps: (1) determine the required competencies, (2) assess the experts and potential experts against the required competencies, and (3) close the gaps.

1. Determine the required competencies.

   a. *Identify the customers of the experts and potential experts.* Customers are the recipients of the outputs of the processes or tasks performed (or to be performed) by the experts. Procurement managers, design engineers, operations managers and technicians, process owners, customer service, and senior management are examples of customers.

   b. *Determine customer needs and prioritize them.* As a reminder, needs should be expressed as benefits (not as features). For example, process owners and managers require the benefit of having capable processes (or the benefit of improved process capability). A customer-needs matrix cross-referencing customers and needs is developed. The needs are then prioritized based on the relevance and criticality of each need to the customer base.

   c. *Translate customer needs into competencies and capabilities.* Continuing with the example, translate the benefit (need) of the improved process capability into competencies and capabilities. So the expert(s) would need the following competencies and capabilities: (1) be able to conduct process capability studies and to understand and interpret metrics such as C_p and C_{pk}, and (2) be able to carry out a DMAIC project, including design of experiments (DOE) to determine the X's in the $Y = f(x)$ of that process in order to improve and optimize the process. In many established organizations, the job description and job profiles state the duties, competencies, and capabilities required of the expert. However, it would be wise to verify that those are indeed what are required and that they are adequately described and that the list is complete.

2. Assess the experts (and potential experts) against the required competencies. (The assessment may be done in a variety of ways. A combination of written and oral assessments has been found to be effective.)

   a. *Written assessment.* Questions for the written assessment should be developed based on the required competencies and capabilities. The assessment consists of multiple-choice questions, essay questions, and problem-solving questions requiring computations using formulas that the expert should know. With the process capability example, questions on when and how to conduct a process capability study, what is control versus capability, and calculating C_p and C_{pk} would be appropriate to evaluate whether the expert is able to evaluate process capability.

b. *Oral assessment.* Organization-specific scenarios can be created and described to generate interview questions on how best the scenario or situation might be handled by the expert. For example, if one of the required competencies is being able to deal with resistance when implementing change, specific change management questions based on the scenario might be asked, such as, "What steps would you take if you are faced with personnel resisting and not wanting to implement the solutions your improvement project selected for implementation?"

3. Close the gaps.

From the written and oral assessments, gaps in knowledge, competencies, and capabilities for each individual are identified. A development plan for each individual can be developed, which may include training (self-study, instructor-led, and/or on-the-job), assignment of tasks or project work to provide a means for him or her to demonstrate capability, and/or undergoing a development protocol for internal certification or external certification to a professional certification entity [such as those from the Juran Institute, American Society for Quality, Software Enterprise Institute (CMMI), Project Management Institute, and others].

The assessment results can also be summarized at the organization level to determine organizational gaps and training needs. Identification of the gaps can be conducted for each job grade level or job function. For example, questions such as the following may be of interest: Do we have capable quality engineers? Do our quality managers have the required competencies to be effective in their jobs? Are our Lean Six Sigma Black Belts capable of coaching others to drive process improvement? Are the Green Belts knowledgeable enough to be effective after they have been certified for over a year? Are our quality engineers still capable for 5 years after first attaining CQE certification? An example of an executive summary of a skills assessment conducted by the Juran Institute is shown in the appendix.

Training and development strategies for the organization can be developed by the Office in collaboration with HR and other departments, accordingly. The necessary budgets can be developed and plans put in place for the coming months or fiscal year. The Office and HR can work to deploy the plans, which may include the use of outside training providers and consultants.

References

Anderson, L. W., and Krathwohl, D. R. (eds.) (2001). *A Taxonomy for Learning, Teaching and Assessing: A Revision of Bloom's Taxonomy of Educational Objectives*, complete edition, Longman, New York.

Bloom, B. S. (ed.) (1956). *Taxonomy of Educational Objectives: The Classification of Educational Goals—Handbook I: Cognitive Domain*, McKay, New York.

Crosby, P. B. (2000). "Creating a Useful and Reliable Organization: The Quality Professional's Role." *Annual Quality Congress*, vol. 54, May, pp. 720–722. American Society for Quality, Milwaukee, WI.

De Feo, J. A. (2008). "Enterprise Assurance." *Juran Institute White Paper*, Juran Institute, Southbury, Connecticut.

Gryna, F. M. (2002), "Interview: The Role of Quality and Teams in the 21st Century." *AQP News for a Change*, vol. 6, no. 7, July, pp. 1–3.

Gryna, F. M., Chua, R. C. H., and De Feo, J. A. (2007). "Organization for Quality." Chapter 7 in *Juran's Quality Planning and Analysis for Enterprise Quality*, McGraw-Hill, New York.

Imler, K. (2006). Core Roles in a Strategic Quality System. *Quality Progress*, June, American Society for Quality, Milwaukee, WI. pp. 57–62.

Juran, J. M., and Blanton Godfrey, D. A. (1984). *Managerial Breakthrough: The Classic Book on Improving Management Performance*. McGraw-Hill, New York.

Spichiger, J. (2002). The Changing Role of Quality Professionals. *Quality Progress*, vol. 35, no. 11, November, American Society for Quality, Milwaukee, WI, pp. 31–35.

Watkins, D. K. (2005). Quality Management's Role in Global Sourcing. *Quality Progress*, vol. 38. no. 4, April, American Society for Quality, Milwaukee, WI, pp. 24–31.

Westcott, R. (2004). Metamorphosis of the Quality Professional. *Quality Progress*, October, American Society for Quality, Milwaukee, WI, pp. 22–32.

CHAPTER 4

Quality Planning and Design of New Goods and Services

John Early and Joseph A. De Feo

High Points of This Chapter

1. Designing for quality and innovation is one of the three universal processes of the Juran Trilogy. It is required to achieve breakthroughs, new products, services, and processes.
2. An effective design process requires a robust method and structure to create new products (goods, services, information) and ensure that these together with key operational processes—including process controls—are developed prior to the introduction of the products to the marketplace.

3. The Juran Quality by Design Model consists of following simple steps, primarily leading to a much better understanding of the customers that will benefit from the new product. It is not a statistical design method as Design for Six Sigma is considered. It is often used to design new services and processes. The steps are as follows:
   - Establish the design targets and goals.
   - Define the market and customers that will be targeted.
   - Discover the market, customers, and societal needs.
   - Develop the features of the new design that will meet the needs.
   - Develop or redevelop the processes to produce the features.
   - Develop process controls to be able to transfer the new designs to operations.

4. The Design for Six Sigma model, often called DMADV, consists of a statistical approach to design applicable to manufactured goods. It follows similar steps and incorporates some of the tools in Juran's model:
   - Define the project and the targets.
   - Measure what is critical to customers and quality (CTQs) to establish the required features.
   - Analyze the information and create a high-level design incorporating the CTQs.
   - Design by creating detailed designs, evaluate them, and optimize them before transferring them to operations.
   - Verify the design requirements and execute the final product.

Tackling the First Process of the Trilogy: Designing Innovative Products

An organization's ability to satisfy its customers depends on the robustness of the design processes because the goods you sell and the services you offer originate there.

The design process is the first of the three elements of the Juran Trilogy. It is one of the three basic functions by which management ensures the survival of the organization. The design process enables innovation to happen by designing products (goods, services, or information) together with the processes—including controls—to produce the final outputs. When design is complete, the other two elements—control and improvement—kick in to continuously improve upon the design as customer needs and technology change.

This handbook addresses two versions of the design process. In this chapter, we will discuss the first version, Juran's universal Quality by Design model. It has been in place since 1986 and provides a structure that can be incorporated into an organization's new product development function, or it can be used independently to be carried out project by project as needed.

The second version, Design for Six Sigma (DFSS), which is referred to by the steps in the process DMADV (define, measure, analyze, design, and verify), is the most recent adaption to Juran's model. It builds upon the Six Sigma Improvement or DMAIC (define, measure, analyze, improve, and control) methodology to improve performance. DMADV was first introduced by GE. It uses elements of the Juran model and incorporates many of the statistical tools common to improvement. DFSS is covered in detail in Chap. 17, Continuous Innovation Using Design for Six Sigma.

The Juran model is especially useful for designing products and redesigning processes simply and economically. The authors have witnessed the design of superb products, processes, and services using this model.

Examples include a prize-winning safety program for a multiple-plant manufacturer; an information system that enables both sales and manufacturing to track the procession of an order throughout the entire order fulfillment process so customers can be informed—on a daily basis—of the exact status of their order; and a redesigned accounts receivable system much faster and more efficient than its predecessor.

The DFSS model is the classic model enhanced by the addition of computers and statistical software packages, which permit the utilization of numerous design tools not easily used without a computer. The Six Sigma model is suitable for designing even complex products and for achieving extraordinary levels of quality. Although it is time consuming and expensive in the short term, when executed properly, it produces a healthy return on investment.

The Juran Quality by Design Model

Modern, structured quality design is the methodology used to plan both features that respond to customers' needs and the process to be used to make those features. "Quality by Design" refers to the product or service development processes in organizations. Note the dual responsibility of those who plan: to provide the features to meet customer needs and to provide the process to meet operational needs. In times past, the idea that product design stopped at understanding the features that a product should have was the blissful domain of marketers, salespeople, and research and development people. But this new dual responsibility requires that the excitement generated by understanding the features and customer needs be tempered in the fire of operational understanding.

That is, can the processes make the required features without generating waste? To answer this question requires understanding both the current processes' capabilities and customer specifications. If the current processes cannot meet the requirement, modern design must include finding alternative processes that are capable to meet the need.

The Juran Trilogy points out that the word "quality" incorporates two meanings: first, the presence of features that create customer satisfaction; second, freedom from failures regarding those features is also needed. In short, failures in features create dissatisfactions.

1. Removing failures is the purpose of quality improvement.
2. Creating features is the purpose of Quality by Design.

Kano, Juran, and others have long ago agreed that the absence of failures, that is, no customer dissatisfaction, may not lead us to the belief that satisfaction is thus in hand. We can readily conclude that dissatisfaction goes down as failures are removed. We cannot conclude that satisfaction is therefore going up, because the removal of irritants does not lead to satisfaction—it leads to less dissatisfaction.

It is only the presence of features that creates satisfaction. Satisfaction and dissatisfaction are not co-opposite terms. It is amazing how many organizations fail to grasp this point. Let's take, for example, the typical "bingo card" seen in many hotels. These are replete with "closed-ended" questions. For example, they ask, "How well do you like this on a scale of 1 to 5?" They do not ask, "How well do you like this?" This is the exact opposite of the question "How well don't you like it?" Therefore, any so-called satisfaction rating that does not allow for open-ended questioning such as "What should we do that we are not already doing?" or "Is there someone who provides a service we do not offer?" will always fall into a one-sided dimension of quality understanding. What, then, does a composite score of 3.5 for one branch in a chain of hotels really mean compared to another branch scoring 4.0? It means little. Their so-called satisfaction indices are really dissatisfaction indices.

So we arrive at the basic fundamental of what quality really is. As stated in Chap. 1, Universal Principles of Quality Management, the authors adopted a definition that Juran had postulated long before: "quality" means fitness for use, and we now have extended it to "fitness for purpose." Let's explore this concept.

First, the definition of "fitness for use" takes into account both dimensions of quality—the presence of features and the absence of failures. The sticky points are these: Who gets to decide what "fitness" means? Who decides what "purpose" means? The user decides what "use" means, and the user decides what "fitness" means. Any other answer is bound to lead to argument and misunderstanding. Providers rarely win here. Users, especially society at large, generally always win. For example, take yourself as a consumer. Did you ever use a screwdriver as a pry bar to open a paint can? Of course you did. Did you ever use it to punch holes into a jar lid so your child could catch bugs? Of course you did. Did you ever use it as a chisel to remove some wood, or metal that was in the way of a job you were doing around the house? Of course you did. Now wait just a moment ... a screwdriver's intended use is to drive screws!

So the word "use" has two components, *intended* use and *actual* use. When the user utilizes it in the intended way, both the provider and the user are satisfied. Conformance to specification and fitness for purpose match. But what about when the user uses it in the nonintended way, as in the screwdriver example? What, then, regarding specifications and fitness?

To delve even deeper, how does the user actually use the product? What need is it meeting for the user? Here we find another juncture: the user can create artful new uses for a product. For example:

> *2000 Uses for WD-40.* WD-40 was formulated years ago to meet the needs of the U.S. space program. Not many know the origins of the brand name. "WD" refers to water displacement, and 40 is simply the 40th recipe the company came up with. But as the product moved into the consumer market, all kinds of new uses were uncovered by the users. People claimed it was excellent for removing scuff marks from flooring. They claimed it could easily remove price stickers from lamps, inspection stickers from windshields, and bubble gum from children's hair. The company delighted in all this. But the company didn't release all those clever new uses for public consumption. People also claimed that if they sprayed bait or lures with it, they caught more fish. Those with arthritis swore that a quick spray on a stiff elbow gave them relief. Let's not go too far. What about use where the product obviously cannot work? In Latin there is a word for this: ab-use (abuse), where the prefix "ab" simply means "not."

Some examples will help: back to the screwdriver. You could argue that using the screwdriver as a pry bar, chisel, or punch is abuse of its original designed purpose. But clearly many manufacturers have provided a product that can withstand this abuse, and so use then falls back into the "intended" column (whether this came as a result of lawsuits or from some other source). Further, a look at commercial aircraft "black boxes" (which are orange, by the way), show that they clearly survive in circumstances where the aircraft do not survive. Understanding of use in all its forms is what modern design seeks to achieve.

Last, modern design and planning, as we see over and over, seeks to create features in response to understanding customer needs. We are referring to customer-driven features. The sum of all features is the new product, service, or process.

A different type of product planning in which features meeting no stated need are put out for users to explore is beyond the scope of this chapter. 3M's Post-it Notes and the Internet are examples where we collectively did not voice needs, but which we cannot imagine life without them, once we embraced their features.

The Quality by Design Problem

The Quality by Design model and its associated methods, tools, and techniques have been developed because in the history of modern society, organizations rather universally have demonstrated a consistent failure to produce the goods and services that unerringly delight

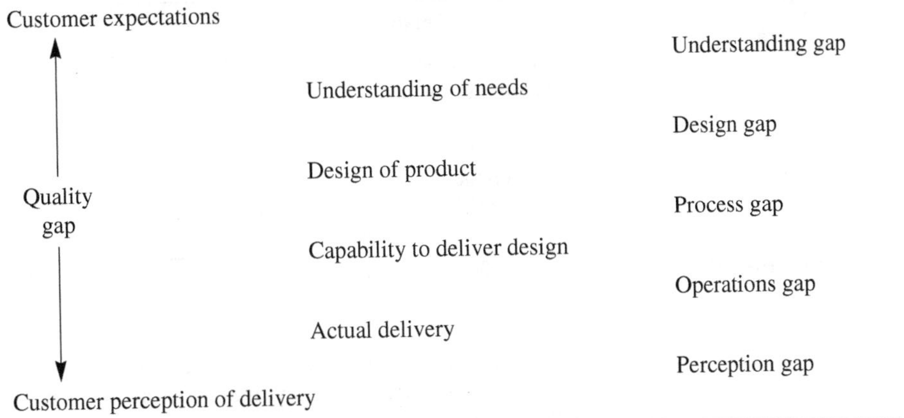

Figure 4.1 The quality gap. (Inspired by A. Parasuraman, Valarie A. Zeithaml, and Leonard L. Berry, "A Conceptual Model for Service Quality and Its Implications for Further Research," *Journal of Marketing*, Fall 1985, pp. 41–50.)

their customers. As a customer, everyone has been dismayed time and again when flights are delayed, radioactive contamination spreads, medical treatment is not consistent with best practices, a child's toy fails to function, a new piece of software is not as fast or user-friendly as anticipated, government responds with glacial speed (if at all), or a home washing machine with the latest high-tech gadget delivers at higher cost clothes that are no cleaner than before. These frequent, large quality gaps are really the compound result of a number of four smaller gaps, illustrated in Fig. 4.1.

The first component of the quality gap is the *understanding gap*, that is, lack of understanding of what the customer needs are. Sometimes this gap is wider because the producer simply fails to consider who the customers are and what they need. More often the gap is there because the supplying organization has erroneous confidence in its ability to understand exactly what the customer really needs. The final perception gap in Fig. 4.1 also arises from a failure to understand customer needs. Customers do not experience a new suit of clothes or the continuity in service from a local utility simply based on the technical merits of the product. Customers react to how they perceive the good or service provides them with a benefit.

The second constituent of the quality gap is a *design gap*. Even if there were perfect knowledge about customer needs and perceptions, many organizations would fail to create designs for their goods and services that are fully consistent with that understanding. Some of this failure arises from the fact that the people who understand customers and the disciplines they use for understanding customer needs are often systematically isolated from those who actually create the designs. In addition, designers—whether they design sophisticated equipment or delicate human services—often lack the simple tools that would enable them to combine their technical expertise with an understanding of the customer needs to create a truly superior product.

The third gap is the *process gap*. Many splendid designs fail because the process by which the physical product is created or the service is delivered is not capable of conforming to the design consistently time after time. This lack of process capability is one of the most persistent and bedeviling failures in the total quality gap.

The fourth gap is the *operations gap*. The means by which the process is operated and controlled may create additional failures in the delivery of the final good or service.

Quality Planning
1. Establish the project and design goals.
2. Identify the customers.
3. Discover the customer needs.
4. Develop the product or service features.
5. Develop the process features.
6. Develop the controls and transfer to operations.

Figure 4.2 Quality by Design steps. (Copyright 1994, Quality by Design, Juran Institute, Inc.)

Quality by Design provides the process, methods, tools, and techniques for closing each of these component gaps and thereby ensuring that any final gap is at a minimum. Figure 4.2 summarizes at a high level the basic steps of Quality by Design. The remainder of this section will provide the details and examples for each of these steps.

Juran Quality by Design Model

We look at each of these as we step through the sequence at a high level.

Step 1: Establish the Project and Design Goals

All design should take place project by project. There is no such thing as design in general; there is only design in specific. In strategic planning, we set out the vision, mission, strategies, objectives, and so on. Each is a specific thing. In product planning, we start with a project, that is, something to plan. We might design a new training room, a new car, a wedding, a customer toll-free hotline, or a new Internet process for bidding on travel booking (such as Priceline .com, Expedia.com). Note that each is a specific thing, and each can be clearly differentiated from anything else. A training room is not a cafeteria, a new car is not a Howitzer, a hotline is not long-distance service, and the travel booking process is not a bookstore online. This is a significant point. Without being able to differentiate what we are designing from anything else, everything collapses into vagueness. So a project is our starting point.

Step 2: Identify the Customers

Going back to the 1980s Total Quality Management (TQM) days, we learned that those who receive the product are customers in some way. If we were designing a training room, the trainees would be an important customer segment. So, too, would the custodians, because they have to clean the room, set it up in different ways, and so on. Customers of the new car include the purchasers, the insurance organizations, the dealers, the carriers, etc. Customers of the hotline include our clients, our service agents, etc. We can include as customers for the travel process the travelers, airlines, and the Web server entity. From all this emerges the basic understanding: A customer is a cast of characters, and each has unique needs that must be met.

Step 3: Discover the Customers' Needs

Wants, needs, perceptions, desires, and other emotions are all involved in our discovery of customer needs. We need to learn how to separate things and prioritize them. But at this point, we need to emphasize that not all high-priority customers (such as the car buyer) are the only ones with high-priority needs. We also stress that just because some customer entity is lower in priority doesn't mean at all that it automatically has lesser-priority needs. We need to understand the "voice of the customer" and the "voice of the market."

Take, for example, the automobile carriers; we simply cannot overlook their needs for the car to be only so high and only so wide. If we ignored their needs, they could stop the product from reaching the cash-paying ultimate customer, our buyer. So, too, could regulators (the various states, the National Highway Transportation Safety Board, the Environmental Protection Agency, etc., impose "needs" that if unmet, could stop the process from going forward at all). So from all this, we reach another point: Customers have to be prioritized in an agreed upon way.

Step 4: Develop the Product or Service Features

The word "feature," as used in product planning, means what the product does, its characteristics, or its functionality. In structured product planning, we adopt a different definition: A feature is the thing that the customer employs to get her or his needs met. For example, in our training room, the trainees need to take notes as they learn. A feature might then be a flip chart, a white board, or a desk. Our custodians might need to move things around quite a bit, so features might include portability, size, weight, and modularity.

As our list of features grows, we soon realize that we cannot possibly have all features at the same priority level. So we need a way to put things in order, once again, and in an agreed upon way. We finalize by optimizing and agreeing on the list of features and the goals for them as well. Note what optimization means: Not all features survive product planning.

Step 5: Develop the Process Features

Because we know that the process is the thing that creates the features, we need to examine current and alternative processes to see which ones will be used to create the features. We need to be sure that the product feature goals can be accomplished via the processes we choose. In other words:

> Process capability must reconcile with product requirements. That statement is very important. No process knows its product goals; product goals come from humans. Ideal product goals would naturally reflect the various customers. But the key issue is this: Variation comes from processes; goals come from humans.

In the example of the training room, process goals might be to reset the room in 20 minutes, keep a supply of flip charts in a closet, certify the trainees to a standard, and so on. As before, we need to list all the possible routes to making the product, select the ones we will use based on some rationale, establish goals for the processes, and reach an optimum.

Step 6: Develop Process Controls and Transfer to Operations

Develop Process Controls

Control is basic to all human activity, from how the body regulates itself as to temperature and metabolism, to financial controls in how we run our organizations or homes. Control consists of three fundamentals:

> In product planning, we need to ensure that the processes work as designed within their capabilities. In the training room, for example, controls might take the form of a checklist for resetting the room and a minimum inventory of flip charts. Control makes use of the concept of the feedback loop.

Here's an example you might keep in mind:

Did you ever check the oil in your car? The dipstick is a form of control point. Note that we begin with a control subject (volume of oil), a unit of measure (quarts or liters), a sensor (you and the dipstick), and a goal (keep the oil somewhere between "full" and "add"—inside those hash marks). Then we move on to sample the process (clean the dipstick, put it back in, remove it, and observe the oil level). Next we adjust when adjustment is called for (oil levels below the add demarcation require us to add oil until we bring the oil up to somewhere between add and full, the agreed goal). If the oil is already within the hash marks, the control activity is to replace the dipstick, shut the hood, and drive on until another checkpoint is reached (perhaps next month). Note that the control activity must reflect the agreed upon goal for control. In the engine oil example, the control point was "inside the hash marks," so the control action is to bring the oil to somewhere "inside the hash marks." Many people miss this point; for example, they add oil until the stick reads "full." This is overcontrol. Control actions must reflect control goals.

Transfer to Operations

Transfer to operations winds up the whole design process. As used here, "operations" means those who run the process, not "manufacturing." To continue the examples used earlier, operations for the training room is the activity of the trainers, the custodians, and the purchasing department. For the new car, operations include manufacturing, transport, dealer relations, and the legal department. For the hotline, operations means the customer service agents who answer the phone. In the travel bidding process, operations include those who shop the bid or reject it and those who maintain the software that interfaces the prospect with the carriers. From the lessons of the era of productivity, the Industrial Revolution, and into the twentieth century, we have learned that the involvement of the operators is key to any well-running process.

With the development of the Ford Taurus came solid understanding of the value of a "platform" team. Designers, engineers, workers, purchasing agents, salespeople, and managers all sat under one roof to develop the car. The concept of platform teams is well ingrained in many car organizations today. The Chrysler Technical Center in Auburn Hills, Michigan, is a later example of such broad collaboration. Thus, successful transfer to operations must include the operators in the design process as early as possible.

The remainder of this section will provide details, practical guidance, and examples for each of these steps.

Juran Quality by Design Model Substeps

Step 1: Establish the Goals and the Project Team

A Quality by Design project is the organized work needed to prepare an organization to deliver a new or revised product, service, or process. The following steps or activities are associated with establishing a Quality by Design project:

1. Identify which projects are required to fulfill the organization's sales or revenue generation strategy.
2. Prepare a goal statement for each project.
3. Establish a team to carry out the project.

Identification of Projects

Deciding which projects to undertake is usually the outgrowth of the strategic and business design of an organization. (See Chap. 7, Strategic Planning and Performance Excellence, for a discussion of how specific projects are deployed from an organization's vision, strategies, and goals.) Typically, design for quality projects create new or updated products that are needed to reach specific strategic goals, to meet new or changing customer needs, to fulfill legal or customer mandates, or to take advantage of a new or emerging technology.

Upper management must take the leadership in identifying and supporting the critical Quality by Design projects. Acting as a design council, council, or similar body, management needs to fulfill the following key roles:

1. *Setting design goals.* Marketing, sales, and similar management functions identify market opportunities and client needs currently not being met. By setting these goals, management is beginning the process to create new products, services, or processes to meet these unmet needs.
2. *Nominating and selecting projects.* The management or council selects the appropriate design projects critical to meeting strategic business and customer goals.
3. *Selecting teams.* Once a project has been identified, a team is appointed to see the project through the remaining steps of the design for quality process. A team may be defined by a project manager in the product development function.
4. *Supporting project team.* New technologies and processes are generally required to meet the new design goals. It is up to management to see that each design team is well prepared, trained, and equipped to carry out its goals. The support may include the following:
   a. Provide education and training in design tools.
   b. Provide a trained project leader to help the team work effectively and learn the design for quality process.
   c. Regularly review team progress.
   d. Approve revision of the project goals.
   e. Identify or help with any issues that may hinder the team.
   f. Provide resource expertise in data analysis.
   g. Furnish resources for unusually demanding data collection such as market studies.
   h. Communicate project results.
5. *Monitoring progress.* The council is responsible for keeping the Quality by Design process on track, evaluating progress, and making midcourse corrections to improve the effectiveness of the entire process. Once the council has reviewed the sources for potential projects, it will select one or more for immediate attention. Next, it must prepare a goal statement for the project.

Prepare Goal Statement

Once the council has identified the need for a project, it should prepare a goal statement that incorporates the specific goal(s) of the project. The goal statement is the written charter for the team that describes the intent and purpose of the project. The team goal describes

- The scope of the project, that is, the product and markets to be addressed
- The goals of the project, that is, the results to be achieved (sales targets)

Writing goal statements requires a firm understanding of the driving force behind the project. The goal helps to answer the following questions:

- Why does the organization want to do the project?
- What will the project accomplish once it is implemented?

A goal statement also fosters a consensus among those who either will be affected by the project or will contribute the time and resources necessary to plan and implement the project goal.

Examples include the following:

- The team goal is to deliver to market a new low-energy, fluorocarbon-free refrigerator that is 25 percent less expensive to produce than similar models.
- The team will create accurate control and minimum cost for the inventory of all stores.

While these goal statements describe what will be done, they are still incomplete. They lack the clarity and specificity required of a complete Quality by Design goal statement that incorporates the goal(s) of a project. Well-written and effective goal statements define the scope of the project by including one or more of the following:

Inherent Performance How the final product will perform on one or more dimensions, for example, 24-hour response time, affects the scope of the project.

Comparative Performance How the final product will perform vis-a-vis the competition, for example, the fastest response time in the metropolitan area, is relevant.

Customer Reaction How will customers rate the product compared with others available? For example, one organization is rated as having a better on-time delivery service than its closest rival.

Voice of Market Who are or will be the customers or target audience for this product, and what share of the market or market niche will it capture, for example, to become the "preferred" source by all business travelers within the continental United States?

Performance Failures How will the product perform with respect to product failure, for example, failure rate of less than 200 for every 1 million hours of use.

Avoidance of Unnecessary Constraints It is important to avoid overspecifying the product for the team; for example, if the product is intended for airline carry-on, specifying the precise dimensions in the goal maybe too restrictive. There may be several ways to meet the carry-on market.

Basis for Establishing Quality Goals In addition to the scope of the project, a goal statement must include the goal(s) of the project. An important consideration in establishing quality goals is the choice of the basis for which the goal(s) are set.

Technology as a Basis In many organizations, it has been the tradition to establish the quality goals on a technological basis. Most of the goals are published in specifications and procedures that define the quality targets for the supervisory and nonsupervisory levels.

The Market as a Basis Quality goals that affect product salability should be based primarily on meeting or exceeding market quality. Because the market and the competition undoubtedly will be changing while the design for quality project is underway, goals should be set so as to meet or beat the competition estimated to be prevailing when the project is completed. Some internal suppliers are internal monopolies. Common examples include payroll preparation, facilities maintenance, cafeteria service, and internal transportation. However, most internal monopolies have potential competitors. There are outside suppliers who offer to sell the same service. Thus the performance of the internal supplier can be compared with the proposals offered by an outside supplier.

Benchmarking as a Basis "Benchmarking" is a recent label for the concept of setting goals based on knowing what has been achieved by others. (See Chap. 18, Benchmarking: Best Practices for Market Leadership.) A common goal is the requirement that the reliability of a new product be at least equal to that of the product it replaces and at least equal to that of the most reliable competing product. Implicit in the use of benchmarking is the concept that the resulting goals are attainable because they have already been attained by others.

History as a Basis A fourth and widely used basis for setting quality goals has been historical performance; that is, goals are based on past performance. Sometimes this is tightened up to stimulate improvement. For some products and processes, the historical basis is an aid to needed stability. In other cases, notably those involving chronically high costs of poor quality, the historical basis helps to perpetuate a chronically wasteful performance. During the goal-setting process, the management team should be on the alert for such misuse of the historical basis.

Goals as a Moving Target It is widely recognized that quality goals must keep shifting to respond to the changes that keep coming over the horizon: new technology, new competition, threats, and opportunities. While organizations that have adopted quality management methods practice this concept, they may not do as well at providing the means to evaluate the impact of those changes and revise the goals accordingly.

Project Goals Specific goals of the project, that is, what the project team is to accomplish, are part of an effective goal statement. In getting the job done, the team must mentally start at the finish. The more focused it is on what the end result will look like, the easier it will be to achieve a successful conclusion.

Measurement of the Goal In addition to stating what will be done and by when, a project goal must show how the team will measure whether it has achieved its stated goals. It is important to spend some time defining how success is measured. Listed below are the four things that can be measured:

1. Quality
2. Quantity
3. Cost
4. Time, speed, agility

An effective Quality by Design project goal must have five characteristics for it to be smart and provide a team with enough information to guide the design process. The goal must be

1. Specific
2. Measurable
3. Agreed to by those affected
4. Realistic—it can be a stretch, but it must be plausible
5. Time-specific—when it will be done

An example of a poorly written goal that is not smart might look something like this: "To design a new life insurance plan for the poor."

Contrast this with the following example: "To design and deliver a whole life plan in less than 90 days that enables poor families to ensure a level of insurance for under $500 per year (at time of introduction). The design also should allow the organization to sell the plans with an average return of between 4 and 6 percent."

The second example is smart—much more detailed, measurable, and time-specific than the first. The target or end result is clearly stated and provides enough direction for the team to plan the features and processes to achieve the goal.

New Product Policies Organizations need to have very clear policy guidance with respect to quality and product development. Most of these should relate to all new products, but specific policies may relate to individual products, product lines, or groups. Four of the most critical policies are as follows.

1. *Failures in new and carryover designs.* Many organizations have established the clear policy that no new product or component of a product will have a higher rate of failures than the old product or component that it is replacing. In addition, they often require that any carryover design have a certain level of performance; otherwise, it must be replaced with a more reliable design. The minimum carryover reliability may be set by one or more of the following criteria: (1) competitor or benchmark reliability, (2) customer requirements, or (3) a stretch goal beyond benchmark or customer requirements.
2. *Intended versus unintended use.* Should stepladders be designed so that the user can stand on the top step without damage, even though the step is clearly labeled "Do Not Step Here?" Should a hospital design its emergency room to handle volumes of routine, nonemergency patients who show up at its doors? These are policy questions that need to be settled before the project begins. The answers can have a significant impact on the final product, and the answers need to be developed with reference to the organization's strategy and the environment within which its products are used.
3. *Requirement of formal Quality by Design process.* A structured, formal process is required to ensure that the product planners identify their customers and design products and processes that will meet those customer needs with minimum failures. Structured formality is sometimes eschewed as a barrier to creativity. Nothing could be more misguided. Formal Quality by Design identifies the points at which creativity is demanded and then encourages, supports, and enables that creativity. Formal design also ensures that the creativity is focused on the customers and that creative designs ultimately are delivered to the customer free of the destructive influences of failures.
4. *Custody of designs and change control.* Specific provision must be made to ensure that approved designs are documented and accessible. Any changes to designs must be validated, receive appropriate approvals, be documented, and be unerringly incorporated into the product or process. Specific individuals must have the assigned authority, responsibility, and resources to maintain the final designs and administer change control.

Establish Team

The cross-functional approach to complete a Quality by Design project is effective for several reasons:

- Team involvement promotes sharing of ideas, experiences, and a sense of commitment to being a part of and helping "our" organization achieve its goal.

- The diversity of team members brings a more complete working knowledge of the product and processes to be planned. Design of a product requires a thorough understanding of how things get done in many parts of the organization.
- Representation from various departments or functions promotes the acceptance and implementation of the new plan throughout the organization. Products or processes designed with the active participation of the affected areas tend to be technically superior and accepted more readily by those who must implement them.

Guidelines for Team Selection When selecting a team, the council identifies those parts of the organization that have a stake in the outcome. There are several places to look:

- Those who will be most affected by the result of the project
- Departments or functions responsible for various steps in the process
- Those with special knowledge, information, or skill in the design of the project
- Areas that can be helpful in implementing the plan

Step 2: Identify the Customers

This step may seem unnecessary; of course, the planners and designers know who their customers are: the driver of the automobile, the depositor in the bank account, the patient who takes the medication. But these are not the only customers—not even necessarily the most important customers. Customers comprise an entire cast of characters that needs to be understood fully.

Generally, there are two primary groups of customers: the external customers—those outside the producing organization—and the internal customers—those inside the producing organization.

Types of External Customers

The term *customer* is often used loosely; it can refer to an entire organization, a unit of a larger organization, or a person. There are many types of customers, some obvious, others hidden. Below is a listing of the major categories to help guide complete customer identification.

The Purchaser This is someone who buys the product for himself or herself or for someone else, for example, anyone who purchases food for his or her family. The end user/ultimate customer is someone who finally benefits from the product, for example, the patient who goes to a health care facility for diagnostic testing.

Merchants These are people who purchase products for resale, wholesalers, distributors, travel agents and brokers, and anyone who handles the product, such as a supermarket employee who places the product on the shelf.

Processors Processors are organizations and people who use the product or output as an input for producing their own product, for example, a refinery that receives crude oil and processes it into different products for a variety of customers.

Suppliers Those who provide input to the process are suppliers, for example, the manufacturer of the spark plugs for an automobile or the law firm that provides advice on the organization's environmental law matters. Suppliers are also customers. They have information needs with respect to product specification, feedback on failures, predictability of orders, and so on.

Potential Customers Those not currently using the product but capable of becoming customers are potential customers; for example, a business traveler renting a car may purchase a similar automobile when the time comes to buy one for personal use.

Hidden Customers Hidden customers comprise an assortment of different customers who are easily overlooked because they may not come to mind readily. They can exert great influence over the product design: regulators, critics, opinion leaders, testing services, payers, the media, the public at large, those directly or potentially threatened by the product, corporate policymakers, labor unions, and professional associations.

Internal Customers

Everyone inside an organization plays three roles: supplier, processor, and customer. Each individual receives something from someone, does something with it, and passes it to a third individual. Effectiveness in meeting the needs of these internal customers can have a major impact on serving the external customers. Identifying the internal customers will require some analysis because many of these relationships tend to be informal, resulting in a hazy perception of who the customers are and how they will be affected. For example, if an organization decides to introduce just-in-time manufacturing to one of its plants, this will have significant effects on purchasing, shipping, sales, operations, and so on.

Most organizations try to set up a mechanism that will allow seemingly competing functions to negotiate and resolve differences based on the higher goal of satisfying customer needs. This might include conducting weekly meetings of department heads or publishing procedure manuals. However, these mechanisms often do not work because the needs of internal customers are not fully understood, and communication among the functions breaks down. This is why a major goal in the design for quality process is to identify who the internal customers are, discover their needs, and plan how those needs will be satisfied. This is also another reason to have a multifunctional team involved in the planning; these are people who are likely to recognize the vested interests of internal customers.

Identifying Customers

In addition to the general guidance just laid out, it is most often helpful to draw a relatively high-level flow diagram of the processes related to the product being planned. Careful analysis of this flow diagram often will provide new insight, identifying customers that might have been missed and refining understanding of how the customers interact with the process. Figure 4.3 is an example of such a diagram. A review of this diagram reveals that the role of "customer" is really two different roles—placing the order and using the product. These may or may not be played by the same individuals, but they are two distinct roles, and each needs to be understood in terms of its needs.

Step 3: Discover Customer Needs

The third step of Quality by Design is to discover the needs of both external customers and internal processors for the product. Some of the key activities required for effective discovery of customer needs include the following:

- Plan to discover customers' needs.
- Collect a list of customers' needs in their language.
- Analyze and prioritize customers' needs.
- Translate their needs into "our" language.
- Establish units of measurement and sensors.

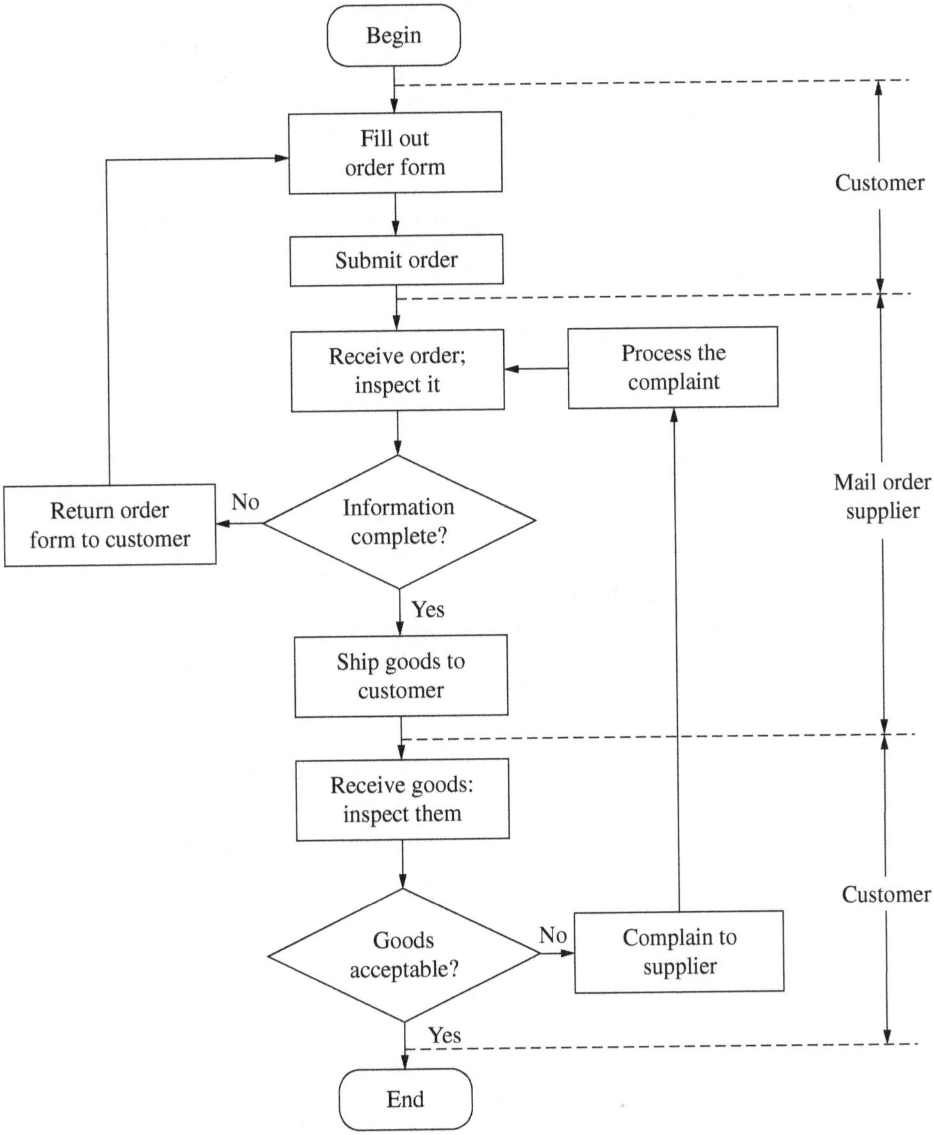

FIGURE 4.3 Flow diagram and customers. (From J. M. Juran, *Quality Control Handbook*, 5th ed., McGraw-Hill, New York, 1999, p. 3.12.)

Our own experience tells us that the needs of human beings are both varied and complex. This can be particularly challenging to a design team because the actions of customers are not always consistent with what they say they want. The challenge for Quality by Design is to identify the most important needs from the full array of those needs expressed or assumed by the customer. Only then can the product delight the customers.

When a product is being designed, there are actually two related but distinct aspects of what is being developed: the technology elements of what the product's features will actually do or how it will function and the human elements of the benefits customers will receive from using the product. The two must be considered together.

Discovering customer needs is a complex task. Experience shows that customers usually do not state, in simple terms, exactly what they want; often they do not even mention some of their most basic needs. Accuracy of bank statements, competence of a physician, reliability of a computer, and grammatical correctness of a publication may be assumed and never stated without probing.

One of the ways customers express their needs is in terms of problems they experience and their expectation that a product will solve their problems. For example, a customer may state, "I cannot always answer my telephone personally, but I do not want callers to be either inconvenienced or disgusted with nonresponsive answering systems." Or the customer may state, "My mother's personal dignity and love of people are very important to me. I want to find an extended care facility that treats her as a person, not a patient." Even when the need is not expressed in such terms, the art and science of discovering needs are to understand exactly the benefit that the customer expects.

When a product's features meet a customer's need, it gives the customer a feeling of satisfaction. If the product fails to deliver the promised feature defect-free, the customer feels dissatisfaction. Even if a product functions the way it has been designed, a competing product, by virtue of superior service or performance, may provide customers with greater satisfaction.

Stated Needs and Real Needs

Customers commonly state their needs as seen from their viewpoint and in their language. Customers may state their needs in terms of the goods or services they wish to buy. However, their real needs are the benefits they believe they will receive.

To illustrate:

Customer wishes to buy:	Benefit customer needs might include:
Fresh pasta	Nourishment and taste
Newest personal computer	Write reports quickly and easily
	Find information on the Web
	Help children learn math
Health insurance	Security against financial disaster
	Access to high-quality health care
	Choice in health care providers
Airline ticket	Transportation, comfort, safety, and convenience

Failure to grasp the difference between stated needs and real needs can undermine the salability of the product in design. Understanding the real needs does not mean that the planners can dismiss the customers' statements and substitute their own superior technical understanding as being the customers' real needs. Understanding the real needs means asking and answering such questions as these:

- Why is the customer buying this product?
- What service does she or he expect from it?
- How will the customer benefit from it?
- How does the customer use it?
- What has created customer complaints in the past?
- Why have customers selected competitors' products over ours?

Perceived Needs

Customers understandably state their needs based on their perceptions. These may differ entirely from the supplier's perceptions of what constitutes product quality. Planners can mislead themselves by considering whether the customers' perceptions are wrong or right rather than focusing on how these perceptions influence customers' buying habits. Although such differences between customers and suppliers are potential troublemakers, they also can be an opportunity. Superior understanding of customer perceptions can lead to competitive advantage.

Cultural Needs

The needs of customers, especially internal customers, go beyond products and processes. They include primary needs for job security, self-respect, respect of others, continuity of habit patterns, and still other elements of what we broadly call the "cultural values"; these are seldom stated openly. Any proposed change becomes a threat to these important values and hence will be resisted until the nature of the threat is understood.

Needs Traceable to Unintended Use

Many quality failures arise because a customer uses the product in a manner different from that intended by the supplier. This practice takes many forms. Patients visit emergency rooms for nonemergency care. Untrained workers are assigned to processes requiring trained workers. Equipment does not receive specified preventive maintenance.

Factors such as safety may add to the cost, yet they may well result in a reduced overall cost by helping to avoid the higher cost arising from misuse of the product. It is essential is to learn the following:

- What will be the actual use (and misuse)?
- What are the associated costs?
- What are the consequences of adhering only to intended use?

Human Safety

Technology places dangerous products into the hands of amateurs who do not always possess the requisite skills to handle them without accidents. It also creates dangerous by-products that threaten human health, safety, and the environment. The extent of all this is so great that much of the effort of product and process design must be directed at reducing these risks to an acceptable level. Numerous laws, criminal and civil, mandate such efforts.

User Friendly

The amateur status of many users has given rise to the term "user friendly" to describe the product feature that enables amateurs to make ready use of technological products. For example, the language of published information should be *simple, unambiguous*, and *readily understood*. (Notorious offenders have included legal documents, owners' operating manuals, administrative forms, etc. Widely used forms such as government tax returns should be field-tested on a sample of the very people who will later be faced with filling out such forms.) The language of published information should also be *broadly compatible*. (For example, new releases of software should be "upward-compatible with earlier releases.")

Promptness of Service

Services should be prompt. In our culture, a major element of competition is promptness of service. Interlocking schedules (as in mail delivery or airline travel) are another source of a growing demand for promptness. Still another example is the growing use of just-in-time manufacturing, which requires dependable deliveries of materials to minimize inventories.

All such examples demonstrate the need to include the element of promptness in design to meet customer needs.

Customer Needs Related to Failures

In the event of product failure, a new set of customer needs emerges—how to get service restored and how to get compensated for the associated losses and inconvenience. Clearly, the ideal solution to all this is to plan quality so that there will be no failures. At this point, we will look at what customers need when failures do occur.

Warranties

The laws governing sales imply that there are certain warranties given by the supplier. However, in our complex society, it has become necessary to provide specific, written contracts to define just what is covered by the warranty and for how long a time. In addition, it should be clear who has what responsibilities.

Effect of Complaint Handling on Sales

While complaints deal primarily with product dissatisfaction, there is a side effect on salability. Research in this area has pointed out the following: of the customers who were dissatisfied with products, nearly 70 percent did not complain. The proportions of these who did complain varied according to the type of product involved. The reasons for not complaining were principally (1) the belief that the effort to complain was not worth it, (2) the belief that complaining would do no good, and (3) lack of knowledge about how to complain. More than 40 percent of the complaining customers were unhappy with the responsive action taken by the suppliers. Again, percentages varied according to the type of product.

Future salability is strongly influenced by the action taken on complaints. This strong influence also extends to brand loyalty. Even customers of popular brands of large-ticket items, such as durable goods, financial services, and automobile services, will reduce their intent to buy when they perceive that their complaints are not addressed.

This same research concluded that an organized approach to complaint handling provides a high return on investment. The elements of such an organized approach may include

- A response center staffed to provide 24-hour access by consumers and/or a toll-free telephone number
- Special training for the employees who answer the telephones
- Active solicitation of complaints to minimize loss of customers in the future

Keeping Customers Informed

Customers are quite sensitive to being victimized by secret actions of a supplier, as the phrase "Let the buyer beware!" implies. When such secrets are later discovered and publicized, the damage to the supplier's quality image can be considerable. In a great many cases, the products are fit for use despite some nonconformances. In other cases, the matter may be debatable. In still other cases, the act of shipment is at least unethical and at worst illegal.

Customers also have a need to be kept informed in many cases involving product failures. There are many situations in which an interruption in service will force customers to wait for an indefinite period until service is restored. Obvious examples are power outages and delays in public transportation. In all such cases, the customers become restive. They are unable to solve the problem—they must leave that to the supplier. Yet they want to be kept informed as to the nature of the problem and especially as to the likely time of solution. Many suppliers are derelict in keeping customers informed and thereby suffer a decline in their quality image. In contrast, some airlines go to great pains to keep their customers informed of the reasons for a delay and of the progress being made in providing a remedy.

Plan to Collect Customers' Needs

Customer needs keep changing. There is no such thing as a final list of customer needs. Although it can be frustrating, design teams must realize that even while they are in the middle of the design process, forces such as technology, competition, social change, and so on can create new customer needs or may change the priority given to existing needs. It becomes extremely important to check with customers frequently and monitor the marketplace. Some of the most common ways to collect customer needs include

1. Customer surveys, focus groups, and market research programs and studies
2. Routine communications, such as sales and service calls and reports, management reviews, house publications
3. Tracking customer complaints, incident reports, letters, and telephone contacts
4. Simulated-use experiments and design processes that involve the customer
5. Employees with special knowledge of the customer: sales, service, clerical, secretarial, and supervisory who come into contact with customers
6. Customer meetings
7. User conferences for the end user
8. Information on competitors' products
9. Personal visits to customer locations; observe and discuss
10. Government or independent laboratory data
11. Changes in federal, state, and local regulations that will identify current need or new opportunity
12. Competitive analysis and field intelligence comparing products with those of competitors
13. Personal experience dealing with the customer and the product (However, it is important to be cautious about giving personal experience too much weight without direct verification by customers. The analysts must remember that looking at customer needs and requirements from a personal viewpoint can be a trap)

Often customers do not express their needs in terms of the benefits they wish to receive from purchasing and using the product.

Collect List of Customers' Needs in Their Language

For a list of customers' needs to have significant meaning in the design of a new product, they must be stated in terms of benefits sought. Another way of saying this is to capture needs in the customer's voice. By focusing on the benefits sought by the customer rather than on the means of delivering the benefit, designers will gain a better understanding of what the customer needs and how the customer will be using the product. Stating needs in terms of the benefits sought also can reveal opportunities for improved quality that often cannot be seen when concentrating on the features alone.

Analyze and Prioritize Customer Needs

The information actually collected from customers is often too broad, too vague, and too voluminous to be used directly in designing a product. Both specificity and priority are needed to ensure that the design really meets the needs and that time is spent on designing for those needs that are truly the most important. The following activities help provide this precision and focus:

- Organizing, consolidating, and prioritizing the list of needs for both internal and external customers
- Determining the importance of each need for both internal and external customers
- Breaking down each need into precise terms so that a specific design response can be identified
- Translating these needs into the supplying organization's language
- Establishing specific measurements and measurement methods for each need

One of the best design tools to analyze and organize customers' needs is the Quality by Design spreadsheet.

Quality by Design Spreadsheets

Designing new products can generate large amounts of information that is both useful and necessary, but without a systematic way to approach the organization and analysis of this information, the design team may be overwhelmed by the volume and miss the message it contains.

Although planners have developed various approaches for organizing all this information, the most convenient and basic design tool is the Quality by Design spreadsheet. The spreadsheet is a highly versatile tool that can be adapted to a number of situations. The Quality by Design process makes use of several kinds of spreadsheets, such as

- Customer needs spreadsheet
- Needs analysis spreadsheet
- Product or service design spreadsheet
- Process design spreadsheet
- Process control spreadsheet

Besides recording information, these tools are particularly useful in analyzing relationships among the data that have been collected and in facilitating the stepwise conversion of customer needs into features and then features into process characteristics and plans. This conversion is illustrated in Fig. 4.4. Analysis of customers and their needs provides the basis for designing the product. The summary of that design feeds the process design, which feeds the control spreadsheet.

For most design projects, simple matrix spreadsheets will suffice. For other projects, more complex quality functional deployment spreadsheets are helpful in computing design tradeoffs. All these spreadsheets are designed to allow the team to record and compare the relationships among many variables at the same time. We will illustrate some of these spreadsheets at the appropriate point in the design process. Figure 4.5 illustrates the generic layout of any one of these spreadsheets. In general, the row headings are the "whats" of the analysis—the customers to be satisfied, the needs to be met, and so on. The columns are the "hows"—the needs that, when met, will satisfy the customer, the features that will meet the needs, and so on. The bottom row of the spreadsheet generally contains specific measurable goals for the how at the top. The body of the spreadsheet expresses with symbols or numerics the impact of the how on the what, for example,

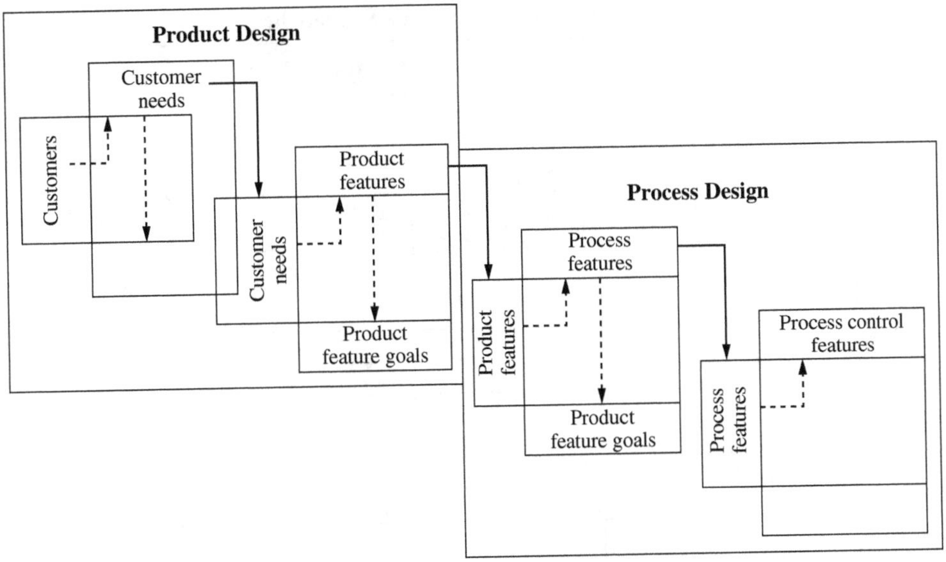

FIGURE 4.4 Sequence of activities. (Juran Institute, Inc. Used by permission.)

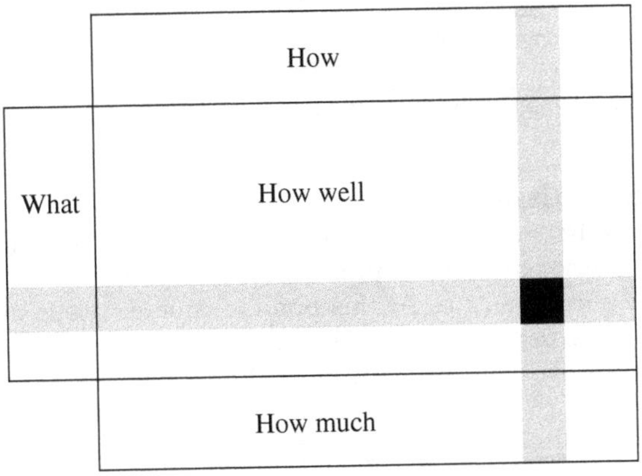

FIGURE 4.5 Planning spreadsheet. (Juran Institute, Inc. Copyright 1994.)

none, moderate, strong, very strong. Other columns can be added to give specific measures of the importance of the respective rows, benchmarks, and so on.

Customer Needs Spreadsheet

Figure 4.6 provides a simple example of a customer needs spreadsheet. The left column lists, in priority order, all the external and internal customers. The column headings are the various needs that have been discovered. By either checking or entering a designation for importance, it is possible to create a simple but comprehensive picture of the importance of meeting each need. All product development must operate within a budget. Prioritizing the customers and their needs ensures that the budget is focused on what is most important.

Customer Needs Spreadsheet

Customers	Customer Needs							
	Attractive	Informative/ well-written articles	Catchy cover lines	Stable circulation	It sells	Enough time	Material complete	No last minute changes
Readers	●	●	●					
Advertisers	●	○	●	●	●			
Printers						●	●	●
Typesetters						●	●	●
Color separators						●	●	●
Newsstand	●	○	●	○	●			

Legend ● Very strong ○ Strong △ Weak

Figure 4.6 Customer needs spreadsheet. (Juran Institute, Inc. Copyright 1994.)

Precise Customer Needs

Once the needs that must be met have been prioritized, they must be described in sufficiently precise terms to design a product based on them. A customer needs spreadsheet helps assemble this analysis. At this point, customer needs are probably a mixture of relatively broad expectations such as "ease of use" and more specific requests such as "access on Saturday." Figure 4.7 illustrates how broad needs (called *primary*) are broken

Primary Need	Secondary Need	Tertiary Need
Convenience	Hours of operation	Open between 5:00 and 9:00 p.m. Saturday hours
	Transportation access	Within three blocks of bus stop Ample parking
	Short wait times	Urgent appointment within 24 hours Routine appointment within 14 days Waiting time at appointment less than 15 minutes
	Complementary services available	Pharmacy on site Lab on site

Figure 4.7 Needs analysis spreadsheet for medical office. (Juran Institute, Inc.)

into succeeding levels of specificity (secondary, tertiary, etc.). Note that primary and secondary do not mean more and less important; they mean, respectively, less specific and more specific. Each need must be broken down to the level at which it can (1) be measured and (2) serve as an unambiguous guide for product design. In some cases two levels of detail may suffice; in others four or five may be required. Figure 4.7 illustrates how this might be done for the primary need "convenience" associated with a group medical practice.

Translate Their Needs into "Our" Language

The precise customer needs that have been identified may be stated in any of several languages, including

- The customer's language
- The supplier's ("our") language
- A common language

An old aphorism claims that the British and Americans are separated by a common language. The appearance of a common language or dialect can be an invitation to trouble because both parties believe that they understand each other and expect to be understood. Failure to communicate because of the unrecognized differences can build additional misunderstanding that only compounds the difficulty. It is imperative, therefore, for planners to take extraordinary steps to ensure that they properly understand customer needs by systematically translating them. The need to translate applies to both internal and external customers. Various organization functions employ local dialects that are often not understood by other functions.

Vague terminology constitutes one special case for translation that can arise even (and often especially) between customers and suppliers who believe they are speaking the same dialect. Identical words have multiple meanings. Descriptive words do not describe with technological precision.

Aids to Translation

Numerous aids are available to clear up vagueness and create a bridge across languages and dialects. The most usual listed are the following: A glossary is a list of terms and their definitions. It is a published agreement on the precise meanings of key terms. The publication may be embellished by other forms of communication, such as sketches, photographs, and videotapes.

Samples can take many forms, such as physical goods (e.g., textile swatches, color chips, audio cassettes) or services (e.g., video recordings to demonstrate "samples" of good service—courtesy, thoughtfulness, etc.). They serve as specifications for features. They make use of human senses beyond those associated with word images.

A special organization to translate communications with external customers may be required because of the high volume of translation. A common example is the order-editing department, which receives orders from clients. Some elements of these orders are in client language. Order editing translates these elements into supplier language, for example, product code numbers, supplier acronyms, and so on.

Standardization is used by many mature industries for the mutual benefit of customers and suppliers. This standardization extends to language, products, processes, and so on. All organizations make use of short design actions for their products, such as code numbers, acronyms, words, phrases, and so on. Such standardized nomenclature makes it easy to communicate with internal customers.

Measurement is the most effective remedy for vagueness and multiple dialects—"Say it in numbers." This is the first, but not the last, point in the design process where measurement is critical. Design for quality also requires measurement of features, process features, process capability, control subjects, and so on.

Establish Units of Measurement and Sensors

Sound Quality by Design requires precise communication between customers and suppliers. Some of the essential information can be conveyed adequately by words. However, an increasingly complex and specialized society demands higher precision for communicating quality-related information. The higher precision is best attained when we say it in numbers.

Quantification Requires a System of Measurement Such a system consists of a unit of measurement, which is a defined amount of some quality feature and permits evaluation of that feature in numbers, for example, hours of time to provide service, kilowatts of electric power, or concentration of a medication.

A sensor, which is a method or instrument of measurement, carries out the evaluation and states the findings in numbers in terms of the unit of measure, for example, a clock for telling time, a thermometer for measuring temperature, or an x ray to measure bone density.

By measuring customer needs, one has established an objective criterion for whether the needs are met. In addition, only with measurement can one answer questions such as these: Is our quality getting better or worse? Are we competitive with others? Which one of our operations provides the best quality? How can we bring all operations up to the level of the best?

Units of Measure for Features

The first task in measurement is to identify the appropriate unit of measurement for each customer need. For features, we know of no simple, convenient, generic formula that is the source of many units of measure. The number and variety of features are simply enormous. In practice, each product feature requires its own unique unit of measure. A good starting point is to ask the customers what their units of measure are for evaluating product quality. If the supplier's units of measure are different, the stage is set for customer dissatisfaction, and the team will need to come up with a unit of measure acceptable to both parties. Even if the customers have not developed an explicit unit of measure, ask them how they would know whether their needs were met. Their response may carry with it an implicit unit of measure.

Application to Goods

Units of measure for quality features of goods make extensive use of "hard" technological units. Some of these are well known to the public: time in minutes, temperature in degrees, or electric current in amperes. Many others are known only to the specialists. There are also "soft" areas of quality for goods. Food technologists need units of measure for flavor, tenderness, and still other properties of food. Household appliances must be "handsome" in appearance. Packaging must be "attractive." To develop units of measure for such features involves much effort and ingenuity.

Application to Services

Evaluation of service quality includes some technological units of measure. A widespread example is promptness, which is measured in days, hours, and so on. Environmental pollutants (e.g., noise, radiation, etc.) generated by service organizations are likewise measured using technological units of measure.

Service quality also involves features such as courtesy of service personnel, decor of surroundings, and readability of reports. Since these features are judged by human beings, the units of measure (and the associated sensors) must be shown to correlate with a jury of customer opinion.

The Ideal Unit of Measure

The criteria for an ideal unit of measure are summarized below. An ideal unit of measure

- Is understandable
- Provides an agreed upon basis for decision making
- Is conducive to uniform interpretation
- Is economical to apply
- Is compatible with existing designs of sensors, if other criteria also can be met

Measuring Abstractions

Some quality features seem to stand apart from the world of physical things. Quality of service often includes courtesy as a significant quality feature. Even in the case of physical goods, we have quality features, such as beauty, taste, aroma, feel, or sound. The challenge is to establish units of measure for such abstractions.

The approach to dealing with abstractions is to break them up into identifiable pieces. Once again, the customer may be the best source to start identifying these components. For example, hotel room appearance is certainly a quality feature, but it also seems like an abstraction. However, we can divide the feature into observable parts and identify those specifics that collectively constitute "appearance," for example, the absence of spots or bare patches on the carpet, clean lavatory, linens free from discoloration and folded to specified sizes, windows free of streaks, bedspreads free of wrinkles and hanging to within specific distances from the floor, and so on. Once units of measure have been established for each piece or component, they should be summarized into an index, for example, number of soiled or damaged carpets to total number of hotel rooms, number of rooms with missing linens to total number of rooms, or number of customer complaints.

Establish the Sensor

To say it in numbers, not only do we need a unit of measure, but also we need to evaluate quality in terms of that unit of measure. A key element in making the evaluation is the sensor.

A *sensor* is a specialized detecting device or measurement tool. It is designed to recognize the presence and intensity of certain phenomena and to convert this sense knowledge into information. In turn, the resulting information becomes an input to decision making because it enables us to evaluate actual performance.

Technological instruments are obviously sensors. So are the senses of human beings. Trends in some data series are used as sensors. Shewhart control charts are sensors.

Precision and Accuracy of Sensors

The *precision* of a sensor is a measure of the ability of the sensor to reproduce its results over and over on repeated tests. For most technological sensors, this reproducibility is high and is also easy to quantify.

At the other end of the spectrum are the cases in which we use human beings as sensors: inspectors, auditors, supervisors, and appraisers. Human sensors are notoriously less precise than technological sensors. Such being the case, planners are well advised to understand the limitations inherent in human sensing before making decisions based on the resulting data.

The *accuracy* of a sensor is the degree to which the sensor tells the truth—the extent to which its evaluation of some phenomenon agrees with the "true" value as judged by an established standard. The difference between the observed evaluation and the true value is the *error*, which can be positive or negative.

For technological sensors, it is usually easy to adjust for accuracy by recalibrating. A simple example is a clock or watch. The owner can listen to the time signals provided over the radio. In contrast, the precision of a sensor is not easy to adjust. The upper limit of precision is usually inherent in the basic design of the sensor. To improve precision beyond its upper limit requires a redesign. The sensor may be operating at a level of precision below that of its capability owing to misuse, inadequate maintenance, and so on. For this reason, when choosing the appropriate sensor for each need, planners will want to consider building in appropriate maintenance schedules along with checklists of actions to be taken during the check.

Translating and Measuring Customer Needs

The customer need for performance illustrates how high-level needs break down into myriad detailed needs. Performance includes all the following detailed, precise needs:

Product Design Spreadsheet All the information on the translation and measurement of a customer need must be recorded and organized. Experience recommends placing these data so that they will be close at hand during product design. The example in Fig. 4.8 shows a few needs all prepared for use in product design. The needs, their translation, and their measurement are all placed to the left of the spreadsheet. The remainder of the spreadsheet will be discussed in the next section.

Step 4: Develop the Product or Service Features

Once the customers and their needs are fully understood, we are ready to design the organization. Most organizations have some process for designing and bringing new products to market. In this step of the Quality by Design process, we will focus on the role of quality in product development and how that role combines with the technical aspects of development and design appropriate for a particular industry. Within product development, product design is a creative process based largely on technological or functional expertise.

The designers of products traditionally have been engineers, systems analysts, operating managers, and many other professionals. In the quality arena, designers can include any whose experience, position, and expertise can contribute to the design process. The outputs of product design are detailed designs, drawings, models, procedures, specifications, and so on.

The two overall quality objectives for this step are:

1. Determine which features and goals will provide the optimal benefit for the customer.
2. Identify what is needed so that the designs can be delivered without failures.

In the case of designing services, the scope of this activity is sometimes puzzling. For example, in delivering health care, where does the product of diagnosing and treating end and the processes of laboratory testing, chart reviews, and so on begin? One useful way to think about the distinction is that the product is the "face to the customer." It is what the customer sees and experiences. The patient sees and experiences the physician interaction, waiting time, clarity of information, and so on. The effectiveness and efficiency of moving blood samples to and around the laboratory have an effect on these features but are really features of the process that delivers the ultimate product to the customer.

Those who are designing physical products also can benefit from thinking about the scope of product design. Given that the customer's needs are the benefits that the customer wants from the product, the design of a piece of consumer electronics includes not only the

Feature Design Spreadsheet

Needs	*Translation*	*Units of Measure*	*Sensors*	Features: Cross resource checking	Auto search for open times	Check resource constraints	FAX info. to scheduling source	Mail instructions to patient	●	●	●		
No double bookings	Double bookings	Yes/No	Review by scheduler	●									
Pt. comes prepared	Pt. followed MD's instructions	Yes/No/Partial	Review by person doing procedure				△	●					
All appointments used	No "holds" used	Yes/No	Review by scheduler		●	○							
All info. easy to find	No "holds" used	Yes/No	Review by scheduler			○							
Quick confirmation	Quick confirmation	Minutes	Software/review by scheduler		○	○							
			Targets and goals	100% of time for all info. entered	One keystroke	Cannot change appt. w/o author from source	Reminder always generated for receiver	For all appointments	●				

Key

- ● Very strong relationship
- ○ Strong relationship
- △ Weak relationship

FIGURE 4.8 Product design spreadsheet. (Juran Institute, Inc. Copyright 1999.)

contents of the box itself but also the instructions for installation and use and the help line for assistance. There are six major activities in this step:

1. Group together related customer needs.
2. Determine methods for identifying features.
3. Select high-level features and goals.
4. Develop detailed features and goals.
5. Optimize features and goals.
6. Set and publish final product design.

Group Together Related Customer Needs

Most Quality by Design projects will be confronted with a large number of customer needs. Based on the data developed in the preceding steps, the team can prioritize and group together those needs that relate to similar functionality. This activity does not require much time, but it can save a lot of time later. Prioritization ensures that the scarce resources of product development are spent most effectively on those items that are most important to the customer. Grouping related needs together allows the design team to "divide and conquer," with subteams working on different parts of the design. Such subsystem or component approaches to design, of course, have been common for years. What may be different here is that the initial focus is on the components of the customers' needs, not the components of the product. The component design for the product will come during the later activities in this step.

Determine Methods for Identifying Features

There are many complementary approaches for identifying the best product design for meeting customers' needs. Most design projects do not use all of them. Before starting to design, however, a team should develop a systematic plan for the methods it will use in its own design. Here are some of the options:

Benchmarking This approach identifies the best in class and the methods behind it that make it best. See Chap. 18, Benchmarking: Best Practices for Market Leadership, for details.

Basic Research One aspect of research might be a new innovation for the product that does not currently exist in the market or with competitors. Another aspect of basic research looks at exploring the feasibility of the product and features. While both these aspects are important, be careful that fascination with the technological abilities of the product does not overwhelm the primary concern of its benefits to the customer.

Market Experiments Introducing and testing ideas for features in the market allows one to analyze and evaluate concepts. The focus group is one technique that can be used to measure customer reactions and determine whether the features actually will meet customer needs. Some organizations also try out their ideas, on an informal basis, with customers at trade shows and association meetings. Still others conduct limited test marketing with a prototype product.

Creativity Developing features allows one to dream about a whole range of possibilities without being hampered by any restrictions or preconceived notions. Design for quality is a proven, structured, data-based approach to meeting customers' needs. But this does not mean it is rigid and uncreative. At this point in the process, the participants in design must be encouraged and given the tools they need to be creative so as to develop alternatives for design. After they have selected a number of promising alternatives, they will use hard analysis and data to design the final product.

Design teams can take advantage of how individuals view the world: from their own perspective. Every employee potentially sees other ways of doing things. The team can encourage people to suggest new ideas and take risks. Team members should avoid getting "stuck" or taking too much time to debate one particular idea or issue. They can put it aside and come back to it later with a fresh viewpoint. They can apply new methods of thinking about customers' needs or problems, such as the following:

- *Change in key words or phrases.* For example, call a "need" or "problem" an "opportunity." Instead of saying, "Deliver on time," say, "Deliver exactly when needed."
- *Random association.* For example, take a common word such as "apple" or "circus" and describe your business, product, or problem as the word. For example, "Our product is like a circus because ... "
- *Central idea.* Shift your thinking away from one central idea to a different one. For example, shift the focus from the product to the customer by saying, "What harm might a child suffer, and how can we avoid it?" rather than "How can we make the toy safer?"
- *Putting yourself in the other person's shoes.* Examine the question from the viewpoint of the other person, your competitor, your customer—and build their case before you build your own.
- *Dreaming.* Imagine that you had a magic wand that you could wave to remove all obstacles to achieving your objectives. What would it look like? What would you do first? How would it change your approach?
- *The spaghetti principle.* When you have difficulty considering a new concept or how to respond to a particular need, allow your team to be comfortable enough to throw out a new idea, as if you were throwing spaghetti against the wall, and see what sticks. Often even "wild" ideas can lead to workable solutions.

The initial design decisions are kept as simple as possible at this point. For example, the idea of placing the control panel for the radio on the steering wheel would be considered a high-level product feature. Its exact location, choice of controls, and how they function can be analyzed later in greater detail. It may become the subject of more detailed features as the design project progresses.

Standards, Regulations, and Policies This is also the time to be certain that all relevant standards, regulations, and policies have been identified and addressed. While some of these requirements are guidelines for how a particular product or product feature can perform, others mandate how they must perform. These may come from inside the organization, and others may come from specific federal, state, or local governments; regulatory agencies; or industry associations. All features and product feature goals must be analyzed against these requirements prior to the final selection of features to be included in the design.

It is important to note that if there is a conflict when evaluating features against any standards, policies, or regulations, it is not always a reason to give up. Sometimes one can work to gain acceptance for a change when it will do a better job of meeting customer needs. This is especially true when it comes to internal policies. However, an advocate for change must be prepared to back up the arguments with appropriate data.

Criteria for Design As part of the preparation for high-level design, the design team must agree on the explicit criteria to be used in evaluating alternative designs and design features. All designs must fulfill the following general criteria:

- Meet the customers' needs
- Meet the suppliers' and producers' needs
- Meet (or beat) the competition
- Optimize the combined costs of the customers and suppliers

In addition to the preceding four general criteria, the team members should agree explicitly on the criteria that they will use to make a selection. (If the choices are relatively complex, the team should consider using the formal discipline of a selection matrix.) One source for these criteria will be the team's vision statement and goals. Some other types of criteria that the team may develop could include

- The impact of the feature on the needs
- The relative importance of the needs being served
- The relative importance of the customers whose needs are affected
- The feasibility and risks of the proposed feature
- The impact on product cost
- The relationship to competitive features uncovered in benchmarking
- The requirements of standards, policies, regulations, mandates, and so on

As part of the decision on how to proceed with design, teams also must consider a number of other important issues regarding what type of product feature will be the best response to customers' needs. When selecting features, they need to consider whether to

- Develop an entirely new functionality
- Replace selected old features with new ones
- Improve or modify existing features
- Eliminate the unnecessary

Select High-Level Features and Goals

This phase of Quality by Design will stimulate the team to consider a whole array of potential features and how each would respond to the needs of the customer. This activity should be performed without being constrained by prior assumptions or notions as to what worked or did not work in the past. A response that previously failed to address a customer need or solve a customer problem might be ready to be considered again because of changes in technology or the market.

The team begins by executing its plan for identifying the possible features. It should then apply its explicit selection criteria to identify the most promising features.

The product design spreadsheet in Fig. 4.8 is a good guide for this effort. Use the right side of the spreadsheet to determine and document the following:

- Which features contribute to meeting which customer needs
- That each priority customer need is addressed by at least one product feature
- That the total impact of the features associated with a customer need is likely to be sufficient for meeting that need

- That every product feature contributes to meeting at least one significant customer need
- That every product feature is necessary for meeting at least one significant customer need (i.e., removing that feature would leave a significant need unmet)

Team Sets Goals for Each Feature In quality terms, a goal is an aimed-at quality target (such as aimed-at values and specification limits). As discussed earlier, this differs from quality standards in that the standard is a mandated model to be followed that typically comes from an external source. While these standards serve as "requirements" that usually dictate uniformity or how the product is to function, product feature goals are often voluntary or negotiated. Therefore, the Quality by Design process must provide the means for meeting both quality standards and quality goals.

Criteria for Setting Product Feature Goals As with all goals, product feature goals must meet certain criteria. While the criteria for establishing product feature goals differ slightly from the criteria for project goals verified in step 1, there are many similarities. Product feature goals should encompass all the important cases and be

- Measurable
- Optimal
- Legitimate
- Understandable
- Applicable
- Attainable

Measuring Features Goals Establishing the measurement for a product feature goal requires the following tasks:

- Determine the unit of measure: meters, seconds, days, percentages, and so on.
- Determine how to measure the goal (i.e., determine what the sensor is).
- Set the value for the goal.

The work done in measuring customer needs should be applied now. The two sets of measurements may be related in one of the following ways:

- Measurement for the need and for the product feature goal may use the same units and sensors. For example, if the customer need relates to timeliness measured in hours, one or more features normally also will be measured in hours, with their combined effects meeting the customer need.
- Measurement for the product feature may be derived in a technical manner from the need measurement. For example, a customer need for transporting specified sizes and weights of loads may be translated into specific engineering measurements of the transport system.
- Measurement for the product feature may be derived from a customer behavioral relationship with the product feature measure. For example, automobile manufacturers have developed the specific parameters for the dimensions and structure of an automobile seat that translate into the customer rating it "comfortable."

Since we can now measure both the customer need and the related product feature goals, it is possible for the Quality by Design team to ensure that the product design will go a long way toward meeting the customers' needs, even before building any prototypes or conducting any test marketing.

For large or complex projects, the work of developing features is often divided among a number of different individuals and work groups. After all these groups have completed their work, the overall Quality by Design team will need to integrate the results. Integration includes

- Combining features when the same features have been identified for more than one cluster
- Identifying and resolving conflicting or competing features and goals for different clusters
- Validating that the combined design meets the criteria established by the team

Develop Detailed Features and Goals

For large and highly complex products, it will usually be necessary to divide the product into a number of components and even subcomponents for detailed design. Each component will typically have its own design team that will complete the detailed design described below. To ensure that the overall design remains integrated, consistent, and effective in meeting customer needs, these large, decentralized projects require

- A steering or core team that provides overall direction and integration
- Explicit charters with quantified goals for each component
- Regular integrated design reviews for all components
- Explicit integration of designs before completion of the product design phase

Once the initial detailed features and goals have been developed, then the technical designers will prepare a preliminary design, with detailed specifications. This is a necessary step before a team can optimize models of features using a number of Quality by Design tools and ultimately set and publish the final features and goals.

It is not uncommon for Quality by Design teams to select features at so high a level that the features are not specific enough to respond to precise customer needs. Just as in the identification of customers' primary needs, high-level features need to be broken down further into terms that are clearly defined and can be measured.

Optimize Features and Goals

Once the preliminary design is complete, it must be optimized. That is, the design must be adjusted so that it meets the needs of both customer and supplier while minimizing their combined costs and meeting or beating the competition.

Finding the optimum can be a complicated matter unless it is approached in an organized fashion and follows quality disciplines. For example, there are many designs in which numerous variables converge to produce a final result. Some of these designs are of a business nature, such as design of an information system involving optimal use of facilities, personnel, energy, capital, and so on. Other such designs are technological, involving optimization of the performance of hardware. Either way, finding the optimum is made easier through the use of certain quality disciplines.

Finding the optimum involves balancing the needs, whether they are multiorganizational needs or within-organization needs. Ideally, the search for the optimum should be done through the participation of suppliers and customers alike. There are several techniques that help achieve this optimum.

Design Review Under this concept, those who will be affected by the product are given the opportunity to review the design during various formative stages. This allows them to use their experience and expertise to make such contributions as

- Early warning of upcoming problems
- Data to aid in finding the optimum
- Challenge to theories and assumptions

Design reviews can take place at different stages of development of the new product. They can be used to review conclusions about customer needs and hence the product specifications (characteristics of product output). Design reviews also can take place at the time of selecting the optimal product design. Typical characteristics of design reviews include the following:

- Participation is mandatory.
- Reviews are conducted by specialists, external to the design team.
- Ultimate decisions for changes remain with the design team.
- Reviews are formal, scheduled, and prepared for with agendas.
- Reviews will be based on clear criteria and predetermined parameters.
- Reviews can be held at various stages of the project.

Ground rules for good design reviews include

- Adequate advance design review of agenda and documents
- Clearly defined meeting structure and roles
- Recognition of interdepartmental conflicts in advance
- Emphasis on constructive—not critical—inputs
- Avoidance of competitive design during review
- Realistic timing and schedules for the reviews
- Sufficient skills and resources provided for the review
- Discussion focus on untried/unproved design ideas
- Participation directed by management

Multifunctional Design Teams Design teams should include all those who have a vested interest in the outcome of the design of the product along with individuals skilled in product design. Under this concept, the team, rather than just the product designers, bears responsibility for the final design.

Structured Negotiation Customers and suppliers are tugged by powerful local forces to an extent that can easily lead to a result other than the optimum. To ensure that these negotiating sessions proceed in as productive a fashion as possible, it is recommended that ground rules be established before the meetings. Here are some examples:

- The team should be guided by a spirit of cooperation, not competition, toward the achievement of a common goal.
- Differences of opinion can be healthy and can lead to a more efficient and effective solution.
- Everyone should have a chance to contribute, and every idea should be considered.
- Everyone's opinions should be heard and respected without interruptions.
- Avoid getting personal; weigh the pros and cons of each idea, looking at its advantages before its disadvantages.
- Challenge conjecture; look at the facts.
- Whenever the discussion bogs down, go back and define areas of agreement before discussing areas of disagreement.
- If no consensus can be reached on a particular issue, it should be tabled and returned to later on in the discussion.

Create New Options Often teams approach a product design with a history of how things were done in the past. Optimization allows a team to take a fresh look at the product and create new options. Some of the most common and useful quality tools for optimizing the design include the following:

- Competitive analysis provides feature-by-feature comparison with competitors' products. (See below for an example.)
- Salability analysis evaluates which features stimulate customers to be willing to buy the product and the price they are willing to pay. (See below for an example.)
- Value analysis calculates not only the incremental cost of specific features of the product but also the cost of meeting specific customer needs and compares the costs of alternative designs. (See below for an example.)
- Critical analysis identifies the "vital few" features that are vulnerable in the design so that they can receive priority for attention and resources.
- Failure mode and effect analysis (FMEA) calculates the combined impact of the probability of a particular failure, the effects of that failure, and the probability that the failure can be detected and corrected, thereby establishing a priority ranking for designing in failure prevention countermeasures.
- Fault-tree analysis aids in the design of preventive countermeasures by tracing all possible combinations of causes that could lead to a particular failure.
- Design for manufacture and assembly evaluates the complexity and potential for problems during manufacture to make assembly as simple and error free as possible. Design for maintainability evaluates particular designs for the ease and cost of maintaining them during their useful life.

Competitive Analysis Figure 4.9 is an example of how a competitive analysis might be displayed. The data for a competitive analysis may require a combination of different approaches such as laboratory analysis of the competitors' products, field testing of those products, or

Product Feature & Goal	Check if Product Feature is Present			Feature Performance vs. Goal (*)			Identify if Significant Risk or Opportunity
	Product A	Product B	Ours	Product A	Product B	Ours	
Retrieve messges from all touch tone phones easily	Yes	Yes	Yes	4	5	4	—
Change message from any remote location	Yes	No	Yes	3	N/A	5	O
2 lines built in	No	No	Yes	N/A	N/A	4	O
Below Add Features in Competitors' Product Not Included in Ours	**Check if Product Feature is Present**			**Feature Performance vs. Goal (*)**			**Identify if Significant Risk or Opportunity**
	Product A	**Product B**	**Ours**	**Product A**	**Product B**	**Ours**	
No cassette used to record message	Yes	Yes		4	N/A		R
Telephone and answering machine in one unit	Yes	Yes		3	4		R

Legend (*)
1 = Poor
2 = Fair
3 = Satisfactory
4 = Good
5 = Excellent

Figure 4.9 Competitive analysis. (Juran Institute, Inc. Copyright 1994.)

in-depth interviews and on-site inspections, where willing customers are using a competitor's product.

Note that by reviewing this analysis, the design team can identify those areas in which the design is vulnerable to the competition as well as those in which the team has developed an advantage. Based on this analysis, the team will then need to make optimization choices about whether to upgrade the product. The team may need to apply a value analysis to make some of these choices.

Salability Analysis An example of salability analysis is shown in Fig. 4.10. This analysis is similar to a competitive analysis, except that the reference point is the response of customers to the proposed design rather than a comparison with the features of the competitors' designs. Note, however, that elements of competitive and salability analyses can be combined, with the salability analysis incorporating customer evaluation of both the proposed new design and existing competitive designs.

Complex products, such as automobiles, with multiple optional features and optional configurations offer a unique opportunity to evaluate salability. Observed installation rates of options on both the existing car line and competitors' cars provide intelligence on both the level of market demand for the feature and the additional price that some segments of the market will pay for the feature, although the other segments of the market may place little or no value on it.

Value Analysis Value analysis has been quite common in architectural design and the development of custom-engineered products, but it also can be applied successfully to other environments as well, as illustrated in Fig. 4.11. By comparing the costs for meeting different

Name of Product: Car Repair Service — Tune-Up	How Do Customers Rate Product? Poor Fair Satisfactory Good Excellent	Basis for Rating Prior Use vs. Opinion	How Do Customers See Differences between Our Products and Competing Products? Positively (+) Negatively (–) No difference (O)	Would Customers Buy if Price Were Not Important? Yes No	Would Customer Buy if Price Were Important?		Of All Products Listed, Prioritize Which Would Customers Buy and Its Basis? Price Features	Identify if Significant Risk or Opportunity
					Price	Yes No		
Ours —	E	U		Y	$175	Y	2-F	
Competitor A —	G	O	+	N	$145	Y	3-P	O
Competitor B —	E	U	O	Y	$175	Y	1-F	R

Product Feature: Pick-Up and Delivery of Car to Be Repaired Product Feature Goal: Same Day Service	How Would Customers Rate Featurs? Poor Fair Satisfactory Good Excellent	Basis for Rating Prior Use vs. Opinion	How Do Customers See Diff. between Our Features against Competing Features? Positively (+) Negatively (–) No Difference (O)	Does the Addition of the Feature Make the Product: More Salable (+) Less Salable (–) No Difference (O)	Identify if Significant Risk or Opportunity
Ours — Offered	G	U		O	
Competitor A — Not offered	S	O	+	—	O
Competitor B — Offered. Also provides loaner car to customer	E	U	—	+	R

Figure 4.10 Salability analysis. (Juran Institute, Inc. Copyright 1994. Used by permission.)

Customer Need (listed in priority order)	Product Feature and Goals						Cost of Meeting Need
	Walk in appointments handled by nurse, 5 days a week	Board certified obstetrician, 2 days a week	Social worker, 5 days a week	Nutritional counselor, 5 days a week	On-site billing clerk takes medicaid insurance from all eligible patients	On-site laboratory—most results under 1 hour	
Convenient to use	60,000	30,000	10,000	10,000	20,000	40,000	170,000
Confidence in staff		70,000	10,000	15,000			95,000
Reasonable cost						25,000	25,000
Sensitivity			15,000	5,000			20,000
Informed choices			5,000	15,000			20,000
Cost for feature	60,000	100,000	40,000	45,000	20,000	65,000	330,000

FIGURE 4.11 Value analysis spreadsheet. (Juran Institute, Inc. Copyright 1994. Used by permission.)

customer needs, the design team can make a number of significant optimization decisions. If the cost for meeting low-priority needs is high, the team must explore alternative ways to meet those needs and even consider not addressing them at all if the product is highly price sensitive. If very important needs have not consumed much of the expense, the team will want to make certain that it has met those needs fully and completely. While low expense for meeting a high-priority need is not necessarily inappropriate, it does present the designers with the challenge of making certain that lower-priority needs are not being met by using resources that could be better directed toward the higher-priority needs. It is not uncommon for products to be overloaded with "bells and whistles" at the expense of the fundamental functionality and performance.

Set and Publish Final Product Design

After the design has been optimized and tested, it is time to select the features and goals to be included in the final design. This is also the stage where the results of product development are officially transmitted to other functions through various forms of documentation. These include the specifications for the features and product feature goals as well as the spreadsheets and other supporting documents. All this is supplemented by instructions, both oral and written. To complete this activity, the team must first determine the process for authorizing and publishing features and product feature goals. Along with the features and goals, the team should include any procedures, specifications, flow diagrams, and other spreadsheets that relate to the final product design. The team should pass along results of experiments, field testing, prototypes, and so on that are appropriate. If an organization has an existing process for authorizing product goals, it should be reexamined in light of recent experience. Ask these questions: Does the authorization process guarantee input from key customers—both internal and external? Does it provide for optimization of the design? If an organization has no existing goal authorization process, now is a good time to initiate one.

Step 5: Develop the Process Features

Once the product is designed and developed, it is necessary to determine the means by which the product will be created and delivered on a continuing basis. These means are, collectively, the *process*. *Process development* is the set of activities for defining the specific means to be used by operating personnel for meeting product quality goals. Some related concepts include

- *Subprocesses.* Large processes may be decomposed into these smaller units for both the development and operation of the process.
- *Activities.* These are steps in a process or subprocess.
- *Tasks.* These comprise detailed step-by-step description for execution of an activity.

For a process to be effective, it must be goal oriented, with specific measurable outcomes; systematic, with the sequence of activities and tasks fully and clearly defined and all inputs and outputs fully specified; and capable, that is, able to meet product quality goals under operating conditions; and legitimate, with clear authority and accountability for its operation.

The 11 major activities involved in developing a process are as follows:

1. Review product goals.
2. Identify operating conditions.
3. Collect known information on alternate processes.
4. Select general process design.
5. Identify process features and goals.
6. Identify detailed process features and goals.
7. Design for critical factors and human error.
8. Optimize process features and goals.
9. Establish process capability.
10. Set and publish final process features and goals.
11. Set and publish final process design.

Review Product Goals

Ideally, this review will be relatively simple. Product quality goals should have been validated with the prior participation of those who would be affected. In many organizations, however, product design and process design often are executed by different teams. There is no real joint participation on either group's part to contribute to the results that both teams are expected to produce. This lack of participation usually reduces the number of alternative designs that could have been readily adopted in earlier stages but become more difficult and more expensive to incorporate later. In addition, those who set the product goals have a vested interest in their own decisions and exhibit cultural resistance to proposals by the process design team to make changes to the product design. If the product and process design efforts are being performed by different groups, then review and confirmation of the product quality goals are absolutely critical.

Review of product quality goals ensures that they are understood by those most affected by the process design. The review helps achieve the optimum. Process designers are able to present product designers with some realities relative to the costs of meeting the quality goals. The review process should provide a legitimate, unobstructed path for challenging costly goals.

Identify Operating Conditions

Seeking to understand operating conditions requires investigation of a number of dimensions.

User's Understanding of the Process By "users," we mean either those who contribute to the processes in order to meet product goals or those who employ the process to meet their own needs. Users consist, in part, of internal customers (organization units or persons) responsible for running the processes to meet the quality goals. Operators or other workers are users. Process planners need to know how these people will understand the work to be done. The process must be designed either to accommodate this level of understanding or to improve the level of understanding.

How the Process Will Be Used Designers always know the intended use of the process they develop. However, they may not necessarily know how the process is actually used (and misused) by the end user. Designers can draw on their own experiences but usually must supplement these with direct observation and interviews with those affected.

The Environments of Use Planners are well aware that their designs must take account of environments that can influence process performance. Planners of physical processes usually do take account of such environmental factors as temperature, vibration, noise level, and so on. Planners who depend heavily on human responses, particularly those in the service areas, should address the impact of the environment on human performance in their process designs. For example, a team designing the process for handling customer inquiries should consider how environmental stress can influence the performance of the customer service representatives. This stress can result from large numbers of customer complaints, abusive customers, lack of current product information, and so on.

Collect Known Information on Alternative Processes Once the goals and environment are clear, the design team needs reliable information on alternative processes available for meeting those goals in the anticipated environment.

Process Anatomy At the highest level, there are some basic process anatomies that have specific characteristics that planners should be aware of. A *process anatomy* is a coherent structure that binds or holds the process together. This structure supports the creation of the goods or the delivery of the service. The selection of a particular anatomy also will have a profound influence on how the product is created and the ability of the organization to respond to customers' needs. Figure 4.12 illustrates these.

The Assembly Tree The *assembly tree* is a familiar process that incorporates the outputs of several subprocesses. Many of these are performed concurrently and are required for final assembly or to achieve an end result at or near the end of the process. This kind of process anatomy is widely used by the great mechanical and electronic industries that build automotive vehicles, household appliances, electronic apparatus, and so on. It is also used to define many processes in a hospital, such as in the case of performing surgery in the operating room. The branches or leaves of the tree represent numerous suppliers or in-house departments making parts and components. The elements are assembled by still other departments.

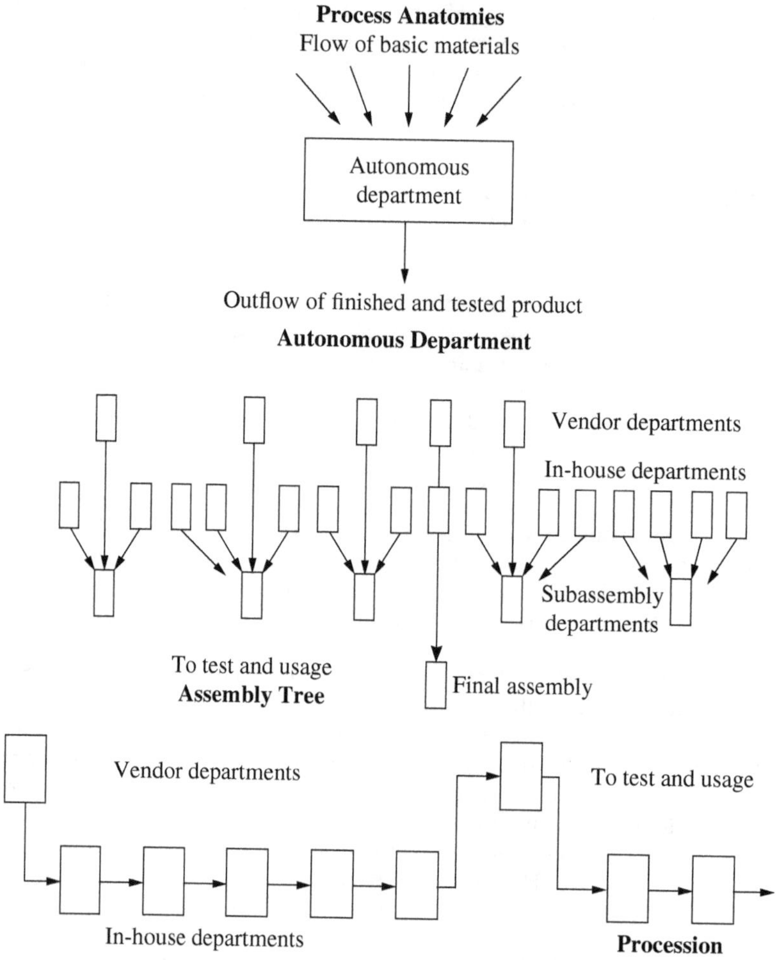

Figure 4.12 Process anatomies. (Juran Institute, Inc. Copyright 1994. Used by permission.)

In the office, certain processes of data collection and summary also exhibit features of the assembly tree. Preparation of major accounting reports (e.g., balance sheet, profit statement) requires assembly of many bits of data into progressively broader summaries that finally converge into the consolidated reports. The assembly tree design has been used at both the multifunctional and departmental levels. In large operations, it is virtually mandatory to use staff specialists who contribute different outputs at various multifunctional levels. An example of this is the budget process. While it is not mandatory to use staff specialists for large departmental processes, this is often the case. This can be illustrated by the design department, where various design engineers contribute drawings of a project that contribute to the overall design.

Business Process Quality Management to Hold the Gains

Increasingly, many planners are applying a fourth, less traditional form of management known as *business process quality management* to their major processes. This new, alternative management form has come about in response to an increased realization that many of today's business goals and objectives are becoming even more heavily dependent on large, complex, cross-functional business processes. Process quality management emphasizes that there are several critical processes that are crucial to an organization if it is to maintain and grow its business. (See Chap. 8, Organization Roles to Support a Quality and Excellence Culture, for a full discussion.)

Measuring the Process

In selecting a specific process design, the team will need to acquire information on the effectiveness and efficiency of alternative designs, including

- Deficiency rates
- Cycle time
- Unit cost
- Output rate

To acquire the needed data, the planners must typically use a number of different approaches, including

- Analyzing the existing process
- Analyzing similar or related processes
- Testing alternative processes
- Analyzing new technology
- Acquiring information from customers
- Simulating and estimating
- Benchmarking

Select General Process Design

Just as product design began with a high-level description expanded to the details, process design should begin by describing the overall process flow with a high-level process flow diagram. From this diagram it will be possible to identify the subprocesses and major activities that can then be designed at a more detailed level. In developing the high-level flow as well as the greater detail later, the team should ensure that it meets the following criteria:

- Deliver the quality goals for the product.
- Incorporate the countermeasures for criticality analysis, FMEA, and fault-tree analysis.
- Meet the project goals.
- Account for actual use, not only intended use.
- Be efficient in consumption of resources.
- Demand no investments that are greater than planned.

While some process designs will largely repeat existing designs and other process designs will represent "green field" or "blank sheet" redesigns, most effective process redesigns are a combination of the tried and true existing processes with some significant quantum changes in some parts of the process.

The preceding criteria should be the guidelines for whether a particular part of the process should be incorporated as it is, improved, or replaced with a fundamentally different approach.

This is the point in process design to think as creatively as possible, using some of the same techniques discussed under product development. Consider the impact of radically different anatomies. Would the customer be served better with dedicated, multispecialty units or with highly specialized expert functionality accessed as needed? What approach is mostly likely to reduce failures? How can cycle time by cut dramatically? Is there a new technology that would allow us to do it differently? Can we develop such a technology?

Once the high-level flow is completed, each activity and decision within the flow diagram needs to be fully documented with a specification of the following for each:

- Inputs
- Outputs
- Goals for outputs
- Cycle time
- Cost
- General description of the conversion of inputs to outputs

Clear specification of these factors makes it possible to divide up the work of detailed design later and still be confident that the final design will be consistent and coordinated.

Once the initial new process flow is completed, it should be reviewed for opportunities to improve it, such as these:

- Eliminate sources of error that lead to rework loops.
- Eliminate or reduce redundant subprocesses, activities, or tasks.
- Decrease the number of handoffs.
- Reduce cycle time.
- Replace tasks, activities, or processes that have outputs with defects.
- Correct sequencing issues in the process to reduce the amount of activity or rework.

Testing Selected Processes

One of the key factors for a successful design is the incorporation of the lessons learned from testing the product, the features, and the overall process and subprocesses to ensure that they meet quality goals. Testing should be conducted throughout the entire Quality by Design process to allow for changes, modifications, and improvements to the plan before it is transferred to operations. Testing is performed at various points to analyze and evaluate alternate designs of the overall process and subprocesses.

Comparisons or Benchmarks

Other units inside and outside the organization may already be using a process similar to the one designed. The process can be validated by comparing it with existing similar processes.

Test Limitations

All tests have some limitations. The following are common limitations that should be understood and addressed:

Differences in Operating Conditions Dry runs and modular testing obviously differ from operating conditions. Even pilot tests and benchmarks will differ in some details from the actual, full implementation. Some common differences between conditions for testing and conditions for full-scale use include

- People operating the process
- Customers of the process
- Extreme values and unusual conditions
- Interactions with other processes and other parts of the organization

Differences in Size Especially with critical failures, such as breakdown of equipment, loss of key personnel, or any other potential failure, as in the case of complications in a surgical procedure, a test might not be large enough to allow these rare failures to occur with any high degree of certainty.

Other Effects Sometimes designing a new process or redesigning an existing process may create or exacerbate problems in other processes. For example, improved turnaround time in approving home loans may create a backlog for the closing department. Such interactions among processes might not occur in an isolated test.

Identify Process Features and Goals

A *process feature* is any property, attribute, and so on that is needed to create the goods or deliver the service and achieve the product feature goals that will satisfy a customer need. A *process goal* is the numeric target for one of the features.

Whereas features answer the question "What characteristics of the product do we need to meet customer needs?" process features answer the question "What mechanisms do we need to create or deliver those characteristics (and meet quality goals) over and over without failures?" Collectively, process features define a process. The flow diagram is the source of many of, but not all, these features and goals.

As the process design progresses from the macro level down into details, a long list of specific process features emerges. Each of these is aimed directly at producing one or more features. For example,

- Creating an invoice requires a process feature that can perform arithmetic calculations so that accurate information can be added
- Manufacturing a gear wheel requires a process feature that can bore precise holes into the center of the gear blank
- Selling a credit card through telemarketing requires a process feature that accurately collects customer information

Most process features fall into one of the following categories:

- Procedures—a series of steps followed in a regular, definite order
- Methods—an orderly arrangement of a series of tasks, activities, or procedures
- Equipment and supplies—"physical" devices and other hard goods that will be needed to perform the process
- Materials—tangible elements, data, facts, figures, or information (these, along with equipment and supplies, also may make up inputs required as well as what is to be done to them)

- People—numbers of individuals, skills they will require, goals, and tasks they will perform
- Training—skills and knowledge required to complete the process
- Other resources—additional resources that may be needed
- Support processes—secretarial support, occasionally other support, such as outsources of printing services, copying services, temporary help, and so on

Just as in the case of product design, process design is easier to manage and optimize if the process features and goals are organized into a spreadsheet indicating how the process delivers the features and goals. Figure 4.13 illustrates such a spreadsheet.

The spreadsheet serves not only as a convenient summary of the key attributes of the process, it also facilitates answering two key questions that are necessary for effective and efficient process design. First, will every product feature and goal be attained by the process? Second, is each process feature absolutely necessary for at least one product feature; that is, are there any unnecessary or redundant process features? Also, verify that one of the other process features cannot be used to create the same effect on the product.

Often high-level process designs will identify features and goals that are required from organization wide macro processes. Examples might include cycle times from the purchasing process, specific data from financial systems, and new skills training. Because the new process will depend on these macro processes for support, now is the time to verify that they are capable of meeting the goals. If they are not, the macro processes will need to be improved as part of the process design, or they will need to be replaced with an alternative delivery method.

		Process Features			
Product Feature	**Product Feature Goal**	Spray delivery capacity	Crew Size	Certified materials	Scheduling forecast on PC to determine to/from and work needed
Time to perform job	Less than one hour 100 percent of time	○	●		●
Guaranteed appointment time	99 percent of jobs within 15 min. of appointment				●
All materials environmentally safe	All naturally occuring/no synthetics			●	
		10 gallons per minute	One person per 10,000 sq. ft. of yd.	100 percent approved by State Dept. of Agriculture	Forecast time always within 10 percent of actual
		Process Feature Goals			

Legend ● Very strong ○ Strong △ Weak

FIGURE 4.13 Process design spreadsheet. (Juran Institute, Inc. Copyright 1994. Used by permission.)

Identify Detailed Process Features and Goals

In most cases, it will be most efficient and effective for individual subteams to carry out the detailed designs of subprocesses and major activities. These detailed designs will have the process features and goals as their objectives and criteria. Each subprocess team will create the design to the level at which standard operating procedures can be developed, software coded, equipment produced or purchased, and materials acquired.

Design for Critical Factors and Human Error

One key element of process design is determining the effect that critical factors will have on the design. "Critical factors" are those aspects which present serious danger to human life, health, and the environment or risk the loss of very large sums of money. Some examples of such factors involve massive scales of operations: airport traffic control systems, huge construction projects, systems of patient care in hospital, and even the process for managing the stock market. Design for such factors should obviously include ample margins of safety as to structural integrity, fail-safe provisions, redundancy systems, multiple alarms, and so on. Criticality analysis and failure-mode and effect analysis (see Chap. 20, Accurate, and Reliable Data and Measurement Systems) are helpful tools in identifying those factors which require special attention at this point.

Workers vary in their capabilities to perform specific tasks and activities. Some workers perform well, whereas others do not perform nearly as well. What is consistent about all workers is that they are a part of the human family, and human beings are fallible. Collectively, the extent of human errors is large enough to require that the process design provides for means to reduce and control human error. Begin by analyzing the data on human errors, and then apply the Pareto principle. The vital few error types individually become candidates for special process design. The human errors that can be addressed by process design fall into these major classes:

- Technique errors arising from individuals lacking specific, needed skills
- Errors aggravated by lack of feedback
- Errors arising from the fact that humans cannot remain indefinitely in a state of complete, ready attention

Principles of Mistake Proofing

Research has indicated that there are a number of different classifications of error proofing methods, and these are spelled out below. *Elimination* consists of changing the technology to eliminate operations that are error prone. For example, in some materials handling operations, the worker should insert a protective pad between the lifting wire and the product so that the wire will not damage the product. Elimination could consist of using nylon bands to do the lifting.

Optimize Process Features and Goals

After the planners have designed for critical factors and made modifications to the plan for ways of reducing human error, the next activity is to optimize first the subprocesses and then the overall process design. In step 4, develop product, the concept of optimization was introduced. The same activities performed for optimizing features and product feature goals also apply to process planning. Optimization applies to both the design of the overall process and the design of individual subprocesses.

Establish Process Capability

Before a process begins operation, it must be demonstrated to be capable of meeting its quality goals. The concepts and methods for establishing process capability are discussed in detail in Chap. 22, Managing Quality in Operations—Manufacturing. Any design project must measure the capability of its process with respect to the key quality goals. Failure to achieve process capability should be followed by systematic diagnosis of the root causes of the failure and improvement of the process to eliminate those root causes before the process becomes operational.

Reduction in Cycle Time

Process capability relates to the effectiveness of the process in meeting customer needs. One special class of needs may relate to subprocess cycle time—the total time elapsed from the beginning of a process to the end. Reducing cycle time has almost become an obsession for many organizations. Pressures from customers, increasing costs, and competitive forces are driving organizations to discover faster ways of performing their processes. Often these targeted processes include launching new products, providing service to customers, recruiting new employees, responding to customer complaints, and so on. For existing processes, designers follow the well-known quality improvement process to reduce cycle time. Diagnosis identifies causes for excessive time consumption. Specific remedies are then developed to alleviate these causes.

Set and Publish Final Process Features and Goals

After the design team has established the flow of the process, identified initial process features and goals, designed for critical processes and human error, optimized process features and goals, and established process capabilities, it is ready to define all the detailed process features and goals to be included in the final design. This is also the stage where the results of process development are officially transmitted to other functions through various forms of documentation. These include the specifications for the features and product feature goals as well as the spreadsheets and other supporting documents. All this is supplemented by instructions, both oral and written.

Filling out the process design spreadsheet is an ongoing process throughout process development. The spreadsheet should have been continually updated to reflect design revisions from such activities as reviewing alternative options, designing for critical factors and human error, optimizing, testing process capability, and so on. After making the last revision to the process design spreadsheet, it should be checked once more to verify the following:

- Each product feature has one or more process features with strong or very strong relation. This will ensure the effective delivery of the product feature without significant defects. Each product feature goal will be met if each process goal is met.
- Each process feature is important to the delivery of one or more features. Process features with no strong relationship to other features are unnecessary and should be discarded.

The completed process design spreadsheet and detailed flow diagrams are the common information needed by managers, supervisors, and workers throughout the process. In addition, the design team must ensure that the following are specified for each task within the process:

- Who is responsible for doing it
- How the task is to be competed

- Its inputs
- Its outputs
- Problems that can arise during operations and how to deal with them
- Specification of equipment and materials to be used
- Information required by the task
- Information generated by the task
- Training, standard operating procedures, job aids that are needed

Step 6: Develop Process Controls and Transfer to Operations

In this step, planners develop controls for the processes, arrange to transfer the entire product plan to operational forces, and validate the implementation of the transfer. There are seven major activities in this step.

1. Identify controls needed.
2. Design feedback loop.
3. Optimize self-control and self-inspection.
4. Establish audit.
5. Demonstrate process capability and controllability.
6. Plan for transfer to operations.
7. Implement plan and validate transfer.

Once design is complete, these plans are placed in the hands of the operating departments. It then becomes the responsibility of the operational personnel to manufacture the goods or deliver the service and to ensure that quality goals are met precisely and accurately. They do this through a planned system of quality control. Control is largely directed toward continuously meeting goals and preventing adverse changes from affecting the quality of the product. Another way of saying this is that no matter what takes place during production (change or loss of personnel, equipment or electrical failure, changes in suppliers, etc.), workers will be able to adjust or adapt the process to these changes or variations to ensure that quality goals can be achieved.

Identify Controls Needed

Process control consists of three basic activities:

- Evaluate the actual performance of the process.
- Compare actual performance with the goals.
- Take action on the difference.

Detailed discussions of these activities in the context of the feedback loop are contained in Chap. 6, Quality Control to Assure Compliance to Customer Requirements.

Process Features

Much control consists of evaluating those process features that most directly affect the features, for example, the state of the toner cartridge in the printer, the temperature of the furnace for smelting iron, or the validity of the formulas used in the researcher's report. Some features become candidates for control subjects as a means of avoiding or reducing failures.

These control subjects typically are chosen from previously identified critical factors or from conducting FMEA, fault-tree analysis (FTA), and criticality analysis. Process controls are associated with the decision: Should the process run or stop?

- Setting the standards for control, that is, the levels at which the process is out of control and the tools, such as control charts, that will be used to make the determination
- Deciding what action is needed when those standards are not met, for example, troubleshooting
- Designating who will take those actions

A detailed process flow diagram should be used to identify and document the points at which control measurements and actions will be taken. Then each control point should be documented on a control spreadsheet similar to Fig. 4.14.

Training

Workers should be trained to make the product conformance decisions and should also be tested to ensure that they make good decisions. Specifications must be unequivocally clear.

The quality audit and audit of control systems are treated elsewhere in detail; see, for example, Chap. 9, Quality Assurance and Audits.While the audit of a control system is a function independent of the design team, the design team does have the responsibility for ensuring that adequate documentation is available to make an effective audit possible and that there are provisions of resources and time for conducting the audit on an ongoing basis.

	Process Controls						
Product feature	Control subject	Sensor	Goal	Measurement frequency	Sample size	Criterion	Responsibility
Process feature 1							
Process feature 2							
– – –							
Wave solder	Solder temperature	Thermo-couple	505°F	Continuous	n/a	≥510°F decrease heat 500°F increase heat	Operator
	Conveyor speed	ft/min meter	4.5 ft/min	1/hour	n/a	≥5 ft/min reduce speed ≤4 ft/min increase speed	Operator
	Alloy purity	Lab chem analysis	1.5% max total contaminants	1/month	15 grams	≥1.5% drain bath, replace solder	Process engineer

FIGURE 4.14 Control spreadsheet. (From J. M. Juran, *Quality Control Handbook*, 5th ed., McGraw-Hill, New York, 1999, p. 3.48.)

Demonstrate Process Capability and Controllability

While process capability must be addressed during the design of the process, it is during implementation that initial findings of process capability and controllability must be verified.

Plan for Transfer to Operations

In many organizations, receipt of the process by operations is structured and formalized. An information package is prepared consisting of certain standardized essentials: goals to be met, facilities to be used, procedures to be followed, instructions, cautions, and so on. There are also supplements unique to the project. In addition, provision is made for briefing and training the operating forces in such areas as maintenance, dealing with crisis, and so on. The package is accompanied by a formal document of transfer of responsibility. In some organizations, this transfer takes place in a near-ceremonial atmosphere.

The Structured Approach Has Value It tends to evolve checklists and countdowns that help ensure that the transfer is orderly and complete. If the organization already has a structure for transfer, project information may be adapted to conform to established practice. If the organization has a loose structure or none at all, the following material will aid in design the transfer of the project.

Regardless of whether the organization has a structure, the team should not let go of the responsibility of the project until it has been validated that the transfer has taken place and everyone affected has all the information, processes, and procedures needed to produce the final product.

Transfer of Know-How During process design, the planners acquire a great deal of know-how about the process. The operating personnel could benefit from this know-how if it were transferred. There are various ways of making this transfer, and most effective transfers make use of several complementary channels of communication, including

- Process specifications
- Briefings
- On-the-job training
- Formal training courses
- Prior participation

Audit Plan for the Transfer As part of the plan for formal transfer, a separate audit plan should be developed as a vehicle for validating the transfer of the plan. This kind of audit is different from the control audits described previously. The purpose of this audit is to evaluate how successful the transfer was. For the audit to have real meaning, specific goals should be established during the design phase of the transfer. Generally, these goals relate to the quality goals established during the development of the product, features, and process features. The team may decide to add other goals inherent to the transfer or to modify newly planned quality goals during the first series of operations. For example, during the first trial runs for producing the product, total cycle time may exceed expected goals by 15 percent. This modification takes into account that workers may need time to adjust to the plan. As they become more skilled, gain experience with the process, and get more comfortable with their new set of responsibilities, cycle time will move closer to targeted quality goals. The audit plan for the transfer should include the following:

- Goals to meet
- How meeting the goals will be measured
- The time phasing for goals, measurement, and analysis
- Who will audit
- What reports will be generated
- Who will have responsibility for corrective action for failure to meet specific goals

Implement Plan and Validate Transfer

The final activity of the Quality by Design process is to implement the plan and validate that the transfer has occurred. A great deal of time and effort has gone into creating the product plan, and validating that it all works is well worth the effort.

Frequently Used Design Tools

- *Affinity diagrams*. This diagram clusters together items of similar type, is a prelude to a cause-effect diagram used in quality improvement, and is used in quality design to group together similar needs or features.
- *Benchmarking*. This technique involves openly sharing and investigating the best practices of organizations, largely for business and internal processes (not for competitive or proprietary manufacturing). In today's world, this has improved from "industrial tourism" to research, largely through participation in online databases.
- *Brainstorming*. This popular technique obtains group ideas as to cause (for improvement) or as to features (for planning).
- *Carryover analyses*. Usually a matrix depicts the degree of carryover of design elements, with particular regard to failure proneness.
- *Competitive analyses*. Usually a matrix depicts a feature-by-feature comparison to the competition, with particular regard to "best-in-class" targets.
- *Control chart*. This is a widely used depiction of process change over time. The most popular is the Shewhart control chart for averages.
- *Criticality analyses*. Usually a matrix depicts the degree of failure of a feature or component against the ranking of customer needs, along with responsibilities detailed for correction.
- *Data collection: focus group*. This popular technique places customers in a setting led by a trained facilitator to probe for the understanding of needs.
- *Data collection: market research*. Any of a variety of techniques aim at answering the three fundamental questions: What is important to the users? What is the order of the items of importance? How well do we do in meeting them in that order, as compared to the competition?
- *Data collection: surveys*. This passive technique elicits answers to preset questions about satisfaction or needs. Usually it is "closed-ended," with meager space for comments or answers to open-ended questions. Poor return rates are a hallmark of

this technique, along with the suspicion that those with dissatisfactions respond at higher rates.

- *Failure mode and effect analyses.* Otherwise called FMEA, the matrix presents the probability of failure, significance of the failure, and ease of detection, resulting in a *risk priority number* (RPN). Higher RPNs are attacked first. This is used in both improvement and design settings, although the chief use is as a design tool.
- *Fault-tree analyses.* A graphical presentation of the modes of failure shows events that must occur together ("and") or separately ("or") in order to have the failure occur. Usually this is shown vertically, with the "ANDed" and "ORed" events cascading as branches on a tree.
- *Flow diagram.* This extremely popular depiction of a process uses standard symbols for activities and flow directions. It originated in software design during the 1950s and evolved into the process mapping widely used today.
- *Glossary.* The glossary is the chief weapon used to remove the ambiguity of words and terms between customers and providers. This is a working dictionary of in-context usage, for example, the meaning of "comfortable" as it applies to an office chair.
- *Design network.* A tree diagram depicts the events that occur either in parallel or sequentially in the design of something. Usually the network is shown with the total time needed to complete the event, along with earliest start and subsequent stop dates. It is used to manage a particularly complex design effort. Like techniques include the program evaluation and review technique (PERT) and critical path method (CPM). Today's spreadsheet-like project management software usually combines the key features of each.
- *Process analysis technique.* This process flowchart technique also shows the time necessary to do each task, the dependencies the task requires (such as access to the computer network), and the time "wasted" in between tasks. Usually it is interview-driven and requires a skilled process expert.
- *Process capability.* This term is given to any number of tools, usually statistical, that thereby reveal the ability of a process to repeat itself and the ability of the process to meet its requirements.
- *Salability analyses.* This is another matrix tool used to depict the price willing to be borne, or the cost needed to deliver, a given feature of a product.
- *Scatter diagram.* This is a graphical technique of plotting one variable against another, to determine co-relationship. It is a prelude to regression analyses to determine prediction equations.
- *Selection matrix.* This matrix tool shows the choices to be made ranked according to agreed upon criteria. It is used in both improvement and design settings.
- *Customer needs spreadsheet.* This spreadsheet tool depicts the relationship between customer communities and the statements of need. Needs strongly relating to a wide customer base subsequently rise in priority when features are considered. Advanced forms of this spreadsheet and others appear as the "house of quality," or quality function deployment (QFD); see the section in this chapter about Design for Six Sigma.
- *Needs analysis spreadsheet.* This spreadsheet tool is used to "decompose" primary statements of need into other levels. Thus, "economical" for a new car purchaser might break down further to purchase price, operating costs, insurance costs, fuel

economy, and resale value. Decomposing needs has the principal benefit of single-point response and measurement if taken to the most elemental level.

- *Product design spreadsheet.* This is a continuation of the customer needs spreadsheet, further developing the features and feature goals that map to the customer needs. The features with the strongest relationship to needs are elevated in priority when considering the processes used to make them.
- *Tree diagram.* Any of a variety of diagrams depicts events that are completed in parallel or simultaneously as branches of a tree. This technique is less refined than the design network, but useful to understand the activities from a "big picture" perspective.
- *Value analysis.* This is a matrix depiction of customer needs and costs required to support or deliver a given feature to meet that need. It is a close cousin to salability analysis.

Design for Six Sigma

Product and service design is the creation of a detailed description for a physical good or service, together with the processes to actually produce that good or service. In quality theory terms, product design means establishing quality goals and putting in place the means to reach those goals on a sustained basis. In Six Sigma terms, product design [Design for Six Sigma (DFSS)] means contemporaneously creating a design for a product and includes the process to produce it in such a way that defects in the product and the process are not only extremely rare, but also predictable. What is more, defects are rare and predictable, even at the point when full-scale production begins. To achieve this level of excellence and its attendant low costs and short cycle times, as well as soaring levels of customer satisfaction, requires some enhancements to traditional design methods. For example, each DFSS design project starts with an identification of customers and a detailed analysis and understanding of their needs. Even "redesign" starts at the beginning because all successful designs are based on customer needs, and in this world of rapid change, customer needs—and even customers—have a way of rapidly changing. Another example is the widespread intensive use of statistical methods in DFSS. The power of the information gained from statistical analyses provides the means to achieve Six Sigma levels of quality, which are measured in parts per million. DFSS is carried out in a series of phases known as DMADV.

DMADV stands for: define, measure, analyze, design, and verify. The discussion that follows does not cover all the details of procedures and tools used in DMADV; that would require many hundreds of pages, and they can be found elsewhere in published form. We will, however, attempt to acquaint the reader with what any manager needs to know about the purpose, the issues, the questions, and the sequence of steps associated with the respective phases of DMADV.

A "new" codification of the process for developing quality products is known as Design for Six Sigma. It combines the concept of quality design with the popular goal of Six Sigma quality. The DFSS process directs the designers of the product to create their designs so that manufacturing can produce them at Six Sigma quality levels. In the case of services, it means developing the service process so that it can be delivered at Six Sigma quality levels.

DFSS is targeted at design activities that result in a new product, a new design of an existing product, or the modification of an existing design. It consists of five phases in the following sequence: define, measure, analyze, design, verify. Figure 4.15 expands on the activities of each phase. (See Chap. 17, Continuous Innovation Using Design for Six Sigma, for more details on DFSS.)

Define	Measure	Analyze	Design	Verify
• Initiate the project • Scope the project • Plan and manage the project	• Discover and prioritize customer needs • Develop and prioritize CTQs • Measure baseline performance	• Develop design alternative • Develop high-level design • Evaluate high-level design	• Optimize detail level design parameters • Evaluate detail level design • Plan detail design verification tests • Verify detail and design of product • Optimize process performance	• Execute pilot/ analyze results • Implement production process • Transition to owners

FIGURE 4.15 Major activities in DFSS.

Reference

Parasuraman, A., Zeithami, V. A., and Berry, L. L. (1985). "A Conceptual Model for Service Quality and Its Implications for Further Research." *Journal of Marketing*, Fall, pp. 41–50.

CHAPTER 5

Quality Improvement and Breakthrough Performance

Joseph M. Juran

High Points of This Chapter

1. A breakthrough in current performance aims to eliminate failures such as excessive number of defects, excessive delays, excessively long time cycles, and the high costs of poor quality due to poorly performing processes.
2. The Juran Universal Sequence for Breakthrough, identified in the 1950s, consists of six steps to achieve superior results. The steps are
   a. Nominate and identify problems. (Management does this.)
   b. Establish a project and team. (Management does this.)
   c. Diagnose the cause(s). (The project team does this.)
   d. Remedy the cause(s). (The project team plus the work group where the cause[s] originate do this.)
   e. Hold the gains. (The project team and affected operating forces do this.)
   f. Replicate results and nominate new projects. (Management does this.)
3. All improvement happens project by project. To achieve breakthrough requires leaders to define goals and projects that are resourced to ensure completion and results.
4. It is upper management's responsibility to mandate breakthrough. Specifically, upper management must
   a. Establish multifunctional councils or steering teams to prioritize projects.
   b. Nominate and select breakthrough projects.
   c. Create project charters that include problem and goal statements.
   d. Provide resources, especially people and time, to carry out the project.
   e. Assign teams, team leaders, and facilitators, "Black Belts" to projects.
   f. Review progress, remove barriers, and manage cultural resistance.
   g. Provide recognition and rewards.
5. Project selection requires expertise and practice on the part of management, so "doable" projects are identified so that the team clearly understands both the problem and the goal.
6. To attain a breakthrough in current performance requires two "journeys": the diagnostic journey and the remedial journey. These journeys represent the application of the fact-based method to solve the performance problems.
7. The diagnostic journey proceeds as follows:
   a. From problem to symptoms of the problem
   b. From symptoms to theories of causes of the symptoms
   c. From theories to testing of the theories
   d. From tests to establishing root cause(s) of the symptoms
8. The remedial journey proceeds as follows:
   a. From root cause(s) to design of remedies of the cause(s)
   b. From design of remedies to testing and proving the remedies under operating conditions
   c. From workable remedies to dealing with predictable resistance to change
   d. From dealing with resistance to establishing new controls on the remedies to hold the gains
9. There have been numerous efforts to create simpler and less intensive improvement methods. Most of them failed to deliver the results. The Six Sigma DMAIC

Improvement Model has gained wide acceptance and is the most widely used. This will be covered in more detail in Chap. 15, Six Sigma: Breakthrough to in-Process Effectiveness. It follows these basic steps:

a. Select the problem and launch a project. (Management does this.)
b. Define the problem. (Champions and management do this.)
c. Measure the magnitude of the symptoms. (The project team does this.)
d. Analyze information to discover the root cause(s). (The project team does this.)
e. Improve by providing a remedy for the cause(s). (Project teams do this.)
f. Control to hold the gains. (Project team and departments do this.)

10. All projects and teams will encounter obstacles when making changes. Objections will be raised by various sources. There may be delaying tactics or rejection by a manager, the work force, or the union. We refer to this as resistance to change. All managers must understand how to overcome this resistance.

The Universal Sequence for Breakthrough

Improvement happens every day, in every organization—even among the poor performers. That is how businesses survive—in the short term. Improvement is an activity in which every organization carries out tasks to make incremental improvements, day after day. Improvement is different from breakthrough improvement. Breakthrough requires special methods and support to attain significant changes and results. It also differs from planning and control. Breakthrough requires taking a "step back" to discover what may be preventing the current level of performance from meeting the needs of its customers. This chapter focuses on attaining breakthrough improvement and how leaders can create a system to increase the rate of improvement. By attaining just a few (the Pareto principle) vital breakthroughs year after year, the organization can outperform its competitors and meet stakeholder needs.

As used here, "breakthrough" means "the organized creation of beneficial change and the attainment of unprecedented levels of performance." Synonyms are "quality improvement" or "Six Sigma improvement." Unprecedented change may require attaining a Six Sigma level (3.4 ppm) or 10-fold levels of improvement over current levels of process performance. Breakthrough results in significant cost reduction, customer satisfaction enhancement, and superior results that will satisfy stakeholders.

The concept of a universal sequence evolved from my experience first in Western Electric Organization (1924–1941) and later during my years as an independent consultant, starting in 1945. Following a few preliminary published papers, a universal sequence was published in book form (Juran 1964). This sequence then continued to evolve based on experience gained from applications by operating managers.

The creation of the Juran Institute in 1979 led to the publication of the videocassette series *Juran on Breakthrough* (Juran 1981). This series was widely received and was influential in launching breakthrough initiatives in many organizations. These organizations then developed internal training programs and spelled out their own versions of a universal sequence. All of these have much in common with the original sequence published in 1964. In some cases, the organizations have come up with welcome revisions or additions.

Breakthrough means change: a dynamic, decisive movement to new, higher levels of performance. In a truly static society, breakthrough is taboo, forbidden. There have been many such societies, and some have endured for centuries. During those centuries, their members either suffered or enjoyed complete predictability. They knew precisely what their station in life was—the same as that lived out by their forebears—but this predictability was, in due course, paid for by a later generation. The price paid was the extinction of the static society through conquest or another takeover by some form of society that was on the move. The threat of extinction may well have been known to the leaders of some of these static societies. Some gambled that the threat would not become a reality until they were gone. It was well stated in Madame de Pompadour's famous letter to Louis XV of France: "After us, the deluge."

History is vital to today's leaders. The threat to the static society stems from basic human drives: the drive for more of everything—knowledge, goods, power, and wealth. The resulting competition is what makes breakthrough important (Juran 1964).

There is an unvarying sequence of events by which we break out of the old levels of performance and into the new. The details of this sequence are important. The starting point is the attitude that a breakthrough is both desirable and feasible. In human organizations, there is no change unless there is first an advocate of change. If someone does not want change, there is a long, hard road before change is finally achieved. The first step on that road is someone's belief that a change—a breakthrough—is desirable and feasible. That change is desirable is mainly an act of faith or belief. Feasibility requires some digging. This leads to the second step.

The second step is to see whether a breakthrough is likely to happen if we mobilize for it—a feasibility study or demonstration project. This study will help separate the problem into major parts, the vital few from the useful many. I call this the Pareto analysis. These vital few problems then become the subject of a drive for new knowledge. But the creation of new knowledge does not just happen—we must organize for it. This leads to the next step.

Organization for breakthrough in knowledge is next. It requires that we appoint or create two systems: one that directs or guides the breakthrough, and one that does the fact-gathering and analysis. We call them the steering arm and the diagnostic arm, respectively. For breakthrough in knowledge, both of these arms are necessary. Neither one alone is sufficient. When both are in place, diagnosis begins. Facts are collected and examined, and new knowledge gained. At this stage, a breakthrough in knowledge has been achieved.

However, a breakthrough in knowledge does not automatically create a breakthrough in performance. Experience has shown that the technical changes needed usually affect the status, habits, beliefs, etc., of the people involved. Anthropologists have given the name "cultural pattern" to this collection of human beliefs, practices, etc.

Breakthroughs in the cultural pattern are in this way an added essential step. Before new levels of performance can be reached, we must discover the effect of the proposed changes on the cultural pattern and find ways to deal with the resistances generated. This turns out at times to be a difficult and important problem.

Finally, a breakthrough in performance can be achieved. This is the result we had set out to attain. To sustain it, we must rely on controls to maintain the status quo until another breakthrough comes along.

Two Kinds of Breakthrough

Breakthrough can be aimed at both sides of quality.

1. *Having higher-quality product and service features* provides customer satisfaction and revenue for the producing organization. These product features drive revenue.
2. *Achieving freedom from failures will reduce* customer dissatisfaction and non value-added waste. To the producing organization, reducing the product failures, which reduce costs, is a target for breakthrough.

Breakthrough is applicable to any industry, problem, or process. To better understand why so many organizations create extensive quality improvement programs such as Lean Six Sigma we must contrast planning versus improvement. In the previous chapter, we discussed the quality planning process to design features.

Breakthrough to reduce excess failures and deficiencies may consist of such actions as

- Increase the yield of production processes
- Reduce error rates of administrative reports
- Reduce field failures
- Reduce claim denials
- Reduce the time it takes to perform critical patient clinical procedures

The result in both cases is performance improvement, which can lead to performance excellence. However, the rate of improvement required to attain market leadership needs to move at a revolutionary rate, and this often eludes most organizations. The methods and tools used to secure superior results are fundamentally different from day-to-day improvement methods, and for subtle reasons.

Creating breakthrough to increase revenue starts by setting strategic goals, such as new product development goals to provide best-in-class features, or reducing cycle times to beat the competition. Meeting such new goals requires a systematic "quality planning" process (Juran 1993). Multiple levels of quality planning are needed. An organization needs to plan new products or to design for quality. Other forms of quality planning are to design for manufacturing, Design for Six Sigma, and even Design for Green and Lean.

Quality planning differs from most product and service development methods in that it is carried out through a universal series of steps focusing on understanding the "voice of the customers" (internal and external) and incorporating it into the design of the product. The best design methods always begin with an identification of who we are designing for. In other words, who are the "customers?" This is often followed by determining the needs of those customers, then developing the product or service features required to meet those needs, and so on. Collectively, this series of steps is the "quality planning or quality by design roadmap." Creating breakthroughs in design is covered in Chap. 4, Quality Planning and Design of New Goods and Services.

Many organizations maintain an organized approach for evolving new products and services, year after year. Under this organized approach

- Product development projects are a part of the business plan.
- A new product development function maintains business surveillance over these projects.
- Full-time product and process development departments are equipped with personnel, laboratories, and other resources to carry out the technological work.
- There is clear responsibility for carrying out the essential technological work.
- A structured procedure is used to process the new developments through the functional departments.
- The continuing existence of this structure favors new product development on a year-to-year basis.

This special organizational structure, while necessary, is not sufficient to ensure good results. In some organizations, the cycle time for getting new products to market is lengthy, the new models compete poorly in the market, or new chronic wastes are created. Such weaknesses usually are traceable to weaknesses in the planning process.

In the case of too many non-value-added tasks or too high a cost associated with chronic waste, the product or service is already in production, the goals are already in place. The processes for meeting those goals and the means to maintain them are being carried out by the workforce. However, the resulting products (goods and services) do not always meet the goals. Consequently, the approach to reducing these non-value-added tasks or chronic waste is different from the design or planning methods. Instead, to attain breakthroughs in current levels of performance, we must first have management commit to a program of quality improvement such as Six Sigma. This program can provide the means to identify the problems and then discover their causes. The organization must make the time to carry out a diagnosis of the current process. Once the causes are uncovered, remedies can be applied to remove the causes. It is this approach—to attain breakthroughs—that is the subject of this chapter.

Continuing to attain breakthrough is needed to meet the changing needs of customers, which are a moving target. Competitive prices are also a moving target. However, breakthroughs in improvement usually lag behind breakthroughs in design. They have progressed at very different rates. The chief reason is that many upper managers give a higher priority to increasing revenue from other means than on focusing resources on attaining breakthroughs by achieving unprecedented levels of performance in this way. This difference in priority is usually reflected in the respective organizational structures. An example is seen in the approach to new product development.

Historically, the efforts to meet the competition and improve performance proceeded along two lines based on two very different philosophies:

- Political leaders focused on traditional political solutions—import quotas, tariffs, legislation on "fair trade," and so on.
- Business leaders increasingly became convinced that the necessary response to competition was to become more competitive. This approach required applying the lessons learned from the role models across the entire national economy. Such a massive scaling up has extended well into the twenty-first century.

The experience of recent decades has led to an emerging consensus that managing for quality (planning, control, and improvement) is one of the most cost-effective means to deal with the threats and opportunities, and provide a means of actions that need to be taken. As it relates to breakthrough, the high points of this consensus include the following:

- Global competition has intensified and has become a permanent unpleasant fact. A needed response is to create a high rate of breakthrough, year after year.
- Customers are increasingly demanding improved products from their suppliers. These demands are then transmitted through the entire supplier chain. The demands may go beyond product breakthrough and extend to improving the system of managing for quality.
- The chronic wastes can be huge in organizations that do not have a strategic program aimed at reducing them. In many organizations during the early 1980s, about a third of all work consisted of redoing what was done previously, due to deficiencies. By the end of the 1990s, this number improved to only 20 to 25 percent (estimated by the authors). The emerging consensus is that such waste should not continue, since it reduces competitiveness and profitability.
- Breakthroughs must be directed at all areas that influence an organization's performance: all business, transactional, and manufacturing processes.

- Breakthroughs should not be left solely to voluntary initiatives; they should be built into the strategic plan and DNA of a system. They must be mandated.
- Attainment of market leadership requires that the upper managers personally take charge of managing for quality. In organizations that did attain market leadership, the upper managers personally guided the initiative. The authors are not aware of any exceptions.

Unstructured Reduction of Chronic Waste

In most organizations, the urge to reduce costs has been much lower than the urge to increase sales. As a result,

- The business plan has not included goals for the reduction of chronic waste.
- Responsibility for such breakthroughs has been vague. It has been left to volunteers to initiate action.
- The needed resources have not been provided, since such breakthroughs have not been a part of the business plan.

The lack of priority by upper managers is traceable in large part to two factors that influence the thinking processes of many upper managers:

- Not only do many upper managers give top priority to increasing sales, but some of them even regard cost reduction as a form of lower-priority work that is not worthy of the time of upper managers. This is especially the case in high-tech industries.
- Upper managers have not been aware of the size of the chronic waste, nor of the associated potential for high return on investment. The "instrument panel or scorecards" available to upper managers have stressed performance measures such as sales, profit, cash flow, and so on, but not the size of chronic waste and the associated opportunities. The managers have contributed to this unawareness by presenting their reports in the language of specialists rather than in the language of management—the language of money.

Breakthrough Models and Methods

Breakthrough addresses the question, How do I reduce or eliminate things that are wrong with my products, services, or processes and the associated customer dissatisfaction? Breakthrough models must address problems that create customer dissatisfaction, products and services of poor quality, and failures to meet the specific needs of specific customers, internal and external.

Based on my research, attaining breakthroughs in current performance by reducing customer-related problems has one of the greatest returns on investment and usually comes down to correcting just a few types of things that go wrong, including

- Excessive number of defects
- Excessive numbers of delays or excessively long time cycles
- Excessive costs of the resulting rework, scrap, late deliveries, dealing with dissatisfied customers, replacement of returned goods, loss of customers and clients, loss of goodwill, etc.
- High costs and ultimately high prices, due to the waste

Effective breakthrough models require that

- Leaders mandate it, project by project year after year
- Projects be assigned to teams that must discover root causes of the problems to sustain the gains
- Teams devise remedial changes to the "guilty" processes to remove or deal with the cause(s)
- Teams work with functions to install new controls to prevent the return of the causes
- Teams look for ways to replicate the remedies to increase the effect of the breakthrough
- All teams must follow a systematic fact-based method, which requires making two journeys:
    - *The diagnostic journey.* From symptoms (evidence a problem exists) to theories about what may cause the symptom(s); from theories to testing of the theories; from tests to establishing root cause(s). Once the causes are found, a second journey takes place.
    - *The remedial journey.* From root causes to remedial changes in the process to remove or deal with the cause(s); from remedies to testing and proving the remedies under operating conditions; from workable remedies to dealing with resistance to change; from dealing with resistance to establishing new controls to hold the gains.
- Regardless of what your organization calls or brands its improvement model, breakthrough results only occur after the completion of both journeys.

It has been more than 50 years since Dr. Juran first published articles on the universal sequence for breakthrough. Over that stretch of time, Dr. Juran had witnessed many models and many organizations trying to simplify, reengineer, and rename this simple method called breakthrough. Some have worked; some have not.

The most recent success is Six Sigma or Six Sigma DMAIC. Six Sigma has become the most effective "brand" of improvement since the Motorola Corporation first began using the quality improvement method Dr. Juran espoused in the late 1970s. Six Sigma methods and tools employ many of these universal principles. They have been combined with the rigor of statistical and technological tools to collect and analyze data.

GE's former chairman, Jack Welch, defined Six Sigma in this way: "Six Sigma is a quality program that, when all is said and done, improves your customers' experiences, lowers your costs and builds better leaders" (Welch 2005).

We will discuss Six Sigma in detail and fill in the blanks on its steps: define, measure, analyze, improve, and control in Chap. 15, Six Sigma: Breakthrough to in-Process Effectiveness.

A Breakthrough Improvement Case

The following is the outline of the anatomy of a breakthrough improvement project. Because this book is written as a guide, we will limit our detailed discussions to some of the more important activities that are carried out by management. Each of the following topics that are outlined contains a large body of technical knowledge, tools, and techniques.

Identify a project (management does this):

- Nominate projects.
- Evaluate projects.
- Select a project.
- Determine if this is a design project, an improvement project, or another type, such as a lean project.

Establish the project (the champions do this):

- Prepare a problem statement and a goal statement.
- Select a facilitator, or in a Six Sigma Program, a Black Belt or expert (see Chap. 15, Six Sigma: Breakthrough to in-Process Effectiveness).
- Select and launch a team.

Diagnose the cause (the project team and Black Belts do this):

- Analyze symptoms.
- Confirm and quantify or modify the goal.
- Formulate theories of causes.
- Test likely theories of causes.
- Identify root cause(s).

Remedy the cause (the project team and the workgroup where the cause(s) originate do this, perhaps with assistance from many others who are affected by, or who contribute to, the remedy):

- Evaluate alternative remedial changes.
- Design the solution, remedy, and changes needed to eliminate the root causes.
- Design new controls to hold the gains.
- Design for the culture (prevent or overcome resistance to the remedial changes).
- Prove the effectiveness of the remedy under operating conditions.
- Implement the remedial changes.

Hold the gains (the project team and the affected operating forces do this):

- Design and implement effective controls.
- Mistake-proof the process, as necessary.
- Audit the controls.

Replicate results and nominate new projects (management does this):

- Replicate the results (clone, perhaps with modifications, the remedy).
- Nominate new projects based on lessons learned from the last project.
- Organize leaders into "performance excellence" or "quality councils."

- Select problems or new goals that need to be improved, and establish projects for them.
- Create project charters: problem and goal statements.
- Provide resources: training, staff, expertise, coaching, and especially time to complete the improvement.
- Assign teams and projects to teams to stimulate remedies and controls.
- Review progress and provide recognition and rewards.

The Mysterious Damage to Linoleum in Manufactured Housing

Here is a brief case of a straightforward, relatively simple (yet valuable) project that illustrates the breakthrough improvement methodology.

Nearly half of the residential single-family dwelling units built in the United States are manufactured on moving production lines. The modular units are transported to remote locations, joined together there, and set upon prepared foundations on the home purchaser's lot. It is hard to tell the difference between an assembled manufactured house and a stick-built house once they are finished and landscaped.

A large manufacturer of modular housing units was dissatisfied with the level of very expensive rework some of its factories in various locations around the country were experiencing. Customer dissatisfaction was rising; profits were eroding. Quality councils consisting of the general manager and all direct reports were formed at each factory. They received training in quality improvement, identified the most expensive rework, formed and trained teams in quality improvement, and set them to reducing the amount of rework. This is the story of one such improvement project. We begin by identifying the problem.

Identify the Problem: One Factory's Quality Council Listed and Prioritized Its Rework Problems Using the Pareto Analysis

The Pareto distribution (arranged in descending order of cumulative percent) of their most costly rework types during the past six months looked as follows (learn more about the Pareto analysis in Chap. 19, Graphical Tools to Improve Process Performance):

- Replacing damaged linoleum: 51 percent
- Repairing cut electrical wires in walls: 15 percent
- Replacing missing fixtures at the site: 14 percent
- Repairing leaks in water pipes: 12 percent
- Repairing cracks in drywall: 8 percent

Based on the Pareto analysis, the quality council selected public enemy number one: replacing damaged linoleum. This problem is expensive to repair. Often, walls had to be removed and new linoleum laid, followed by replacing the wall. The next step was to establish a legitimate project with responsibility to resolve the problem.

Establish the Project

- A problem statement was formulated: *"The excess number of occurrences of replacing damaged linoleum accounts for 51 percent of all rework."*
- The goal statement was provided to the team as to the direction they should take: *"Reduce the number of occurrences of replacing damaged linoleum."*

Note that both the problem and goal are described and the variable and unit of measure in the problem statement and goal statement are identical. This is important because the problem statement tells the team what problem it is trying to solve. The rest of the project focuses on whatever the council selects as the problem; if they do not match, the team may carry out the goal and not solve the problem. The council chartered a project team consisting of representatives of the workstations where the linoleum was installed and where the damaged linoleum was observed. The council appointed a worker in one of those workstations to be the project team leader. The project team leader received training not only in quality improvement, but also in leading a project team. A trained facilitator coached the team in the breakthrough improvement methodology. The team began its diagnostic journey: the journey from symptom to cause.

Diagnose the Cause

The team's first task was to analyze the symptoms. (Symptoms are outward evidence of the problem.) The primary symptom was, of course, the number of occurrences of replacing damaged linoleum. Secondary symptoms were the cost of replacement, the various types of damage, the location where the damage showed up, downtime due to replacement, overtime to do the replacement, and the like. The symptoms were analyzed by *defining* them, *quantifying* them, and *visualizing* them. What follows is an analysis of the primary symptom.

Various types of damage were identified and defined as gouges, scrapes, cuts, gaps, and smears. A flow diagram was constructed showing all operations in all workstations that related to linoleum or replacement of linoleum. The flow diagram also identified the workstations where damage showed up. Several Pareto analyses were performed. The first Pareto was "by type of damage." It showed:

- Gouges (dents): 45 percent
- Scrapes: 30 percent
- Cuts: 21 percent
- Gaps: 4 percent
- Smears: 2 percent

Accordingly, the team now focused temporarily on the top priority: gouges.

A second Pareto analysis of gouges by location in the house was performed. It showed which areas of the home had the most occurrences of damage:

- Kitchen: 38 percent
- Interior hall: 21 percent
- Bathroom 1: 18 percent
- Bathroom 2: 14 percent
- Laundry: 9 percent

Now the team refocused their attention on gouges in kitchens to the temporary exclusion of all the other symptoms. The Pareto principle states that for any given effect (an output of a process or a symptom in this case), there are a number of contributors. These contributors make unequal contributions. By far, a relatively few contributors make the greatest contribution. These are called the *vital few*. Some contributors occur less often and are called the *useful many*. Following the Pareto analysis, the team concentrated on the vital few contributors to the problem to get the greatest return for the least effort.

A third Pareto analysis of gouges in kitchens by work shift showed no difference in occurrences between shifts, indicating that "shift" is not a contributor to gouges in kitchens. Based on the experience of the team, they next generated a list of theories (or hypotheses) about what causes gouges in kitchens. They generated a long list of theories. The most compelling ones were

- Dropping heavy, sharp objects (tools)
- Dragging objects across the floor
- Grit on boots of employees
- Careless employees
- No protection for the new linoleum

In the manufactured housing industry, it is known that the first three theories, if they in fact occur, cause gouges in linoleum. Those theories did not need testing. What about "lack of protection"? The only way to test that theory is first to correlate gouges in kitchens with the presence or absence of protection. The team arranged that all reports of linoleum damage or replacement would include an indication if protection was "present."

When this was done, it was discovered that virtually all cases of gouge damage to linoleum occurred when floor protection was missing. Furthermore, the team discovered that there was no formal control plan for protection to be installed. Consequently, no quality inspections or quality assurance checks revealed that no controls were being exercised and that none even existed! Floor protection was a haphazard phenomenon at best.

Remedy the Cause

Workers, purchasing personnel, and engineers went to work to select and procure material that was strong and economical to lay on freshly installed linoleum. All agreed that the operator would be responsible for laying it immediately after each job, and that supervisors would check to see that it happened. Incidents of gouge damage—and other types of damage—to linoleum went down dramatically. (It seemed several damage types had common causes, one of which was no protection.) For a few weeks, damage to linoleum almost entirely disappeared. Celebrations were held. The plant manager began to look forward to granting bigger bonuses—and getting one himself!

At the weekly meeting of the factory management team a few weeks later, the quality manager reported the mysterious reappearance of gouge damage. This news was greeted with incredulity and disappointment. "We thought we had gouge damage licked!" Indeed they had, except for a couple of "small details."

Hold the Gains

When the team investigated, it discovered that (1) no formal control plan for providing protection had been devised and published; (2) there had been a turnover of workers in the various workstations who had not been trained in the procedure; and (3) the new workers had not been trained because there was no published plan; what's more, there also was no formal training program (with controls to ensure that training actually happened). Consequently, no training could or did take place. It became apparent that the "factory" operated more like a construction site under a roof, with standards upheld by the skill and pride of artisans. A factory, by contrast, is characterized by more formal procedures and controls. All this was a valuable lesson learned for all concerned, and led to a number of additional new improvement and planning projects, new attitudes toward the work, and a maturing of the plant as it evolved from construction site to factory. Controls and training were formalized.

Initiatives of the Past

In response to a crisis or economic downturns, many organizations, especially in the United States, undertake "improvement" initiatives to improve their performance. For various reasons, many of these initiatives fall far short of their goals. Some of the methods selected were doomed from the beginning; however, a few organizations made stunning breakthroughs, improved their performance, and became the role models, the market leaders in best practices.

The methods used by these role models have been analyzed and provide us with some lessons learned—the actions that are needed to attain breakthrough and market leadership and what methods and tools must be used to enable those results to happen.

Breakthrough Lessons Learned

My analysis of the actions taken by the successful organizations shows that most of them carried out many or all of the following tasks or strategies:

1. They enlarged the business plan at all levels to include annual goals for breakthrough and customer satisfaction.
2. They implemented a systematic process for making breakthroughs and set up special infrastructure or organizational machinery to carry out that process.
3. They adopted the big Q concept—they applied the breakthrough methods to all business processes, not just the manufacturing processes.
4. They trained all levels of personnel, including upper management, in the methods and tools to carry out their respective goals.
5. They enabled the workforce to participate in making breakthroughs in their daily work practices.
6. They established measures and scorecards to evaluate progress against the breakthrough goals.
7. The managers, including the upper managers, reviewed progress against the breakthrough goals.
8. They expanded the use of recognition and revised the reward system to recognize the changes in job responsibilities and using the new methods and tools.
9. They renewed their programs every few years to include changes to their programs as their performance improved.
10. They created a "rate of improvement" that exceeded the competition's.

The Rate of Breakthrough Is Most Important

The tenth lesson learned is an important one. Just having a system of breakthrough may not be enough. This lesson learned demonstrated that the annual rate of breakthrough determines which organizations will emerge as the market leaders. Figure 5.1 shows the effect of differing rates of breakthrough.

In this figure, the vertical scale represents product salability, so what goes up is good. The upper line shows the performance of Organization A, which at the outset was the industry leader. Organization A kept getting better, year after year. In addition, Organization A was profitable. Organization A seemed to face a bright future.

The lower line shows that Organization B, a competitor, was at the outset not the leader. However, Organization B is improving at a rate much faster than that of Organization A.

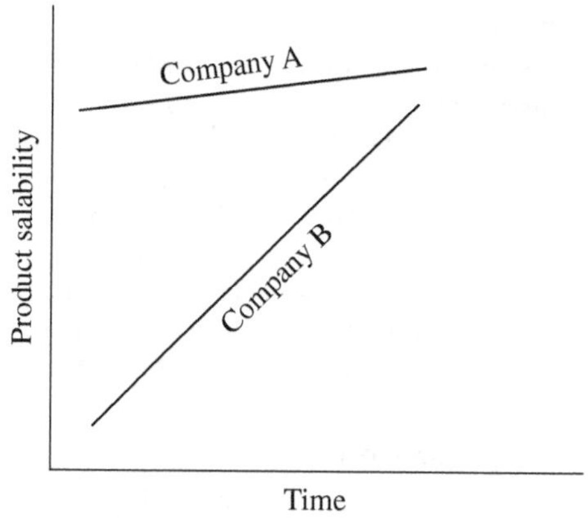

Figure 5.1 Two contrasting rates of improvement. (Juran Institute, Inc., 2009.)

Organization A is now threatened with loss of its leadership when Organization B surpasses them. The lesson is clear:

> The most decisive factor in the competition for market leadership is the rate of breakthrough an organization maintains.
>
> —Joseph M. Juran

The sloping lines of Fig. 5.1 help to explain why Japanese goods attained market leadership through quality in so many products. The major reason was that the Japanese organizations' rate of breakthrough was for decades revolutionary when compared with the evolutionary rate of the West. Eventually, they had to surpass the evolutionary rate of the Western organizations. The result was an economic disaster for many U.S. organizations in the early 1980s. Today, U.S. automobile manufacturers have made great strides in quality while Toyota has had recalls. Figure 5.2 shows my estimate of the rates of breakthrough in the automobile industry from 1950 to 1990.

There are also lessons to be learned from the numerous initiatives to improve competitiveness during the 1980s, some of which failed to produce bottom-line results. The introduction of quality circles, employee involvement teams, TQM, reengineering, and National Quality Awards all were methods used to respond to the Japanese quality revolution. Some were not sustainable and failed. Each of them may have helped the organization that used them at that point in time. An important lesson does stand out. The initiatives showed us that attaining a revolutionary rate of breakthrough is not simple at all. It takes a strategic focus to sustain market leadership. Only the National Quality Awards continue today in most parts of the world. Organizations that made statements like, Quality is dead or TQM did not work, blamed the methodology for their failures. This was only partially true. In some cases, the wrong method was selected and in others, their own management did not deal with the numerous obstacles and cultural resistance that prohibited these methods from working in the first place. These obstacles and the means to manage them will be discussed throughout this chapter.

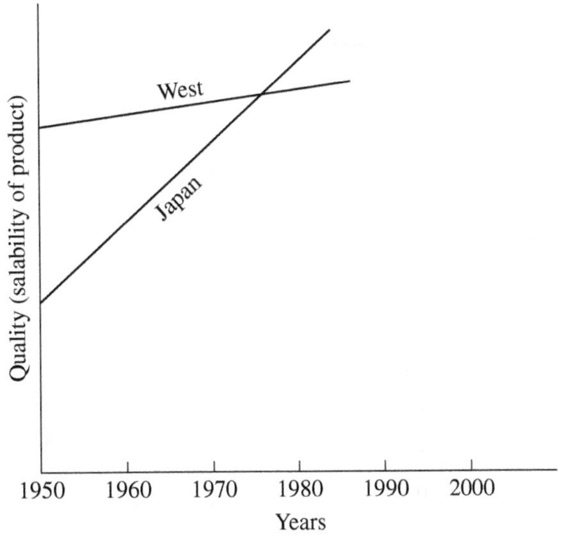

Figure 5.2 Estimate of rates of quality improvement in the automobile industry. (Juran Institute, Inc., 1994. Used by permission.)

The Breakthrough Fundamentals

Creating breakthroughs rests on just a few fundamental concepts. For most organizations and managers, annual breakthrough is not only a new responsibility; it is also a radical change in the style of management—a change in the organization's culture. Therefore, it is important to grasp the basic concepts before getting into the breakthrough process itself.

Breakthrough Distinguished from Design and Control

Breakthrough improvement differs from design (planning) and control. The trilogy diagram (Fig. 5.3) shows this difference. In this figure, the chronic waste level (the cost of poor quality) was originally about 23 percent of the amount produced. This chronic waste was built into the process—"It was planned that way." Later, a breakthrough improvement project reduced this waste to about 5 percent. Under my definition, this reduction in chronic waste is a breakthrough—it attained an unprecedented level of performance.

Figure 5.3 also shows a "sporadic spike"—a sudden increase in waste to about 40 percent. Such spikes are unplanned—they arise from various unexpected sources. The personnel promptly got rid of that spike and restored the previous chronic level of about 23 percent. This action did not meet the definition of a breakthrough. It did not attain an unprecedented level of performance. It removed the spike and returned performance to the planned level. This is referred to as root-cause analysis, taking corrective action, or "firefighting."

Creating breakthroughs in current performance differs from creating breakthroughs in design. Current performance has well-known customer needs and targets. New product or service development is trying to create something new that meets a new need of the customer. It is new, innovative, and requires proper planning.

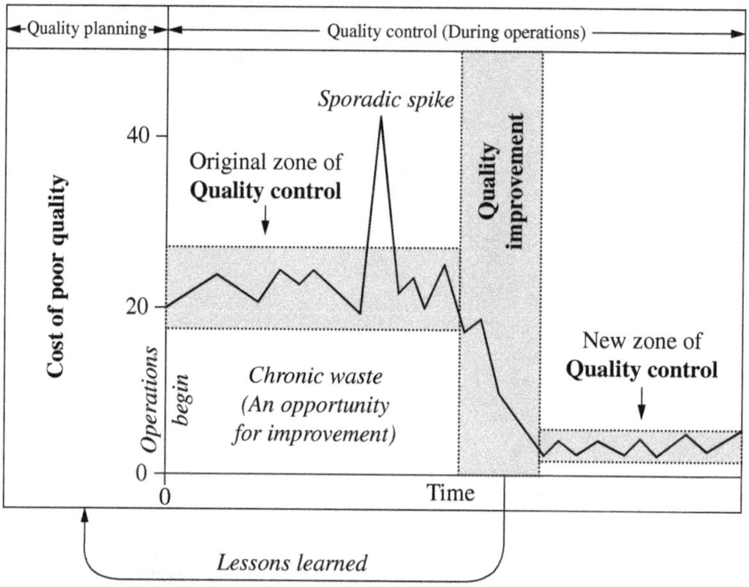

Figure 5.3 The Juran Trilogy. (Juran Institute, Inc., 1986.)

All three—design, control, and improvement—result in better performance, and all use teams to get there. It is only the steps that must be carried out that are different. This is analogous to a carpenter, electrician, and a plumber. Each tradesperson works on building a home or solving a problem (leaky pipes, rotted wood, failed circuit breaker). They have common methods and use similar tools, but at different times and for different purposes.

All Breakthrough Takes Place Project by Project

There is no such thing as breakthrough in a general way. All breakthrough takes place project by project and in no other way.

As used here, "breakthrough" means "the solving of a chronic problem by scheduling (launching a project) to find a solution." Since the word breakthrough has multiple meanings, the organization should create a glossary and educate all employees on what it means. The definition is helped by presenting a few examples that were carried out successfully in your organization.

Breakthrough Is Applicable Universally

The huge numbers of projects carried out during the 1980s and 1990s demonstrated that breakthrough is applicable to all:

- Service industries as well as manufacturing industries
- Business processes as well as manufacturing processes
- Support activities as well as operations
- Software- and information-based industries

During this period, breakthrough was applied to virtually all industries, including government, education, and health. In addition, breakthrough has been applied successfully to the entire spectrum of organization functions: finance, product development, marketing, legal, and so on.

In one organization, the legal vice president doubted that breakthrough could be applied to legal work. Yet within two years, she reduced by more than 50 percent the cycle time of filing for a patent. (For elaboration and many more case examples, see the Resources library at www.juran.com.)

Breakthrough Expands to Many Parameters

Published reports of breakthroughs show that the effects have extended to all parameters:

- *Productivity*. The output per person-hour.
- *Cycle time*. The time required to carry out processes, especially those that involve many steps performed sequentially in various departments.
- *Human safety*. Many projects improve human safety through mistake-proofing, fail-safe designs, and so on.
- *The environment*. Similarly, many projects have been directed at protecting the environment by reducing toxic goals, and so on.

Some projects provide benefits across multiple parameters. A classic example was the color television set. The Japanese Matsushita Organization had purchased an American color television factory (Quasar). Matsushita then made various breakthroughs, including

- Product redesign to reduce field failures
- Process redesign to reduce internal defect rates
- Joint action with suppliers to improve purchased components

The results of these and other changes are set out in the before and after data:

	1974	1977
Fall-off rate, i.e., defects (on assembled set) requiring repair	150 per 100 sets	4 per 100 sets
Number of repair and inspection personnel	120	15
Failure rate during the warranty period	70%	10%
Cost of service calls	$22 million	$4 million

The manufacturer benefited in multiple ways: lower costs, higher productivity, more reliable deliveries, and greater salability. The ultimate users also benefited—the field failure rate was reduced by more than 80 percent.

The Backlog of Breakthrough Projects Is Never Ending

The existence of a huge backlog of problems to solve is evident from the numbers of breakthroughs actually made by organizations that carried out successful initiatives during the 1980s and 1990s. Some reported making breakthroughs by the thousands, year after year. In very large organizations, the numbers are higher still, by orders of magnitude.

The backlog of breakthrough projects exists in part because of internal and external factors. Internally, the planning of new products and processes has long been deficient. In effect, the planning process has been a dual hatchery. It hatched out new plans. It also hatched out new chronic wastes, and these accumulated year after year. Each such chronic waste then became a potential breakthrough project.

A further reason for a huge backlog is the nature of human ingenuity—it seems to have no limit. The world's exemplar for idea systems is certainly Toyota, which, between 1951 and 1991, logged more than 20 million ideas. It is not a coincidence that the company is the world's leading manufacturing firm, as Toyota's idea system has been a major contributor to that success. The story of that system is nicely documented in the book with this compelling title: *40 Years, 20 Million Ideas* by Yuzo Yasuda.

Externally, the constantly changing needs of customers and our society will always challenge the status quo. Targets today are not good enough for tomorrow. This creates a never-ending backlog of projects.

Breakthrough Does Not Come Free

Breakthrough and the resulting reduction of chronic waste do not come free—they require an expenditure of effort in several forms. It is necessary to create an infrastructure to mobilize the organization's resources toward annual breakthrough. This involves setting specific goals to be reached, choosing projects to be tackled, assigning responsibilities, following progress, and so on.

There is also a need to conduct extensive training in the nature of the breakthrough improvement methods and tools, how to serve on breakthrough teams, how to use the tools, and so on.

In addition to all this preparatory effort, each breakthrough improvement project requires added effort to conduct diagnoses to discover the causes of the chronic waste and provide remedies to eliminate the causes. This is the time it takes for all the people involved in the team to solve the problem.

The preceding adds up to a significant front-end outlay, but the results can be stunning. They *have* been stunning in the successful organizations, the role models. Detailed accounts of such results have been widely published, notably in the proceedings of the annual conferences held by the U.S. National Institute for Standards and Technology (NIST), which administers the Malcolm Baldrige National Award.

Reduction in Chronic Waste Is Not Capital Intensive

Reduction in chronic waste seldom requires capital expenditures. Diagnosis to discover the causes usually consists of the time of the breakthrough project teams. Remedies to remove the causes usually involve fine tuning the process. In most cases, a process that is already producing more than 80 percent good work can be raised to the high 90s without capital investment. Such avoidance of capital investment is a major reason why reduction of chronic waste has a high return on investment (ROI).

In contrast, projects to create breakthroughs in product design and development to increase sales can involve costly outlays to discover customer needs, design products and processes, build facilities, and so on. Such outlays are largely classified as capital expenditures and thereby lower the ROI estimates. There is also a time lag between investing in design and receiving revenue from the sale of the new designs.

The Return on Investment for Breakthrough Improvement Is High

This is evident from results publicly reported by national award winners in Japan (Deming Prize), the United States (Baldrige Award), Europe, and elsewhere. More and more organizations have been publishing reports describing their breakthroughs, including the gains made.

It has been noted that the actual return on investment from breakthrough projects has not been well researched. My own research conducted by examining papers published by organizations found that the average breakthrough project yielded about $100,000 of cost reduction (Juran 1985). The organizations were large—sales in the range of over $1 billion per year.

It has also been estimated that for projects at the $100,000 level, the investment in diagnosis and remedy combined runs to about $15,000 or 15 percent. The resulting ROI is among the highest available to managers. It has caused some managers to quip, "The best business to be in is breakthrough." Today, breakthrough projects return many more dollars, but the cost of attaining breakthrough has not changed from the 15 percent investment level.

It is astounding that some of the recent organizations have become world quality leaders using the project-by-project approach of Six Sigma. One of them is Samsung Electronics.

Samsung Electronics Co. (SEC) of Seoul, Korea, has perfected its fundamental improvement approach using Six Sigma as a tool for innovation, efficiency, and quality. SEC was founded in 1969 and sold its first product, a television receiver, in 1971. Since that time, the company has used tools and techniques such as total quality control, total process management, product data management, enterprise resource management, supply chain management, and customer relationship management. Six Sigma was added to upgrade these existing innovations and improve SEC's competitive position in world markets. The financial benefits made possible by Six Sigma, including cost savings and increased profits from sales and new product development, are expected to approach $1.5 billion.

SEC completed 3290 Six Sigma improvement projects in the first 2 years; 1512 of these were Black Belt-level projects. By the third year, 4720 projects are expected to be completed, 1640 of them by Black Belts.

SEC's Six Sigma projects have also contributed to an average of 50 percent reduction in defects. There is no thought of improvement in quality and productivity without Six Sigma. These impressive numbers have certainly played a major role in Samsung's recent growth. Some indications of this include the following:

- By 2001, SEC had earned a net income of $2.2 billion on total revenues of $24.4 billion. Market capitalization stood at $43.6 billion.
- According to SEC's 2001 annual report, SEC now is one of the top 10 electronic and electrical equipment manufacturing companies in the world, with the best operating profit ratios and superior fiscal soundness.
- The report also says the debt-to-equity ratio is lower than that of any top ranking company, and the shareholders' equity-to-net-assets ratio surpasses the average.
- SEC says its technological strengths, Six Sigma quality initiatives, and product marketability helped increase its share of the memory chip market in 2001 to 29 percent, monitors to 21 percent, and microwave ovens to 25 percent of those sold worldwide.

Despite a downturn in the world economy and a reduction in exports to the United States, credit for SEC's current operating profit margin of 8.5 percent is due mostly to quality improvements and Six Sigma deployment.

SEC's quality and innovative strategy helped it reach the number-one position in the *BusinessWeek* 2002 information technology guide. The guide noted SEC's computer monitors, memory chips, telephone handsets, and other digital products, focusing on four Standard & Poor's criteria: shareholder return, return on equity, revenue growth, and total revenues.

The *BusinessWeek* ranking was also due to SEC's employees' belief that quality is the single most important reason for the company's higher sales, lower costs, satisfied customers, and profitable growth. Only a few years ago, SEC's products were virtually unknown by Americans or were known as the cheaper, lower-quality substitute for Japanese brands. This perception is changing. The U.S. market now represents 37 percent of SEC's total sales.

The Major Gains Come from the Vital Few Projects

The bulk of the measurable gains come from a minority of the breakthrough projects—the "vital few." These are multifunctional in nature, so they need multifunctional teams to carry them out. In contrast, the majority of the projects are in the "useful many" category and are carried out by local departmental teams. Such projects typically produce results that are orders of magnitude smaller than those of the vital few.

While the useful many projects contribute only a minor part of the measurable gains, they provide an opportunity for the lower levels of the hierarchy, including the workforce, to participate in breakthrough. We will discuss the useful many projects in Chap. 14, Lean Techniques and the Shingo Prize. In the minds of many managers, the resulting gain in work life is as important as the tangible gains in operating performance.

Breakthrough—Some Inhibitors

While the role-model organizations achieved stunning results through breakthrough, most organizations did not. Some of these failures were due to honest ignorance of how to mobilize for breakthrough, but there are also some inherent inhibitors to establishing breakthrough on a year-to-year basis. It is useful to understand the nature of some of the principal inhibitors before setting out.

Disillusioned by the Failures

The lack of results mentioned earlier has led some influential journals to conclude that breakthrough initiatives are inherently doomed to failure. Such conclusions ignore the stunning results achieved by the role-model organizations. (Their results prove that these are achievable.) In addition, the role models have explained how they got those results, thereby providing lessons learned for other organizations to follow. Nevertheless, the conclusions of the media have made some upper managers wary about going into breakthrough.

Higher Quality Costs More

Some managers hold a mindset that "higher costs more." This mindset may be based on the outmoded belief that the way to improve is to increase inspection so that fewer defects escape to the customer. It also may be based on the confusion caused by the two meanings of the word.

Higher in the sense of improved product features (through product development) usually requires capital investment. In this sense, it does cost more. However, higher in the sense of lower chronic waste usually costs less—a lot less. Those who are responsible for preparing proposals for management's approval should be careful to define the key words—which kind are they talking about?

The Illusion of Delegation

Managers are busy people, yet they are constantly bombarded with new demands on their time. They try to keep their workload in balance through delegation. The principle that "a good manager is a good delegator" has wide application, but it has been overdone as applied to breakthrough. The lessons learned from the role-model organizations show that going into annual breakthrough adds minimally about 10 percent to the workload of the entire management team, including the upper managers.

Most upper managers have tried to avoid this added workload through sweeping delegation. Some established vague goals and then exhorted everyone to do better—"Do it right the first time." In the role-model organizations, it was different. In every such organization, the upper managers took charge of the initiative and personally carried out certain nondelegable roles.

Employee Apprehensions

Going into breakthrough involves profound changes in an organization's way of life—far more than is evident on the surface. It adds new roles to the job descriptions and more work to the job holders. It requires accepting the concept of teams for tackling projects—a concept that is alien to many organizations and that invades the jurisdictions of the functional departments. It requires training on how to do all this. Collectively, the mega change disturbs the peace and breeds many unwanted side effects.

To the employees, the most frightening effect of this profound set of changes is the threat to jobs and/or status. Reduction of chronic waste reduces the need for redoing prior work and hence, the jobs of people engaged in such redoing. Elimination of such jobs then becomes a threat to the status and/or jobs of the associated supervision. It should come as no surprise if the efforts to reduce waste are resisted by the workforce, the union, the supervision, and others.

Nevertheless, breakthrough is essential to remaining competitive. Failure to go forward puts all jobs at risk. Therefore, the organization should go into breakthrough while realizing that employee apprehension is a logical reaction of worried people to worrisome proposals. A communication link must be opened to explain the why, understand the worries, and search for optimal solutions. In the absence of forthright communication, the informal channels take over, breeding suspicions and rumors.

Additional apprehension has its origin in cultural patterns. (The preceding apprehensions do not apply to breakthrough of product features to increase sales. These are welcomed as having the potential to provide new opportunities and greater job security.)

Securing Upper Management Approval and Participation

The lessons learned during the 1980s and 1990s included a major finding: Personal participation by upper managers is indispensable to getting a high rate of annual breakthrough. This finding suggests that advocates for initiatives should take positive steps to convince the upper managers of

- The merits of planning for annual breakthrough
- The need for active upper management to provide resources
- The precise nature of the needed upper management participation

Proof of the Need

Upper managers respond best when they are shown a major threat or opportunity. An example of a major threat is seen in the case of Organization G, a maker of household appliances. Organization G and its competitors, R and T, were all suppliers to a major customer involving four models of appliances (Table 5.1). This table shows that in 2000, Organization G was a supplier for two of the four models. Organization G was competitive in price, on-time delivery, and product features, but it was definitely inferior in the customer's perception of the chief

Model Number	2000	2001	2002	2003
1	G	G	R	R
2	R	R	R	R
3	G	G	G	R
4	T	R	R	R

TABLE 5.1 Suppliers to a Major Customer

Category	Amount, $	Percent of Total
Internal failures	7,279,000	79.4
External failures	283,000	3.1
Appraisal	1,430,000	15.6
Prevention	170,000	1.9
	9,162,000	100.0

TABLE 5.2 Analysis of Cost of Poor Quality

problem being field failures. By 2002, lack of response had cost Organization G the business on model number 1. By 2003, Organization G also had lost the business on model number 3.

Awareness also can be created by showing upper managers other opportunities, such as cost reduction through cutting chronic waste.

The Size of the Chronic Waste

A widespread major opportunity for upper managers is to reduce the cost of poor quality or the costs associated with poorly performing processes. In most cases, this cost is greater than the organization's annual profit, often much greater. Quantifying this cost can go far toward proving the need for a radical change in the approach to breakthrough. An example is shown in Table 5.2. This table shows the estimated cost of poor quality for an organization in a process industry using the traditional accounting classifications. The table brings out several matters of importance to upper managers:

The order of magnitude. The total of the costs is estimated at $9.2 million per year. For this organization, this sum represented a major opportunity. (When such costs have never before been brought together, the total is usually much larger than anyone would have expected.)

The areas of concentration. The table is dominated by the costs of internal failures—they are 79.4 percent of the total. Clearly, any major cost reduction must come from the internal failures.

COPQ versus Cost Reduction

Company X wanted to reduce operating costs by 10 percent. It began with a mission to have each executive identify where costs could be cut in business units. The executives created a list of 60 items, including things like eliminating quality audits, changing suppliers, adding new computer systems, reducing staff in customer services, and cutting back R&D.

The executives removed functions that provide quality and services to meet customer needs. They bought inferior parts and replaced computer systems at great expense. They disrupted their organization, particularly where the customers were most affected, and reduced the potential for new services in the future.

After accomplishing this, most of the executives were rewarded for their achievements. The result? Their cost reduction goal was met, but they had dissatisfied employees, upset customers, and an organization that still had a significant amount of expense caused by poor performance.

The financial benefit to the bottom line of an organization's balance sheet by improving the cost of quality is not always fully appreciated or understood. This misunderstanding stems from old misconceptions that improving quality is expensive.

However, this misconception is partially true. For example, if an organization provides a service to clients for a given price and a competitor provides the same basic service with enhanced features for the same price, it will cost your organization more to add those features that the competitor already provides.

If your organization does not add those features, it may lose revenue because customers will go to a competitor. If you counteract by reducing the price, you may still lose revenue. In other words, the quality of your competitor's service is better.

For your organization to remain competitive, it will have to invest in developing new features. This positively affects revenue. To improve quality, features have to be designed in—or in today's terminology, a new design must be provided at high Sigma levels.

Because of this historical misconception, organizations do not always support the notion that improving quality will affect costs and not add to them. They overlook the enormous costs associated with poor performance of products, services, and processes—costs associated with not meeting customer requirements, not providing products or services on time, or reworking them to meet the customer needs. These are the costs of poor quality (COPQ) or the cost of poorly performing processes (COP^3).

If quantified, these costs will get immediate attention at all management levels. Why? When added together, costs of poor quality make up as much as 15 to 30 percent of all costs. Quality in this complete sense, unlike the quality that affects only income, affects costs. If we improve the performance of products, services, and processes by reducing deficiencies, we will reduce these costs. To improve the quality of deficiencies that exist throughout an organization, we must apply breakthrough improvements.

A Six Sigma program focused on reducing the costs of poor quality due to low Sigma levels of performance and on designing in new features (increasing the Sigma levels) will enable management to reap increased customer satisfaction and bottom-line results. Too many organizations reduce costs by eliminating essential product or service features that provide satisfaction to customers, while ignoring poor performance that costs the bottom line and shareholders millions of dollars.

A Better Approach

Company Y approached its situation differently than did Company X, as described at the beginning of this section. The executives identified all costs that would disappear if everything worked better at higher Sigma levels. Their list included costs associated with credits or allowances given to customers because of late delivery, inaccuracy or errors in billings, scrap and rework, and accounts payable mistakes caused by discount errors and other mistakes.

When this company documented its costs of poor quality, the management team was astounded by the millions of dollars lost due to poor quality of performance within the organization.

This total cost of poor quality then became the target. The result? Elimination of waste and a return to the bottom line from planned cost reductions and more satisfied customers. Why? Because the company eliminated the reasons these costs existed in the first place. There were process and product deficiencies that caused customer dissatisfaction. Once these deficiencies were removed, the quality was higher and the costs were lower.

While responding to customer demands for improved quality in everything an organization does is becoming essential, organizations should not overlook the financial impact of poor performance. In fact, these costs should be the driver of the project selection process for Six Sigma.

In other words, the cost of poor quality provides proof of why changes must be made. The need to improve an organization's financial condition correlates directly with the process of making and measuring quality improvements. Regardless of the objective you start with, enhancing features as well as reducing costs of poor quality will affect the continuing financial success of an operation.

While there is a limit to the amount quality can be improved when cost effectiveness and savings are measured against the costs of achieving them, it's not likely this will occur until you approach Five or Six Sigma levels. A business must pursue the next level of quality based on what is of critical importance to its customers. If customers demand something, chances are it must be done to keep the business. If they do not, there's time to plan.

Driving Bottom-Line Performance

If you accept the reality that customers and the marketplace define quality, then your organization must have the right product or service features and lower your deficiencies to create loyal customers.

With a competitive price and market share strongly supported by fast cycle time, low warranty costs, and low scrap and rework costs, revenue will be higher and total cost lower. The substantial bonus that falls to the profit column comes, in effect, from a combination of enhancing features and reducing the costs of poor quality.

Before getting into specific ways to identify, measures, and account for the impact of costs of poor quality on financial results, look at what to do first if you are trying to understand how the costs of quality can drive a financial target.

For example, if your organization sets a cost reduction target to save $50 million, there is a simple methodology to determine how many improvement projects it will take to reach that goal. The organization can then manage the improvement initiative more effectively if it puts some thought behind how much activity it can afford. The answer will help determine how many experts or Black Belts are needed to manage the improvements and how much training will be required.

The methodology includes the following six steps:

1. Identify your cost reduction goal of $50 million over the next 2 years—$25 million per year.
2. Using an average return of $250,000 for each improvement, calculate how many projects are needed to meet the goal for each year. For this example, we would need an incredible 200 projects—100 per year.
3. Calculate how many projects per year can be completed and how many experts will be required to lead the team. If each project can be completed in 4 months, that means one Black Belt on two projects per 4 months. Hence, one Black Belt can complete six projects in 1 year. We will then need about 17 Black Belts.

4. Estimate how many employees will be involved on a part-time basis to work with the Black Belts to meet their targets. Assume four per Black Belt per 4 months. We would need about 200 employees involved at some level each year, possibly as little as 10 percent of their time.
5. Identify the specific costs related to poor performance, and select projects from this list that are already causing your organization to incur at least $250,000 per deficiency. If you haven't created this list, use a small team to identify the costs and create a Pareto analysis prior to launching any projects.
6. Use this method and debate each variable among the executive team to ensure the right amount of improvement can be supported. All organizations make improvements, but world-class organizations improve at a faster rate than their competition.

Where to Find Costs of Poor Performance

To put targets of opportunity into perspective, look at the traditional costs of poor quality and, even more critically, the hidden costs of poor quality, as shown in Figs. 5.2 and 5.3. The hidden costs must be quantified to get a complete picture of losses due to poor performance. These costs of poor quality could disappear entirely if every activity were performed without deficiency every time.

Three major categories of costs of poor quality exist in organizations. You can focus your efforts better if you put them into the following three categories:

1. Appraisal and inspection costs
2. Internal failure costs
3. External failure costs

Appraisal and Inspection Costs

Appraisal and inspection costs are costs associated with inspection—checking or assuring that deficiencies are discovered before customers are affected.

Examples include

- Testing products or checking documents before providing them to customers
- Reviewing documents and correcting errors before mailing
- Inspecting equipment or supplies
- Proofreading reports or correspondence
- Auditing customer bills prior to sending invoices
- Retooling due to poor design

Discovering deficiencies at this stage avoids serious failure costs later and helps develop more effective and efficient inspection methods. There will always be some costs in this category because some level of auditing will be needed to ensure consistent performance. The point is to avoid excessive costs.

Internal Failure Costs

Failure costs within an organization are attributed to the repair, replacement, or discarding of defective work the customer does not see.

Examples include

- Replacing metal stampings that do not meet specifications during production
- Repainting scratched surfaces
- Making up for unplanned computer downtime
- Replacing components damaged when being moved from one station to another
- Rewriting parts of a proposal
- Working overtime to make up for slippage
- Correcting database errors
- Stocking extra parts to replace defective components
- Scrapping products that do not meet specifications
- Spending excess accounts-payable time to correct supplier invoice errors
- Engineering change notices to correct errors in specifications or drawings

These costs may affect customer service indirectly.

External Failure Costs

External failure affects customers directly; these usually are the most expensive failures to correct. External failure costs may result from

- Satisfying warranty claims
- Investigating complaints
- Offsetting customer dissatisfaction with a recovery strategy
- Collecting bad debts
- Correcting billing errors
- Processing complaints
- Expediting late shipments by purchasing more expensive means of transportation
- Replacing or repairing damaged or lost goods
- Housing stranded passengers from cancelled flights
- Paying interest or losing discounts for late payments to vendors
- Providing on-site assistance to customers when field problems occur
- Providing credits and allowances to clients for lack of performance or late deliveries

Efforts to correct external failures usually focus on regaining customer confidence or lost sales. Both are debatable costs that may or may not be fully calculated.

Interpreting the Costs of Poor Quality

The costs of poor quality at this stage are determined by educated estimates used to guide organizational decisions. They should not be part of a monthly financial analysis, although understanding these costs may affect the way financial and cost accounting data are compiled and interpreted.

The precision required to identify the costs of poor quality varies depending on how data are used. When used to help select an improvement project, data need not be as precise as those used in developing new budgets for a process after it has been approved.

When you are evaluating projects, data on poor quality help identify, charter, and support projects with the greatest potential for reducing costs. Black Belts and teams may select some projects because of the impact on customers or internal culture, but data must show where costs are highest so that focus can be concentrated on the vital few.

The amount of cost reduction provided by a remedy is another indicator of project effectiveness. When planning for a remedy, a task force should develop supportable estimates of costs that will be eliminated by the remedy and use those estimates to develop a budget for the revised process.

There are four major steps in measuring the costs of poor quality:

1. Identify activities resulting from poor quality.
2. Decide how to estimate costs.
3. Collect data and estimate costs.
4. Analyze results and decide on the next steps.

Identify Activities Resulting from Poor Quality

Activities are categorized as resulting from poor quality only if they exist solely because of deficiencies assessed when doing appraisals, inspections, and internal or external cost estimates.

A project team usually begins by measuring the obvious costs of a problem's primary symptom, such as discarded supplies, customer complaints, or erroneous shipments. After a flow diagram of the process in question has been created and further analysis has been conducted, additional activities are usually identified as those required, for example, to dispose of and replace returned items.

Efforts to identify remedial activities are generally more global since the focus is on costs of poor quality throughout an organization. This effort is best undertaken by one or a small number of analysts working with a team of midlevel and senior managers experienced in key areas.

The task force usually launches its efforts by identifying major organizational processes and their customers. For each process, the task force brainstorms major activities associated with poor quality and expands the list through carefully constructed interviews with individuals representing different levels within the most critical functions. At this point, the objective is to prepare a list of activities related to poor quality, not estimate costs.

Project teams and task forces find it easier to explain what they are looking for if they have a full list of typical examples associated with poor quality. The examples described earlier fall into major categories of poor quality costs. Using key words such as rework, waste, fix, return, scrap, complaint, repair, expedite, adjust, refund, penalty, waiting, and excess usually stimulates a healthy response, too.

Decide How to Estimate Costs

When a specific activity related to poor quality is identified, two strategies help estimate its costs: total resources and unit costs. These strategies can be used individually or together.

An example of the total resource approach is how an operational unit calculated the human resource time to process customer complaints and the dollar value of that time.

This approach requires two pieces of data: total resources consumed in a category and the percentage of those resources consumed for activities associated with poor quality.

An example of the unit cost approach is when a project team calculates the annual cost of correcting erroneous shipments. To find that cost, the team should estimate the cost of correcting an average erroneous shipment and how many errors occurred in one year, and then multiply the average cost by the annual number of errors.

Data for calculating the total resources used in a category might come from a variety of sources, such as accounting, time reporting, other information systems, informed judgment, special time reporting, special data collections, and unit costs. These sources are described in the section "Calculating Resources Used."

Collect Data and Estimate Costs

Procedures for collecting data on costs of poor quality are generally the same as those for any good data collection:

- Formulate questions to be answered.
- Know how data will be used and analyzed.
- Determine where data will be collected.
- Decide who will collect it.
- Understand data collectors' needs.
- Design a simple data collection form.
- Prepare clear instructions.
- Test forms and procedures.
- Train data collectors.
- Audit results.

To estimate the costs of poor quality, it is sometimes necessary to collect personal opinions and judgments about relative magnitudes of time spent or cost. Even though precise numeric data is not required for such estimates, it is important to plan carefully. The manner in which opinions are solicited affects responses.

Sampling works when the same activity is performed often in different parts of an organization. All field sales offices, for example, perform similar functions. If a company has 10 field sales offices, estimates from one or two would provide a reasonable value for calculating overall costs of poor quality.

Analyze Results and Decide on the Next Steps

Collecting data on costs of poor quality helps make decisions such as

- Selecting the most important quality improvement projects
- Identifying the most costly aspects of a specific problem
- Identifying specific costs to be eliminated

The Results

Of note is the fact that every organization that has adopted Six Sigma and integrated the discipline throughout its operations has produced impressive savings that were reflected on the bottom line. More customers were satisfied and became loyal, and revenues, earnings, and operating margins improved significantly.

For example, Honeywell's cost savings have exceeded $2 billion since implementing Six Sigma in 1994. At General Electric, the Six Sigma initiative began in 1996 and produced more than $2 billion in benefits in 1999. Black & Decker's Six Sigma productivity savings rose to about $75 million in 2000, more than double the prior year's level, bringing the total saved since 1997 to over $110 million.

A more revealing insight into the cost of poor quality as a function of Six Sigma performance levels is the following:

- When ±3 Sigma of the process that produces a part is within specification, there will be 66,807 defects per million parts produced. If each defect cost $1000 to correct, then the total COPQ would be $66,807,000.
- When an organization improves the process to within ±4 Sigma, there will be only 6210 defects per million at a COPQ of $6,210,000.
- At ±5 Sigma, the cost of defects declines to $233,000 per million, a savings of $66,574,000 more than the savings at a process capability of ±3 Sigma.
- At the near perfection level of ±6 Sigma, defects are almost eliminated at $3400 per million parts produced.

After all data are collected and tabulated, and decisions are made, no study of the cost of poor quality should end without a continuing action plan to eliminate a major portion of the costs that have been identified. There is no need to use a complex accounting method for measuring costs because it would be expensive and waste valuable effort. Simple methods are sufficient.

The most important step in developing useful COPQ data is simply to identify activities and other factors that affect costs. Any consistent and unbiased method for estimating costs will yield adequate information that will identify key targets for quality improvement. More refined estimates may be needed for specific projects when diagnosing the cause of a specific problem or identifying specific savings.

Calculating Resources Used

Data for calculating the total resources used in an expense category come from a variety of sources.

Accounting Categories

Financial and cost accounting systems often contain specific categories that can be allocated partly or totally to costs of poor quality. Typical examples include scrap accounts, warranty costs, professional liability, discarded inventory, and total department operating costs.

Time Reporting

Many organizations routinely ask employees to report how much time they spend on specific activities. This makes it possible to assign some or all of the time in a category to a specific cost of poor quality.

Other Information Systems

Other information systems include cost accounting, activity-based cost accounting, materials management, sales, or similar reports.

Data for calculating the percentage of resources used for cost of poor quality activities can be obtained through a variety of techniques, including

- *Informed judgment.* Supervisors and experienced employees can make adequate judgments about what proportion of a department's time is spent on an activity. This is especially true if the unit performs very few distinct functions or if the effort consumes a very large or small portion of total time.
- *Special time reporting.* This method has been used to calculate costs for processing computer complaints. A special short-term collection of time distribution data may be appropriate if a department performs many different functions, activity is neither unusually small nor large, or there is uncertainty or significant disagreement among informed individuals as to the percentage of time or money allocated to a specific activity. A significant disagreement would typically be one of more than 10 percent of the total amount allocated.
- *Special data collections.* Besides collecting data on how much employee time is spent on an activity, an organization might also collect data on the amount of time a computer network is inoperative, the volume of items consumed or discarded, or the amount of time special equipment or other resources are not used.

In all these examples, the general calculation to determine costs of poor quality is

$$\text{Cost of poor quality} = (\text{cost of total resources in a category}) \times \\ (\text{percentage of resources in category used for activities related to poor quality})$$

Unit Cost

An example of this strategy occurs when a project team calculates the annual cost of correcting erroneous shipments. To find out the cost, the team should estimate the cost of correcting an average erroneous shipment, estimate how many such errors occurred in 1 year, and then multiply the average cost by the annual number of errors.

Focusing on unit cost requires two pieces of data: the number of times a particular deficiency occurs and the average cost for correcting and recovering from that deficiency when it does occur.

This average cost, in turn, is computed from a list of resources used to make corrections, on the amount used of each resource, and on the cost of each resource unit.

Unit cost is often the most appropriate strategy when deficiencies occur rarely and may be costly, when deficiencies are complex and require the participation of many departments to correct, or when deficiencies occur frequently and correcting them is so routine that those involved may not realize their pervasiveness.

Data on frequency of a deficiency may come from any of the following:

- Quality assurance
- Warranty data
- Customer surveys
- Field service reports
- Customer complaints
- Management engineering studies
- Internal audit reports

- Operational logs
- Special surveys

Estimating the cost of a single occurrence usually requires some analysis. A flowchart showing various rework loops associated with a deficiency can often help identify all-important resources used.

When searching for resources, consider hours worked by occupation and level, contracted services, materials and supplies, capital equipment and facilities, and cost of money for borrowed or uncollected funds.

To find out how much of each resource is used, check the following sources:

- Time reporting systems
- Cost accounting systems
- Various administrative logs
- Management engineering studies
- Informed judgment
- Special data collections

When a team has identified the amount of each resource used, it is ready to calculate the cost for each and add up costs for all resources. The finance or engineering functions typically will have standard methods for calculating the unit costs a team might require.

Here are hints to remember when calculating unit costs:

- Include benefits as well as wages and salaries.
- Include allocated capital costs for major equipment and facilities. While this is a minor consideration for many activities that can be safely ignored, it is vital for some activities.

Do not be misled by the argument that capital costs are fixed and would exist even if deficiencies did not occur. This is a typical example of the cost of poor quality being hidden by standard practices. If computers were used more efficiently, it would be possible to process more jobs without buying additional equipment. Idle capital or misused capital resources are a cost of poor quality just as surely as discarded paper from a faulty print job.

Be sure to include penalties or misused discounts for late payments and premium prices paid for rush orders or shipments.

Other Methods

Still other methods can be developed for special projects. For example, with regard to lost supplies, the organization should calculate the cost that would have been consumed if there had been no defects and the cost of supplies had actually been consumed. The difference between the two is the cost of poor quality. This type of approach might also be applied in comparing actual outcomes with the best outcomes others have achieved.

Special circumstances may lead a team to develop still other approaches that are appropriate to the specific problem. For example, a greater investment in prevention would be cost-effective.

The Potential Return on Investment

A major responsibility of upper managers is to make the best use of the organization's assets. A key measure of judging what is best is return on investment (ROI). In general terms, ROI is the ratio of (1) the estimated gain to (2) the estimated resources needed. Computing ROI for projects to reduce chronic waste requires assembling estimates such as

- The costs of chronic waste associated with the projects
- The potential cost reductions if the projects are successful
- The costs of the needed diagnosis and remedy

Many proposals to go into breakthrough have failed to gain management support because no one has quantified the ROI. Such a goal is a handicap to the upper managers—they are unable to compare (1) the potential ROI from breakthrough with (2) the potential ROI from other opportunities for investment.

Managers and others who prepare such proposals are well advised to prepare the information on ROI in collaboration with those who have expertise in the intricacies of ROI. Computation of ROI can be complicated because two kinds of money are involved—capital and expenses. Each is money, but in some countries (including the United States) they are taxed differently. Capital expenditures are made from after-tax money, whereas expenses are paid out of pretax money.

This difference in taxation is reflected in the rules of accounting. Expenses are written off promptly, thereby reducing the stated earnings and hence, the income taxes on earnings. Capital expenditures are written off gradually—usually over a period of years. This increases the stated earnings and hence, the income taxes on those earnings. This means it is advantageous for proposals to go into breakthrough because breakthrough is seldom capital intensive. (Some upper managers tend to use the word *investment* as applying only to capital investment.)

Getting Cost Figures

Organization accounting systems typically quantify only a minority of the costs of poor quality. The majority are scattered throughout the various overheads. As a result, specialists have looked for ways to supply what is missing. Their main efforts toward solution have been as follows:

- *Make estimates.* This is the "quick and dirty" approach. It is usually done by sampling, and involves only a modest amount of effort. It can, in a few days or weeks, provide (a) an evaluation of the approximate cost of chronic waste and (b) indicate where this is concentrated.
- *Expand the accounting system.* This is much more elaborate. It requires a lot of work from various departments, especially accounting, and it runs into a lot of calendar time, often 2 or 3 years.

> In my experience, estimates involve much less work, can be prepared in far less time, and yet are adequate for managerial decision making.
>
> —J.M. Juran

Note that the demand for "accuracy" of the cost figures depends on the use to which the figures will be put. Balancing the books demands a high degree of accuracy. Making managerial decisions sometimes can tolerate a margin of error. For example, a potential breakthrough project has been estimated to incur about $300,000 in annual cost of poor quality. This figure is challenged. The contesting estimates range from $240,000 to $360,000—quite a wide range. Then someone makes an incisive observation: "It doesn't matter which estimate is correct. Even at the lowest figure, this is a good opportunity for breakthrough, so let's tackle it." In other words, the managerial decision to tackle the project is identical despite a wide range of estimate.

Languages in the Hierarchy

A subtle aspect of securing upper management approval is the choice of language. Industrial organizations make use of two standard languages—the language of money and the language of things. (There are also local dialects, each peculiar to a specific function.) However, as seen in Fig. 5.4, use of the standard languages is not uniform.

Figure 5.4 shows the use of standard languages in different levels of a typical hierarchy. At the apex, the principal language of the top management team is the language of money. At the base, the principal language of the first-line supervisors and the workforce is the language of things. In between, the middle managers and the specialists need to understand both principal languages—*the middle managers should be bilingual.*

It is quite common for chronic waste to be measured in the language of things: percent errors, process yields, hours of rework, and so on. Converting these measures into the language of money enables upper managers to relate them to the financial measures that have long dominated the management "instrument panel."

Years ago, Dr. Juran was invited to visit a major British manufacturer to study its approach to managing for quality and to provide a critique. He found that the organization's cost of poor quality was huge, that it was feasible to cut this in half in 5 years, and that the resulting return on investment would be much greater than that of making and selling the organization's products. When he explained this to the managing director, he was most impressed—it was the first time that the problem of chronic waste had been explained to him in the language of return on investment. He promptly convened his directors (vice presidents) to discuss what to do about this opportunity.

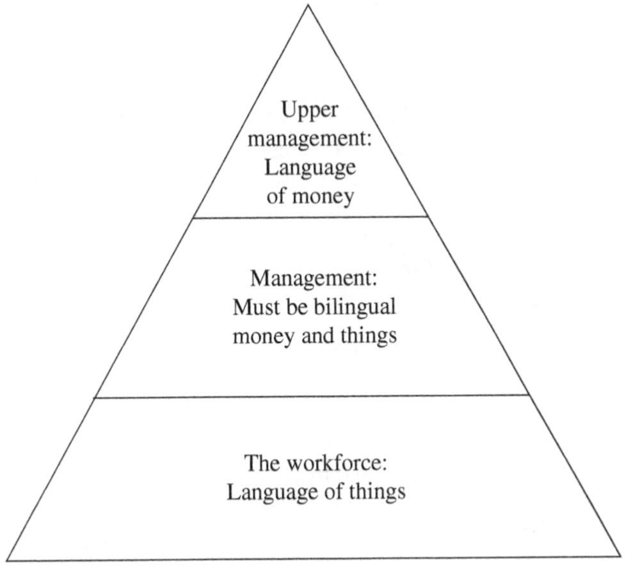

FIGURE 5.4 Common languages in the hierarchy. (Juran Institute, Inc., 1994.)

Presentations to Upper Managers

Presentations to upper managers should focus on the goals of the upper managers, not on the goals of the advocates. Upper managers are faced with meeting the needs of various stakeholders: customers, owners, employees, suppliers, the public (e.g., safety, health, environment), and so on. It helps if the proposals identify specific problems of stakeholders and estimate the benefits to be gained.

Upper managers receive numerous proposals for allocating the organization's resources: invade foreign markets, develop new products, buy new equipment to increase productivity, make acquisitions, enter joint ventures, and so on. These proposals compete with each other for priority, and a major test is ROI. It helps if the proposal to go into breakthrough includes estimates of ROI.

Converting the supporting data into units of measure that are already familiar to upper managers sometimes helps an explanation of proposals. For example,

- Last year's cost of poor quality was five times last year's profit of $1.5 million.
- Cutting the cost of poor quality in half would increase earnings by 13 cents per share of stock.
- Thirteen percent of last year's sales orders were canceled due to poor quality.
- Thirty-two percent of engineering time was spent in finding and correcting design weaknesses.
- Twenty-five percent of manufacturing capacity is devoted to correcting problems.
- Seventy percent of the inventory carried is traceable to poor quality.
- Twenty-five percent of all manufacturing hours were spent in finding and correcting defects.
- Last year's cost of poor quality was the equivalent of our operation making 100 percent defective work during the entire year.

Experience in making presentations to upper management has provided some useful dos and don'ts:

- *Do* summarize the total of the estimated costs of poor quality. The total will be big enough to command upper management's attention.
- *Do* show where these costs are concentrated. A common grouping is in the form of Table 5.2. Typically (as in that case), most of the costs are associated with failures, internal and external. Table 5.2 also shows the fallacy of trying to start by reducing inspection and test. The failure costs should be reduced first. After the defect levels come down, inspection costs can be reduced as well.
- *Do* describe the principal projects that are at the heart of the proposal.
- *Do* estimate the potential gains, as well as the return on investment. If the organization has never before undertaken an organized approach to reducing related costs, then a reasonable goal is to cut these costs in half within a space of 5 years.
- *Do* have the figures reviewed in advance by those people in finance (and elsewhere) to whom upper management looks for checking the validity of financial figures.
- *Do not* inflate the present costs by including debatable or borderline items. The risk is that the decisive review meetings will get bogged down in debating the validity of the figures without ever discussing the merits of the proposals.

- *Do not* imply that the total costs will be reduced to zero. Any such implication will likewise divert attention from the merits of the proposals.
- *Do not* force the first few projects on managers who are not really sold on them or on unions who are strongly opposed. Instead, start in areas that show a climate of receptivity. The results obtained in these areas will determine whether the overall initiative will expand or die out.

The needs for breakthrough go beyond satisfying customers or making cost reductions. New forces keep coming over the horizon. Recent examples have included growth in product liability, the consumerism movement, foreign competition, legislation, and environmental concerns of all sorts. Breakthrough has provided much of the response to such forces.

Similarly, the means of convincing upper managers of the need for breakthrough go beyond reports from advocates. Conviction also may be supplied by visits to successful organizations, hearing papers presented at conferences, reading reports published by successful organizations, and listening to the experts, both internal and external. However, none of these is as persuasive as results achieved within one's own organization.

A final element of presentations to upper managers is to explain their personal responsibilities in launching and perpetuating breakthrough.

Mobilizing for Breakthrough

Until the 1980s, breakthrough in the West was not mandated—it was not a part of the business plan or a part of the job descriptions. Some breakthrough did take place, but on a voluntary basis. Here and there, a manager or a nonmanager, for whatever reason, elected to tackle some breakthrough project. He or she might persuade others to join an informal team. The result might be favorable, or it might not. This voluntary, informal approach yielded few breakthroughs. The emphasis remained on inspection, control, and firefighting.

The Need for Formality

The crisis that followed the Japanese revolution called for *new strategies*, one of which was a much higher rate of breakthrough. It then became evident that an informal approach would not produce thousands (or more) breakthroughs year after year. This led to experiments with structured approaches that in due course helped some organizations become role models.

Some upper managers protested the need for formality: "Why don't we just do it?" The answer depends on how many breakthroughs are needed. For just a few projects each year, informality is adequate; there is no need to mobilize. However, making breakthroughs by the hundreds or the thousands requires a formal structure.

As it has turns out, mobilizing for breakthrough requires two levels of activity, as shown in Fig. 5.5. The figure shows the two levels of activity. One of these mobilizes the organization's resources to deal with the breakthrough projects collectively. This becomes the responsibility of management. The other activity is needed to carry out the projects individually. This becomes the responsibility of the breakthrough teams.

The Executive "Quality Council"

The first step in mobilizing for breakthrough is to establish the organization's council (or similar name). The basic responsibility of this council is to launch, coordinate, and "institutionalize" annual breakthrough. Such councils have been established in many organizations. Their experiences provide useful guidelines.

Activities by Management	Activities by Project Teams
Establish infrastructure: quality councils	Verify problem
Select problems; determine goals and targets	Analyze symptoms
Create project charters and assign teams	Theorize as to causes
Launch teams and review progress	Test theories
Provide recognition and rewards	Discover causes
	Stimulate remedies and controls

Figure 5.5 Mobilizing for breakthrough.

Membership and Responsibilities

Council membership is typically drawn from the ranks of senior managers. Often, the senior management committee is also the council. Experience has shown that councils are most effective when upper managers are personally the leaders and members of the senior councils.

In large organizations, it is common to establish councils at the divisional level as well as at the corporate level. In addition, some individual facilities may be so large as to warrant establishing a local council. When multiple councils are established, they are usually linked together—members of high-level councils serve as chairpersons of lower-level councils. Figure 5.6 is an example of such linkage.

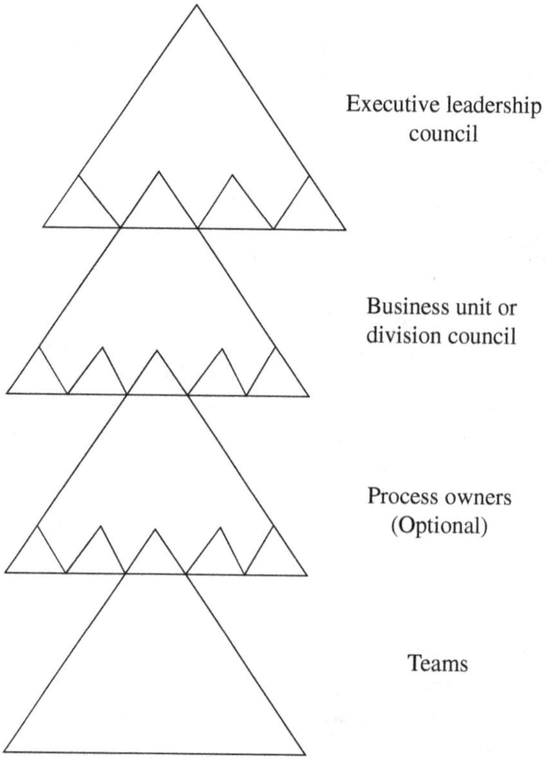

Figure 5.6 Quality councils are linked together. (Juran Institute, Inc., 1994.)

Experience has shown that organizing councils solely in the lower levels of management is ineffective. Such organization limits breakthrough projects to the "useful many" while neglecting the "vital few" projects—those that can produce the greatest results. In addition, councils solely at lower levels send a message to all: "Breakthrough is not high on upper management's agenda."

It is important for each council to define and publish its responsibilities so that (1) the members agree on their goal and (2) the rest of the organization can become informed relative to upcoming events.

Many councils have published their statements of responsibility. Major common elements have included the following:

- Formulate the policies, such as focus on the customer has top priority, breakthrough must go on year after year, participation should be universal, or the reward system should reflect performance on breakthrough.
- Estimate the major dimensions, such as the status of the company's quality compared with its competitors, the extent of chronic waste, the adequacy of major business processes, or the results achieved by prior breakthroughs.
- Establish processes for selecting projects, such as soliciting and screening nominations, choosing projects to be tackled, preparing goal statements, or creating a favorable climate for breakthrough.
- Establish processes for carrying out the projects, such as selecting team leaders and members or defining the role of project teams.
- Provide support for the project teams, such as training time for working on projects, diagnostic support, facilitator support, or access to facilities for tests and tryouts.
- Establish measures of progress, such as effect on customer satisfaction, effect on financial performance, or extent of participation by teams.
- Review progress, assist teams in the event of obstacles, and ensure that remedies are implemented.
- Provide for public recognition of teams.
- Revise the reward system to reflect the changes demanded by introducing annual breakthrough.

Councils should anticipate the troublesome questions and, to the extent feasible, provide answers at the time of announcing the intention to go into annual breakthrough. Some senior managers have gone to the extent of creating a videotape to enable a wide audience to hear the identical message from a source of undoubted authority.

Leaders Must Face Up to the Apprehensions about Elimination of Jobs

Employees not only want dialogue on such an important issue, they also want assurance relative to their apprehensions, notably the risk of job loss due to improvements. Most upper managers have been reluctant to face up to these apprehensions. Such reluctance is understandable. It is risky to provide assurances when the future is uncertain.

Nevertheless, some managers have estimated in some depth the two pertinent rates of change:

- The rate of creation of job openings due to attrition: retirements, offers of early retirement, resignation, and so on. This rate can be estimated with a fair degree of accuracy.

- The rate of elimination of jobs due to reduction of chronic waste. This estimate is more speculative—it is difficult to predict how soon the breakthrough rate will get up to speed. In practice, organizations have been overly optimistic in their estimates.

Analysis of these estimates can help managers judge what assurances they can provide, if any. It also can shed light on the choice of alternatives for action: retrain for jobs that have opened up, reassign to areas that have job openings, offer early retirement, assist in finding jobs in other organizations, and/or provide assistance in the event of termination.

Assistance from the Quality and/or Performance Excellence Functions

Many councils secure the assistance of the performance excellence and quality departments. These are specialists that are skilled in the methods and tools to attain high quality. They are there to

- Provide inputs needed by the council for planning to introduce breakthrough
- Draft proposals and procedures
- Carry out essential details such as screening nominations for projects
- Develop training materials
- Develop new scorecards
- Prepare reports on progress

It is also usual for the quality directors to serve as secretaries of the council.

Breakthrough Goals in the Business Plan

Organizations that have become the market leaders—the role models—all adopted the practice of enlarging their business plan to include quality goals. In effect, they translated the threats and opportunities faced by their organizations into goals, such as

- Increase on-time deliveries from 83 to 100 percent over the next 2 years.
- Reduce the cost of poor quality by 50 percent over the next 5 years.

Such goals are clear—each is quantified, and each has a timetable. Convincing upper managers to establish such goals is a big step, but it is only the first step.

Deployment of Goals

Goals are merely a wish list until they are deployed—until they are broken down into specific projects to be carried out and assigned to specific individuals or teams who are then provided with the resources needed to take action. Figure 5.7 shows the anatomy of the deployment process. In the figure, the broad (strategic) goals are established by the council and become a part of the organization's business plan. These goals are then divided and allocated to lower levels to be translated into action. In large organizations, there may be further subdivision before the action levels are reached. The final action level may consist of individuals or teams.

In response, the action levels select breakthrough projects that collectively will meet the goals. These projects are then proposed to the upper levels along with estimates of the resources needed. The proposals and estimates are discussed and revised until final decisions are reached. The end result is an agreement on which projects to tackle, what resources to provide, and who will be responsible for carrying out the projects.

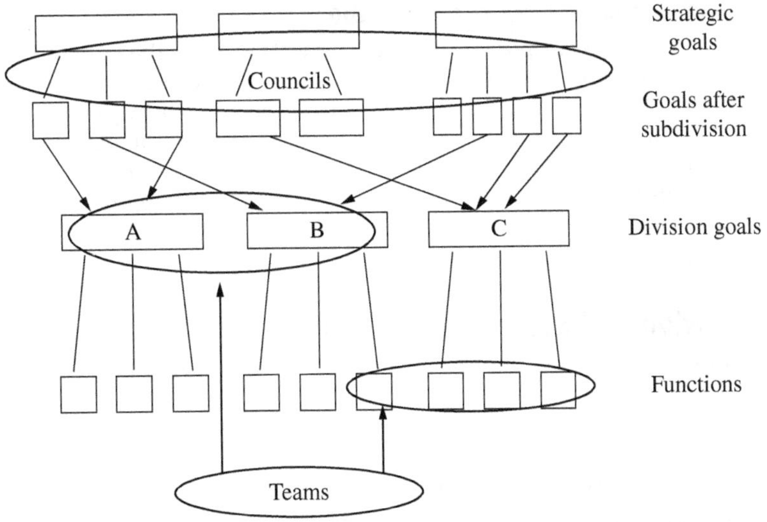

FIGURE 5.7 Anatomy of the deployment process. (Juran Institute, Inc.)

This approach of starting at the top with strategic goals may seem like a purely top-down activity. However, the deployment process aims to provide open discussion in both directions before final decisions are made, and such is the way it usually works out.

The concept of strategic goals involves the vital few matters, but it is not limited to the corporate level. Goals also may be included in the business plans of divisions, profit centers, field offices, and still other facilities. The deployment process is applicable to all of these. (For added discussion of the deployment process, see Chap. 7, Strategic Planning and Performance Excellence.)

The Project Concept

As used here, a project is a chronic problem scheduled for solution. The project is the focus of actions for breakthrough. All breakthrough takes place project by project and in no other way.

Some projects are derived from the goals that are in the organization's business plan. These are relatively few in number, but each is quite important. Collectively, these are among the vital few projects (see "Use of the Pareto Principle"). However, most projects are derived not from the organization's business plan but from the nomination-selection process, as discussed later.

Use of the Pareto Principle

A valuable aid to the selection of projects during the deployment process is the Pareto principle. This principle states that in any population that contributes to a common effect, a relative few of the contributors—the vital few—account for the bulk of the effect. The principle applies widely in human affairs. Relatively small percentages of the individuals write most of the books, commit most of the crimes, own most of the wealth, and so on.

Presentation of data in the form of a Pareto diagram greatly enhances communication of the information, most notably in convincing upper management of the source of a problem and gaining support for a proposed course of action to remedy the problem. (For an account of how Dr. Juran came to name the Pareto principle, see Appendix in this Handbook.)

The Useful Many Problems and Solutions

Under the Pareto principle, the vital few projects provide the bulk of the breakthrough, so they receive top priority. Beyond the vital few are the useful many problems. Collectively, they contribute only a minority of the breakthrough, but they provide most of the opportunity for employee participation. The useful many projects are made through the application of workplace improvement teams, quality circles, the Lean 5S tools, or self-directed work teams. See Chap. 8, Organization Roles to Support a Quality and Excellence Culture.

The Nomination and Selection Process

Most projects are chosen through the nomination and selection process, involving several steps:

- Project nomination
- Project screening and selection
- Preparation and publication of project goal statements

Sources of Nominations

Nominations for projects can come from all levels of the organization. At the higher levels, the nominations tend to be extensive in size (the vital few) and multifunctional in their scope. At lower levels, the nominations are smaller in size (the useful many) and tend to be limited in scope to the boundaries of a single department.

Nominations come from many sources. These include

- *Formal data systems*, such as field reports on product performance, customer complaints, claims, returns, and so on; accounting reports on warranty charges and on internal costs of poor quality; and service call reports. (Some of these data systems provide for analyzing the data to identify problem areas.)
- *Special studies*, such as customer surveys, employee surveys, audits, assessments, benchmarking against competitors, and so on.
- *Reactions from customers* who have run into product dissatisfactions are often vocal and insistent. In contrast, customers who judge product features to be not competitive may simply (and quietly) become ex-customers.
- *Field intelligence* derived from visits to customers, suppliers, and others; actions taken by competitors; and stories published in the media (as reported by sales, customer service, technical service, and others).
- *The impact on society*, such as new legislation, extension of government regulation, and growth of product liability lawsuits.
- *The managerial hierarchy*, such as the council, managers, supervisors, professional specialists, and project teams.
- *The workforce* through informal ideas presented to supervisors, formal suggestions, ideas from circles, and so on.
- *Proposals* relating to business processes.

Effect of the Organization-wide or Big Q Concept

Beginning in the 1980s and continuing for the near future, the scope of nominations for projects broadened considerably under the big Q concept. The breadth of the big Q concept is evident from the wide variety of projects that have already been tackled:

- Improve the precision of the sales forecast.
- Reduce the cycle time for developing new products.
- Increase the success rate in bidding for business.
- Reduce the time required to fill customers' orders.
- Reduce the number of sales cancellations.
- Reduce the errors in invoices.
- Reduce the number of delinquent accounts.
- Reduce the time required to recruit new employees.
- Improve the on-time arrival rate (for transportation services).
- Reduce the time required to file for patents.

The Nomination Process

Nominations must come from human beings. Data systems are impersonal—they make no nominations. Various means are used to stimulate nominations for breakthrough projects:

- *Call for nominations*. Letters or bulletin boards are used to invite all personnel to submit nominations, either through the chain of command or to a designated recipient, such as the secretary of the council.
- *Make the rounds*. In this approach, specialists (such as engineers) are assigned to visit the various departments, talk with the key people, and secure their views and nominations.
- *The council members themselves*. They become a focal point for extensive data analyses and proposals.
- *Brainstorming meetings*. These are organized for the specific purpose of making nominations.

Whatever the method used, it will produce the most nominations if it urges use of the big Q concept—the entire spectrum of activities, products, and processes.

Nominations from the Employees at All Levels

The workforce is potentially a source of numerous nominations. Workers have extensive residence in the workplace. They are exposed to many local cycles of activity. Through this exposure, they are well poised to identify the existence of problems and to theorize about their causes. As to the details of goings-on in the workplace, no one is better informed than the workforce. "That machine hasn't seen a maintenance man for the last six months." In addition, many workers are well poised to identify opportunities and to propose new ways.

Workforce nominations consist mainly of local useful many projects along with proposals of a human relations nature. For such nominations, workers can supply useful theories of causes as well as practical proposals for remedies. For projects of a multifunctional nature, most workers are handicapped by their limited knowledge of the overall process and of the interactions among the steps that collectively make up the overall process.

In some organizations, the solicitation of nominations from the workforce has implied that such nominations would receive top priority. The effect was that the workforce was deciding which projects the managers should tackle first. It should have been made clear that workers' nominations must compete for priority with nominations from other sources.

Joint Projects with Suppliers and Customers

All organizations buy goods and services from suppliers; over half the content of the finished product may come from suppliers. In earlier decades, it was common for customers to contend. The supplier should solve his problems. Now there is growing awareness that these problems require a partnership approach based on

- Establishing mutual trust
- Defining customer needs as well as specifications
- Exchanging essential data
- Direct communication at the technical level as well as the commercial level

Project Screening

Calls for nominations can produce large numbers of responses—numbers that are beyond the digestive capacity of the organization. In such cases, an essential further step is screening to identify those nominations that promise the most benefits for the effort expended.

To start with a long list of nominations and end up with a list of agreed-upon projects requires an organized approach—an infrastructure and a methodology. The screening process is time-consuming, so the council usually delegates it to a secretariat, often the department. The secretariat screens the nominations—it judges the extent to which the nominations meet the criteria set out below. These judgments result in some preliminary decision-making. Some nominations are rejected. Others are deferred. The remainder is analyzed in greater depth to estimate potential benefits, resources needed, and so on.

The councils and/or the secretariats have found it useful to establish criteria to be used during the screening process. Experience has shown that there is a need for two sets of criteria:

- Criteria for choosing the first projects to be tackled by any of the project teams
- Criteria for choosing projects thereafter

Criteria for Projects

During the beginning stages of project-by-project breakthrough, everyone is in a learning state. Projects are assigned to project teams, who are in training. Completing a project is a part of that training. Experience with such teams has evolved a broad set of criteria:

- The project should deal with a *chronic problem*—one that has been awaiting a solution for a long time.
- The project should be *feasible*. There should be a good likelihood of completing it within a few months. Feedback from organizations suggests that the most frequent reason for failure of the first project has been failure to meet the criterion of feasibility.
- The project should be *significant*. The end result should be sufficiently useful to merit attention and recognition.

- The results should be *measurable*, whether in money or in other significant terms.
- The first projects should be winners.

Additional criteria to select projects are aimed at what will do the organization the most good:

- *Return on investment*. This factor has great weight and is decisive, all other things being equal. Projects that do not lend themselves to computing return on investment must rely for their priority on managerial judgment.
- *The amount of potential breakthrough*. One large project will take priority over several small ones.
- *Urgency*. There may be a need to respond promptly to pressures associated with product safety, employee morale, and customer service.
- *Ease of technological solution*. Projects for which the technology is well developed will take precedence over projects that require research to discover the needed technology.
- *Health of the product line*. Projects involving thriving product lines will take precedence over projects involving obsolescent product lines.
- *Probable resistance to change*. Projects that will meet a favorable reception take precedence over projects that may meet strong resistance, such as from the labor union or from a manager set in his or her ways.

Most organizations use a systematic approach to evaluate nominations relative to these criteria. This yields a composite evaluation that then becomes an indication of the relative priorities of the nominations. (For more detail and an example of a project selection matrix, see Chap. 12, Six Sigma: Breakthrough to in-Process Effectiveness.)

Project Selection

The result of the screening process is a list of recommended projects in their order of priority. Each recommendation is supported by the available information on compatibility with the criteria and potential benefits, resources required, and so on. This list is commonly limited to matters in which the council has a direct interest.

The council reviews the recommendations and makes the final determination on which projects are to be tackled. These projects then become an official part of the organization's business. Other recommended projects are outside the scope of the direct interest of the council. Such projects are recommended to appropriate sub-councils, managers, and so on. None of the preceding prevents projects from being undertaken at local levels by supervisors or by the workforce.

Vital Few and Useful Many

Some organizations completed many projects. Then, when questions were raised—"What have we gotten for all this effort?"—they were dismayed to learn that there was no noticeable effect on the bottom line. Investigation then showed that the reason could be traced to the process used for project selection. The projects actually selected had consisted of

- *Firefighting projects*. These are special projects for getting rid of sporadic "spikes." Such projects did not attack the chronic waste and hence, could not improve financial performance. (See Chap. 16, Root Cause Analysis Methods.)

- *Useful many projects.* By definition, these have only a minor effect on financial performance but have great effect on human relations.
- *Projects for improving human relations.* These can be quite effective in their field, but the financial results are usually not measurable.

To achieve a significant effect on the bottom line requires selecting the "vital few" projects as well as the "useful many." It is feasible to work on both, since different people are assigned to each.

There is a school of thought emerging that contends that the key to market leadership is "tiny breakthroughs in a thousand places"—in other words, the useful many. Another school urges focus on the vital few. In my experience, neither of these schools has the complete answer. Both are needed—at the right time.

The vital few projects are the major contributors to leadership and to the bottom line. The useful many projects are the major contributors to employee participation and to the quality of work life. Each is necessary; neither is sufficient.

The vital few and useful many projects can be carried out simultaneously. Successful organizations have done just that by recognizing that while there are these two types of projects, they require the time of different categories of organization personnel.

The interrelation of these two types of projects is shown in Fig. 5.8. In this figure, the horizontal scale is time. The vertical scale is chronic waste. What goes up is bad. The useful many breakthroughs collectively create a gradually sloping line. The vital few breakthroughs, though less frequent, contribute the bulk of the total breakthrough.

Cost Figures for Projects

To meet the preceding criteria (especially that of return on investment) requires information on various costs:

- The cost of chronic waste associated with a given nomination
- The potential cost reduction if the project is successful
- The cost of the needed diagnosis and remedy

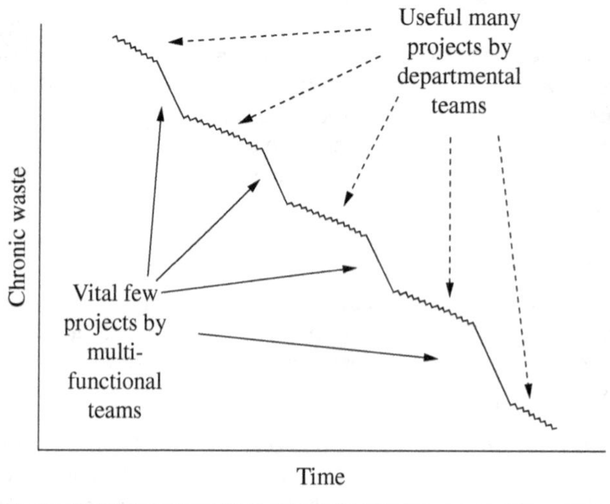

FIGURE 5.8 Interrelation of projects. (Juran Institute, Inc., 1994.)

Costs versus Percent Deficiencies

It is risky to judge priorities based solely on the percentage of deficiencies (errors, defects, and so on). On the face of it, when this percentage is low, the priority of the nomination also should be low. In some cases this is true, but in others it can be seriously misleading.

Elephant-Sized and Bite-Sized Projects

There is only one way to eat an elephant: bite by bite. Some projects are "elephant-sized"; that is, they cover so broad an area of activity that they must be subdivided into multiple "bite-sized" projects. In such cases, one project team can be assigned to "cut up the elephant." Other teams are then assigned to tackle the resulting bite-sized projects. This approach shortens the time to complete the project, since the teams work concurrently. In contrast, use of a single team stretches the time out to several years. Frustration sets in, team membership changes due to attrition, the project drags, and morale declines.

A most useful tool for cutting up the elephant is the Pareto analysis. For an application, see the paper mill example earlier, under "Use of the Pareto Principle." For elephant-sized projects, separate goal statements are prepared for the broad coordinating team and for each team assigned to a bite-sized project.

Replication and Cloning

Some organizations consist of multiple autonomous units that exhibit much commonality. A widespread example is the chains of retail stores, repair shops, hospitals, and so on. In such organizations, a breakthrough project that is carried out successfully in one operating unit logically becomes a nomination for application to other units. This is called cloning the project.

It is quite common for the other units to resist applying the breakthrough to their operation. Some of this resistance is cultural in nature (not invented here, and so on). Other resistance may be due to real differences in operating conditions. For example, telephone exchanges perform similar functions for their customers. However, some serve mainly industrial customers, whereas others serve mainly residential customers.

Upper managers are wary of ordering autonomous units to clone breakthroughs that originated elsewhere. Yet cloning has advantages. Where feasible, it provides additional breakthroughs without the need to duplicate the prior work of diagnosis and design of remedy.

What has emerged is a process as follows:

- Project teams are asked to include in their final report their suggestions as to sites that may be opportunities for cloning.
- Copies of such final reports go to those sites.
- The decision of whether to clone is made by the sites.

However, the sites are required to make a response as to their disposition of the matter. This response is typically in one of three forms:

1. We have adopted the breakthrough.
2. We will adopt the breakthrough, but we must first adapt it to our conditions.
3. We are not able to adopt the breakthrough for the following reasons.

In effect, this process requires the units to adopt the breakthrough or give reasons for not doing so. The units cannot just quietly ignore the recommendation.

A more subtle but familiar form of cloning is done through projects that have repetitive application over a wide variety of subject matter.

A project team develops computer software to find errors in spelling. Another team evolves an improved procedure for processing customer orders through the organization. A third team works up a procedure for conducting design reviews. What is common about such projects is that the result permits repetitive application of the same process to a wide variety of subject matter: many different misspelled words, many different customer orders, and many different designs.

Project Charters: Problem and Goal Statements for Projects

Each project selected should be accompanied by a written problem and goal statement that sets out the intended focus and the intended result of the project. Upon approval, this statement defines the actions required of the team assigned to carry out the project.

The Purpose of the Project Charter

The problem and goal statement serves a number of essential purposes:

- It defines the problem and the intended result and so helps the team know when it has completed the project.
- It establishes clear responsibility—the goal becomes an addition to each team member's job description.
- It provides legitimacy—the project becomes official organization business. The team members are authorized to spend the time needed to carry out the goal.
- It confers rights—the team has the right to hold meetings, ask people to attend and assist the team, and request data and other services germane to the project.

Perfection as a Goal

There is universal agreement that perfection is the ideal goal—complete freedom from errors, defects, failures, and so on. The reality is that the absence of perfection is due to many kinds of such deficiencies and that each requires its own breakthrough project. If an organization tries to eliminate all of them, the Pareto principle applies

- The vital few kinds of deficiencies cause most of the trouble but also readily justify the resources needed to root them out. Hence, they receive high priority during the screening process and become projects to be tackled.
- The remaining many types of deficiencies cause only a small minority of the trouble. As one comes closer and closer to perfection, each remaining kind of deficiency becomes rarer and rarer and hence, receives lower and lower priority during the screening process.

All organizations tackle those rare types of failure that threaten human life or that risk significant economic loss. In addition, organizations that make breakthroughs by the thousands year after year tackle even the mild, rare kinds of deficiency. To do so, they enlist the creativity of the workforce.

Some critics contend that publication of any goal other than perfection is proof of a misguided policy—a willingness to tolerate defects. Such contentions arise out of a lack of experience with the realities. It is easy to set goals that demand perfection now. Such goals, however, require organizations to tackle failure types so rare that they do not survive the screening process.

Nevertheless, there has been progress. During the twentieth century, there was a remarkable revision in the unit of measure for deficiencies. In the first half of the century, the usual measure was in percent defective, or defects per hundred units. By the 1990s, many industries had adopted a measure of defects per million units and use Sigma metrics and calculations. The leading organizations now do make thousands of breakthroughs year after year. They keep coming closer to perfection, but it is a never-ending process.

While many nominated projects cannot be justified solely on their return on investment, they may provide the means for employee participation in the breakthrough process, which has value in its own right.

The Project Team

For each selected project, a team is assigned. This team then becomes responsible for completing the project. Why a team? The most important projects are the vital few, and they are almost invariably multifunctional in nature. The symptoms typically show up in one department, but there is no agreement on where the causes lie, what the causes are, or what the remedies should be. Experience has shown that the most effective organizational mechanisms for dealing with such multifunctional problems are multifunctional teams.

Some managers prefer to assign problems to individuals rather than to teams. ("A camel is a horse designed by a committee.") The concept of individual responsibility is in fact quite appropriate if applied to control. ("The best form of control is self-control.") However, breakthrough, certainly for multifunctional problems, inherently requires teams. For such problems, assignment to individuals runs severe risks of departmental biases in the diagnosis and remedy.

A process engineer was assigned to reduce the number of defects coming from a wave soldering process. His diagnosis concluded that a new process was needed. Management rejected this conclusion on the grounds of excess investment. A multifunctional team was then appointed to restudy the problem. The team found a way to solve the problem by refining the existing process (Betker 1983).

Individual biases also show up as cultural resistance to proposed remedies. However, such resistance is minimal if the remedial department has been represented on the project team.

Appointment of Teams/Sponsors

Project teams are not attached to the chain of command on the organization chart. This can be a handicap in the event that teams encounter an impasse. For this reason, some organizations assign council members or other upper managers to be sponsors (or "champions") of specific projects. These sponsors follow team progress (or lack thereof). If the team does run into an impasse, the sponsor may be able to help the team get access to the proper person in the hierarchy.

Teams are appointed by sponsors of the projects, process owners, local managers, or others. In some organizations, workforce members are authorized to form teams (circles and so on) to work on breakthrough projects. Whatever the origin, the team is empowered to make the breakthrough as defined in the goal statement.

Most teams are organized for a specific project and are disbanded on completion of the project. Such teams are called ad hoc, meaning "for this purpose." During their next project, the members will be scattered among several different teams. There are also "standing" teams that have continuity—the members remain together as a team and tackle project after project.

Team Responsibilities

A project team has responsibilities that are coextensive with the goal statement. The basic responsibilities are to carry out the assigned goal and to follow the universal breakthrough process. In addition, the responsibilities include

- Proposing revisions to the goal statement
- Developing measurements as needed
- Communicating progress and results to all who have a need to know

Membership

The team is selected by the sponsor after consulting with the managers who are affected. The selection process includes consideration of (1) which departments should be represented on the team, (2) what level in the hierarchy team members should come from, and (3) which individuals in that level.

The departments to be represented should include

- *The ailing department*. The symptoms show up in this department, and it endures the effects.
- *Suspect departments*. They are suspected of harboring the causes. (They do not necessarily agree that they are suspect.)
- *Remedial departments*. They will likely provide the remedies. This is speculative, since in many cases, the causes and remedies come as surprises.
- *Diagnostic departments*. They are needed in projects that require extensive data collection and analysis.
- *On-call departments and subject matter experts (SMEs)*. They are invited in as needed to provide special knowledge or other services required by the team.

This list includes the usual sources of members. However, there is need for flexibility.

Choice of level in the hierarchy depends on the subject matter of the project. Some projects relate strongly to the technological and procedural aspects of the products and processes. Such projects require team membership from the lower levels of the hierarchy. Other projects relate to broad business and managerial matters. For such projects, the team members should have appropriate business and managerial experience.

Finally comes the selection of individuals. This is negotiated with the respective supervisors, giving due consideration to workloads, competing priorities, and so on. The focus is on the individual's ability to contribute to the team project. The individuals need

- *Time* to attend the team meetings and to carry out assignments outside the meetings—"the homework."
- A *knowledge base* that enables the individual to contribute theories, insights, and ideas, as well as job information based on his or her hands-on experience.
- *Training* in the breakthrough process and the associated tools. During the first projects, this training can and should be done concurrently with carrying out the projects.

Most teams consist of six to eight members. Larger numbers tend to make the team unwieldy as well as costly. (A convoy travels only as fast as the slowest ship.)

Should team members all come from the same level in the hierarchy? Behind this question is the fear that the biases of high-ranking members will dominate the meeting. Some of this no doubt takes place, especially during the first few meetings. However, it declines as the group dynamics take over and as members learn to distinguish between theory and fact.

Once the team is selected, the members' names are published, along with their project goal. The act of publication officially assigns responsibility to the individuals as well as to the team. In effect, serving on the project team becomes a part of the individuals' job descriptions. This same publication also gives the team the legitimacy and rights discussed earlier.

Membership from the Workforce

During the early years of using breakthrough teams, organizations tended to maintain a strict separation of team membership. Teams for multifunctional projects consisted exclusively of members from the managerial hierarchy plus professional specialists. Teams for local departmental projects (such as quality circles and employee involvement teams) consisted exclusively of members from the workforce. Figure 5.9 compares the usual features of these teams with those of multifunctional teams.

Experience then showed that as to the details of operating conditions, no one is better informed than the workforce. Through residence in the workplace, workers can observe local changes and recall the chronology of events. This has led to a growing practice of securing such information by interviewing the workers. The workers can be "on call" or full-time

Feature	Department Teams or Quality Circles	Breakthrough Teams
Primary purpose	To improve departmental processes and human relations	To improve performance by creating breakthroughs across multiple departments
Secondary purpose	To improve quality	To improve teamwork and participation
Scope of project	Within a single department	Across multiple departments
Size of project	One of the useful many	One of the vital few
Membership	From a single department	From multiple departments
Basis of membership	Voluntary or mandatory	Mandatory
Hierarchical status of members	The manager, staff in any department	Management, subject matter experts, and the workforce
Continuity	Team remains intact, project after project	Team is ad hoc, disbands after project is completed

FIGURE 5.9 Contrast departmental teams and quality circles to multifunctional teams. (From Making Quality Happen, Juran Institute, Inc., 1998.)

team members. In a hospital, a doctor can be considered in the same way. Removing a worker tied directly to the production of a product must minimize their time away from their work or patients.

One result of all this experience has been a growing interest in broadening worker participation generally. This has led to experimenting with project teams that make no distinction as to rank in the hierarchy. These teams may become the rule rather than the exception. (For further discussion on the trends in workforce participation, see Chap. 8, Organization Roles to Support a Quality and Excellence Culture.)

Upper Managers on Teams

Some projects, by their very nature, require that the team include members from the ranks of upper management. Here are some examples of breakthrough projects actually tackled by teams that included upper managers:

- Shorten the time to put new products on the market.
- Improve the accuracy of the sales forecast.
- Reduce the carryover of prior failure-prone features into new product models.
- Establish a teamwork relationship with suppliers.
- Develop the new measures needed for strategic planning.
- Revise the system of recognition and rewards for breakthrough.

There are some persuasive reasons urging all upper managers to personally serve on some project teams. Personal participation on project teams is an act of leadership by example. This is the highest form of leadership. Personal participation on project teams also enables upper managers to understand what they are asking their subordinates to do, what kind of training is needed, how many hours per week are demanded, how many months it takes to complete the project, and what kinds of resources are needed. Lack of upper management understanding of such realities has contributed to the failure of some well-intentioned efforts to establish annual breakthrough.

Model of the Infrastructure

There are several ways to show in graphic form the infrastructure for breakthrough—the elements of the organization, how they relate to each other, and the flow of events. Figure 5.10 shows the elements of the infrastructure in pyramid form. The pyramid depicts a hierarchy consisting of top management, the autonomous operating units, and the major staff functions. At the top of the pyramid are the corporate council and the subsidiary councils, if any. Below these levels are the multifunctional breakthrough teams. (There may be a committee structure between the councils and the teams.)

At the intradepartmental level are teams from the workforce—circles or other forms. This infrastructure permits employees in all levels of the organization to participate in breakthrough projects, the useful many as well as the vital few.

Team Organization

Breakthrough teams do not appear on the organization chart. Each "floats"—it has no personal boss. Instead, the team is supervised *impersonally* by its goal statement and by the breakthrough roadmap.

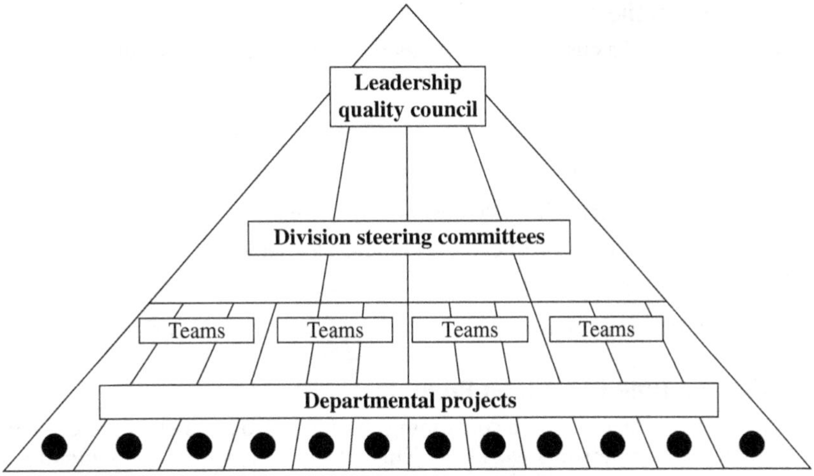

Figure 5.10 Model of the infrastructure for breakthrough quality improvement. (Juran Institute, Inc.)

The team does have its own internal organizational structure. This structure invariably includes a team *leader* (chairperson and so on) and a team *secretary*. In addition, there is usually a *facilitator*.

The Team Leader

The leader is usually appointed by the sponsor—the council or other supervising group. Alternatively, the team may be authorized to elect its leader.

The leader has several responsibilities. As a team member, the leader *shares* in the responsibility for completing the team's goal. In addition, the leader has administrative duties. These are *unshared* and include

- Ensuring that meetings start and finish on time
- Helping the members attend the team meetings
- Ensuring that the agendas, minutes, reports, and so on are prepared and published
- Maintaining contact with the sponsoring body

Finally, the leader has the responsibility of *oversight*. This is met not through the power of command—the leader is not the boss of the team—it is met through the power of leadership. The responsibilities include

- Orchestrating the team activities
- Inspiring all members to contribute
- Helping to resolve conflicts among members
- Assigning the homework to be done between meetings

To meet such responsibilities requires multiple skills, which include

- A trained capability for leading people
- Familiarity with the subject matter of the goal
- A firm grasp of the breakthrough process and the associated tools

The Team Members

"Team members" as used here includes the team leader and secretary. The responsibilities of any team member consist mainly of the following:

- Arranging to attend the team meetings
- Representing his or her department
- Contributing job knowledge and expertise
- Proposing theories of causes and ideas for remedies
- Constructively challenging the theories and ideas of other team members
- Volunteering for or accepting assignments for homework

Finding the Time to Work on Projects

Work on project teams is time consuming. Assigning someone to a project team adds about 10 percent to that person's workload. This added time is needed to attend team meetings, perform the assigned homework, and so on. Finding the time to do all this is a problem to be solved, since this added work is thrust on people who are already fully occupied.

No upper manager known to me has been willing to solve the problem by hiring new people to make up for the time demanded by the breakthrough projects. Instead, it has been left to each team member to solve the problem in his or her own way. In turn, the team members have adopted such strategies as

- Delegating more activities to subordinates
- Slowing down the work on lower-priority activities
- Improving time management on the traditional responsibilities
- Looking for ongoing activities that can be terminated. (In several organizations, there has been a specific drive to clear out unneeded work to provide time for breakthrough projects.)

As projects begin to demonstrate high returns on investment, the climate changes. Upper managers become more receptive to providing resources. In addition, the successful projects begin to reduce workloads that previously were inflated by the presence of chronic wastes.

Facilitators and Black Belts

Most organizations make use of internal consultants, usually called *facilitators* or *Black Belts*, to assist teams. A facilitator like a Black Belt does not have to be a member of the team and may not have any responsibility for carrying out the team goal. (The literal meaning of the word facilitate is "to make things easy.") The prime role of the facilitator is to help the team to carry out its goal. The usual roles of facilitators consist of a selection from the following:

Explain the organization's intentions. The facilitator usually has attended briefing sessions that explain what the organization is trying to accomplish. Much of this briefing is of interest to the project teams.

Assist in team building. The facilitator helps the team members to learn to contribute to the team effort: propose theories, challenge the theories of others, and/or propose lines of investigation. Where the team concept is new to an organization, this role may require working directly with individuals to stimulate those who are unsure about how to contribute and to restrain the overenthusiastic ones. The facilitator also may evaluate the progress in team building and provide feedback to the team.

Assist in training. Most facilitators have undergone training in team building and in the breakthrough process. They usually have served as facilitators for other teams. Such experiences qualify them to help train project teams in several areas: team building, the breakthrough roadmap, and/or use of the tools.

Relate experiences from other projects. Facilitators have multiple sources of such experiences:

- Project teams previously served on
- Meetings with other facilitators to share experiences in facilitating project teams
- Final published reports of project teams
- Projects reported in the literature

Assist in redirecting the project. The facilitator maintains a detached view that helps him or her sense when the team is getting bogged down. As the team gets into the project, it may find itself getting deeper and deeper into a swamp. The project goal may turn out to be too broad, vaguely defined, or not doable. The facilitator usually can sense such situations earlier than the team and can help guide it to a redirection of the project.

Assist the team leader. Facilitators provide such assistance in various ways:

- Assist in planning the team meetings. This may be done with the team leader before each meeting.
- Inspire attendance. Most nonattendance is due to conflicting demands made on a team member's time. The remedy often must come from the member's boss.
- Improve human relations. Some teams include members who have not been on good terms with each other or who develop friction as the project moves along. As an "outsider," the facilitator can help to direct the energies of such members into constructive channels. Such action usually takes place outside the team meetings. (Sometimes the leader is part of the problem. In such cases, the facilitator may be in the best position to help.)
- Assist on matters outside the team's sphere of activity. Projects sometimes require decisions or actions from sources that are outside the easy reach of the team. Facilitators may be helpful due to their wider range of contacts.

Support the team members. Such support is provided in multiple ways:

- Keep the team focused on the goal by raising questions when the focus drifts.
- Challenge opinionated assertions by questions such as, Are there facts to support that theory?
- Provide feedback to the team based on perceptions from seeing the team in action.

Report progress to the councils. In this role, the facilitator is a part of the process of reporting on progress of the projects collectively. Each project team issues minutes of its meetings. In due course, each also issues its final report, often including an oral presentation to the council.

However, reports on the projects collectively require an added process. The facilitators are often a part of this added reporting network.

The Qualifications of Facilitators and Black Belts

Facilitators undergo special training to qualify them for these roles. The training includes skills in team building, resolving conflicts, communication, and management of quality change; knowledge relative to the breakthrough processes, for example, the breakthrough roadmap and the tools and techniques; and knowledge of the relationship of breakthrough to the organization's policies and goals. In addition, facilitators acquire maturity through having served on project teams and providing facilitation to teams. This topic is covered in more detail in Chap. 12, Six Sigma: Breakthrough to in-Process Effectiveness.

This prerequisite training and experience are essential assets to the facilitator. Without them, he or she has great difficulty winning the respect and confidence of the project's team.

Most organizations are aware that to go into a high rate of breakthrough requires extensive facilitation. In turn, this requires a buildup of trained facilitators. However, facilitation is needed mainly during the startup phase. Then, as team leaders and members acquire training and experience, there is less need for facilitator support. The buildup job becomes a maintenance job.

This phased rise and decline has caused most organizations to avoid creating full-time facilitators or a facilitator career concept. Facilitation is done on a part-time basis. Facilitators spend most of their time on their regular job.

In many larger organizations, Black Belts are full-time specialists. Following intensive training in the breakthrough process, these persons devote all their time to the breakthrough activity. Their responsibilities go beyond facilitating project teams and may include

- Assisting in project nomination and screening
- Conducting training courses in the methods and tools
- Coordinating the activities of the project team with those of other activities in the organization, including conducting difficult analyses
- Assisting in the preparation of summarized reports for upper managers

A team has no personal boss. Instead, the team is supervised impersonally. Its responsibilities are defined in:

- *The project charter*. This goal statement is unique to each team.
- *The steps or universal sequence for breakthrough*. This is identical for all teams. It defines the actions to be taken by the team to accomplish its goal.

The project team has the principal responsibility for the steps that now follow—taking the two "journeys." (See p. 162 for a description of the two journeys.)

Diagnosis is based on the factual approach and requires a firm grasp of the meanings of key words. It is helpful to define some of these key words at the outset.

Leaders Must Learn Key Breakthrough Terminology

A "defect" is any state of unfitness for use or nonconformance to specification. Examples are illegible invoices, scrap, and low mean time between failures. Other names include "error," "discrepancy," and "nonconformance."

A "symptom" is the outward evidence that something is wrong or that there is a defect. A defect may have multiple symptoms. The same word may serve as a description of both defect and symptom.

A "theory" or "hypothesis" are unproven assertions as to reasons for the existence of defects and symptoms. Usually, multiple theories are advanced to explain the presence of defects.

A "cause" is a proved reason for the existence of a defect. Often there are multiple causes, in which case they follow the Pareto principle—the vital few causes will dominate all the rest.

A "dominant cause" is a major contributor to the existence of defects and one that must be remedied before there can be an adequate breakthrough.

"Diagnosis" is the process of studying symptoms, theorizing as to causes, testing theories, and discovering causes.

A "remedy" is a change that can eliminate or neutralize a cause of defects.

Diagnosis Should Precede Remedy

It may seem obvious that diagnosis should precede remedy, yet biases or outdated beliefs can get in the way.

For example, during the twentieth century, many upper managers held deep-seated beliefs that most defects were due to workforce errors. The facts seldom bore this out, but the belief persisted. As a result, during the 1980s, many of these managers tried to solve their problems by exhorting the workforce to make no defects. (In fact, defects are generally over 80 percent management-controllable and under 20 percent worker-controllable.)

Untrained teams often try to apply remedies before the causes are known. ("Ready, fire, aim.") For example,

- An insistent team member "knows" the cause and pressures the team to apply a remedy for that cause.
- The team is briefed on the technology by an acknowledged expert. The expert has a firm opinion about the cause of the symptom, and the team does not question the expert's opinion.
- As team members acquire experience, they also acquire confidence in their diagnostic skills. This confidence then enables them to challenge unproved assertions.
- Where deep-seated beliefs are widespread, special research may be needed.

In a classic study, Greenridge (1953) examined 850 failures of electronic products supplied by various organizations. The data showed that 43 percent of the failures were traceable to product design, 30 percent to field operation conditions, 20 percent to manufacture, and the rest to miscellaneous causes.

Institutionalizing Breakthrough

Numerous organizations have initiated breakthrough, but few have succeeded in institutionalizing it so that it goes on year after year. Yet many of these organizations have a long history of annually conducting product development, cost reduction, productivity breakthrough, and so on. The methods they used to achieve such annual breakthrough are well known and can be applied to breakthrough. They are

- Enlarge the annual business plan to include goals for breakthrough.
- Make breakthrough a part of everyone's job description. In most organizations, the activity of breakthrough has been regarded as incidental to the regular job of meeting the goals for cost, delivery, and so on. The need is to make breakthrough a part of the regular job.
- Establish upper management audits that include review of progress on breakthrough.
- Revise the merit rating and reward system to include a new parameter—performance on breakthrough—and give it proper weight.
- Create well-publicized occasions to provide recognition for performance on breakthrough.

Review Progress

Scheduled, periodic reviews of progress by upper managers are an essential part of maintaining annual breakthroughs. Activities that do not receive such review cannot compete for priority with activities that do receive such review. Subordinates understandably give top priority to matters that are reviewed regularly by their superiors.

There is also a need for regular review of the breakthrough process. This is done through audits that may extend to all aspects of managing for quality. (Refer to Chap. 11, ISO 9000 Quality Management System.)

Much of the database for progress review comes from the reports issued by the project teams. However, it takes added work to analyze these reports and to prepare the summaries needed by upper managers. Usually, this added work is done by the secretary of the council with the aid of the facilitators, the team leaders, and other sources such as finance.

As organizations gain experience, they design standardized reporting formats to make it easy to summarize reports by groups of projects, by product lines, by business units, by divisions, and for the corporation. One such format, used by a large European organization, determines for each project:

- The original estimated amount of chronic waste
- The original estimated reduction in cost if the project were to be successful
- The actual cost reduction achieved
- The capital investment
- The net cost reduction
- The summaries are reviewed at various levels. The corporate summary is reviewed quarterly at the chairperson's staff meeting (personal communication to the author)

Evaluation of Performance

One of the objectives of progress review is evaluation of performance. This evaluation extends to individuals as well as to projects. Evaluation of individual performance on breakthrough projects runs into the complication that the results are achieved by teams. The problem then becomes one of evaluating individual contribution to team efforts. This new problem has as yet no scientific solution. Thus, each supervisor is left to judge subordinates' contributions based on inputs from all available sources.

At higher levels of an organization, the evaluations extend to judging the performance of supervisors and managers. Such evaluations necessarily must consider results achieved on multiple projects. This has led to an evolution of measurement (metrics) to evaluate managers' performance on projects collectively. These metrics include

- Numbers of breakthrough projects: initiated, in progress, completed, and aborted
- Value of completed projects in terms of breakthrough in product performance, reduction in costs, and return on investment
- Percentage of subordinates active on project teams
- Superiors then judge their subordinates based on these and other inputs

Training for Breakthrough

Throughout this chapter, there have been numerous observations on the needs for training employees. These needs are extensive because all employees must understand the methods and tools employed to attain breakthrough. Project-by-project breakthrough may be new to the organization, turnover may be high, or employees may be assigned new responsibilities. To carry out these new responsibilities requires extensive training.

So far in this decade, many organizations made significant investments in training their workforces in the methods and tools to attain performance excellence. According to *Six Sigma* and the American Society for Quality (ASQ), more than 100,000 people were trained as Black Belts. Another 500,000 may have been trained as Green Belts. A new certification process has been added at the ASQ and at many firms like the Juran Institute to ensure that these experts are qualified and competent to drive results. A Black Belt training program may consist of up to six weeks of training plus time to be certified.

This trend has been reversed from the 1990s. Training budgets were cut to reduce costs. Today, training is an investment in the future. This will benefit us as organizations move into the future.

References

Betker, H. A. (1983). "Breakthrough Program: Reducing Solder Defects on Printed Circuit Board Assembly." *The Juran Report*, No. 2, November, pp. 53–58.

Greenridge, R. M. C. (1953). "The Case of Reliability vs. Defective Components et al." *Electronic Applications Reliability Review*, No. 1, p. 12.

Juran, J. M. (1964). *Managerial Breakthrough*. McGraw-Hill, New York. Revised edition, 1995.

Juran, J. M. (1981). "Juran on Breakthrough," a series of 16 videocassettes on the subject. Juran Institute, Inc., Wilton, CT.

Juran, J. M. (1985). "A Prescription for the West—Four Years Later." European Organization for Quality, 29th Annual Conference. Reprinted in *The Juran Report*, No. 5, Summer 1985.

Juran, J. M. (1993), "Made in USA, a Renaissance in Quality." *Harvard Business Review*, July-August, pp. 42–50.

Juran, J. M. "Juran on Quality Leadership," (1988) A video package, Juran Institute, Inc., Wilton, CT.

Welch, J. (2005). *Winning*. Harper Collins, New York, NY.

CHAPTER 6

Quality Control to Assure Compliance to Customer Requirements

Joseph M. Juran and Joseph A. De Feo

High Points of This Chapter

1. The quality control process is a universal managerial process for conducting operations so as to provide stability—to prevent adverse change and to "maintain the status quo." Quality control takes place by use of the feedback loop.
2. Each feature of the product or process becomes a control subject—a center around which the feedback loop is built. As much as possible, human control should be done by the workforce—the office clerical force, factory workers, salespersons, etc.
3. The flow diagram is widely used during the planning of quality controls. The weakest link in facilities control has been adherence to schedule.
4. To ensure strict adherence to schedule requires an independent audit.
5. Knowing which process variable is dominant helps planners during allocation of resources and priorities.
6. The design for process control should provide the tools needed to help the operating forces distinguish between real change and false alarms. It is most desirable to provide umpires with tools that can help to distinguish between special causes and common causes. An elegant tool for this purpose is the Shewhart control chart (or just control chart). The criteria for self-control are applicable to processes in all functions and all levels, from general manager to nonsupervisory worker.
7. Responsibility for results should be keyed to controllability. Ideally the decision of whether the process conforms to process quality goals should be made by the workforce.
8. To make use of self-inspection requires meeting several essential criteria: Quality is number one; mutual confidence, self-control, training, and certification are the others. Personnel who are assigned to make product conformance decisions should be provided with clear definitions of responsibility as well as guidelines for decision making.
9. The proper sequence in managing is first to establish goals and then to plan how to meet those goals, including the choice of the appropriate tools. The planning for quality control should provide an information network that can serve all decision makers.

Compliance and Control Defined

Compliance or quality control is the third universal process in the Juran Trilogy. The others are quality planning in and quality improvement. The Juran Trilogy diagram (Fig. 6.1) shows the interrelation of these processes.

Figure 6.1 is used in several other chapters in this handbook to describe the relationships between planning, improvement, and control—the fundamental managerial processes in quality management. What is important for this chapter is to concentrate on the two "zones of control."

In Fig. 6.1, we can easily see that although the process is in control in the middle of the chart, we are running the process at an unacceptable level of performance and "waste." What is necessary here is not more control, but improvement—actions to change the level of performance.

After the improvements have been made, a new level of performance has been achieved. Now it is important to establish new controls at this level to prevent the performance level

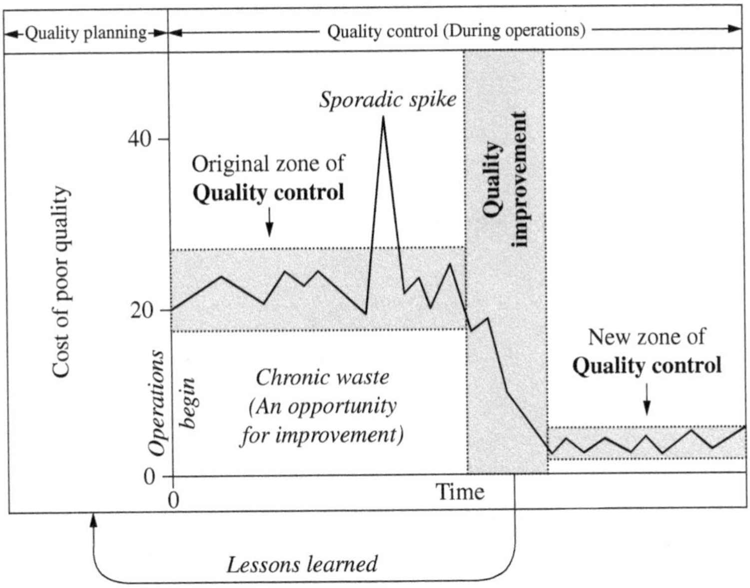

FIGURE 6.1 Juran Trilogy diagram. (Juran Institute, Inc., 1986.)

from deteriorating to the previous level or even worse. This is indicated by the second zone of control.

The term "control of quality" emerged early in the twentieth century (Radford 1917, 1922). The concept was to broaden the approach to achieving quality, from the then-prevailing after-the-fact inspection (detection control) to what we now call "prevention (proactive control)." For a few decades, the word "control" had a broad meaning, which included the concept of quality planning. Then came events that narrowed the meaning of "quality control." The "statistical quality control" movement gave the impression that quality control consisted of using statistical methods. The "reliability" movement claimed that quality control applied only to quality at the time of test but not during service life.

In the United States, the term "quality control" now often has the meaning defined previously. It is a piece of a "performance excellence, operational excellence, business excellence, or total quality program," which are now used interchangeably to comprise the all-embracing term to describe the methods, tools, and techniques to manage the quality of an organization.

In Japan, the term "quality control" retains a broad meaning. Their "total quality control" is equivalent to our term "business excellence." In 1997, the Union of Japanese Scientists and Engineers (JUSE) adopted the term Total Quality Management (TQM) to replace Total Quality Control (TQC) to more closely align themselves with the more common terminology used in the rest of the world. Figure 6.2 shows the input-output features of this step.

In Fig. 6.2, the input is operating process features, or key control characteristics, developed to produce the product features, or key product characteristics, required to meet customer needs. The output consists of a system of product and process controls, which can provide stability to the operating process.

A key product characteristic is a product characteristic for which reasonably anticipated variation *could significantly affect* a product's safety, compliance to government regulations, performance, or fit.

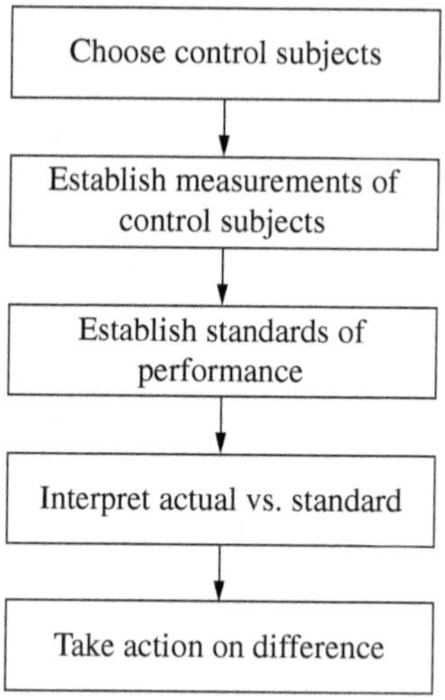

FIGURE 6.2 Input-output diagram.

Key product characteristics (KPCs) are *outputs from a process that are measurable* on, within, or about the product itself. They are the outputs perceived by the customer. Examples of KPCs include

- KPCs "On:" the product: width, thickness, coating adherence, surface cleanliness, etc.
- KPCs "Within:" the product: hardness, density, tensile strength, mass, etc.
- KPCs "About:" the product: performance, weight, etc.

In general, key control characteristics (KCCs) are *inputs that affect the outputs* (KPCs). They are unseen by the customer and are measurable only when they occur. A KCC is

- A process parameter for which variation must be controlled around some target value to ensure that variation in a KPC is maintained around its target values during manufacturing and assembly.
- A process parameter for which reduction in variation will reduce the variation of a KPC.
- Directly traceable to a KPC.
- Particularly significant in ensuring a KPC achieves target value.
- Not specified on a product drawing or product documentation.

The Relation to Quality Assurance

Quality control and quality assurance have much in common. Each evaluates performance. Each compares performance to goals. Each acts on the difference. However, they also differ from each other. Quality control has as its primary purpose maintaining control. Performance is evaluated during operations, and performance is compared to targets during operations.

In the process, metrics are utilized to monitor adherence to standards. The resulting information is received and used by the employees.

The main purpose of quality assurance is to verify that control is being maintained. Performance is evaluated after operations, and the resulting information is provided to both the employees and others who have a need to know. Results metrics are utilized to determine conformance to customer needs and expectations. Others may include leadership, plant, functional; corporate staffs; regulatory bodies; customers; and the general public.

The Feedback Loop

Quality control takes place by use of the feedback loop. A generic form of the feedback loop is shown in Fig. 6.3.

The progression of steps in Fig. 6.3 is as follows:

1. A sensor is "plugged in" to evaluate the actual quality of the control subject—the product or process feature in question. The performance of a process may be determined directly by evaluation of the process feature, or indirectly by evaluation of the product feature—the product "tells" on the process.
2. The sensor reports the performance to an umpire.
3. The umpire also receives information on the quality goal or standard.
4. The umpire compares actual performance to standard. If the difference is too great, the umpire energizes an actuator.
5. The actuator stimulates the process (whether human or technological) to change the performance so as to bring quality into line with the quality goal.
6. The process responds by restoring conformance.

Note that in Fig. 6.3 the elements of the feedback loop are functions. These functions are universal for all applications, but responsibility for carrying out these functions can vary widely. Much control is carried out through automated feedback loops. No human beings are involved. Common examples are the thermostat used to control temperature and the cruise control used in automobiles to control speed.

Another form of control is self-control carried out by employees. An example of such self-control is the village artisan who performs every one of the steps of the feedback loop. The artisan chooses the control subjects based on understanding the needs of customers, sets the quality targets to meet the needs, senses the actual quality performance, judges conformance, and becomes the actuator in the event of nonconformance.

This concept of self-control is illustrated in Fig. 6.4. The essential elements here are the need for the employee or work team to know what they are expected to do, to know how

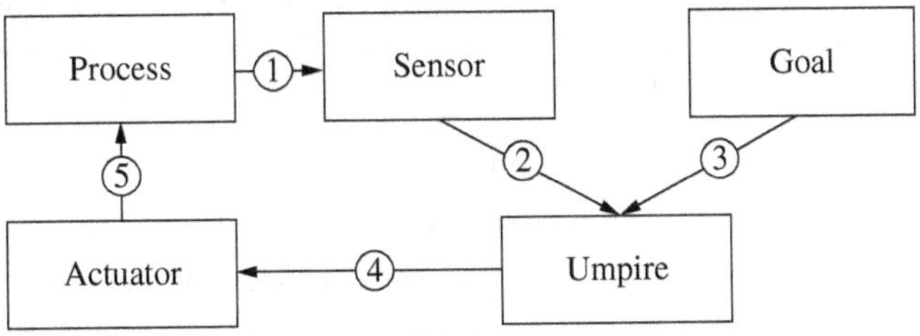

FIGURE 6.3 Feedback loop.

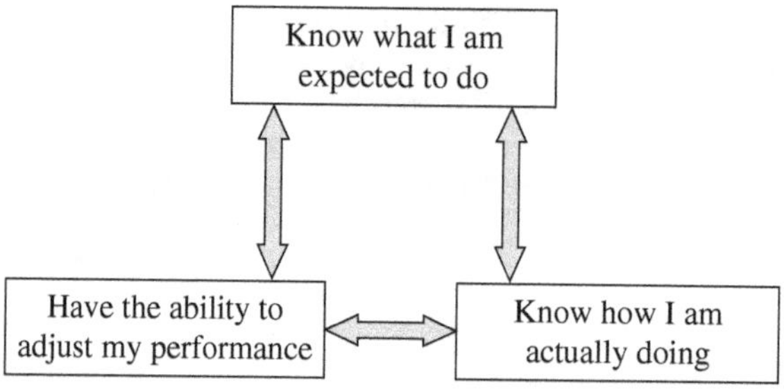

Figure 6.4 Concept of self-control. (The Juran Institute, Inc.)

they are actually doing, and to have the means to regulate performance. This implies that they have a capable process and have the tools, skills, and knowledge necessary to make the adjustments and the authority to do so.

A further common form of feedback loop involves office clerks or factory workers whose work is reviewed by umpires in the form of inspectors. This design of a feedback loop is largely the result of the Taylor Management System adopted in the early twentieth century. It focused on the separation of planning for quality from the execution or operations. The Taylor Management System emerged a century ago and contributed greatly to increasing productivity. However, the effect on quality was largely negative. The negative impact resulted in large costs associated with poor quality, products and services that have higher levels of failure, and customer dissatisfaction.

The Elements of the Feedback Loop

The feedback loop is a universal. It is fundamental to maintaining control of every process. It applies to all types of operations, whether in service industries or manufacturing industries, whether for profit or not. The feedback loop applies to all levels in the hierarchy, from the chief executive officer to the members of the workforce. However, there is wide variation in the nature of the elements of the feedback loop.

In Fig. 6.5, a simple flowchart is shown describing the control process with the simple universal feedback loop imbedded.

The Control Subjects

Each feature of the product (goods and services) or process becomes a control subject (the specific attribute or variable to be controlled)—a center around which the feedback loop is built. The critical first step is to choose the control subject. To choose control subjects, you should identify the major work processes and products, define the objectives of the work processes; succinctly define the work processes; identify the customers of the process, and then select the control subjects (KPCs and/or KCCs). Control subjects are derived from multiple sources, which include

- Stated customer needs for product features
- Translated "voice of the customer" needs into product features
- Defined process features that create the product or service features
- Industry and government standards and regulations (i.e., Sarbanes Oxley, ISO 9000, etc.)

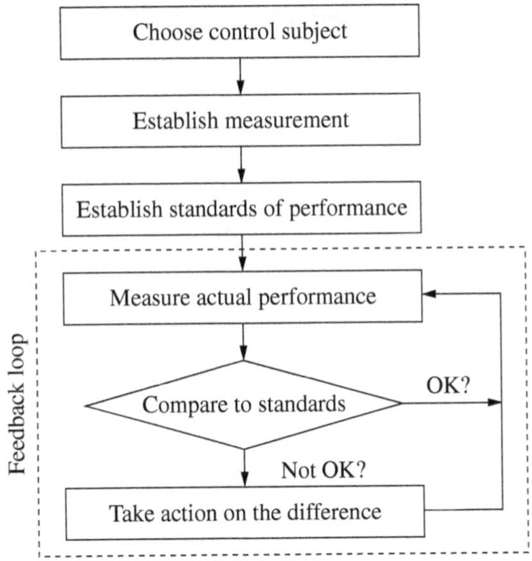

Figure 6.5 Simple flowchart describing the control process.

- Need to protect human safety and the environment (i.e., OSHA, ISO 14000)
- Need to avoid side effects such as irritations to stakeholders, employees, or to a neighboring community
- Failure mode and effects analyses
- Control plans
- Results of design of experiments

At the staff level, control subjects consist mainly of product and process features defined in technical specifications and procedures manuals. At managerial levels, the control subjects are broader and increasingly business oriented. Emphasis shifts to customer needs and to competition in the marketplace. This shift in emphasis then demands broader control subjects, which, in turn, have an influence on the remaining steps of the feedback loop.

Establish Measurement

After choosing the control subjects, the next step is to establish the means of measuring the actual performance of the process or the quality level of the goods or services being created. Measurement is one of the most difficult tasks of management and is discussed in almost every chapter of this handbook. In establishing the measurement, we need to clearly specify the means of measuring (the sensor), the accuracy and precision of the measurement tool, the unit of measure, the frequency of measuring, the means by which data will be recorded, the format for reporting the data, the analysis to be made on the data to convert it to usable information, and who will make the measurement. In establishing the unit of measure, one should select a unit of measure that is understandable, provides an agreed-upon basis for decision making, is customer focused, and can be applied broadly.

Establish Standards of Performance: Product Goals and Process Goals

For each control subject it is necessary to establish a standard of performance—a target or goal (also metrics, objectives, etc.). A standard of performance is an aimed-at target toward which work is expended. Table 6.1 gives some examples of control subjects and the associated goals.

The prime goal for products and services is to meet customer needs. Industrial customers often specify their needs with some degree of precision. Such specified needs then become goals for the producing company. In contrast, consumers tend to state their needs in vague terms. Such statements must then be translated into the language of the producer in order to become product goals.

Other goals for products that are also important are those for reliability and durability. Whether the products and services meet these goals can have a critical impact on customer satisfaction, loyalty, and overall costs. The failures of products under warranty can seriously affect the profitability of a company through both direct and indirect costs (loss of repeat sales, word of mouth, etc.).

The processes that produce products have two sets of goals:

- To produce products and services that meet customer needs. Ideally, each and every unit produced should meet customer needs (meet specifications)
- To operate in a stable and predictable manner. In the dialect of the quality specialist, each process should be "in a state of control." We will later elaborate on this, in the section "Process Conformance"

Quality targets may also be established for functions, departments, or people. Performance against such goals then becomes an input to the company's scorecard, dashboard, and reward system. Ideally such goals should be

- *Legitimate.* They should have undoubted official status.
- *Measurable.* They can be communicated with precision.
- *Attainable.* As evidenced by the fact that they have already been attained by others.
- *Equitable.* Attainability should be reasonably alike for individuals with comparable responsibilities.

Control Subject	Goal
Vehicle mileage	Minimum of 25 mi/gal highway driving
Overnight delivery	99.5% delivered prior to 10:30 A.M. next morning
Reliability	Fewer than three failures in 25 years of service
Temperature	Minimum 505°F; maximum 515°F
Purchase-order error rate	No more than 3 errors/1000 purchase orders
Competitive performance	Equal or better than top three competitors on six factors
Customer satisfaction	90% or better rate, service outstanding or excellent
Customer retention	95% retention of key customers from year to year
Customer loyalty	100% of market share of over 80% of customers

TABLE 6.1 Control Subjects and Associated Quality Goals

Quality goals may be set from a combination of the following bases:

- Goals for product and service features and process features are largely based on technological analysis
- Goals for functions, departments, and people should be based on the need of the business and external benchmarking rather than historical performance

In recent years, quality goals used at the highest levels of an organization have become commonplace. Establishing long-term goals such as reducing the costs of poor quality or becoming best in class have become a normal part of strategic business plans. The emerging practice is to establish goals on "metrics that matter," such as meeting customers' needs, exceeding the competition, maintaining a high pace of improvement, improving the effectiveness of business processes, and setting stretch goals to avoid failure-prone products and processes.

Measure Actual Performance

A critical step in controlling quality characteristics is to measure the actual performance of a process as precisely as possible. To do this requires measuring with a "sensor." A sensor is a device or a person that makes the actual measurement.

The Sensor

A "sensor" is a specialized detecting device. It is designed to recognize the presence and intensity of certain phenomena and to convert the resulting data into "information." This information then becomes the basis of decision making. At the lower levels of an organization, the information is often on a real-time basis and is used for daily control. At higher levels, the information is summarized in various ways to provide broader measures, detect trends, and identify the vital few problems.

The wide variety of control subjects requires a wide variety of sensors. A major category is the numerous technological instruments used to measure product features and process features. Familiar examples are thermometers, clocks, and weight scales. Another major category of sensors is the data systems and associated reports, which supply summarized information to the managerial hierarchy. Yet another category involves the use of human beings as sensors. Questionnaires, surveys, focus groups, and interviews are also forms of sensors.

Sensing for control is done on an organization level. Information is needed to manage for the short and long term. This has led to the use of computers to aid in the sensing and in converting the resulting data into information.

Most sensors provide their evaluations in terms of a unit of measure—a defined amount of some feature—that permits evaluation of that feature in numbers or pictures. Familiar examples of units of measure are degrees of temperature, hours, inches, and tons. Human beings do a considerable amount of sensing. Such sensing is subject to numerous sources of error. The use of pictures as a standard to comparison can help reduce human errors. Also of vital importance to alleviate human errors is the application of detailed instructions.

Compare to Standards

The act of comparing to standards is often seen as the role of an umpire. The umpire may be a person or a technological device. Either way, the umpire may be called on to carry out any or all of the following activities:

- Compare the actual process performance to the targets
- Interpret the observed difference (if any); determine if there is conformance to the target

- Decide on the action to be take
- Stimulate corrective action
- Record the results

These activities require elaboration and will be examined more closely in an upcoming section.

Take Action on the Difference

In any well-functioning control system we need a means of taking action on any difference between desired standards of performance and actual performance. For this we need an actuator. This device (human or technological or both) is the means for stimulating action to restore conformance. At the operations or employee level, it may be a keypad for giving orders to a centralized computer database, a change in a new procedure, a new specification document, or a new setting of a dial to adjust a machine to the right measure. At the management level, it may be a memorandum to subordinates, a new company policy, or a team to change a process.

The Key Process

In the preceding discussion we have assumed a process. This may also be human or technological or both. It is the means for producing the product and service features, each of which requires control subjects to ensure conformance to specifications. A process does all work. A process consists of inputs, labor, technology, procedures, energy, materials, and outputs.

Taking Corrective Action

There are many ways of taking corrective action to troubleshoot a process and return to the "status quo." A popular example of a root cause and corrective action method is the so-called PDCA or PDSA Cycle (first popularized by Walter Shewhart and then by Dr. Deming as the Deming Wheel) as shown in Fig. 6.6. Deming (1986) referred to this as the Shewhart Cycle, which is the name many still use when describing this version of the feedback loop.

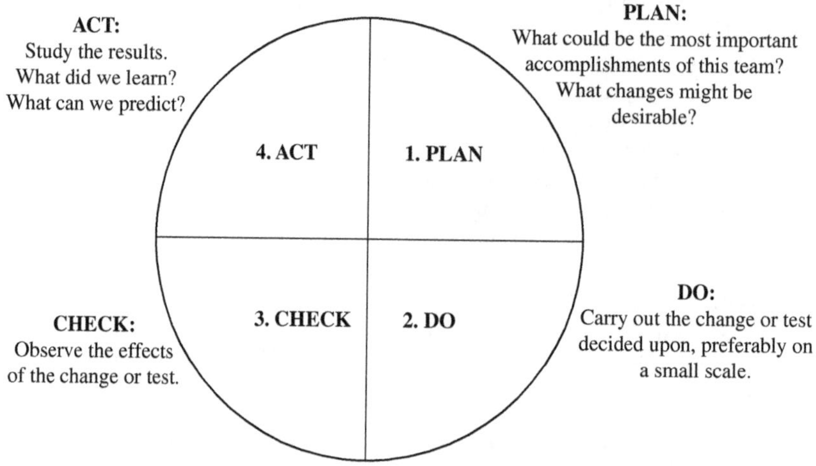
ACT:
Study the results.
What did we learn?
What can we predict?
PLAN:
What could be the most important accomplishments of this team?
What changes might be desirable?
4. ACT
1. PLAN
3. CHECK
2. DO
CHECK:
Observe the effects of the change or test.
DO:
Carry out the change or test decided upon, preferably on a small scale.

FIGURE 6.6 The PDCA Cycle. (Shewhart and Deming, 1986.)

In this example, the feedback loop is divided into four steps labeled Plan, Do, Check, and Act (PDCA) or Plan, Do, Study, Act (PDSA). This model is used by many health care and service industries. These steps correspond roughly to the following:

- "Plan" includes choosing control subjects and setting goals.
- "Do" includes running and monitoring the process.
- "Check" or "Study" includes sensing and umpiring.
- "Act" includes stimulating the actuator and taking corrective action.

An early version of the PDCA cycle was included in W. Edwards Deming's first lectures in Japan (Deming 1950). Since then, additional versions have been used, like PDSA, PDCA, RCCA, and so on.

Some of these versions have attempted to label the PDCA cycle in ways that make it serve as a universal series of steps for both control and improvement. The authors feel that this confuses matters, since two very different processes are involved. Our experience is that all organizations should define two separate methods. One is to take corrective action on a "sporadic change" in performance.

RCCA, PDSA, and PDCA differ from improvement methods like Six Sigma in that the scope of the problem lends itself to a simpler, less complex analysis to find the root cause of a "sporadic problem." RCCA analytical and communication tools contribute to the reduction of day-to-day problems that plague processes. Tools utilized for analysis and diagnosis of sporadic spikes typically take the form of graphical tools with less emphasis on statistical applications. Often many organizations that have been trained in RCCA and the like do not have the right tools and methods to solve chronic problems. It is best to use the Six Sigma D-M-A-I-C improvement methods.

The Pyramid of Control

Control subjects run to large numbers, but the number of "things" to be controlled is far larger. These things include the published catalogs and price lists sent out, multiplied by the number of items in each; the sales made, multiplied by the number of items in each sale; the units of product produced, multiplied by the associated numbers of quality features; and so on for the numbers of items associated with employee relations, supplier relations, cost control, inventory control, product and process developments, etc.

A study in one small company employing about 350 people found that there were more than a billion things to be controlled (Juran 1964, pp. 181–182).

There is no possibility for upper leaders to control huge numbers of control subjects. Instead, they divide up the work of control using a plan of delegation similar to that shown in Fig. 6.7.

This division of work establishes three areas of responsibility for control: control by nonhuman means, control by the workforce, and control by the managerial hierarchy.

Control by Technology (Nonhuman Means)

At the base of the pyramid are the automated feedback loops and error-proofed processes, which operate with no human intervention other than maintenance of facilities (which, however, is critical). These nonhuman methods provide control over a great majority of things. The control subjects are exclusively technological, and control takes place on a real-time basis.

The remaining controls in the pyramid require human intervention. By a wide margin, the most amazing achievement in quality control takes place during a biological process that

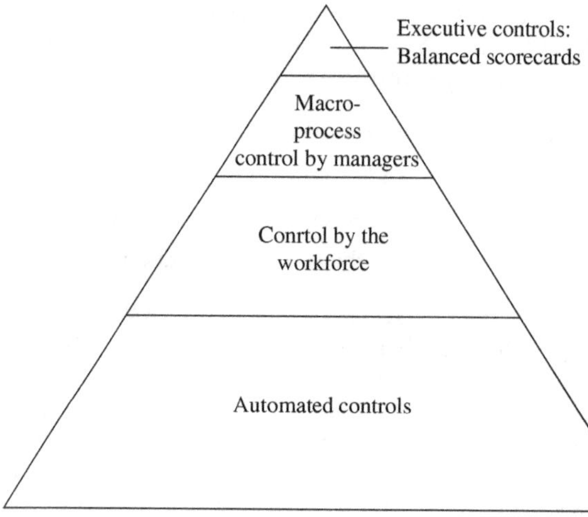

Figure 6.7 The pyramid of control. (Making Quality Happen, Juran Institute, Inc., Senior Executive Workshop, p. F-5.)

is millions of years old—the growth of the fertilized egg into an animal organism. In human beings the genetic instructions that program this growth consist of a sequence of about three billion "letters." This sequence—the human genome—is contained in two strands of DNA (the double helix), which "unzip" and replicate about a million billion times during the growth process from fertilized egg to birth of the human being.

Given such huge numbers, the opportunities for error are enormous. (Some errors are harmless, but others are damaging and even lethal.) Yet the actual error rate is of the order of about 1 in 10 billion. This incredibly low error rate is achieved through a feedback loop involving three processes (Radman and Wagner 1988):

- A high-fidelity selection process for attaching the right "letters," using chemical lock-and-key combinations
- A proofreading process for reading the most recent letter, and removing it if incorrect
- A corrective action process to rectify the errors that are detected

Control by the Employees (Workforce)

Delegating such decisions to the workforce yields important benefits in human relations and in conduct of operations. These benefits include shortening the feedback loop; providing the workforce with a greater sense of ownership of the operating processes, often referred to as "empowerment"; and liberating supervisors and leaders to devote more of their time to planning and improvement.

It is feasible to delegate many quality control decisions to the workforce. Many organizations already do. However, to delegate process control decisions require meeting the criteria of "self-control" or "self-management."

Control by the Managerial Hierarchy

The peak of the pyramid of control consists of the "vital few" control subjects. These are delegated to the various levels in the managerial hierarchy, including the upper leaders.

	At Workforce Levels	At Managerial Levels
Control goals	Product and process features in specifications and procedures	Business-oriented, product salability, competitiveness
Sensors	Technological	Data systems
Decisions to be made	Conformance or not?	Meet customer needs or not?

(*Making Quality Happen*, Juran Institute, Inc., Senior Executive Workshop, p. F-4, Southbury, CT.)

TABLE 6.2 Contrast of Quality Control and Two Levels—Workforce and Upper Management

Leaders should avoid getting too deep into making decisions on quality control. Instead, they should

- Make the vital few decisions
- Provide criteria to distinguish the vital few decisions from the rest. For an example of providing such criteria see Table 6.2
- Delegate the rest under a decision-making process that provides the essential tools and training

The distinction between vital few matters and others originates with the control subjects. Table 6.2 shows how control subjects at two levels—workforce and upper management—affect the elements of the feedback loop.

Planning for Control

Planning for control is the activity that provides the system—the concepts, methodology, and tools—through which company personnel can keep the operating processes stable and thereby produce the product features required to meet customer needs. The input-output features of this system (also plan, process) were depicted in Fig. 6.2.

Critical to Quality (CTQ): Customers and Their Needs

The principal customers of control systems are the company personnel engaged in control—those who carry out the steps that enable the feedback loop. Such personnel require (1) an understanding of what is critical to quality (CTQ), customers' needs, and (2) a definition of their own role in meeting those needs. However, most of them lack direct contact with customers. Planning for control helps to bridge that gap by supplying a translation of what customers' needs are, along with defining responsibility for meeting those needs. In this way, planning for quality control includes providing operating personnel with information on customer needs (whether direct or translated) and defining the related control responsibilities of the operating personnel. Planning for quality control can run into extensive detail.

Who plans for control? Planning for control has in the past been assigned to

- Product development staff
- Quality engineers and specialists
- Multifunctional design teams
- Departmental leaders and supervisors
- The workforce

Planning for control of critical processes has traditionally been the responsibility of those who plan the operating process. For noncritical processes, the responsibility was usually assigned to quality specialists from the quality department. Their draft plans were then submitted to the operating heads for approval.

Recent trends have been to increase the use of the team concept. The team membership includes the operating forces and may also include suppliers and customers of the operating process. The recent trend has also been to increase participation by the workforce.

Compliance and Control Concepts

The methodologies of compliance and control are built around various concepts, such as the feedback loop, process capability, self-control, etc. Some of these concepts are of ancient origin; others have evolved in this and the last centuries. During the discussion of planning for control, we will elaborate on some of the more widely used concepts.

The Process Map or Flow Diagram

The usual first step in planning for control is to map out the flow of the operating process. Figure 6.8 is an example of a flow diagram.

The flow diagram is widely used during the planning of quality controls. It helps the planning team to

- Understand the overall operating process. Each team member is quite knowledgeable about his or her segment of the process, but less so about other segments and about the interrelationships.
- Identify the control subjects around which the feedback loops are to be built. The nature of these control subjects was discussed previously in the section "The Control Subjects."

The Control Spreadsheet

The work of the planners is usually summarized on a control spreadsheet. This spreadsheet is a major planning tool. An example can be seen in Fig. 6.9.

In this spreadsheet the horizontal rows are the various control subjects. The vertical columns consist of elements of the feedback loop, plus other features needed by the operating forces to exercise control so as to meet the quality goals.

Some of the contents of the vertical columns are unique to specific control subjects. However, certain vertical columns apply widely to many control subjects. These include unit of measure, type of sensor, quality goal, frequency of measurement, sample size, criteria for decision making, and responsibility for decision making.

Who Does What?

The feedback loop involves multiple tasks, each of which requires a clear assignment of responsibility. At any control station there may be multiple people available to perform those tasks. For example, at the workforce level, a control station may include setup specialists, operators, maintenance personnel, inspectors, etc. In such cases it is necessary to agree on who should make which decisions and who should take which actions. An aid to reaching such agreement is a special spreadsheet similar to Fig. 6.9.

In this spreadsheet, the essential decisions and actions are listed in the leftmost column. The remaining columns are headed up by the names of the job categories associated with the

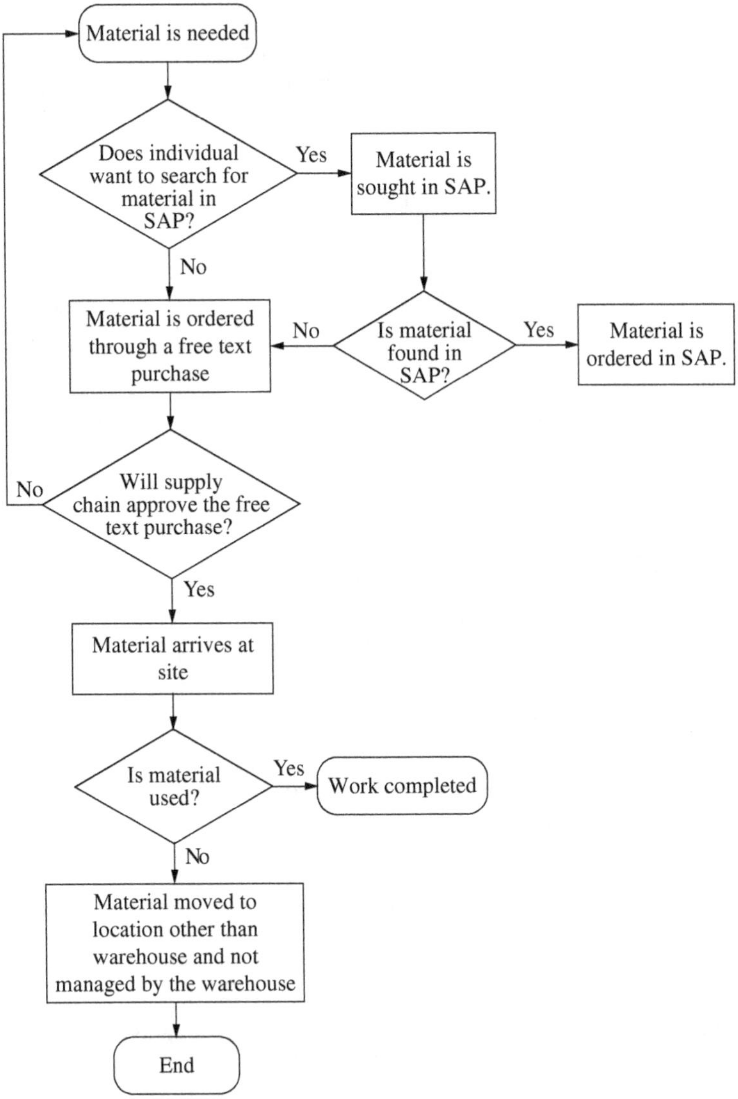

Figure 6.8 The flow diagram.

control station. Then, through discussion among the cognizant personnel, agreement is reached on who is to do what.

This spreadsheet is a proven way to find answers to the long-standing, but vague, question, who is responsible for quality? This question has never been answered because it is inherently unanswerable. However, if the question is restated in terms of decisions and actions, the way is open to agree on the answers. This clears up the vagueness.

Test and Control Stations

A "control station" is an area in which quality control takes place. In the lower levels of an organization, a control station is usually confined to a limited physical area. Alternatively,

	Process Control Features							
Control Subjects	Unit of Measure	Type of Sensor	Goal	Frequency of Measurement	Sample Size	Criteria for Decision Making	Responsibility for Decision Making	...
Wave solder conditions solder temperature	Degree (°F)	Thermo-couple	505°F	Continuous	N/A	510°F reduce heat 500°F increase heat	Operator	...
Conveyor speed	Feet per minute (ft/min)	ft/min	4.5 ft/min	1 per hour	N/A	5 ft/min reduce speed 4 ft/min increase speed	Operator	...
Alloy purity	% Total contaminates	Lab chemical analysis	1.5% max	1 per month	15 grams	At 1.5%, drain bath, replace solder	Process engineer	...

Figure 6.9 The control spreadsheet.

the control station can take such forms as a patrol beat or a "control tower." At higher levels, control stations may be widely dispersed geographically, as in the scope of a manager's responsibility.

A review of numerous control stations shows that they are usually designed to provide evaluations and/or early warnings in the following ways:

- At changes of jurisdiction, where responsibility is transferred from one organization to another
- Before embarking on some significant irreversible activity, such as signing a contract
- After creation of a critical quality feature
- At the site of dominant process variables
- At areas ("windows") that allow economical evaluation to be made

Stages of Control

The flow diagram not only discloses the progression of events in the operating process, it also suggests which stages should become the centers of control activity. Several of these stages apply to the majority of operating processes.

Setup (Startup) Control

The end result of this form of control is the decision of whether or not to "push the start button." Typically setup control involves a number of steps:

- A *countdown* listing the preparatory steps needed to get the process ready to produce. Such countdowns sometime come from suppliers. Airlines provide checklists to help travelers plan their trips; electric power organizations provide checklists to help householders prepare the house for winter weather.
- *Evaluation* of process and/or product features to determine whether, if started, the process will meet the goals.
- *Criteria* to be met by the evaluations.

- *Verification* that the criteria have been met.
- *Assignment* of responsibility. This assignment varies, depending largely on the criticality of the quality goals. The greater the criticality, the greater is the tendency to assign the verification to specialists, supervisors, and "independent" verifiers rather than to nonsupervisory workers.

Running Control

This form of control takes place periodically during the operation of the process. The purpose is to make the "run or stop" decision—whether the process should continue to produce a product or whether it should stop.

Running control consists of closing the feedback loop, over and over again. The process and/or product performance is evaluated and compared with goals. If the product and/or process conform to goals, and if the process has not undergone some significant adverse change, the decision is to "continue to run." If there is nonconformance, or if there has been a significant change, corrective action is in order.

The term "significant" has meanings beyond those in the dictionary. One of these meanings relates to whether an indicated change is a real change or is a false alarm due to chance variation. The design for process control should provide the tools needed to help the operating forces distinguish between real changes and false alarms. Statistical process control (SPC) methodology is aimed at providing such tools.

Product Control

This form of control takes place after some amount of product has been produced. The purpose of the control is to decide whether the product conforms to the product quality goals. Assignment of responsibility for this decision differs from company to company. However, in all cases, those who are to make the decision must be provided with the facilities and training that will enable them to understand the product quality goals, evaluate the actual product quality, and decide whether there is conformance.

Since all this involves making a factual decision, it can, in theory, be delegated to anyone, including members of the workforce. In practice, this delegation is not made to those whose assigned priorities might bias their judgment. In such cases, the delegation is usually to those whose responsibilities are free from such bias, for example, "independent" inspectors. Statistical quality control (SQC) is a methodology frequently employed to yield freedom from biases.

Facilities Control

Most operating processes employ physical facilities: equipment, instruments, and tools. Increasingly, the trend has been to use automated processes, computers, robots, etc. This same trend makes product quality more and more dependent on maintenance of the facilities. Leading organizations have moved to elements of total productive maintenance (TPM). The extent of application varies by company, but TPM and reliability-centered maintenance (RCM) support sound facilities control. The elements of design for facilities control are well known:

- Establish a schedule for conducting facilities maintenance
- Establish a checklist—a list of tasks to be performed during a maintenance action
- Train the maintenance forces to perform the tasks
- Assign clear responsibility for adherence to schedule
- Enhance management of critical spares

- Standardize preventive maintenance tasks and frequency on equipment
- Optimize efficiency related to maintenance staffing and organization
- Increase mechanical interface with equipment from operator

The weakest link in facilities control has been adherence to schedule. To ensure strict adherence to schedule requires an independent audit.

In cases involving introduction of a new technology, a further weak link is training the maintenance forces (White 1988).

During the 1980s, the automakers began to introduce computers and other electronics into their vehicles. It soon emerged that many repair shop technicians lacked the technological education base needed to diagnose and remedy the associated field failures. To make matters worse, the automakers did not give high priority to standardizing the computers. As a result, a massive training backlog developed.

Concept of Dominance

Control subjects are so numerous that planners are well advised to identify the vital few so that they will receive appropriate priority. One tool for identifying the vital few is the concept of dominance.

Operating processes are influenced by many variables, but often one variable is more important than all the rest combined. Such a variable is said to be the "dominant variable." Knowledge of which process variable is dominant helps planners during allocation of resources and priorities. The more usual dominant variables include

- *Setup dominant.* Some processes exhibit high stability and reproducibility of results over many cycles of operation. A common example is the printing process. The design for control should provide the operating forces with the means for precise setup and validation before operations proceed.
- *Time dominant.* Here, the process is known to change progressively with time, for example, depletion of consumable supplies, heating up, and wear of tools. The design for control should provide means for periodic evaluation of the effect of progressive change and for convenient readjustment.
- *Component dominant.* Here, the main variable is the quality of the input materials and components. An example is the assembly of electronic or mechanical equipment. The design for control should be directed at supplier relations, including joint planning with suppliers to upgrade the quality of the inputs.
- *People dominant.* In these processes, quality depends mainly on the skill and knack possessed by the workers. The skilled trades are well-known examples. The design for control should emphasize aptitude testing of workers, training and certification, quality rating of workers, and error-proofing to reduce employee errors.
- *Information dominant.* Here, the processes are of a "job-shop" nature so that there is frequent change in what product is to be produced. As a result, the job information changes frequently. The design for control should concentrate on providing an information system that can deliver accurate, up-to-date information on just how this job differs from its predecessors.

Process Capability

One of the most important concepts in the quality planning process is "process capability." The prime application of this concept is during planning of the operating processes.

This same concept also has applications in quality control. To explain this, a brief review is in order. All operating processes have an inherent uniformity for producing products. This uniformity can often be quantified, even during the planning stages. The process planners can use the resulting information for making decisions on adequacy of processes, choice of alternative processes, need for revision of processes, and so forth, with respect to the inherent uniformity and its relationship to process goals.

Applied to planning for quality control, the state of process capability becomes a major factor in decisions on frequency of measuring process performance, scheduling maintenance of facilities, etc.; the greater the stability and uniformity of the process, the less the need for frequent measurement and maintenance.

Those who plan for quality control should have a thorough understanding of the concept of process capability and its application to both areas of planning—planning the operating processes as well as planning the controls.

Process Conformance

Does the process conform to its quality goals? The umpire answers this question by interpreting the observed differences between process performance and process goals. When current performance does differ from the quality goals, the question arises, what is the cause of this difference?

Special and Common Causes of Variation

Observed differences usually originate in one of two ways: (1) the observed change is caused by the behavior of a major variable in the process (or by the entry of a new major variable) or (2) the observed change is caused by the interplay of multiple minor variables in the process.

Shewhart called (1) and (2) "assignable" and "nonassignable" causes of variation, respectively (Shewhart 1931). Deming later coined the terms "special" and "common" causes of variation (Deming 1986). In what follows we will use Deming's terminology.

"Special" causes are typically sporadic, and often have their origin in single variables. For such cases, it is comparatively easy to conduct a diagnosis and provide remedies. "Common" causes are typically chronic and usually have their origin in the interplay among multiple minor variables. As a result, it is difficult to diagnose them and to provide remedies. This contrast makes clear the importance of distinguishing special causes from common causes when interpreting differences. The need for making such distinctions is widespread. Special causes are the subject of quality control; common causes are the subject of quality improvement.

The Shewhart Control Chart

It is most desirable to provide umpires with tools that can help to distinguish between special causes and common causes. An elegant tool for this purpose is the Shewhart control chart (or just control chart) shown in Fig. 6.10.

In Fig. 6.10, the horizontal scale is time and the vertical scale is quality performance. The plotted points show quality performance as time progresses.

The chart also exhibits three horizontal lines. The middle line is the average of past performance and is, therefore, the expected level of performance. The other two lines are statistical "limit lines." They are intended to separate special causes from common causes, based on some chosen level of probability, such as 1 chance in 100.

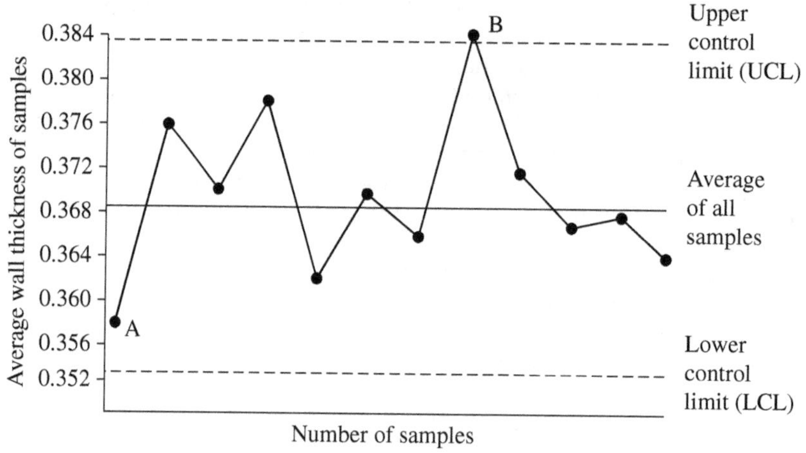

FIGURE 6.10 Shewhart control chart. ("Quality Control," Leadership for the Quality Century, Juran Institute, Inc.)

Points within Control Limits

Point A on the chart differs from the historical average. However, since point A is within the limit lines, this difference could be due to common causes (at a probability of more than 1 in 100). Hence, we assume that there is no special cause. In the absence of special causes, the prevailing assumptions include

- Only common causes are present
- The process is in a state of "statistical control"
- The process is doing the best it can
- The variations must be endured
- No action need be taken—taking action may make matters worse (a phenomenon known as "hunting" or "tampering")

The preceding assumptions are being challenged by a broad movement to improve process uniformity. Some processes exhibit no points outside of control chart limits, yet the interplay of minor variables produces some defects.

In one example, a process in statistical control was nevertheless improved by an order of magnitude. The improvement was by a multifunctional improvement team, which identified and addressed some of the minor variables. This example is a challenge to the traditional assumption that variations due to common causes must be endured (Pyzdek 1990).

In other cases, the challenge is subtler. There are again no points outside the control limits, but in addition, no defects are being produced. Nevertheless, the customers demand greater and greater uniformity. Examples are found in business processes (precision of estimating), as well as in manufacturing (batch-to-batch uniformity of chemicals, uniformity of components going into random assembly). Such customer demands are on the increase, and they force suppliers to undertake projects to improve the uniformity of even the minor variables in the process. There are many types of control charts.

Points Outside of Control Limits

Point B also differs from the historical average, but is outside of the limit lines. Now the probability is against this being the result of common causes, less than 1 chance in 100. Hence, we assume that point B is the result of special causes. Traditionally such "out-of-control" points become nominations for corrective action.

Ideally, all such nominations should stimulate prompt corrective action to restore the status quo. In practice many out-of-control changes do not result in corrective action. The usual reason is that the changes involving special causes are too numerous—the available personnel cannot deal with all of them. Hence, priorities are established based on economic significance or on other criteria of importance. Corrective action is taken for the high-priority cases; the rest must wait their turn. Some changes at low levels of priority may wait a long time for corrective action.

A further reason for failure to take corrective action is a lingering confusion between statistical control limits and quality tolerances. It is easy to be carried away by the elegance and sensitivity of the control chart. This happened on a large scale during the 1940s and 1950s. Here are two examples from my personal experience:

- A large automotive components factory placed a control chart at every machine
- A viscose yarn factory created a "war room" of more than 400 control charts

In virtually all such cases the charts were maintained by the quality departments but ignored by the operating personnel. Experience with such excesses has led leaders and planners to be wary of employing control charts just because they are sensitive detectors of change. Instead, the charts should be justified based on value added. Such justifications include

- Customer needs are directly involved
- There is risk to human safety or the environment
- Substantial economics are at stake
- The added precision is needed for control

Statistical Control Limits and Tolerances

For most of human history, targets and goals consisted of product features or process features, usually defined in words. Words such as "the color is red" and "the length is long enough" are targets, but are open to too much interpretation. The growth of technology stimulated the growth of measurement, plus a trend to define targets and goals in precise numbers. In addition, there emerged the concept of limits, or "tolerances," around the targets and goals. For example,

- Ninety-five percent of the shipments shall meet the scheduled delivery date.
- The length of the bar shall be within 1 mm of the specified number.
- The length of time to respond to customers is 10 minutes, plus or minus 2 minutes.

Such targets had official status. They were set by product or process designers, and published as official specifications. The designers were the official quality legislators—they enacted the laws. Operating personnel were responsible for obeying the quality laws—meeting the specified goals and tolerances.

Statistical control limits in the form of control charts were virtually unknown until the 1940s. At that time, these charts lacked official status. They were prepared and published by quality specialists from the quality department. To the operating forces, control charts were a mysterious, alien concept. In addition, the charts threatened to create added work in the form of unnecessary corrective action. The operating personnel reasoned as follows: It has always been our responsibility to take corrective action whenever the product becomes nonconforming. These charts are so sensitive that they detect process changes that do not result in nonconforming products. We are then asked to take corrective action even when the products meet the quality goals and tolerances.

So there emerged a confusion of responsibility. The quality specialists were convinced that the control charts provided useful early-warning signals that should not be ignored. Yet the quality departments failed to recognize that the operating forces were now faced with a confusion of responsibility. The latter felt that so long as the products met the quality goals there was no need for corrective action. The upper leaders of those days were of no help—they did not involve themselves in such matters. Since the control charts lacked official status, the operating forces solved their problem by ignoring the charts. This contributed to the collapse in the 1950s of the movement known as "statistical quality control."

The 1980s created a new wave of interest in applying the tools of statistics to the control of quality. Many operating personnel underwent training in "statistical process control." This training helped to reduce the confusion, but some confusion remains. To get rid of the confusion, leaders should

- Clarify the responsibility for corrective action on points outside the control limits
- Establish guidelines on actions to be taken when points are outside the statistical control limits but the product still meets the quality tolerances

The need for guidelines for decision making is evident from Fig. 6.11. The guidelines for quadrants A and C are obvious. If both process and product conform to their respective goals, the process may continue to run. If neither process nor product conforms to their respective goals, the process should be stopped and remedial action should be taken. The guidelines for quadrants B and D are often vague, and this vagueness has been the source of a good deal of confusion. If the choice of action is delegated to the workforce, the leaders should establish clear guidelines.

		Product	
		Conforms	Does not conform
Process	Does not conform	B vague	C clear
	Conforms	A clear	D vague

Figure 6.11 Areas of decision making. (Making Quality Happen, Juran Institute, Inc. Used by permission.)

Numerous efforts have been made to design control chart limits in ways that help operating personnel detect whether product quality is threatening to exceed the product quality limits.

Self-Control and Controllability

Workers are in a state of self-control when they have been provided with all the essentials for doing good work. These essentials include

- Means of knowing what the goals are.
- Means of knowing what their actual performance is.
- Means for changing their performance in the event that performance does not conform to goals. To meet this criterion requires an operating process that (1) is inherently capable of meeting the goals and (2) is provided with features that make it possible for the operating forces to adjust the process as needed to bring it into conformance with the goals.

These criteria for self-control are applicable to processes in all functions and at all levels, from general manager to nonsupervisory worker.

It is all too easy for leaders to conclude that the above criteria have been met. In practice, however, there are many details to be worked out before the criteria can be met. The nature of these details is evident from checklists, which have been prepared for specific processes in order to ensure that the criteria for self-control are met. Examples of these checklists include those designed for product designers, production workers, and administrative and support personnel. Examples of such checklists can be found by referring to the subject index of this handbook.

If all the criteria for self-control have been met at the worker level, any resulting product nonconformance are said to be worker-controllable. If any of the criteria for self-control have not been met, then management's planning has been incomplete—the planning has not fully provided the means for carrying out the activities within the feedback loop. The nonconforming products resulting from such deficient planning are then said to be management-controllable. In such cases it is risky for leaders to hold the workers responsible for quality.

Responsibility for results should, of course, be keyed to controllability. However, in the past, many leaders were not aware of the extent of controllability as it prevailed at the worker level. Studies conducted by Juran during the 1930s and 1940s showed that at the worker level, the proportion of management-controllable to worker-controllable nonconformance was of the order of 80 to 20. These findings were confirmed by other studies during the 1950s and 1960s. That ratio of 80 to 20 helps to explain the failure of so many efforts to solve the organizations' quality problems solely by motivating the workforce.

Effect on the Process Conformance Decision

Ideally, the workforce should make the decision of whether the process conforms to process quality goals. There is no shorter feedback loop. For many processes, this is the actual arrangement. In other cases, the process conformance decision is assigned to nonoperating personnel—independent checkers or inspectors. The reasons include

- The worker is not in a state of self-control
- The process is critical to human safety or to the environment
- Quality does not have top priority
- There is a lack of mutual trust between the leaders and the workforce

Product Conformance: Fitness for Purpose

There are two levels of product features, and they serve different purposes. One of these levels serves such purposes as

- Meeting customer needs
- Protecting human safety
- Protecting the environment

Product features are said to possess "fitness for use" if they are able to serve the above purposes.

The second level of product features serves purposes such as

- Providing working criteria to those who lack knowledge of fitness for use
- Creating an atmosphere of law and order
- Protecting innocents from unwarranted blame

Such product features are typically contained in internal specifications, procedures, standards, etc. Product features that are able to serve the second list of purposes are said to possess conformance to specifications, etc. We will use the shorter label "conformance."

The presence of two levels of product features results in two levels of decision making: Is the product in conformance? Is the product fit for use? Figure 6.12 shows the interrelation of these decisions to the flow diagram.

The Product Conformance Decision

Under prevailing policies, products that conform to specification are sent on to the next destination or customer. The assumption is that products that conform to specification are also fit for use. This assumption is valid in the great majority of cases.

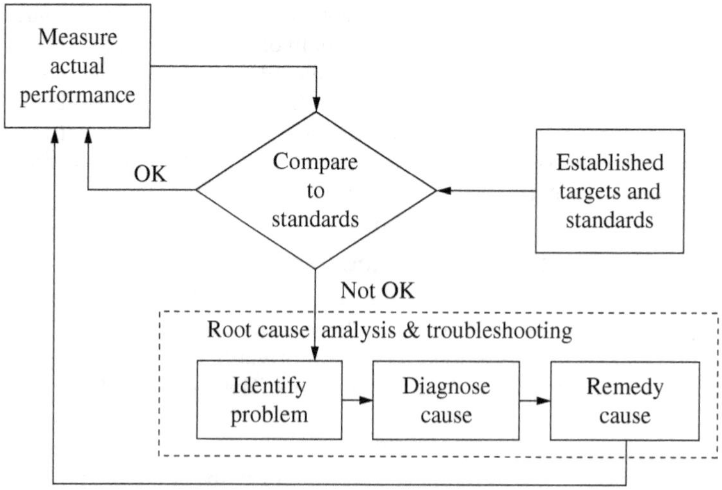

FIGURE 6.12 Interrelation of QC and RCA.

The combination of large numbers of product features, when multiplied by large volumes of product, creates huge numbers of product conformance decisions to be made. Ideally, these decisions should be delegated to the lowest levels of organization—to the automated devices and the operating workforce. Delegation of this decision to the workforce creates what is called "self-inspection."

Self-Inspection

We define "self-inspection" as a state in which decisions on the product are delegated to the workforce. The delegated decisions consist mainly of: Does product quality conform to the quality goals? What disposition is to be made of the product?

Note that self-inspection is very different from self-control, which involves decisions on the *process*.

The merits of self-inspection are considerable:

- The feedback loop is short; the feedback often goes directly to the actuator—the energizer for corrective action.
- Self-inspection enlarges the job of the workforce—it confers a greater sense of job ownership.
- Self-inspection removes the police atmosphere created by use of inspectors, checkers, etc.

However, to make use of self-inspection requires meeting several essential criteria:

- *Quality is number one.* Quality must undoubtedly be the top priority.
- *Mutual confidence.* The leaders must have enough trust in the workforce to be willing to make the delegation, and the workforce must have enough confidence in the leaders to be willing to accept the responsibility.
- *Self-control.* The conditions for self-control should be in place so that the workforce has all the means necessary to do good work.
- *Training.* The workers should be trained to make the product conformance decisions.
- *Certification.* The recent trend is to include a certification procedure. Workers who are candidates for self-inspection undergo examinations to ensure that they are qualified to make good decisions. The successful candidates are certified and may be subject to audit of decisions thereafter.

In many organizations, these criteria are not fully met, especially the criterion of priority. If some parameter other than quality has top priority, there is a real risk that evaluation of product conformance will be biased. This problem happens frequently when personal performance goals are in conflict with overall quality goals. For example, a chemical company found that it was rewarding sales personnel on revenue targets without regard to product availability or even profitability. The salespeople were making all their goals, but the company was struggling.

The Fitness for Purpose Decision

The great majority of products do conform to specifications. For the nonconforming products there arises a new question: Is the nonconforming product nevertheless fit for use?

A complete basis for making this decision requires answers to questions such as

- Who are the user(s)?
- How will this product be used?
- Are there risks to structural integrity, human safety, or the environment?
- What is the urgency for delivery?
- How do the alternatives affect the producer's and the user's economics?

To answer such questions can involve considerable effort. Organizations have tried to minimize the effort through procedural guidelines. The methods in use include

- *Treat all nonconforming products as unfit for use.* This approach is widely used for products that can pose risks to human safety or the environment—products such as pharmaceuticals or nuclear energy.
- *Create a mechanism for decision making.* An example is the material review board so widely used in the defense industry. This device is practical for matters of importance, but is rather elaborate for the more numerous cases in which little is at stake.
- *Create a system of multiple delegation.* Under such a system, the "vital few" decisions are reserved for a formal decision-making body such as a material review board. The rest are delegated to other people.

Table 6.3 is an example of a table of delegation used by a specific company.

	Amount of Product or Money at Stake Is:	
Effect of Nonconformance Is on	**Small**	**Large**
Internal economics only	Department head directly involved, quality engineer	Plant managers involved, quality manager
Economic relations with supplier	Supplier, purchasing agent, quality engineer	Supplier, manager
Economic relations with client	Client, salesperson, quality engineer	Client (for marketing, manufacturing, technical, quality)
Field performance of the product	Product designer, salesperson, quality engineer	Client (managers for technical, manufacturing, marketing, quality)
Risk of damage to society or of nonconformance to government regulations	Product design manager, compliance officer, lawyer, quality managers	General manager and team of upper managers

*For those industries whose quality mission is really one of conformance to specification (for example, atomic energy, space), the real decision maker on fitness for use is the client or the government regulator.

Table 6.3 Multiple Delegations of Decision Making on Fitness for Purpose*

Disposition of Unfit Product

Unfit product is disposed of in various ways: scrap, sort, rework, return to supplier, sell at a discount, etc. The internal costs can be estimated to arrive at an economic optimum. However, the effects go beyond money: schedules are disrupted, people are blamed, etc. To minimize the resulting human abrasion, some organizations have established rules of conduct, such as

- Choose the alternative that minimizes the total loss to all parties involved. Now there is less to argue about, and it becomes easier to agree on how to share the loss.
- Treat the loss as an opportunity for quality improvement and avoid looking for blame.
- Use "charge backs" sparingly. Charging the vital few losses to the departments responsible has merit from an accounting viewpoint. However, when applied to the numerous minor losses, this is often uneconomic as well as detrimental to efforts to improve quality.

Failure to use products that meet customer needs is a waste. Sending out products that do not meet customer needs is worse. Personnel who are assigned to make product conformance decisions should be provided with clear definitions of responsibility as well as guidelines for decision making. Leaders should, as part of their audit, ensure that product conformance decisions would be made using a process appropriate to company needs.

Corrective Action

The final step in closing the feedback loop is to actuate a change that restores conformance with quality goals. This step is popularly known as "troubleshooting" or "firefighting."

Note that the term "corrective action" has been applied loosely to two very different situations, as shown in Fig. 6.1. The feedback loop is well designed to eliminate sporadic nonconformance, like that "spike" in Fig. 6.1; the feedback loop is not well designed to deal with the area of chronic waste shown in the figure. Instead, the need is to employ the quality improvement process. We will use the term "corrective action" in the sense of troubleshooting—eliminating sporadic nonconformance.

Corrective action requires the journeys of diagnosis and remedy. These journeys are simpler than for quality improvement. Sporadic problems are the result of adverse change, so the diagnostic journey aims to discover what has changed. The remedial journey aims to remove the adverse change and restore conformance.

Diagnosing Sporadic Change

During the diagnostic journey, the focus is on what has changed. Sometimes the causes are not obvious, so the main obstacle to corrective action is diagnosis. The diagnosis makes use of methods and tools such as

- Forensic autopsies to determine with precision the symptoms exhibited by the product and process
- Comparison of products made before and after the trouble began to see what has changed; also comparison of good and bad products made since the trouble began
- Comparison of process data before and after the problem began to see what process conditions have changed

- Reconstruction of the chronology, which consists of logging on a time scale (of hours, days, etc.) (1) the events that took place in the process before and after the sporadic change—that is, rotation of shifts, new employees on the job, maintenance actions, etc., and (2) the time related to product information—that is, date codes, cycle time for processing, waiting time, move dates, etc.

Analysis of the resulting data usually sheds a good deal of light on the validity of the various theories of causes. Certain theories are denied. Other theories survive to be tested further.

Operating personnel who lack the training needed to conduct such diagnoses may be forced to shut down the process and request assistance from specialists, the maintenance department, etc. They may also run the process "as is" in order to meet schedules and thereby risk failure to meet the quality goals.

Corrective Action—Remedy

Once the cause(s) of the sporadic change is known, the worst is over. Most remedies consist of going back to what was done before. This is a return to the familiar, not a journey into the unknown (as is the case with chronic problems). The local personnel are usually able to take the necessary action to restore the status quo.

Process designs should provide means to adjust the process as required to attain conformance with quality goals. Such adjustments are needed at startup and during running of the process. This aspect of design for process control ideally should meet the following criteria:

- There should be a known relationship between the process variables and the product results.
- Means should be provided for ready adjustment of the process settings for the key process variables.
- A predictable relationship should exist between the amount of change in the process settings and the amount of effect on the product features.

If such criteria are not met, the operating personnel will, in due course, be forced to cut corners in order to carry out remedial action. The resulting frustrations become a disincentive to putting high priority on quality.

The Role of Statistical Methods in Control

An essential activity within the feedback loop is the collection and analysis of data. This activity falls within the scientific discipline known as "statistics." The methods and tools used are often called "statistical methods." These methods have long been used to aid in data collection and analysis in many fields: biology, government, economics, finance, management, etc.

Statistical Process Control

Statistical process control (SPC) has multiple meanings, but in most organizations, it is considered to include basic data collection; analysis through such tools as frequency distributions, Pareto principle, Ishikawa (fish bone) diagram, Shewhart control chart, etc.; and application of the concept of process capability.

Advanced tools, such as design of experiments and analysis of variance, are a part of statistical methods but are not normally considered part of statistical process control.

The Merits

These statistical methods and tools have contributed in an important way to quality control and to the other processes of the Juran Trilogy—quality improvement and quality planning. For some types of quality problems, the statistical tools are more than useful—the problems cannot be solved at all without using the appropriate statistical tools.

The SPC movement has succeeded in training a great many supervisors and workers in basic statistical tools. The resulting increase in statistical literacy has made it possible for them to improve their grasp of the behavior of processes and products. In addition, many have learned that decisions based on data collection and analysis yield superior results.

The Risks

There is a danger in taking a tool-oriented approach to quality instead of a problem-oriented or results-oriented approach. During the 1950s, this preoccupation became so extensive that the entire statistical quality control movement collapsed; the word "statistical" had to be eliminated from the names of the departments.

The proper sequence in managing is first to establish goals and then plan how to meet those goals, including choosing the appropriate tools. Similarly, when dealing with problems—threats or opportunities—experienced leaders start by first identifying the problems. They then try to solve those problems by various means, including choosing the proper tools.

During the 1980s, numerous organizations did, in fact, try a tool-oriented approach by training large numbers of their personnel in the use of statistical tools. However, there was no significant effect on the bottom line. The reason was that no infrastructure had been created to identify which projects to tackle, to assign clear responsibility for tackling those projects, to provide needed resources, to review progress, etc.

Leaders should ensure that training in statistical tools does not become an end in itself. One form of such assurance is through measures of progress. These measures should be designed to evaluate the effect on operations, such as improvement in customer satisfaction or product performance, reduction in cost of poor quality, etc. Measures such as numbers of courses held or numbers of people trained do not evaluate the effect on operations and hence should be regarded as subsidiary in nature.

Information for Decision Making

Quality control requires extensive decision making. These decisions cover a wide variety of subject matter and take place at all levels of the hierarchy. The planning for quality control should provide an information network that can serve all decision makers. At some levels of the hierarchy, a major need is for real-time information to permit prompt detection and correction of nonconformance to goals. At other levels, the emphasis is on summaries that enable leaders to exercise control over the vital few control subjects. In addition, the network should provide information as needed to detect major trends, identify threats and opportunities, and evaluate the performance of organization units and personnel.

In some organizations, the quality information system is designed to go beyond control of product features and process features; the system is also used to control the quality performance of organizations and individuals, such as departments and department heads. For example, many organizations prepare and regularly publish scoreboards showing summarized quality performance data for various market areas, product lines, operating functions, etc. These performance data are often used as indicators of the quality performance of the personnel in charge.

To provide information that can serve all those purposes requires planning that is directed specifically at the information system. A multifunctional team whose mission is focused on the quality information system best does such planning. That team properly includes the customers as well as the suppliers of information. The management audit of the quality control system should include assurance that the quality information system meets the needs of the various customers.

The Quality Control System and Policy Manual

A great deal of quality planning is done through "procedures," which are really repetitive-use plans. Such procedures are thought out, written out, and approved formally. Once published, they become the authorized ways of conducting the company's affairs. It is quite common for the procedures relating to managing for quality to be published collectively in a "quality manual" (or similar title). A significant part of the manual relates to quality control.

Quality manuals add to the usefulness of procedures in several ways:

- *Legitimacy.* The manuals are approved at the highest levels of organization.
- *Accessibility.* The procedures are assembled into a well-known reference source rather than being scattered among many memoranda, oral agreements, reports, minutes, etc.
- *Stability (or consistency).* The procedures survive despite lapses in memory and employee turnover.

Study of company quality manuals shows that most of them contain a core content, which is quite similar from company to company. Relative to quality control, this core content includes procedures for

- Applying the feedback loop to process and product control
- Ensuring that operating processes are capable of meeting the quality goals
- Maintaining facilities and calibration of measuring instruments
- Relating to suppliers on quality matters
- Collecting and analyzing the data required for the quality information system
- Training the personnel to carry out the provisions of the manual
- Auditing to ensure adherence to procedures

The need for repetitive-use quality control systems has led to an evolution of standards at industry, national, and international levels. For an example of developing standard operating procedures, including the use of videocassettes, see Murphy and McNealey (1990). Workforce participation during the preparation of procedures helps to ensure that the procedures will be followed.

Format of Quality Manuals

Here, again, there is much commonality. The general chapters of the manual include

1. An official statement by the general manager. It includes the signatures that confer legitimacy.
2. The purpose of the manual and how to use it.

3. The pertinent company (or divisional, etc.) quality policies.
4. The organizational charts and tables of responsibility relative to the quality function.
5. Provision for audit of performance against the mandates of the manual.

Leaders are able to influence the adequacy of the quality control manual in several ways:

- Participate in defining the criteria to be met by the manual
- Approve the final draft of the manual to make it official
- Periodically audit the up-to-date-ness of the manual as well as conformance to the manual

Provision for Audits

Experience has shown that control systems are subject to "slippage" of all sorts. Personnel turnover may result in loss of essential knowledge. Entry of unanticipated changes may result in obsolescence. Shortcuts and misuse may gradually undermine the system until it is no longer effective.

The major tool for guarding against deterioration of a control system has been the audit. Under the audit concept, a periodic, independent review is established to provide answers to the following questions: Is the control system still adequate for the job? Is the system being followed?

The answers are obviously useful to the operating leaders. However, that is not the only purpose of the audit. A further purpose is to provide those answers to people who, though not directly involved in operations, nevertheless have a need to know. If quality is to have top priority, those who have a need to know include the upper leaders.

It follows that one of the responsibilities of leaders is to mandate establishment of a periodic audit of the quality control system.

Tasks for Leaders

1. Leaders should avoid getting too deeply involved in making decisions on quality control. They should make the vital few decisions, provide criteria to distinguish the vital few from the rest, and delegate the rest under a decision-making process.
2. To eliminate the confusion relative to control limits and product quality tolerance, leaders should clarify the responsibility for corrective action on points outside the control limits and establish guidelines on action to be taken when points are outside the statistical control limits but the product still meets the quality tolerances.
3. Leaders should, as part of their audit, ensure that product conformance decisions would be made using a process appropriate to company needs. They should also ensure that training in statistical tools does not become an end in itself. The management audit of the quality control system should include assurance that the quality information system meets the needs of the various customers.
4. Leaders are able to influence the adequacy of the quality control manual in several ways: participate in defining the criteria to be met, approve the final draft to make it official, and periodically audit the currentness of the manual as well as the state of conformance.

References

Deming, W. E. (1950). "Elementary Principles of the Statistical Control of Quality." Nippon Kagaku Gijutsu Renmei (Japanese Union of Scientists and Engineers), Tokyo.

Deming, W. E. (1986). "Out of the Crisis." MIT Center for Advanced Engineering Study. Cambridge, MA.

Juran, J. M. (1964). *Managerial Breakthrough*. McGraw-Hill, New York.Murphy, R. W., and McNealey, J. E. (1990). "A Technique for Developing Standard Operating Procedures to Provide Consistent Quality." *1990 Juran IMPRO Conference Proceedings*, pp. 3D1–3D6.

Pyzdek, T. (1990). "There's No Such Thing as a Common Cause." *ASQC Quality Congress Transactions*, pp. 102–108.

Radford, G. S. (1917). "The Control of Quality." *Industrial Management*, vol. 54, p. 100.

Radford, G. S. (1922). *The Control of Quality in Manufacturing*. Ronald Press Company, New York.

Radman, M., and Wagner, R. (1988). "The High Fidelity of DNA Duplication." *Scientific American*, August, pp. 40–46.

Shewhart, W. A. (1931). *Economic Control of Quality of Manufactured Product*. Van Nostrand, New York, 1931. Reprinted by ASQC, Milwaukee, 1980.

White, J. B. (1988). "Auto Mechanics Struggle to Cope with Technology in Today's Cars." *The Wall Street Journal*, July 26, p. 37.

CHAPTER 7

Strategic Planning and Performance Excellence

Joseph A. De Feo, Rick Edgeman, Gerhard Plenert, and Kenneth Snyder

High Points of This Chapter

1. Strategic planning is the systematic approach to defining long-term business goals and planning the means to achieve them. An organization transformation based on the management of quality should be integrated with the strategic plans of the organization.
2. The strategic plan enables organizations to deploy all goals, including quality improvement goals, to the organization. It provides the basis for senior management to make sound strategic choices and prioritize the organization's focus and other change activities.
3. Activities not aligned with the organization's strategic goals should be changed or eliminated.
4. In this chapter, we define the strategic planning process and its deployment tasks and describe the systematic approach to deploying "quality goals."
5. This chapter also explains the specific roles of leaders when implementing and ensuring the success of the strategic plan and its deployment.

Strategic Planning and Quality: The Benefits

Strategic planning is the systematic approach to defining long-term business goals and planning the means to achieve them. Once an organization has established its long-term goals, effective strategic planning enables it, year by year, to create an annual business plan, which includes the necessary annual goals, resources, and actions needed to move toward those goals.

Many organizations have created a vision to be the best performers by creating and producing high-quality products and services for their customers. By doing so, they have outperformed those that did not. This performance is not just related to the quality of their goods and services, but to the business itself: more sales, fewer costs, and better culture through employee satisfaction and ultimately better market success for its stakeholders.

It is necessary to incorporate these goals into the strategic planning process and into the annual business plans. This will ensure that the new focus becomes part of the plan and does not compete with the well-established priorities for resources. Otherwise, the best-intended desired changes will fail.

Many leaders understand the meaning of strategic planning as it relates to the creation of the strategic plan and the financial goals and targets to be achieved. Often, they do not include the deployment of strategic "quality" goals, sub goals, and annual goals or the assignment of the resources and actions to achieve them. We will try to highlight this difference and use the term "strategic planning and deployment" throughout this chapter. Many organizations have overcome failures of change programs and have achieved long-lasting results through strategic deployment.

Six Sigma, Lean Six Sigma, and in prior years Total Quality Management all became pervasive change processes and were natural candidates for inclusion in the strategic plan of many organizations. The integration of these "quality and customer-driven" methods with strategic planning is important for their success.

Organizations have chosen different terms for this process. Some have used the Japanese term "hoshin kanri." Others have partially translated the term and called it "hoshin planning" (discussed in the following section). Still others have used a rough translation of the term and called it *policy deployment*. In an earlier version of the United States Malcolm Baldrige National Quality Award, this process was called *strategic quality planning*. Later this award criterion was renamed "strategic planning."

Whether the upper managers should align quality with the plan is a decision unique to each organization. What is decisive is the importance of integrating major change initiatives or quality programs into the strategic plan. The potential benefits of strategic planning and deployment are clear:

- The goals become clear—the planning process forces clarification of any vagueness
- The planning process then makes the goals achievable
- The monitoring process helps to ensure that the goals are reached
- Chronic wastes are "scheduled" to be reduced through the improvement process
- Creation of new focus on the customers and quality is attained as progress is made

Hoshin Kanri (Hoshin Planning)

The Hoshin Kanri developed in Japan integrated the work of W. Edwards Deming and Joseph M. Juran with Management by Objectives (MBO) elements out of Peter Drucker's, *The Practice of Management*. *The Balanced Scorecard*, by Robert S. Kaplan and David P. Norton reinforces the Plan, Do, Study, Act nature of the process by providing focus on the four key perspectives of Financial, Customer, Internal Processes, and Learning and Growth, along with the development of a scorecard that is used to track progress and support the governance system.

Roots of *Hoshin Kanri* may be traced to *A Book of Five Rings* written in 1645 by Miyamoto Musashi (Harris 1982). This book, the essence of which is captured by the word *heiho* or strategy, was a resource intended to provide instruction to samurai warriors, including instruction in what is perhaps the quintessential samurai skill—kendo, or precision swordsmanship. Relative to kendo, *A Book of Five Rings*, asserts that those thoroughly conversant with strategy will recognize the intentions of their enemies and through preparation and recognition will have many opportunities to cultivate and execute strategies capable of thwarting the objectives of their adversaries and positioning themselves to be victorious.

Like heiho, the word *hoshin* is comprised of two Chinese characters: ho—which means method or form, and shin, which is often translated as "shiny metal—the glint from the spear that leads the way" (Lee and Dale 1998) or, in a more contemporary form, an aim. When assembled, the word hoshin can be taken to mean "a methodology for strategic direction setting." The word *kanri* is commonly interpreted as "management" so that hoshin kanri becomes "management of the strategic direction setting process." Given this interpretation, in the West, hoshin kanri is commonly referred to as either policy deployment or strategy deployment or by the East / West hybrid term that we will henceforth use: hoshin planning.

Generally speaking, a given hoshin is mission and vision critical to an enterprise and is stated in terms of a goal or objective (a policy or a strategy) that is intended to elevate associated business processes and outcomes to a target performance level. The underlying structure of hoshin planning implies that it can be applied at essentially any level of the enterprise, ranging from the senior executive level to the day-to-day operational level.

Often, a high level (senior executive) hoshin is of such foundational importance to the enterprise that failure to attain or fulfill it within an appropriate timeframe will place the organization at risk. As such, a high level hoshin can be thought of as representing "big (enterprise) vision." Organizations that practice enterprise level hoshin planning ordinarily have a limited number of hoshins—typically three to five—that must be realized within a specified time span that, in the West, will ordinarily range from 1 to 5 years, with specified mileposts and periodic stage gate reviews along the way.

At the enterprise level, hoshin planning begins with "big vision" that is progressively unfolded by cascading the various hoshins from one level of the enterprise to the next to the next and so on—beginning with the executive level and ending with the operational level. Thus, from one level to the next to the next until the bottom of the waterfall an increasingly detailed scheme emerges. In this way hoshin planning begins with strategy or policy, is progressively transformed into plan, is progressively executed, leading to full strategy / policy implementation. Hoshin planning beginning at the operations level is executed in like manner, but with generally less far-reaching strategic implications and nearer-term fulfillment needs. In its high-level incarnation hoshin planning is highly strategic and focused on breakthrough improvement (Witcher 2003) whereas at the operations level it is ordinarily more incremental and focused on continuous improvement (Hutchins 2008).

Benefits of Hoshin Planning

We can conclude that a key benefit of hoshin planning is its ability to create consensus (Watson 2003) and facilitate enterprise alignment through significant workforce participation (Kondo 1998) which requires extensive communication both lateral and multi-level in nature. Such communication assures that each individual involved in the hoshin planning process is conversant with the "big goals and objectives" or hoshins of those both immediately before them (their direct supervisor) and immediately following them (their direct reports) as well with those of their immediate colleagues. This occurs because their own hoshins and related activities are driven by hoshins received from their direct supervisor and in turn inform the hoshins and related activities of their direct reports so that all involved in the process are familiar with three or more levels. This communication process is fundamentally a negotiated dialogue that is often referred to as "catchball" (Tennant and Roberts 2001) and "connects the planners and the doers" (Sussland 2002). Successful hoshin planning implementation is often associated with complementary and skilled use of effective performance management and measurement approaches such as the balanced scorecard (Kapland and Norton 1996; Witcher and Chau 2007). Together these approaches provide an exceptional means of rationally applying management of objectives as developed by the father of modern strategic management—Peter Drucker (Greenwood 1981).

The value of hoshin planning, as with most approaches, is bounded by the value and timeliness of the strategy or policy being deployed, not to mention the quality of the "plan" as it unfolds through the organization. Although hoshin planning may begin at any level of an organization and cascade downward through relevant other levels until sufficient execution is attained, we will provide the high level view that emerges by beginning at the senior executive level (CEO) of the enterprise.

Hoshin Kanri Policy Deployment

Figure 7-1 depicts hoshin planning from an enterprise perspective. To explain Fig. 7-1 we use the increasingly common scenario wherein organizations must produce not only acceptable financial performance and impacts to satisfy key stakeholders, but also socially equitable

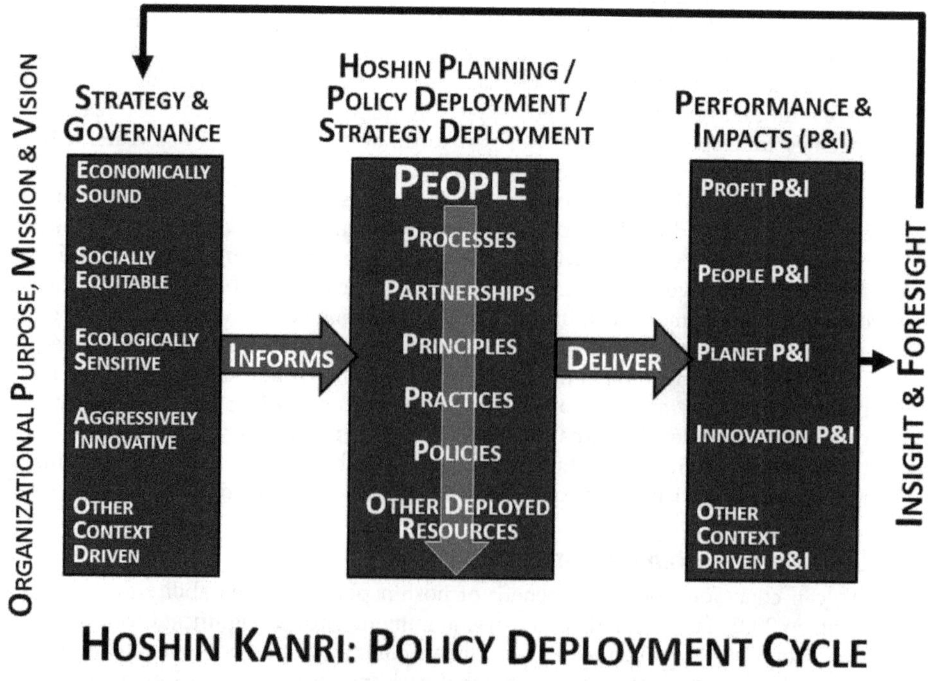

Figure 7.1 Hoshin planning from an enterprise perspective.

and environmentally sensitive performance. Impacts are demanded by citizens and regulatory agencies if not by our own consciences.

Examining Fig. 7-1 from left-to-right we see that most enterprises have a clearly defined purpose, mission and vision. The function of enterprise strategy and governance is to serve this purpose, vision and mission. Contemporary enterprises increasingly strive to be economically sound, socially equitable, and ecologically sensitive and hence formulate their strategy accordingly, with many organizations also needing to incorporate other context driven strategy elements such as being aggressively innovative in order to compete or to remain or become relevant in the marketplace. Although a typical organization will have numerous strategies, the Pareto Principle (Juran 2005) of separating the "vital few" (strategies) from the "trivial many" suggests that a few of these will be primary—that is—hoshins that populate the vital few, while the others will be relatively less important and will constitute the trivial many. Given the growth of triple top line approaches, and the importance of innovation, many organizations may have one or two hoshins that emerge from each of these categories.

Once executive level hoshins are determined, those executives will communicate these "what to" priorities on to the subsequent organizational level. Those responsible at the next level are provided with these hoshins or whats generally with little to no guidance as to "how to" fulfill them: determination of how is up to those at that level as is the selection of which hoshins are relevant to their span of influence. Those responsible at this next level will then determine the relevant how to elements and these become the hoshins or whats that are cascaded to the following level.

This process continues, with the hoshins or whats at one level translated into hows at the next level until the plan is fully elaborated, transforming in the process from "big vision" to "execution." Relative to Fig. 7-1, this process begins with strategy at the executive level seen on the left side of the Figure, and is unfolded through various levels—with people doing the unfolding through progressive translation of whats into hows into whats into hows …as represented by the center portion of Fig. 7-1, ultimately delivering performance and impacts along the way as seen in the rightmost box. The mechanisms of the transformation are portrayed in the center portion of Fig. 7-1: people, processes, partnerships, principles, practices, policies, and whatever other resources might be deployed / applied.

It is important to note that this is a living or cyclical process in that performance and impacts resulting from hoshin implementation are intended to provide both insight into recent enterprise performance and foresight into future enterprise priorities. Of course it is also important for the organization to be externally aware so that future priorities might be influenced by new, pending or likely legislation; by technological changes; by economic cycles; by emerging megatrends; or by other things not herein cited, but yet highly relevant to the enterprise's competitive landscape.

Hoshin kanri is known by many names, including policy deployment, strategy deployment, and hoshin planning. Originating in Japan, the primary intention of hoshin planning is to translate strategy into actions that ultimately yield relevant performance and impacts. A number of tools and methods are available to support this process but it is critical not to place undue focus on the tools, numerous variations and adaptations of which can be found. Equally, it is important not to "fall in love" with a given strategy and to recognize that there is no perfect strategy—only better and worse ones; relevant, less relevant, and irrelevant ones.

What Is Strategic Planning and Deployment?

Strategic planning is the systematic process by which an organization defines its long-term goals with respect to quality and customers, and integrates them—on an equal basis—with financial, human resources, marketing, and research and development goals into one cohesive business plan. The plan is then deployed throughout the entire organization.

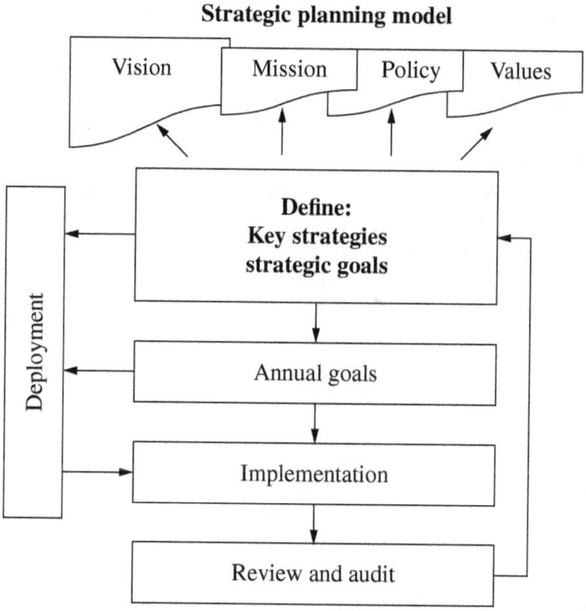

FIGURE 7.2 Strategic planning model.

As a component of an effective business management system, strategic planning enables an organization to plan and execute strategic organizational breakthroughs. Over the long term, the intended effect of such breakthroughs is to achieve competitive advantage or to attain a status of "quality leadership."

Strategic planning has evolved during the past decades to become an integral part of many organizational change processes, like Six Sigma or Operational Excellence (OpEx). It is now part of the foundation that supports the broader system of managing the business of the organization. A simple strategic planning and deployment model is shown in Fig. 7.2. This is what will be used throughout this chapter.

Strategic planning and deployment is also a key element of the U.S. Malcolm Baldrige National Quality Award and the European Foundation for Quality Management (EFQM) Award, as well as other international and state awards. The criteria for these awards stress that customer-driven quality and operational performance excellence are key strategic business issues, which need to be an integral part of overall business planning. A critical assessment of the Malcolm Baldrige National Quality Award winners demonstrates that those organizations that won the award outperformed those that did not (Fig. 7.3).

	1988–1996 investments	Value on 12/1/97	Percent change
All recipients	$7,496.54	$33,185.69	342
Standard & Poor's 500	$7,496.54	$18,613.28	148

Data: National institute of standards and technology.

FIGURE 7.3 Malcolm Baldrige National Quality Award winner performance. (*BusinessWeek*, March 16, 1998, p. 60.)

Quality is profitable, from 1995 to 2002, the "Baldrige Index" outperformed the S&P 500 stock index for 8 straight years, in certain years beating the S&P by wide margins of 4:1 or 5:1. The index was discontinued in 2004 when Baldrige began to recognize and award small businesses and educational entities along with their normal categories for National Quality Awards. The additions of smaller organizations skewed the "Baldrige Index," yet the results from the original study, when the playing fields were level, speak volumes; quality pays off.

Godfrey (1997) has observed that to be effective, strategic deployment should be used as a tool, a means to an end, not as the goal itself. It should be an endeavor that involves people throughout the organization. It must capture existing activities, not just add to already overflowing plates. It must help senior managers face difficult decisions, set priorities, and not just start new initiatives but also eliminate many current activities that add no value.

Strategic Planning Today

The approach used to establish organization-wide financial goals has evolved into a more robust strategic plan. To be effective in the global marketplace, large organizations must create a strategic plan that includes the elements discussed in the following sections.

Quality and Customer Loyalty Goals

These major goals are incorporated and supported by a hierarchy of goals at lower levels: subgoals, projects, etc. Improvement goals are goals aimed at creating a breakthrough in performance of a product, serving process, or people by focusing on the needs of customers, suppliers, and shareholders. The plan incorporates the "voice of the customer" and aligns them to the plan. This alignment enables the goals to be legitimate and balances the financial goals (which are important to shareholders) with those of importance to the customers. It also eliminates the concern that there are two plans, one for finance and one for quality.

A systematic, structured methodology for establishing annual goals and providing resources must include the following:

- *A provision of rewards.* Performance against improvement goals is given substantial weight in the system of merit rating and recognition. A change in the structure is required to include rewarding the right behaviors.
- *Required and universal participation.* The goals, reports, reviews, etc. are designed to gain participation from within the organization's hierarchy. This participation involves every employee at every level, providing support for the change initiative and helping achieve the desired results.
- *A common language.* Key terms, such as quality, benchmarking, and strategic quality deployment, acquire standard meanings so that communication becomes more and more precise.
- *Training.* It is common for all employees to undergo training in various concepts, processes, methods, tools, etc. Organizations that have so trained their workforce, in all functions, at all levels, and at the right time, are well poised to outperform organizations in which such training has been confined to the quality department or managers.

Why Strategic Deployment? The Benefits

The first question that often arises in the beginning stages of strategic planning in an organization is, Why do strategic planning in the first place? To answer this question requires a look at the benefits other organizations have realized from strategic planning. They report that it

- Focuses the organization's resources on the activities that are essential to increasing customer satisfaction, lowering costs, and increasing shareholder value (see Fig. 7.3)
- Creates a planning and implementation system that is responsive, flexible, and disciplined
- Encourages interdepartmental cooperation
- Provides a method to execute breakthroughs year after year
- Empowers leaders, managers, and employees by providing them with the resources to carry out the planned initiatives
- Eliminates unnecessary and wasteful initiatives that are not in the plan
- Eliminates the existence of many potentially conflicting plans—the finance plan, the marketing plan, the technology plan, and the quality plan
- Focuses resources to ensure financial plans are achievable

Why Strategic Deployment? The Risks

Different organizations have tried to implement total quality management systems as well as other change management systems. Some organizations have achieved stunning results; others have been disappointed by their results, often achieving little in the way of bottom-line savings or increased customer satisfaction. Some of these efforts have been classified as failures. One of the primary causes of these disappointments has been the inability to incorporate these "quality programs" into the business plans of the organization.

Other reasons for failure are include the following:

- Strategic planning was assigned to planning departments, not to the upper managers themselves. These planners lacked training in concepts and methods, and were not among the decision makers in the organization. This led to a strategic plan that did not include improvement goals aimed at customer satisfaction, process improvement, etc.
- Individual departments had been pursuing their own departmental goals, failing to integrate them with the overall organizational goals.
- New products or services continued to be designed with failures from prior designs that were carried over into new models, year after year. The new designs were not evaluated or improved and hence, were not customer driven.
- Projects suffered delays and waste due to inadequate participation and ended before positive business results were achieved.
- Improvement goals were assumed to apply only to manufactured goods and manufacturing processes. Customers became irritated not only by receipt of defective goods; receiving incorrect invoices and late deliveries also irritated them. The business processes that produce invoices and deliveries were not subject to modern quality planning and improvement because there were no such goals in the annual plan to do so.

The deficiencies of the past strategic planning processes had their origin in the lack of a systematic, structured approach to integrate programs into one plan. As more organizations became familiar with strategic quality deployment, many adopted its techniques, which treat managing for change on the same organization wide basis as managing for finance.

Launching Strategic Planning and Deployment

The Strategic Deployment Process

The strategic deployment process requires that the organization incorporate customer focus into the organization's vision, mission, values, policies, strategies, and long- and short-term goals and projects. Projects are the day-to-day, month-to-month activities that link quality improvement activities, reengineering efforts, and quality planning teams to the organization's business objectives. Juran (1988) has stated, "You need participation by the people that are going to be impacted, not just in the execution of the plan but in the planning itself. You have to be able to go slow, no surprises, use test sites in order to get an understanding of what are some things that are damaging and correct them."

The elements needed to establish strategic deployment are generally alike for all organizations. However, each organization's uniqueness will determine the sequence and pace of application and the extent to which additional elements must be provided.

There exists an abundance of jargon used to communicate the strategic deployment process. Depending on the organization, one may use different terms to describe similar concepts. For example, what one organization calls a vision; another organization may call a mission (Fig. 7.4).

The following definitions of elements of strategic planning are in widespread use and are used in this chapter:

Vision. A desired future state of the organization or enterprise. Imagination and inspiration are important components of a vision. Typically, a vision can be viewed as the ultimate goal of the organization, one that may take 5 or even 10 years to achieve.

Mission. The purpose of or the reason for the organization's existence and usually states, for example, what we do and whom we serve.

Selected definitions	
Mission	What business we are in
Vision	Desired future state of organization
Values	Principles to be observed to meet vision or principle to be served by meeting vision
Policy	How we will operate and our commitment to customers and society

Figure 7.4 Organizational vision and mission. (Juran Institute, Inc.)

- The presence of JetBlue at JFK International is unmatched. Measured by number of passengers booked, JetBlue carries almost the equivalent of every other airline conducting business at JFK. With their entrenchment in the United States' largest travel market, JetBlue ensures itself profitability even in difficult markets. "Our mission is to bring humanity back to air travel."

Strategies. The means to achieve the vision. Strategies are few and define the key success factors, such as price, value, technology, market share, and culture that the organization must pursue. Strategies are sometimes referred to as "key objectives" or "long-term goals."

Annual goals. The aim or end to which work effort is directed. Goals are referred to as "long term" (2 to 3 years) and "short term" (1 to 2 years). Achievement of goals signals the successful execution of the strategy.

- JetBlue aims to preserve the core JetBlue experience of unique, low-cost, high-quality flights while adding optional product offerings for all customers.

Ethics and values. The beliefs of the organization.

- For the fourth year in a row, JetBlue was ranked number one in customer service for low-cost carriers by J.D. Power & Associates. It is this exceptional customer service that continues to drive JetBlue and set it apart. Partnerships with Sirius XM, and Direct TV, and improved legroom all make the flight experience for every customer a more enjoyable experience.

Policies. A guide to managerial action. An organization may have policies in a number of areas: quality, environment, safety, human resources, etc. These policies guide day-to-day decision making.

Initiatives and projects. An initiative or project implies assignment of selected individuals to a team, which is given the responsibility, tools, and authority to achieve the specific goal or goals. These should be multifunctional teams launched to address a deployed goal, and whose successful completion ensures that the strategic goals are achieved.

- After 6 years of planning and 3 years of construction, JetBlue's Terminal 5 opened at JFK. Terminal 5 offers JetBlue customers their own parking lot and road for improved access to the airliner. It comprises 26 gates, affords the highest in modern amenities and concession offerings, and due to its proximity to the runway allows JetBlue to be more efficient in their processes. Terminal 5 advances the company's stake in the New York travel market.

Deployment plan. The vision must be broken apart and translated into successively smaller and more specific parts—key strategies, strategic goals, etc.—to the level of projects and even departmental actions. The detailed plan for decomposition and distribution throughout the organization is called the *deployment plan*. It includes the assignment of roles and responsibilities, and the identification of resources needed to implement and achieve the project goals (Fig. 7.5).

Scorecards and key performance indicators. Measurements that are visible throughout the organization for evaluating the degree to which the strategic plan is being achieved.

- By the end of 2008, JetBlue was the seventh largest passenger carrier in the United States and conducted 600 flights daily.

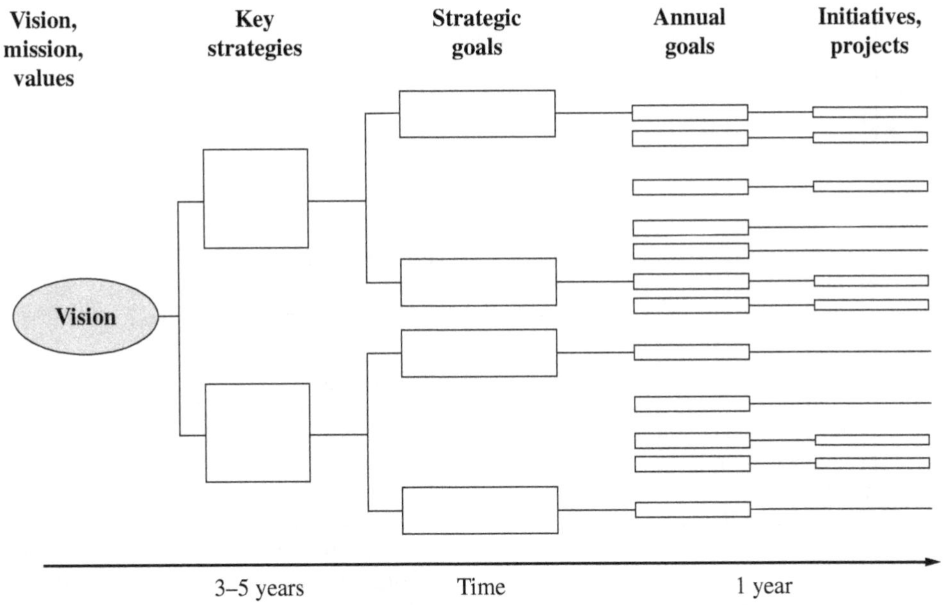

FIGURE 7.5 Deploying the vision. (Juran Institute, Inc.)

Developing the Elements of Strategic Planning and Deployment

Establish a Vision

Strategic deployment begins with a vision that is customer-focused. In the organizations we know that are successfully making the transition to a more collaborative organization, the key to success is developing and living by a common strategic vision. When you agree on an overall direction, you can be flexible about the means to achieve it (Tregoe and Tobia 1990).

Really powerful visions are simply told. The Ten Commandments, the Declaration of Independence, and a Winston Churchill World War II speech—all present messages that are so simple and direct you can almost touch them. Our corporate strategies should be equally compelling.

A vision should define the benefits a customer, an employee, a shareholder, or society at large can expect from the organization:

Here are a few examples:

- Samsung, the world's largest manufacturer of high-quality digital products is guided by a singular vision: "to lead the digital convergence movement."

  Samsung believes that through technology innovation today, we will find the solutions we need to address the challenges of tomorrow. From technology comes opportunity—for businesses to grow, for citizens in emerging markets to prosper by tapping into the digital economy, and for people to invent new possibilities. It's our aim to develop innovative technologies and efficient processes that create new markets, enrich people's lives, and continue to make Samsung a trusted market leader.

- Sentara Health (based in the mid-Atlantic states): We have commitment to grow as one of the nation's leading health care organizations by creating innovative systems of care that help people achieve and maintain their best possible state of health.
- Kaiser Permanente (a large U.S.-based health care system): We are committed to providing our members with quality, cost-effective health care. Our physicians and managers work together to improve care, service, and the overall performance of our organization.

Each of the preceding visions offers a very different view of the direction and character of the organization. Each conveys a general image to customers and employees of where the organization is headed. For the organization, the vision provides, often for the first time in its history, a clear picture of where it is headed and why it is going there.

Good vision statements should also be compelling and shared throughout the organization. It is often a good idea to make the vision a stretch for the organization but possible of being achieved within 3 to 5 years, and to state a measurable achievement (e.g., being the best). In creating the vision, organizations should take into account its customers, the markets in which it wants to compete, the environment within which the organization operates, and the current state of the organization's culture.

Vision statements, by themselves, are little more than words. Publication of such a statement does not inform the members of an organization what they should do differently from what they have done in the past. The deployment process and the strategic plan become the basis for making the vision a reality. The words of the vision are just a reminder of what the organization is pursuing. The vision must be carried out through deeds and actions.

Some common pitfalls when forming a vision include

- Focusing the vision exclusively on shareholders as customers
- Thinking that once a strategic plan is written it will be carried out with no further work
- Failing to explain the vision as a benefit to customers, employees, suppliers, and other stakeholders
- Creating a vision that is either too easy or too difficult to achieve
- Failing to consider the effects of rapid changes taking place in the global economy will have 3 to 5 years into the future
- Failing to involve key employees at all levels in creating the vision
- Failing to benchmark competitors or to consider all possible sources of information on future needs, internal capabilities, and external trends

Agree on Your Mission

Most organizations have a mission statement. A mission statement is designed to address the question, What business(es) are we in? A mission is often confused with a vision and even published as one. A mission statement should clarify the organization's purpose or reason for existence. It helps clarify what your organization is.

The following are some examples:

- **Samsung:** Everything we do at Samsung is guided by our mission: to be the best "digital-e-company."
- **Amazon.com:** Our vision is to be earth's most customer-centric company, to build a place where people can come find and discover anything they might want to buy online.

- **Dell:** To be the most successful computer company in the world and delivering the best customer experience in the markets they share.
- **eBay:** Pioneers communities built on commerce, sustained by trust, and inspired by opportunity. eBay brings together millions of people every day on a local, national, and international basis through an array of websites that focus on commerce, payments, and communications.
- **Facebook:** Is a social utility that helps people communicate more efficiently with their friends, family members, and coworkers. The company develops technologies that facilitate the sharing of information through the social graph, the digital mapping of people's real-world social connections. Anyone can sign up for Facebook and interact with the people they know in a trusted environment.
- **Google:** It's mission is to organize the world's information and make it universally accessible and useful.
- **The Ritz-Carlton Hotel:** Is a place where the genuine care and comfort of our guests is our highest mission.
- **Sentara Health:** We will focus, plan, and act on our commitments to our community mission, to our customers, and to the highest quality standards of health care to achieve our vision for the future.

In the Sentara example, the references to leadership and the future may lead the reader to confuse this mission statement (what business we are in) with a vision statement (what we aim to become). Only the organization itself can decide whether these words belong in its mission statement. It is in debating such points that an organization comes to a consensus on its vision and mission.

Together, a vision and a mission provide a common agreed-upon direction for the entire organization. This direction can be used as a basis for daily decision making.

Develop Long-Term Strategies or Goals

The first step in converting the vision into an achievable plan is to break the vision into a small number of key strategies (usually four or five). Key strategies represent the most fundamental choices that the organization will make about how it will go about reaching its vision. Each must contribute significantly to the overall vision. For example,

- Xerox initiated their Leadership Through Quality program as part of a broader corporate focus on quality. More than 100,000 employees were trained over a 3-year period in a six-step process. Empowered employees started a number of initiatives, many involving environmental and quality improvements, yielding millions of dollars in added profits each year. Xerox management credits the success of new environmental initiatives primarily to employees using quality management practices. Cross-function teams are formed to focus *on a variety of issues.*

Responsibility for executing these key strategies is distributed (or deployed) to key executives within the organization, the first step in a succession of subdivisions and deployments by which the vision is converted into action.

In order to determine what the key strategies should be, one may need to assess five areas of the organization and obtain the necessary data on

- Customer loyalty and customer satisfaction
- Costs related to poor quality or products, services, and processes

- Organization culture and employee satisfaction
- Internal business processes (including suppliers)
- Competitive benchmarking

Each of these areas, when assessed, can form the basis for a balanced business scorecard. Data must be analyzed to discover specific strengths, weaknesses, opportunities, and threats as they relate to customers, quality, and costs. Once complete, the key strategies can be created or modified to reflect measurable and observable long-term goals.

Develop Annual Goals

An organization sets specific, measurable strategic goals that must be achieved for the broad strategy to be a success. These quantitative goals will guide the organization's efforts toward achieving each strategy. As used here, a goal is an aimed-at target. A goal must be specific. It must be quantifiable (measurable) and is to be met within a specific period. At first, an organization may not know how specific the goal should be. Over time, the measurement systems will improve and the goal setting will become more specific and more measurable.

Despite the uniqueness of specific industries and organizations, certain goals are widely applicable. There are seven areas that are minimally required to ensure that the proper goals are established. They are

Product performance. Goals in this area relate to product features that determine response to customer needs, for example, promptness of service, fuel consumption, mean time between failures, and courteousness. These product features directly influence product salability and affect revenues.

Competitive performance. This has always been a goal in market-based economies, but seldom a part of the business plan. The trend to make competitive performance a long-term business goal is recent but irreversible. It differs from other goals in that it sets the target relative to the competition, which, in a global economy, is a rapidly moving target. For example, all of our products will be considered the "best in class" within 1 year of introduction as compared to products of the top five competitors.

Business improvement. Goals in this area may be aimed at improving product deficiencies or process failures, or reducing the cost of poor quality waste in the system. Improvement goals are deployed through a formal structure of quality improvement projects with assignment of associated responsibilities. Collectively, these projects focus on reducing deficiencies in the organization, thereby leading to improved performance.

Cost of poor quality. Goals related to quality improvement usually include a goal of reducing the costs due to poor quality or waste in the processes. These costs are not known with precision, though they are estimated to be very high. Nevertheless, it is feasible, through estimates, to bring this goal into the business plan and to deploy it successfully to lower levels. A typical goal is to reduce the cost of poor quality by 50 percent each year for 3 years.

Performance of business processes. Goals in this area have only recently entered the strategic business plan. These goals relate to the performance of major processes that are multifunctional in nature, for example, new product development, supply-chain management, and information technology, and subprocesses, such as accounts receivable and purchasing. For such macro processes, a special problem is to decide who should have the responsibility for meeting the goal? We discuss this later under "Deployment to Whom?"

Customer satisfaction. Setting specific goals for customer satisfaction helps keep the organization focused on the customer. Clearly, deployment of these goals requires a good deal of sound data on the current level of satisfaction/dissatisfaction and what factors will contribute to increasing satisfaction and removing dissatisfaction. If the customers' most important needs are known, the organization's strategies can be altered to meet those needs most effectively.

Customer loyalty and retention. Beyond direct measurement of customer satisfaction, it is even more useful to understand the concept of customer loyalty. Customer loyalty is a measure of customer purchasing behavior between customer and supplier. A customer who buys solely from a single supplier is said to display a loyalty with respect to A of 100 percent. A study of loyalty opens the organization to a better understanding of product salability from the customer's viewpoint and provides the incentive to determine how to better satisfy customer needs. The organization can benchmark to discover the competition's performance, and then set goals to exceed that performance (Fig. 7.6).

The goals selected for the annual business plan are chosen from a list of nominations made by all levels of the hierarchy. Only a few of these nominations will survive the screening process and end up as part of the organization wide business plan. Other nominations may instead enter the business plans at lower levels in the organization. Many nominations will be deferred because they fail to attract the necessary priority and, therefore, will get no organization resources.

Upper managers should become an important source of nominations for strategic goals, since they receive important inputs from sources such as membership on the executive council, contacts with customers, periodic reviews of business performance, contacts with upper managers in other organizations, shareholders, and employee complaints.

Goals that affect product salability and revenue generation should be based primarily on meeting or exceeding marketplace quality. Some of these goals relate to projects that have a long lead time, for example, a new product development involving a cycle time of several years, computerizing a major business process, or a large construction project that will not be commissioned for several years. In such cases, the goal should be set so as to meet the competition estimated to be prevailing when these projects are completed, thereby "leap-frogging" the competition.

Product performance (customer focus). This relates to performance features that determine response to customer needs, such as promptness of service, fuel consumption, MTBF, and courtesy. (Product includes goods and services.)

Competitive performance. Meeting or exceeding competitive performance has always been a goal. What is new is putting it into the business plan.

Performance improvement. This is a new goal. It is mandated by the fact that the rate of quality improvement decides who will be the quality leader of the future.

Reducing the cost of poor quality. The goal here relates to being competitive as to costs. The measures of cost of poor quality must be based on estimates.

Performance of business processes. This relates to the performance of major multifunctional processes such as billing, purchasing, and launching new products.

Figure 7.6 Quality goals in the business plan. (Juran Institute, Inc.)

In industries that are natural monopolies (e.g., certain utilities), the organizations often are able to make comparisons through use of industry databanks. In some organizations there is internal competition as well—the performances of regional branches are compared with each other.

Some internal departments may also be internal monopolies. However, most internal monopolies have potential competitors—outside suppliers who offer the same services. The performance of the internal supplier can be compared with the proposals offered by an outside supplier.

A third and widely used basis for setting goals has been historical performance. For some products and processes, the historical basis is an aid to needed stability. For other cases, notably those involving high chronic costs of poor quality, the historical basis has done a lot of damage by helping to perpetuate a chronically wasteful performance. During the goal-setting process, upper managers should be on the alert for such misuse of the historical data. Goals for chronically high cost of poor quality should be based on planned breakthroughs using the breakthrough improvement process.

Articulate Ethics and Values

Corporate values reflect an organization's culture.

Simply said: "Culture is a set of habits and beliefs that a group of people have in common—for example, facing similar questions and problems because they operate in a similar business. This is the internal element of culture; the other one is external-oriented: In what environment does this group of people operate and how does this affect them? How do they interact with the environment?" (Bool 2008).

Social responsibility is obviously a value that is focused on this second element: the interaction of the group (corporation) with its environment.

Some organizations create value statements to further define themselves. Values are what an organization stands for and believe in. A list of values must be supported with actions and deeds from management, lest its publication create cynicism in the organization. Training and communication of values for all employees becomes a prerequisite to participation in the planning process. Organization-published values are policies that must be changed to support the values of the organization.

Samsung's value statements are an example of this:

- We will devote our human resources and technology to create superior products and services, thereby contributing to a better global society.
- Our management philosophy represents our strong determination to contribute directly to the prosperity of people all over the world. The talent, creativity, and dedication of our people are key factors to our efforts, and the strides we have made in technology offer endless possibilities to achieving higher standards of living everywhere.
- At Samsung we believe that the success of our contributions to society and to the mutual prosperity of people across national boundaries truly depends on how we manage our company.
- Our goal is to create the future with our customers.

Communicate Organization Policies

Policy declarations are a necessity during a period of major change, and organizations have acted accordingly. Since the 1980s we have seen an unprecedented surge of activity in publishing "quality policies." "Policy" as used here is a guide to managerial action. Published policy statements are the result of a good deal of deliberation by management, followed by approval at the highest level. The senior executive team or quality council plays a prominent role in this process.

While the details vary, the published policies have much in common from organization to organization. For instance, most published quality policies declare the intention to meet the needs of customers. The wording often includes identification of specific needs to be met, for example, "The organization's products should provide customer satisfaction."

Most published policies include language relative to competitiveness in quality, for example, "Our organization's products shall equal or exceed the competition."

A third frequent area of published quality policy relates to quality improvement, declaring, for example, the intention to conduct improvement annually.

Some quality policy statements include specific reference to internal customers or indicate that the improvement effort should extend to all phases of the business. For example,

> Helix Energy Solutions Group, Inc., (Helix) is fully committed to being the leading provider of select life-of-field solutions. A primary goal of Helix is to achieve the highest standards of quality in all business units' practices and operations without compromise. Our objective is to continually improve our organization performance, while offering our customers a safe, cost-effective, and professional service.

Enforcement of policies is a new problem due to the relative newness of documented quality policies. In some organizations, provision is made for independent review of adherence to policies. ISO 9000, the international standard for quality assurance, requires a quality policy as a declaration of intent to meet the needs of customers. An audit process is mandated to ensure that the policy is carried out.

Leadership

A fundamental step in the establishment of any strategic plan is the participation of upper management acting as an executive "council." Membership typically consists of the key executives. Top-level management must come together as a team to determine and agree upon the strategic direction of the organization. The council is formed to oversee and coordinate all strategic activities aimed at achieving the strategic plan. The council is responsible for executing the strategic business plan and monitoring the key performance indicators. At the highest level of the organization, an executive council should meet monthly or quarterly.

The executives are responsible for ensuring that all business units have a similar council at the subordinate levels of the organization. In such cases, the councils are interlocked, that is, members of upper-level councils serve as chairpersons for lower-level councils.

If a council is not in place, the organization should create one. In a global organization, processes are too complex to be managed functionally. A council ensures a multifunctional team working together to maximize process efficiency and effectiveness. Although this may sound easy, in practice it is not. The senior management team members may not want to give up the monopolies they have enjoyed in the past. For instance, the manager of sales and marketing is accustomed to defining customer needs, the manager of engineering is accustomed to sole responsibility for creating products, and the manager of manufacturing has enjoyed free rein in producing products. In the short run, these managers may not easily give up their monopolies to become team players. The creation of a council would help ensure better collaboration among the various divisions of the organization.

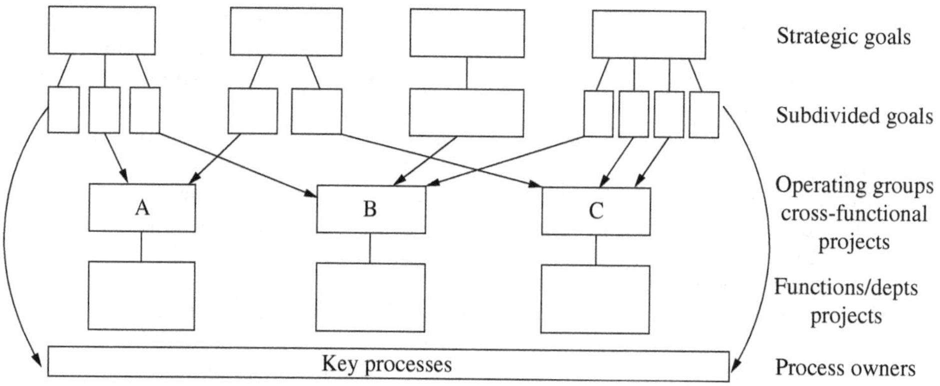

Figure 7.7 Deployment of strategic goals. (Juran Institute, Inc.)

Deployment of Goals

The deployment of long- and short-term goals is the conversion of goals into operational plans and projects. "Deployment" as used here means subdividing the goals and allocating the sub goals to lower levels. This conversion requires careful attention to such details as the actions needed to meet these goals, who is to take these actions, the resources needed, and the planned timetables and milestones. Successful deployment requires establishment of an infrastructure for managing the plan. Goals are deployed to multifunctional teams, functions, and individuals (Fig. 7.7).

Subdividing the Goals

Once the strategic goals have been agreed upon, they must be subdivided and communicated to lower levels. The deployment process also includes dividing up broad goals into manageable pieces (short-term goals or projects). For example,

- An airline's goal of attaining 99 percent on-time arrivals may require specific short-term (8 to 12 months) initiatives to deal with such matters as
  - The policy of delaying departures in order to accommodate delayed connecting flights
  - The decision making of gate agents at departure gates
  - The availability of equipment to clean the plane
  - The need for revisions in departmental procedures to clean the plane
  - The state of employee behavior and awareness
- A hospital's goal of improving the health status of the communities they serve may require initiatives that
  - Reduce incidence of preventable disease and illness
  - Improve patient access to care
  - Improve the management of chronic disease conditions
  - Develop new services and programs in response to community needs

Such deployment accomplishes some essential purposes:

- The subdivision continues until it identifies specific deeds to be done.
- The allocation continues until it assigns specific responsibility for doing the specific deeds.

Those who are assigned responsibility respond by determining the resources needed and communicating this to higher levels. Many times, the council must define specific projects, complete with team charters and team members, to ensure goals are met (Fig. 7.8). (For more on the improvement process, see Chapter 5, Quality Improvement and Breakthrough Performance.)

Deployment to Whom?

The deployment process starts by identifying the needs of the organization and the upper managers. Those needs determine what deeds are required. The deployment process leads to an optimum set of goals through consideration of the resources required. The specific projects to be carried out address the subdivided goals. For example, in the early 1980s, the goal of having the newly designed Ford Taurus/Sable become "Best in Class" was divided into more than 400 specific sub goals, each related to a specific product feature. The total planning effort was enormous and required more than 1500 project teams.

To some degree, deployment can follow hierarchical lines, such as corporate to division and division to function. However, this simple arrangement fails when goals relate to cross-functional business processes and problems that affect customers.

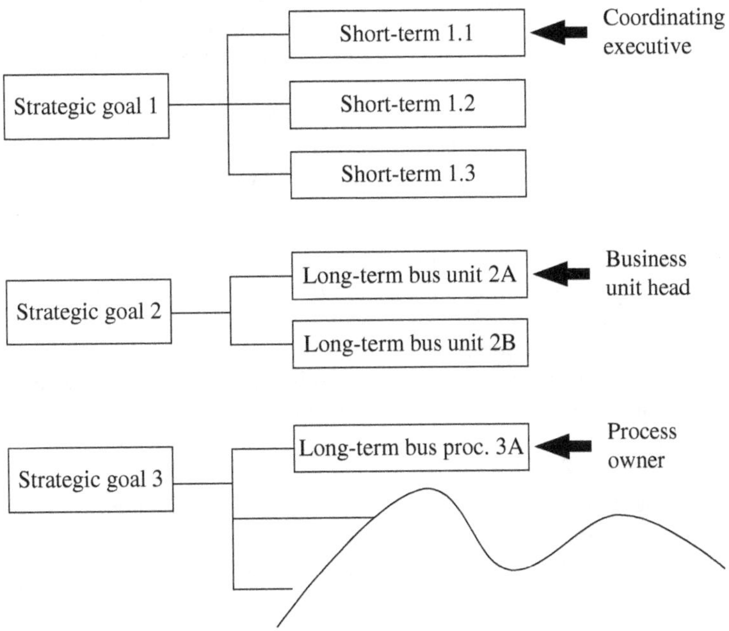

FIGURE 7.8 Sub goals. (Juran Institute, Inc.)

Major activities of organizations are carried out by the use of interconnecting networks of business processes. Each business process is a multifunctional system consisting of a series of sequential operations. Since it is multifunctional, the process has no single "owner"; hence, there is no obvious answer to the question, Deployment to whom? Deployment is thus made to multifunctional teams. At the conclusion of the team project, an owner is identified. The owner (who may be more than one person) then monitors and maintains this business process.

Communicating the Plan: "Catch Ball"

Once the goals have been established, the goals are communicated to the appropriate organization units. In effect, the executive leadership asks their top management, What do you need to support this goal? The managers at this level discuss the goal and ask their subordinates a similar question, and so on. The responses are summarized and passed back up to the executives. This process may be repeated several times until there is general satisfaction with the final plan.

This two-way communication process is called "catch ball," a term coined by the Japanese. Catch ball includes the following:

- Clear communication of what top management proposes as the key focus areas of the strategic plan for the coming business year
- Identification and nomination by managers at various lower levels of other areas for organization attention
- Decisions as to what departments and functions should do about the areas that have been identified in the plan

This two-way communication requires that the recipients be trained in how to respond. The most useful training is prior experience in quality improvement. Feedback from organizations using catch ball suggests that it outperforms the process of unilateral goal setting by upper managers.

For example, Boeing Aerospace Systems has been very successful in introducing its strategic quality plan, with its mission, vision, key strategies, and strategic goals. To review and refine mission statements, strategies, and the overall vision of the organization, Boeing conducts yearly assessments drawing from customer satisfaction assessments, human resource assessments, supplier assessments, risk assessments, and financial assessments. By essentially taking feedback from all facets of their business (customers, workforce, suppliers, community, and shareholders), Boeing is able to enact improved implementation plans and better manage their allocation of resources. The identification of needs within the infrastructure, addressing problems within the culture/training of the workforce, and modifying institutionalized processes for the better are all results of continually deploying assessments and consistent communication between the management and its workforce.

A Useful Tool for Deployment

The tree diagram is a graphic tool that aids in the deployment process (see Fig. 7.8). It displays the hierarchical relationship of the goals, long-term goals, short-term goals, and projects, and indicates where each is assigned in the organization. A tree diagram is useful in visualizing the relationship between goals and objectives or teams and goals. It also provides a visual way to determine if all goals are supported.

Measure Progress with KPI

There are several reasons why measurement of performance is necessary and why there should be an organized approach to it:

- Performance measures indicate the degree of accomplishment of objectives and, therefore, quantify progress toward the attainment of goals.
- Performance measures are needed to monitor the continuous improvement process, which is central to the changes required to become competitive.
- Measures of individual, team, and business unit performance are required for periodic performance reviews by management.

Once goals have been set and broken down into sub goals, key measures (performance indicators) need to be established. A measurement system that clearly monitors performance against plans has the following properties:

- Indicators that link strongly to strategic goals and to the vision and mission of the organization.
- Indicators that include customer concerns; that is, the measures focus on the needs and requirements of internal and external customers.
- A small number of key measures of key processes that can be easily obtained on a timely basis for executive decision-making.
- The identification of chronic waste or cost of poor quality.

For example, Poudre Valley Health Systems (PVHS) established measures of their processes early in the implementation of their business plan and were able to monitor and quantify the following:

- Improve and maintain employee satisfaction to the top 10 percent of vacancy rate in all U.S. organizations.
- Strengthen overall service area market share by establishing market strategies specific to service area needs. By breaking down service areas to primary/local and total/national market shares, PVHS aims to control 65 percent of their primary market share and 31.8 percent of total market share by 2012.
- Support facility development by opening a cancer center.
- Enhance physician relations by initiating a physician engagement survey tool and reaching a goal of 80 percent satisfaction.
- Strengthen the company's financial position by achieving a financial flexibility unit of 11 and meeting a 5-year plan.

The best measures of the implementation of the strategic planning process are simple, quantitative, and graphical. A basic spreadsheet that describes the key measures and how they will be implemented is shown in Fig. 7.9. It is simply a method to monitor the measures.

As goals are set and deployed, the means to achieve them at each level must be analyzed to ensure that they satisfy the objective that they support. Then the proposed resource

Annual quality goals	Specific measurements	Frequency	Format	Data source	Name

FIGURE 7.9 Measurement of quality goals. (Juran Institute, Inc.)

expenditure must be compared with the proposed result and the benefit/cost ratio assessed. Examples of such measures are

- Financial results
  - Gains
  - Investment
  - Return on investment
- People development
  - Trained
  - Active on project teams
- Number of projects
  - Undertaken
  - In process
  - Completed
  - Aborted
- New product or service development
  - Number or percentage of successful product launches
  - Return on investment of new product development effort
  - Cost of developing a product versus the cost of the product it replaces
  - Percent of revenue attributable to new products
  - Percent of market share gain attributable to products launched during the last 2 years
  - Percent of on-time product launches
  - Cost of poor quality associated with new product development
  - Number of engineering changes in the first 12 months of introduction

- Supply-chain management
  - Manufacturing lead times—fill rates
  - Inventory turnover
  - Percent on-time delivery
  - First-pass yield
  - Cost of poor quality

The following is an example of measures that one bank used to monitor teller quality:

- Speed
  - Number of customers in the queue
  - Amount of time in the queue (timeliness)
  - Time per transaction
  - Turnaround time for no-wait or mail transactions
- Accuracy
  - Teller differences in adding up the money at the end of the day
  - Amount charged off/amount handled

Once the measurement system is in place, it must be reviewed periodically to ensure that goals are being met.

Reviewing Progress

A formal, efficient review process will increase the probability of reaching the goals. When planning actions, an organization should look at the gaps between measurement of the current state and the target it is seeking. The review process looks at gaps between what has been achieved and the target (Fig. 7.10).

Frequent measurements of strategic deployment progress displayed in graphic form help identify the gaps in need of attention. Success in closing those gaps depends on a formal

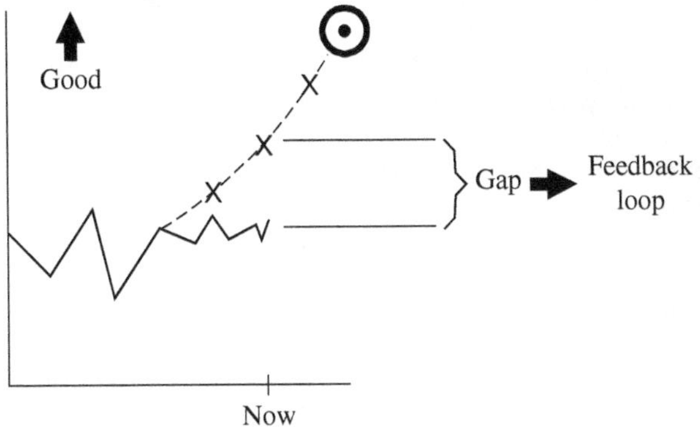

FIGURE 7.10 Review. (Juran Institute, Inc.)

Projects	Project leaders	Baseline measurements	Targets	Initial plan	Review points				Review leader
					Resources	Analysis	Plan	Results	

FIGURE 7.11 Progress review plan. (Juran Institute, Inc.)

feedback loop with clear responsibility and authority for acting on those differences. In addition to the review of results, progress reviews are needed for projects under way to identify potential problems before it is too late to take effective action. Every project should have specific, planned review points, much like those in Fig. 7.11.

Organizations today include key performance indicators as discussed in the following sections.

Product and Service Performance

There may be several product or service features. For the great majority of product features, there exist performance metrics and technological sensors to provide objective product evaluation.

Competitive Quality

These metrics relate to those qualities that influence product salability—for example, promptness of service, responsiveness, courtesy of pre-sale and after-sale service, and order fulfillment accuracy. For automobiles, qualities include top speed, acceleration, braking distance, and safety. For some product features, the needed data must be acquired from customers through negotiation, persuasion, or purchase. For other product features, it is feasible to secure the data through laboratory tests. In still other cases, it is necessary to conduct market research.

Trends must now be studied so that goals for new products can be set to correspond to the state of competition anticipated at the time of launch.

Some organizations operate as natural monopolies, for example, regional public utilities. In such cases, the industry association gathers and publishes performance data. In the case of internal monopolies (e.g., payroll preparation, transportation), it is sometimes feasible to secure competitive information from organizations that offer similar services for sale.

Performance on Improvement

This evaluation is important to organizations that go into quality improvement on a project-by-project basis. Due to lack of commonality among the projects, collective evaluation is limited to the summary of such features as

- *Number of projects.* Undertaken, in process, completed, and aborted.
- *Financial results.* Amounts gained, amounts invested, returns on investment.
- *Persons involved as project team members.* Note that a key measure is the proportion of the organization's management team that is actually involved in improvement projects. Ideally, this proportion should be over 90 percent. In the great majority of organizations, the actual proportion has been less than 10 percent.

Cost of Poor Quality

As stated in prior chapters, the "cost of poor quality" is those costs that would disappear if our products and processes were perfect and generated no waste. Those costs are huge. Our research indicates that 15 to 25 percent of all work performed consisted of redoing prior work because products and processes were not perfect.

The costs are not known with precision. In most organizations, the accounting system provides only a minority of the information needed to quantify this cost of poor quality. It takes a great deal of time and effort to extend the accounting system so as to provide full coverage. Most organizations have concluded that such effort is not cost-effective.

The gap can be filled somewhat by estimates that provide upper managers with approximate information as to the total cost of poor quality and the major areas of concentration. These areas of concentration then become the target for quality improvement projects. Thereafter, the completed projects provide fairly precise figures on quality costs before and after the improvements.

Product and Process Failures

Even though the accounting system does not provide for evaluating the cost of poor quality, much evaluation is available through measures of product and process deficiencies, either in natural units of measure or in money equivalents—for example, cost of poor quality per dollar of sales, dollar of cost of sales, hour of work, or unit shipped. Most measures lend themselves to summation at progressively higher levels. This feature enables goals in identical units of measure to be set at multiple levels: corporate, division, and department.

Performance of Business Processes

Despite the wide prevalence and importance of business processes, they have been only recently controlled as to performance. A contributing factor is their multifunctional nature. There is no obvious owner and hence, no clear, sole responsibility for their performance. Responsibility is clear only for the subordinate micro processes. The system of upper management controls must include control of the macro processes. That requires establishing goals in terms of cycle times, deficiencies, etc., and the means for evaluating performances against those goals.

The Scorecard

To enable upper managers to "know the score" relative to achieving strategic quality deployment, it is necessary to design a report package, or scorecard. In effect, the strategic plan dictates the choice of subjects and identifies the measures needed on the upper management scorecard.

The scorecard should consist of several conventional components:

- Key performance indicators (at the highest levels of the organization)
- Quantitative reports on performance, based on data
- Narrative reports on such matters as threats, opportunities, and pertinent events
- Audits conducted (see "Business Audits" later in this chapter)

These conventional components are supplemented as required to deal with the fact that each organization is different. The end result should be a report package that assists upper managers to meet the quality goals in much the same way as the financial report package assists the upper managers to meet the financial goals.

The council has the ultimate responsibility for designing such a scorecard. In large organizations, design of such a report package requires inputs from the corporate offices and divisional offices alike. At the division level, the inputs should be from multifunctional sources.

The report package should be specially designed to be read at a glance and to permit easy concentration on those exceptional matters that call for attention and action. Reports in tabular form should present the three essentials: goals, actual performances, and variances. Reports in graphic form should, at the least, show the trends of performances against goals. The choice of format should be made only after learning the preferences of the customers, that is, the upper managers.

Managerial reports are usually published monthly or quarterly. The schedule is established to coincide with the meetings schedule of the council or other key reviewing body. The editor of the scorecard is usually the director of quality (quality manager, etc.), who is usually also the secretary of the council.

Scorecards have become an increasing staple in corporations across the globe, so much so that they have moved beyond their initial purpose. Scorecards have now been created not just to document an organization's bottom line but also to judge how green an organization actually is. A "Climate Counts" Organization Scorecard rates organizations across different industry sectors in their practices to reduce global warming and create greener business practices. Organizations that are making concerted efforts to alleviate these causes receive higher scores. Like regular scorecards, the information is available to the public, and the opportunity to further a positive public image is at hand. Items include

- Leading indicators (e.g., quality of purchased components)
- Concurrent indicators (e.g., product test results, process conditions, and service to customers)
- Lagging indicators (e.g., data feedback from customers and returns)
- Data on cost of poor quality

The scorecard should be reviewed formally on a regular schedule. Formality adds legitimacy and status to the reports. Scheduling the reviews adds visibility. The fact that upper managers personally participate in the reviews indicates to the rest of the organization that the reviews are of great importance.

Many organizations have combined their measurements from financial, customer, operational, and human resource areas into "instrument panels" or "balanced business scorecards."

Business Audits

An essential tool for upper managers is the audit. By "audit," we mean an independent review of performance. "Independent" signifies that the auditors have no direct responsibility for the adequacy of the performance being audited. (See Chapter 9, Quality Assurance and Audits.)

The purpose of the audit is to provide independent, unbiased information to the operating managers and others who have a need to know. For certain aspects of performance, those who have a need to know include the upper managers.

To ensure quality, upper management must confirm that

- The systems are in place and operating properly.
- The desired results are being achieved.

Growing to encompass a broad range of fields, quality audits are now utilized in a plethora of industries, including science. The Royal College of Pathologists implements quality audits on a number of their research reports. The quality audit ensures that individuals and teams are meeting the procedures and standards expected of them and that their work is in line with the mission of the study.

These audits may be based on externally developed criteria, on specific internal objectives, or on some combination of both. Three well-known external sets of criteria to audit organization performance are those of the United States' Malcolm Baldrige National Award for Excellence, the European Foundation Quality Management Award (EFQM), and Japan's Deming Prize. All provide similar criteria for assessing business excellence throughout the entire organization.

Traditionally, quality audits have been used to provide assurance that products conform to specifications and that operations conform to procedures. At upper-management levels, the subject matter of quality audits expands to provide answers to such questions as

- Are our policies and goals appropriate to our organization's mission?
- Does our quality provide product satisfaction to our clients?
- Is our quality competitive with the moving target of the marketplace?
- Are we making progress in reducing the cost of poor quality?
- Is the collaboration among our functional departments adequate to ensure optimizing organization performance?
- Are we meeting our responsibilities to society?

Questions such as these are not answered by conventional technological audits. Moreover, the auditors who conduct technological audits seldom have the managerial experience and training needed to conduct business-oriented quality audits. As a consequence, organizations that wish to carry out quality audits oriented to business matters usually do so by using upper managers or outside consultants as auditors.

Juran (1998) has stated:

> One of the things the upper managers should do is maintain an audit of how the processes of managing for achieving the plan is being carried out. Now, when you go into an audit, you have

> three things to do. One is to identify what are the questions to which we need answers. That's no delegable; the upper managers have to participate in identifying these questions. Then you have to put together the information that's needed to give the answers to those questions. That can be delegated, and that's most of the work, collecting and analyzing the data. And there are the decisions of what to do in light of those answers. That's no delegable. That's something the upper managers must participate in.

Audits conducted by executives at the highest levels of the organization where the president personally participates are usually called "The President's Audit" (Kondo 1988). Such audits can have major impacts throughout the organization. The subject matter is so fundamental in nature that the audits reach into every major function. The personal participation of the upper managers simplifies the problem of communicating to the upper levels and increases the likelihood that action will be forthcoming. The very fact that the upper managers participate in person sends a message to the entire organization relative to the priority placed on quality and to the kind of leadership being provided by the upper managers—leading, not cheerleading (Shimoyamada 1987).

Lessons Learned

There are some important lessons learned about the risks in implementing strategic deployment:

- Pursuing too many objectives, long term and short term, at the same time will dilute the results and blur the focus of the organization.
- Excessive planning and paperwork will drive out the needed activities and demotivate managers.
- Trying to plan strategically without adequate data about customers, competitors, and internal employees can create an unachievable plan or a plan with targets so easy to achieve that the financial improvements are not significant enough.
- If leaders delegate too much of the responsibility, there will be a real and perceived lack of direction.
- For an organization to elevate quality and customer focus to top priority creates the impression that it is reducing the importance of finance, which formerly occupied that priority. This perceived downgrading is particularly disruptive to those who have been associated with the former top-priority financial goals.

When embarking on strategic planning, imposing a structured approach on those who prefer not to have it creates the biggest disruption. Resistance to the structured approach will be evident at the outset. The single most important prerequisite for embarking on a long-term, effective, organization wide improvement effort is the creation of an environment conducive to the many changes that are necessary for success. Organizations have aggressively sought to eliminate these barriers that have taken years or decades to establish. The process of change takes time, however, and change will occur only as an evolutionary process.

References

Bool, H. (2008). Social Responsibility & The Corporate Values Statement, http://ezinearticles.com/?id=1871566, 2008.

Godfrey, A. B. (1997). "A Short History of Managing Quality in Health Care." In Chip Caldwell, ed., *The Handbook for Managing Change in Health Care*. ASQ Quality Press, Milwaukee, WI.

Greenwood, R. G. (1981). "Management by Objectives: As Developed by Peter Drucker, Assisted by Harold Smiddy." *The Academy of Management Review*, Vol. 6, No. 2, pp. 225–230.

Harris, V. (1982). *A Book of Five Rings* by Miyamoto Musashi (1645), translated by V. Harris, Overlook Press, Woodstock, New York.

Hutchins, D. (2008). Hoshin Kanri: *The Strategic Approach to Continuous Improvement*, Gower Publishing Limited, Hampshire, UK.

JetBlue Airways, Terminal 5, JFK International Airport, http://phx.corporate-ir.net/External.File?item=UGFyZW50SUQ9MzMzODAzfENoaWxkSUQ9MzE2NTMwfFR5cGU9MQ==&t=1, 2008.

Juran, J. M. (1988). *Juran on Planning for Quality*. Free Press, New York.

Juran, J. M. (2005). "Pareto, Lorenz, Cournot, Bernoulli, Juran and Others." *Joseph M. Juran: Critical Evaluations in Business and Management*, Vol. 1, No. 4, p. 47.

Kaplan, R. S., and Norton, D. P. (1996). *The Balanced Scorecard: Translating Strategy into Action*, Harvard Business School Press, Boston, Massachusetts.

Kondo, Y. (1988). "Quality in Japan." In J. M. Juran, ed., *Juran's Quality Control Handbook*, 4th ed. McGraw-Hill, New York. (Kondo provides a detailed discussion of quality audits by Japanese top managers, including The President's Audit. See Chapter 35F, "Quality in Japan," under "Internal QC Audit by Top Management.")

Kondo, Y. (1998). "Hoshin Kanri: A Participative Way of Quality Management in Japan." *The TQM Magazine*, Vol. 10, No. 6, pp. 425–431.

Lee, R. G., and Dale, B. G. (1998). "Policy Deployment: An Examination of the Theory." *International Journal of Quality and Reliability Management*, Vol. 15, No. 5, pp. 520–540.

Shimoyamada, K. (1987). "The President's Audit: QC Audits at Komatsu." *Quality Progress*, January, pp. 44–49. (Special Audit Issue.)

Sussland, W. A. (2002). "Connecting the Planners and the Doers." Quality Progress, Vol. 35, No. 6, pp. 55–61.

Tennant, C., and Roberts, P. (2001). "Hoshin Kanri: Implementing the Catchball Process." Long Range Planning, Vol. 34, No. 3, pp. 287–308.

Tregoe, B., and Tobia, P. (1990). "Strategy and the New American Organization." *Industry Week*. August 6.

Watson, G. H. (2003). "Policy Deployment: Consensus Method of Strategy Realization." *Quality into the 21st Century: Perspectives on Quality and Competitiveness for Sustained Performance*, pp. 191–218, T. Conti, Y. Kondo, and G. H. Watson (Eds.), ASQ Quality Press, Milwaukee, Wisconsin.

Witcher, B. J. (2003). "Policy Management of Strategy (Hoshin Kanri)." Strategic Change, Vol. 12, No. 2, pp. 83–94.

Witcher, B. J., and Chau, V. S. (2007). "Balanced Scorecard and Hoshin Kanri: Dynamic Capabilities for Managing Strategic Fit." *Management Decision*, Vol. 45, No. 3, pp. 518–538.

CHAPTER 8

Organization Roles to Support a Quality and Excellence Culture

Joseph A. De Feo and Peter Robustelli

High Points of This Chapter

1. Coordination of quality activities throughout an organization requires two efforts: coordination for control and coordination for creating change.
2. Coordination for control is often the focus of a quality department; coordination for creating change often involves "parallel organizations" such as a quality council and quality project teams.
3. New forms of organization aim to remove the barriers, or walls, between functional departments.
4. To achieve quality excellence, upper management must lead the quality effort. The roles in this leadership can be identified.
5. A quality council is a group of upper managers who develop quality strategy and guide and support its implementation.

6. Middle management executes the quality strategy through a variety of roles.
7. Inputs from the workforce are essential to identify the causes of quality problems and to design work systems for self-control.

Role of Upper Management

A company president said this: "My people have disappointed me. I clearly told them that quality is our first priority; I provided training; now two years later there is little evidence of improvement." He missed the point of active leadership. Of all the ingredients for successfully achieving quality superiority, one stands out: active leadership by upper management. Commitment to quality is assumed but is not enough.

Certain roles can be identified:

- Establish and serve on a quality council
- Establish quality strategies
- Establish, align, and deploy quality goals
- Provide the resources
- Provide training in quality methodology
- Serve on upper management quality improvement teams that address chronic problems of an upper management nature
- Review progress and stimulate improvement
- Provide for reward and recognition

In brief, upper management develops the strategies for quality and ensures their implementation through personal leadership.

One effective example is the action taken by the head of a manufacturing division. He personally chairs an annual meeting at which improvement projects are proposed and discussed. By the end of the meeting, a list of approved projects for the coming year is finalized, and responsibility and resources are assigned for each project.

Unfortunately, there is a price to be paid for this active leadership. The price is time. Upper managers will need to spend at least 10 percent of their time on quality activities—with other managers, with frontline employees, with suppliers, and with customers.

Providing Resources for Quality Activities

The modern approach to quality requires an investment of time and resources throughout the entire organization—for many people, the price is about 10 percent of their time. In the long run, this investment yields time savings that then become available for quality activities or other activities; in the short run, the investment of resources can be a problem.

Upper management has the key role in providing resources for quality activities. One alternative is to add resources, but in highly competitive times, this approach may not be feasible. Often, time and resources can be found only by changing the priorities of both line and staff work units. Thus some work must be eliminated or delayed to make personnel available for quality activities.

People who are assigned to quality teams should be aware of the amount of time that will be required while they are on a team. If time is a problem, they should be encouraged to propose changes in their other priorities (before the team activity starts). Resources for project teams will be made available only if the pilot teams demonstrate benefits by achieving tangible results—thus the importance of nurturing those pilot teams to yield results. As a track record builds for successful teams, the resource issue becomes less of a problem. One outcome can be for the annual budgeting process of an organization to include a list of proposed projects and necessary resources on the agenda for discussion.

In a dramatic action to provide resources for a comprehensive quality effort, upper management at a modem manufacturer assigned 3 percent of the total workforce (of 3000 people) to work full time on quality. The team included six upper level managers who were responsible for designing and overseeing the effort. These managers returned to their original jobs after 1 year.

Quality Leadership Council

A quality council (sometimes called a "leadership team") is a group of upper managers who develop the quality strategy and guide and support the implementation. Councils may be established at several levels—corporate, division, business unit. When multiple councils are established, they are usually linked together, that is, members of high-level councils serve as chairpersons of lower level councils. For any level, membership consists of upper managers—both line and staff. The chairperson is the manager who has overall responsibility and authority for that level, for example, the president for the corporate council and the division and site managers for their levels. One member of the council is the quality director whose role in the company is discussed later. Each council should prepare a charter that includes responsibilities such as

- Formulating quality strategies and policies
- Estimating major dimensions of the quality issue
- Establishing an infrastructure for selecting quality projects and assigning project team leaders and members
- Providing resources including support for teams
- Planning for training for all levels
- Establishing strategic measures of progress
- Reviewing progress and remove any obstacles to improvement
- Providing for public recognition of teams
- Revising the reward system to reflect progress in quality improvement

At Kelly Services, Inc. (McCain 1995), the quality council identified its key responsibilities as

- Setting and deploying vision, mission, shared values, quality policy, and quality goals
- Reviewing progress against goals
- Integrating quality goals into business plans and performance management plans

Upper managers often ask, Isn't the quality council identical in membership to the regular top-management team? Usually, yes. If so, instead of having a separate council, why not add quality issues to the agenda of the periodic upper management meetings? Eventually (when quality has become "a way of life"), the two can be combined, but not at the start. The seriousness and complexity of quality issues require a focus that is best achieved by meetings that address quality alone.

As a council works on its various activities, it often designates one or more full-time staff people to assist the council by preparing draft recommendations for council review. In another approach, several council members serve on ad hoc task forces to investigate various issues for the council.

In addition, individual council members often serve as advocates ("champions") for key quality-related projects. In this role, they continually monitor the project, use their executive position to remove obstacles to completing the project, and recognize project teams for their efforts.

Providing evidence of upper management leadership is clearly important in establishing a positive quality culture.

Finally, note that leadership is the first of the Baldrige criteria. The leadership criterion covers organizational leadership, public responsibility, and citizenship. Examination of the complete criteria reveals that leadership should be linked to other Baldrige criteria such as strategic planning, customer and market focus, information and analysis, and human resources.

Role of the Quality Director

The quality director of the future is likely to have two primary roles—administering the quality department and assisting upper management with strategic quality management (Gryna, 1993).

The Quality Department of the Future

What will the future role of a quality department be? The major activities are indicated in Table 8.1. The table indicates some traditional activities of the quality department and some important departures from the current norm.

Note, for example, "transferring activities to line departments." Recent decades have shown that the best way by far to implement quality methodologies is through line organizations rather than through a staff quality department. (Isn't it a shame that it took us so long to understand this point?) A few organizations have stressed this approach for several years, and some have been eminently successful in transferring many quality activities to the line organization.

To achieve success in such a transfer, the line departments must clearly and fully understand the activities for which they are responsible. In addition, the line departments must be trained to execute these newly acquired responsibilities.

Examples of such transfer include shifting the sentinel (sorting) type inspection activity from a quality department to the workforce itself, transferring reliability engineering work

Companywide quality planning
Setting up quality measurement at all levels
Auditing outgoing quality
Auditing quality practices
Coordinating and assisting with quality projects
Participating in supplier partnerships
Training for quality
Consulting for quality
Developing new quality methodologies
Transferring activities to line departments

TABLE 8.1 Functions of the Quality Department of the Future

from a quality department to the design engineering department, and transferring supplier quality activities from a quality department to the purchasing department. A key role for the quality department is to help its internal customers achieve their quality objectives. In most organizations, the key internal customer is the operations department (called *operations* in the service sector and *manufacturing* or *production* in the manufacturing sector). Clearly, quality and operations must be partners, not adversaries. As a quality director, the author each year had to obtain 50 percent of his departmental budget from operations departments. This experience vividly causes a manager to become customer oriented. If a quality director truly wants to assist internal customers, a constructive step is to conduct internal "market research" on these customers. Bourquin (1995) describes how a unit of AT&T conducted a survey of internal customers (mainly manufacturing but also marketing, design, and other departments). The survey contained 10 questions reflecting both the relative importance of certain services and the degree of satisfaction with the services. The survey required little time or money but provided valuable quantitative feedback from the main internal customers.

The author believes that there will always be a need for a quality department to provide an independent evaluation of product quality and to furnish services to internal customers (see Table 8.1). Changing business conditions, however, make it essential to review the role of the department periodically and make appropriate changes. These changing business conditions include company mergers, changes in customer (external and internal) expectations, outsourcing, the information explosion, new communication technologies, the impact of global business and cultures, and the maturity of the quality effort within the organization. In addition, the emphasis on continuous improvement within organizations has resulted in internal competitors to a quality department for coordinating improvement activities. Such competitors include industrial engineering, human resources, financial audit, and information technology.

Assisting Upper Management with Strategic Quality Management (SQM)

A wonderful opportunity exists for a quality director to assist upper management to plan and execute the many activities of SQM. Some of these activities are shown in Table 8.2. Can a quality director achieve such an exalted role? It is useful to cite an analogy to the area of finance (Table 8.3). Many organizations today have a CFO. This officer is concerned with broad financial planning, addressing questions such as, Where should our company be going with respect to finance? Other financial managers direct and manage detailed financial processes, such as accounts payable, accounts receivable, cash management, acquisitions, and budgeting. These roles are also vital in the organization but different from the broader role of the CFO. Finally, of course, line managers throughout the entire organization have

Assessing quality
Formulating goals and policies
Developing quality strategies to increase sales revenue and reduce internal costs
Delegating organizational responsibilities for quality
Carrying out reward and recognition
Reviewing progress
Determining personal roles for upper ma nagement
Acting as facilitator to the quality council
Integrating quality during strategic business planning cycle

Table 8.2 Assisting Upper Management with Strategic Quality Management

Finance (today)	Quality (the future)
Chief financial officer	Quality director
Other financial managers	Other quality managers
Line managers	Line managers

TABLE 8.3 A Contrast of Roles

specific activities that help meet the financial objectives of the company. Note that the quality function comprises both technical quality activities and strategic quality management activities. Provision must be made to staff a quality department with skills in these two areas. Larger organizations may benefit from having both a quality engineering function and a quality management function (just as accounting and financial planning are separate functions). These functions would then report to the quality director.

The quality director of the future could act as the right hand of upper management on quality in the same way that the chief financial officer acts as the right hand of upper management on finance. (Note that other essential quality activities such as inspection, audit, and quality measurement must be directed and managed by other quality managers.)

Some of the scars of experience that have been accumulated within financial circles apply also to the quality function. Years ago, there was no such role as chief financial officer. The detailed financial processes were handled by one or several managers. As time went on, the need for a person with a broader financial viewpoint became apparent. In some companies, the person who was then the "controller" was promoted to become the chief financial officer. Other companies felt that the controller did not have the breadth of vision and scope to assist upper management in broad financial planning, even though the controller was excellent in administering some of the detailed financial processes. Thus not every controller became a chief financial officer.

A similar issue arises in regard to the broad scope necessary for the quality director of the future. Is the present quality director prepared, or is that person willing to become prepared, for the broader business role with respect to quality? Quality directors who wish to assume this broader role in the future need to learn from the lessons of the financial controllers.

Table 8.4 shows some ingredients for success as a quality director of the type indicated here.

Focus on customer orientation and customer advocacy
Ability to build collaborative relationships across functions
Strong oral and written communication skills to encourage sharing of information
Goal orientation
Ability to analyze complex issues and generate innovative solutions
Initiative, persistence, and self-confidence in gai ning acceptance of new ideas
Ability to organize activities
Ability to provide for self-development of subordinates

Source: "Ingredients for success as a quality director" adapted from G. Watson, "The Emancipation of Quality: Building Bridges and Closing Gaps" as appeared in *Quality Progress*, August, 1998, Vol. 31, No. 8. Reprinted by permission of Gregory H. Watson.

TABLE 8.4 Ingredients for Success as a Quality Director

Clearly, this list goes far beyond the current scope of quality directors in many companies. In larger companies, quality managers, including those who will manage some of the technical quality activities, will be needed at various levels. A quality director is likely to be involved in administering technical activities but will increasingly be called on to assist upper management as upper management—not the quality director—leads the company on quality. For elaboration of the future role of the quality department, see Gryna (1993).

Cinelli and Schein (1994) present the results of a survey on the profile of the U.S. senior quality executive. The survey examines the professional background, day-to-day activities, and responsibilities of quality executives at 223 Fortune 500 firms. For the results of a survey of 38 U.K. companies on organizing for quality, see Groocock (1994).

Role of Management

Middle managers, supervisors, professional specialists, and the workforce are the people who execute the quality strategy developed by upper management.

The roles of middle managers, supervisors, and specialists include

- Nominating quality problems for solutions.
- Serving as leaders of various types of quality teams.
- Serving as members of quality teams.
- Serving on task forces to assist the quality council in developing elements of the quality strategy.
- Leading the quality activities within their own area by demonstrating a personal commitment and encouraging their employees.
- Identifying customers and suppliers and meeting with them to discover and address their needs.

Increasingly, middle managers are asked to serve as team leaders as a continuing part of their job. For many of them, their roles as team leaders require special managerial skills. For a manager of a department directing the people in that department, a traditional hierarchical approach is common. The leader of a cross-functional quality improvement team faces several challenges, for example, the leader usually has no hierarchical authority over anyone on the team because members come from various departments, serve part time, and have priorities in their home departments. The success of a team leader depends on technical competence, ability to get people to work together as a team, and a personal sense of responsibility for arriving at a solution to the assigned problem. Leading a team calls for quite an array of talents and willingness to assume responsibility. For some middle managers, the required change in managerial style is too much of a burden; for others, the role presents an opportunity. Dietch et al. (1989) identified and studied 15 characteristics of team leaders at the Southern California Edison Company. The research concluded that team leaders (as compared to team members) have a higher tolerance for handling setbacks, believe that they have great influence over what happens to them, exhibit a greater tolerance for ambiguity, are more flexible, and are more curious.

Role of the Workforce

By workforce, we mean all employees except those in management and professional specialists.

Recall that most quality problems are management or system controllable. Therefore management must (1) direct the steps necessary to identify and remove the causes of quality problems and (2) provide the system that places workers in a state of self-control.

Inputs from and cooperation by the workforce are essential. The roles of the workforce include

- Nominating quality problems for solution
- Serving as members of various types of quality teams
- Identifying elements of their own jobs that do not meet the three criteria of self-control
- Becoming knowledgeable as to the needs of their customers (internal and external)

At last we are starting to tap the potential of the workforce by using its experience, training, and knowledge. One plant manager says this: "No one knows a workplace and a radius of 20 feet around it better than the worker." Quality goals cannot be achieved unless we use the hands and the heads of the workforce. Some of the workforce roles on teams are discussed next.

Role of Teams

The "organization of the future" will be influenced by the interaction of two systems that are present in all organizations: the technical system (equipment, procedures, etc.) and the social system (people, roles, etc.); thus the name "sociotechnical systems" (STS).

Much of the research on sociotechnical systems has concentrated on designing new ways of organizing work, particularly at the workforce level. Team concepts play an important role in these new approaches. Some organizations now report that, within a given year, 40 percent of their people participate on a team; some organizations have a goal of 80 percent. A summary of the most common types of quality teams is given in Table 8.5.

Aubrey and Gryna (1991) describe the experiences of more than 1000 quality teams during four years at 75 banking affiliates of Banc One. This effort yielded significant results: $18 million in cost savings and revenue enhancement; a 10 to 15 percent improvement in customer satisfaction; and a 5 to 10 percent reduction in costs, defects, and lost customers. On some teams, membership was assigned; on other teams, membership was voluntary. A summary of some of the organizational results is given in Table 8.6. The focus is to improve customer satisfaction, reduce cost and increase revenue, and improve communication between frontline employees and management.

Next we examine three types of teams: quality project teams, workforce teams, and self-directed teams (see Fig. 8.1 for reference).

Training and Certification of Personnel

The principles of selection, training, and retention of personnel are known but are not always practiced with sufficient intensity in many functions.

Selection of Personnel

Norrell, a human resource company, provides client companies with traditional temporary help, managed staffing, and outsourcing services. A survey of more than 1000 of their clients clarified the client definition of quality as "excellence of personnel." Excellence was defined in terms of eight criteria for clerical positions and nine criteria for technical-industrial positions. These criteria are used in selecting and training personnel assigned to the client companies. In another example based on a survey of five service organizations and nine manufacturers, Jeffrey (1995) identified 15 competencies that these organizations and their

	Improvement Quality Project Team	Workforce Team	Business Process Quality Team	Self-Directed Team
Purpose	Solve cross-functional quality problems	Solve problems within a department	Plan, control, and improve the quality of a key cross-functional process	Plan, execute, and control work to achieve a defined output
Membership	Combination of managers, professionals, and workforce from multiple departments	Primarily workforce from one department	Primarily managers and professionals from multiple departments	Primarily workforce from one work area
Basis and size of membership	Mandatory; 4–8 members	Voluntary; 6–12 members	Mandatory; 4–6 members	Mandatory; all members in the work area (6–18)
Continuity	Team disbands after project is completed	Team remains intact, project after project	Permanent	Permanent
Other names	Quality improvement team	Employee involvement group	Business process management team; process team	Self-supervising team; semiautonomous team

Table 8.5 Summary of Types of Quality Teams

customers viewed as important in customer service activities by frontline employees. Many of these competencies apply to both frontline personnel and back-office personnel.

A human resources firm that provides temporary personnel to clients performed research to select a series of questions for prospective employees. A battery of questions was developed and given to all employees. Employees who are rated superior by clients (based on market research studies) answer certain questions differently from employees who are

Feature	Results of Research
Team size	Average of 7 members, with a range of 2–11
Project selection	75% by management, 15% by quality council, 10% by individual team
Average savings related to project selection	Projects selected by management or the quality council achieved savings about twice as high as projects selected by the team
Project duration	A verage of 3 months; 24 worker-hours per team member (excluding time spent outside team meetings)
Factors to maximize team success	Ideal team size of 4–5 employees; 75% officer/staff level, 25% nonexempt employees; members elected by management; project selected by management or quality council; project duration of 3–4 months with weekly 90-minute team meetings

Table 8.6 Observations on Organization of Quality Teams at a Bank

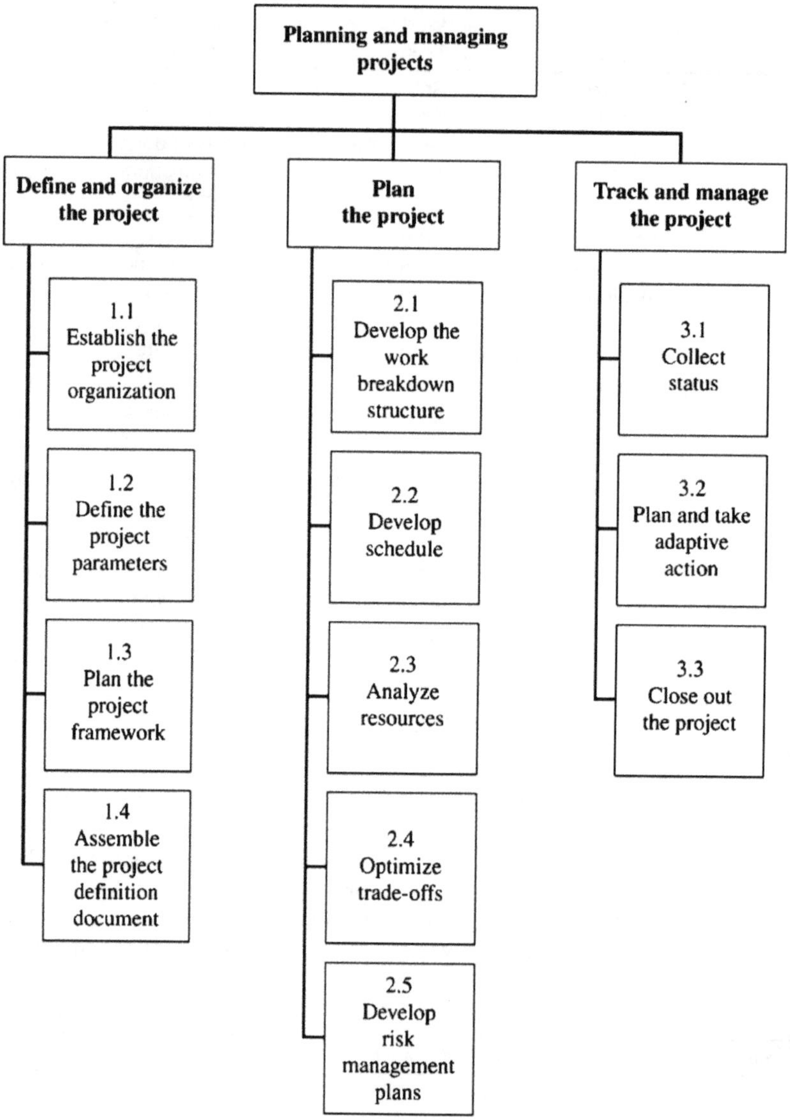

FIGURE 8.1 Project management process model. From Project Management Manual. Boston, MA, 1996. Copyright © 1996 by the Harvard Business School Publishing Corporation; all rights reserved. Reprinted by permission of Harvard Business School Press.

not superior. (Other questions result in the same response from most employees.) Responses to these "differentiating questions" help to select new employees.

Personality is one important attribute for many (but not all) positions in the operations function. This attribute is increasingly important as organizing by teams becomes more prevalent. A chemical manufacturer even places job applicants in a team problem-solving situation as part of the selection process. Larry Silver of Raymond James Financial Services Inc. states the case well: "We need to recruit people who play well in the sandbox with others, that is, don't throw sand in people's face and do work together to build castles."

One tool for evaluating personality types is the Myers-Briggs Type Indicator. This personality test describes 16 personality types that are based on four preference scales: extrovert or introvert, sensing or intuition, thinking or feeling, and judgment or perception. Thus one personality type is an extrovert, sensing, thinking, judgment person. Analyzing responses to test questions from prospective or current employees helps to determine the personality types of individuals. Organizations need many personality types, and the Myers-Briggs approach describes the contributions of each of the 16 types to the organization. By understanding the types and making job assignments accordingly, an organization can take advantage of all personality types to achieve high performance in the workplace. McDermott (1994) explains the 16 types and how the tool can help in recruiting new personnel and assigning current personnel. Anderson and Anderson (1997) describe a candidate assessment tool that uses four areas of abilities, three areas of motivation and interests, and 13 areas of personality.

Training

One essential ingredient of a broad quality program is extensive training. Table 8.7 identifies constituencies and subject matter.

Experience in training has identified the reasons that some training programs fail:

- Failure to provide training when it will be used. In too many cases, training is given to employees who have little or no opportunity to use it until many months later (if ever). A much better approach schedules training for each group when it is needed—"just in time" training.
- Lack of participation by line managers in designing training. Without this participation, training is often technique oriented rather than problem and results oriented.
- Reliance on the lecture method of training. Particularly in the industrial world, training must be highly interactive, that is, it must enable a trainee to apply the concepts during the training process.
- Poor communication during training. The technology of quality, particularly statistical methodology, can be mystifying to some people. Many benefits are possible if we emphasize simple language and graphical techniques.

Training programs are a failure if they do not result in a change in behavior. Applying these lessons can help to prevent such failures.

Retention

Investing increased resources in selection and training helps to retain skilled employees. Compensation, of course, is also an essential contributor to employee retention. Other factors, however, are also essential, including

1. Providing for career planning and development
2. Designing jobs for self-control
3. Providing sufficient empowerment and other means for personnel to excel
4. Removing sources of job stress and burnout
5. Providing continuous coaching for personnel
6. Providing for participation in departmental planning
7. Providing the opportunity to interact with customers (both external and internal)
8. Providing various forms of reward and recognition

Subject Matter	Top Management	Quality Managers	Other Middle Managers	Specialists	Facilitators	Workforce
Quality awareness	X		X	X	X	X
Basic concepts	X	X	X	X	X	X
Strategic quality management	X	X				
Personal roles	X	X	X	X	X	X
Three quality processes	X	X	X	X	X	
Problem-solving methods		X	X	X	X	X
Basic statistics	X	X	X	X	X	X
Advanced statistics		X		X		
Quality in functional areas		X	X	X		
Motivation for quality	X	X	X		X	

TABLE 8.7 Who Needs to Be Trained in What

Performance appraisal has become a subject of intense debate (and confusion). The author believes that, when properly conducted, performance appraisal is a useful tool.

Retention of key employees depends on many issues. An information technology services firm conducted a survey of more than 200 employees to determine their reasons for leaving—with some surprising results. In rank order (based on frequency and intensity of their feeling), employees gave the following reasons for leaving: communicate with me more, train my boss, train me more, help me set goals and give me feedback, provide a higher salary, and provide better benefits.

Morgan and Smith (1996) is a useful reference for selection, training, and retention of personnel.

References

Anderson, D. N., and T. Anderson (1997). "Finding the Right People with the Right Tools," IIE Solutions Conference Proceedings, Institute of Industrial Engineers, Norcross, GA.

Bourquin, C. R. (1995). "A Quality Department Surveys Its Customers (or, Shoes for the Cobbler's Children)," Annual Quality Congress Proceedings, ASQ, Milwaukee, pp. 970–976.

Cinelli, D. L., and L. Schein (1994). "A Profile of the U.S. Senior Quality Executive," The Conference Board, New York.

Dietch, R., S. Tashian, and H. Green (1989). "Leadership Characteristics and Culture Change: An Exploratory Research Study," Impro Conference Proceedings, Juran Institute, Inc., Wilton, CT, pp. 3C-21 to 3C-29.

Groocock, J. M. (1994). "Organizing for Quality—Including a Study of Corporate-Level Quality Management in Large U.K.-Owned Companies," *Quality Management Journal*, January, pp. 25–35.

Gryna, F. M. (1993). "The Role of the Quality Director—Revisited," Impro Conference Proceedings, Juran Institute, Inc., Wilton, CT, pp. 11.12–11.14.

Gryna F. M. (2004). *Work Overload!: Redesigning Jobs to Minimize Stress and Burnout*. ASQ Quality Press, Milwaukee, Wisconsin.

Jeffrey, J. R. (1995). "Preparing the Front Line," *Quality Progress*, February, pp. 79–82.

McCain, C. (1995). "Successfully Solving the Quality Puzzle in a Service Company," Impro Conference Proceedings, Juran Institute, Inc., Wilton, CT, pp. 6A.1-1 to 6A.1-13.

McDermott, R. (1994). "The Human Dynamics of Total Quality," Quality Congress Transactions, ASQ, Milwaukee, pp. 225–233.

Morgan, R. B. and J. E. Smith (1996). *Staffing the New Workplace*, ASQ Quality Press, Milwaukee, and CCH Inc., Chicago.

CHAPTER 9

Quality Assurance and Audits

Joseph A. De Feo

High Points of This Chapter

1. Quality assurance is the activity of providing evidence to establish confidence that quality requirements will be met.
2. A quality audit is an independent review conducted to compare some aspect of quality performance with a standard for that performance. It is a process conducted by either an internal or external auditor that helps to ensure that an organization's systems are in place and are being followed.
3. Experienced auditors often can discover opportunities for improvement as a by-product of their search for nonconformance to stated needs. These opportunities may even be known to the operations personnel so that the auditor is only making a rediscovery.
4. Audits are conducted by individuals or by a team. A team usually has a lead auditor who plans the audit, creates the audit schedule, assembles or creates checklists or aid memoirs, conducts the meetings, reviews the findings and comments of the

auditors, prepares the audit report, evaluates corrective action, and presents the audit report.

5. The final phase of the audit is follow-up to confirm that corrective action has been taken by the audited activity and that the corrective action is effective.

Definition and Concept of Assurance

In this book, quality assurance is the activity of providing evidence to establish confidence that quality requirements will be met. ISO defines quality assurance as all the planned and systematic activities implemented within the quality system, and demonstrated as needed, to provide adequate confidence that an entity will fulfill requirements for quality. Readers are warned that other meanings are common. For example, "quality assurance" is sometimes the name of a department concerned with many quality management activities such as quality planning, quality control, quality improvement, quality audit, and reliability.

Many quality assurance activities provide protection against quality problems through early warnings of trouble ahead. The assurance comes from evidence—a set of facts. For simple products, the evidence is usually some form of inspection or testing of the product. For complex products, the evidence is inspection and test data and also reviews of plans and audits of the execution of plans. A family of assurance techniques is available to cover a wide variety of needs.

Quality assurance is similar to the concept of the financial audit, which provides assurance of financial integrity by establishing, through "independent" audit, that the plan of accounting is (1) such that, if followed, it will correctly reflect the financial condition of the company and (2) is actually being followed. Today, independent financial auditors (certified public accountants) have become an influential force in the field of finance.

Many forms of assurance previously discussed in this book are performed within functional departments (Table 9.1). This chapter discusses three forms of company-wide quality assurance: quality audits, quality assessments, and product audits.

Concept of Quality Audits

A quality audit is an independent review conducted to compare some aspect of quality performance with a standard for that performance. It is a process conducted by either an internal or external auditor that helps to ensure that an organization's systems are in place and are being followed. The objective of the audit is to draw attention to needed improvements and ensure legal and regulatory requirements are being followed in order to bring consistent quality goods and services to the consumer. A successful quality audit concentrates on the needs of the organization. The term independent is critical and is used in the sense that the reviewer (called the *auditor*) is neither the person responsible for the performance under review nor the immediate supervisor of that person. An independent audit provides an unbiased picture of performance. The terms "quality assessment" (or quality evaluation) and "quality audit" have similar meanings, but in common usage, assessment refers to a total spectrum of quality activities often including managerial matters such as the cost of poor quality, standing in the market place, and quality culture.

The ISO 19011:2002 First Edition definition spells out some additional aspects: A quality audit is a systematic and independent examination to determine whether quality activities and related results comply with planned arrangements and whether these arrangements are implemented effectively and are suitable for achieving objectives. (A product audit, discussed later in this chapter, is a review of physical product; a quality audit is a review of an activity.)

Department	Assurance Activity
Marketing	Product evaluation by a test market Controlled use of product Product monitoring Captive service activity Special surveys Competitive evaluations
Product development	Design review Reliability analysis Maintainability analysis Safety analysis Human factors analysis Manufacturing, inspection, and transportation analysis Value engineering Self-control analysis
Supplier relations	Qualification of supplier design Qualification of supplier process Evaluation of initial samples Evaluation of first shipments
Production	Design review Process capability analysis Preproduction trials Preproduction runs Failure mode, effect, and criticality analysis for processes Review of manufacturing planning (checklist) Evaluation of proposed process control tools Self-control analysis Audit of production quality
Inspection and test	Interlaboratory tests Measuring inspector accuracy
Customer service	Audit of packaging, transportation, and storage Evaluation of maintenance services

Source: JQH4, p. 9.3.

TABLE 9.1 Examples of Departmental Assurance Activities

Internal audits, sometimes called *first-party audits*, are conducted by, or on behalf of, the organization itself for management review and other internal purposes and may form the basis for an organization's self-declaration of conformity. In many cases, particularly in smaller organizations, independence can be demonstrated by the freedom from responsibility for the activity being audited. External audits include those generally termed second- and third-party audits. Second-party audits are conducted by parties having an interest in the organization, such as customers, or by other persons on their behalf. Third-party audits are conducted by external, independent auditing organizations, such as those providing registration or certification of conformity to the requirements of ISO 9001 or ISO 14001.

Companies use quality audits to evaluate their own quality performance and the performance of their suppliers, licensees, agents, and others; regulatory agencies use quality audits to evaluate the performance of organizations they regulate.

The specific purpose of quality audits is to provide independent assurance that

- Plans for attaining quality are such that, if followed, the intended quality will be attained.
- Products are fit for use and safe for the user.
- Standards and regulations defined by government agencies, industry associations, and professional societies are being followed.
- There is conformance to specifications.
- Procedures are adequate and are being followed.
- The data system provides accurate and adequate information on quality to all concerned.
- Deficiencies are identified, and corrective action is taken.
- Opportunities for improvement are identified, and the appropriate personnel are alerted.

A key question in establishing an audit program is whether the audits should be compliance oriented or effectiveness oriented or both. In practice, many quality audits are compliance oriented; the audits compare quality-related activities to some standard or requirement for those activities (e.g., do written work instructions exist for production and service operations?). The emphasis is on determining conformance to the requirement of written work instructions and maintenance of procedures for those instructions as evidence of that conformance. These audits have matured to "process-centered" audits, whereby the auditor will follow a product or particular service through its process life cycle, checking for conformance to requirements. Effectiveness audits evaluate whether the requirement is achieving the desired result (for external and internal customers) and whether the activity is making efficient use of resources (for elaboration, see Russell and Regel 1996).

At first glance, it seems that audits should be both compliance oriented and effectiveness oriented—and sometimes they can be both. When audits are conducted internally, they can and should be both compliance and effectiveness oriented. But when audits are conducted by external parties, the companies audited can have serious and reasonable issues if an audit concerns matters of effectiveness of operations including use of resources. Effectiveness evaluations of internal operations by external auditors open a broad range of considerations about customer satisfaction and requirements and internal management processes that make it difficult to conduct such audits in a fair and useful way.

Principles of a Quality Audit Program

Five principles are essential to a successful quality audit program:

1. An uncompromising emphasis on conclusions based on facts. Any conclusions lacking a factual base must be so labeled.
2. An attitude on the part of auditors that the audits provide assurance to management and also a useful service to line managers in managing their departments. Thus audit

reports must provide sufficient detail on deficiencies to facilitate analysis and action by line managers.

3. An attitude on the part of auditors to identify opportunities for improvement (this is not usually the case with third-party audits because independence must be respected, but it is important for first- and second-party audits). Such opportunities include highlighting good ideas used in practice that are not part of formal procedures. Sometimes an audit can help to overcome deficiencies by communicating through the hierarchy the reasons for deficiencies that have a source in another department.
4. Addressing the Human Relations: Issues Discussed.
5. Competence of Auditors. The basic education and experience of the auditors should be sufficient to enable them to learn in short order the technological aspects of the operations they are to audit. Lacking this background, they will be unable to earn the respect of the operations personnel. In addition, they should receive special training in the human relations aspects of auditing. The American Society for Quality provides a program for the certification of quality auditors.

These five essentials for a successful quality audit activity were responsible for a dramatic tribute to an audit activity within one company. Line managers voluntarily give up part of their own budget each year to provide funds for a quality audit group.

Subject Matter of Audits

For simple products, the range of audits is also simple and is dominated by product audits (discussed later). For complex products, the audit is far more complex. In large companies, even the division of the subject matter is a perplexing problem. For such companies, the programs of audit use one or more of the following approaches to divide up the subject matter:

Organizational Units

Large companies comprise several layers of organization, each with specific assigned missions: corporate office, operating divisions, plants, etc. Such companies commonly use multiple teams of quality auditors; each reviews its specialized subject matter and reports the results to its own "clientele."

Product Lines

Here the audits evaluate the quality aspects of specific product lines (e.g., printed circuit boards, hydraulic pumps) all the way from design through field performance.

Quality Systems

Here the audits are directed at the quality aspects of various segments of the overall systematic approach to quality such as design, manufacturing, supplier quality, and other processes. A system-oriented audit reviews any such system over a whole range of products. Table 9.2 provides an example from Mallinckrodt Inc., a medical products manufacturer.

Product and Process Control Systems

These type audits evaluate whether current product and process controls (including measurement) are in place and able to meet the needs of customers with regard to quality.

Specific Activities

Audits may also be designed to single out specific procedures that have special significance to the quality mission: disposition of nonconforming products, documentation, instrument calibration, software (Table 9.3).

A. Organizational design
   1. Management responsibility
   2. Job descriptions

B. Customer management practices
   1. Corrective action
   2. Servicing
   3. Complaint and inquiry handling
   4. Recall and field correction

C. Organizational and individual development practices
   1. Training
   2. Personnel hygiene

D. Product development practices
   1. Device design control
   2. Concept generation
   3. Device development
   4. Transfer to operations
   5. Life-cycle maintenance/postmarket surveillance

E. Product and process control practices
   1. Process control
   2. Special processes
   3. Process capability
   4. Facilities and equipment
   5. Control of contamination
   6. Recovered material

F. Procurement practices
   1. Purchasing
   2. Contract review

G. Warehousing and distribution practices
   1. Handling, storage, distribution, and installation

H. Quality assurance practices
   1. Product identification and traceability
   2. Acceptance activities
   3. Nonconforming goods
   4. Labeling
   5. Internal quality audits
   6. Electronic data processing

I. Information analysis practices
   1. Inspection, measuring, and test equipment
   2. Statistical techniques
   3. Analytical methods and laboratories

J. Document management practices
   1. Document control
   2. Quality records
   3. Product registration and approval dossiers

TABLE 9.2 Quality Systems Evaluation—Components and Elements

Scope or Activity	Examples of Specific Tasks Audited
Engineering documentation	Use of latest issue of specifications by operators; time required for design changes to reach shop
Job instructions	Existence and adequacy of written job instructions
Machines and tools	Use of specified machines and tools; adequacy of preventive maintenance
Calibration of measuring equipment	Existence of calibration procedures and degree to which calibration intervals are met
Production and inspection	Adequacy of certification program for critical skills; adequacy of training
Production facilities	General cleanliness and control of critical environmental conditions
Inspection instructions	Existence and adequacy of written instructions
Documentation of inspection results	Adequacy of detail; feedback and use by production personnel
Material status	Identification of inspection status and product configuration; segregation of defective product
Materials handling and storage	Procedure for handling critical materials; protection from damage during handling; control of in-process storage environments

Table 9.3 Examples of Audits of Tasks

Audits of quality systems as well as specific activities may take the form of (1) audit of the plans or (2) audit of the execution versus the plans. Further, the subject matter may include internal activities or external activities such as those conducted by suppliers.

Identifying Opportunities

Experienced auditors often can discover opportunities for improvement as a by-product of their search for nonconformance to stated needs. These opportunities may even be known to the operations personnel so that the auditor is only making a rediscovery. However, these personnel may have been unable to act due to any of a variety of handicaps: preoccupation with day-to-day control, inability to communicate through the layers of the hierarchy, and lack of diagnostic support.

The auditor's relatively independent status and lack of preoccupation with day-to-day control may enable him or her to prevail over these handicaps. In addition, the auditor's reports go to multiple layers of the hierarchy and thereby have a greater likelihood of reaching the ear of someone who has the power to act on the opportunity. For example, the auditor may find that the quality cost reports are seriously delayed owing to backlogs of work in the accounting department. His or her recommendation to expedite the reports may reach the person who can act, whereas the same proposal by the operations personnel may never reach that level.

Structuring the Audit Program

Audits of individual tasks or systems of tasks are usually structured. For example, they are designed to carry out agreed upon purposes and are conducted under agreed rules of conduct.

	Audit Department	Line Department	Upper Management
Discussion of purposes to be achieved by audits and general approach for conducting audits	X	X	X
Draft of policies, procedures, and other rules to be followed	X	X	
Final approval			X
Scheduling of audits	X	X	
Conduct of audits	X		
Verification of factual findings		X	
Publication of report with facts and recommendations	X		
Discussion of reports	X	X	X
Decisions on action to be taken		X	
Subsequent follow-up	X		

Table 9.4 Steps in Structuring an Audit Program

Reaching agreement on these rules and purposes requires collaboration among three essential participating groups:

- The heads of the activities which are to be the subject of audit
- The heads of the auditing department(s)
- The upper management, which presides over both

Without such collective agreements, the audit program may fail. The usual failure modes are (1) an abrasive relationship between auditors and line managers or (2) a failure of line managers to heed the audit reports.

Table 9.4 depicts the typical flow of events through which audit programs are agreed on and audits are carried out. A published statement of purposes, policies, and methods becomes the charter that legitimizes audits and provides continuing guidelines for all concerned.

Audits are often done by full-time auditors who are skilled in both technical and human relations aspects. Audit teams of upper managers, middle managers, and specialists can also be effective.

Planning Audits of Activities

The main steps in performing an audit are planning, opening meeting, performing, reporting, follow-up on corrective action, and closure. The flowchart in Fig. 9.1 describes these steps in some detail. An excellent reference for the auditing process is ASQ Quality Audit Division (2013).

Behind the steps in Fig. 9.1 are a number of important policy issues:

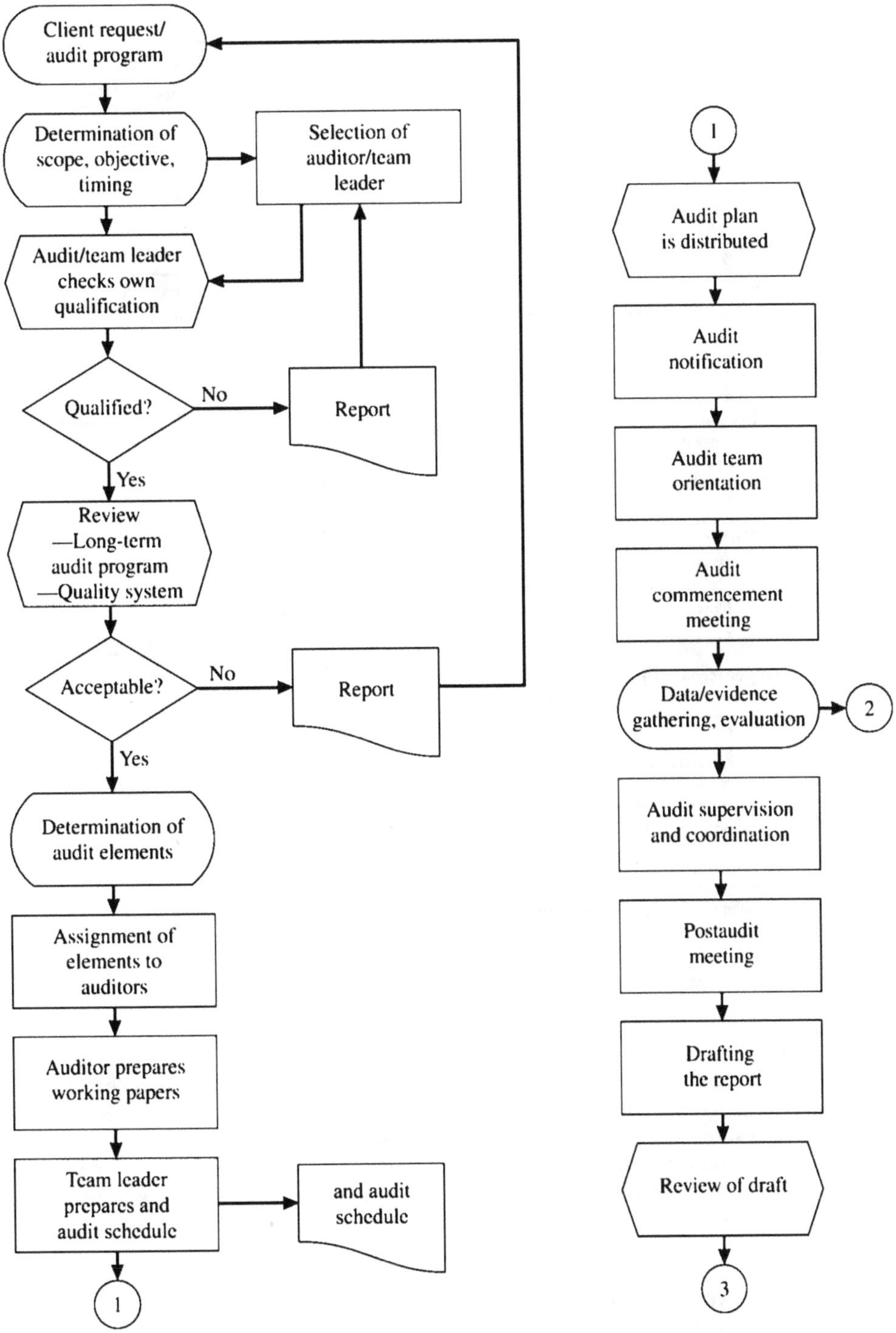
Client request/ audit program
Determination of scope, objective, timing
Selection of auditor/team leader
Audit/team leader checks own qualification
Qualified?
No
Report
Yes
Review —Long-term audit program —Quality system
Acceptable?
No
Report
Yes
Determination of audit elements
Assignment of elements to auditors
Auditor prepares working papers
Team leader prepares and audit schedule
and audit schedule
1
1
Audit plan is distributed
Audit notification
Audit team orientation
Audit commencement meeting
Data/evidence gathering, evaluation
2
Audit supervision and coordination
Postaudit meeting
Drafting the report
Review of draft
3

FIGURE 9.1

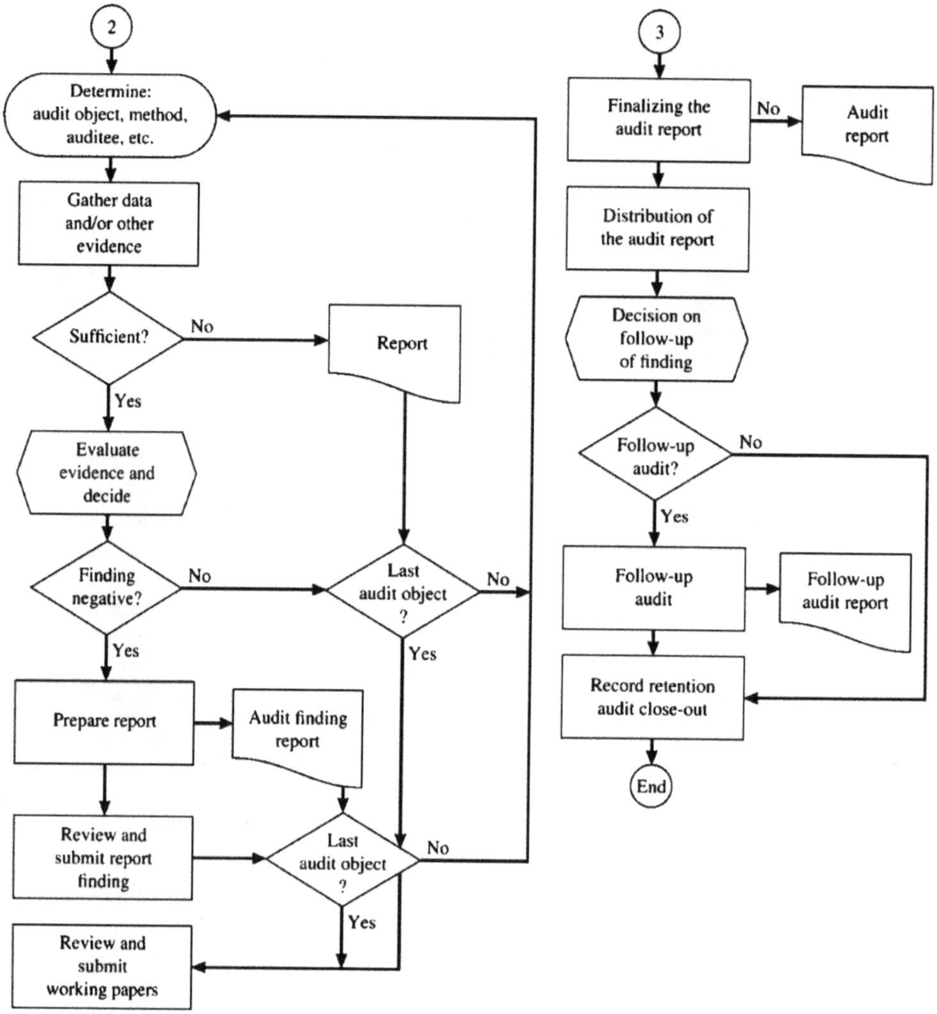

FIGURE 9.1 *(Continued)*

Legitimacy

The basic right to conduct audits is derived from the "charter" that has been approved by upper management, following participation by all concerned. Beyond this basic right are other questions of legitimacy: What is the scope and objective? What shall be the subject matter for audit? Should the auditor be accompanied during the tour? Whom may the auditor interview? The bulk of auditing practice provides for legitimacy—the auditor acts within the provisions of the charter plus supplemental agreements reached after discussion with all concerned.

Scheduled versus Unannounced

Most auditing is done on a scheduled basis. "No surprises, no secrets." This practice enables all concerned to organize workloads, assign personnel, etc., in an orderly manner. It also minimizes the irritations that are inevitable when audits are unannounced.

Customer

The customer of the audit is anyone who is affected by the audit. The key customer is the person responsible for the activity being audited. Other customers include upper management and functions affected by the activity. Each customer has needs that should be recognized during the planning of the audit. Note that this orientation of service to the activity audited means that the audit must go beyond compliance with a requirement. Such an orientation is not practiced (or even accepted) by all auditors, but the author believes that the concept is basic to useful audits.

Audit Team

Audits are conducted by individuals or by a team. A team usually has a lead auditor who plans the audit, creates the audit schedule, assembles or creates checklists or aid memoirs, conducts the meetings, reviews the findings and comments of the auditors, prepares the audit report, evaluates corrective action, and presents the audit report.

Clearly, auditors must be open-minded and possess sound judgment, have the trust and respect of line management, and be knowledgeable in the area audited. ISO 19011:2002 First Edition recommends other qualifications for auditors, including education, training, experience, personal attributes, and management capabilities. Auditing is a sensitive task. A survey of auditors and auditees in the financial services industry investigated five attributes of the audit and auditor: professionalism, business knowledge, risk perspective, audit planning and conduct, and reporting audit results. Several surprises: Auditees viewed professionalism (objectivity, knowledge of the area being audited) as three times more important than auditors did; auditors thought risk perspective (coverage of key risk areas, audit in sufficient detail) as three times more important than auditees did. Notice how this disparity illustrates the importance of understanding the needs of audit customers.

Use of Reference Standards and Checklists

As far as possible, the auditor is expected to compare activities as they are with some objective standard of what they should be. Where such standards are available, there is less need for the auditor to make a subjective judgment and thereby less opportunity for wide differences of opinion. However, provision should be made for challenge of the standard itself. The reference standards normally available include

- Written policies of the company as they apply to quality
- Stated objectives in the budgets, programs, contracts, etc.
- Customer and company quality specifications
- Pertinent government specifications and handbooks
- Company, industry, and other pertinent quality standards on products, processes, and computer software
- Published guides for conduct of quality audits
- Pertinent quality departmental instructions
- General literature on auditing

One type of checklist identifies areas of subject matter that are to be checked, leaving it to the auditor to supply the detailed checklist. Typical examples of such areas are maintenance of machines and tools or control of engineering change orders. Having a standard for comparison in audits is an important matter. Steven Ehrhardt of Mallinckrodt Inc. uses a perceptive principle. When his audit group is asked to make an audit, he first determines

Stage at Which Product Auditing is Conducted	Pros and Cons of Using This Stage
After acceptance by inspectors	Most economical, but does not reflect effects of packing, shipping, storage, or usage
After packing but before shipment to field	Requires unpacking and repacking, but evaluates effect of original packing
Upon receipt by dealers	Difficult to administer at such multiple locations, but reflects effects of shipping, storage
Upon receipt by users	Even more difficult to administer, but evaluates the added effects of dealer handling and storage plus effects of shipment to user and unpacking
Performance in service	The ideal, but also the most difficult to administer because of the number and variety of usages; can be simplified through sampling

TABLE 9.5 Potential Stages of Product Auditing

whether the company has a standard or clear job specification for the activity. If not, then before making the audit, he asks those responsible for the activity to define a standard. Without a standard, Ehrhardt believes, people do not know what they are supposed to do, and thus an audit is not appropriate. Clarification of standards may be more important than the audit itself. Imagine how this initiative to provide service by an auditor helps to build a trusting relationship with operations people.

Some checklists go into great detail, requiring the auditor to check numerous items of operational performance (and to record the fact that such items were checked). For example, an auditor checking a test performed by an inspector might be required to check the work of the inspector for the correctness of the specification issue number used, the list of characteristics checked, the type of instruments employed, the sample size, data entries, etc. In a hospital, an audit checklist could include questions such as, "Are applicable intravenous solutions stored under refrigeration prior to delivery?" and "Are all drugs, chemicals, and biologicals clearly, accurately, and appropriately labeled?" For an example, see Table 9.5.

Audit Performance

Several policy issues affect audit performance.

Verification of Facts

Auditors are universally expected to review with the line supervision the facts (outward symptoms) of any deficiencies discovered during the audit. The facts should be agreed on before the item enters a report that will go to higher management.

Discovery of Causes

Many companies expect the auditor to investigate major deficiencies to determine their causes. This investigation then becomes the basis of the auditor's recommendation.

Other companies expect the auditor to leave such investigations to the line people; audit recommendations will then include proposals for such investigations. As mentioned earlier, typically recommendations are not given for third-party audits.

Recommendations and Remedies

Auditors are invariably expected to make recommendations to reduce deficiencies and improve performance. In contrast, auditors are commonly told to avoid becoming involved in designing remedies and making them effective. However, auditors are expected to follow up recommendations to ensure that something specific is done, that is, that the recommendation is accepted or else considered and rejected.

Policy issues are often incorporated into a "quality audit manual." Such a manual also includes details on the subject matter to be covered in audits; checklists of items to be checked and questions to be asked; classification of the seriousness of deficiencies observed; use of software for entry, processing, storage, and retrieval of audit data; and guidelines for audit reports.

Status of the Audit

The key customer should be kept informed about progress of the audit—what has been covered and what remains to be done. The status of lengthy audits can be reported through debriefing meetings, informal discussions, and electronic mail. Status reporting includes explaining what deficiencies or problems have been detected—even before preparing a draft of the audit report. This is usually accomplished during the report close-out meeting at the conclusion of the audit or could be a daily report close-out meeting. Status reports enable the company to check the accuracy of the auditors' observations and give the people responsible for the activity audited a chance to explain their plan to correct the deficiency.

Audit Reporting

Audit results should be documented in a report, and a draft should be reviewed (preferably at the postaudit meeting) with the management of the activity that was audited. Auditors and the activity audited should agree in advance on the distribution of the audit report. If desired, the report may be issued by the auditor and the auditee. All members of an audit team and the auditee should sign the report.

The report should include the following items:

- Executive summary.
- Purpose and scope of the audit.
- Details of the audit plan, including audit personnel, dates, the activity that was audited (personnel contacted, material reviewed, number of observations made, etc.). Details should be placed in an appendix.
- Standards, checklist, or other reference documents that were used during the audit.
- Audit observations, including supporting evidence, conclusions, and recommendations—using the audit customer's terminology.
- Recommendations, if applicable, for improvement opportunities.
- Recommendations, if applicable, for follow-up on the corrective action that is to be proposed and implemented by line management, along with subsequent audits if necessary.
- Distribution list for the audit report.

Summarizing Audit Data

In an audit most elements of performance are found adequate, whereas some are found in a state of nonconformance to established standards. Reporting of these findings requires two levels of communication:

- Reports of nonconformance to secure corrective action. These reports are made promptly to the responsible operating personnel, with copies to some of the managerial levels.
- A report of the overall status of the subject matter under review. To meet these requirements, the report should:
  - Evaluate overall quality performance in ways that provide answers to the major questions raised by upper managers, for example, Is the product safe? Are we complying with legal requirements? Is the product fit for use? Is the product marketable? Is the performance of the department under review adequate?
  - Provide evaluations of the status of the major subdivisions of the overall performance—the quality systems and subsystems, the divisions, the plants, the procedures, etc.
  - Provide some estimate of the frequency of inadequacies in relation to the number of opportunities for inadequacies (see "Units of Measure").
  - Provide some estimate of the trend of this ratio (of inadequacies found to inadequacies possible) and of the effectiveness of programs to control the frequency of occurrence of inadequacies.

Seriousness Classification

Some audit programs use seriousness classification of inadequacies. This approach is quite common in product audits, where defects are classified in terms such as critical, major, and minor, each with some "weight" in the form of demerits. These systems of seriousness classification are highly standardized.

Some audit programs also apply seriousness classification to discrepancies found in planning, in procedures, in decision making, in data recording, and so on. The approach parallels that used for product audits. Definitions are established for such terms as "serious," "major," and "minor"; demerit values are assigned; and total demerits are computed.

Units of Measure

For audits of plans, procedures, documentation, etc., it is desirable to compare the inadequacies found against some estimate of the opportunities for inadequacies. Some companies provide an actual count of the opportunities, such as the number of criteria or check points called out by the plans and procedures. Another form is to count the inadequacies per audit with a correction factor based on the length of time consumed by the audit. The obvious reason is that more time spent in auditing means more ground covered and more inadequacies found.

Distribution of Audit Report

Traditionally, copies of the audit report are sent to upper management for notification, review, and possible follow-up. Clearly, managers of audited activities are not happy with audit reports listing various nonconformances that are sent to their superiors. To promote harmony and a constructive viewpoint on audits, some organizations have adopted a different policy. The audit report is sent only to the manager whose activity is audited, and a follow-up audit is scheduled. If the nonconformances are corrected in time for the follow-up audit, the audit file is closed; otherwise, a copy of both audit reports is sent to upper management.

Another approach is to establish prior to the audit the single point of accountability for the audited function. This individual is recognized as the point of contact for the lead auditor and the individual singularly responsible in the audited function for report receipt and internal distribution.

In the spirit of ongoing improvement, after the report is issued, the auditees should be asked about the value they received from the audit and the report.

Regel (2000) reports on a survey of quality auditors. Of their seven primary concerns in achieving closure on their audits, the one ranked first in importance was "report issues" that accounted for 25.3 percent of the total concerns. Examples of reasons included "insufficient detail," "not linked to business goals," "not stated clearly," and "not stated in management language."

Corrective Action Follow-Up

The final phase of the audit is follow-up to confirm that corrective action has been taken by the audited activity and that the corrective action is effective. The steps are shown in Fig. 9.2. ASQ Quality Audit Division provides details on this process.

It is important to remember the key purpose of an audit: improvement. If corrective action is not implemented for some reason, the auditor should first verify that the conclusions in the report are correct and agreed upon by the audited area. If this is not the case, then the lack of agreement must be resolved. If there is agreement with the report but the audited area has not been able to implement corrective action because of lack of resources or other reasons, the auditor should determine whether he or she can somehow help the audited area. One possible approach is to review how the seriousness of the deficiency is presented to management. Restating the deficiency in monetary or other terms that will have an impact on management can help to obtain the necessary resources or remove obstacles to implementation of the corrective action.

Human Relations in Auditing

In theory, the audit is an instrument plugged into operations to secure an independent source of information. Where it is a physical instrument, for example, the propeller speed indicator on the bridge of a ship, there is no clash of personalities. However, auditors are human beings, and in practice, their relationships with those whose work is being audited can become quite strained. Deficiencies turned up in the audit may be resented because of the implied criticism.

Recommendations in the audit may be resented as an invasion of responsibilities. In the reverse direction, auditors may regard slow responses to requests for information as a form of grudging cooperation. These and other human relations problems are sufficiently important to warrant extensive discussion plus indoctrination of both auditing personnel and operations personnel with respect to the following issues:

- *The reasons behind the audits.* These reasons may have been well discussed during the basic formulation of the audit program. However, that discussion was held among the managers. There is also a need to explain to both supervisors and nonsupervisors the "why" of the audits. (It is not enough to explain that upper management wants audits done.) Obviously, all employees are also customers, consumers, and concerned citizens, so it is easy to point out the benefits they derive from audits conducted in other companies. In addition, it can be made clear that the managers, customers, regulators, etc., of this company likewise require added assurance.
- *Avoiding an atmosphere of blame.* A sure way to cause a deterioration in human relations is to look for someone to blame rather than how to achieve improvement. Line managers as well as auditors can fall into this trap. An atmosphere of blame

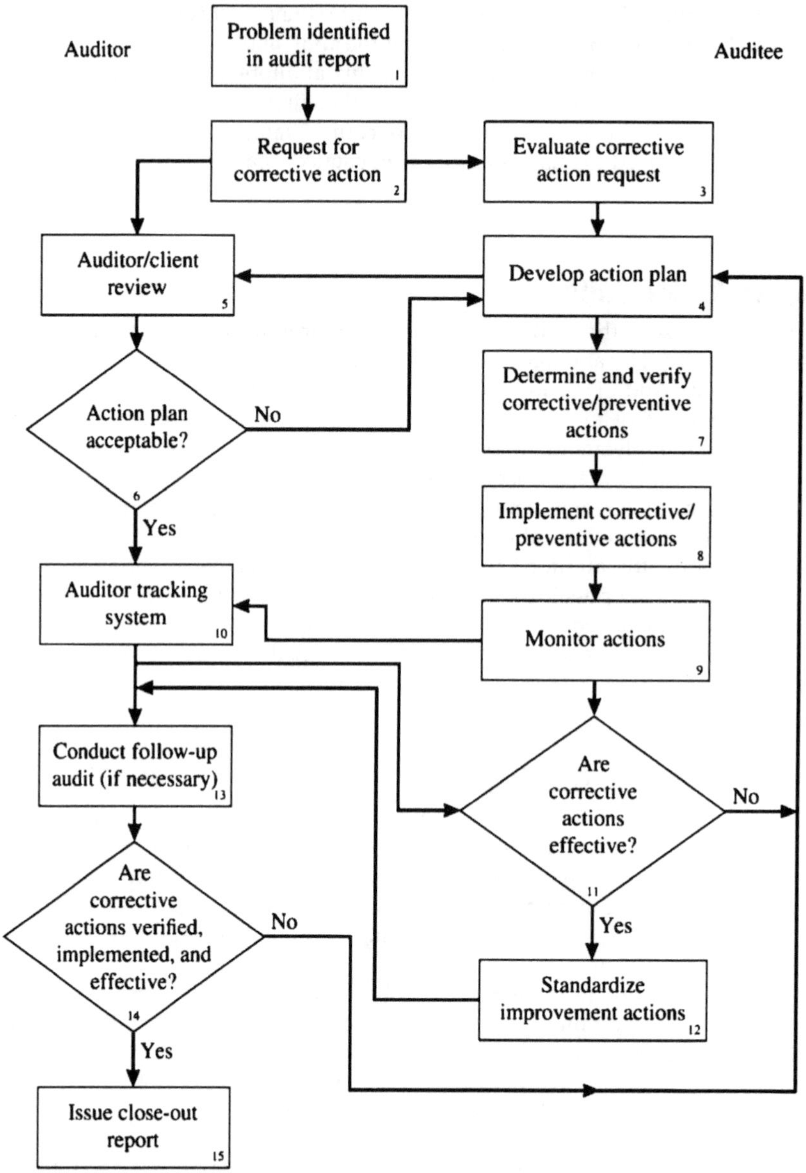

FIGURE 9.2 The audit function improvement process. (From Russell and Regel, 1996; reprinted by permission of the ASQ.)

breeds resentment and also dries up the sources of information. Audit reports and recommendations should be problem oriented rather than person oriented.

- *Balance in reporting.* An audit that reports only deficiencies may be factual as far as it goes. Yet it will be resented because nothing is said about the far greater number of elements of performance that are done well. ("Even a broken clock is correct twice a day"—Anonymous.) Some companies require that the auditors start their reports with "commendable observations." Others have evolved overall summaries or ratings that consider deficiencies and also the opportunities for deficiencies.

- *Depersonalizing the report.* In many companies, auditors derive much influence from the fact that their reports are reviewed by upper management. Auditing departments should be careful to avoid misusing this influence. The ideal is to depersonalize the reports and recommendations. The real basis of the recommendations should be the facts rather than the opinion of the auditor. Where there is room for a difference of opinion, auditors have a right and a duty to give their opinions as input to the decision-making process. However, any position of undue advocacy should be avoided because this tends to reduce the auditor's credibility as an objective observer. (The ultimate responsibility for results rests on the line managers, not on the auditors.)
- *Postaudit meeting.* An important part of the implementation phase is the postaudit meeting that is held with the manager and his or her team of the audited activity. At this meeting, the audit observations are presented so that the manager and the team can plan for corrective action. In addition, the manager and team can point out to the auditor any mistakes with respect to the facts that have been collected.

A self-audit and an independent audit can be combined to provide a two-tier audit. Each audit has an audit plan, execution, and report. The advantages include using the expertise of the person responsible for the function, assuring objectivity with an independent auditor, and minimizing some of the human relationship issues.

The aim of both the self-audit and the independent audit is to build an atmosphere of trust based on the prior reputation of the auditors, the approach used during the audit, and an emphasis on being helpful to the activity audited. Even such small matters as the title of the audit process should be carefully considered. Occasionally, people try to avoid the use of the term "audit" when what will be done is observation and evaluation. Also, audits may be hidden in a company education program. Such subterfuges detract from the trust that must be developed for audits to be effective and useful.

Auditors must function within the culture of an organization. Anthropology is the study of the origin; behavior; and physical, social, and cultural development of human beings. For the astute comments of a quality manager (who is a certified quality auditor) with a formal background in anthropology.

Quality auditors must, of course, adhere to the highest standards of ethics and professional conduct. For a discussion of the specifics including the American Society for Quality Code of Ethics and the Institute of Internal Auditors Code of Ethics.

Product Audit

A product audit is an independent evaluation of a product's quality to determine its fitness for use and conformance to specification. Product auditing takes place after inspections have been completed. The purposes of product auditing include

- Estimating the quality level delivered to customers
- Evaluating the effectiveness of the inspection decisions in determining conformance to specifications
- Providing information useful in improving the outgoing product quality level and improving the effectiveness of inspection
- Providing additional assurance beyond routine inspection activities

There is a good deal of logic behind such product audits. In many cases, the inspection and testing department is subordinate to a manager who is also responsible for meeting other standards (schedules, costs, etc.). In addition, there is a value in reviewing the performance of the entire quality control function, which includes inspection and test planning as well as the conduct of the tests themselves. Finally, the more critical the product, the greater the need for some redundancy as a form of assurance.

Stage of Evaluation

Ideally, the product audit should compare actual service performance with users' service needs. This ideal is so difficult and costly to administer that most product auditing consists of an approximation (see Table 9.5).

For many simple, stable products, the approximation of test results versus specifications is a useful, economical way of conducting a product audit. Even for products not so simple, most quality characteristics identifiable by the user are also completely identifiable while the product is still at the factory. Thus product characteristics that are essential to use are properly evaluated at some appropriate stage, whether in the factory or at some more advanced stage.

As products become increasingly complex, product auditing is increasingly conducted at several of the stages shown in Table 9.5. The bulk of the characteristics may be evaluated at the most economical stage, that is, shortly after factory inspection. However, the remaining (and usually more sophisticated) characteristics may be evaluated at other stages.

Scope of the Product Audit

The scope of some product audits completely misses the mark in measuring customer reaction.

As one example, the plant manager of an electronics manufacturing firm received a rating of 98 percent on a product audit from the plant. For this rating, the plant received an award for quality. When the mean time between failures of that same product was measured in the field, the value was only 200 hours. This problem was a known reason for customer complaints, but such matters had not been evaluated by the product audit.

In another case, a vehicle manufacturer had a system of taking a weekly product audit sample from production. A comparison of separate market research results with the internal product audit was devastating. Only 18 percent of the characteristics that customers claimed were important to them were being checked in the product audit.

For simple products, a representative sample of finished goods may be bought on the open market. These samples are then checked for fitness for use and conformance to specification. Some companies conduct such audits annually as part of broad annual planning for the product line. Such audits may include a review of competitive product as well.

For complex consumer products, for example, household appliances, it is feasible to secure product audit data at multiple stages of the product progression shown in Table 9.5. The most extensive product audit takes place immediately following factory inspection and testing. Additional audit data are then secured from selected distributors and dealers under a special joint "open and test" audit. Similar arrangements are made to secure data from selected servicing dealers. In addition, the data from consumer "arrival cards" are used. When properly arranged with due regard to time lags, all of these data sources can be charted to show trends as well as levels.

Audit plans must spell out, or give guidance on, the selection of detailed product dimensions or properties that are to be checked. Provision should be made for two types of audit—random and focused. The former is based on a random selection of product characteristics to yield an unbiased picture of quality status. A focused audit, on the other hand, concentrates on a specific area of the product that experience suggests needs to be studied. Many companies use audit manuals to spell out the design of the audit for the auditor, almost to the last level of detail. For example, the manual may specify particular categories of dimensions to be audited (i.e., length) but may rely on the auditor to select which length dimension to audit.

Sampling for Product Audit

For products manufactured by mass production, sample sizes for product audit can often be determined by using conventional statistical methods. These methods determine the sample size required for stated degrees of risk. Sample sizes for product audit determined by these methods when applied to mass production still represent a small fraction of the product that needs to be sampled. In contrast, for products manufactured as large units or in small quantities, the conventional concepts of statistical sampling are prohibitively costly. In such cases, sample sizes are often arbitrary, and they seem small from the viewpoint of probability. For example, a vehicle manufacturer uses a product audit sample consisting of 2 percent of production per shift with a minimum of five vehicles—whichever number is larger. Even though the number of vehicles sampled may be small, the total number of characteristics that is sampled may be quite large. Auditors check 380 items on each vehicle, and the product audit test includes a 17-mile road test. In some cases of highly homogeneous production, a sample of one unit taken from batch production can be adequate for a product audit.

Reporting the Results of Product Audit

The results of a product audit appear in the form of the presence or absence of defects, failures, etc. A continuing score or "rating" of quality is then prepared based on the audit results.

Product audit programs often use a seriousness classification of defects. Defects are classified in terms such as critical, major, minor A, minor B, and incidental at times, each with some "weight" in the form of demerits. In product audits, the usual unit of measure is demerits per unit of product.

EXAMPLE 9.1 *A product audit system uses four classes of seriousness of defects. During 1 month, the product auditors inspected 1200 finished units of product with the following results:*

Type of Defects	Number Found	Demerits per Defect	Total Demerits
Critical	1	100	100
Major	5	25	125
Minor A	21	5	105
Minor B	64	1	64
Total	91		394

Although the 91 defects found represented many defect types and four classes of seriousness, the total of 394 demerits, when divided by the 1200 units inspected, gives a single number, that is, 0.33 demerit per unit.

The actual number of demerits per unit for the current month is often compared against historical data to observe trends. (Sometimes it is compared with competitors' products to judge the company's quality versus market quality.) A major value of a measure such as

Class of Defect	Probability	Number of Defects Revealed by Audit	Cost per Service Call, $	Expected Costs, $	Expected Number of Service Calls
V1	1.00	1	15.00	15.00	1.00
V2	0.60	3	15.00	27.00	1.80
E1	1.00	3	30.00	90.00	3.00
E2	0.60	4	30.00	72.00	2.40
P1	1.00	1	25.00	25.00	1.00
P2	0.60	2	25.00	30.00	1.20
P3	0.20	2	25.00	10.00	0.40
Totals				269.00	10.80

TABLE 9.6 Audit Data of 50 Units

demerits per unit is that it compares discrepancies found with the opportunity for discrepancies. Such an index appeals to operating personnel as eminently fair.

A scoreboard in demerits per unit is by no means universally accepted. Managers in some industries want ready access to the figures on critical and major defects. These managers believe that such figures represent the real problems regardless of demerits per unit.

It is often useful to summarize the product results in other languages. A manufacturer of consumer products classifies defects in a product audit as visual (V), electrical (E), and performance (P) and then predicts service costs for products in the field. This is done by first establishing classes for each type of defect in terms of the probability of receiving a field complaint (e.g., a class 2 visual defect has a 60 percent probability). Service call costs are then combined with audit data. For example, Table 9.6 shows the results of an audit of 50 units. The expected cost is the product of the probability, the number of defects, and the cost per service call. The expected service cost per unit is then estimated as $269/50 = $5.38. Alternatively, as indicated in Table 9.6, the expected number of service calls is the product of the probability and the number of defects. The expected number of service calls per unit is then estimated as 10.8/50 = 0.22, or about 22 of every 100 products delivered to the field can be expected to have service calls.

In addition to summarizing the defects found (in both number and relative seriousness), the audit results can be tallied by functional responsibility (i.e., design, purchasing, production).

Audit results can also be summarized to show the effectiveness of the previous inspection activities. Typically, a simple ratio is used, such as the percentage of total defects detected by inspection. For example, if the previous inspection revealed a total of 45 defects in a sample of N pieces and if the product audit inspection revealed five additional defects, the inspection effectiveness would be (45/50)(100), or 90 percent.

Stravinskas (1989), Lane (1989), and Williams (1989) describe how AT&T Microelectronics changed from a traditional audit approach of reinspection of product (product audit) to an approach that uses system audits and process audits combined with a reduced amount of product audit. During one period, inspection costs were reduced by 12 percent, and the savings were then used to provide additional prevention activity, which resulted in a $2 saving in failure costs for each $1 added in prevention cost.

References

ASQ Quality Audit Division (2013). *The Quality Audit Handbook*, 2nd ed., ASQ, Milwaukee. Hunt, J. R. ed. (1997). "The Quality Auditor: Helping Beans Take Root," Quality Progress, December.

ISO 19011:2002: A New Auditing Standard for QMS and EMS, pp. 27–33.

Lane, P. A. (1989). "Continuous Improvement—AT&T QA Audits," ASQC Quality Congress Transactions, Milwaukee, pp. 772–775.

Regel, T. (2000). "Management Audit and Compliance Audit Compatibility," Annual Quality Congress Proceedings, ASQ, pp. 606–609.

Russell, J. P. and T. Regel (1996). *After the Quality Audit: Closing the Loop on the Audit Process*, Quality Press, ASQ, Milwaukee.

Stravinskas, J. M. (1989). "Manufacturing System and Process Audits," ASQC Quality Congress Transactions, Milwaukee, pp. 91–94.

Williams, C. A. (1989). "Improving Your Quality Auditing Systems," ASQC Quality Congress Transactions, Milwaukee, pp. 797–799.

CHAPTER 10

The Board of Directors and Organizational Excellence

Marcos E.J. Bertin

High Points of This Chapter

1. The role of the board of directors is to guide management in developing plans to maximize the stakeholder's satisfaction in line with sound organization sustainability and control that these plans are properly implemented.
2. What a board of directors does, or does not, do has an impact on the organization and on the organization's performance, including quality results. Unfortunately, this is not common knowledge to all boards, but it will become clear when analyzing the evolution of corporate governance.
3. This chapter addresses the quality of the board of directors (or equivalent governance body) and provides guidelines on
   a. Planning corporate governance implementation
   b. Improving the performance of the board of directors
   c. Becoming an effective director

4. Effective corporate governance provides proper incentives for the board and management to pursue objectives that are in the interests of the organizations and its stakeholders.
5. The most successful global organizations today recognize the key guidance and control role the board has to balance the alignment of the interests of all stakeholders to ensure sustainable growth and profitability.

Corporate Governance Evolves and Has an Impact on Organizations

Corporate governance is a dynamic issue. Here we discuss its evolution and its influence not only on organization practices but, more importantly, results, starting a few millenniums in the past all the way to the present role of the board.

For example, governing boards of health care organizations increasingly are called to respond to important new developments—clinical, operational and regulatory—associated with quality of care. Important new policy issues are arising with respect to how quality of care affects matters of reimbursement and payment, efficiency, cost controls, collaboration between organizational providers and individual and group practitioners. These new issues are so critical to the operation of health care organizations that they require attention and oversight, as a matter of fiduciary obligation, by the governing board. This oversight obligation is based upon the application of the fiduciary duty of care board members owe the organization and, for nonprofit organizations, the duty of obedience to charitable mission. It is additive to the traditional duty of board members in the hospital setting to be responsible for granting, restricting, and revoking privileges of membership in the organized medical staff.

The Role of the Board in Excellence and Quality

Corporate governance has not been a stranger to quality for quite some time:

- The International Academy for Quality (IAQ) started a quality-in-corporate governance (QiCG) project in 1996 during their triennial assembly in Yokohama, Japan. In 2007 the IAQ published the book *Corporate Governance Quality at the Top*, with major conclusions of the QiCG committee continuing to be one of the major IAQ activities.
- The National Malcolm Baldrige National Quality Award for Excellence, United States, includes corporate governance criteria in the leadership category, as do other national quality award organizations.
- The Secretary of State's Award for Corporate Excellence (ACE), established by the State Department in 1999, emphasizes the important role U.S. organizations play to advance best practices, good corporate governance, and democratic values overseas.

Corporate governance is much about leadership by example; as such, the board's commitment toward organizational excellence and quality should begin at home by becoming an effective board. The concepts and guidelines discussed in this chapter should lead in the right direction.

Effective or *quality* boards require great directors. The directors should become literate in excellence principle and quality methods as well as other major subjects such as IT, and any other subjects directly related to the particular organization. Training programs should keep directors updated.

The most successful global organizations today recognize the key guidance and control role the board has to balance the alignment of the interests of all stakeholders to ensure sustainable growth and profitability. To perform a quality job, directors should be aware of following developments:

- Significant increase of government regulations gradually becoming global.
- More demanding consumers.
- Global customers requesting suppliers to meet standards that include transparency and governance.
- Investment and pension funds corporate governance requirements.
- Rating agencies increasing requirements.
- Society and customers awareness requiring that organizations have sound social responsibility programs.
- Corporate governance standards are in the process of development.
- Certification programs available for directors in organizations such as the National Association of Corporate Directors, Washington, D.C.
- Corporate governance institutes in universities have research programs regarding the establishment of metrics and variability of corporate governance processes related to organizations results.

Organizations are gradually recognizing the need to adopt scientific quality tools in corporate governance, adapting those that were successfully applied in factories, service units, government, and health care, for example.

We all agree that metrics are important. However, the numbers are only as good as the help metrics provide the board in performing its functions. *The board's role is to guide management in the development of plans to maximize the stakeholders' satisfaction in line with sound organization sustainability and control that these plans are properly implemented.*

The processes of corporate governance are interdependent and complex involving all stakeholders. Therefore, although customer loyalty is still one of the key pillars of success, we must expand the vision that *quality is what the customer thinks it is* to: *quality is what the stakeholders think it is.*

The board has a significant role in driving quality of goods and services by ensuring there is a clear organization mission and vision with respect to quality in place. For nonprofit boards—in the context of what is generally referred to as the fiduciary duty of obedience to the corporate purpose and mission of health care organizations. Nonprofit corporations are formed to achieve a specific goal or objective (e.g., the promotion of health), as recognized under state nonprofit corporation laws. This is in contrast to the typical business corporation, which often is formed to pursue a general corporate purpose. It is often said of nonprofits that "the means and the mission are inseparable." The fundamental nature of the duty of obedience to corporate purpose is that the nonprofit director is charged with the obligation to further the purposes of the organization as set forth in its articles of incorporation or bylaws.

As Warren G. Bennis, the distinguished management teacher and author, said, "Leaders keep their eyes on the horizon, not just on the bottom line." To gain this broader, more future-oriented perspective, I believe that corporate boards should look at more than the rules to be followed; they should also carefully consider their role as leaders and the responsibilities that this leadership brings.

Ivey research studies have shown, that the Board must articulate a clear set of organizational values embracing integrity, fairness and compassion. Values are not the same as ethical guidelines or codes of conduct. Guidelines and codes are specific rules about what a board member can and cannot do. Values, on the other hand, are more profound and enduring. Values communicate beliefs.

During their extensive research, Ivey professor Pratima Bansal and Ivey MBA student Sonia Kandola discovered that organizations need a strong and consistent set of values to ensure all employees act with integrity. "These values establish a framework for what is considered to be acceptable within an organization." They further concluded that these values must be lived by every employee, especially board members and executives. Every employee looks to the corporate leadership team to show the way.

Ivey also states that boards of directors must have a vision for their organizations—a vision that recognizes that a company has a purpose as part of a larger community. The best companies are passionate about contributing to society—from the products they sell, to the examples they set, to the charities they support. The best companies have a vision that employees are proud of, that customers believe in, and that investors understand.

The Historical Responsibility to Shareholders

Modern organizations developed in the middle of the nineteenth century after a long history since its birth as far back as 3000 BC in Mesopotamia. However, it was not until the beginning of the twentieth century that the gradual separation of ownership from control took place: the shareholders who own the organizations from their agents who run them; this was the birth of the professional manager. The most important characteristics were

- The adoption of Frederick Taylor's first scientific rationalistic approach to management that resulted in a remarkable improvement in productivity.
- The introduction of the humanistic school of management thinkers led by Elton Mayo.
- A new class of management consultants such as Arthur D. Little, James McKinsey, and Peter Drucker that contributed to the development of the Organizations Man following Watson's IBM and Sloan's General Motors, successful models at the time.
- The impact of the research, courses, conferences, and publications organized the American Management Association founded in 1926.

No reference is found on the board of directors and its members in the huge amount of management literature generated during this period. Also, there were no other forms of professional developments for corporate directors as it was available for financial, commercial, human resources, or other management activities, including American societies for security analysts, quality control, and many other specialities. The National Association for Corporate Directors was founded in 1977. This confirms that in this "Organization Man" period, corporate directors believed that success depended only on the ability of management and that the role of the board of directors was solely to protect the interests of the owners. We review the limited view of the responsibilities of boards in the 1960s.

The 1960s: Juran and Louden Board Professionalizing Initiatives

In September to October 1961, The *Harvard Organization Review* published the first Code of Conduct for Executives. The President's Association of the American Management Association starts also in the 1960s to organize seminars for board members. A significant amount of consulting work and coaching is made with boards. These experiences resulted in the first book ever published on what we today call corporate governance by authors Dr. Joseph Juran and J. Keith Louden in 1966. This remarkable book was *The Corporate Director*. It is interesting to note that Dr. Juran was a member of several boards.

One of the two authors of this chapter received this book from Dr. Juran in 1969; this was the beginning of a lifetime interest in corporate governance and a valuable guide for the "what" and "how" of a long "hands-on" experience.

It is amazing to see that this book, most probably one of the "vital few" at the time, includes full chapters on subjects that today are required by government regulations and organizations: codes of best corporate governance practices worldwide. This includes, for example

- Composition of the boards
- Formalizing the jobs of directors and boards
- The professional director
- Organizing the boards and their committees
- Board meetings
- The CEO and the board
- Maintaining a healthy board

The final conclusion of Juran and Louden (1966) has become a reality today more than 40 years later:

> The present practice is mainly empirical. The men work from experience and instinct and many of them do a good job on it. But the growing importance of the board and the resulting need for ever improving performance by directors, suggest that empiricism has had its day. The board of tomorrow will work from a base of professionalism.

Today: Learning from the Consequences of Short-Term Profit and Other Fashions

We see the revival of corporate governance as Juran and Louden visualized it back in the 1960s. The question of how to align the interests of those who ran the organizations with the interests of those that owned has returned. Therefore, the need to enforce proper checks and balances, taking in consideration all stakeholders by developing and implementing sound and effective governance practices and regulations, has become paramount.

The revival of corporate governance started due to two well-known economic crises with the preparation and publication in 1999 of the *OECD Principles of Corporate Governance*. The first revision was in 2004 based on 5 years of applied experience. These principles assist governments in evaluating and improving a legal, institutional, and regulatory framework for corporate governance in their countries and to provide guidance and suggestions for stock exchanges, investors, corporations, and other parties that have a role in the process of developing good corporate governance. The principles represent a common basis essential for the development of good governance practices in all types of organizations.

Corporate governance provides the structure through which the objectives of the organizations are set and the means of attaining those objectives and monitoring performance are determined. Good corporate governance should provide proper incentives for the board and management to pursue objectives that are in the interests of the organizations and its shareholders and should facilitate effective monitoring. The presence of an effective corporate governance system, within individual organizations and across an economy as a whole, helps to provide a degree of confidence that is necessary for a market economy to function properly. As a result, the cost of capital is lower and firms are encouraged to use resources more efficiently, thereby underpinning growth.

The OECD Principles of corporate governance involve six basic areas:

1. Ensuring the basis for an effective corporate governance framework
2. Rights of shareholders and key ownership functions
3. Equitable treatment of shareholders
4. Role of stakeholders in corporate governance
5. Disclosure and transparency
6. Responsibilities of the board

There is a strong tendency to increase and enforce new regulations at the national and global level. It is interesting to observe the gradual globalization of regulations and the need the organizations have to update their corporate governance practices to comply with the increasing demands of investment funds, banks, major customers, and, of course, governments regulatory bodies.

On the other hand, research made by corporate governance institutes in an increasing number of universities, for example, Yale in the United States, St. Gallen in Switzerland, and Nankai in China and corporate directors associations such as the National Association of Corporate Directors (NACD) in Washington, D.C. and the Institute of Directors in London definitely prove the significant contributions that the application of corporate governance best practices have on organizations' results:

- Better corporate governance at both the firm and the country level results in higher valuations.
- Better corporate governance increases the variety of financing instruments available.
- Better corporate governance increases the effectiveness of management improving their performance.

Planning to Implement Corporate Governance

The approach to implement corporate governance should be based in identifying *what is good for the organization's organization*, which definitively includes complying with regulations. There are no recipes. Also, there are many different situations. Thus, after presenting a descriptive model and a couple of useful governance bodies we will provide guidelines from two different angles:

- Family organizations issues and requirements
- Stock market issues and requirements

Ancient Symbols for a Current Governance Model

In the previous section, it should have become clear about the relevance stakeholders currently have and how critical the board's job is in aligning interests. Here we discuss a pertinent model to continue exploring these and other relevant concepts.

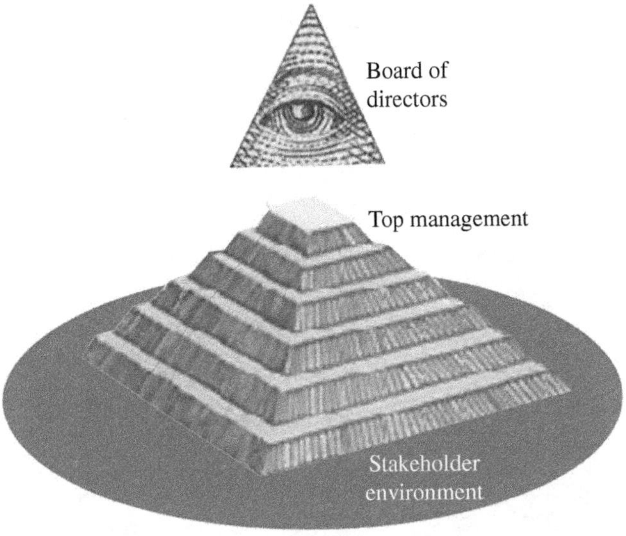

Figure 10.1 Step pyramid in perspective.

Again going back a few millennia in time, a step pyramid was built in Saqqarah, Egypt. Unlike the triangular shapes that were commonly used, this pyramid clearly differentiated the layers within the structure (see the comparison in Fig. 10.1).

As for the board, what better than to use another ancient symbol, one embedded in dollar bills, the All-Seeing Eye and the stakeholder environment? It simply surrounds the organization.

Now let us take a look from another direction (Fig. 10.2). At its center is the board (the triangle), then senior management, and then the lower levels within the organization. Finally, on the outside there's the stakeholder environment.

The arrows pointing from the top down refer to simple policies and rules that originate at board level and are passed to senior management for subsequent implementation and communication to the lower levels of the pyramid. The arrows pointing from the bottom up represent the feedback traveling back.

What about the outside-in arrow connecting the stakeholders and the board? It is very important. Directors must become aware about stakeholder expectations, evaluate them, and then determine which will be considered in the organization's strategy. Stakeholders then become part to the top-down/bottom-up process.

Board actions must be top-down to ensure that strategies are clear, easy to understand, budgeted, and in line with corporate objectives. The board does not implement; its main functions are to

- Provide strategic guidance of the organization
- Monitor management

Now, let us not forget that implementing corporate governance is about "what is good for the organization's organization." The lack of proven infallible recipes does not mean that there are no useful resources available, such as the following.

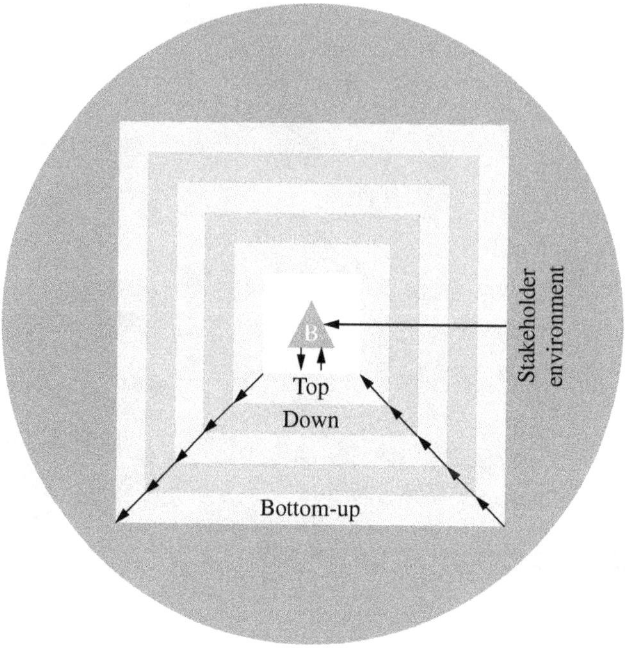

Figure 10.2 Step pyramid view from above.

Governance Bodies for Medium and Small Organizations

Among the most outstanding corporate governance characteristics in faster-growing medium and small organizations, we can highlight that

- It is not unusual to find an organization's CEO also acting as chairman of the board.
- Among the most important functions the board performs is advising senior management, right behind strategic planning.
- Boards do not have committees, or if they do, it is just the auditing committee.
- Many boards have a "hands-on" approach (absolutely not recommended; see Fig. 10.1), where directors complement management functions and get involved in running the organizations.

Hence, it becomes obvious that to achieve corporate governance, at the very least boards will need to go through a transition. Something similar can be said regarding many family organizations. The entities we describe here can help this and other purposes.

Advisory Boards

"Advise" is the keyword. Advisory boards act as an informal think tank and have no legal responsibilities (Fig. 10.3). Advisory boards have the same requirements as professional boards, and their effectiveness depends on the development and application of best practices based on well-documented processes that add value as explained in other sections in the chapter. Advisory boards can also be a first step for those organizations in transition to integrate a formal legal board.

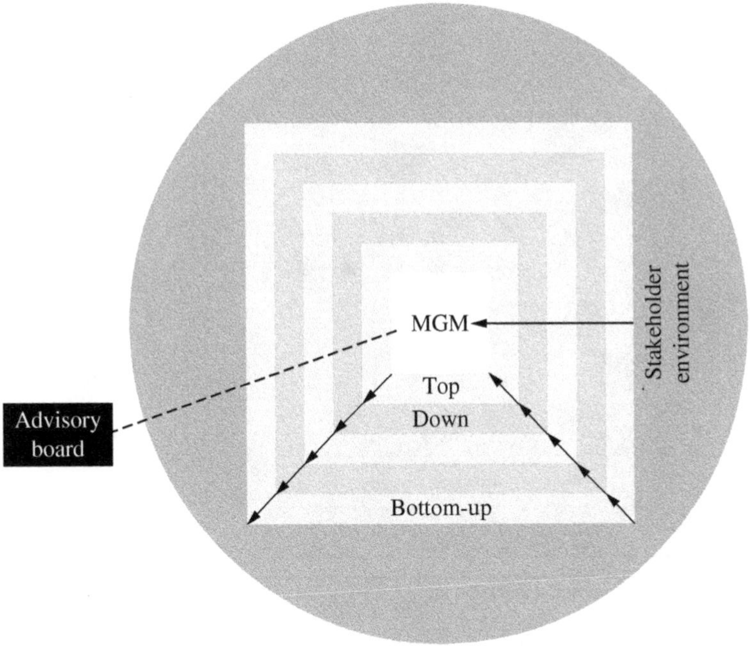

Figure 10.3 Advisory board relationship.

Seeking good advisory boards early in an organization's development can both ease and quicken the planning process. In the case of family organizations, an advisory board has the advantage of being detached from the family and can be more objective when it comes to who within the next generation is more capable of running the organization.

Family Councils

As we address in the following section, the organization's governance in family-owned businesses becomes more complex when more family members are directly or indirectly involved in the organization. Successful family organizations in many countries have, in addition to a board of directors, a family council that represents the family's shareholders (Fig. 10.4), which have the following functions and responsibilities:

- Holds formal meetings to discuss family and family-owned organization issues
- Establishes rules for family participation and guidelines for the board
- Becomes the only means of communication between the family-owned organization and the family shareholders
- Knows about the course of the family-owned organization/organizations
- Avoids individual contacts with management

Planning for Family-Owned Organizations

A family organization is one of the foundations of the world organization community. They represent over 70 percent of the total registered organizations. The moment to evaluate the

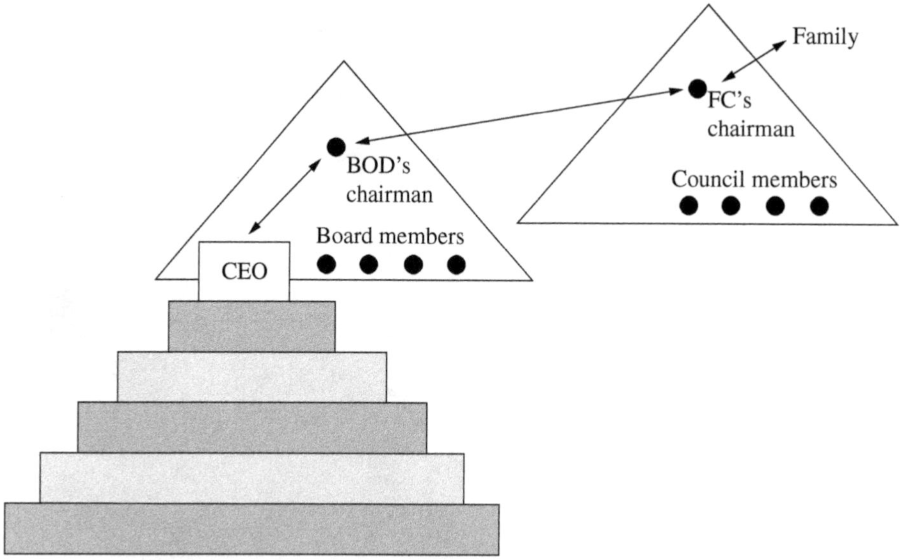

FIGURE 10.4 Family council relationship.

implementation of corporate governance is when the successful founders start planning retirement and the family members decide to continue with the organization. Steps that successful family organizations worldwide usually take are as follows:

- Initially with the collaboration of an outside consultant, the owner of the organization prepares a master plan that includes new members, a family council (FC), a professional board of directors, and potential members (family, key managers, independent external directors), including desired qualifications and needed training. If there is an existing board usually is a "rubber stamp" type that meets the minimum law requirements for organizations not listed in the stock market.
- Organize an FC with well-defined family directives to the board and formal operating procedures. This FC should be the only mean of communications between the family shareholders and the board.
- It is frequently preferred to have an advisory board before appointing a formal professional board. This facilitates the transition with family and management and the search for professional independent directors who might not be willing to become a board member with all the legal responsibilities involved in an organization they do not know much about.
- The right people are the key to the success of this project. In the case of the family members, they have to realize they can be managers, board members, FC members, or just shareholders. This depending on their background, preferences, capabilities, and training. Some family members might be delighted to become marketing managers and others might prefer to remain as shareholders in the organization. The number of family members tends to increase with each generation. When that's the case, it is also important to prune the family tree by buying the stock of members not interested in the organization.

To select external independent directors for the initial advisory board or the formal legal board in addition to the necessary experience and knowledge requires that independent directors have the following:

- A personality that earns the respect of his or her peers, family directors, and executive managers
- The ability to work effectively as a team
- The ability to challenge management while avoiding confrontation and refraining from invading their functions
- The ability to improve the chances to correct the sources of family conflicts

Systematic Corporate Governance Improvement

Improve by Focusing on Results

Having already discussed the impact that corporate governance has on an organization, the logical step is to figure out how to improve that governance. Actually, there are several viable diagnostic alternatives, each with its particular strengths and weaknesses, such as self-assessments or 360° evaluations. However, we felt that a TQM environment may require a special emphasis over results, which are a board's product, in a way.

To that extent, we have based ours on the method described in *An Approach to the Evaluation of a Board of Directors* by Hugo Strachan and Marcos E. J. Bertin, published in 2005 by The International Academy for Quality (IAQ). Although it was originally intended as a director's self-assessment, we have already successfully applied this concept to improve the quality of boards of several organizations.

It is important to note, that among its itemized categories, unlike others, this method includes board contributions (to results), which is probably the most relevant and the key to our scope. It covers two angles:

1. *Contributions on organizations results.* Successful critical strategies generated and controlled by the board that should be evaluated as well as key performance indicators, including intangibles. That is, the following items:
   - Organizations financial results
   - Competitive access to capital
   - Performance indicators
   - Brand value/organization image
   - Organization's intellectual capital value
   - Risk management
2. *Stakeholder's evaluation of the board.* Stakeholders include management, personnel, major suppliers, customers, government, society, and controlling and minority shareholders. It is the following two items:
   - Stakeholders' evaluation of the Board
   - Community perception of the organizations as a whole

Note that this is "outside-in" input; thus, the board must define how the stakeholders' opinions are determined and evaluated.

The Four-Level Scoring Criteria Applied to Results

The chosen method encourages scoring according to the level of evolution of corporate governance implementation that has been reached, that is

Level 1. Understanding (qualification 0 to 1): The board understands the need to improve corporate governance in the respective point or area.

Level 2. First steps (qualification: >1 to 3): The board has taken concrete steps toward establishing best practices in the respective point or area.

Level 3. Implementation (qualification: >3 to 7): The board has implemented improvements to corporate governance in the respective point or area.

Level 4. Leadership (qualification: >7 to 10): The board has reached the best achievable improvements to corporate governance in the industry for the respective point or area.

This scheme helps to narrow the variance in estimating values. The following questions are also designed to reinforce that effect:

1. *Organizations financial results.* Are they consistent with shareholders'/owners' expectations? If not, how proactive is the board in aligning them with their needs?
2. *Competitive access to capital.* Does the board assist in this regard? If not, how aware is the board of its role?
3. *Performance indicators.* Are they already in the desired state? Is the board helping the organization to further improve performance?
4. *Brand value/organization image.* Have goals been met yet? Is the board proactively working on improving them? How close are they to align the organization's image with industry standards?
5. *Organizations' intellectual/intangible capital value.* Are these capitals a board's concern? Is it active in generating/preserving such capital? Are results good enough?
6. *Risk management.* Is risk management a key factor in an organization's success? If so, does the board have any merit?
7. *Stakeholders' evaluation of the board.* Do they perceive that critical/needed strategies are implemented? At least improving? How relevant is the board in generating them?
8. *Community perception of the organization's as a whole.* Is the organization evaluated? If so, is it perceived as being good enough? If not, is it at least improving?

Any board of directors that evaluates these or similar questions regularly enough has definitively taken a step in the right direction.

As for metrics, a form such as the one in Fig. 10.5 could be used. In this case for board contributions, note that on the right side, we include an example for clarification purposes. Qualifications are represented by the *Q* column in Fig. 10.5 and their respective levels by the *I–V* column.

Although a resulting level of *III* in Fig. 10.5 may seem fine, being one the board's most relevant functions, a level of *I* for stakeholder evaluation would be inadmissible. This shows why each and every board contribution item should be seriously considered throughout the improvement method we describe as follows.

	Board contributions	Q	I–IV	Example Q	Example I–IV
a	Company financial results			7	III
b	Competitive access to capital			3	II
c	Performance indicators			5	III
d	Brand value company image			8	IV
e	Intellectual capital value			4	III
f	Risk management influence			7	III
g	Stakeholders' evaluation of the board			1	I
h	Company's community perception			4	II
	Max	80		80	
	Total			39	
	Result			**4,88**	**III**

Q = Qualification
IV = Level

FIGURE **10.5** Calculations form.

The Result-Oriented Improvement Method

The processes of corporate governance are interdependent and complex and involve all stakeholders. Although a handful of organizations have research programs regarding establishing their metrics and variability, it may take a while and it is still yet to be seen how practical they will be, especially to boards of small- and medium-sized organizations. Thus, we take a different approach, "Improvement through Awareness."

Further along this section we supply a list of categorized subjects for frequent review. The idea is to identify issues that could potentially influence a board's capability to contribute to results, the eight points we discussed earlier. Although experienced directors may find this easier, the less experienced would gain familiarity. Given the divergence some board matters tend to have, such as stakeholder interests, it is important to refocus periodically.

A form like the one in Fig. 10.6 should prove useful to

- List the potentially relevant subjects (left rows)
- List respective board contributions (middle columns)
- Mark identified relationships between subjects and contributions (crosses in row intersections with columns)
- Quantify using the four-level scoring criteria described earlier

				Board contributions										
				Company financial results	Competitive access to capital	Performance indicators	Brand value company image	Intellectual capital value	Risk management influence	Stakeholders' evaluation of the board	Company's community perception	Goal (I–V)	Priority	Goal met?
	Identified relevant items	Q	I–IV	III	II	III	IV	III	III	I	II			
6.3	Sample item: Third in category 6	3	II		x			x				III		
6.4	Sample item: Fourth in category 6	5	III					x		x		IV		
5.2	Sample item: Second in category 5	4	III											
4.1	Sample item: First in category 4	1	I	x					x			III		
4.4	Sample item: Fourth in category 4	4	II											
			Goal	IV	III	III	IV	IV	IV	III	III	Goals to reach during the following period		
			Priority											
			Met?											

Q = Qualification
IV = Levels

FIGURE 10.6 Cross-reference improvement form.

- The assembled information would then allow for:
  - Disregarding subjects that ended up not being that relevant (remove crosses)
  - Adding newly identified cause-effect relationships (add crosses)
  - Establishing improvement goals and priorities (right and bottom bars)

This would provide a full and single view of many of the probably most important practical board matters.

Categorized Subjects

Purposely, to avoid biasing, sample rows do not mention any real issues. In the example in Fig. 10.6, items are chosen from hypothetical categories 4 to 6. In the first row, it is the third item of category 6.

Because every organization has different objectives and needs, each one should eventually be able to select what serves their own requirements most in this chapter. These categorized subjects are intended to serve that purpose by inspiring useful topics for the board's agenda, or for custom self-assessment questionnaires or to use in the improvement process described earlier (see Fig. 10.6) and are chosen from the following categories:

1. Mission and principles
2. Board and stakeholders
3. Board and shareholders
4. Board and management
5. Operating procedures
6. Board structure

There are also board subjects regarding IT governance, but, for didactic matters, they are treated separately in the "Corporate and IT Governance" section of this chapter, where in

					Board contributions							
			Q	I-IV	Company financial results	Competitive access to capital	Performance indicators	Brand value company image	Intellectual capital value	Risk Management influence	Stakeholders' evaluation of the board	Company's community perception
IT governance	a	Commitment to IT										
	b	IT alignment										
	c	IT added value										
	d	IT cost management										
	e	IT projects success										
	f	Information availability										
	g	IT related risks										
	h	Operational IT framework										
		Max	80									
		Total										
		Result										

Q = Qualification
IV = Levels

Figure 10.7 IT governance sample form.

addition to the pertinent subjects, we supply a sample form (Fig. 10.7) and suggest a few guiding questions for each. This should give an idea on processing whole categories. A thorough description of all categories is available in Bertin and Watson (2007).

Categories here are numbered in reverse order, not just to catch the reader's attention, but because it represents moving from outside the pyramid (see Figs. 10.1 and 10.2)—stakeholders, all the way to the core, the mission, principles and values of the organizations. This means that the depth of any analysis would depend on category choices, the subjects of which are listed as follows.

Mission and Principles

Proactive establishment of mission and principles, their deployment and employee commitment, liabilities awareness, risk anticipation and mitigation, risks and crisis management policies, integrated risks and compliance vision, and code of best practices.

Board Structure

Adequate size, board member profiles and selection, independent chairman versus CEO and lead director, influence of independent directors, board committee functions, and integration.

Board Operating Procedures

Member selection policy, independent director policy, function descriptions, training and orientation, professional management meetings, members' contribution to agenda, board compensation and review, nondirectors' participation in meetings, and board and director assessment.

Board and Management

CEO evaluations, senior management compensation, access to management, CEO succession, relevant information supply, consideration of risk and crisis management, management training, and development.

Board and Shareholders

Remunerations disclosure, one share one vote, director, committee and board assessments, organizations information delivery, ownership structure, compliance disclosure, organizations profits, extraordinary transactions, member election, and dismissal.

Board and Stakeholders

Stakeholder expectations and strategy, organizations formal disclosures (including institutional investors, customers, and press) and communications, and board activity regarding community issues.

References

Bertin, M. E. J. (1996). "Quality on the Board of Directors." *Proceedings of the JUSE International Conference on Quality*, Yokohama, Japan.

Bertin, M. E. J. (2004). "The Impact of an Effective Board of Directors in Organizations Results." *Proceedings of the 48th European Organization for Quality*, Moscow, Russia.

Bertin, M. E. J., and Strachan, H. (2005). *An Approach to the Evaluation of a Board of Directors*. ASQ World Conference on Quality and Improvement. Seattle, Washington.

Bertin, M. E. J., and Watson, G. H., eds. (2007). *Corporate Governance: Quality at the Top*. International Academy for Quality, Salem, New Hampshire, pp. 83–126.

Bertin, M. E. J. (2007). "Quality in the Board of Directors—What the Board Should Do—How to Measure Board Performance." *Proceedings of the 1st Middle East Quality Organization Annual Conference*, Dubai, UAE.

Bertin, M. E. J., and Bertin Schmidt, M. (2009). "Where Were the Boards? Thoughts Regarding the Global Financial Crisis." *Proceedings of the 3rd Middle East Quality Organization Annual Conference*, Dubai, UAE.

Corporate Social Responsibility: Why Good People Behave Badly in Organizations. By Bansal, Pratima, Kandola, Sonia. Ivey Business Journal Online. March/April 2003.

Juran, J. M., and Louden, J. K. (1966). *The Corporate Director*. American Management Association, New York.

OECD—Organisation for Economic Co-Operation and Development (2004). OECD Principles of Corporate Governance. Paris, France.

Stephenson, C. (2004). "A Board's Role in Fostering Vision, Values, and Integrity." *Ivey Business Journal (Web)*.

Strachan, H., and Bertin, M. E. J. (2005). "An Approach to the Evaluation of a Board of Directors." *ASQ World Conference on Quality and Improvement*, Seattle, WA.

CHAPTER 11

ISO 9000 Quality Management System

Joseph A. De Feo

High Points of This Chapter

1. ISO 9000 standards have had great impact on the implementation of international trade and quality systems by organizations worldwide. The standards have been applied in a wide range of industry/economic sectors and government regulatory areas. The ISO 9000 standards deal with the management systems used by organizations to ensure quality in: design, production, delivery, and support products.
2. The new ISO 9001:2015 standard includes several important changes for Quality Management Systems, including modifications in terminology, the introduction of new context-based clauses, emphasis on management's role in quality, and a focus on risk-based approach.
3. To maintain its registered status, the supplier organization must pass periodic surveillance audits by a registrar. Surveillance audits are often conducted semiannually. The audits may be less comprehensive than a full audit. If so, a full audit is performed every few years.

4. The ISO 14000 is a standard for an environmental management system. It is applicable to any business, regardless of size, location, or industry. The purpose of the standard is to reduce the environmental footprint of a business and to decrease the pollution and the waste a business produces.
5. cGMP refers to the Current Good Manufacturing Practice regulations enforced by the U.S. Food and Drug Administration (FDA). cGMPs provide for systems that ensure proper design, monitoring, and control of manufacturing processes and facilities.
6. AS9100 is a widely adopted and standardized quality management system for the aerospace industry.

International Standards Overview

Standards exist principally to facilitate international trade and to avoid harming customers and society. In the prestandardization era (before 1980), there were various national and multinational standards. Standards for electrical, mechanical, and chemical process compatibility have been around for decades. Other standards such as military standards were developed for the military and other groups for the nuclear power industry, and, to a lesser extent, for commercial and industrial use. These standards have commonalities and historical linkages. However, they were often not consistent in terminology or content for widespread use in international trade. As a result, organizations were left to recreate their own standards or adapt the existing ones. This only led to even less commonality. In the 1980s as most of the organizations in the industrialized world began to improve quality and safety at record paces there became a need to fill a void. That void was a common quality management system that would be a nonbinding "contract" between the customer and the supplier. This void was filled by the ISO 176 Technical Committee in the form of the ISO 9000 set of standards. This was later followed by filling a similar void for environmental standards with ISO 14000. Many organizations globally began using these standards as a "certified" standard for performance. Although their intent was important, the standards became more of an opportunity to get a certificate of compliance that could be used to impress customers, rather than a set of requirements that ensured that customer needs are met.

Certain industry/economic sectors then began developing industrywide quality system standards, based upon the verbatim adoption of ISO 9000, together with industrywide supplemental requirements. The automotive industry (QS 9000), the pharmaceutical and medical devices industry (cGMPs), government regulatory agencies, and military procurement agencies (AS9100 and the Mission Assurance Provisions, MAP), are adopting this approach in many places worldwide. Even software development uses the CMMI standard of software quality systems created in the early 1990s at Carnegie Mellon University to ensure a common approach to manage software quality. The standards play an important—but not always understood—role in managing for superior quality.

We will include a brief discussion on the following standards and or industry practices:

- ISO 9000 for Quality Management Systems
- ISO 14000 for Environmental Managements Systems
- cGMPs for Pharmaceutical and Medical Devices
- ISO/TS 16949: Automotive Industry
- AS9100 and MAP in the U.S. defense industry

ISO 9000 Quality Management System Standard

The ISO 9000 standards have had great impact on international trade and quality systems implementation by organizations worldwide. The international standards have been adopted as national standards by over 70 countries. They have been applied in a wide range of industry/economic sectors and government regulatory areas. ISO 9000 standards deal with management systems used by organizations to ensure quality in: design, production, delivery, and support products. The standards apply to all generic product categories: hardware, software, processed materials, and services. The complete set of ISO 9000 family of standards provides quality management guidance, quality assurance requirements, and supporting technology for an organization's quality management system. The standards provide guidelines or requirements on what features are to be present in the management system of an organization but do not prescribe how the features are to be implemented. This nonprescriptive character gives the standards their wide applicability for various products and situations. Upon implementing ISO 9000, an organization can be registered as a Certified Quality Management System.

The standards in the ISO 9000 family were created and are produced and maintained by Technical Committee 176 of the International Organization for Standardization (ISO). The first meeting of ISO/TC176 was held in 1980. ISO 8402, the vocabulary standard, was first published in 1986. The initial ISO 9000 series was published in 1987, consisting of the following:

- Fundamental concepts and road map guideline standard ISO 9000
- Three alternative requirements standards for quality assurance (ISO 9001, ISO 9002, or ISO 9003)
- Quality management guideline standard ISO 9004

Since 1987, additional standards have been published. The ISO 9000 family now contains a variety of standards supplementary to the original series. In particular, revisions of the basic ISO 9000 series, ISO 9000 through ISO 9004, were published in 1994, 2000, 2008, and, most recently, in 2015 under the name ISO 9000:2015. This section is written in relation to the 2015 revisions after an initial introduction to the original standard.

ISO 9000 has been adopted and implemented worldwide for quality assurance purposes in both two-party contractual situations and third-party certification/registration situations. Periodic surveillance audits that are part of the third-party certification/registration arrangements worldwide provide continuing motivation for supplier organizations to maintain their quality systems in complete conformance and to improve the systems to continually meet their objectives for quality.

The market for quality management and quality assurance standards itself grew rapidly, partly in response to trade agreements such as the European Union (EU), the General Agreement on Tariffs and Trade (GATT), and the North American Free Trade Association (NAFTA). These agreements all depend upon standards that implement the reduction of nontariff trade barriers. The ISO 9000 family occupies a key role in implementing such agreements.

ISO 9001:2015

Certificates for the ISO 9001 Standard have been issued to over one million organizations in over 170 countries. After 25 years since the implementation of the first ISO 9001 Standard, it became necessary to introduce a revised version. While the scope of the 9001 Standard did not change substantially, the structure and terms have been updated to allow for easier

ISO 9001:2008	ISO 9001:2015
Products	Products and services
Exclusions	Not used
Documentation, records	Documented information
Work environment	Environment for the operation of processes
Purchased product	Externally provided products and services
Supplier	External provider

TABLE 11.1 Major Differences in Terminology from ISO 9001:2008 to ISO 9001:2015

alignment with other management systems standards. ISO 9001:2015 now follows the same *High-Level Structure* as other ISO Standards, which makes integration easier for organizations using multiple management systems. As far as new terminology, organizations may choose which terms best suit their operations and are not required to adapt the new terms. Table 11.1 displays the major differences in terminology from the ISO 9001:2008 to the ISO 9001:2015 standards.

One of the key terminology changes is the replacement of *products* with *products and services;* this was done so that the standard is more applicable to a wider range of industries, and specifically service industries. *Documents* and *records* were replaced with *documented information* in order to allow for more flexibility in organizations' documentation processes. This leaves the decision to the organization to determine which procedures need to be documented to what extent.

In addition to changes in terminology, two new clauses have been introduced regarding the context of the organization: clause 4.1 Understanding the organization and its context and clause 4.2 Understanding the needs and expectations of interested parties. These new clauses require the organization to identify a wide range of issues and requirements that may impact the planning of their quality management system. The clauses assume a stakeholder approach and focus on Stakeholder Relationship Management (SRM), which goes beyond the previous emphasis on Customer Relationship Management (CRM). With this approach, stakeholders include all interested parties, but only requires the organization to fulfil the needs and expectations of those interested parties who are deemed relevant to the Quality Management System.

The 2015 revision also places a more significant focus on leadership for being accountable for quality. The standard requires that management assigns, communicates, and ensures understanding of quality roles and responsibilities. It also requires that management widen their scope of review to the strategic level, including the interests of *relevant interested parties* when considering risks and opportunities for the organization.

Another significant change in the 9001:2015 Standard is the emphasis on a risk-based approach. While there is risk inherent in all aspects of a quality management system, risk-based thinking ensures that these risks are identified, considered and controlled throughout the design and implementation of the process. Risk-based thinking was introduced in this version of the standard in several clauses. See Table 11.2 for where risk-based thinking is addressed in the new standard. Organizations are required to understand risk assessment and identify and act on both risks and opportunities. However, with the new standard there is no formal requirement for a documented risk management process.

ISO 9001:2015 Clause	Application of Risk-Based Thinking
Clause 4	The organization is required to determine its QMS processes and to address its risks and opportunities
Clause 5	Top management is required to • Promote awareness of risk-based thinking • Determine and address risks and opportunities that can affect product/service conformity
Clause 6	The organization is required to identify risks and opportunities related to QMS performance and take appropriate actions to address them
Clause 7	The organization is required to determine and provide necessary resources (risk is implicit whenever "suitable" or "appropriate" is mentioned)
Clause 8	The organization is required to manage its operational processes (risk is implicit whenever "suitable" or "appropriate" is mentioned)
Clause 9	The organization is required to monitor, measure, analyze and evaluate effectiveness of actions taken to address the risks and opportunities
Clause 10	The organization is required to correct, prevent or reduce undesired effects and improve the QMS and update risks and opportunities

TABLE 11.2 How Risk-Based Thinking Is Addressed in ISO 9001:2015

External Driving Forces

The driving forces that have resulted in widespread implementation of the ISO 9000 standards can be summed up in one phrase: the globalization of business. Expressions such as the "postindustrial economy" and "the global village" reflect profound changes in recent decades. These changes include the following:

- New technology in virtually all industry/economic sectors
- Worldwide electronic communication networks
- Widespread worldwide travel
- Dramatic increase in world population
- Depletion of natural resource reserves, arable land, fishing grounds, and fossil fuels
- More intensive use of land, water, energy, and air
- Widespread environmental problems/concerns
- Downsizing of large organizations and other organizations, flattened organizational structure and outsourcing of functions outside the core functions of the organization
- Number and complexity of language, culture, and legal and social frameworks encountered in the global economy
- Diversity a permanent key factor
- Developing countries becoming a larger proportion of the total global economy; there are new kinds of competitors and new markets

These changes have led to increased economic competition, increased customer expectations for quality, and increased demands upon organizations to meet more stringent requirements for quality of their products.

Globalization of business is a reality even for many small- and medium-size organizations. These smaller organizations, as well as their large counterparts, now find that some of their prime competitors are likely to be based in another country. Fewer and fewer businesses are able to survive by considering only competition within the local community. This affects the strategic approach and the product planning of organizations of all sizes.

Internal Response to the External Forces

Organizations everywhere are dealing with the need to change. There is now greater focus on human resources and organizational culture and on empowering and enabling people to do their jobs. ISO 9000 implementation involves establishing policy, setting objectives for quality, designing management systems, documenting procedures, and training for job skills. All of these elements are parts of clarifying what people's jobs are.

Organizations have adopted performance excellence programs that include business process management as a means of adapting to changing customer needs. This concept is emphasized in the ISO 9000 standards. Metrics are being used increasingly to characterize product quality and customer satisfaction more effectively.

Organizations are implementing better product design and work-process design procedures, and improved production strategies. Benchmarking and competitive assessment are used increasingly.

An important question that is often asked is "In this world of rapid change, how can a single family of standards, ISO 9000, apply to all industry and economic sectors, all products, and all sizes of organizations?"

ISO 9000 standards are founded on the concept that the assurance of consistent product quality is best achieved by simultaneous application of two kinds of standards:

- Product standards (technical specifications)
- Quality system (management system) standards

Product standards provide the technical specifications that apply to the characteristics of the product and, often, the characteristics of the process by which the product is produced. Product standards are specific to the particular product: both its intended functionality and its end-use situations that the product may encounter.

The management system is the domain of the ISO 9000 standards. It is by means of the distinction between product specifications and management system features that the ISO 9000 standards apply to all industry/economic sectors, all products, and all sizes of organizations.

Distinctions between Organizational Performance Excellence Programs and ISO Standards

The ISO 9000 family standards contain requirements and guidelines. It is a quality management system model to be used for quality assurance purposes for providing confidence in product and service quality. A requirements standard becomes binding upon an organization wherever the organization:

- Is explicitly called up in a contract between the organization and its customer
- Seeks and earns third-party certification and registration

All of the ISO 9000 family standards are generic, in the sense that they apply to any product or any organization. All of the ISO 9000 family standards are nonprescriptive in the sense that they describe what management system functions shall or should be in place; but they do not prescribe how to carry out those functions.

ISO 9004 is similar to many National Awards for Excellence in that it provides a model for organizational performance excellence and quality management. The major difference is most National Awards Criteria are business outcomes focused and the ISO 9000 standards are not. Why? Because the ISO 9000 was not developed to include many of the enablers and influencers that will assure that all processes (production and nonproduction) in an organization are continuously improved. In fact, ISO 9000 is focused on effectiveness of processes, and National Awards and most Organization Performance Excellence Systems are focused on effectiveness and efficiency.

The ISO 9000 standards do not include the full scope of managing for quality. They were not designed to be. As a result of complaints, we often hear ISO 9000 did not do what we expected it to do, whereas others said it was great for them. The ones that stated it did not work had an expectation that the standard alone, once implemented, would guarantee improved quality and better financial performance. They were not satisfied. They also did not know that the standard does not include provisions for these other tasks that must happen beyond the product and service production processes.

If ISO 9004 were the registration standard, more organizations would see the benefit of registrations because ISO 9004 can be used as an organizational performance excellence system. ISO 9001 is only an assurance system focused on processes that only impact customer requirements being met (Table 11.3). This is a subset of what is needed to for organizational performance excellence programs; the organizations that stated that ISO 9004 worked for them used the standard as a building block to a better system. They filled in the gaps where the standard was not designed to do. As a result, these organizations saw ISO Standards as an important part of their performance excellence program.

We define organizational performance excellence systems with the inclusion of planning, control, and improvement methods and applied to all processes in an enterprise, as essential to manage for quality. Quality Assurance is an important part of this system since it provides information on how our system is performing to predetermine product specs and plans.

The Prime Focus of	
Quality Management	**Quality Assurance**
• *Achieving* results that satisfy the requirements for quality • Motivated by stakeholders *internal* to the organization, especially the organization's management • Goal is to satisfy *all stakeholders* • Effective, efficient, and continually improving overall quality-related *performance* is the intended result • Scope covers all activities that affect the total quality-related *business results* of the organization	• *Demonstrating* that the requirements for quality have been (and can be) achieved • Motivated by stakeholders, especially customers, *external* to the organization • Goal is to satisfy all *customers* • *Confidence* in the organization's products is the intended result • Scope of demonstration covers activities that directly affect quality-related *process and product results*

TABLE 11.3 Quality Management and Quality Assurance

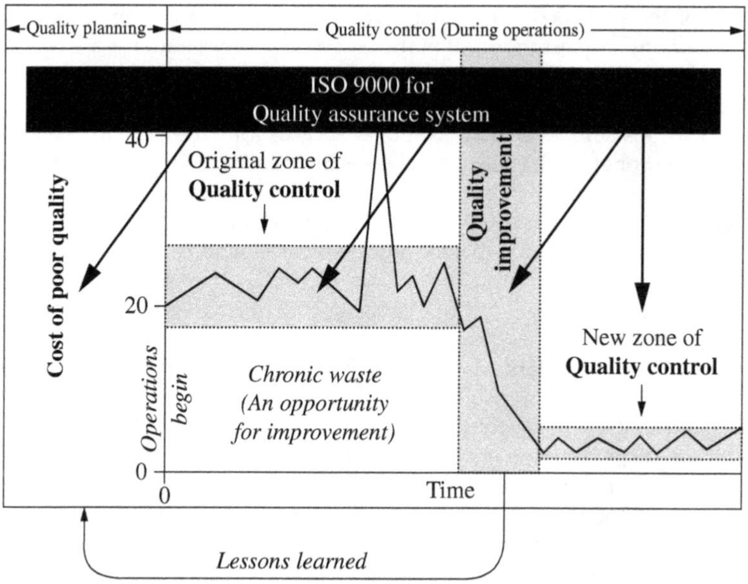

Figure 11.1 ISO 9000 and the Juran Trilogy. (Juran Institute, Inc., Southbury CT.)

Quality control is different from quality assurance. Control is about what to monitor to ensure requirements are met while assurance is about reviewing and auditing your system to prove that your organization is indeed holding the gains on improvements or changing customer needs (Fig. 11.1).

One of the most pressing needs in the early years of ISO/TC176 work was to internationally harmonize the meanings of terms such as "quality control" and "quality assurance." These two terms, in particular, were used with diametrically different meanings among various nations, and even within nations. The term "quality management" was introduced into the ISO 9000 standards as the umbrella term for quality control and quality assurance. The term "quality management" was defined, included in ISO 8402, and adopted internationally. This, in turn, enabled agreement on harmonized definitions of the meanings of each of the terms "quality control" and "quality assurance."

According to ISO 9000:2015, the fundamental concepts and principles of quality management in the International Standard are applicable to the following:

- Organizations seeking sustained success through the implementation of a quality management system
- Customers seeking confidence in an organization's ability to consistently provide products and services conforming to their requirements
- Organizations seeking confidence in their supply chain that product and service requirements will be met
- Organizations and interested parties seeking to improve communication through a common understanding of the vocabulary used in quality management
- Organizations performing conformity assessments against the requirements of ISO 9001

- Providers of training, assessment, or advice in quality management
- Developers of related standards

Quality System Certification/Registration

The earliest users of quality assurance requirements standards were large customer organizations such as electric power providers and military organizations. These customers often purchase complex products to specific functional design. In such situations, quality assurance requirements are called up in a two-party contract where the providing organization (i.e., the supplier) is referred to as the "first party" and the customer organization is referred to as the "second party." Such quality assurance requirements typically include provisions for the providing organization to have internal audits sponsored by its management to verify that its quality system meets the contract requirements. These are first-party audits. Such contracts typically also include provisions to have external audits sponsored by the management of the customer organization to verify that the supplier organization's quality system meets the contract requirements. These are second-party audits. Within a contractual arrangement between two such parties, it is possible to tailor the requirements, as appropriate, and to maintain an ongoing dialogue between customer and supplier.

When such assurance arrangements become a widespread practice throughout the economy, the two-party, individual-contract approach becomes burdensome. There develops a situation where each organization in the supply chain is subject to periodic management system audits by many customers and is itself subjecting many of its subsuppliers to such audits. There is a lot of redundant effort throughout the supply chain because each organization is audited multiple times for essentially the same requirements. The conduct of audits becomes a significant cost element for both the organizations performing the audit and the organizations being audited.

Certification/Registration-Level Activities

The development of quality system certification/registration is a means to reduce the redundant, non-value-adding effort of these multiple audits. A third-party organization, which is called a "certification body" in some countries, or a "registrar" in other countries (including the United States), conducts a formal audit of a supplier organization to assess conformance to the appropriate quality system standard, say, ISO 9001 or ISO 9002. When the supplier organization is judged to be in complete conformance, the third party issues a certificate to the supplying organization and registers the organization's quality system in a publicly available register. Thus, the terms "certification" and "registration" carry the same marketplace meaning because they are two successive steps signifying successful completion of the same process.

To maintain its registered status, the supplier organization must pass periodic surveillance audits by the registrar. Surveillance audits are often conducted semiannually. They may be less comprehensive than the full audit. If so, a full audit is performed every few years.

In the world today, there are hundreds of certification bodies/registrars. Most of them are private, for-profit organizations. Their services are valued by the supplier organizations they register, and by the customer organizations of the supplier organizations, because the registration service adds value in the supply chain. It is critical that the registrars do their work competently and objectively and that all registrars meet standard requirements for their business activities. They are, in fact, supplier organizations that provide a needed service product in the economy.

Accreditation-Level Activities

To ensure competence and objectivity of the registrars, systems of registrar accreditation have been set up worldwide. Accreditation bodies audit the registrars for conformity to standard international guides for the operation of certification bodies. The quality system of the registrar comes under scrutiny by the accreditation body through audits that cover the registrar's documented quality management system, the qualifications and certification of auditors used by the registrar, the record keeping, and other features of the office operations. In addition, the accreditation body witnesses selected audits done by the registrar's auditors at the facility of the client supplier organization.

Mutual International Acceptance

Various other countries have also implemented these three areas of activity:

1. Accreditation of certification bodies/registrars
2. Certification of auditors
3. Accreditation of auditor training courses

Various bilateral mutual recognition agreements are in place between certain countries whereby, for example, the certification of an auditor in one country carries over into automatic recognition of that certification in another country. In other situations, a memorandum of understanding has been negotiated between, say, the accreditation bodies in two countries, whereby they enter into a cooperative mode of operation preliminary to entering into a formal mutual recognition agreement. Under a memorandum of understanding, the accreditation bodies may jointly conduct the audit of a registrar, and the auditors may jointly document the results of the audit. However, each of the accreditation bodies would make its own decision whether to grant or continue the accreditation, as the case may be.

In principle, there should be no need for a supplier organization to obtain more than one certification/registration. A certificate from a registrar accredited anywhere in the world should, in principle, be accepted by customer organizations anywhere else in the world. In practice, it takes time to build infrastructure comparable in any country. It takes additional time (measured in years) for that infrastructure to mature in its operation and for confidence to build in other countries. Of course, not all countries decide to set up their own infrastructure but may choose to have their supplier organizations who wish to become registered do so by employing the services of an accredited registrar from another country.

Indeed, many registrar organizations have established operations internationally and provide services in many countries. Such registrars often seek accreditation in multiple countries because their customers (supplier organizations) look for accreditation under a system with which they are familiar and have developed confidence.

At the present time, there is a multiplicity of arrangements involving single or multiple accreditations of registrars, single or multiple certifications of auditors, and single or multiple accreditations of training courses. The overall system is moving toward widespread mutual recognition, but the ultimate test of credibility is the marketplace willingness to accept a single certification and a single accreditation.

The International Organization for Standardization (ISO), in January 1995 reaffirmed its support for the Quality System Assessment Recognition (QSAR) and approved a plan of action for setting the program in motion. This effectively laid the foundation for a voluntary system aimed at encouraging worldwide acceptance of ISO 9000 certificates.

The current status where registrars and course providers may have multiple accreditations, and auditors may have multiple certifications, may seem to have more redundancy than is necessary. If we step back and compare the current situation to the alternative of widespread second-party auditing of the quality systems of supplier organizations, it must be acknowledged that the present situation is better because there is

- Much less redundancy of auditing
- Much improved consistency of auditing
- The potential for even less redundancy and further improved consistency through the use of international standards and guides as criteria and through mutual harmonization efforts driven by the marketplace

Formal International Mutual Recognition

For the United States, there is one further complication. Almost alone among the countries of the world, the U.S. standards system is a private sector activity. The American National Standards Institute (ANSI), a private sector organization, is the coordinating body for standards in the United States. Under the ANSI umbrella, many organizations produce and maintain numbers of American national standards. Most of these standards relate to product technical specifications. Among the largest U.S. producers of standards are such organizations as the American Society of Testing and Materials (ASTM), the American Society of Mechanical Engineers (ASME), and the Institute of Electrical and Electronics Engineers (IEEE), but there are many other organizations that produce American national standards applicable to specific products or fields of activity. The ANSI system provides a consistent standards development process that is open, fair, and provides access to all parties that may be materially affected by a standard. The success of the U.S. system is attested to by the predominance of the U.S. economy internationally and the widespread adoption of U.S. standards for multinational or international use.

However, there are three levels of activities and infrastructure in relation to conformity assessment in international trade. Two levels have already been discussed: the certification/registration level and the accreditation level. The third level is recognition. At the recognition level, the national government of country A affirms to the government of country B that A's certification and accreditation infrastructure conforms to international standards and guides. In most countries of the world where the standards system is run by a government or semigovernment agency and the accreditation activities are carried out by that agency, the recognition level is virtually automatic. In the United States, various government agencies may be called upon to provide the formal recognition.

For example, in dealing with the European Union (EU) on products that fall under one of the EU directives that regulate products that have health, safety, and environmental risks, the EU insists upon dealing through designated government channels. The relevant U.S. government agency varies from one EU directive to another. In many areas, the recognition responsibility falls under the National Voluntary Conformity Assessment System Evaluation (NVCASE) program. NVCASE is a program run by the Department of Commerce through the National Institute of Standards and Technology (NIST). To be evaluated by NVCASE, an applicant must provide NIST with sufficient information to allow thorough assessment. The applicant's management system is thoroughly reviewed based on established internationally accepted criteria.

Conformity Assessment and International Trade

The conformity assessment approach of the EU typifies what is happening in many parts of the world. For a regulated product to be sold in any EU country, it must bear the "CE" mark.

Under the EU's modular approach, to qualify to be able to use the mark, the supplier organization must produce evidence of conformity in four areas:

1. Technical documentation of product design
2. Type testing
3. Product surveillance (by samples, or by each product)
4. Surveillance of quality assurance

Depending on the directive, the EU will offer suppliers various routes (modules) to satisfy the requirements. These routes range from "Internal Control of Production," which focuses on the product surveillance aspects, to "Full Quality Assurance," which typically focuses on certification/registration to ISO 9001 and relies upon the ISO 9001 requirements for capability in product design. In most modules, the manufacturer must submit product units, and/or product design technical information, and/or quality system information to a certification body that has been designated by the government as a "notified body." In some modules, the notified body must also provide for product tests where required. Several modules involve certification to ISO 9001, ISO 9002, or ISO 9003.

Implementing this modular approach to conformity assessment for regulated products by the European Union (then called the European Community) was the largest, single, early impetus to the rapid spread of certification/registration to ISO 9001 or ISO 9002 worldwide. For example, about half of the dollar volume of U.S. trade with Europe is in regulated products. Nevertheless, global trends in technology and in requirements for quality, and the cost savings of third-party versus widespread second-party auditing, as discussed previously, are powerful additional incentives and staying power for sustained international use and growth of third-party quality system certification/registration.

Moreover, for a supplier organization it is not effective to attempt to have two quality management systems, one for regulated products and another for nonregulated products. Consequently, there are multiple incentives for large numbers of supplier organizations, engaged directly or indirectly in international trade, to operate a quality management system that conforms to ISO 9001 or ISO 9002, as appropriate.

Guiding Principles

There are many registrars; each is registering many supplier quality systems. Each supplier is dealing with many customers. It is impractical to adequately monitor the operations of such a system solely by periodic audits conducted by an accreditation body. Consequently, the guiding principle should be that primary reliance must be placed on the concept of "truth in labeling," by means of which every customer has routine, ready access to the information upon which to judge all four elements of the scope of a supplier's registered quality system.

Industry-Specific Adoptions and Extensions of ISO 9000 Standards

Medical Device Industry

Circumstance 1 relates to the medical device manufacturing industry. For example, in the United States, the Food and Drug Administration (FDA) developed and promulgated the Good Manufacturing Practice (GMP) regulations. The GMP operates under the legal imprimatur of the FDA regulations, which predate ISO 9000 standards. The FDA regularly inspects medical device manufacturers for their compliance with the GMP requirements. Many of these requirements are quality management system requirements that parallel the subsequently published ISO 9002:1987 requirements. Other GMP regulatory requirements relate more specifically to health, safety, or environmental aspects. Many other nations have similar regulatory requirements for such products.

In the United States, the FDA has created revised GMPs that parallel closely the ISO 9000 standard plus specific regulatory requirements related to health, safety, or the environment. Expanding the scope of ISO 9000 to include quality system requirements related to product design reflects the recognition of the importance of product design and the greater maturity of quality management practices in the medical device industry worldwide. Similar trends are taking place in other nations, many of which are adopting ISO 9001 verbatim for their equivalent of the GMP regulations.

Current Good Manufacturing Practices (cGMPs) for human pharmaceuticals affect every American. Consumers expect that each batch of medicines they take will meet quality standards so that they will be safe and effective. Most people, however, are not aware of cGMPs, or how FDA assures that drug manufacturing processes meet these basic objectives. Recently, FDA has announced a number of regulatory actions taken against drug manufacturers based on the lack of cGMPs.

What Are cGMPs?

cGMP refers to the Current Good Manufacturing Practice regulations enforced by the FDA. cGMPs provide for systems that ensure proper design, monitoring, and control of manufacturing processes and facilities. Adherence to the cGMP regulations ensures the identity, strength, quality, and purity of drug products by requiring that manufacturers of medications adequately control their manufacturing operations. This includes establishing strong quality management systems, obtaining appropriate quality raw materials, establishing robust operating procedures, detecting and investigating product quality deviations, and maintaining reliable testing laboratories. This formal system of controls at a pharmaceutical organization, if adequately put into practice, helps to prevent instances of contamination, mix-ups, deviations, failures, and errors. This ensures that drug products meet their quality standards.

cGMP requirements were established to be flexible in order to allow each manufacturer to decide individually how to best implement the necessary controls by using scientifically sound design, processing methods, and testing procedures. The flexibility in these regulations allows companies to use modern technologies and innovative approaches to achieve higher quality through continual improvement. Accordingly, the "C" in cGMP stands for "Current," requiring companies to use technologies and systems that are up-to-date in order to comply with the regulations. Systems and equipment that may have been "top-of-the-line" to prevent contamination, mix-ups, and errors 10 or 20 years ago may be less than adequate by today's standards.

It is important to note that cGMPs are minimum requirements. Many pharmaceutical manufacturers are already implementing comprehensive, modern quality systems and risk management approaches that exceed these minimum standards.

Why Are cGMPs Important to Software Development?

A consumer usually cannot detect (through smell, touch, or sight) that a drug product is safe or if it will work. Although cGMPs require testing, testing alone is not enough to ensure quality. In most instances, testing is done on a small sample of a batch (e.g., a drug manufacturer may test 100 tablets from a batch that contains 2 million tablets) so that most of the batch can be used for patients rather than be destroyed by testing. Therefore, it is important that drugs are manufactured under conditions and practices required by cGMP regulations to ensure that quality is built into the design and manufacturing process at every step. Facilities that are in good condition, equipment that is properly maintained and calibrated, employees who are qualified and fully trained, and processes that are reliable and reproducible are a few examples of how cGMP requirements help to ensure the safety and efficacy of drug products.

How Does the FDA Determine if an Organization Is Complying with cGMP Regulations?

The FDA inspects pharmaceutical manufacturing facilities worldwide using scientifically and cGMP-trained individuals whose job it is to evaluate whether the organization is following cGMP regulations. The FDA also relies upon reports of potentially defective drug products from the public and from the industry. The FDA will often use these reports to identify sites for which an inspection or investigation is needed. Most companies that are inspected are found to be fully compliant with the cGMP regulations.

In August 2002, the FDA announced the pharmaceutical cGMPs for the twenty-first Century Initiative. In that announcement, the FDA explained the agency's intent to integrate quality systems and risk management approaches into its existing programs to encourage industry to adopt modern and innovative manufacturing technologies. The cGMP initiative was spurred by the fact that since 1978, when the last major revision of the cGMP regulations was published, there have been many advances in manufacturing science and in our understanding of quality systems. In addition, many pharmaceutical manufacturers are already implementing comprehensive, modern quality systems and risk management approaches. This guidance is intended to help manufacturers implementing modern quality systems and risk management approaches to meet the requirements of the agency's cGMP regulations. The agency also saw a need to harmonize cGMPs with other non-U.S. pharmaceutical regulatory systems and with FDA's own medical device quality systems regulations. This guidance supports these goals. It also supports the objectives of the Critical Path Initiative, which intends to make the development of innovative medical products more efficient so that safe and effective therapies can reach patients sooner.

cGMPs for the twenty-first century initiative steering committee created a Quality System Guidance Development working group (QS working group) to compare current cGMP regulations, which call for specific quality management elements, to other existing quality management systems. The QS working group mapped the relationship between cGMP regulations parts 210 and 211 and the 1978 Preamble to the cGMP regulations and various quality system models, such as the Drug Manufacturing Inspections Program (i.e., systems-based inspectional program), and the Environmental Protection Agency's Guidance for Developing Quality Systems for Environmental Programs, ISO Quality Standards, other quality publications, and experience from regulatory cases. The QS working group determined that, although the cGMP regulations do provide great flexibility, they do not incorporate explicitly all of the elements that today constitute most quality management systems.

cGMP regulations and other quality management systems differ somewhat in organization and in certain constituent elements; however, they are very similar and share some underlying principles. For example, cGMP regulations stress quality control. More recently developed quality systems stress quality management, quality assurance, and the use of risk management tools, in addition to quality control. The QS working group decided that it would be very useful to examine exactly how the cGMP regulations and the elements of a modern, comprehensive quality system fit together in today's manufacturing world. This guidance is the result of that examination.

In ISO, a new technical committee, ISO/TC210, has been formed specifically for medical device systems. TC210 has developed standards that provide supplements to ISO 9001 clauses.

These supplements primarily reflect the health, safety, and environment aspects of medical devices and tend to parallel regulatory requirements in various nations.

ISO/TS 16949: Automotive Industry

In the years preceding publication of the 1987 ISO 9000 standards, various original equipment manufacturers (OEMs) in the automotive industry had developed organization-specific proprietary quality system requirements documents. These requirements were part of OEM contract arrangements for purchasing parts, materials, and subassemblies from the thousands of organizations in their supply chain. The OEMs had large staffs of second-party auditors to verify that these OEM-specific requirements were being met.

Upon publication of ISO 9001:1994, the major U.S. OEMs began implementation of an industrywide common standard—QS-9000—that incorporates ISO 9001 verbatim plus industry-specific supplementary requirements. Some of the supplementary requirements are really prescriptive approaches to some of the generic ISO 9001 requirements; others are additional quality system requirements that have been agreed on by the major OEMs; a few are OEM specific.

On December 14, 2006, all QS 9000 certifications were terminated. With QS 9000, the middle certification between ISO 9001 and ISO/TS 16949 were no longer valid; businesses had a choice between either ISO 9001 or TS 16949. QS 9000 is considered to have been superseded by ISO/TS 16949.

ISO/TS 16949:2009, in conjunction with ISO 9001:2008, defines quality management system requirements for design and development, production and, when relevant, installation and service of automotive-related products.

ISO/TS 16949:2009 applies to sites of the organization where customer-specified parts are manufactured for production and/or service.

Supporting functions, whether on-site or remote (such as design centers, corporate headquarters and distribution centers), form part of the site audit as they support the site, but they cannot obtain stand-alone certification to ISO/TS 16949:2009. ISO/TS 16949:2009 can be applied throughout the automotive supply chain.

Computer Software

The global economy has become permeated with electronic information technology (IT). The IT industry now plays a major role in shaping and driving the global economy. As in past major technological advances, the world seems fundamentally very different, and paradoxically, fundamentally the same. Computer software development occupies a central position in this paradox.

First, note that computer software development is not so much an industry as it is a discipline.

Second, many IT practitioners emphasize that computer software issues are complicated by the multiplicity of ways that computer software quality may be critical in a supplier organization's business. For example,

- The supplier's product may be complex software whose functional design requirements are specified by the customer.
- The supplier may actually write most of its software product, or may integrate off-the-shelf packaged software from subsuppliers.
- The supplier may incorporate computer software/firmware into its product, which may be primarily hardware and/or services.
- The supplier may develop and/or purchase from subsuppliers software that will be used in the supplier's own design and/or production processes of its product.

However, it is important to acknowledge that hardware, processed materials, and services often are involved in a supplier organization's business in the same multiple ways.

What, then, are the issues in applying ISO 9001 to computer software development? There is general consensus worldwide that

- The generic quality management system activities and associated requirements in ISO 9001 are relevant to computer software, just as they are relevant in other generic product categories (hardware, other forms of software, processed materials, and services).
- There are some things that are different in applying ISO 9001 to computer software.

There is at this time no worldwide consensus as to which things, if any, are different enough to make a difference and what to do about any things that are different enough to make a difference.

ISO/TC176 developed and published ISO 9000-3:1991 as a means of dealing with this important, paradoxical issue. ISO 9000-3 contains guidelines for applying ISO 9001 to the development, supply, and maintenance of (computer) software and has been useful and widely used. ISO 9000-3 offers guidance that goes beyond the requirements of ISO 9001, and it makes some assumptions about the life cycle model for software development, supply, and maintenance. In the United Kingdom, a separate certification scheme (TickIT) for software development has been operating for several years, using the combination of ISO 9001 and ISO 9003. The scheme has received both praise and criticism from various constituencies worldwide. Those who praise the scheme claim that it

- Addresses an important need in the economy to provide assurance for customer organizations that the requirements for quality in software they purchase (as a separate product, or incorporated in a hardware product) will be satisfied.
- Includes explicit provisions beyond those for conventional certification to ISO 9001 to ensure competency of software auditors, their training, and audit program administration by the certification body.
- Provides a separate certification scheme and logo to exhibit this status publicly. Critics claim that the scheme.
- Is inflexible and attempts to prescribe a particular life cycle approach to computer software development that is out of tune with current best practices for developing many types of computer software.
- Includes unrealistically stringent auditor qualifications in the technology aspects of software development, qualifications whose technical depth is not necessary for effective auditing of management systems for software development.
- Is almost totally redundant with conventional third-party certification to ISO 9001, under which the certification body/registrar already is responsible for competency of auditors. Accreditation bodies verify the competency as part of accreditation procedures.
- Adds substantial cost beyond conventional certification to ISO 9001 and provides little added value to the supply chain.

In the United States, a proposal to adopt a TickIT-like software scheme was presented to the ANSI/RAB (Registrar Accreditation Board) accreditation program. The proposal was rejected, primarily on the basis that there was not consensus and support in the IT industry and the IT-user community.

Standardization Is Here to Stay

Standards are here to stay. Many industries are working together with various standards bodies to periodically improve their standards and mandate as many systems as possible to ensure the safety and quality of our products. For instance, a new International Standard—ISO 31000:2009, *Risk Management—Principles and Guidelines*—was developed to help organizations manage risk effectively. ISO 31000 provides principles, framework, and a process for managing any form of risk in a transparent, systematic, and credible manner within any scope or context.

In addition, the ISO has published a standard to facilitate implementation of quality management systems, based on ISO 9001:2000, by the medical device industry. The key objectives of ISO 13485:2003 are to maximize the probability that a medical device organization will meet regulatory quality management system requirements worldwide, will provide safe and effective medical devices, and will meet customer requirements—Ed Kimmelman, convener of the working group that developed the new standard.

ISO 13485:2003, Medical Devices—Quality Management Systems—Requirements for Regulatory Purposes, is based on quality management system requirements currently contained in medical device regulations around the world as well as those appropriate requirements contained in ISO 9001:2000. The new standard is used by organizations involved in the design, production, installation, and servicing of medical devices as well as in the design, development, and provision of related services. It can also be used by external certification bodies to assess an organization's ability to meet requirements. The new standard, which replaces ISO 13485:1996, is the work of ISO technical committee ISO/TC 210, Quality Management and Corresponding General Aspects for Medical Devices, working group WG 1, Application of Quality Systems to Medical Devices, in conjunction with members of the Global Harmonization Task Force (Study Group 3), conceived in 1992 in an effort to achieve greater uniformity between national medical device regulatory systems.

> Standardization will embrace common operational sequences, part-dimensional strategies, and guidelines for equipment use.
>
> —Barney and De Feo (2008)

Standards are becoming a way of life for global organizations. Get ready; there is more to come.

References

Barney M., and De Feo, J. A. (2008). "The Future of Manufacturing." Quality Digest, Chico CA.

Carnegie Mellon Software Engineering Institute. (2006). "CMMI Executive Overview." Sponsored by U.S. Department of Defense, Carnegie Mellon University. Pittsburgh, PA.

Department of Health and Human Services (2004). "Final Report: Pharmaceutical CGMPS for the 21st Century—Risk-Based Approach." U.S. Food and Drug Administration, September.

Gasiorowski-Denis, E. (2003). Quality Management Systems for Medical Device Industry, ISO Management Systems, November–December 2003, Available at http://www.iso.org/iso/medical_device_ims6_2003.pdf. Accessed January 22, 2015.

Humphrey, W. S. (1989). *Managing for Software Process*. Addison-Wesley, Reading, MA.

International Organization for Standardization (2015). "ISO 9000—Quality management." Available at http://www.iso.org/iso/home/standards/management-standards/iso_9000.htm Accessed January 15, 2015.

International Organization for Standardization. (2015). "Risk-Based Thinking in ISO 9001:2015." Available at www.iso.org/tc176/sc02/public Accessed January 15, 2015.

International Standard ISO 9001 (2008). 4th ed., November 15, 2008. Available at www.iso.org

Liu, X. (1993). Technical Report CMU/SEI-93-TR-024 ESC-TR-93-177, Capability Maturity Model SM for Software, Version 1.1.

National Institute of Standards and Technology. (2009). "National Voluntary Conformity Assessment Systems Evaluation." U.S. Department of Commerce, June. Available at http://gsi.nist.gov/global/index.cfm/L1-4/L2-38/L3-97 Accessed January 20, 2015.

CHAPTER 12

The Role of Statistics and Probability

John Early

High Points of This Chapter

1. Statistical methods are essential in the modern approach to quality.
2. Variation is a fact of nature and a fact of business life.
3. In summarizing data, useful tabular and graphical tools include the frequency distribution, histogram, boxplot, and probability plots, and useful numerical indexes include the average, median, range, and standard deviation.
4. A sample is a limited number of items taken from a larger source called the population.

5. A probability distribution function relates the values of a characteristic to their probability of occurrence in the population.
6. The important continuous probability distributions are the normal, exponential, and Weibull; important discrete distributions are the Poisson and binomial.
7. Three theorems of probability are basic in analyzing the probability of specific events.

Statistical Tools in Quality

Statistics or statistical inference is the science of making decisions under uncertainty. We must confront uncertainty in our decisions because the real world is subject to random variation (discussed later). Statistical methods help us make good decisions—and avoid bad ones—in the face of that variation-induced uncertainty. At a deeper level, quality, in the sense of freedom from deficiencies, requires statistical tools because most deficiencies arise as the result of random variation.

The term statistics is also used in everyday speech to mean any systematic collection, tabulation, analysis, or interpretation of data. It can even mean just data.

Probability is a measure that describes the chance that an event will occur based on the underlying variation in the phenomenon being analyzed. Measures of probability allow us to make quantitative statements about the risks associated with the decisions we must make based on the data available to us.

Of course, statistical tools are only a part of the capabilities one must master to achieve outstanding quality, but they are a critical part.

The Concept of Variation

The concept of variation states that no two items will be perfectly identical. Variation is a fact of nature and a fact of business life. For example, even "identical" twins vary slightly in height and weight at birth.

The dimensions of a large-scale, integrated chip vary from chip to chip; the proportions of the various ingredients in cans of tomato soup vary slightly from can to can; the time required to check in at an airline check-in counter varies from passenger to passenger. To disregard the existence of variation (or to rationalize falsely that it is small) can lead to seriously incorrect decisions on major problems. Statistics helps to analyze data properly and draw conclusions, taking into account the existence of variation.

Statistical variation—variation due to random causes—is much greater than most people think. Often we decide what action to take based on the most recent data point, and we forget that the data point is part of a history of data. Malcolm Roberts once said: "Many managers run their systems by the last data point."

Data summarization can take several forms: tabular, graphical, and numerical. Sometimes one form will provide a useful, complete summarization. In other cases, two or even three forms are needed for complete clarity.

Tabular Summarization of Data: Frequency Distribution

A frequency distribution is a tabulation of data arranged according to magnitude. The raw data of the electrical resistance of 100 coils are given in Table 12.1. Table 12.2 shows the frequency distribution of these data with all measurements tabulated at their actual values. For example, there were 14 coils each of which had a resistance of 3.35 ohms (Ω); there were 5 coils each of which had a resistance of 3.30 Ω. The frequency distribution spotlights where most of the data are grouped (the data are centered about a resistance of 3.35) and how much

3.37	3.34	3.38	3.32	3.33	3.28	3.34	3.31	3.33	3.34
3.29	3.36	3.30	3.31	3.33	3.34	3.34	3.36	3.39	3.34
3.35	3.36	3.30	3.32	3.33	3.35	3.35	3.34	3.32	3.38
3.32	3.37	3.34	3.38	3.36	3.37	3.36	3.31	3.33	3.30
3.35	3.33	3.38	3.37	3.44	3.32	3.36	3.32	3.29	3.35
3.38	3.39	3.34	3.32	3.30	3.39	3.36	3.40	3.32	3.33
3.29	3.41	3.27	3.36	3.41	3.37	3.36	3.37	3.33	3.36
3.31	3.33	3.35	3.34	3.35	3.34	3.31	3.36	3.37	3.35
3.40	3.35	3.37	3.35	3.32	3.36	3.38	3.35	3.31	3.34
3.35	3.36	3.39	3.31	3.31	3.30	3.35	3.33	3.35	3.31

Table 12.1 Resistances of 100 Coils, Ω

Resistance, Ω	Tabulation	Frequency	Cumulative Frequency
3.45			
3.44	|	1	1
3.43			
3.42			
3.41	||	2	3
3.40	||	2	5
3.39	||||	4	9
3.38	~~||||~~ |	6	15
3.37	~~||||~~ |||	8	23
3.36	~~||||~~ ~~||||~~ |||	13	36
3.35	~~||||~~ ~~||||~~ ||||	14	50
3.34	~~||||~~ ~~||||~~ ||	12	62
3.33	~~||||~~ ~~||||~~	10	72
3.32	~~||||~~ ||||	9	81
3.31	~~||||~~ ||||	9	90
3.30	~~||||~~	5	95
3.29	|||	3	98
3.28	|	1	99
3.27	|	1	100
3.26			
Total		100	

Table 12.2 Tally of Resistance Values of 100 Coils

variation there is in the data (resistance runs from 3.27 to 3.44 Ω). Table 12.2 shows the conventional frequency distribution and the cumulative frequency distribution in which the frequency values are accumulated to show the number of coils with resistances equal to or less than a specific value. The particular problem determines whether the conventional, cumulative, or both distributions are required. Note, of course, that resistance is measured here only to the nearest hundredth of an ohm. The five 3.30 values are not likely equal if measured to additional significant figures.

When there are large amounts of highly variable data, the frequency distribution can become too large to serve as a summary of the original data. The data may be grouped into cells to provide a better summary. Table 12.3 shows the frequency distribution for these data grouped into six cells, each 0.03 Ω wide. Grouping the data into cells condenses the original data, and therefore some detail is lost.

The following is a common procedure for constructing a frequency distribution:

1. Decide on the number of cells. Table 12.4 provides a guide.
2. Calculate the approximate cell interval i. The cell interval equals the largest observation minus the smallest observation divided by the number of cells. Round this result to some convenient number (preferably the nearest odd number with the same number of significant digits as the actual data).

Resistance, Ω	Frequency
3.415–3.445	1
3.385–3.415	8
3.355–3.385	27
3.325–3.355	36
3.295–3.325	23
3.265–3.295	5
Total	100

Table 12.3 Frequency Table of Resistance Values

Number of Observations	Recommended Number of Cells
20–50	6
51–100	7
101–200	8
201–500	9
501–1000	10
Over 1000	11–20

Table 12.4 Number of Cells in Frequency Distribution

3. Construct the cells by listing cell boundaries.
   - Each cell boundary should be to one more significant digit than the actual data and should end in a 5.
   - The cell interval should be constant throughout the entire frequency distribution.
4. Tally each observation in the appropriate cell and then list the total frequency f for each cell.

This procedure should be adjusted when necessary to provide a clear summary of the data and to reveal the underlying pattern of variation.

Graphical Summarization of Data: The Histogram

A histogram is a vertical bar chart of a frequency distribution. Figure 12.1 shows the histogram for the electrical resistance data. Note that as in the frequency distribution, the histogram highlights the center and amount of variation in the sample of data. The simplicity of construction and interpretation of the histogram makes it an effective tool in the elementary analysis of data.

Graphical methods are essential to effective data analysis and clear presentation of results. Many of these methods are used throughout this book. More are available. Experience dictates that the first step in data analysis is to plot the data using an appropriate graphical tool. While most graphical analysis will also require statistical support, simply calculating the various statistics without examining the data graphically can lead to faulty analysis.

One variation of the histogram is the stem-and-leaf plot. Heyes (1985) presents data on wire break strength in grams (see Table 12.5) for supplier A. The corresponding stem-and-leaf

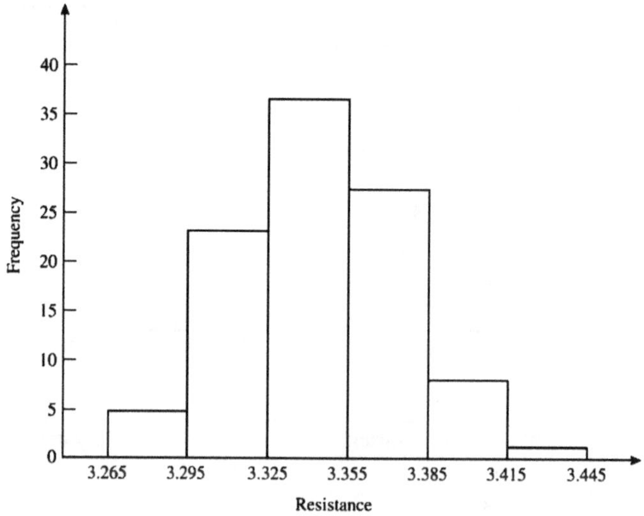

FIGURE 12.1 Histogram of resistance.

1. 346	6. 402	11. 368
2. 338	7. 635	12. 376
3. 323	8. 281	13. 311
4. 438	9. 431	14. 379
5. 398	10. 390	15. 216

TABLE 12.5 Original Data on Wire Break Strength

plot is shown in Fig. 12.2. Note that the stem is the first digit(s) of each value and the leaf is the remaining digits, for example, for a value of 216, the stem is 2 and the leaf is 16. Note that this plot reveals the shape of the histogram and also makes it possible to see the original values of the data. The great thing about this simple tool is that it can be quickly and easily constructed without a computer. With minimal technology or mathematics, the analyst can visually assess variation and determine the five-number summary of the boxplot presented later.

Stem	Leaf
2	16, 81
3	11, 23, 38, 46, 68, 76, 79, 90, 98
4	02, 31, 38
5	–
6	35

FIGURE 12.2 Stem-and-leaf plot. (From Heyes, 1985.)

1. 216	6. 346	11. 398
2. 281	7. 368	12. 402
3. 311	8. 376	13. 431
4. 323	9. 379	14. 438
5. 338	10. 390	15. 635

TABLE 12.6 Ordered Data on Wire Break Strength

Boxplots

A simple, clever, and effective way to summarize data is a boxplot. The boxplot is a graphical five-number summary of the data. In the basic boxplot, the five values are the median, maximum value, minimum value, first quartile, and third quartile. The quartiles are the values below which one-fourth and three-fourths of the observations lie.

Using the wire break strength data, the data are first arranged in rank order (see Table 12.6). The median is the middle value (the eighth rank, or 376). The maximum and minimum values are 216 and 635. The quartiles are 323 and 402 because those values divide the data into quarters. This can also be described as the medians of each half. Figure 12.3 shows the resulting boxplot. The box, bounded by the two quartiles with the median inside the box, summarizes the middle half of the data. The lines extending out to the extreme values are the "whiskers." The longer whisker on the right suggests that the data include some values that are much larger than the other values. Also, the location of the median indicates that the values above the median are, as a group, closer to the median than the values below the median.

Innovative methods of graphical analysis and display of data are discussed in a now classic text by Tukey (1977). A good summary of graphical methods including the boxplot is presented in Wadsworth et al. (2001).

In another example, a large service organization had a problem of poor reliability of copy machines. The copiers were repaired by two different firms (A and B). An improvement team conducted a study to see which contractor provided the faster and more consistent response to a repair call. The response time, in minutes, was recorded for a sample of 10 repair calls for each contractor. The resulting boxplots are shown in Fig. 12.4. The team concluded that contractor B was superior because B was usually faster and more consistent than A.

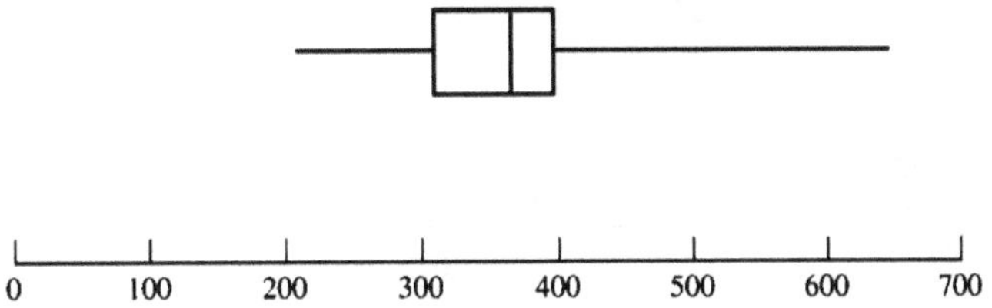

FIGURE 12.3 Boxplot. (From Heyes, 1985.)

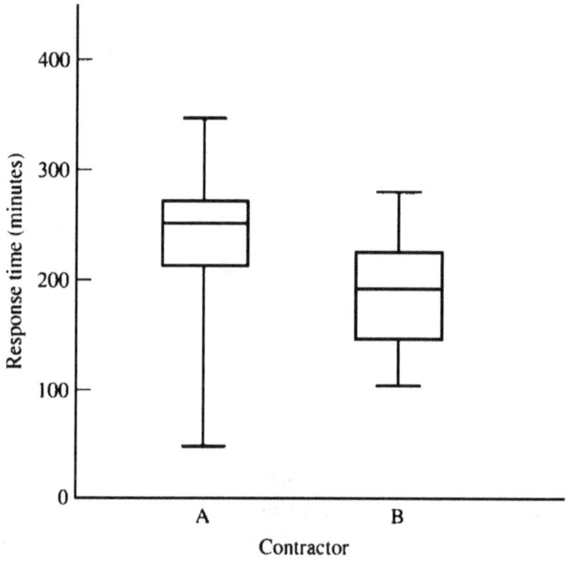

FIGURE 12.4 Basic boxplots of copier repair response time. (*Source*: Juran Institute, Inc.)

Graphical Summarization of Time-Oriented Data: The Run Chart

The histogram is a simple and effective way to summarize data variation according to magnitude as a snapshot of a specific time period. The run chart is a dynamic plot of the data versus time. Such plots can reveal trends, cycles, and other changes over time. An example is presented in Fig. 12.5.

Readers should be cautious in employing certain graphical features of popular software. Some of these (e.g., three-dimensional bar graphs) sacrifice clarity for glitz. A summary of graphical methods for quality is presented in Wadsworth et al. (2001). Harris (1996) provides a comprehensive reference on "information graphics." For unusually creative ideas on displaying all types of information, see Tufte (1997).

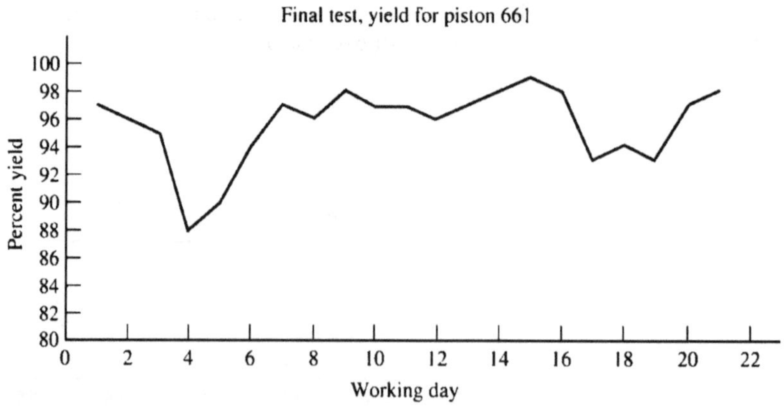

FIGURE 12.5 Run chart. (*Source*: Juran Institute, Inc.)

Methods of Summarizing Data: Numerical

Data can also be summarized by computing (1) a measure of central tendency to indicate where most of the data are centered and (2) the measure of dispersion to indicate the amount of scatter in the data. Often these two measures provide an adequate summary.

The key measure of the central tendency is the arithmetic mean, or average. The definition of the mean is

$$\bar{X} = \frac{\sum_{i=1}^{n} X_i}{n}$$

where $\bar{X}$ = sample mean
X_i = individual observations
n = number of observations
$\sum_{i=1}^{n} X_i$ = summation of the X_i

Another measure of central tendency is the median—the middle value when the data are arranged according to magnitude. The median is useful for asymmetric distributions, data with outliers, or data that can be ranked but are not measured on a continuous metric, such as color or visual appearance.

Two measures of dispersion are commonly calculated.

In general, the standard deviation is the most useful measure of dispersion. Like the mean, the definition of the standard deviation is a formula:

$$s = \sqrt{\frac{\sum_{i=1}^{n} (X_i - \bar{X})^2}{n = 1}}$$

where s is the sample standard deviation. The square of the standard deviation, s^2, is called the *variance*.

The standard deviation is the square root of the average of squared deviations of the observations from their mean. There is usually difficulty in understanding the "meaning" of the standard deviation. The only definition is the formula. There is no hidden meaning to the standard deviation, and it is best viewed as an index that shows the amount of variation in a set of data. Later, applications of the standard deviation to predictions will help clarify its meaning.

One useful technique is to calculate a relative measure of variation as the standard deviation divided by the mean (the coefficient of variation). The coefficient of variation is sometimes useful when comparing the spread between data sets with different means.

When the amount of data is small (10 or fewer observations), the range is a more useful measure of dispersion. The range is the difference between the maximum value and the minimum value in the data. Because the range is based on only two values, it contains less information about the data. For data sets of more than 10, the standard deviation is usually the best choice.

A problem that sometimes arises in the summarization of data is that one or more extreme values are far from the rest of the data. A simple (but not necessarily correct) solution is available: drop such values. The reasoning is that a measurement error or some other unknown factor makes the values "unrepresentative." Unfortunately, this approach may be rationalizing to eliminate an annoying problem of data analysis and may seriously mislead the investigation. Extreme values should be excluded only if evidence can be developed that, in fact, the outlier arises from some special cause such as a measurement error. Remember that we are using statistics to help us make better decisions. If we change the data to fit our priors, then we have defeated the whole purpose of using data.

Probability Distributions: General

A sample is a limited number of items taken from a larger source. A population is a large source of items from which the sample is taken. Usually it is not possible to describe a population of data without an estimation from a sample. We take measurements from a sample and calculate a sample statistic, for example, the mean. A statistic is a quantity computed from a sample to estimate a population parameter. Statistical inference requires that the sample be random, that is, each possible sample of n items has an equal chance of being selected.

A probability distribution function is a mathematical formula that relates the values of the characteristic with their probability of occurrence in the population. The collection of these probabilities is called a probability distribution. The mean (μ) of a probability distribution is often called the expected value. Some distributions and their functions are summarized in Fig. 12.6. Distributions are of two types:

1. *Continuous* (or *"variable" data*). When the characteristic being measured can take on any value (subject to the fineness of the measuring process), its probability distribution is called a *continuous probability distribution*. For example, the probability distribution of the resistance data of Table 12.1 is an example of a continuous probability distribution because the resistance could have any value, limited only by the fineness of the measuring instrument. Most continuous characteristics follow one of several common probability distributions: the normal distribution, the exponential distribution, and the Weibull distribution. These distributions find the probabilities associated with occurrences of the actual values of the characteristic. Other continuous distributions (e.g., t, F, and chi square) are important in data analysis but are not helpful in directly predicting the probability of occurrence of actual values.
2. *Discrete* (or *"attribute" data*). When the characteristic being measured can take on only certain specific values (e.g., integers 0, 1, 2, 3), its probability distribution is called a *discrete probability distribution*. For example, the distribution of the number of defectives r in a sample of five items is a discrete probability distribution because r can be only 0, 1, 2, 3, 4, or 5. The most common discrete distributions are the Poisson and binomial (see Fig. 12.6).

The following sections explain how probability distributions can be used with a sample of observations to make predictions about the larger population. Such predictions assume that the data come from a process that is stable over time, which is not always the case. Plotting the data points in order of production provides a rough test for stability; plotting the data on a statistical control chart provides a rigorous test.

Distribution	Form	Probability function	Comments on application
Normal	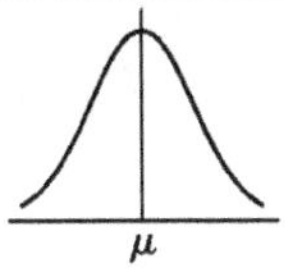	$y = \frac{1}{\sigma\sqrt{2\pi}} e^{-\frac{(X-\mu)^2}{2\sigma^2}}$ μ = Mean σ = Standard deviation	Applicable when there is a concentration of observations about the mean and it is equally likely that observations will occur above and below the mean. X can take any real positive or negative number.
Exponential	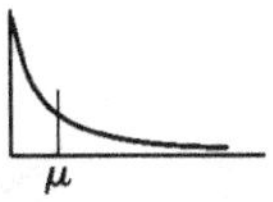	$y = \frac{1}{\mu} e^{-\frac{x}{\mu}}$	Applicable when it is likely that more observations will occur below the average than above $X > 0$.
Weibull	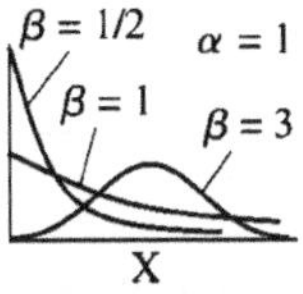	$y = \alpha\beta(X - \gamma)^{\beta-1}e^{-\alpha(X-\gamma)^{\beta}}$ α = Scale parameter β = Shape parameter γ = Location parameter	Applicable in describing a wide variety of patterns in variation $X > -\gamma$
Poisson*	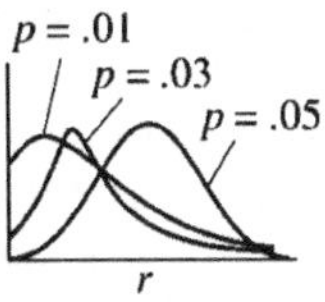	$y = \frac{(np)^r e^{-np}}{r!}$ n = Number of trials r = Number of occurrences p = Probability of occurrence	Applicable in defining the probability of occurrence per unit time or unit space. np = average occurrences per unit and can be less than 1.
Binomial*	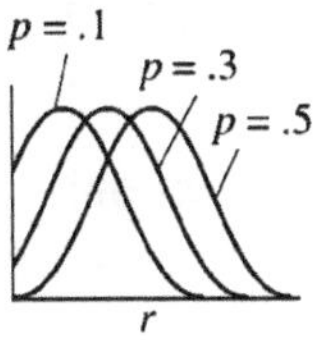	$y = \frac{n!}{r!(n-r)!} p^r q^{n-r}$ n = Number of trials r = Number of occurrences p = Probability of occurrence $q = 1 - p$	Applicable in defining the probability of r occurrences in n trials of an event that has a constant probability of occurrence on each independent trial.

FIGURE 12.6 Summary of common probability distributions. (Asterisks indicate that these are discrete distributions, but the curves are shown as continuous for ease of comparison with the continuous distributions. Strictly, the actual plots are discrete, not continuous.)

The Normal Probability Distribution

Many quality characteristics can be approximated by the normal density function:

$$y = \frac{1}{\sigma\sqrt{2\pi}} e^{-(X-\mu)^2/2\sigma^2}$$

where $e = 2.7183$ (approximate)
$\pi = 3.1416$ (approximate)
μ = population mean
σ = population standard deviation

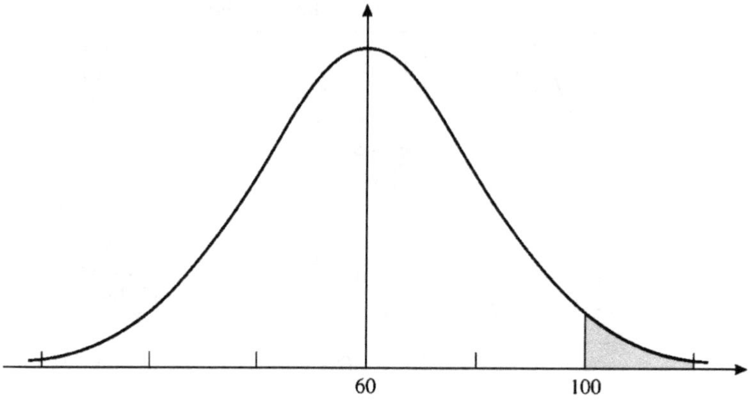

FIGURE 12.7 Distribution of light bulb life.

Note that the distribution is fully defined by the mean μ and standard deviation σ of the population.[1] The curve for the normal probability distribution is related to a frequency distribution and its histogram. As the sample becomes larger and larger and the width of each cell becomes smaller and smaller, the histogram approaches a smooth curve. If the entire population were measured and if it were normally distributed, the result would be as shown in Fig. 12.6. Thus the shape of a histogram of sample data provides some indication of the probability distribution for the population. If the histogram resembles[2] the "bell" shape, shown in Fig. 12.6; this is a basis for suspecting that the population follows a normal probability distribution. There are statistical tests to support determining the underlying distribution when that is necessary.

Using the Normal Probability Distribution for Predictions

Predictions require just two estimates and a table. The estimates are

$$\text{Estimate of } \mu \text{ is } \bar{x} \qquad \text{Estimate of } \sigma \text{ is } s$$

The calculations of the sample $\bar{x}$ and s are made by the methods previously discussed. For example, from past experience, a manufacturer concludes that the burnout time of a particular light bulb follows a normal distribution. A sample of 50 bulbs has been tested, and the average life is 60 days with a standard deviation of 20 days. How many bulbs in the entire population of light bulbs can be expected to be still working after 100 days of life?

The problem is to find the area under the curve beyond 100 days (Fig. 12.7).

The area under a distribution curve between two stated limits represents the probability of an occurrence between those limits. Therefore, the shaded area under the curve is the probability that a bulb will last more than 100 days. To find the area, calculate the difference Z between a particular value and the average of the curve in units of standard deviation. Every conceivable combination of means and standard deviation defines a unique normal curve, but these unique curves are all related to each other in a fairly simple way. Every set of data (Xs) can be transformed standardized into Zs using the following:

$$Z = \frac{X - \mu}{\sigma}$$

[1]Unless otherwise indicated, Greek symbols will be used for population values and Roman symbols for sample values.

[2]The sample histogram may not look as if it came from a normal population. Small deviations from exact normality are expected in random samples. Statistical tests are available to determine whether the deviations are significant.

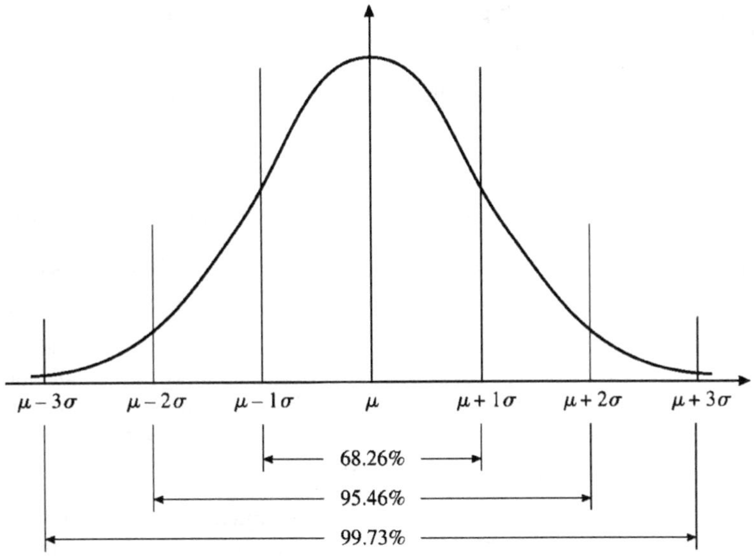

FIGURE 12.8 Areas of the normal curve.

This Z statistic has a mean of 0 and a standard deviation of 1.0. So, a table to describe the areas under this standard normal distribution is all we need to compute the probabilities for any normal distribution.

In this problem $Z = (100 - 60) \div 20 = +2.0$. Table A in App. II shows a probability of .9773 for $Z = 2$. The statistical distribution tables provide probabilities that cover the span from $-\infty$ up to and including the value of Z included in the formula (i.e., cumulative probabilities). Thus .9773 is the probability that a bulb will last 100 days or less. The normal curve is symmetrical about the average, and the total area is 1.000. The probability of a bulb lasting more than 100 days then is 1.0000 − .9773, or .0227, or 2.27 percent of the bulbs in the population will still be working after 100 days.

Similarly, if a characteristic is normally distributed and if estimates of the average and standard deviation of the population are obtained, this method can estimate the total percentage of production that will fall within specification limits.

Figure 12.8 shows representative areas under the normal distribution curve. Thus 68.26 percent of the population will fall between the average of the population plus or minus 1 standard deviation of the population, 95.46 percent of the population will fall between the average and 62s, and finally, 63s will include 99.73 percent of the population.

Probability Curves and Histogram Analysis

We can combine the histogram concept and the probability curve concept to yield a practical working tool known as histogram analysis.

A random sample is selected from the process, and measurements are made for the selected quality characteristics. A histogram is prepared and specification limits are added. Knowledge of the production process is then combined with insights provided by the histogram to draw conclusions about the ability of the process to meet the specifications.

Figure 12.9 shows 16 typical histograms. Readers are encouraged to interpret each of these pictures by asking two questions:

1. Can the process meet the specification limits? This is the question of process capability.
2. What action on the process, if any, is appropriate?

These questions can be answered by analyzing the following characteristics:

1. *The centering of the histogram.* This defines the aim of the process.
2. *The width of the histogram.* This defines the variability about the aim.
3. *The shape of the histogram.* When a normal or bell-shaped curve is expected, then any significant deviation or other aberration is an alert that either normality does not apply to the process or there are quality problems. For example, histograms with two or more peaks may reveal that several "populations" have been mixed together or that several processes are at work.

An example of a histogram from the service industry is presented in Fig. 12.10. Data were collected from four bank tellers on the time to conduct a transaction during the busy lunch-hour period. Note the similarity of the histogram to Fig. 12.9*g*. The two peaks suggest a combination of two bell-shaped distributions, which in turn suggests two different work processes. Even when separate histograms by teller were plotted, the same double-peaked histograms resulted. Then the types of transactions were examined. Developing histograms by type of transaction revealed the reason for the double-peaked histograms. Simple transactions such as withdrawals and deposits resulted in short transaction times; complex transactions such as opening retirement accounts and obtaining certificates of deposit resulted in longer transaction times. Originally, management believed that two inexperienced tellers were the primary reason for delays in serving customers. Histogram analysis revealed the true causes as the extra steps required for complex transactions.

Histograms illustrate how variable (or continuous) data provide much more information than attribute (or categorical) data. For example, Fig. 12.9*b*, *d*, *g*, and *i* warn of potential trouble, even though all units in the sample are within the specification limits. With attribute measurement, all units would simply be classified as acceptable, and the inspection report would have stated "50 inspected, 0 defective"—therefore no problem. One customer had a dramatic experience based on a lot that yielded a sample histogram similar to Fig. 12.9*i*. Although the sample indicated that the lot met quality requirements, the customer realized that the supplier must have made much scrap and screened it out before delivery. A rough calculation indicated that full production must have been about 25 percent defective. The histogram enabled the customer to deduce this without ever having been inside the supplier's plant. Note how the "product tells on the process." As the customer would eventually pay for this scrap (in the selling price), he wanted the situation corrected. The supplier was contacted, and advice was offered in a constructive manner.

As a general rule, at least 40 measurements are needed for the histogram to reveal the basic pattern of variation. Histograms based on too few measurements can lead to incorrect conclusions because the shape of the histogram may be incomplete without the observer realizing it.

Histograms have limitations. Because histograms do not display the order of production, the time-to-time process variations during production are not disclosed. Hence the seeming central tendency of a histogram may be illusory—the process may have drifted substantially.

The histogram is an important analytical tool. The key to its usefulness is its simplicity. It speaks a language that everyone understands—comparison of product measurements

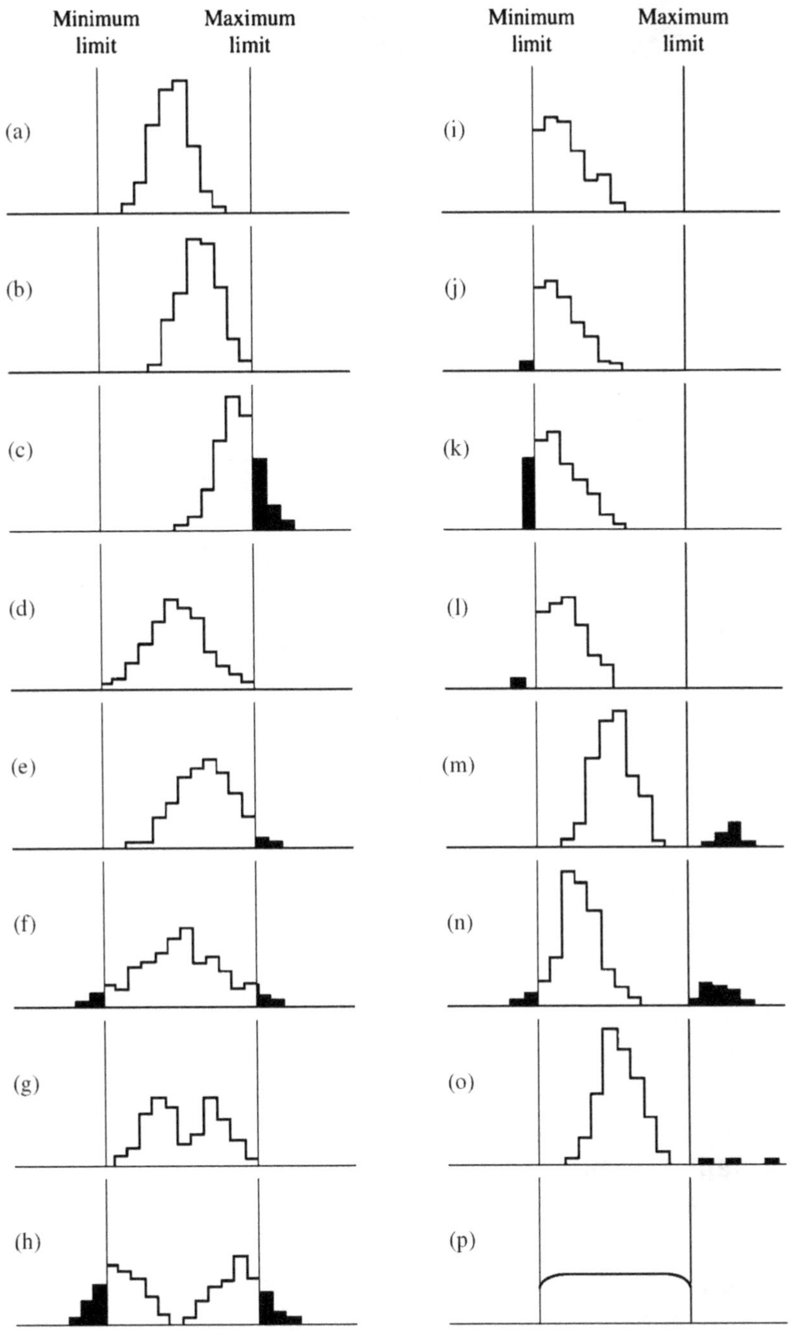

FIGURE 12.9 Distribution patterns related to tolerances. (Adapted from Armstrong and Clarke, 1946.)

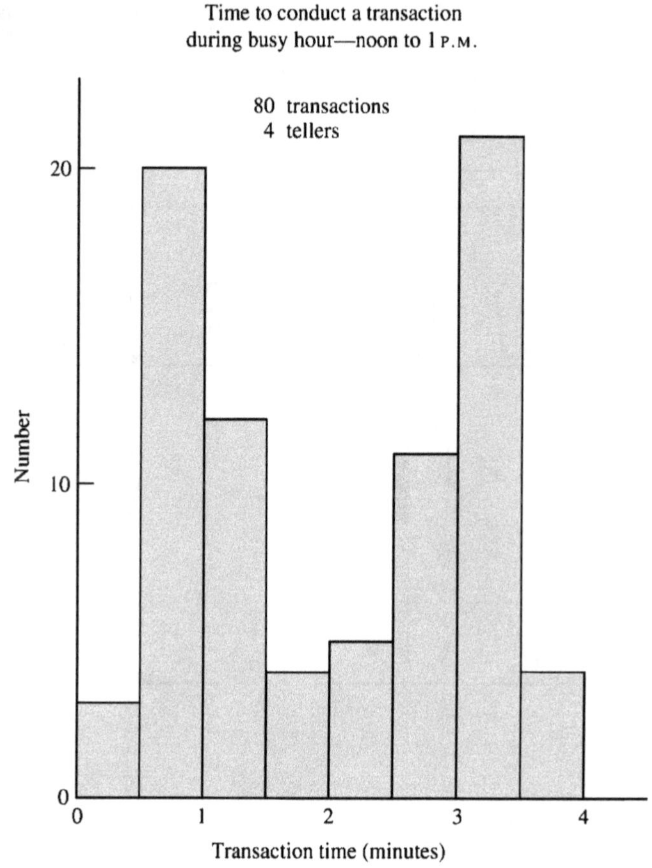

FIGURE 12.10 Histogram of bank teller transaction times.

against specification limits. To draw useful conclusions from this comparison requires little experience in interpreting frequency distributions and no formal training in statistics. The experience soon expands to include applications in development, manufacturing, supplier relations, and field data.

The Exponential Probability Distribution

The exponential probability function is

$$y = \frac{1}{\mu} e^{-X/\mu} \quad (x > 0)$$

Figure 12.6 shows the shape of an exponential distribution curve. Note that the normal and exponential distributions have distinctly different shapes. An examination of the tables of areas shows that 50 percent of a normally distributed population occurs above the mean value and 50 percent below. In an exponential population, 36.8 percent is above the mean and 63.2 percent below the mean. This refutes the intuitive idea that the mean is always

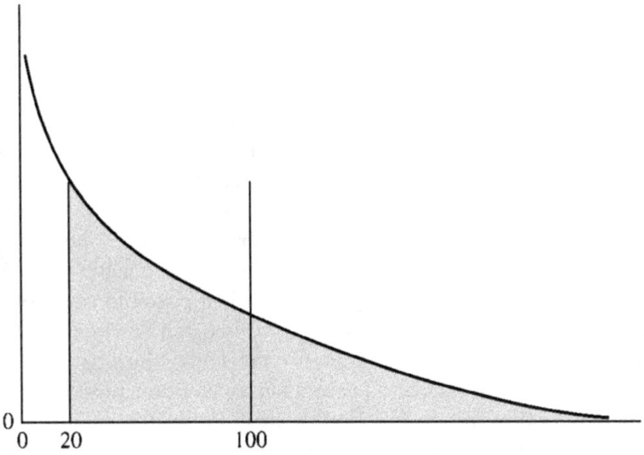

Figure 12.11 Distribution of time between failures.

associated with a 50 percent probability. The exponential curve can be used to describe the loading pattern for some structural members because smaller loads are more numerous than larger loads. The exponential curve is also useful in describing the distribution of failure times of complex equipment.

A fascinating property of the exponential distribution is that the standard deviation equals the mean.

Using the Exponential Probability Distribution for Predictions

Predictions based on an exponentially distributed population require only an estimate of the population mean. For example, the time between successive failures of a complex piece of repairable equipment is measured, and the resulting histogram is found to resemble the exponential probability curve. For the measurement made, the mean time between failures (commonly called MTBF) is 100 hours. What is the probability that the time between two successive failures of this equipment will be at least 20 hours?

The problem is finding the area under the curve beyond 20 hours (Fig. 12.11). Table B in App. II gives the area under the curve beyond any particular value X that is substituted in the ratio X/μ. In this problem:

$$\frac{X}{\mu} = \frac{20}{100} = 0.20$$

From Table B, the area under the curve beyond 20 hours is 0.8187. The probability that the time between two successive failures is greater than 20 hours is .8187, that is, there is about an 82 percent chance that the equipment will operate continuously without failure for 20 or more hours. Similar calculations would give a probability of .9048 for 10 or more hours.

The Weibull Probability Distribution

The Weibull distribution is a family of distributions having the general function

$$y = \alpha\beta(X-\gamma)^{\beta-1}e^{-\alpha(X-\gamma)^{\beta}} \quad (\text{for } x > 0)$$

where α = scale parameter
β = shape parameter
γ = location parameter

The curve of the function (Fig. 12.6) varies greatly depending on the numerical values of the parameters. Most important is the shape parameter β, which reflects the pattern of the curve. Note that when β is 1.0, the Weibull function approximates the exponential. When b is about 3.5 (and $\alpha = 1$ and $\gamma = 0$), the Weibull closely approximates the normal distribution. In most practical applications, β varies from about 1/3 to 5. The scale parameter a is related to the peakedness of the curve, that is, as a becomes larger the curve becomes stretched out and flatter. The location parameter γ is the smallest possible value of X. Increasing the location parameter slides the curve to the right. It is often 0, thereby simplifying the equation to two parameters. For example, datasets concerning time usually have a natural location parameter equal to zero. It is often unnecessary to determine the values of these parameters because predictions are made directly from Weibull probability paper, but King (1981) gives procedures for graphically finding α, β, and γ. Table J in App. II provides a sample of Weibull paper. With this sample paper, b can be estimated by drawing a line parallel to the line of best.

The Weibull covers many shapes of distributions. This feature makes the Weibull popular in practice because it reduces the problems of examining a set of data and deciding which of the common distributions (e.g., normal or exponential) fits best. Computer software such as Excel or MINITAB is useful in performing Weibull analyses.

Using the Weibull Probability Distribution for Predictions

Consider the case of seven heat-treated shafts that were stress tested until each of them failed. The fatigue life (in terms of number of cycles to failure) was as follows:

11,251	40,122
17,786	46,638
26,432	52,374
28,811	

The problem is to predict the percentage of failure of the population for various values of fatigue life. The solution is to plot the data on a probability plot grid, observe whether the points fall approximately in a straight line, and if so, read the probability predictions (percentage of failure) from the graph or compute them from the fitted parameters. Before the ready availability of laptop computing power, a paper and pencil were required for the application of a Weibull probability paper.

In a Weibull plot, the original data are usually[3] plotted against mean ranks. (Thus the mean rank for the ith value in a sample of n ranked observations refers to the mean value of the percentage of the population that would be less than the ith value in repeated experiments of size n.) The mean rank is calculated as $i/(n + 1)$. The mean ranks necessary for this example are based on a sample size of seven failures and are as shown in Table 12.7.

The cycles to failure are now plotted on the probability graph against the corresponding values of the mean rank (see Fig. 12.12). Note that the horizontal scale is logarithmic. The vertical scale reflects the cumulative percentages of the data expected in the fitted Weibull

---

[3]There are other plotting positions such as $(i - 0.5)/n$ and $(i - 0.3)/(n + 0.4)$. The latter is used in most software.

Failure Number (*i*)	Mean Rank
1	0.125
2	0.250
3	0.375
4	0.500
5	0.625
6	0.750
7	0.875

TABLE 12.7 Table of Mean Ranks

distribution. The vertical axis gives the cumulative percentage of failures in the population corresponding to the fatigue life shown on the horizontal axis. For example, about 50 percent of the population of shafts will fail in fewer than 31,000 cycles. About 80 percent of the population will fail in fewer than 44,000 cycles. By appropriate subtractions, predictions can be made of the percentage of failures between any two fatigue life limits. The Minitab software permits reading the exact values from the graph.

It is tempting to extrapolate particularly to predict life. For example, suppose that the minimum fatigue life were specified as 8000 cycles and the seven measurements displayed earlier were from tests conducted to evaluate the ability of the design to meet 8000 cycles. Because all seven tests exceeded 8000 cycles, the design seems adequate and might therefore be released for production. However, extrapolation on the fitted distribution predicts that about 2.3 percent of the population of shafts would fail in less than 8000 cycles. The available

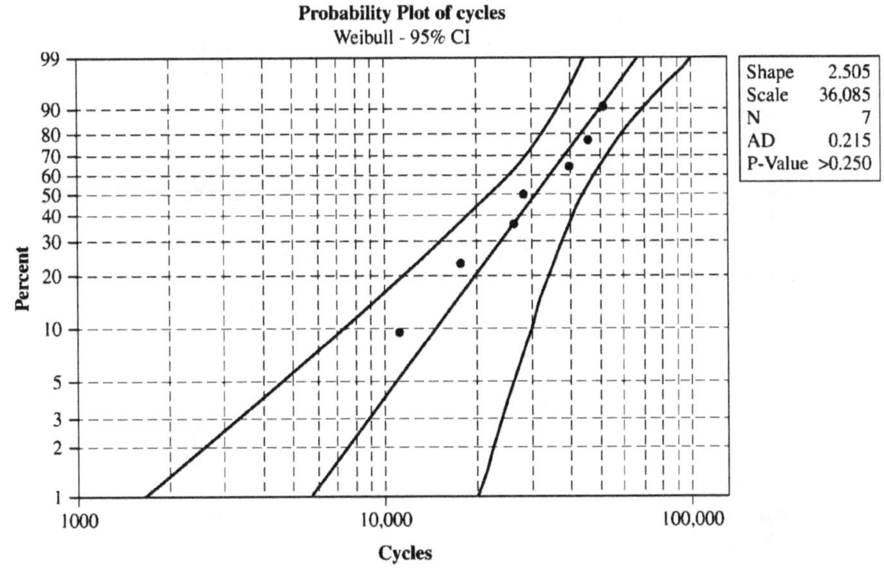

FIGURE 12.12 Probability plot of fatigue life in cycles.

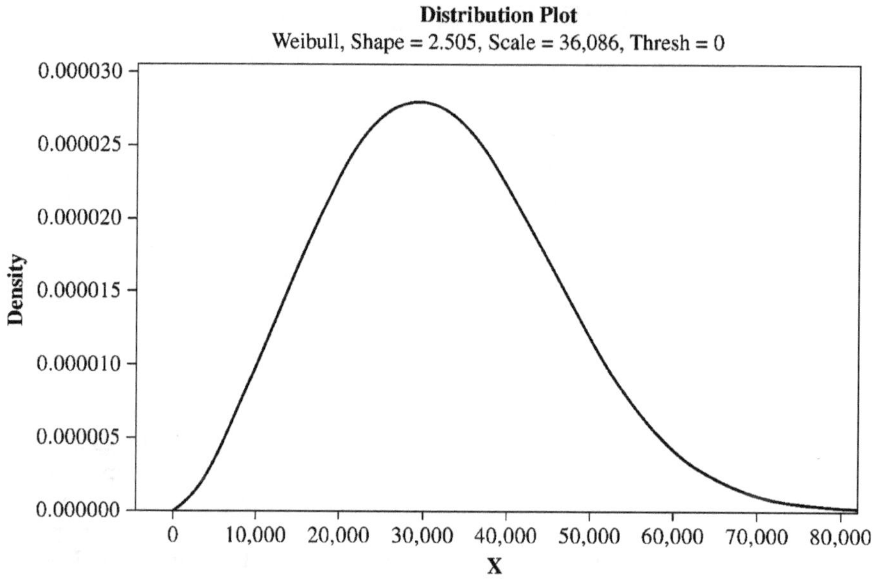

Figure 12.12b Distribution of fatigue life in cycles.

confidence intervals around the fitted distribution suggest that at a 95 percent level of confidence, the 8000 cycle failure rate could be as high as 11 percent. This information suggests a review of the design before release to production. Thus the small sample (all within specifications) gives a deceptive result, but the Weibull plot acts as an alarm signal by highlighting a potential problem and the risks associated with it.

Extrapolation can go in the other direction. Note that a probability plot of life-test data does not require completing all tests before the plotting starts. As each unit fails, the failure time can be plotted against the mean rank. If the early points appear to be following a straight line, it is tempting to draw in the line before all tests are finished. The line can then be extrapolated beyond the actual test data, and the life predictions can be made without accumulating a large amount of test time. The approach has been applied to predicting, early in a warranty period, the "vital few" components of a complex product that will be most troublesome. However, extrapolation has dangers. It requires the judicious melding of statistical theory, process experience, and judgment.

To make a valid Weibull plot, at least seven points are needed. Such small samples can have significant sampling variability, and it is wise to calculate and evaluate the confidence limits. Any fewer data points casts doubt on the ability of the plot to reveal the underlying pattern of variation. The Anderson-Darling (A-D) statistic on the graph is a measure of how well the actual data fit the proposed distribution. If the P-value is less than .05, then the fit is too poor to be used. Other distributions should be evaluated or outlying values evaluated for indications of special cause variation.

Probability distributions for normal, exponential, Weibull, and other probability functions can also be tested as appropriate against life data. Although the mathematical functions and tables provide the same information, the graph reveals relationships between probabilities and values of X that are not as readily apparent from the calculations. For example, the reduction in percentage defective in a population as a function of wider and wider specification limits can be easily portrayed by the graph.

Other continuous distributions include the continuous uniform (roughly speaking, all values having equal probabilities), log normal (logarithms of the original values are normally distributed), and the multinormal (e.g., a product with two measurement parameters, each normally distributed, is called a bivariate normal).

The Poisson Probability Distribution

The fundamental Poisson distribution provides the probability of a specific number of occurrences for some unit of time or space, as follows: If the probability of occurrence p of an event is constant in each of n independent trials of the event, the probability of r occurrences in n trials is

$$\frac{(np)^r e^{-np}}{r!}$$

where n = number of trials
p = probability of occurrence
r = number of occurrences

Using the Poisson Distribution for Predictions

The Poisson distribution is most useful for analyzing where the probability of occurrences is small and the total number of trials can vary. Poisson distributions are widely used in estimating and predicting defects per unit of time or for defects per unit calculations, especially for complex electronic components. The Poisson is also useful as an approximation in calculating probabilities associated with sampling procedures. Table C in App. II gives cumulative Poisson probabilities directly, that is, the probability of r or fewer occurrences in n trials of an event having probability p. For example, suppose that a lot of 300 units of product is submitted by a vendor whose past quality has been about 2 percent defective. A random sample of 40 units is selected from the lot. Table C in App. II provides the probability of r or fewer defectives in a sample of n units. Entering the table with a value of np equal to 40(0.02), or 0.8, for various values of r results in Table 12.8. Individual probabilities can be found by subtracting cumulative probabilities. Thus the probability of exactly two defectives is .953 – .809, or .144. Of course, the probabilities in Table 12.8 could also be found by substituting the formula six times ($r = 0, 1, 2, 3, 4, 5$).

r	Probability of r or Fewer in Sample with 0.8 Expected per Sample
0	.449
1	.809
2	.953
3	.991
4	.999
5	1.000

Table 12.8 Table of Poisson Probabilities

The Poisson, in this case, is an approximation of the more exact, but complex, binomial distribution and applies when the sample size is at least 16, the population size is at least 10 times the sample size, and the probability of occurrence p in each trial is less than .1. These conditions are often met.

The Binomial Probability Distribution

If the conditions of the Poisson distribution are not met, the binomial distribution may be applicable. The Binomial distribution is the general condition with a fixed number of tri. The Binomial distribution is the general condition with a fixed number of trials and constant probability of success. The Poisson is a special usage when we are studying rare events. If the probability of occurrence p of an event is constant in each of n independent trials of the event, then the probability of r occurrences in n trials is

$$\frac{n!}{r!(n-r)!}p^r q^{n-r}$$

where $q = 1 - p$.

In practice, the assumption of constant probability of occurrence is considered reasonable when the population size is at least 10 times the sample size.[4]

Tables for the binomial are available, but today it is usually easiest to use the calculation capability of software like Minitab or Excel.

Using the Binomial Probability Distribution for Predictions

A lot of 100 units of product is submitted by a vendor whose past quality has been about 5 percent defective. A random sample of six units is selected from the lot. The probabilities of various sample results are given in Table 12.9.

In using the formula, note that $0! = 1$.

Other discrete distributions include the hypergeometric (used when the probability of an occurrence can change as in sampling without replacement in a small universe),

r	P (Exactly r Defectives in 6) = $[6!/r!(6-r)!](0.05)^r(0.95)^{6-r}$
0	.7351
1	.2321
2	.0306
3	.0021
4	.0001
5	.000
6	.000

TABLE 12.9 Table of Binomial Probabilities

[4] Under this condition, the change in probability from one trial to the next is negligible. If this condition is not met, the hypergeometric distribution should be used.

discrete uniform (all values have equal probabilities), and the multinomial (when two or more parameters are observed in a sample).

Basic Theorems of Probability

Probability is expressed as a number that lies between 1.0 (certainty that an event will occur) and .0 (impossibility of occurrence).

A convenient definition of probability is one based on a frequency interpretation: If an event A can occur in s cases out of a total of n possible and equally probable cases, the probability that the event will occur is

$$P(A) = \frac{s}{n} = \frac{\text{Number of successful cases}}{\text{Total number of possible cases}}$$

Example 12.1 A lot consists of 100 parts. A single part is selected at random, and thus each of the 100 parts has an equal chance of being selected. Suppose that a lot contains a total of eight defectives. Then the probability of drawing a single part that is defective is 8/100, or .08.

The following theorems are useful in solving problems:

Theorem 12.1 If $P(A)$ is the probability that an event A will occur, then the probability that A will not occur is $1 - P(A)$. This theorem is sometimes called the complementary rule of probability.

Theorem 12.2 If A and B are two events, then the probability that either A or B or both will occur is

$$P(A \text{ or } B) = P(A) + P(B) - P(A \text{ and } B)$$

A special case of this theorem occurs when A and B cannot occur simultaneously (i.e., A and B are mutually exclusive). Then the probability that either A or B will occur is

$$P(A \text{ or } B) = P(A) + P(B)$$

Example 12.2 The probability of r defectives in a sample of six units from a 5 percent defective lot was previously found by the binomial (sampling is with replacement). The probability of zero defectives was .7351; the probability of one defective was .2321. The probability of zero or one defective is then .7351 + .2321, or .9672.

Theorem 12.3 If A or B are two events, then the probability that events A and B occur together is

$$P(A \text{ and } B) = P(A) \times P(B \mid A)$$

where $P(B \mid A)$ = probability that B will occur assuming that A has already occurred.

A special case of this theorem occurs when the two events are independent, that is, when the occurrence of one event has no influence on the probability of the other event. If A and B are independent, then the probability of both A and B occurring is

$$P(A \text{ and } B) = P(A) \times P(B)$$

Example 12.3 A complex system consists of two major subsystems that operate independently. The probability of successful performance of the first subsystem is .95; the corresponding probability for the second subsystem is .90. Both subsystems must operate successfully to achieve total system success. The probability of the successful operation of the total system is therefore .95 × .90 = .855.

The theorems preceding have been stated in terms of two events but can be expanded for any number of events.

Computer Software for Statistical Analysis

With the advent of statistical software packages, the practitioner can now use many statistical techniques that were not previously considered because of the difficulty in doing the calculations. Currently, most software packages provide a basic explanation of a technique, define the inputs required, and then present the results. But such accessibility has a danger. The practitioner must understand the assumptions behind the methods and what the final results do and do not mean. In the haste to obtain an answer and avoid tedious detail, there is a danger of applying a technique incorrectly or misunderstanding a result. Beware of these serious consequences.

References

Armstrong, G. R., and P. C. Clarke (1946). "Frequency Distribution vs Acceptance Table," *Industrial Quality Control*, vol. 3, no. 2, pp. 22–27.

Harris, R. L. (1996). *Information Graphics*. Management Graphics, Atlanta, GA.

Heyes, G. B. (1985). "The Box Plot." *Quality Progress*, December, pp. 12–17.

Juran Institute, Inc., (1989). *Quality Improvement Tools—Box Plots*. Wilton, CT, p. 7.

King, J. R. (1981). *Probability Charts for Decision Making*, rev. ed., TEAM, Tamworth, NH.

Tufte, E. R. (1997). *Visual Explanations*. Graphics Press, Cheshire, CT.

Tukey, J. W. (1977). *Exploratory Data Analysis*. Addison-Wesley, Reading, MA.

Wadsworth, H. M., K. S. Stephens, and A. B. Godfrey (2001). *Modern Methods for Quality Control and Improvement*, 2nd ed., John Wiley and Sons, New York.

CHAPTER 13

Managing Risk

Leonard W. Heflich

High Points of This Chapter

1. Identifying risks
2. Assessing risks and loss potential
3. Communicating to build awareness of risks
4. Performing a risk assessment
5. Using risk assessment to focus our activities and projects
6. Managing to reduce, mitigate or eliminate loss due to risk

Introduction

Risk is anything that can go wrong and result in a loss. Operational risk is defined as "the direct or indirect loss resulting from inadequate or failed internal processes, people and systems or from external events." (Power, 2004) Under good conditions, risk can be invisible, making it difficult to see or plan for (Marks, 2014). Under adverse conditions our people, processes and systems are stressed and may fail resulting in loss. Risk assessment and risk management are critical capabilities that we must develop in ourselves, our teams, our organizations and our companies in order to prepare for what goes wrong under adverse conditions and minimize the loss.

This chapter focuses on the practical versus classical. There are many excellent books and publications on the classical methods including ORM (operational risk management), which was originally developed by the U.S. Air Force and NASA. These classical methods are required reading and are helpful in developing a risk management process. I will attempt to take us beyond the theoretical approach as defined in ORM to identify practical methods that will work in today's business environment. ORM may be beyond our expertise, and we may not have the same resources as NASA.

Risk management is applicable to all human activities as all human activities are inherently risky. The fact that our activities are risky, with the potential for loss under adverse conditions, means that we need a framework for assessing the risk, the potential loss/benefits and for identifying steps we can take to manage, minimize or even eliminate the risk and the resulting loss.

Risk management is an excellent decision tool that we can use to plan our work and projects. Basing our activities on risk will focus our efforts on the work that can add the most value. Today risk management is being successfully applied to every field of work.

Risk management is complicated. Risk encompasses every facet of our business. It involves multiple parties including colleagues, competitors, consumers, customers, regulators, and the media and is international in scope. Risk is constrained by many factors; it is driven by change in technology and social forces and is only partially predictable or controllable. Risk management involves tradesoffs but is not a zero sum situation.

Get used to the idea that every situation has two sides like a coin and we must examine both when assessing risk. It is very easy for a team to develop groupthink and head down a path together in apparent consensus without stopping to consider the other side of the coin. Edward de Bono's book *Six Thinking Hats* is an excellent treatise on the subject (De Bono, 1999). He assigns different color hats to each team member and specific goals to each. One of the members must wear the black hat and their goal is to find everything wrong with the idea. This is a great way to prevent groupthink. In a risk assessment exercise, we especially want to be thinking of the dark side to find all that can go wrong with our processes, people and our systems. This is not the time for blind optimism. We want balanced, intelligent pessimism. Another great book to read by Andy Grove, former CEO of Intel, is *Only The Paranoid Survive* (Grove, 1996). I tell people, you don't need to read the book, just the title!

What Is a Risk?

A risk is something that can go wrong and result in a loss. It is wise to be aware of risks so that we can take prudent steps to prepare for the possibility of the occurrence so as to minimize or eliminate the potential loss. Preventing loss is the goal. Risks are difficult to predict, or they wouldn't be risky. We are not omniscient of course, but we don't have to be, just smart and diligent. Risks are often hidden, especially when things are going smoothly. A risk may only make itself known under adverse conditions, which of course can vary as well. A risk may not be evident under mild adverse conditions because our systems and people have the capability to recognize the situation and manage it effectively. The risk may only become evident and cause loss under more extreme adverse conditions. This is difficult to predict. The other problem with risks is that all our efforts to prepare for them will appear to be a waste of time and money until adverse conditions expose the risk and the resulting loss. Selling a program to reduce risk can be difficult, as it may appear to be unnecessary. This further underscores the need for awareness. If management is unaware of a risk or the potential loss that can occur under adverse conditions, they will be unwilling to spend time or money on systems to minimize, manage, or eliminate the risk.

Identifying Risks

The first step is to identify the potential risks. This is not a trivial activity, and is critically important, as it will be impossible for us to prepare for or prevent a loss due to a risk unless we are aware of it. I suggest that the first step to identify the risks that are relevant to your business or operation is to form a multifunctional crisis team with broad knowledge of your business, your systems, your people and your processes.

1. Start with a list of activities, systems and processes that are used in your business.
2. Go through the list and ask the question "What can possibly go wrong?"
3. Make a list of significant failures that have occurred in the past 3 years and use this to identify additional potential risks.

Assessing the Risk

Once we have a list of potential risks, we need to assess the importance of each risk to allow us to prioritize. We cannot work on everything and it will be best to focus our efforts on the most important risks. On the other hand, in order to be diligent, we must do what is possible and reasonable to reduce or eliminate risk when it is possible, regardless of the importance rating. Assessing the importance of the risks is a great learning and team development exercise. The team will learn a lot from each other and learn about the risks facing the business, especially those risks that originate in another function but could have impact on them as well.

Risks vary in the extent of the potential loss to our business. Classical ORM assesses risk on two factors: severity and probability. I suggest that there is benefit to consider a few others including

- The impact of speed
- Consumer outrage
- Complexity, scope
- Uncertainty

In the classical method, we assign a numerical rating (1, 2, or 3) to each of the factors to denote the magnitude of the potential loss, basically low, medium, and high. We can use a 3- or 5-point scale or higher if desired. I suggest that the point of the exercise is to separate and prioritize the risks and a 3-point scale is adequate. Typically we multiply the values assigned to each risk for each factor. If we include multiple factors other than severity and probability, it may be better to add the values assigned to each risk rather than multiply, to keep the resulting numbers manageable. The result is a list of the potential risks and a numerical value assigned to each that is proportional to the importance of the risk to our business.

In assessing the risks, we want our estimates to be grounded in reality. We don't want to speculate wildly. On the other hand, data is usually lacking so some judgment will be required. Historical data may be a poor indicator of future risk, so we should not depend solely on history. The world is changing, our business is changing and our systems are changing. We need to extrapolate but not in a linear fashion. We need to imagine, based on our experience, what could go wrong and then assess each risk accordingly.

We will use this list of prioritized risks to create awareness in our organization of the risks and the importance of each. We will use the list and numeric values to assess our options for prevention, elimination or management of each risk. There are costs associated with prevention and spending a lot of money to prepare for a risk that will never occur is a waste of money. Of course it's a waste of money only until the day that the risk situation occurs, in which case our planning and investment were brilliant and prudent.

Severity

Severity is an estimate of how great the loss could be if a risk situation were to occur. We need to consider potential extreme adverse conditions. A good question to ask is "What is the worst that can happen?" If we are aware of the worst that can happen, build our systems to prevent or at least manage a worst-case scenario, then we have done a good job to prepare our company and ourselves. For example, when assessing the severity of damage that could result from a storm, we typically consider a one hundred year time frame and assess what is the worst storm we have experienced or expect to experience in a one hundred year time frame? Communities typically write building codes based on the severity of this potential situation.

Probability

Probability is the likelihood that a risk situation could occur. Consider how often the risk situation is likely to occur. In the example of a storm we are using a 100-year time frame. One hundred years sounds like a long time but in reality the buildings that we are constructing today will likely still be in use 100 years from now. Historical data is a good place to start, but is insufficient. The world is changing and the past may not be a good indicator of frequency of occurrence in the future.

Impact of Speed

The impact of speed is an assessment of how fast a loss could occur in the event of a risk situation and whether our systems are capable of responding quickly enough to prevent a loss. Speed is becoming a more important factor in risk assessment due to the impact of social media and the fact that the world is moving and changing faster today than ever before. If our systems are obsolete, complicated or cumbersome, speed could have a severe multiplier effect on the potential loss in a risk situation. The speed at which a loss occurs demonstrates the value of being aware, prepared and practiced when an event happens. There will be little or no time to consider the options or to make plans once the event has occurred. We need to have at least a framework of a plan ready to execute at a moment's notice in order to prevent loss due to speed.

Outrage

Consumers become outraged at some failures but not others. We need to assess if the potential risks we are considering will result in outrage in the event of a failure or not. If there is outrage, a seemingly innocuous situation could become a catastrophic loss. For example, consumers are not outraged by the risks associated with driving a car as is evidenced by the fact that in the United States alone, 34,000 people die each year in automobile crashes. Worldwide the number is 1.23 million. Yet public outrage is very low. On the other hand, at most a few hundred people die each year in commercial airplane crashes, yet the outrage level is very high. Consumers do not accept failure in commercial flight so prevention efforts must be of the highest level. Outrage is not rational, but it can be important if the potential failures you are considering invoke consumer outrage.

Complexity

Complexity hits us from several different angles. One perspective we need to consider is the complexity of the systems and processes in our company. Complex systems that are inflexible and delay our response to a potential crisis can make the difference between a minor event and a major loss. Our systems need to be flexible and capable of rapid identification of the situation, enable rapid and good decisions and allow for rapid and effective response to the situation. Rapid and effective response will minimize the loss. Complex systems that are difficult to navigate will slow down the response and increase the potential loss. We need to honestly assess the capability of our systems to gather the right information, make a good decision and take effective action.

The other perspective we need to consider is the complexity of the risk situation. A complex situation with multiple parties involved can make it difficult, or at least slow, to get the information we need after a risk situation has occurred. It can make it more difficult to assess the situation and decide what actions need to be taken. Ultimately, a complex situation can paralyze our organization and prevent us from taking any action. If a risk situation is known to be complex, we can take steps up front to identify the issues and parties involved and prepare a plan in advance with buy-in from all. We could agree on roles and responsibilities in the event of a failure so that everyone knows what to do and gathers the needed information quickly. We could assign decision authority to one person to assure that actions are taken quickly and effectively. If we consider the situation rationally and calmly beforehand, we can create a protocol to manage it. If we wait for the crisis situation to occur and then try to navigate a complex situation, we will delay our response and increase the loss.

Scope

The scope of a potential loss is defined by how broad the impact could be. Is the loss isolated to a specific product, brand, customer, plant, country, etc. or could the impact spread more broadly? Scope is an especially critical factor to consider today where the Internet will link an event that occurs in a division in Brazil to our businesses in the United States or Europe. We are all linked and within seconds the Internet will disperse information about a failure in one country with no regard for borders. Scope could also involve competitors. When peanuts from one company are at risk, all peanut producers are impacted without regard for the fact that our product has had no similar failure. A broad scope can expand the potential loss due to a risk situation to other producers or even other industries. Keep in mind that nothing ever really disappears from the Internet. Even if we are able to remove videos, they will always keep popping up and be active for years or decades. The scope of impact can greatly expand the loss due to a failure.

Uncertainty

We need to assess the degree to which uncertainty can impact a potential loss situation. Some situations are inherently more uncertain than others. The information that we need to make a good decision may be lacking because of our inability to quantify or even assess the potential loss. The systems that exist inside our company or outside may be inadequate to provide accurate, timely information that we will need to make a good decision.

One aspect of assessing uncertainty is to assess the adequacy of the information that will be available to us in a crisis situation. For example, in the event of a fire in a plant, the local fire department will evacuate the building and not allow our people to enter the building until they decide that it is safe to do so. During this time, when we are not allowed to enter the building, it is impossible for us to assess the condition of the facility or even to estimate when we will be allowed back into the building. Uncertainty is high. Another example is one of our vendors could have a foreign material contamination failure. If we don't know our vendors or what kind of systems they have in place and have no historical records on their performance then it will be difficult for us to assess the risk of a failure. There is even an element of trust to consider. If we trust another party to be honest and complete in their communication of a risk event it gives us some confidence in their performance. If we don't trust them, or worse, know that we cannot trust them, then we will be second-guessing them and wasting valuable time.

When in a crisis situation, we are almost always operating with incomplete or wrong information. And yet, we need to assess the situation and take effective action as soon as we are able. This puts us in a difficult situation. If we act too soon, we may miss important information, which will at best require that we reassess the situation and make a better decision sometime later, or at worst, result in a mistake that compounds the loss. On the other hand, the information we have will never be complete or totally accurate, so delaying the decision and action will increase the loss.

Another aspect of uncertainty is excessive certainty! Remember that every coin has two sides. So just when we think we have all the information we need and are certain of the its validity, that is a good time to step back and ask if the information we have is really as good as it looks on paper. If we think the information is incorruptible and fail to be adequately critical, we could make a big mistake and overlook a serious risk. One of my favorite bosses used to say "You're really in trouble when you start to believe your own BS!"

Risk Is Risky

Now we have identified the potential risks and assessed each on several factors. We know a lot more about our business than we did before. We are starting to see where losses could occur in the event of a failure and how the loss could be compounded under extreme conditions. Let's turn this coin over and look at the other side; meaning how good a job did we do in identifying and assessing the risks? Uncertainty, and therefore risk, will creep into our assessment in several ways.

Unknown Unknowns

At the bottom of the list are the unknown unknowns—we don't know what we don't know. This is a bad position to be in, as we don't even realize that we are in a potential loss situation. We need at least to be aware. This is where self-education and outreach to others can be so helpful. Industry trade groups are excellent at helping companies share their expertise in a noncompetitive environment. Many companies do not consider their safety programs to be a competitive advantage, recognizing that a failure in one company impacts all, so there is value in sharing best practices in risk prevention. A great example is the automobile airbag. Mercedes Benz spent millions of dollars developing a workable airbag for cars and started installing them in their cars. Then instead of patenting it, they made copies of the blueprints and sent them to all the other major automobile manufacturers and told them that they needed to do the same. They never asked for a penny of compensation. The airbag has become standard safety equipment in almost all cars today and is credited with saving hundreds of thousands of lives.

Known Unknowns

Next we can be aware of a risk but not have sufficient data to assess the potential loss. This is a known unknown. At least we are aware of the existence of a risk. Not having adequate data with which to assess it puts us in a difficult and potentially indefensible position. We need to get more information. The EU has enlisted the precautionary principle in these situations to elevate these risks to the level of a hazard, and therefore, to be avoided. This could be a good approach, but the downside is that avoidance means losing the benefits as well.

Known Knowns

Then at the top of the pile, we are aware of a situation and have good data to assess the potential loss and feel confident in the data. The risk, of course, is that we are wrong. The adverse conditions experienced in the past may have been inadequate to stress the situation sufficiently to expose the full extent of potential loss. This situation can be as bad as or worse than the others as we have some confidence in our assessment, but the confidence could be unfounded. The data that is available could be incomplete or wrong. In fact, we should expect it to be incomplete and wrong.

Predictability

Classical risk management methods use algorithms and models to assess the risk due to certain situations. I suggest that if a situation can be effectively modeled or described by an algorithm, then it is predictable and therefore is not a risk. Risky situations are situations that cannot be predicted or modeled. That is why they are risky. We don't have the data or an effective model to predict the potential loss. We only know that if something were to go wrong a loss could occur.

Strict Controls

Strict, inflexible controls, defined systems and quantified risks can in themselves create risk. This sounds like an oxymoron, but the tighter the process the more likely it is to fail in the event of a crisis situation. Every crisis situation is different. The risk may be the same but the event is different due to the conditions and other unanticipated and uncontrollable factors. Our controls and systems must be flexible enough to adapt to the change in conditions. First the system must be capable of identifying the change in conditions and then be capable of adapting to it. There is no substitute for thinking in a crisis. Developing a well-defined system could be excellent practice for the team to prepare them for a real crisis, but there must be the realization that once the real thing is in progress, the well-defined system is only a reference that guides our actions and is not a strait jacket that limits our actions.

Failure Is Necessary

This is really going to surprise you—failure is actually necessary (Power, 2004). It is clear that we need to do everything we can to anticipate, prepare for and prevent a potential risk situation from occurring and causing loss. However things will go wrong and crises will occur; therefore, we need to be excellent at managing them in order to minimize the loss. Some failures are outside of our control, such as those caused by weather, earthquakes, floods, fires and even strikes or regulatory events. These are real risks and we need to assess what we do can to minimize the impact and loss in advance while the sun is shining, but then be prepared to manage them effectively and quickly when the rain starts to fall.

In a real world, perfection is not possible and not affordable. The best quality systems aim for Six Sigma or less than 3.4 defects per million units produced. That is a very low rate of failure but it is not zero and therefore not perfect. Your organization may not be able to afford a failure rate of Six Sigma. Maybe the cost of failure is lower and the cost of avoidance higher, so you can only afford a failure rate of Three Sigma or one in 99 units produced. You need to assess the cost of failure versus the cost of a quality system to prevent it to find out where the two curves intersect. This will help you to find out how much failure you can afford or not afford.

All human systems fail eventually. In fact if we are not failing, then we are probably not trying hard enough. New products, changes to existing products, competitive pressures, a failure that occurs at a vendor or a customer, etc. will result in failures that we need to respond to. These can and should be part of our risk assessment process. But the expectation that we can effectively anticipate and eliminate risk or the losses associated with all risk events is beyond our ability. We need to accept that failure will occur in spite of, or even because of, our efforts and then build systems to detect and manage the situation to minimize the loss. If we think that our systems and processes are so good that failure cannot occur, we will be inviting failure, because almost certainly we are mistaken. If you have not observed a failure in some time in your organization, you need to examine what is going on. What is your definition of failure? Are people hiding failures? Are losses occurring but not being considered a failure? Failing to fail could itself be a failure.

What Can We Do?

In a risk situation, we need to act quickly and effectively. Some simple guidelines up front can make this easier.

1. Define what constitutes a crisis situation for your organization. I work in the food industry, so our definition of a crisis event is when we have produced food that has the potential to cause harm to a consumer.
2. Demand rapid communication. When a crisis event occurs our expectation is that notice is communicated to the crisis team within 30 minutes. This means that we cannot wait for an investigation or for the team to gather all of the facts. As soon as a crisis event is identified it is communicated.
3. There must be two people in the box. This means that no one has the right to put the company at risk by themselves. They must include someone who is outside of their "box" in making the decision on what to do next. Can they manage the situation themselves or do they need to elevate the decision? This is a difficult decision as no one likes to call his or her boss at 3 AM in the morning with bad news. We need to thank people for being brave enough to elevate a concern, even if we find afterwards that it was a false alarm. Never punish the messenger or the system will not work.
4. Ask simple questions and take simple, effective actions
   - What do we know? (Or better yet—what do we think we know?)
   - What don't we know that we need in order to make a good decision?
   - Who is going to get that information and when?
   - What is a reasonable explanation of the events that occurred? If the pieces fit and there are no outlying or missing facts, then we have reason to feel confident in the facts and the description of the situation. If any key fact does not fit or is missing, then we need to dig further.
   - Repeat until we have enough information to make a good decision.
5. Make the necessary decisions and communicate to the team for rapid execution.
6. Execute quickly and flawlessly. Do not allow failure. If there is an indication of weakness or hesitation in any person or group, immediately reinforce it to ensure success.

7. Regroup the team as often as needed to assess the progress to ensure success.
8. Debrief at the conclusion of the event to assess what we did well and what to do better next time.

It is absolutely critical to document everything during a crisis event—what we know, the time when we know it and the source. Time and source is critical during debrief. In the heat of the moment we will quickly forget what we know, when and how we know it. We must keep an accurate and complete log of the events *during* the crisis event.

Managing Risk

Once we are aware and know something about the potential risks that we face, then we are ready to develop plans for managing, minimizing and perhaps even eliminating those risks and the losses that can occur under adverse conditions. One of the biggest failures that I see commonly in risk management is that we fail to think big enough. We aim for a 10 percent reduction in some loss in the next year. It sounds like a reasonable goal and perhaps a greater reduction is not feasible in a short time. First, don't assume that it is impossible to eliminate a risk. Perhaps a longer time frame will help expand our thinking enough to allow us to find a way. Challenge your team to find a way to eliminate the risk in 3 or 5 years and you will be surprised to find out that they can.

Concerns

Make a list of your concerns about the risk. Brainstorm it. Be expansive and don't leave anything out at this point. Your concerns are all the aspects of the risk that make it risky. Include all the factors in your list: severity, probability, impact of speed, outrage, complexity, scope, and uncertainty.

Options

Make a list of all of your options to minimize or eliminate the risk. Again, brainstorm it and be expansive and all-inclusive. Leave nothing out. What could you do if you had the time and resources to do it?

Consequences

Consider the consequences of action and inaction. Taking no action is an action. There are always consequences to our actions, both good and bad. Often we must accept some bad consequences that come along with the good. A revenge effect is an unanticipated bad result that happens because we did something we thought was good. We require football players to wear padded helmets in order to prevent head injuries reducing minor injuries. But the unanticipated result is that serious head injuries increase, because the helmet makes the players feel safe to hit even harder with their heads.

Future Impact

When we consider the consequences of our actions, we must consider the possibility that our decision today could have an impact on what we can do in the future, resulting in a loss of flexibility or even a loss. Deciding that we will not produce a certain product because of an inherent risk can cause a loss in the future if one of our competitors figures out how to safely manufacture the product and manage the risk.

Cost and Benefit

There is a cost to our actions both in time and money. We do not have unlimited amounts of either, so we will need to focus our efforts to keep within our budgets for time and money. And we are a business, so we need to assess the cost/benefit tradeoff. If the benefit of our action or project doesn't justify the expense, then we need to reconsider. Perhaps there is a more efficient way to get the benefit with less cost. Perhaps the benefit can be enhanced at the same cost. Perhaps we need to put this action or project on hold for now and further develop the proposal to improve the cost/benefit ratio. Or, perhaps the risk is such that we need to work to reduce or manage it at any cost. These are difficult but necessary decisions to make.

What Will You Do?

Last, we need to decide what we will do, who will do it and when. Now is the time for practical actions that we can execute. Use the list of concerns and options tempered by consideration of the consequences, the cost/benefit and future impact to develop a list of real actions and projects that we can and will do. This list of work is the real purpose of doing a risk assessment. I say jokingly that the purpose of risk assessment is not to do a risk assessment! It's no joke. If we do a risk assessment and stop short of identifying real, concrete action steps and projects to minimize or eliminate the risk, then we have wasted our efforts.

Making Risk Management Work

Finally, after we have identified the potential risks, assessed the importance of each and identified some effective actions to take to manage, minimize or eliminate the risks and the losses associated with them, we must execute our plan.

Risk Avoidance

All human activities involve some level of risk. Some activities are hazardous and should be avoided, as the risks are not worth accepting. The potential loss is too great, the cost of managing it too high, our ability to effectively manage it too limited or inadequate, or the benefit too low to justify spending to avoid the potential loss. In these cases, it may be better to completely avoid the risky practice altogether. It is an option to consider. If we decide to continue the practice in spite of the identified risk and potential loss, then we do so with awareness and hopefully commitment from our management to do what is necessary to manage the risk effectively. Some hazardous activities are worth the benefit even if it comes with a risk of severe loss. Chemotherapy is a hazardous practice with significant risk of loss to the patient, but also with significant potential benefit that motivates a person with cancer to accept the risk and the potential loss.

Anticipate and Prepare

The objective of risk management is to allow us to anticipate and prepare effectively when we are engaging in risky practices or activities. Making everyone aware of the risk and potential loss is the first step. They simply may not be aware of the risk. Then, once we have their attention, we need to identify the options that we have to manage, minimize and even eliminate the risk and the associated potential loss. Anticipation of what could go wrong and what losses could occur is itself a risky practice with lots of uncertainty. But it puts us in a far better position than being ignorant or negligent where we do not know or know but choose to do nothing about it. We want to be diligent regarding risk and educate ourselves as best we can about the existence of risk and then take appropriate and prudent actions to manage, minimize or eliminate the risk and the potential loss. How well we can do this depends on how well we understand the risks. There is no substitute for thinking and hard work. This effort will often take us beyond the current state of the art into areas that are either unknown or partially known at best. This is cutting edge work. Try to fit what you learn into a model to assess the validity and whether new information is consistent with what you already know. A good model can help us to understand and predict in areas of uncertainty. We said earlier that risk management should not be a competitive advantage, but the fact is that those who are better able to manage risk in their businesses will have a competitive advantage. If we can build systems and teams of people that are more competent and more capable of managing risk than our competition, it will be a significant competitive advantage.

Managing under Uncertainty

Risk is uncertain and risky. We can improve our success rate by building an effective early detection system. A healthy dose of paranoia is useful. Teach your team to always examine both sides of the coin—after identifying all the good reasons to do a project, stop and examine if there are any potential bad effects or revenge effects. What can go wrong? Remember that Murphy was an optimist! Murphy's law is "If something can go wrong, it will go wrong." That is an optimistic perspective. The pessimistic paranoid perspective is "Even when we think it cannot possibly go wrong, it will still go wrong." So when you think that you have done "everything" to manage and prevent a risk from occurring or resulting in a loss, check again and look for gaps and for what could go wrong anyway. Reassess and reexamine your risk assessment at least annually to update it. The world is changing, our business is changing and the risks we face every day are changing. Assume that your plan is obsolete almost as soon as you finish writing it. See Figure 13-1 for a visual representation of the universe of risk in which we operate. The inner elipse are the risks that we can identify and manage. The larger box includes the manageable risks and risks that we can identify but

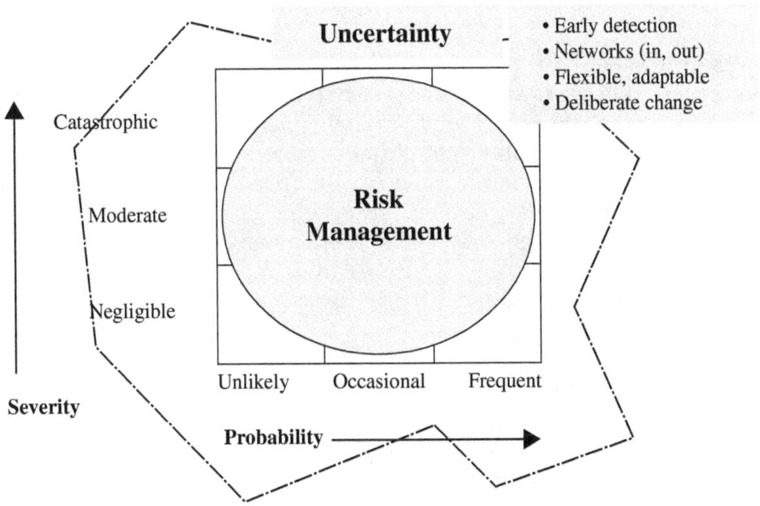

FIGURE 13.1 Risk management with uncertainty.

not manage. The outer irregular dashed line represents all the risks that exists, including those that are unknown and unmanageable.

Reassess and Reevaluate

When you complete the activities or projects that were identified in the risk assessment process, it is time to review it and make a new list of activities and projects. It is a never-ending process. You will find that over time your team gets better, your activities get more focused and results improve. You may also find that the opportunities get harder to find as you pick the low hanging fruit. Then it's time to expand the time frame and make the goal bigger. Instead of small incremental improvements in risk reduction, look for bigger, bolder opportunities to eliminate the risk. Perhaps with a 3 or 5 year horizon the team can find a way to do this.

Networking

Get outside of your organization to see what people in other companies or industries are doing to reduce or eliminate risk. Who has a best practice that you can copy or leverage? Benchmarking with other companies is a great way to trade best practices to the benefit of both parties. Using a consultant or being active in a trade group can expose us to what others are doing. And of course, read about failures and successes. It may be better to study the failures, as a good failure investigation will reveal more than the flowery description of a success.

Flexible Systems

Examine your systems and test them. Write a good crisis drill and use it to reveal weak spots in your processes, people and systems. Developing a meaningful drill is itself instructive. Make it as real as possible. I like to deliver the drill in stages and purposely include wrong and irrelevant information in the early stages. After the drill, review the documentation, the decisions, the speed and the effectiveness of the team. If your systems are slow, cumbersome, obsolete, inadequate or needlessly complex, then this is an important area to work on. Developing or changing systems is a difficult and painful process. Better to start now than wait for a loss situation to uncover the issues.

Change Management

Your organization needs an effective system for managing change. Launching a new product, modifying an existing product, changing vendors, changing packaging, complying with new regulatory requirements, etc. all require that we can manage change effectively and quickly. Mistakes or delays can be costly. There is a natural tradeoff between process and results and between systems that are effective and ones that are complex and burdensome. People will use a good system and learn how to use it effectively to manage their area of responsibility. People will avoid a bad or unnecessarily complex system, creating a major risk for the organization. Change is happening all the time. We must be good at managing it.

Change as a Competitive Advantage

Having the ability to manage change can itself be a competitive advantage. If our organization is able to make and manage change more quickly or more effectively than our competitor, this could be a significant competitive advantage. It is not just about avoiding losses that come with delays or failures, but also the benefits in the marketplace of being first or being agile.

Communicating Risk

We must communicate appropriately about risk and risk management to the people in our organization as well as to our customers, consumers and vendors. Making everyone aware of the existence of a risk is a critical first step. We cannot act without awareness. Communicate, communicate and communicate. Then communicate again! We cannot over communicate on a topic as important as risk management. Drills, webinars, company websites, policies, guidelines, best practice definition and the risk assessment process itself are all good tools for communication of risk. We should use them all and repeat often. Finally, use real crisis situations as a tool for training and practice. Always debrief after a crisis situation is over to identify areas for improvement. A good practice is to use the crisis team and the crisis response process to handle minor failures or even near miss situations as practice. Real failures may occur too infrequently (we hope) to be useful for training, so use minor events. In this manner the team can handle a real situation, gather real data and make real decisions to assess their performance and become excellent at managing crises.

Roles and Responsibilities

We need well-trained teams where all members know their roles and responsibilities in the event of a crisis situation. We cannot have people trying to figure out who to call or who is supposed to do what during a crisis. We want our teams to know what to do and to do it quickly and effectively. Failure or delay during a crisis will compound the potential loss. We must do everything we can to minimize the impact of a crisis on our consumers, customers, employees and our business. Be careful that the communication is not excessively complicated or inflexible. Then perform a test to make sure that the communication was effective.

Common Mistakes in Risk Management (Taleb et al., 2009)

- We believe that we can predict extreme events.
- We extrapolate past events to predict the future.
- We make decisions based on experience and intuition over data (sample size is too small).
- We think the world is normally distributed.
- We don't heed negative advice.
- We think we cannot afford redundancy.

Make Risk Management Part of the Culture

We can require that all functions use risk assessment and risk management as part of their normal practices to identify activities and projects. We can insist that a risk assessment be performed for every project or decision. In this way, we can make risk management part of the culture in our organization.

An Example of a Simple Risk Assessment

I find that too often we make risk assessment a difficult and mysterious process. The result is paralysis and inactivity. I think we are better off with a simple process that enables us to assess the situation and then take effective action to manage, minimize or eliminate the identified risks. We want to make risk assessment and risk management easy enough so people will make it part of their normal activities. Every project or decision should start with a risk assessment.

Here is an example of a simple approach to assessing the risk level in a manufacturing plant. In a situation where we have many manufacturing plants, many contract manufacturers or many vendors; we will need a mechanism to help us focus our efforts on the most important or riskiest plants. This will improve our success rate and allow us to focus our limited resources on the plants that will benefit the most. It is not a perfect process and we need to keep in mind that there is risk in performing the risk assessment, but there are also benefits. We will use the risk assessment process to make our team aware of the risks associated with the manufacturing plants and help them to decide how to apply their limited resources in the most effective manner to reduce risk.

In this example, my team has made a list of the criteria that we think are relevant to assess the risk in a food manufacturing plant. Figure 13.2 contains a list of criteria to assess the risks and a scoring system to quantify the relative risk for each criterion. Quantifying the relative risk due to each criterion allows us to prioritize the relative risk for all the facilities. When resources are limited, as they always are, prioritizing allows us to focus our efforts where we can have the most impact to reduce risk.

The idea is that the team will assess each manufacturing plant on each of the 16 criteria in the table and assign a score for each as defined in the table. At the end of the assessment, we will add up the scores and then rank the facilities from high to low. The plants with the highest scores will have the highest risk. We will then pick the 10 or 20 percent of the plants with the highest risk score and focus our efforts in the next year on helping them reduce their risk. We repeat this process at least annually. The result is a structured, focused effort to

Plant	Presence of allergens	Performance on allergens	Food safety management	Workforce turnover	HACCP	Self inspection program	Food defense program	GFSI certification
A								
B								
C								
	Presence of allergens	Performance on allergens	Food safety management	Workforce turnover	HACCP	Self inspection program	Food defense program	GFSI certification
	0 = None or all	0 = Excellent	0 = Experienced	0 = < 2%	0 = Excellent	0 = Excellent	0 = Excellent	0 = Yes
	1 = Soy, milk	1 = Good	1 = < 5 years	1 = < 4%	1 = Good	1 = Good	1 = Good	3 = Partial
	3 = Tree nuts, seeds	3 = Fair	3 = < 2 years	3 = < 6%	3 = Fair	3 = Fair	3 = Fair	5 = No
	5 = Peanuts	5 = Poor	5 = < 1 year	5 = > 8%	5 = Poor	5 = Poor	5 = Poor	
Highest possible	5	5	5	5	5	5	5	5

FM consumer complaints	Mold consumer complaints	Environmental pathogen program	Vendor management	Training program	Number of copackers	Copacker management	Performance on last audit	Total score
FM consumer complaints	Mold consumer complaints	Environmental pathogen program	Vendor management	Training program	Copackers	Copacker management	Performance on last audit	
0 = < 1 CPM	0 = < 1 CPM	0 = > 1 year	0 = Solid	0 = Strong	0 = None	2 = Defined	0 = > 90	
1 = < 2 CPM	1 = < 2 CPM	3 = < 6 months	3 = Some	3 = Some	1 = < 5	5 = Undefined	5 = > 80	
3 = < 3 CPM	3 = < 3 CPM	5 = No program	5 = Weak	5 = Weak	3 = < 10		15 = > 70	
5 = > 5 CPM	5 = > 5 CPM				5 = > 10		25 = < 70	
5	5	5	5	5	5	5	25	100

Figure 13.2 Manufacturing facility food safety risk assessment.

reduce risk under a constraint of limited resources. It is also a great way to educate the team about the manufacturing plants. If there is someone on the team who is familiar with a plant and can assess the criteria in the table, then we all learn about that plant. If no one on the team knows how to assess a plant, then we learned that we have a gap in our knowledge about that plant and we need to fix it.

An Example of an Organization Risk Assessment

Assessing the risks involved in an organization, business or function is a different and more open process. We don't have a well-defined list of criteria. We want a simple process that is effective and that allows us to develop a list of activities and projects based on the risks. Here is an example of such a process that I use to lead my team.

1. We are responsible for food safety and quality in our company, so we start out with a list of programs and potential risks that could be a source of risk based on our past experience. We keep this list updated as new areas of risk emerge:

   - Allergens
   - Cleaning practices
   - Facility condition
   - Equipment condition
   - Consumer complaints
   - Sales complaints
   - Findings of the Internal Food Safety Inspection
   - Findings of external auditors
   - Employee practices
   - Ingredient storage
   - Pest presence in the facility
   - Food safety skill level of associates and management
   - Workforce training
   - Workforce turnover rate
   - Current status of food safety programs
   - Test results for yeast/mold and environmental pathogens
   - Vendor management program
   - Vendor performance issues

   We use this list as part of our brainstorming process to identify potential risks.

2. Next, we consider recent failures and near miss situations that have occurred in our organization or in the industry as a source of potential risks.
3. Finally we come up with a list of the potential risks.
4. We rate each for severity and probability. Multiply the values and arrive at a risk score for each identified risk. We can add in other factors, such as: the impact of speed, consumer outrage, complexity, scope and uncertainty. If we include all these factors, it would be better to add the scores to arrive at a risk score for each identified risk.

1	Unlabeled allergen
2	Allergen cross contact
3	FS inspection failure
4	Bakery pest control
5	FM contamination
6	Post bake contamination
7	Depot pest control
8	GB product quality failure
9	Copacker quality failure
10	Lack of qualified FS Mgt
11	Bakery security
12	Depot security
13	Dirty trays

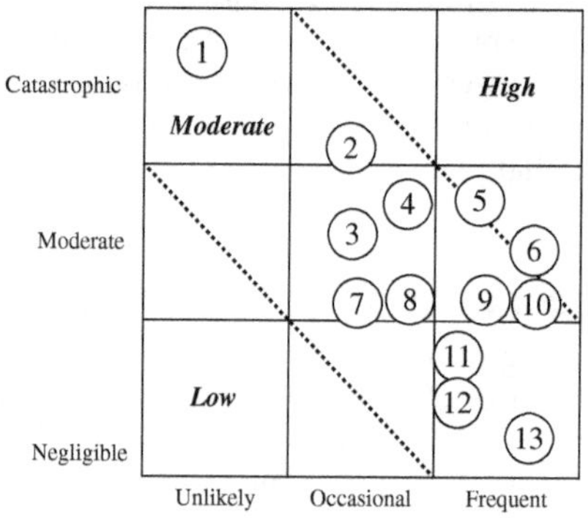

FIGURE 13.3 Global safety risk assessment.

5. We graph the risks and the risk scores as shown in Fig. 13.3 to allow us to see how the risks are distributed. It also makes it easier to see progress over time as the graph changes.
6. Next, we select each risk and identify our concerns, options and next steps as shown in Fig. 13.4. When considering concerns and options we brainstorm without constraint or limits. Next steps, on the other hand, are real actions or projects that are feasible, cost effective, and effective at reducing the risk. We identify the action or project, the name of the person who is responsible and the date it is due. We hold

Risk assessment–allergen changeover practices

Concerns:
- Allergen cross contamination will occur if cleaning and separation practices are not effective
- A changeover checklist is a good reminder of steps to take along the entire line
- Sign-off by the associate performing the work documents that the work was done
- Review and sign-off by a supervisor confirms that the work was done properly
- It is important that associates and supervisors take ownership for verification
- The cleaning checklist verification needs to be supported by a validation process

Options:
- Review current changeover checklists in facilities to assure adequacy
- Review current validation processess–update as needed
- Is the checklist verification supported by the validation process?

Next steps:
- Review current checklist verification and validation processes (AB, June 15)
- Develop a model changeover checklist for facilities to use for verification (DA, July 30)
- Develop guidelines for validating allergen changeover processes (GJ, June 30)

FIGURE 13.4 Risk assessment—allergen changeover practices.

ourselves accountable to deliver the result on time and in budget. Here is an example of a risk assessment performed for risks that can occur during allergen changeovers in a manufacturing plant. Note the three stages of the risk assessment: concerns, options and next steps. We will not get all the concerns addressed or work on all of the options this time. We will use the concerns and options next time we review the risk assessment to expand our action steps to go further in reducing or eliminating the risk.

7. Last, do it again. When the action steps are completed, it is time to review the risk assessment again and come up with a new list of actions and projects to continue the process of risk reduction and elimination.

Some Final Thoughts on Risk Management (Taleb, 2008)

Andy Grove says that only the paranoid survive (Grove, 1996). I would change that to only those who effectively manage risk live a life worth living. We can attempt to avoid or eliminate risk from our lives but that will only expose us to other risks that may actually be less pleasant and we lose the benefits of the risky activity. If we decide to sit at home and never leave the house, we are trading one set of risks for another and missing many of the benefits of the risky activity. If we eat candy, we risk getting cavities. The consequences and the remediation are not so terrible, so we brush our teeth to reduce the risk, visit the dentist twice a year to detect and take corrective action and proceed to eat some reasonable amount of candy in spite of the risk in order to gain the benefit of the experience. Justifying smoking cigarettes may be more difficult as the consequences and remediation are significantly more costly, but again we if we chose not to smoke we lose the benefits that may be highly enjoyable to us. So the answer is to rationally assess the risk, the consequences, and the cost of a failure, the benefits and the steps we can take to reduce the risk; then choose for our situation the best decision. If we have done the assessment properly, we will maximize the benefits while minimizing the costs to achieve an optimal balance and result. I suggest that the goal is not surviving, but living.

Things that can be mistaken for the other:

Luck	Skill
Randomness	Determinism
Probability	Certainty
Belief	Knowledge
Conjecture	Certitude
Theory	Reality
Coincidence	Causality
Lucky idiot	Skilled practitioner

"In theory there is no difference between theory and practice, but in practice there is."

—Yogi Berra

"Nothing is easier than self-deceit. For what each man wishes, that he also believes to be true."

—Demosthenes

References

de Bono, E., *Six Thinking Hats*, Little, Brown and Company, New York, New York, September 1999.

Grove, A., *Only the Paranoid Survive*, Broadway Books, New York, New York, September 1996.

Marks, H., *The Most Important Thing: Uncommon Sense for the Thoughtful Investor*, Colombia University Press, New York, Chichester, West Sussex, January 2014.

Power, M., *The Risk Management of Everything, Rethinking the Politics of Uncertainty*, Demos, London, 2004.

Taleb, N. N., *Fooled by Randomness*, Random House Publishing Group, October 2008.

Taleb, N. N., Goldstein, D. G., and M. W. Spitznagel, M. W., "The Six Mistakes that Executives Make in Risk Assessment," *Harvard Business Review*, October 2009.

CHAPTER 14

Lean Techniques and the Shingo Prize

Kevin Caldwell, Rick Edgeman, Kenneth Snyder

High Points of This Chapter

1. Lean is based on creating a "pull system" to produce faster rather than the traditional "push" systems used by most organizations. One of the main goals of Lean is to always pull from the customer demand, not push to the customer.
2. The Shingo Prize program publicly recognizes enterprises at various stages of their enterprise excellence journey via differing levels: *Shingo Prize, Shingo Silver Medallion,* and the *Shingo Bronze Medallion.*
3. Value Stream Mapping is an important Lean tool. It maps and documents all the tasks (material and information flow) and the metrics associated with them (cycle time, costs) within a process, including inherent waste. This provides the guidance to select the right problems and solve them as process improvement projects.
4. There is a standardized approach and set of tools, such as rapid improvement events, or *kaizens* (Japanese word for "improvement") to attack embedded wastes and increase the velocity of a process. Improving velocity exposes the problems—waste—and eliminates them, thereby making the processes faster, better, and cheaper.

5. 6S (sort, set in order, shine, standardize, sustain, and safety) is a Lean method to achieve a highly effective workplace that is clean and well organized. The benefits of an efficient workplace include prevention of defects, prevention of accidents, and elimination of time wasted searching for tools, documentation, and other ingredients to produce goods or services.
6. The integration of Lean and Six Sigma has become known as Lean Six Sigma. Lean focuses on efficiency and Six Sigma focuses on how effectiveness can lead to faster results than either method applied independent of the other.

A Truly Lean Introduction

Lean is the process of optimizing systems to reduce costs and improve efficiency by eliminating product and process waste. The emphasis is on eliminating non-value-added activities such as producing late services, defective products, excess inventory charges and excess finished goods inventory, excess internal and external transportation of products, excessive inspection, and idle time of equipment or workers due to poor balance of work steps in a sequential process. The goal of Lean has long been a goal of industrial engineering—to improve the efficiency of all processes.

As Shuker states in his article "The Leap to Lean (2000)," creating a lean organization encompasses the delivery of goods and services using less of everything: less waste, less human effort, less manufacturing space, less investment in tools, less inventory, and less engineering time to develop a new product, and less motion, for example. Lean manufacturing was a process management philosophy derived mostly from the War Manpower Commission, a World War II U.S. agency, which led to the Toyota Production System (TPS) and from other sources. The War Manpower Commission is renowned for its focus on reducing the original Toyota seven deadly wastes: overproduction, wait time, transportation, processing methods, inventory, motion, and defects (sometimes called the eight deadly wastes) in order to improve overall customer satisfaction. The eighth deadly waste was the waste of people's unused creativity. Lean is often linked with Six Sigma because of that methodology's emphasis on reduction of process variation (or its converse smoothness) and Toyota's combined usage (with TPS). Although Lean concepts began in manufacturing operations, it has been successfully applied in many industries as diverse as hospital patient care, internal auditing, and insurance customer service. Lean principles can be applied in most processes because mostly all contain waste that a customer is not willing to pay for, nor is the business willing to accept higher costs because of them. For additional information on TPS please reference Spear and Bowen's (1999) article in the Harvard Business Review entitled *Decoding the DNA of the Toyota Production System (TPS)*.

For many, Lean is the set of TPS "tools" that assist in the identification and steady elimination of waste (*muda* in Japanese terminology), the improvement of quality in production time, and costs. This and other Japanese terms used by Toyota are strongly represented in the Lean vernacular. To solve the problem of waste, Lean has several tools at its disposal, including continuous process improvement (*kaizen*), a method to achieve a highly effective workplace that is clean and well organized (6S), and mistake proofing (*poka-yoke*). In this way, Lean can be seen as taking a very similar approach to other improvement methodologies.

The second, and complementary approach to Lean, which is also promoted by the TPS, is the focus upon improving the "flow" or smoothness of work (thereby steadily eliminating *mura*, unevenness) through the system and not upon waste reduction per se. Techniques to improve flow include "production leveling," "pull production" (by means of *kanban*, signboard, or billboard), and the *Heijunka* box (achieving smoother production flow).

Lean implementation and the TPS are therefore focused on getting the right things to the right place, at the right time, and in the right quantity to achieve perfect work flow while minimizing waste and being flexible and able to change. More importantly, all of these concepts have to be understood, appreciated, and embraced by the actual employees who build the products and therefore own the processes that deliver the value. The cultural and managerial aspects of a Lean organization are just as, and possibly more, important than the actual tools or methodologies of production itself.

Lean and the Shingo Prize

That Japan's post-World War II rise to recognition of superior products and services is due in no small way to the influence and activity of American quality luminaries such as Dr. Joseph M. Juran and Dr. W. Edwards Deming is well known. What Drs. Juran and Deming and others found in Japan was a highly talented, highly motivated collection of business and engineering leaders who embraced quality methods and philosophy with near "tent revival" zeal. In addition to broad and expert adaptation and deployment of these methods, Japanese businesses and professionals added new and highly pragmatic approaches. This was done with the sort of efficiency that extreme resource scarcity can motivate, augmented by the effectiveness that dedication to precision births.

Just as Drs. Juran and Deming, and other American quality luminaries wielded significant influence in Japan, a new constellation of Japanese "quality stars" arose, the names and developments of whom have been and remain integral to contemporary expressions of quality in its many forms that include but are not limited to lean manufacturing and enterprise methods. Among these have been Taiichi Ohno who was co-primary developer of the famed Toyota Production System (TPS); Masaaki Imai and *kaizen* (continuous improvement); Kauro Ishikawa and cause-and-effect diagrams; Yoji Akao and quality function deployment; Genichi Taguchi and robust product design; Noriaki Kano and the Kano customer needs model; and *Hoshin Kanri*—development of which is not attributed to any single individual, but to Japan's Bridgestone Tire company in 1965 (Watson, 2003).

Arguably, the most influential of Japanese quality professionals was Dr. Shigeo Shingo (1909 to 1990). Dr. Shingo is credited with development or co-development of numerous methods that are today central to quality practice in general and lean practice in particular. Among these methods are *just-in-time manufacturing* (JIT) that is also referred to as the *Toyota Production System* (White et al. 1999, White and Prybutok 2001), *poka-yoke* or "error proofing" (Chase and Stewart 1994), *gemba* walks, and "quick changeover" or *single minute exchange of dies* or SMED (McIntosh et al. 2007) where the name of this technique derives from a suggestion by Dr. Shingo's son, Ritsuo who is former president of Toyota China and Hino Motors—Toyota's heavy truck division. Each of these approaches has been critical to the founding and development of the theory and practices that are collectively referred to as lean manufacturing or lean enterprise. Dr. Shingo's books addressing these and other approaches were published in Japanese with some, but not all translated into English (Shingo 1981, Shingo 1985, Shingo 1986, Shingo 1988, Shingo and Dillon 1989). Translation of his other books heretofore published only in Japanese is being undertaken by the Shingo Institute. Dr. Shingo's influence has reverberated around the world, not only directly through his own efforts and publications, but through the organizations implementing his work and through widely read books that have become "must read" classics such as *The Machine That Changed the World* that introduced the term "lean manufacturing" (Womack et al. 1990) and *World Class Manufacturing: The Lessons of Simplicity Applied* (Schonberger 1986).

Dr. Shingo's quality quest has been championed since the late 1980s by an institute bearing his name, the Shingo Institute, that is a part of the Jon M. Huntsman School of Business at Utah

State University (USA). Among other significant efforts, the Shingo Institute administers the *Shingo Prize for Enterprise Excellence*, a global prize that in recognizing excellence shares a kinship with America's Malcolm Baldrige National Quality Award and the EFQM Excellence Award—formerly known as the European Quality Award. Differentiating aspects of the Shingo Prize program are its emphasis on lean approaches and multiple recognition levels.

After briefly discussing the founding and history of the Shingo Institute and Shingo Prize program, attention will shift to the Shingo Enterprise Excellence Model. Discussed will be insights into enterprise excellence, guiding principles derived from these insights and upon which the Shingo model is founded, the model itself, relationships among primary model elements, and assessment approaches.

The Shingo Institute and Shingo Prize

Dr. Shingo envisioned collaboration with an organization that would advance his life's work through research, practical yet rigorous education, and a program for recognizing the best in enterprise excellence throughout the world. Fulfillment of his vision began in 1988 and is both ongoing and evolving. Soon after being awarded an honorary Doctorate of Management from Utah State University that year, the Shingo Prize was established and—later—the Shingo Institute was formed.

The Shingo Institute promotes systematic enterprise adoption, cultivation, and exercise of timeless excellence principles proven to drive and sustain enhanced financial, environmental, and social performance and impacts. The Shingo Institute supports this through cutting edge research, relevant education, insightful enterprise assessment, and public recognition of organizations committed to achieving sustainable world-class results. While the Shingo Prize is integral to the activities and priorities of the Shingo Institute, the scope of the Institute has expanded to include a focus on research, educational offerings among which is the "Shingo MBA" program initiated in 2015, and a large and growing international network of Shingo Institute Licensed Affiliates—companies that provide training in areas relevant to the Shingo model and principles. The Shingo Institute also acknowledges relevant and impactful research and practitioner publications through its Shingo Research and Professional Publication Award initiative.

Most well-known of Shingo Institute activities is the Shingo Prize program—a program that, unlike the Baldrige Award Program or the EFQM Excellence Award—is global in scope. Equally, the Shingo Prize program publicly recognizes enterprises at various stages of their enterprise excellence journey via differing levels: *Shingo Prize*, *Shingo Silver Medallion*, and the *Shingo Bronze Medallion*. As a further distinction, there is no specific limitation on the number of recipients during a given year or for any of the *Shingo Prize*, *Silver Medallion*, or *Bronze Medallion* levels. Each of these points is illustrated by the list of 2013 through 2015 recipients that is provided in Table 14.1.

Shingo Prize recipient enterprises are ones recognized as world-class for successful establishment of a culture anchored on principles of enterprise excellence that are deeply embedded into the thinking and behavior of all leaders, as well as attainment of world-class results. Performance is measured both in terms of business results and the degree to which business, management, improvement and work systems are driving appropriate and ideal behavior at all levels. Leadership is strongly focused on ensuring that enterprise excellence principles are deeply embedded in enterprise culture and that this is routinely assessed and subject to improvement. Managers are focused on continuously improving systems to drive behavior that is closely aligned with enterprise excellence principles. Associates are assuming responsibility for improving not only their own work systems, but other systems that are within their value stream. Understanding of why things are being done in the way they are has saturated the associate level of the enterprise. Improvement activity has begun to focus on the enterprise

Recipient Level	Year	Enterprise/Location
Shingo Prize	2015	• Envases Universales Rexam de Centroamerican, S.A./Amatitlan, Guatemala
	2014	• NewsUK – Newsprinters Ltd./Holytown, Motherwell, UK • Barnes Aerospace OEM Strategic Business/Ogden, Utah, USA • DuPuy Synthes Ireland/Cork, Ireland • Abbott Vascular Clonmell/Tipperary, Ireland
Silver Medallion	2014	• Rexam Beverage Can South America, Jacareí/Jacareí, São Paulo, Brazil • PyMPSA Plásticos y Materias Primas/Guadalajara, Jalisco, Mexico • Rexam Healthcare, Neuenburg/Neuenburg am Rhein, Germany • Rexam Beverage Can, Enzesfeld/Enzesfeld, Vienna, Austria • Boston Scientific, Maple Grove/Maple Grove, Minnesota, USA • Vale Europe Ltd., Clydach Refinery/Clydach, Swansea, UK
	2013	• MEI Queretaro/El Marques, Querétaro, Mexico • Pentair Water Pool and Spa/Moorpark, California, USA • Rexam Beverage Can South America, Manaus Ends/Manaus, Amazonas, Brazil
Bronze Medallion	2015	• Boston Scientific, Costa Rica/El Coyol, Alajuela, Costa Rica • Carestream Health, Yokneam/Yokneam, Israel
	2014	• Lundbeck Pharmaceuticals Italy S.p.A./Padova, Italy • Vistaprint Deer Park Australia/Derriut, Victoria, Australia • Corporation Steris Canada/Québec, QC, Canada • Autoliv Inflator Co., Ltd./Shanghai, China • Rexam Beverage Can South America/Rio de Janeiro, Brazil
	2013	• Covidien/Athlone, Ireland • Letterkenny Army Depot, Force Provider/Chambersburg, Pennsylvania, USA

TABLE 14.1 *Shingo Prize, Silver Medallion, Bronze Medallion* Recipients: 2013 through 2015

as a whole and the enterprise scorecard has clearly defined performance measures that include behavioral measures. Key measures are stable, predictable and mature with positive trends and few anomalies. Goals that are challenging, while also realistic are present in most areas and there is solid understanding of world-class performance. Most measures are aligned to corporate goals and cascade to the lowest levels. Silos are difficult to identify.

Shingo Silver Medallion recipient enterprises are ones demonstrating strong use of tools and techniques that also have mature systems that drive improvement, and that are beginning to align their thinking and behavior with principles of enterprise excellence. Enterprise leadership is involved in improvement efforts and supports the alignment of enterprise excellence principles with systems. Managers are deeply involved and focused on driving behaviors through the design of systems. Associates are routinely involved in use of improvement tools to drive both continuous and breakthrough improvement in their areas of responsibility. Understanding of why this is done has begun to penetrate the associate level of the enterprise. Improvement activity is focused on multiple business systems. The business scorecard uses a broad and flexible spectrum of measures and is beginning to include behavioral elements. Key measures are stable and trending in mostly positive directions and people at all levels of the enterprise understand how to affect the measures appropriately for their areas of responsibility. Performance goals are being set for most business systems and alignment is clear and obvious in most of these systems, with plans set to align the remaining ones.

Shingo Bronze Medallion recipient enterprises are ones demonstrating strong use of tools and techniques for business improvement and are working to develop effective systems to create continuity and consistency of tools applied throughout the enterprise. Enterprise leadership is setting the direction for improvement and supports the efforts of others while managers are involved in developing systems and helping others use tools and techniques effectively. Associates are trained and actively participate on improvement projects. Not fully mature at the associate level is understanding of why things are being done the way they are so that improvement activity is generally focused heavily at the operations level and has only begun in support level. Measures are beginning to communicate cause-and-effect while key measures have begun to stabilize and are trending in mostly positive directions.

Further differentiating the Shingo Prize program are its emphases on lean culture and practice, as well as behaviors and principles.

These differentiators have historically led to a relatively lower consideration of performance and impacts during assessment of *Shingo Prize* applicants than either America's Malcolm Baldrige Award or Europe's EFQM Excellence Award. Although this has been the historical case, greater emphasis on performance and impacts—results—is rapidly evolving.

Getting There: Insights of Enterprise Excellence

Enterprise excellence is a term with various aliases. Among common aliases for *enterprise excellence* are *performance excellence*—the term used relative to America's Malcolm Baldrige National Quality Award principles, model and criteria; *business excellence* as associated with the European Quality Award and the European Foundation for Quality Management (EFQM) model upon which it is based; *organizational excellence*; and the less encompassing term "operational excellence" historically used relative to the Shingo Prize. The contemporary Shingo model is a comprehensive enterprise excellence model and is now referred to as such, it is also a model that attends—more than other excellence models—to the operational strategies, processes, activities, and performance of organizations—hence its historic reference to the term "operational excellence."

Principles upon which the Shingo model is founded are derived in part from the following three insights of enterprise excellence:

- *Ideal results require ideal behaviors.* Results, that is, performance and impacts, are the aim of every enterprise, but there are varied methods by which these may be attained—some better, some less so. Ideal results are positive outcomes that are sustainable over the long term. Simply learning or acquiring new tools or systems does not lead to ideal results, though of course these may be supportive of ideal results. Consistent achievement of ideal results requires leaders and managers to create an environment or culture where ideal behaviors are manifest in each enterprise associate. Exceptional leaders understand that a cause-and-effect relationship exists between results and behavior with behavior serving as the cause and results as the effect.
- *Beliefs and systems drive behaviors.* The beliefs that we hold collectively and as individuals profoundly impact our behavior. Commonly overlooked is the substantial effect that systems have on behavior. People's work in enterprises is largely governed by systems designed to create specific business results, without

due attention dedicated to the behavior that system generates. Many enterprises have de facto systems that have evolved due to specific needs for specific results. As such leaders and managers are confronted by the necessary and often massive challenge of creating or realigning management and work systems that drive the ideal behaviors that are essential to producing ideal business results.

- *Principles inform ideal behaviors.* Principles are foundational rules that help to reveal both positive and negative consequences of our behaviors. Such rules inform our decisions, especially concerning our behaviors. When our decisions (behaviors) align with principles the result is cognitive consonance, whereas decisions and actions contrary to principles create cognitive dissonance that we often refer to as moral or ethical dilemmas. The more deeply leaders, managers and associates understand the principles of enterprise excellence and the more perfectly aligned that systems are aligned to reinforce ideal behavior, the greater the likelihood is that a sustainable culture of enterprise excellence will be established. In such cultures achievement of ideal performance and impacts becomes the normative, rather than aspirational.

These insights imply that apart from enabling tools, enterprise excellence is advanced through soft technologies that include policies, practices, partnerships, leadership, governance and—more generally—the behaviors these either enable or impede. The Shingo model emphasizes ideal behaviors as critical to fulfillment of its 10 core principles and, in turn, relentless advancement toward enterprise excellence.

Enabling Principles: Timeless, Universal, and Self-Evident

Principles upon which the Shingo model is based are classified into four related categories: *cultural enablers, continuous* (and breakthrough) *improvement, enterprise alignment,* and *results* (performance and impacts) with each principles category populated by a number of *supporting (sub-)principles* that are elaborated in the *Shingo Prize for Operational Excellence Guidelines and Application* (2010). While these principles may be independently pursued, they are intended to work collaboratively to create synergy that leads to a whole that is more than its constituent elements. The four categories and their respective principles are provided and elaborated in the following.

Cultural enabler principles speak to the issue that knowledge, insight, and foresight are found in abundance and are pervasive across individuals in essentially any organization. As such enterprise leaders should demonstrate both humility and wisdom through exercising two principles:

1. *Lead with humility.* Organizational and personal growth is enabled when a leader thinks less about his/herself and more about others. They seek out and value the ideas of others and are willing to change when they learn something new.
2. *Respect every individual.* Every person in any organization has the right of being respected. When people feel respected, they contribute to organizational purpose not with their hands only, but with their minds and hearts.

Supporting (sub-)principles behind cultural enablers include the nurture of long-term relationships, empowering and involving the whole of enterprise human capital, developing human capital, and assuring that this is done in a safe environment.

Continuous improvement principles suggest that enterprises should relentlessly pursue beneficial change. Such change may range from incremental to breakthrough in magnitude and may hence involve innovation that is incremental, radical or disruptive. Further, the ongoing search for improvement is both central to enterprise culture and a clear expectation throughout the entirety of the enterprises' human ecology. Most real improvement is accomplished through process change and innovation so that processes should be subject to ongoing review and reexamination, always seeking improvement as far upstream in processes as possible. In short, processes should be subject to constant scrutiny. It is not only improvement of enterprise processes that should be pursued, but constant improvement of the enterprise's human capital relative to knowledge, skills, and both anticipated and unexpected changes on the horizon. These ideas suggest the widely acknowledged tenet that the pursuit of perfection is likely a never-ending one as "perfection" is a moving target in a dynamic environment. In such environments variation is—if not inevitable—then at least the norm. The *Continuous Improvement Principles* of the Shingo model are as follow:

- *Focus on process.* All outcomes are the consequence of a process. It is nearly impossible for even competent and highly motivated people to consistently produce ideal results with a poor process.
- *Embrace scientific thinking.* Innovation and improvement are the consequence of repeated cycles of experimentation and learning. Failure leads to more learning through a relentless and systematic exploration of new ideas.
- *Flow and pull value.* Value for customers is maximized when it is created in response to real demand and a continuous and uninterrupted flow. Anything that disrupts the continuous flow of value is waste.
- *Ensure quality at the source.* Perfect quality can only be achieved when every element of work is performed perfectly the first time and every time. When and if errors occur, they must be detected and corrected at the point and time of their creation. Solutions to problems are best found where the work is actually done by the people performing the work.
- *Seek perfection.* Perfection is an aspirational goal that is not likely to be achieved, but the pursuit of which creates a mindset and culture of both continuous and radical improvement. The realization of what is possible is only limited by the paradigms through which we see and understand the world.

Subprinciples supporting continuous (and breakthrough) improvement include those of stabilizing and standardizing processes, data-driven decision making, insistence on direct observation, focus on value streams, keeping things as simple and visual as reasonable possible, identifying and eliminating waste, and integration of improvement with work.

Relative to *waste*, it is perhaps too formulaic to focus only on the standard categories, discussed in lean manufacturing environments as *muda*, *mura*, and *mudi*. It can be additionally valuable to think of the goal of waste identification and elimination as being to

> ... seek to prevent, recover, or reclaim value sacrificed to poor practices, poor purposes, poor policies, poor processes, poor partnerships, poor strategies, lack of focus, etc. ...

Enterprise alignment principles suggest that anything less yields an enterprise not performing to its fullest potential. This idea is simply enough illustrated by using the analogy of an automobile that is "out of alignment": when that is the case, its tires will wear unevenly, it will persistently wander off its intended path, and it will consume more energy and produce more waste than it will when it is "perfectly aligned." In other words, its parts and processes will work at cross-purposes and its operator will engage in non-value-added activities that prevent them from pursuing ideal ones in ideal ways. Should we then expect that something as inherently more complex as an organization should perform any better when it is misaligned?

Enterprises are complex with many people, processes and motivations so that less than holistic consideration will *almost surely* in the mathematical sense and *always* in the practical sense lead to compromised, suboptimal results. This implies that there exist relationships between and among enterprise people, processes and organizational units that form both singular and interrelated systems and that such systems should not be thought of in isolation. There is a progression in both the order and way that actions are performed or decisions are made and these deliver consequences that are either degraded or enhanced by antecedent actions and decisions. As such, it behooves the human ecology of the enterprise to think and act from the perspective of "cause and effect" as well as sequence or linkage.

The leadership of an aligned enterprise is well-aware of its identity and purpose, the values and beliefs upon which it was founded, its path forward, and is confident of not only surviving, but of prospering both in and beyond its present circumstances. This requires that leadership and a critical mass of the enterprise's human ecology to move forward with unity of purpose.

Together these ideas lead to the following *enterprise alignment principles*:

- *Create constancy of purpose.* When leaders communicate with unwavering clarity why an organization exists, where it is going, how it will get there, and the principles upon which it acts, people become empowered to innovate, adapt and take risks with greater confidence.
- *Think systematically.* Solutions to enterprise problems must embrace the connectedness and impact of the one on the many. The broader the perspective of any individual or team, the greater their ability to conceive and implement solutions that create a positive outcome for the enterprise.

Supportive subprinciples of enterprise alignment include realism relative to both the condition and potential of the enterprise, focus on long-term change rather than short-term fixes, alignment of systems, and alignment of strategy.

There is a final Shingo model principle: *create value for the customer*. This principle is classified into a category referred to as *results* where results should be more broadly construed and both *performance* and *impacts* where impacts are direct and indirect consequences of performance and may be of either intended or unintended natures. The argument in favor of *performance and impacts* derives from the concept of a chain reaction wherein the initial results spread the reaction or impacts, which under certain conditions may dramatically accelerate. Thinking of *results* as *performance* and *impacts* allows extension of enterprise thinking into the realm of its intended beneficiaries (customers and other direct stakeholders) ... and the beneficiaries of their beneficiaries ... and the beneficiaries of their beneficiaries of their beneficiaries ... *ad infinitum*.

Though often cast in a negative light, it is instructive to think of this in a way similar to the way that we might think of contagion or of something "going viral." Use of contagion as an analogy for impacts from antecedent results should serve to heighten our awareness that "what is good for the goose is *not always* good for the gander" and that in fact that which is beneficial for one party may in turn create detrimental effects or impacts for a subsequent downstream party. This implies that an organization should endeavor to be cognizant of secondary and ancillary uses customers may have for products and services the organization provides and, in so far as possible, what may happen beyond those immediate uses.

It is with the foregoing comments in mind that we note that creating value for customers should always occur within the larger context of thinking, behaving, creating, and providing value not only legally, ethically, and morally, but also systematically, always with the awareness that an enterprise ecosystem may extend well beyond the organization, its suppliers and its *intended* beneficiaries to unforeseen parties and that such parties may incur unintentional harm. In other words, we take as a learning point that *performance* and *impacts* must be concurrently considered. As such, *systematic thinking*—previously cited as core to *enterprise alignment*—must occur with full cognizance of cause-and-effect relationships that exist between *performance and impacts*.

- *Create value for customers.* Value must ultimately be determined through the lens of what customers want and for which they are willing to pay. Enterprises that fail to deliver effectively and efficiently on this fundamental outcome are unsustainable in the longer run.

Supporting subprinciples include identifying and using measures that matter, aligning behavior with performance, and identifying and leveraging cause-and-effect relationships.

The Shingo Model

As can be seen from the list of 2013 to 2015 *Shingo Prize, Silver Medallion*, and *Bronze Medallion* recipients provided in Table 14.1, the *Shingo Enterprise Excellence Model* has global reach. Examination of this list also reveals that, as with other major excellence programs, such as America's Baldrige Award Program or the EFQM Excellence Award, there has been increasing penetration of excellence principles and practices into non-manufacturing enterprises. Acknowledgement of this trend supports more balanced usage of the term "operational excellence" commonly linked to manufacturing enterprises and historically used in reference to the Shingo model, and the more omnibus term "enterprise excellence" that is seen as immediately relevant to any enterprise.

The *Shingo Enterprise (e.g., Operational/Performance/Business) Excellence Model* is presented in Fig. 14.1. Enterprise culture lies at the center of the model. Heeding the quote by Peter Drucker, the father of modern strategic management, that "culture eats strategy for breakfast" (Bider and Kowalski 2014), the model places enterprise *culture* at its heart, or center. Surrounding culture

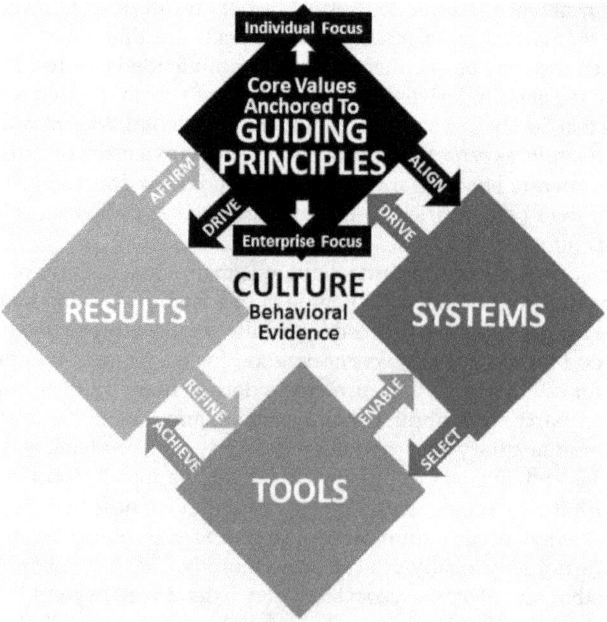

Figure 14.1 The Shingo Enterprise Excellence Model.

we find the four areas of emphasis of the model. Moving in a clockwise direction and beginning at the top of the model we see *guiding principles* that are anchored to core values, expectations of which are that these should be not only for the entity we call the enterprise, but also manifested in its people at individual levels. *Guiding principles* must be aligned with *systems* where, in turn, *systems* are supported by use of appropriate *tools* that must be selected (Doggett 2005). Through tools we achieve *results* that affirm our *guiding principles*. Next, beginning with *guiding principles* and moving in a counterclockwise direction we see that *principles* drive *results* and that once attained, *results* can be used to refine our *tools*. *Tools* then enable better functioning *systems* that drive *guiding principles*. In this sense we see that enterprise excellence is pursued on a continuous basis, which can be enabled by self-reinforcing cycles.

The model is further elaborated in the *Shingo Prize for Operational Excellence Model & Application Guidelines* (2010).

Enterprise Assessment

At this writing, *Shingo Prize* applicant enterprises are assessed on a 1000 points possible scale relative to the four dimensions of the *Shingo Model*: *Cultural Enablers* (250 points), *Continuous Process Improvement* (350 points), *Enterprise Alignment* (200 points), and *Results* (200 points). As previously noted, future rebalancing of these is anticipated, with a shift toward greater emphasis on results. That being said, it is the long-standing perspective of the Shingo Institute that *cultural enablers*, *continuous process improvement*, and *enterprise alignment* drive *results* (i.e., performance and impacts) so that any rebalancing reflects only fine tuning and the effort to make the importance of *results* more overt. Tables provided in this section can be found in the *Shingo Prize for Operational Excellence Model & Application Guidelines* (2010) available at the Shingo Institute website: www.shingoprize.org.

Table 14.2 provides the standard scoring matrix used for assessing applicant enterprises. In this matrix it should be noted that with respect to *cultural enablers*, *continuous process*

Scoring Matrix		Senior Leadership	Operations		Support	
			Managers	Associates	Managers	Associates
Cultural Enablers (250 points)	Weight	40%	20%	10%	20%	10%
Continuous Process Improvement (350 points)	Weight	15%	35%	15%	25%	10%
Enterprise Alignment (200 points)	Weight	50%	20%	5%	20%	5%
Results (200 points)	**Quality**	Measures (20%)				
	Cost/ Productivity	Measures (20%)				
	Delivery	Measures (20%)				
	Customer Satisfaction	Measures (20%)				
	Safety/ Environment/ Morale	Measures (20%)				

Table 14.2 Shingo Prize Applicant Assessment Scoring Matrix

improvement, and *enterprise alignment* that the assessment process allocates differing percentages of points to senior leadership, operations managers, operations associates, support service managers, and support service associates. The results category assesses results relative to the five categories of quality, cost/productivity, delivery, customer satisfaction, and safety/environment/morale with each category equally weighted—however, assessment of results does not differentiate with respect to leaders, managers, and associates.

It delineating scoring in this way, each dimension is associated with specific principles emphasized by the Shingo model. As such, assessment seeks to identify the presence, practice, and relative effectiveness of these principles in the enterprise at behavioral and performance levels. Simultaneously, however, this is done through examination of key processes that are contextually relevant to the enterprise. Table 14.3 provides a *representative example* of how this might be done, with the key processes actually examined for a specific enterprise being contextually driven.

The *cultural enablers* dimension is associated with the principles of lead with humility and respect every individual. Similarly, the *continuous* (and breakthrough) *process improvement* dimension is related to the principles of focus on process, embrace scientific thinking, flow, and pull value, ensuring quality and the source, and seeking perfection.

The *enterprise alignment* dimension of the Shingo model is related to the principles of creating constancy and unity of purpose, and thinking systemically. Finally, the *results* dimension is associated with creating value for the customer.

The first three of these dimensions—*cultural enablers, continuous improvement,* and *enterprise alignment* are assessed using the behavior assessment scale provided in Table 14.4. The *results* dimension is assessed using the assessment scale provided in Table 14.5.

The general assessment approach taken in each case is to use what are commonly referred to as (graduated) maturity scales. Such scales are descriptive with each step in the scale providing typical findings of behaviors or performance at progressively more mature states. Effective maturity scale use requires significant expertise of the area being assessed. This is fundamentally the case because expert judgment is needed to expand reasonably beyond what are fundamentally examples, performance at specific levels of maturity.

The percentages provided in Table 14.2 represent a "level playing field" in the sense that enterprises applying for a prize are assessed in a consistent manner for comparability purposes. Many enterprises using excellence models, however, do not formally pursue a prize or award such as the Shingo Prize, Baldrige National Quality Award, or EFQM Excellence Award or the recognition that come with being named a recipient—that is—"a winner." Instead, many enterprises use such models as a self-assessment tool and do so routinely—typically annually—for the purpose of ongoing improvement. In such cases it is often reasonable for those organizations to alter or customize weights such as the percentages provided in Table 14.2 to better fit the context of their competitive landscape.

One Vision of the Future

There are a number of trends to which it seems the Shingo and other models will likely gradually adapt to. It is clear that a rapidly increasing number of enterprises are concerned about their social and environmental performance and impacts. Such concerns maybe driven by conscience, regulatory demands, or market pressures but whatever the driving force behind such concerns, business strategy, practice, and performance emphases are shifting. We see that significant components of enterprise strategy are concerned not only with being financially prudent, but with social equity and environmental sensitivity—the three components of triple top line strategy (McDonough and Braungart, 2002). It is not only enterprise

Assessment Areas		Senior Leadership	Customer Relations	Product/Service Development	Operations	Supply	Management Support Processes
		Business Processes					
Cultural Enablers (150 Points)	People Development: Education, Training and Coaching (50 Points)						
	People Development Empowerment and Involvement (50 Points)						
	People Development Environmental and Satety (50 Points)						
	Weight	40%	Weighting in this dimension for the five business processes is determined by the percentage of total associates in each area,multiplied by 60%				
Continuous Improvement (400 Points)	Weight	10%	10%	5%	40%	15%	20%
Enterprise Alignment (200 Points)	Weight	60%	5%	5%	15%	5%	10%
Results (250 Points)	Quality	Measures (50 Points)					
	Cost I Productivity	Measures (50 Points)					
	Delivery	Measures (50 Points)					
	Customer Satisfaction	Measures (50 Points)					
	Morale	Measures (50 Points)					

TABLE 14.3 Example Assessment Matrix for an Example Enterprise

policies and practices that must be aligned with these, but processes, systems, and partnerships. It is through these mechanisms that performance and impacts result.

Further, given the now well-known criticality of needs, it is clear that significant innovation leading not so much to continuous improvement, but rather to breakthrough advances in areas that—among others—are socially or environmentally relevant, timely, and impactful.

Lenses	Level 1 0% to 20%	Level 2 21% to 40%	Level 3 41% to 60%	Level 4 61% to 80%	Level 5 81% to 100%
Role	Leaders are focused mostly on fire-fighting and largely absent from improvement efforts.	Leaders are aware of other's initiatives to improve but are largely uninvolved.	Leaders set direction for improvement and support efforts of others.	Leaders are involved in improvement efforts and support the alignment of principles of enterprise excellence with systems.	Leaders are focused on ensuring that enterprise excellence principles are driven deeply into the culture and regularly assessed for improvement.
	Managers are oriented toward getting results "at all costs."	Managers mostly look to specialists to create improvement through a project orientation.	Managers are involved in developing systems and helping others to use tools effectively.	Managers focus on driving behaviors through the design of systems.	Managers are primarily focused on continuously improving systems to drive behavior that is more closely aligned with enterprise excellence principles.
	Associates focus on doing their jobs and are generally treated like an expense.	Associates are occasionally asked to participate on an improvement team that is usually led by someone outside their natural work team.	Associates are trained and participate in improvement projects.	Associates are routinely involved in using tools to drive continuous improvement in their own areas of responsibility.	Associates understand principles, the "why" behind the tools, and are leaders for improving their own work systems and other systems within their value stream.
Frequency	Infrequent or rare.	Event-based and irregular.	Frequent and common.	Consistent and predominant.	Constant and uniform.
Duration	Initiated or undeveloped.	Experimental and formative.	Repeatable and predictable.	Established and stable.	Culturally integrated and mature.
Intensity	Apathetic or indifferent.	Apparent with individual commitment.	Moderate local commitment.	Persistent, wide commitment.	Tenacious with full commitment.
Scope	Isolated with point solutions.	Silos, internal value stream only.	Predominantly in operations with functional value stream consideration.	Multiple business processes with integrated value stream consideration.	Enterprise-wide with extended value stream consideration.

Table 14.4 Behavior Assessment Scale

Lenses	Level 1 0 to 20%	Level 2 21 to 40%	Level 3 41 to 60%	Level 4 61 to 80%	Level 5 81 to 100%
Stability	Little or no evidence of stability. Little to no predictability. Beginning to implement. Unpredictable. 0-1 year.		Has begun to stabilize. Initiating predictability. Building maturity. All levels have become comfortable with measures. 2 to 3 years.		Stable. Predictable. Long-term. Mature. 4+ years.
Trend/Level	Level is low. Trend is poor. Little to no evidence of goals. Little to no evidence of benchmarking.		Moderate improvement in level. Trends are mostly positive to flat with some backsliding. Benchmarking is industry-focused.		High level of attainment considered world-class. Benchmarks constantly raise the bar and are a function of process, not industry. Positive trend with very few anomalies to explain. Trend is well above expectations.
Alignment	Isolated with inconsistent usage of measures. Little alignment. Strong silos.		Some areas are aligned. Performance measures aligned in operations. Silos are beginning to fall. Working toward enterprise-wide alignment.		All measures align to corporate goals and down to the lowest level. Enterprise-wide extended value stream. No silos.
Improvement	Little to no systematic feedback. Sporadic feedback. Little evidence of goal setting, some evidence in operations.		Regular feedback in some areas. All areas do not address feedback systematically. Many areas beyond operations have a process to set goals.		Routine feedback to appropriate party. Evidence of feedback in all areas. Almost all areas have realistic and challenging goals.

TABLE 14.5 Results Assessment Scale

When these are combined, it is reasonable to speculate that the sorts of results—that is, performance and impacts—that will need to be considered will be ones in the so-called triple bottom line areas of people, planet and profit (Elkington 1997), as well as innovation. Performance is driven partially be processes and systems however, so that processes and systems supporting innovation will be of increasing concern.

There are other enterprise consequences of the sort of social volatility that is increasingly common, as well as the strains we see in the natural environment. As such it is likely that while pursuing excellence, enterprises will also pursue sustainability, resilience and robustness. An enterprise is (Edgeman and Williams 2014):

> ... *sustainable* to the extent that it is able to create and maintain economics, ecological, and social value for itself, its stakeholders, society at large, and policy makers.
>
> ... *resilient* to the extent that it possesses the capacity to self-renew through innovation by adapting its responses to negative shocks and challenges over time.
>
> ... *robust* to the degree that the enterprise is highly resistant or immune to a critical subset of such shocks and challenges.
>
> ... *excellent* when its leadership and strategy, as deployed through people, processes, partnerships, and policies deliver superior performance and impacts in specified areas.

It is the "specified areas" in which enterprises must deliver superior performance may change over time. A current and future list of areas might include, for example, the following domains: enterprise human capital, innovation, financial, environmental, social, data analytics and intelligence, marketplace, supply chain, operations, and various, more context-specific ones.

Lean in Nonmanufacturing-Based Industries

Lean methods and tools have made their way into most industries. A method that was used in manufacturing to reduce waste is now used to improve cycle time, flow, and velocity, improve workplace department performance and, yes, reduce waste in hospitals, insurance companies, financial services, and more. Here is one example from a hospital (Volland 2005):

Adapted from ***A Case Study: Now That's Lean***

Jennifer Volland
Reprinted with Permission from *Medical Imaging Magazine.*

In the hopes of improving workflow and patient throughput, the Nebraska Medical Center (Omaha) began implementation of Lean Six Sigma in December 2002. As a 735-bed nonprofit hospital, the center is the largest teaching hospital in Nebraska with both academic and private practice physicians. One of the first Six Sigma projects for the organization was in the Interventional Radiology (IR) department, where such invasive procedures are performed.

A project team—which included the lead nurse scheduler, lead technologist, and department manager—was assembled to address patient throughput problems. Physician involvement was initiated early with ongoing input and information sharing for process improvements.

The project team defined physicians who referred patients into the IR department as their primary customer. They quickly realized that current volumes supported by the department did not fully meet the needs of referring physicians. Patients were lost to other healthcare systems that could accommodate the additional patients within the community, resulting in loss of revenue and market share.

The project team measured the cycle time of each step to determine where to best focus improvement efforts. Reducing holding room (HR) time quickly became evident as an area of opportunity. A patient's HR time averaged 151 minutes with a standard deviation of 242.4 minutes (February 4–19, 2003). Upon further examination, however, many more problems were identified. First, patient flow coordination from the HR into one of three procedure rooms was problematic because of different equipment in the rooms. Often, the nurse scheduler was pulled to function as the department appointment scheduler as well as the person coordinating patient flow. The duality of tasks created problems for timeliness in appointment scheduling with the referring clinics and flow of patients through the HR.

Changes made during the Lean Six Sigma implementation had a significant impact on the amount of time patients spent in the HR. The amount of time a patient spent in the HR, after the improvements, averaged 32.7 minutes with a standard deviation of 37.71 minutes (March 17–24, 2003). Follow-up monitoring during the control phase showed sustained improvements; with the HR time leveraging 31.02 minutes and a standard deviation of 24.86 minutes (October 29–December 16, 2003).

Lean techniques applied within the IR department resulted in improved processes and ability to better meet customer expectations. As a result of the project, referring clinics were successfully able to feel the impact of changes for improved interventional radiologists within the department. Not only were the changes significant, but, post-project, the department has been able to successfully sustain the gains made in the HR.

Reducing Waste Alone Is Not Lean

It is not enough to just believe "if I eliminate the no valued waste we will be Lean." This is only one aspect of a Lean organization. Although the elimination of waste may seem like a simple and clear subject, it is noticeable that waste is often very conservatively identified. This then hugely reduces the potential of an organization. Although the elimination of waste is the goal of Lean, the TPS defines three types of waste: *muri* or overburden, *mura* or unevenness, and *muda* or non-value-added work.

Muri is all the unreasonable work that management imposes on workers and machines because of poor organization, such as carrying heavy weights, moving things around, dangerous tasks, and even working significantly faster than usual. *Muri* is pushing a person or a machine beyond its natural limits.

Mura focuses on implementing and eliminating fluctuation at the scheduling or operations level, such as quality and volume.

Muda is discovered after the process is in place and is dealt with reactively rather than proactively with *muri* and *mura*. It is seen through variation in output (which as mentioned earlier) and can blend well with Six Sigma applications. It is the role of management to examine the *muda*, or waste, in the processes and eliminate the deeper causes by considering the connections to the *muri* and *mura* of the system. The *muda* (waste) and *muri* (overburden) must be fed back to the *mura* planning stage for the next project.

More often than not, most organizations improperly only focus on *muda* or non-value-added waste and fail to understand this approach is reactive and will only partially position the organization for success (if at all). One must ensure that all three waste types are addressed.

Muri can be avoided through standard work disciplines. To achieve this, a standard condition or output must be defined. Then every process and function must be reduced to its simplest elements for examination and later recombination. This is done by taking simple work elements and combining them, one by one, into standard work sequences.

Mura is avoided by using Just-in-Time (JIT) systems that are based on little or no inventory by supplying the production process with the right part, at the right time, in the right amount, and first in, first out component flow. JIT systems create a "pull system" in which each subprocess withdraws its needs from the preceding sub processes, and ultimately from an outside supplier. When a preceding process does not receive a request or withdrawal, it does not make more parts.

To properly manage outcomes in a Lean organization, you must ensure that all three types of waste are managed and controlled. Demand and capacity must be balanced to that demand and must be fully understood. Current state conditions must be understood in order to move to future state pull production and the elimination of non-value-added activities creating waste. Standard work must be institutionalized, which alleviates overburdening associates as they perform activities. These activities will create the model for cultural transformation from a batch-and-queue operation to an operation with synchronous flow, team-based activities, and a true focus on the customer mindset.

Lean Manufacturing Case Study

AGC Flat Glass North America, a wholly owned subsidiary of the world's second-largest glass producer, Asahi Glass Company, operates 45 facilities throughout North America, and all were experiencing pressure to provide the lowest total cost product with rapid order fulfillment in a highly competitive market. In September 2006, AGC launched an initiative to drive operational excellence and improve profitability. This initiative was coined JPI (Jikko Process Improvement) by AGC and is based on the principles of the TPS and Lean enterprise.

One of the first facilities to implement the JPI process was AGC Hebron, a fabrication facility located near Columbus, Ohio. Hebron serves the Ohio market and neighboring states. Hebron receives glass from one of AGC's primary glass facilities and transforms these raw materials into a number of end products, including single-pane products, sealed insulated units for window manufacturing, and tempered (heat-treated) glass for safety applications. The Hebron fabrication processes include cutting, tempering, and insulating unit assembly. An initial assessment of the facility was performed, and the results indicated that manufacturing lead times were exceeding seven days with wide swings up to weeks in some cases. Excess inventory made it nearly impossible to quickly find a specific job or determine what to fabricate next. There was also a concern for employee safety, specifically increased risk of injury attributed to the large cut-glass inventory. Wide swings in product demand placed on manufacturing also served to complicate the business. Some days, the plant capacity was underutilized while other days customer demand exceeded capacity by twofold.

A cross-functional team was formed to drive the improvement efforts. Team members included sales, production control, purchasing, production employees, corporate JPI members, and a transformation coach. In the first days, the team was introduced to the concepts of the TPS and Lean manufacturing.

One of the first things the team quickly developed was a "Current State Map," a valuable tool to understand the actual situation on the production floor and in-order fulfillment activities. Once completed, the current state map clearly told the present story and set a firm direction for future improvement.

The first step to improve the efficiency of the workplace focused on implementing the 6S (sort, set-in-place, sweep/shine, standardize, self-discipline and safety) process. After the initial training, the team began to attack waste; sorting unnecessary items from needed items, implementing visual control for tools and materials, cleaning everything, and putting in place a robust auditing system to sustain the gains. From there, the team focused on their "Current State Map." Points of delay and inventory builds were addressed and, in most cases, eliminated. Equipment was relocated to aid product flow, which reduced movement and product queues. To further consolidate inventory, over half the material-handling racks used to store glass were removed. The reduction in inventory in a matter of days translated to improved lead times to the customer. At this point, the Hebron team adopted the motto, "There is no tomorrow." A key to a Lean enterprise and the TPS inherent in this philosophy is the idea that customer delivery requirements will be met and that all products can and will be produced in a single day to customer demand and pull. This expectation was well within the plant capabilities for cycle times. The team also studied demand patterns compared to the demonstrated capacity. Once this relationship was understood and lead times were reduced, the plant could successfully be level loaded, thus further solidifying delivery reliability to levels above 99 percent on time. This percentage was well above historic levels. The improved product flow quickly identified quality issues that were previously hidden by excess work in process. In the weeks that followed, a number of other enhancements were included such as improved equipment maintenance to ensure reliability, mistake proofing methods, *kanbans* for supply replenishment, and a focus on faster changeovers. During the time the physical changes were occurring, another important transformation took place—the culture slowly changed. The plant began running differently. Employees knew what the customer needed by the hour and produced accordingly. Orders moved seamlessly through the operations without heroic efforts, making work life easier and, more importantly, safer.

Within weeks, the customers began to see and feel the changes. The new Hebron customer complaints turned to customer compliments. Overall demand steadily increased as past customers lost due to service issues began to return and new customers began to come to Hebron for their glass needs. The financial results followed as Hebron experienced a turnaround in profitability. Commenting on profitability, Jerry Hackler, Hebron's Operations Manager remarked, "The effect of the bottom line came quickly. Even in the early months the facility generated more operating income on fewer sales, a clear indication of the cost improvement impact."

History of Lean

The history of manufacturing and the introduction of Lean are summarized in Fig. 14.2. The Lean mission is to have the following throughout the entire supply chain to win the marketplace:

- Shortest possible lead time
- Optimum level of strategic inventory
- Highest practical customer service levels
- Highest possible quality (low defect rate)
- Lowest possible waste (low cost of poor quality)

This is accomplished by synchronizing the flow of work (both internal and external to the organization) to the "drumbeat" of the customer's requirements. All kinds of waste are driven out (time, material, labor, space, and motion). The overall intent is to reduce variation and drive out waste by letting customers pull value through the entire value stream (or supply chain).

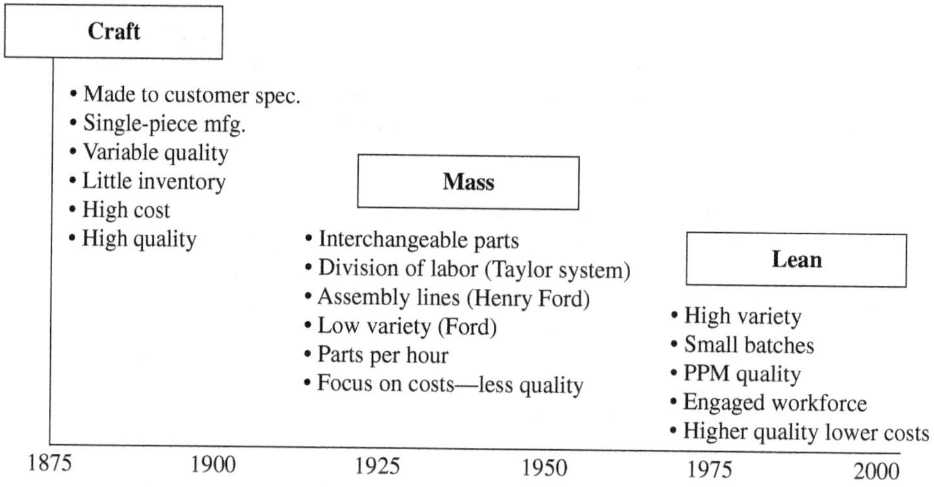

FIGURE 14.2 History of manufacturing.

In their book Lean Thinking, Womack and Jones (2003) state that the key principles of Lean are to

- Specify value in the eyes of the customer; the voice of the customer
- Identify the value stream for each product
- Make value flow without interruptions
- Reduce defects in products and deficiencies in processes
- Let customers pull value
- Pursue perfection—Six Sigma levels
- Drive out variation (short and long term)

The Relationship of Lean to Managing for Quality

One key component of being a Lean organization is the need to create "value" as seen from the eyes of the customers. The operational definition of value is the benefit the customer gains from using the product or service. The customer *creates* value. Providing value to the customer is why the producer exists. Lean starts with defining value in terms of products/services and benefits provided to the customer at the right time at an appropriate price. Anything that does not provide value to the customer can be considered waste (Fig. 14.3).

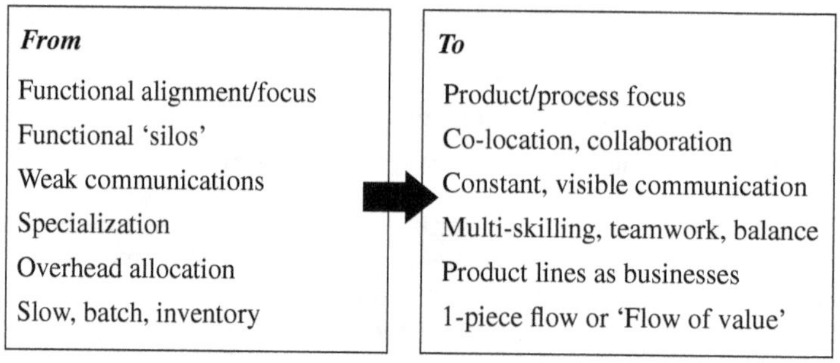

FIGURE 14.3 Lean characteristics.

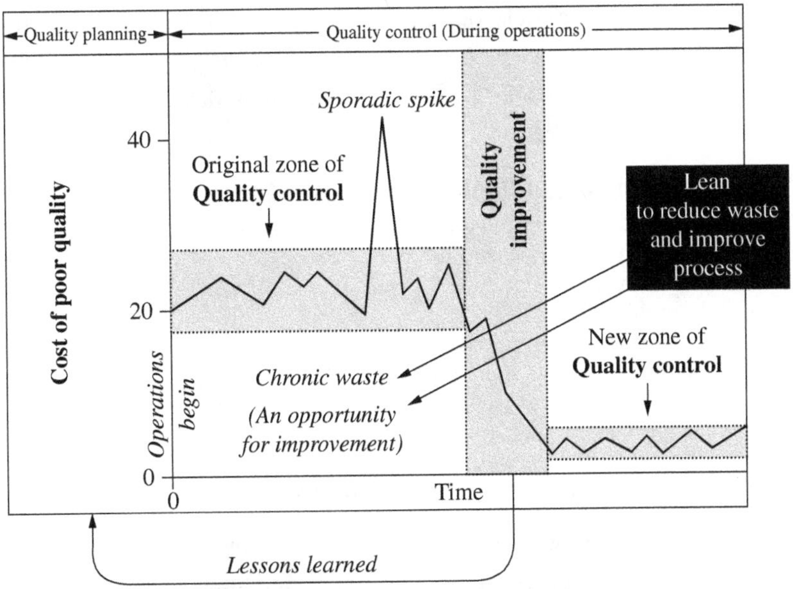

FIGURE 14.4 Lean and the Juran Trilogy. (Juran Institute, Inc., Southington, CT.)

If we review the Juran Trilogy in Fig. 14.4, we can see that Lean supports the definition of quality in that all products and services must be "fit for purpose." Customers define quality as both the features and freedom from failures. Therefore, because Lean is about creating value by eliminating nonvalue, it is important to include in Lean the management of quality. Lean is used in quality control (daily routine practices) because it enables work to be standardized, leading to better compliance. Lean is used in improvement to decrease the costs of nonperforming processes in the form of waste reduction. Most recently, Lean methods are being used in quality planning to design for Lean. Designing for Lean is similar to designing for quality. An organization now must design a product or service so that it can flow easily with little disruption from customer need to customer use.

The Eight Wastes

Taiichi Ohno (1988) identified seven types of waste that exist in most processes and organizational systems. These identifiable wastes lead to the poorly performing processes if they are not dealt with and removed. Lean practitioners and experts must focus on reducing or eliminating these wastes, part of a *kaizen* or Rapid Improvement Event.

The following includes Ohno's seven types of waste, which were focused on production in addition to the eighth waste (which seems to have no origin) directed at all processes:

1. *Overproduction*—making or doing more than is required or earlier than needed
2. *Waiting*—for information, materials, people, and maintenance
3. *Transport*—moving people or goods around or between sites
4. *Poor process design*—too many/too few steps, no standardization, and inspection rather than prevention
5. *Inventory*—raw materials, work in progress, finished goods, papers, and electronic files

6. *Motion*—inefficient layouts at workstations, in offices, poor ergonomics
7. *Defects*—errors, scrap, rework, nonconformance
8. *Underutilized personnel resources and creativity*—ideas that are not listened to, skills that are not used

The Lean Roadmap and Rapid Improvement Events

Six Sigma and Lean have both evolved over decades as part of the continuing revolution of quality, excellence, and breakthrough performance. Motorola created the term "Six Sigma" as it worked to raise the standard for improvement to new heights. Lean grew out of the experiences of the TPS.

Now Lean and Six Sigma have evolved to reflect today's core business challenges: the challenge to execute and to maximize value, as well as respond to "nanosecond customer" needs. Dr. Joseph De Feo of Juran Global, refers to the speed at which today's demanding customers expect results. Lean and Six Sigma are now used for sustainable competitive advantage across all industries and cultures.

Every organization wants to be Lean and have

- The shortest possible process lead times for providing products and services
- The optimum level of strategic inventory and human resources
- The highest practical customer service level
- The highest possible quality (low defect rate)
- The lowest possible waste (low COPQ, cost of poor quality) throughout the entire value chain

Although there have been numerous techniques and tools utilized in Lean implementation, most Lean practitioners did not have a Lean model until the collaboration with Six Sigma DMAIC (define, measure, analyze, improve, control). The Juran Lean Roadmap in Fig. 14.5 is an example of a model designed to carry out "Lean projects or events." It provides the five DMAIC steps as in Six Sigma and includes the Lean tasks. This set of steps provides a Lean or "Lean Six Sigma" practitioner with a reminder to focus both on efficiency and effectiveness.

Figure 14.6 provides a tool grid to demonstrate tools that can be used at every step in the method. Each of the tools in this grid can be found in this chapter as well as in Chap. 19, Graphical Tools to Improve Process Performance and Chap. 20, Accurate, Reliable Data and Measurement Systems.

Rapid Improvement Events or *Kaizens*

Rapid Improvement Events (RIE) or *kaizens* are typically 1-week focused efforts that are facilitated and conducted by Lean Experts or Black Belts to enable Lean teams to analyze the value streams and quickly develop/implement solutions in a short time frame. These events have application in offices, service organizations, health care arenas, and manufacturing operations and consistently yield tremendous, real-time improvement. *Kaizen* is the Japanese word for incremental improvement (or "good change"). It has become associated with the use of small teams carrying out improvements on a regular basis. It is often used as a name for all encompassing continuous improvement methods. We have chosen to use it as it is defined: a small improvement that is made on a regular basis. RIE or *kaizen* teams are multifunctional so that all aspects of the process and problems associated with them are considered and solutions developed will be understood and accepted by all. Rapid improvement teams

Define value

1. Define stakeholder value and critical to quality (CTQ).
2. Map high-level process.
3. Assess for 6S.

Measure value

1. Measure customer demand.
2. Plan for data collection.
3. Create a value stream attribute map.
4. Determine pace, Takt Time and manpower.
5. Identify replenishment and capacity constraints.
6. Implement 6S (S1–S3).

Analyze process—flow

1. Analyze the value stream attribute map.
2. Analyze the process load and capacity.
3. Perform value added/non-value added analysis.
4. Apply Lean problem-solving.

Improve process—pull

1. Conduct rapid improvement events (RIE).
2. Design the process changes and flow.
3. Feed, balance, and load the process.
4. Standardize work tasks.
5. Implement new process.

Maintain control

1. Stabilize and refine value stream.
2. Complete process and visual controls.
3. Identify mistake-proofing opportunities.
4. Implement 6S (S4–S6).
5. Monitor results and close out project.

Figure 14.5 Lean Six Sigma Roadmap and substeps.

are fast because Lean is easier than, say, Six Sigma. Rapid improvement teams are fast because they tackle focused projects bit by bit. They also tackle problems where the data are typically readily available.

This technique is a good tool to involve all levels of the workforce. It can help build an empowered and engaged workforce. RIEs can be used to identify and solve departmental problems as well.

What Do RIE and *Kaizen* Teams Do?

A Lean Expert or a Black Belt works with management to select the area to focus the improvement on. They then carry out the following preparations for the events:

1. One to three weeks prior to conducting the event the expert assembles the team, facilitates development of a charter and gathers as much data as possible surrounding the area to be improved. The type of data depends on the area selected but typically includes a manufacturing area of focus:
    - Process flow diagrams for each product or product family (if available)
    - Yields by operation
    - Setup time by operation

LEAN	LEAN TOOLS																																			
	6S	Basic Statistics/Graphs	Brainstorming/Affinity	Cause-Effect Diagram\5 WHYS	Control Plans	CTQ Matrix	Data Collection	FMEA	Kanban Calculator	Load Charting	Mistake Proofing	MSA	Pareto Analysis	Poka Yoke	Process Load Calculator	Process Modeling	Production Control Board	Product Quantity Analysis	Product Routing Analysis	Project Charter	Reliability Centered Maintenance	Rapid Improvement Events	S1-S3 Planning Document	Selection Matrix	SIPOC	SMED	Sort Sheet	SPC/Control Charts	Standard Work Diagrams	Stratification	Supermarket	Takt Time	Theory of Constraints	TPM	VNVA	Value Stream Mapping
Select Projects		■	■										■					■	■					■				■		■						
Define Demand																				■																
a. Verify/ Refine Project Charter			■				■																	■						■						
b. Confirm the Value Stream Improvement Focus																		■	■																	
c. Determine Stakeholder Value & CTQs						■	■						■																							
d. Map High-Level Process																									■											
Measure Value Stream																																				
a. Define Customer Demand							■																							■		■				
b. Plan for Data Collection							■																							■						
c. Validate Measurement System		■										■																								
d. Create a Value Stream Map																									■											■
e. Identify Replenishment and Capacity Constraints							■			■					■														■				■			
f. Implement 6S (S1-S3)																							■				■									
Analyze Process – Flow																																				
a. Analyze the Value Stream Attribute Map																																				■
b. Analyze the Load Chart							■			■																						■				
c. Analyze Capacity							■								■	■																				■
d. Perform VA/NA Decomposition Analysis							■																												■	
e. Apply Lean Problem Solving to Special Cause				■																																
Improve Process – Pull																																				
a. Conduct the RIE								■														■														
b. Design the Process Changes and Flow																																				■
c. Feed, Balance, Load the Process									■	■																■					■					
d. Standardize Work Tasks																													■							
e. Implement New Processes																	■				■													■		
Maintain Control																																				
a. Stabilize and Refine Value Stream																	■				■													■		
b. Complete Process and Visual Controls					■			■																				■	■							
c. Identify Mistake-proofing Opportunities											■			■	■																					
d. Implement S4-S6	■																																			
e. Monitor Results & Closeout Project																												■								

FIGURE 14.6 Lean methods and tools.

- Changeover time by operation
- Average WIP (work in progress) inventory levels between operations
- Average materials inventory
- Average finished goods inventory
- Cycle times by operation
- Average daily customer demand by end item
- Monthly customer demand by end item
- List of suppliers including items supplied, amounts, annual dollar value, and delivery frequency
- Material move/store times
- Material move distances
- Inspection frequencies and sample sizes
- DPMO (Defects per Million Opportunities) or Sigma levels of each process

2. One week prior to the event, the team is trained in basic methods and tools of Lean.
3. Event week—the team begins by validating the current state Value Stream Maps and develop "Future State" maps, define customer demand, pace, balance the work, define standard work, and implement improvements.
4. After event—ensure controls are in place; monitor progress.

During the event the teams may conduct multiple small assignments. Some of the more important ones are

- Begin current state Value Stream Map
- Understand the data that is available and collect as much needed data as possible
- Ensure the availability of equipment
- Implement S1, S2, and S3 of 6S (sort, set in order, shine, standardize, sustain, and safety)
- Validate Value Stream Maps—understand the "before" values
- Study current conditions
- Complete the following:
  - VA/NVA decomposition analysis
  - Current state load charts, spaghetti diagrams, standard worksheets
  - Review current state analyses
  - Design the future state and design control sheets
- Develop future state standard work
- Implement changes (big moves)
- Implement control boards
- Review standard work, standard work in process, needed fixtures, etc.
- Finalize flow, procedures, standard work, and production control board
- Present results to management and celebrate

Pull versus Push Systems

Traditional operations have worked within a push system. A push system computes start times and then pushes products into operations based on demand. This approach ignores constraints or bottlenecks within the process and can cause unbalanced flow and excess WIP inventories. A pull system, by contrast, only produces when authorized to do so and based on the process status and customer demand.

Pull systems produce faster than push systems, and, by nature, pull production controls and enhances flow. The goal should always be to pull to customer demand.

Lean Value Stream Management

Lean focuses on finding value streams. These value streams consist of all activities required to bring a product from conception to commercialization. They can include all key business processes such as design, order taking, scheduling, production, sales, marketing, and delivery. Understanding the value stream allows one to see value-added steps, non-value-added but needed steps, and non-value-added steps. Value-added activities *transform* or shape material or information into something that meets customer requirements. Non-value-added activities take time or resources, but they do not add value to the customer's requirement (but they may meet the organization's requirements). The value stream improvement journey typically starts with training the team on key concepts of Lean and mapping the *current state* using value stream maps that document materials and information flow as well as any pertinent information on the process (such as cycle times, downtime, capacity, wait times, yield, and inventory levels). The goal is to identify all the necessary components to bring a product to commercialization, as well as all waste inherent in the process. Improvements are identified from here. The desired future state is then documented as a *future state value stream map*, and the improvements are implemented to drive toward the desired future state goal.

Value streams can be mapped for a single product or service but, more often, a process supports more than one single-ended item. When products share the same design and fabrication processes, they are called a *product family*. In practice, value stream maps are frequently developed around a product family. It is not uncommon for maps to commingle with other product families as they progress through the process.

As mentioned above, a value stream comprises all the tasks currently required to move the product family though its process. There are three typically mapped cycles: concept to launch (the design cycle), raw materials to customer (the build cycle), and delivery to recycling (the sustain cycle). The build cycle is the most commonly mapped.

An example of a value stream map for a paint line showing both the current state and future state are shown in Figs. 14.7 and 14.8. There are a number of excellent sources for the techniques of mapping the value stream such as Learning to See (Rother and Shook 2003), Value Stream Management (Tapping et al. 2002), and Creating Mixed Model Value Streams (Duggan 2002). To be most effective, mapping should include all process steps involved, including suppliers and customers. Specific attributes, including information flow, for each step should be well documented and verified. These data should be as realistic as possible and show variation within the attributes if it exists. These data will be the starting point developing the future state map, which incorporates improvements and waste reduction.

Impact of Demand

The impact of demand on an operation cannot be understated. A key component to satisfying the customer is understanding their demands of the product. This is one of the single most important elements within the value stream. It is important to understand the pattern of demand as well, whether growing or declining, seasonal, or stationary. The producer must

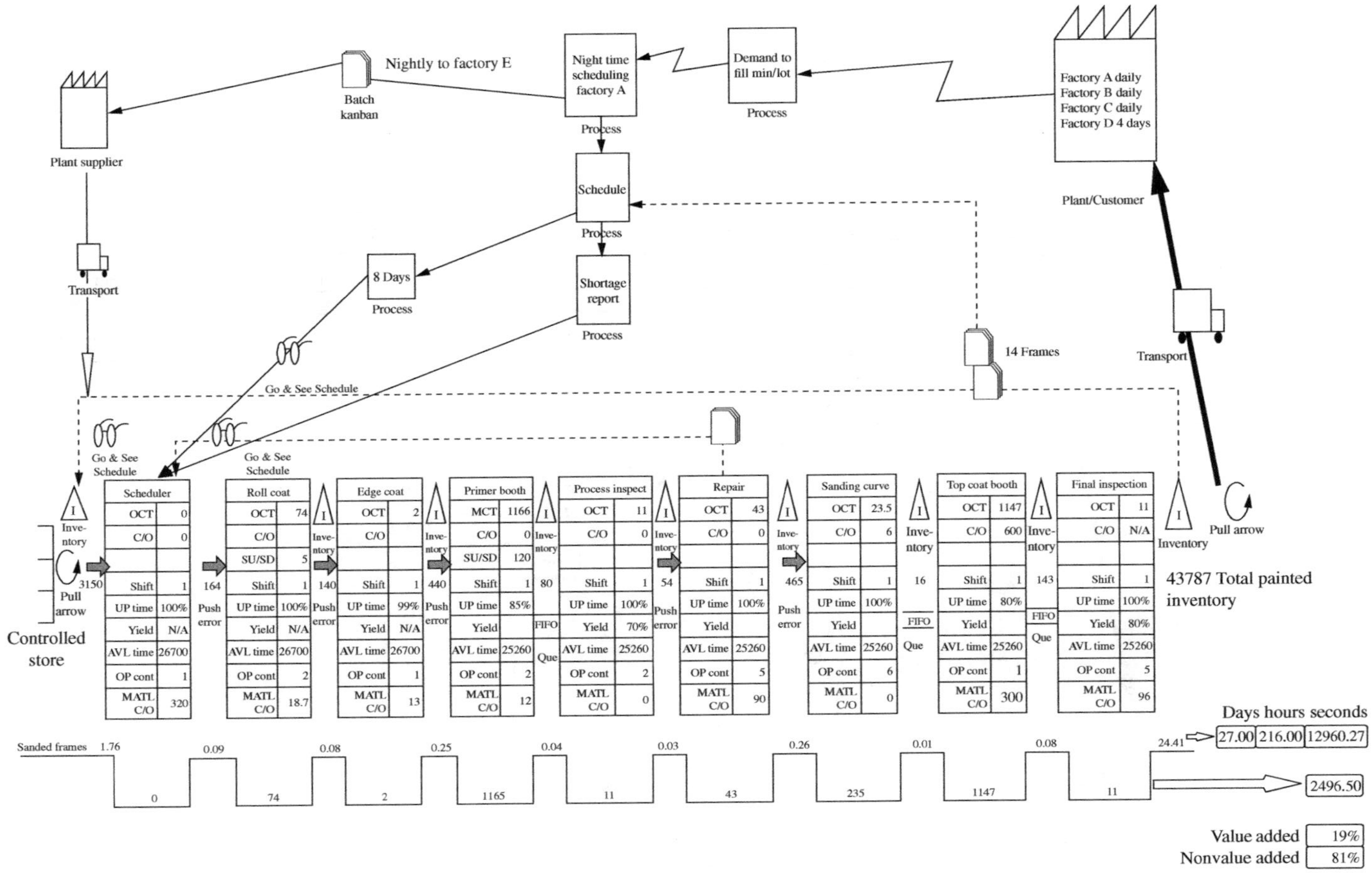

FIGURE 14.7 Value stream map for paint line—current state.

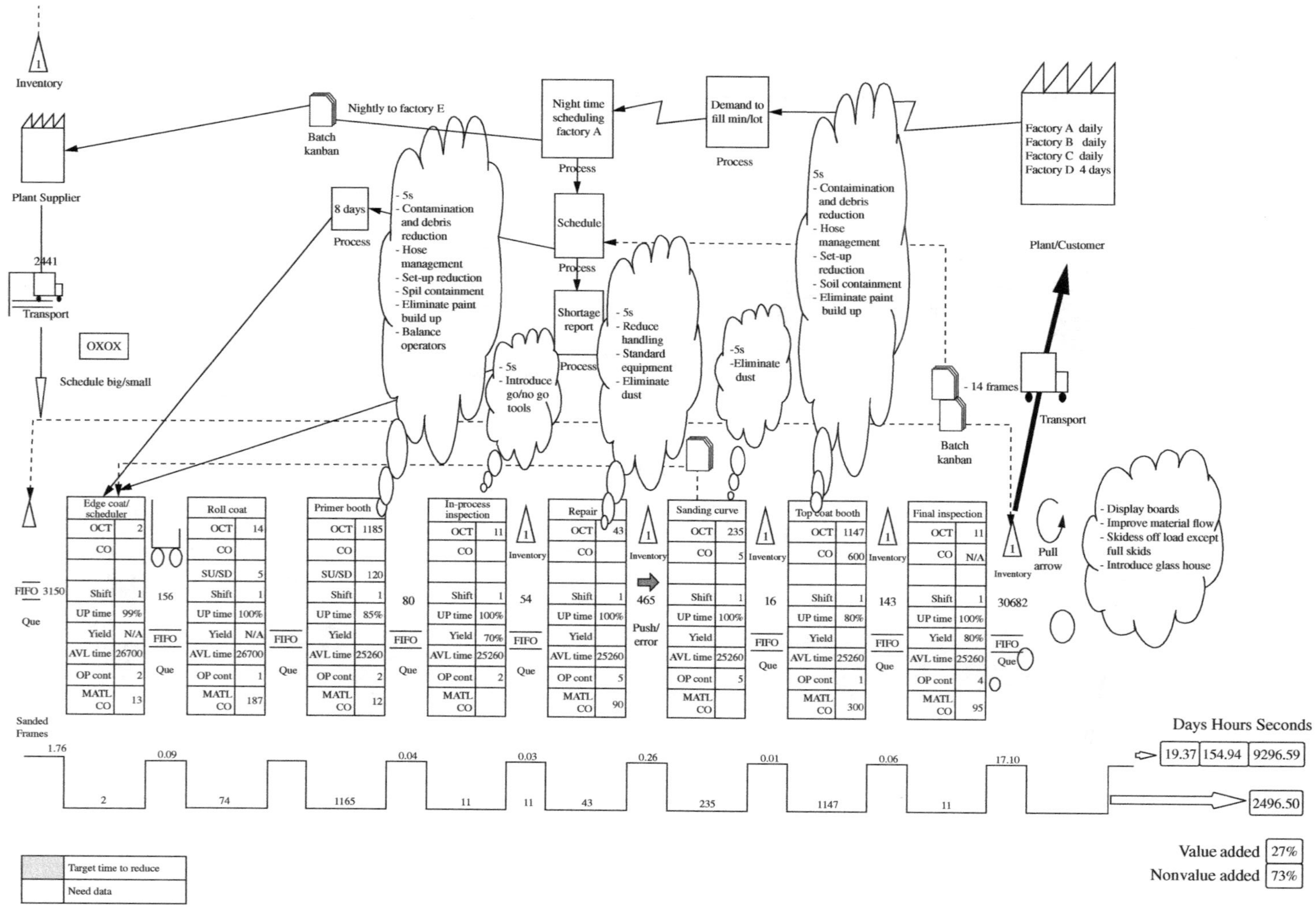

Figure 14.8 Value stream map for paint line—future state.

react quickly and effectively to changing demand to assure delivery reliability and cost effective operations. Demand variability can be mix driven, quantity driven, or as often the case, both. Demand variability can adversely affect delivery reliability, product quality, inventory costs, and total cost, among others, all with negative consequence to the customer. Demand is also utilized to determine takt time (from the German word *Taktzeit*, meaning meter, as in music, which establishes the pace, or beat, of the music), the rate at which customers buy a single unit. Takt time is discussed later in the chapter. Changing demand causes changes in takt time, which causes changes to required resources. If this flux is not understood and managed correctly, many of the adverse effects mentioned will quickly become a reality. It is recommended that if demand varies significantly, multiple value streams be developed; each with the specific takt time and specific resources to match customer expectations.

Capacity and Demand

Capacity and demand must balance to ensure proper flow. With too little capacity, you have unhappy customers; with too much capacity, you have waste. Capacity is the amount of output that a system is capable of sustaining over a given time. It is loosely calculated as available time divided by the longest cycle time. Theoretical capacity (also called engineered capacity or maximum capacity) can be thought of as output at the ideal state. This may be nameplate output information of a machine. It operates under perfect conditions, which are not realized in most facilities. On the other hand, demonstrated capacity can be calculated based on current, real-life situations. The difference between theoretical capacity and demonstrated capacity is improvement opportunity.

Demand should not be confused with capacity. Demand is the customer's requirements and is independent of the producer's abilities.

Value/Non-Value-Added Decomposition Analysis

The main goal of Lean is to identify and eliminate waste. This can be accomplished once we have a solid understanding of the process as it currently is. This is the first step to improvement; determine what is of value to the customer and what is not. As mentioned above, anything that does not provide value to the customer can be considered waste. If constructed carefully, the current state map will provide a wealth of opportunity for improvement. The basic premise of value/non-value-added decomposition analysis simply is to ask the question, is the customer willing to pay for this? This should be performed for each process step. If not, what can be done to reduce the waste or completely eliminate the waste all together? In some cases, due to the current capability of the process, a non-value-added activity is still required, at least for the time being. An example of a non-value-added, but necessary, task would include inspections or other quality checks. This activity will remain in place to ensure customer satisfaction until the process can be made robust enough not to require the non-value-added activity.

Flow and Takt Time

The concept of flow requires the rearrangement of mental thoughts regarding "typical" production processes. One must not think of just "functions" and "departments." We need to redefine how functions, departments, and organizations work to make a positive contribution to the value stream. Flow production requires that we produce at the customer's purchase rate and if necessary, make every product every day to meet customer's orders, that is, to meet the pace or "drumbeat." The pace or drumbeat is determined by takt time.

$$\text{Takt time} = \frac{\text{available time (in a day)}}{\text{average daily demand}}$$

Determine pace

Over 10 days	Demand
1	30
2	40
3	50
4	60
5	10
6	30
7	40
8	20
9	60
10	40
10	**380**

Per day:

$$\frac{\text{Time available in period (840 min.)}}{\text{Average demand (38)}} = \textbf{22.1 minutes}$$

Based on 2 shifts of 7 hours

Figure 14.9 Takt time calculation example.

For example, in Fig. 14.9, the pace or takt time is calculated for the demand shown during a 10-day period.

Takt Time Calculation Example

To be practical, takt time may need to be modified, depending on the variability of the process. When modifying takt time beyond the simple equation, another name should be used, such as cell takt, machine takt, or practical takt. Although modifiers may be planned, they are still waste, or planned waste. Manpower staffing requirements can then be determined as follows:

$$\text{Minimum staffing required} = \frac{\text{total cycle time in process}}{\text{takt time}}$$

6S—A Plan for Neat and Clean Workplaces

Many workplace departments are dirty and disorganized. The benefits of an efficient and effective workplace include the means to prevent defects; accidents; and the elimination of time wasted searching for tools, documentation, and other important items to complete a work process. By focusing on the removal of the dirtiness and organizing the workplace departments, they will perform work safer, faster, and cheaper.

A simple tool called 6S now provides us with a framework to create a neat and clean workplace. Its steps are as follows:

- *Sort.* Remove all items from the workplace that are not needed for current operations.
- *Set in order.* Arrange workplace items so that they are easy to find, to use, and to put away.
- *Shine.* Sweep, wipe, and keep the workplace clean.
- *Standardize.* Make "shine" become a habit.
- *Sustain.* Create the conditions (e.g., time, resources, rewards) to maintain a commitment to the 6S approach.
- *Safety.* Ensure employee safety in operations.

Decades ago, industries producing critical items (e.g., health care, aerospace) learned that clean and neat workplaces are essential in achieving extremely low levels of defects. The quality levels demanded by the Six Sigma approach now provide the same impetus.

Perhaps the significance of the 6S approach is its simplicity. The benefits are obvious: they are the simplest work-simplification tools and are easy to understand and apply. Simple tools sometimes get dramatic results, and that is what has happened with 6S. For elaboration of the five steps (excluding safety), see The Productivity Press Development Team (1996); Fig. 14.10.

6S should be implemented throughout the improvement process and sustained into the future, adjusting as needed. 6S provides a solid foundation for most all Lean tools and techniques.

A note on safety: Once the first 5Ss are firmly in place, a remarkable thing happens, the workplace becomes safer. Very often, no additional effect is required to achieve this benefit. With the work area sustaining organization and cleanliness, a 50 percent reduction in work related safety incidences could occur. Combining the 5S with a formal safety program can deliver amazing results, and it is called 6S for "success."

Inventory Analysis

Inventory is the amount of stock of any item or resource in an organization. In manufacturing inventory normally includes raw materials, finished goods, component parts, supplies, and work in progress (WIP). The purpose of inventory is to manage variation (demand,

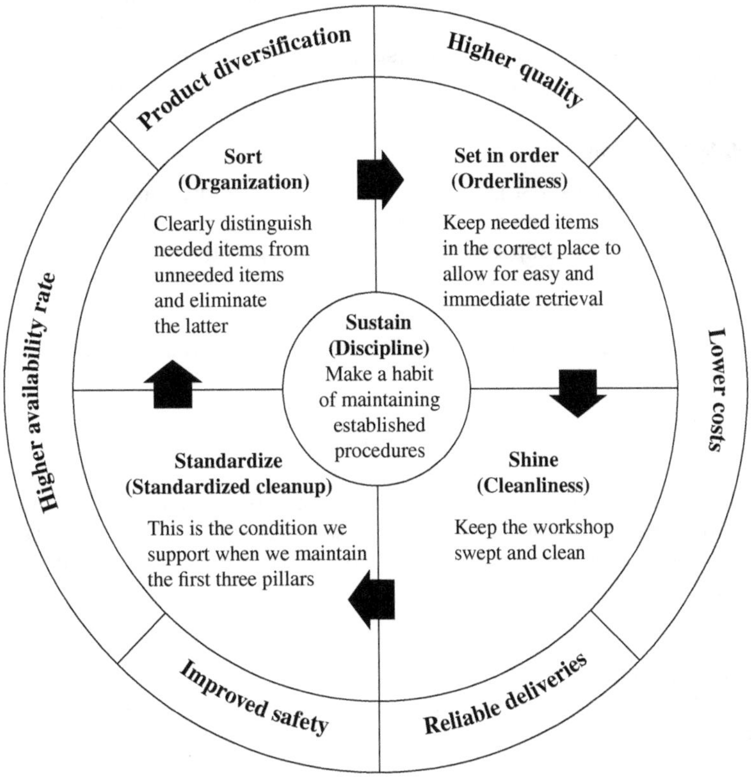

Figure 14.10 The 5S concept. (The productivity press development team (1996). Reprinted with permission of productivity press.)

delivery, and the process itself), ease production scheduling, reduce setups, and balance the quantity of the economic order. Although a certain volume of inventory can have strategic value, inventory is most often viewed as waste. Waste is the cash tied up in the materials and labor, and waste in storage and movement. Inventory is also open to damage, theft, and obsolescence. The aim of Lean is to reduce, if not eliminate, inventory.

There is a place for inventory besides in the hands of the customer. Inherent variation occurs in every process daily. Strategic inventories can compensate for process efficiencies and buffer customer demand fluctuations. Inventory is strategically placed and is set with calculated minimum-maximum stocking levels to ensure optimum flow through the process. When calculating stocking levels, one should consider customer demand (and variation), quantity consumed during replenishment, cycle time intervals for replacement, and impact of flow disruptions.

A regularly overlooked source of waste related to inventory is inventory inaccuracies. The differences between actual counts and recorded counts (commonly known as "book to actual") can be costly to both the producer and the customer. Measuring this difference can be the first step in improving accuracy. Another approach to improvement is cycle counting. Cycle counting is a physical inventory-taking task in which inventory is counted frequently rather than once or twice a year. Benefits of a more perpetual approach include more accurate inventory records, less overproduction, and less stock-outs and can be prioritized based on value.

Inventory in all its forms should be eliminated, or at least minimized. When developing the process improvements, the Lean practitioner should review each point of inventory and ensure continuous flow, and, if necessary, set a countermeasure inventory against variation. The educational society American Production & Inventory Control Society (APICS, now the Educational Society for Resource Management) provides an excellent source of information supporting resource management. Using the Lean Inventory Analysis Tool can reduce the inventory by matching it to the level of demand that occurs in your supply chain.

Little's Law

In our quest to achieve a Lean environment, we are fortunate to have a very simple, yet powerful, relationship known as Little's law. Simply stated, Little's law is a straightforward mathematical relationship among WIP, lead-time, and the process' throughput. Little's law:

$$\mathrm{WIP} = \mathrm{TP} \times \mathrm{LT}$$

where WIP = work-in-process, TP = throughput, and LT = lead-time.

Rewritten:

$$\mathrm{LT} = \frac{\mathrm{WIP}}{\mathrm{TP}}$$

This relationship shows that by reducing WIP, we can directly improve time to the customer through reduce lead-time. It also states that if WIP inventories are allowed to vary, so will lead times. In other words, if WIP is held constant so will lead times (Fig. 14.11).

Managing and Eliminating Constraints

A constraint is anything that limits a system from achieving higher performance or throughput. Constraints can come in many forms, including

- *Equipment*. Capacity, speed, capability
- *Labor*. Supply, skills

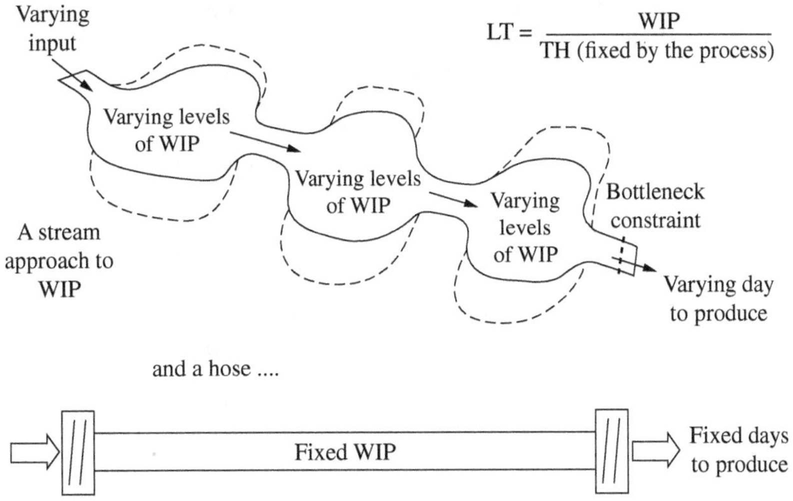

Figure 14.11 Little's law.

- *Information*. Speed, accuracy
- *Suppliers*. Reliability, quality

This is an important concept when evaluating the current state value stream. When evaluating the value stream, special attention should be paid to the constraint (sometimes referred to as the Monument). An improvement in any other area is, by definition, a waste; improvement should occur at the constraint. Once this resource is no longer a constraint, another resource will be the rate-limiting step. Focus should then move to the new constraint. The goal for a manufacturing organization is to drive the constraint to sales.

Goldratt's theory of constraints (Goldratt 1992) offered a five-step process for addressing constraints, involving the following:

- Identifying the constraint
- Deciding how to exploit the constraint
- Subordinating all else to the above decision
- Elevating the performance of the constraint
- Moving to the next constraint and go back to step 1

As we can see, this is an ongoing process to drive continuous improvement.

Improving the Process and Implementing Pull Systems

Once takt time has been calculated, each constraint (such as long setup times) should be identified and managed (or eliminated) to enable smaller batch sizes. Ideally, this leads to single-piece flow. If this reduction can be achieved, it will eliminate overproduction and excess inventories. Pull production scheduling techniques are used so that customer demand pulls demand through the value stream (from supplier to production to the customer). In pull production, materials are staged at the point of consumption. As they are consumed, a signal is sent back to previous steps in the production process to pull forward sufficient materials to replenish only what has been consumed.

The steps for improvement teams (or *kaizen* teams) to Lean out an operation are as follows:

- Determine the pace (takt time and manpower)
- Establish the sequence and replenishment (product family turnover and setup/changeover required)
- Design the line or process (proximity, sequence, interdependence)
- Feed the line or process (strategic inventory, standard WIP)
- Balance the line or process (load, standard work)
- Stabilize and refine (6S, continuous improvement)

Competitive pressures to reduce lead-time are now a driving force to analyze processes for improvement. A flow diagram or preferably a Value Stream Map can reveal a wealth of sources for improvement such as

- The number of functions and how they interact
- The extent to which the same macro process is used for the vital few customers and the useful many
- The existence of rework
- The extent and location of bottlenecks, such as numerous needs for signatures
- The location and amount of inventory

Numerous ways have been found to shorten the cycle time for processes. These include

- Providing a simplified process for the useful many applications
- Reducing the number of steps and handoffs
- Eliminating wasteful "loops"
- Reducing changeover time
- Managing the constraint or bottleneck resource
- Reducing inventory

Physical Design and Proximity

As the Lean practitioner continues to evaluate the value stream for opportunities, it is not uncommon to find movement to be a waste. This is due to sequential operations not being in close physical proximity, as is often the case with departmentalized facilities. Simply moving processes closer together can improve flow and reduce waste of all types. When we expand this idea and group all the interdependent assets into a "cell," the benefits can be become even more significant. The cellular design will minimize space; a 50 percent reduction is common.

Cells should also be designed so that the steps are interdependent and run to the same takt time or pace. This approach will reduce inventory, reduce cycle times, and provide immediate quality feedback.

Another approach to aligning resources is the idea of group technology. Group technology is the process of examining all items produced by an organization to identify those with sufficient similarity that common design or manufacturing plans can be used. This would reduce the number of new designs or new manufacturing plans. In addition to the savings in resources, group technology can improve both the quality of design and the quality of conformance by using proven designs and manufacturing plans. In many companies, only 20 percent of the parts initially thought to require a new design actually need it; of the remaining new parts, 40 percent could be built from an existing design, and modifying an existing design could create the other 40 percent. Relocating production machines can also benefit from the group technology concept. Machines are grouped according to the parts they make and can be sorted into cells of machines, each cell producing one or several part families.

Balancing the Process

When designing improvements into a future state, smooth and sequenced flow is critical. The design should be balanced from step to step. Make process steps interdependent, and run to the same takt time with minimum inventory and the smallest lot sizes possible. In addition to reduced lead time as calculated by Little's law, this approach provides immediate quality feedback. As operations approach a continuous flow and single-piece processing, wastes will be quickly eliminated. Allocation of resources (people and equipment) to accomplish a series of tasks is minimized toward the idle point. Often, by combining work, the process can reduce the required resources by balancing new combined cycle times as close as possible to one another.

Kanbans: Signal to Produce

As mentioned earlier, nonstrategic inventory, excessive transporting, waiting, and overproduction are all forms of waste. An effective way to control these wastes is to use a signaling system to authorize production and motion within the value stream. This is sometimes, but not always, a card. The signaling device, whatever its type, is called a *kanban*. The device is used to control strategic inventory levels, standard WIP and is the trigger for a pull process. Some producers use marked-up floors to identify where the materials should be stored and in what quantity. When the space is empty, the supplying operation is approved to replenish the inventory. Containers can also be used as signaling tools; for example, when a container is empty, this triggers production of the upstream operation. Hopp and Spearman (2000) provide a detailed explanation of the design and applications of *kanban* systems.

Setup Reduction or SMED

In some processes, the waste associated with changeover from one product (process) type to the next scheduled can be sizeable. This was the case at Toyota, which promoted the work of Shigeo Shingo (1989) to reduce the changeover time for stamping presses from four hours to three minutes. The methods for reducing changeover were called "single minute exchange of die" (SMED). SMED is a set of techniques used to perform equipment setup and changeover operations in fewer than 10 minutes, or dramatically reduced from current levels. These principles can be applied to all types of changeovers.

The benefits of SMED include decreased inventory, improved capacity and throughput, and improved on-time delivery to the customer. The longer the setup time, the more likely the operation is to store inventory. Like equipment maintenance breakdowns, changeovers cost productivity that cannot be recouped. Faster changeovers also improve flexibility to produce wider ranges of products at reduced costs (scrap, labor, and skills).

The primary steps to faster changeovers include

- Move as much of the work of change over from *internal* activity (which requires production to stop) to *external* activity (which can be completed without stopping production)

- Streamline the internal activity with the same principles as production: minimizing motion and travel, adjacency, and balancing. Then streamline external activity
- Eliminate the need for adjustments and trial runs
- Streamline external activity

Although originally developed for changing capital equipment configurations for different product runs, the same principles have been applied to improving lead times for service and knowledge work—for example, staging the data for insurance underwriters so that they can began a new case immediately rather than having to retrieve the needed data, minimizing the time for a customer service representative to open a new case by prepopulating key fields in the case documentation, or organizing all audit data in a standard format to facilitate switching from one study to another.

Reliability and Maximizing Equipment Performance

Reliability is the ability to supply a product or service on or before it is promised. Within operations, this normally directly ties to a resource being able to consistently produce the quantity and quality demanded by the customer. To ensure quantity, the asset must be available when called upon. Maintenance excellence is the mindset to maximize resources through the highest levels of equipment consistency and dependability. Maintenance excellence is based on a sound philosophy of guiding performance, combined with a strong tactical approach for implementation. The overall philosophy is called total productive maintenance (TPM) and the tactical approach, reliability-centered maintenance (RCM).

Maintaining equipment is generally recognized as being essential, but pressures for production can result in delaying scheduled maintenance. Sometimes, the delay is indefinite, the equipment breaks down, and maintenance becomes reactive instead of preventive.

The planning should determine how often maintenance is necessary, what form it should take, and how processes should be audited to ensure that maintenance schedules are followed. Prioritizing maintenance activities is discussed as follows in RCM.

In the event of objections to the proposed plan for maintenance on the grounds of high cost, data on the cost of poor quality from the process can help to justify the maintenance plan.

Total Productive Maintenance

Equipment maintenance used to be carried out by the operator. After work was organized and more specialized, maintenance was turned over to specialists. This was typically a small group of highly trained individuals who could fix nearly any problem with the equipment. It has become imperative to return as much of the routine maintenance responsibilities to operators. Total productive maintenance (TPM) looks into the value stream for improvements. TPM identifies the sources of losses and drives toward the elimination of all of them and focuses on zero losses (including quality losses) for productivity.

The operator forms the core of TPM and is the process expert. They are in the best position to help drive improvement in accidents, defects, and breakdowns. TPM is a philosophy based on total employee involvement, which is called autonomous maintenance. Operators are trained to stop abnormalities and other sources of accelerated deterioration. Operators will also perform daily checks for cleanliness, carry out routine lubrication, and tighten fasteners. Training is the key and should be incorporated with 6S mentioned earlier.

Reliability-Centered Maintenance

TPM sets the overall philosophy and standards for maintenance. To complement this, a planning method is needed, a way to prioritize resources and actions. This is called reliability-centered maintenance (RCM). The goal of RCM is to ensure process reliability through data collection, analysis, and detailed planning. Like TPM, if properly deployed, RCM will drive down inventories, shorten lead times, provide more stable operations, and improve job satisfaction.

Prioritization is the foundation of RCM. The basic premise is to allocate resources as effectively as possible to eliminate unplanned downtime, reduce deteriorating quality, or ensure planned output. Assets are prioritized into one of three categories: reactive, preventive, and predictive. The reactive maintenance approach is to run to failure. These assets could include noncritical components, redundant equipment, small simple items, and assets with low failure rates. Examples would include electric solenoids, relay coils, lamps, and all breakdowns. The priority for this class is low; allow for running to failure. The next step is preventive maintenance. This set of assets has a known failure pattern and is often a time-based relationship. Consumables also fall into this group. Motor brushes, bearings and gears, filters, and most normal planned maintenance actions are some examples. Here, a planned schedule can be generated based on the number of cycles or a time interval, performing maintenance activities (hopefully) before failure. The final class is predictive maintenance. This category is the highest priority in terms of planning and assigning of resources. These resources are the most critical to the operations and the ones required to provide customer satisfaction. This group also includes assets with random failure patterns, assets not normally subject to wear, and replacement components with long lead times for replenishment. The group is analyzed based on condition. Methods such as vibration analysis, lubrication analysis, temperature, current signature, and high-speed videos can determine machine conditions. If successfully implemented, RCM can deliver significant business benefits. Experience with Juran's principles has shown that reactive maintenance costs are two to three times higher than preventative; and preventative is two to three times higher than predictive.

Measuring improvement in reliability should include several dimensions. The most encompassing is overall equipment effectiveness (OEE). This measures the cumulative effect of all losses due to equipment condition–machine availability, machine efficiency, and machine quality performance. Figure 14.12 shows a calculation OEE. Other measures for maximizing equipment performance include those found in Fig. 14.13*a* and *b*.

OEE calculation:

$$\text{Machine availability (MA)} = \frac{\text{Actual running time}}{\text{Planned running time}}$$

$$\text{Machine efficiency (ME)} = \frac{\text{Cycle time} \times \text{units produced}}{\text{Uptime}}$$

$$\text{Machine quality performance (MQ)} = \frac{\text{Number of good units}}{\text{Total units produced}}$$

$$\textbf{OEE} = \textbf{MA} \times \textbf{ME} \times \textbf{MQ}$$

Figure 14.12 Calculation of overall equipment effectiveness.

Maintainability – Mean time to repair (MTTR)

$$\text{MTTR} = \frac{\text{Sum of downtime for repair}}{\text{Number of repairs}}$$

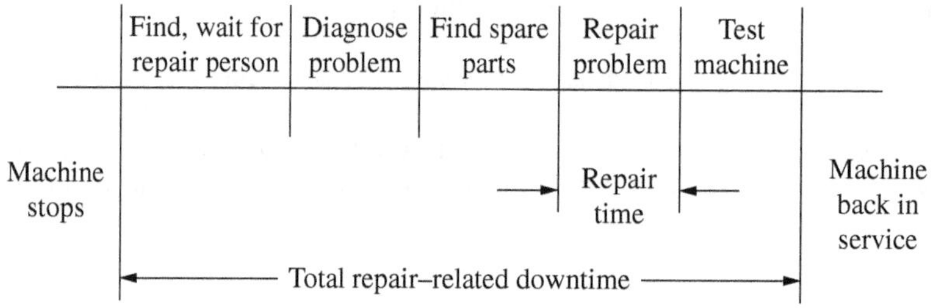

(a)

Reliability – Simple measures

$$\text{Machine availability (MA)} = \frac{\text{Actual running time}}{\text{Planned running time}}$$

$$\text{Mean time between failure (MTBF)} = \frac{\text{Total running time}}{\text{Number of failures}}$$

(b)

FIGURE 14.13 (*a*) Maintainability: mean time to repair (MTTR), (*b*) Simple measures of reliability.

Mistake Proofing the Process

An important element of prevention is designing the process to be error free through "mistake proofing" (the Japanese call it *poka-yoke*).

A widely used form of mistake proofing is the design (or redesign of the machines and tools, the "hardware") to make human error improbable, or even impossible. For example, components and tools may be designed with lugs and notches to achieve a lock-and-key effect, which makes it impossible to misassemble them. Tools may be designed to sense the presence and correctness of prior operations automatically or to stop the process on sensing depletion of the material supply. For example, in the textile industry, a break in a thread releases a spring-loaded device that stops the machine. Protective systems (e.g., fire detection) can be designed to be "fail safe" and to sound alarms as well as all-clear signals.

In a classic study, Nakajo and Kume (1985) discuss five fundamental principles of mistake proofing developed from an analysis of about 1000 examples collected mainly from assembly lines: elimination, replacement, facilitation, detection, and mitigation (Table 14.6).

Mistake proofing is both a proactive and reactive tool. As Fig. 14.14 shows, the upper portion of the chart (prevent defects) highlights a proactive effort, whereas the lower part of the chart (mitigate errors) assumes a reactive effort because a problem already exists. It is better to use mistake proofing in a proactive mode and stop defects from ever occurring by mistake-proofing products and processes at the design stage. However, the next best alternative is to prevent defects from passing along to the next operation, reactive mode.

Principle	Objective	Example
Elimination	Eliminating the possibility of error	Redesigning the process or product so that the task is no longer necessary
Replacement	Substituting a more reliable process for the work	Using robotics (e.g., in welding or painting)
Facilitation	Making the work easier to perform	Color coding parts
Detection	Detecting the error before further processing	Developing computer software that notifies the worker when a wrong type of keyboard entry is made (e.g., alpha versus numeric)
Mitigation	Minimizing the effect of the error	Using fuses for overload circuits

TABLE 14.6 Summary of Mistake-Proofing Principles

Mistake proofing can, of course, result in defect-free work. The advantage can also include eliminating many inspection operations and requiring an immediate response when problems do arise. For more information on mistake proofing reference Mistake Proofing for Operators from The Productivity Press Development Team (1997).

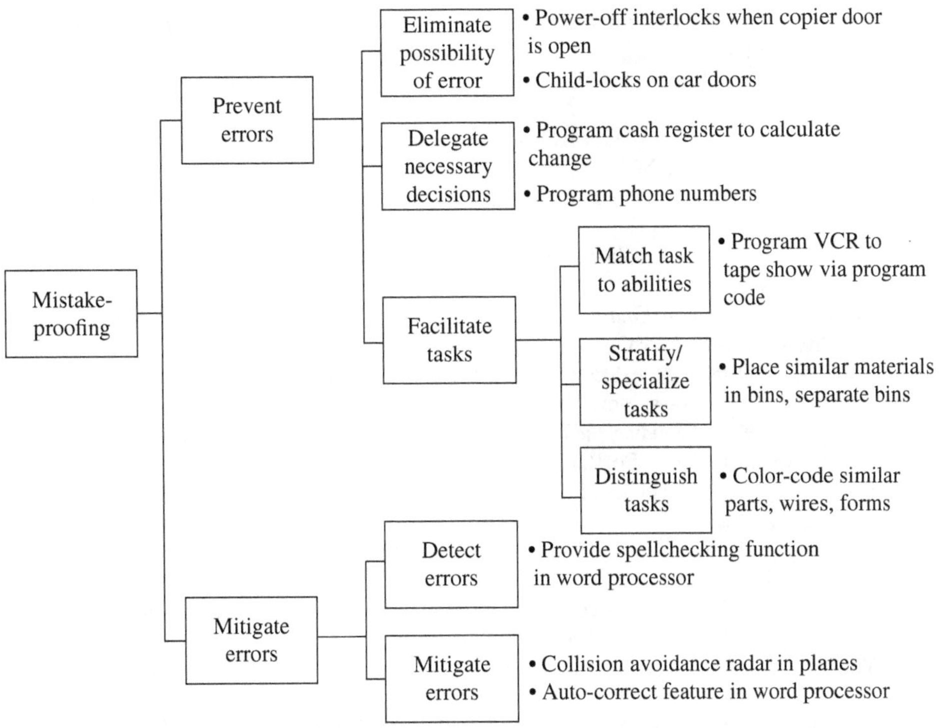

FIGURE 14.14 Mistake-proofing guidelines.

References

Bider, L., and Kowalski, S. (2014). "A Framework for Synchronizing Human Behavior, Processes and Support Systems Using a Socio-Technical Approach," *Enterprise, Business-Process and Information Systems Modeling*, Springer, Berlin, Heidelberg, pp. 109–123.

Chase, R. B., and Stewart, D. M. (1994). "Make Your Service Fail-Safe," *MIT Sloan Management Review*, Vol. 35, No. 3, pp. 35–45.

Doggett, A. M. (2005). "Root Cause Analysis: A Framework for Tool Selection," *Quality Management Journal*, Vol. 12, No. 4, pp. 34–45.

Duggan, K. (2002). *Creating Mixed Model Value Streams*. Productivity Press, New York.

Edgeman, R. L., and Williams, J. A. (2014). "Enterprise Self-Assessment Analytics for Sustainability, Resilience and Robustness," *TQM Journal*, Vol. 26, No. 4, pp. 368–381.

Elkington, J. (1997). *Cannibals with Forks: The Triple Bottom Line of 21st Century Business*. Capstone Publishing, Oxford, UK.

Goldratt, E. M. (1992). *The Goal*, 2nd ed., North River Press, Great Barrington, MA.

Hopp, W., and Spearman, M. (2000). *Factory Physics*, 2nd ed., Irwin McGraw-Hill, New York.

McDonough, W. and Braungart, M. (2002). "Design for the Triple Top Line: New Tools for Sustainable Commerce," *Corporate Environmental Strategy*, Vol. 9, No. 3, pp. 251–258.

McIntosh, R., Owen, G., Culley, S. and Mileham, T. (2007). "Changeover Improvement: Reinterpreting Shingo's "SMED" Method," *IEEE Transactions on Engineering Management*, Vol. 54, No. 1, pp. 98–111.

Nakajo, T., and Kume, H. (1985). "A Case History: Development of a Foolproofing Interface Documentation System," *IEEE Transactions on Software Engineering*, Vol. 19, No. 8, pp. 765–773.

Ohno, T. (1988). *Toyota Production System: Beyond Large-Scale Production*, Productivity Press, Portland, OR.

Rother, M., and Shook, J. (2003). *Learning to See*, The Lean Enterprise Institute, Cambridge, MA.

Schonberger, R. I. (1986). *World Class Manufacturing: The Lessons of Simplicity Applied*, The Free Press, New York, NY.

Shingo, S. (1981). *The Toyota Production System*, Japanese Management Association, Tokyo, Japan.

Shingo, S. (1985). *A Revolution in Manufacturing: the SMED System*, Productivity Press, New York, NY.

Shingo, S. (1986). *Zero Quality Control: Source Inspection and the Poka-Yoke System*, Productivity Press, New York, NY.

Shingo, S. (1988). *Non-Stock Production: the Shingo System of Continuous Improvement*, Productivity Press, New York, NY.

Shingo, S. (1989). *A Study of the Toyota Production System from an Industrial Engineering Viewpoint*. Productivity Press, Portland, OR.

Shuker, T. J. (2000). "The Leap to Lean," *Annual Quality Congress Proceeding*, ASQ, Milwaukee, pp. 105–112.

Tapping, D., Luyster, T., and Shuker, T. (2002). *Value Stream Management*, Productivity Press, New York, NY.

The Productivity Press Development Team (1996). *5S for Operators—5 Pillars of the Visual Workplace*, Productivity Press, Portland, OR.

The Productivity Press Development Team (1997). *Mistake Proofing for Operators: The ZQC System*, Productivity Press, Portland, OR.

Volland, J. (2005). "Case Study: Now That's Lean," *Medical Imaging Magazine*, January 2005. Available at http://www.imagingeconomics.com/issues/articles/mi_2005-01_07.asp.

Watson, G. H. (2003). "Policy Deployment: Consensus Method of Strategy Realization," in *Quality into the 21st Century: Perspectives on Quality and Competitiveness for Sustained Performance*, pp. 191–218, T. Conti, Y. Kondo, and G.H. Watson (Eds.), ASQ Quality Press, Milwaukee, WI, USA.

White, R. E., Pearson, J. N., and Wilson, J. R. (1999). "JIT Manufacturing: A Survey of Implementations in Small and Large US Manufacturers", *Management Science*, Vol. 45, No. 1, pp. 1–15.

White, R. E., and Prybutok, V. (2001). "The Relationship Between JIT Practices and Type of Production Systems", *Omega*, Vol. 29, No. 2, pp. 113–124.

Womack, J. P., Jones, D. T., and Roos, D. (1990), *The Machine that Changed the World*, The Free Press, New York, NY.

Womack, J. P., and Jones, D. T. (2003). *Lean Thinking* (revised and updated), The Free Press, New York, NY.

CHAPTER 15

Six Sigma: Breakthrough to In-Process Effectiveness

Joseph A. De Feo and John F. Early

High Points of This Chapter

1. Six Sigma and Lean Six Sigma have developed into one of the most widely recognized and effective methods for creating breakthrough improvement. Both have evolved from the basis of Juran's Universal on Quality Improvement.
2. Six Sigma methods focus on identifying and meeting the needs of customers first and the business second. In this way, revenues increase and costs decrease, improving results.
3. Many large organizations like Samsung Electronics, General Electric, and Honeywell have experienced great success employing Six Sigma and Lean Six Sigma methods since its inception at Motorola in the 1980s. Today, organizations like Naples Community Hospital Florida, The Mayo Clinic, Bank of America, Telefónica in Spain, and hundreds of others have adopted Lean Six Sigma as their improvement method of choice.

4. Six Sigma and Lean Six Sigma methods help both traditional manufacturers of goods as well as producers of services and information to improve their bottom line and increase customer satisfaction.
5. The two primary Six Sigma methods are DMAIC (define, measure, analyze, improve, control) to improve processes and products (the focus of this chapter) and DMADV (define, measure, analyze, design, verify: Design for Six Sigma) to help ensure that products and processes function well from the voice of the customer (VOC) through the delivery of goods.
6. The five steps to carry out a Six Sigma DMAIC project are discussed in detail.
7. A successful Six Sigma deployment depends on a clear understanding of roles, responsibilities, structures, and training requirements of the employee.

Six Sigma: A New Global Standard for Improvement

Six Sigma and Lean Six Sigma (which adds Lean tools to the basic methodology) are quality improvement methods with value-added enhancements of computers and an increasing array of statistical and other software packages. For simplicity, we will refer to the full range of quality improvement methods and tools simply as Six Sigma for this chapter (Fig. 15.1).

If solutions to your problems are elusive, or if you must attain quality levels measured in parts per million or approaching perfection, Six Sigma will place your ailing process under a microscope to find solutions. Figure 15.2 presents the Six Sigma or DMAIC steps and tools most often used with it. The DMAIC steps are

1. *Define* the problem as clearly as one can in words.
2. *Measure* the current level of performance and voice of the customers.

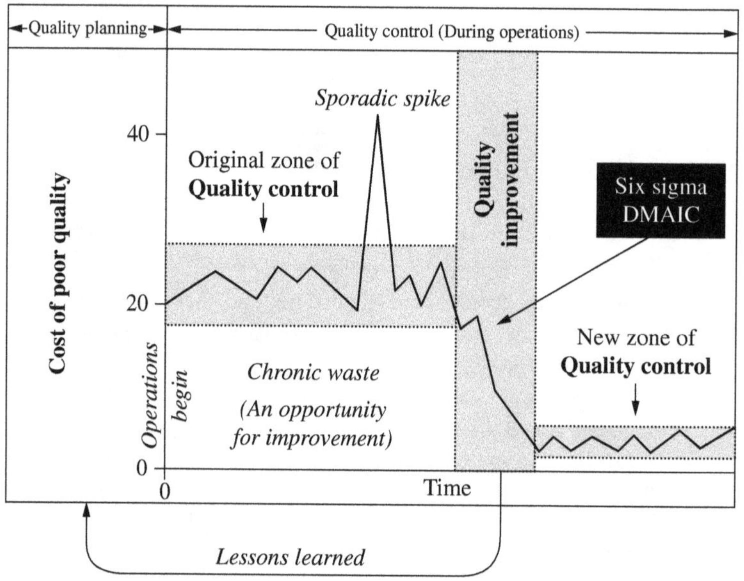

Figure 15.1 Six Sigma and the Juran Trilogy. (Juran Institute, Inc., Southington, CT.)

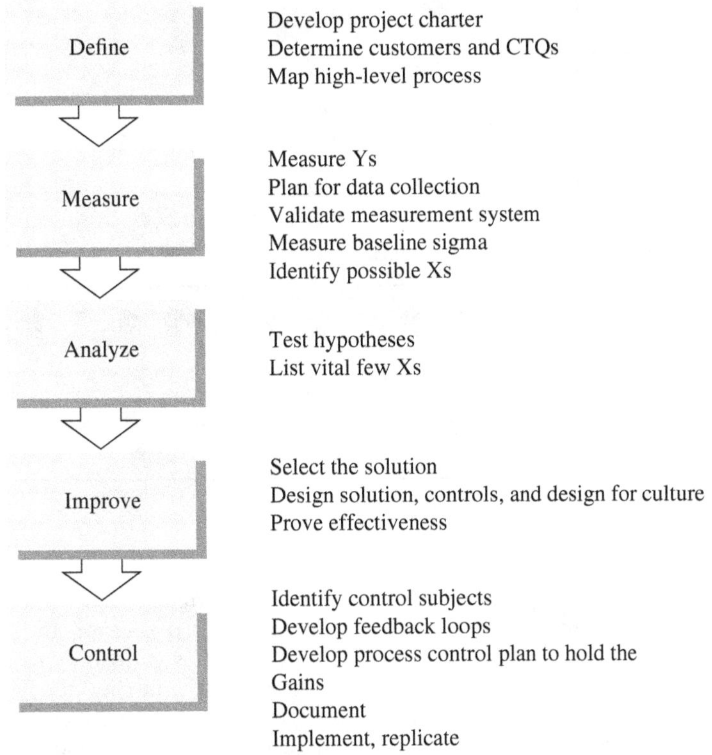

Figure 15.2 Six Sigma phases and steps.

3. *Analyze* collected data to determine the cause(s) of the problem.
4. *Improve* by selecting the right solutions to solve the problem.
5. *Control* to hold the gains.

With these fundamental steps, Six Sigma is enabling many organizations around the world to succeed in achieving performance breakthroughs where they had failed before. The smart companies recognize this as not simply a "fix" to one-time problems, but truly a new way of doing business. Business challenges do not go away in a free marketplace; rather, they continually change in degree and form. Organizations worldwide are under continuing pressure to control costs, maintain high levels of safety and quality, and meet growing customer expectations. This breakthrough improvement process of Six Sigma has been adopted by many companies, including Samsung Electronics, General Electric, Honeywell, and other organizations, as the most effective method for achieving these and other goals.

More than just a formal program or discipline, Six Sigma is an operating philosophy that can be shared beneficially by everyone: customers, shareholders, employees, and suppliers. Fundamentally, it is also a customer-focused methodology that drives out waste, raises levels of quality, and improves the financial and time performance of organizations to breakthrough levels. Six Sigma's target for perfection is to achieve no more than 3.4 defects, errors, or mistakes per million opportunities, whether it involves the design and production of a product or a customer-oriented service process.

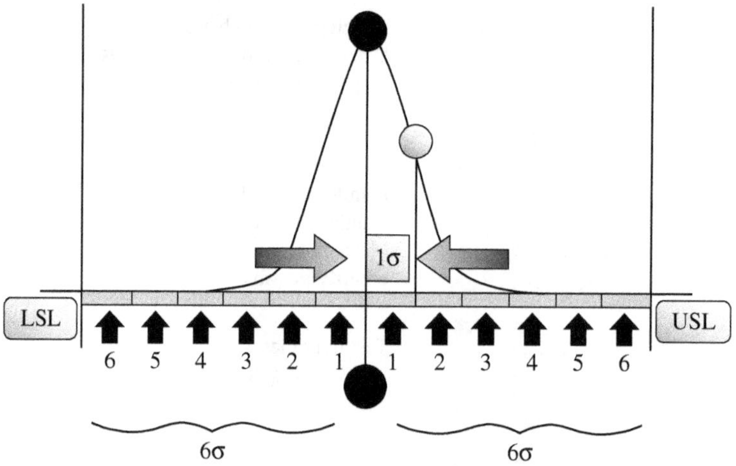

Figure 15.3 Six Sigma level of performance.

It is from this target that the "Six Sigma" name originated. Usually written as a small sigma in the Greek alphabet, sigma (σ) is the symbol used to denote the standard deviation or measure of variation in a process. A process with less variation will be able to fit more standard deviations, or "sigmas," between the process center and the nearest specification limit than a process that is highly variable. The greater the number of sigmas within the specifications, the fewer the defects. The smaller the variation, the lower the cost. A higher sigma level means the process of delivering a good, product, or customer service has greater consistency. Figure 15.3 demonstrates a Six Sigma level of performance. This means that one can fit in six standard deviations, or six sigmas, between the process center and the nearest specification limit.

Most organizations operate at the Three Sigma level, or about 66,800 defects per million opportunities (DPMO) for most of their processes and at a Four or Five Sigma level in some of the mission-critical processes. Comparisons of Sigma levels, yields, and the corresponding defect rates are shown in Table 15.1. It would be foolish, however, to try to achieve Six Sigma levels of performance for every process in the organization. This is because not all processes are equally important. For example, the process for requesting time off for vacation is not as critical as the order fulfillment process. What really counts is significant improvement in the mission-critical areas—that is, critical as defined by the customer.

Six Sigma Is Customer Focused—Organization Examples of Success

Why does Six Sigma work as well as it does? In large part, it is because of a strong emphasis on the customer. While the saying "the customer is always right" is not literally true, customers hold the key that can unlock unrealized potential in your business. Basically, the DMAIC process translates a customer's needs into actionable, operational terms and defines the critical processes and tasks that must be done well to meet the customer needs. Although the details vary, depending on the analysis and improvement interventions that follow. Six Sigma consistently will drive the performance of products, services, and processes to breakthrough levels, that is, to new and sustained levels of performance. Breakthroughs are achieved not by massive teams or flashy initiatives, but by using a steady and

Process Sigma (Short Term)	Long-Term Yield	Defects Per Million
6	99.99966%	3.4
5.5	99.9968%	32
5	99.9767%	230
4.5	99.8650%	1340
4	99.3790%	6200
3.5	97.725%	22,700
3	93.319%	66,800
2.5	84.13%	158,000
2	69.15%	308,000
1.5	50%	499,000
1	31%	691,000
0.5	16%	841,000

TABLE 15.1 Sigma Level, Yield, and Defect Level

concerted project-by-project approach. In this manner, the Six Sigma approach will help organizations:

- Improve cycle times, quality, and cost
- Improve effectiveness and efficiency of processes, including e-commerce
- Design products and services that will sell well
- Reduce chronic waste, or the cost of poor quality (COPQ)
- Grow profits by improving revenue and reducing costs

In short, Six Sigma is financially rewarding. Our experience indicates returns on investment (ROI) are achievable ranging from 10:1 to more than 100:1.

Samsung Electronics

When the decision was made by Samsung Electronics Company, Ltd., Vice Chairman and CEO Jong-Yong Yun to position the company for the future, the catalyst was Six Sigma. Samsung Electronics began its journey with training as the first essential step to prepare for implementing the methodology. Starting initially in manufacturing operations and R&D in 2000, the company expanded to transactional business processes and the entire supply chain, ultimately obtaining significant savings and financial benefits in all 16 of its business units in South Korea and internationally. The methodology's philosophy and methods continue to be integrated still more deeply throughout the company by developing the internal specialists needed to teach, implement, maintain, and grow this competence in the future. No single person, nor any operation in Samsung Electronics, is exempted from the process, and the company is not looking back.

General Electric

Mr. Jack Welch, General Electric's retired CEO, was one of the first high visibility executives who became a Six Sigma leader and advocate. As an international business role model, he was vocal in expressing his views as to what leaders must do to achieve superior results. GE became an early adopter of Six Sigma, and through its demonstrated success and bottom-line results, enabled Mr. Welch to vault Six Sigma from the mailroom to the boardroom. In his book *Winning*, he said "Six Sigma, originally focused on reducing waste and elevating the quality in our products and processes, has delivered billions of dollars to GE's bottom line in savings. Six Sigma has grown from an internally focused activity to an outside focus—also improving the productivity and efficiency of our customers' operations. Increasing the intimacy between GE and its customer base is making everyone more productive and helps all of us grow through tough economic environments." "Today," Mr. Welch explained, "Six Sigma has evolved to an even larger role in GE. Its rigorous process discipline and relentless customer focus has made it the perfect training ground and vehicle for the future leadership of GE. Our best and brightest employees are moving into Six Sigma assignments. I'm confident that when the board picks a successor to Jeffrey Immelt 20 years from now, the man or woman chosen will be someone with Six Sigma in his or her blood. Six Sigma has become the language of leadership in our company in GE. Its rigorous process discipline and relentless customer focus has made it the perfect training ground and vehicle for the future leadership of GE."

Six Sigma Works for Production, Service, and Transactional Processes

The Six Sigma movement gained interest in health care, financial services, legal services, engineering, consulting, and almost all organizations. In addition to achieving major improvement in manufacturing goods, managing inventory, delivering products, and managing repetitive processes, the Six Sigma methods have migrated to transactional processes. Processes that avoided continuous improvement because, as many stated, "the tools did not apply to us" have joined the Six Sigma bandwagon. Processes like completing an invoice, writing a contract, and boarding passengers on an airline, banking, hospitals, insurance, government, and other service organizations have tried Six Sigma. Most succeeded in

- Optimizing equipment usage
- Experiencing fewer rejects or errors
- Cutting response times to customer inquiries
- Reducing inspection, maintenance, inventory, and supply chain costs
- Creating more satisfied customers, external as well as internal

When implemented strategically, Six Sigma also

- Helps turn over working capital faster
- Reduces capital spending
- Makes existing capacity available and new capacity unnecessary
- Fosters an environment that motivates employees
- Improves morale, teamwork, and career potential

Telefónica

One of the biggest names in business in Spain and in the Spanish- and Portuguese-speaking world has a long tradition of quality management practices and achievements. So when the company embarked on a pilot Six Sigma program toward the end of 2000, the scale and ambition of the effort reflected the company's experience of business improvement initiatives.

Between March and July 2001, some 21 first-phase projects were completed. Efficiency savings from these projects amounted to more than 22 million euros; customer satisfaction levels were at all-time highs. Telefónica committed itself to 300 Six Sigma projects for the following year and estimated that it will have conducted 3000 projects during the next 3 years after that (European Quality, 2002).

The Six Sigma Model for Improvement has been widely used to address repetitive production-like processes and ones that address repetitive transactional processes.

We need to clearly establish the difference between production (aka, manufacturing) and service or transactional processes. All processes are transformations that result in the change of state of one or more things that can be physical objects or services. *Production processes* directly transform raw materials or semi-finished goods into a final physical product (aka, goods). The output of production processes is a transformed physical product; these processes are deterministic, workflow-oriented, highly procedural, and, therefore, highly repeatable. Because of this, production processes are well suited for representation by the traditional, workflow-based triple role of input-process-output (IPO) or supplier-input-process-output-customer (SIPOC) models.

A process to produce goods is a series of work activities performed by people and other resource-consuming assets in order to transform given input(s) into output(s).

A service process or transactional process (sometimes also called people or paper processes) directly transforms one state or condition of one or more things (objects, abstractions such as information, data, symbolic representations, etc.) into another. One execution of a transactional process results in a transformation, the outcome of which, in turn, may be a change of state in a number of things (physical objects such as inventories, data and information, people, etc.). Examples of transactional processes include

- Value-added service processes related to production (transporting, installing, storing, repairing, maintaining, etc.)
- Support or back-office processes in manufacturing and service organizations (selling, purchasing, subcontracting, warehousing, billing, human resources, etc.)
- Value-added processes in service industries (banking, insurance, transportation, health care, hospitality, education, etc.)
- Value-added processes in the public sector (including the military) and the not-for-profit sector (legislative and administrative processes, planning, command and control, fundraising, etc.)

The output of transactional processes is a change of state or condition, defined by the transaction. These processes are information (communication)-driven in that successive executions of a transactional process depend on the informational inputs (requests, offers, etc.) received at the outset of each execution. Accordingly, successive executions may be different with different results. Therefore, these processes are not always repeatable in the same sense as stamping out millions of identical parts, but are self-regulating and highly adaptable. A transactional process is a logical set of customer-supplier tasks that drive work activities performed by people.

Transactional process characteristics that differentiate them from production processes may include

- Scarcity of measurement data; available measurements are primarily discrete (attribute)
- Measurement system is partially or entirely I/T defined (e.g., reporting)

- The definition of quality includes information quality
- Dominant variables: people and information
- High-cost labor
- Disproportionately large financial leverage

DMADV (Design) versus DMAIC (Improvement)

As stated in the "High Points of This Chapter" section, there is another Six Sigma methodology for *designing and developing* a new product, service, or process with no defects.

Design for Six Sigma, DFSS for short, follows the DMADV steps. DMADV is different from DMAIC as follows:

1. *Define.* Provides the goals and direction to design a new product or service with development of a team charter.
2. *Measure.* Collects and translates customer needs into CTQs. A CTQ is what is critical to quality in the eyes of the customer. DMADV may deal with many CTQs in one design project. Six Sigma DMAIC typically focuses on only one CTQ that is creating customer dissatisfaction or related to the problem at hand.
3. *Analyze.* Understand the information collected from the voice of the customers and define the design features that collectively will be developed into a concept and then into one or more high-level designs. DMAIC focuses on identifying the root causes of the customer dissatisfaction and the problem at hand.
4. *Design.* In this step, the final product or service design is developed. A detailed design with associated design elements is completed and the critical-to-process variables are identified, from which the process for creating and delivering the good or service is developed.
5. *Verify.* The new design plans are implemented and the organization prepares for full-scale rollout and puts control mechanisms in place. In DMAIC, we control the process to hold the gains. In DMADV, we verify that the project goals are met, that the customer receives the value expected, and assure that control is effective to deliver on the CTQs and product design.

Key Roles to Deploying Six Sigma Successfully

Deploying a Six Sigma program requires building a suitable infrastructure, as described in Chap. 5, Quality Improvement Breakthrough Performance. A number of key roles are important, as shown in Fig. 15.4. Each role is essential, yet, by itself, insufficient to produce the improvement an organization expects from Six Sigma. Each role requires knowledge of the methods and tools. In addition, the Six Sigma community led by the American Society for Quality has established a standard curriculum and certification process for the roles of Green, Black, and Master Black Belts. Certification is granted upon completing subject matter training, carrying out a number of significant projects, and passing written and oral reviews.

The key roles to drive Six Sigma are

- Leadership
- Champion
- Master Black Belt

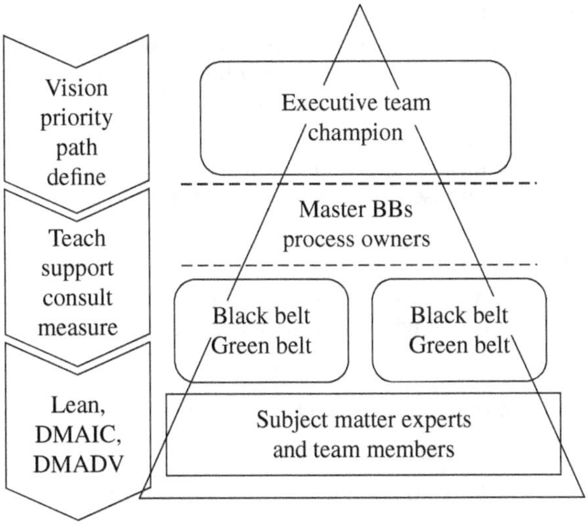

FIGURE 15.4 Key roles for Six Sigma. (Juran Institute, Inc.)

- Black Belt
- Green Belt
- Project Team and Subject Matter Experts (SMEs)
- Process Owner

Leadership's Role

The roles of *all* the members of the organization leadership team to create annual breakthrough when acting as a steering team are

- *Setting improvement goals.* Identify the best opportunities to improve performance and set strategic and annual goals for the organization. Establish accountability for meeting goals.
- *Establish infrastructure to enable Six Sigma Projects to happen.* Establish or revise management systems for selecting and assigning projects, organizational reporting of project progress, and accountability of the various roles, performance appraisal, reward, and recognition.
- *Appoint Champions.* They can sponsor projects and ask the right questions at each phase of DMAIC of the Six Sigma project.
- *Support projects and monitor progress.* Enable project teams to carry out their project goals. Provide the necessary training, resources, facilities, budgets, time, and most importantly, management support. Monitor progress of projects and keep them on track.
- *Provide organizational support to deal with resistance to change that occurs when implementing breakthroughs.*
- *Become educated and receive training in the methods of Six Sigma to be able to support and evaluate the work of all the other roles.*

All members of the executive team and managers at all levels should be committed to the Six Sigma effort, agree to support it, and act with unified focus and consistency to facilitate the gradual cultural changes that will inevitably be required. A fractured executive and management team can, and usually does, wreak havoc and confusion on a Six Sigma effort, drains the energy out of those trying to make it succeed, and leaves in its wake disillusionment and meager results. If the executive team fails to maintain unified focus and transform the culture, it loses its credibility and ability to lead.

Role of Champions

Champions are usually members of management (or at least folks with organizational clout). The ideal Champion is one who wants to sponsor a project and likes change.

The Champion

- Identifies improvement projects that meet strategic goals
- Is responsible for creating a project charter
- Identifies and selects competent Belts and team members
- Mentors and advises on prioritizing, planning, and launching Six Sigma projects
- Removes organizational obstacles that may impede the work of the Belts or project teams
- Provides approval and support to implement improvements designed by the project teams
- Provides recognition and rewards to the Black Belts and teams upon successful completion of their projects
- Communicates with executive management and peers as to the progress and results associated with the Six Sigma efforts
- Removes barriers the teams encounter
- Understands and upholds the Six Sigma methodology

In general, Champions manage, support, defend, protect, fight for, maintain, uphold, and function as an advocate for Six Sigma. Usually, a strong Champion can be found behind every successful project. Weaker Champions are usually associated with weaker results.

After helping the steering team select projects, the Champions mentor and support the overall process. Once criteria are established and business unit managers and Champions are identified, projects are selected for their potential in breakthrough improvement. This means evaluating opportunities for strategic relevance, operational efficiency, product and service quality related to customer satisfaction or dissatisfaction, and bottom-line savings.

The Champions and leadership of each business unit support Six Sigma project teams. As influential members of management, they are expected to promote the application, acceptance, and evolution of the process within their business units in the following ways:

- Project selection
- Leadership reviews
- Project support

- Resource allocation
- Career development

Role of Master Black Belts

A Master Black Belt receives training and coaching beyond that of a Black Belt. Master Black Belts are qualified to train Black Belts. The role of a Master Black Belt includes

- Acting as internal Six Sigma consultant, trainer, and expert on Six Sigma
- Managing and facilitating multiple projects—and their Black Belts
- Supporting and advising Champions and executive management
- Providing technical support and mentoring as needed

Everyone else in the organization—those who are not Champions, Master Black Belts, or Black Belts—becomes either a Green Belt or a team member (some organizations call them Yellow Belts). Suffice it to say that the different colored belts vary according to the amount of skill they will need, the formal training received, and the active roles each takes in participating in Six Sigma activities. In an ideal situation, all organization members receive training at some minimal level and are awarded the appropriate belt. Everyone feels included, and everyone understands what Six Sigma is all about, and just as important, what it is not about. No one is left to wonder what Six Sigma is all about or to resent or resist it. This unifies the organization behind the Six Sigma effort and significantly reduces pockets of resistance.

Role of Black Belts

Black Belts are on-site implementation experts with the ability to develop, coach, and lead cross-functional process improvement teams. They mentor and advise management on Six Sigma issues. Black Belts have an in-depth understanding of Six Sigma philosophy, theory, strategy, tactics, and Six Sigma tools. Each project is targeted to save at least $250,000 ROI per project. Black Belts are expected to guide three to six projects per year, which increases further the ROI of Six Sigma.

The training required to be certified as a Black Belt is rigorous and demanding. An illustrative list of topics would include

Critical team leadership and facilitation skills	Correlation and regression
Six Sigma methodology	Hypothesis testing using attribute and variables data
Core improvement tools	ANOVA: Analysis of variance
Use of an appropriate statistical software package	DOE: Design of experiments
Measurement system analysis	EVOP: Evolutionary operations
Determining process capability	Lean enterprise principles and tools
Process mapping	Mistake-proofing
Quality function deployment	SPC: Statistical process control
FMEA: Failure mode, effect, and criticality analysis	Process control plans
Basic statistical methods	Transfer to operations

Armed with this training—usually delivered in four weeklong sessions with 4- to 5-week intervening intervals—the Black Belt is full-time and devoted to carrying out real Six Sigma projects. When Black Belt training has been completed, employees are able to

- Develop, coach, and lead cross-functional teams
- Mentor and advise management on prioritizing, planning, and launching projects
- Disseminate tools and methods to team members
- Achieve results that match the company's business strategies with a positive benefit to financial performance

Role of Green Belts

Employees who become members of each project team often enter the process by becoming Green Belts. A Green Belt requires about 8 days of training in the overall Six Sigma improvement methods and tools. They become key team members on a Black Belt-level project or can be leaders of smaller-scope projects.

Each week in the classroom is followed by four to five weeks of practical application on the same projects back in their business units. If properly selected, these initial projects will produce significant bottom-line savings and, typically, return more than the entire training investment. Each project is targeted to save at least $100,000 to $250,000 ROI per project.

The total number of employees trained in Six Sigma throughout the world must be in the hundreds of thousands by now. More and more companies, like Samsung and GE, are planning for these employees to move up the ranks to top management levels. In the final analysis, success in achieving results with this process depends on whether top management, particularly CEOs, accept responsibility for their nondelegable roles.

Mr. Bob Galvin at Motorola, Mr. Larry Bossidy at AlliedSignal—now Honeywell—and Mr. Jack Welch at GE were role models for making Six Sigma and opportunities for Black Belt employees a vital part of the culture during their tenure as CEOs. Top management can overcome the powerful forces in any organization that may resist unity of direction. The answer is to find a universal improvement process like Six Sigma that fits all functions in an organization. Six Sigma is an extremely healthy and productive cultural change that takes time to complete. It is not free. It requires resources and training, but customer satisfaction, quality products and services, and a highly competitive organization produce a significant return on investment, satisfaction all employees have from being on a winning team, and pride in being part of such an organization.

Roles of Project Team Members and Subject Matter Experts

The members of the Six Sigma team can come from throughout the organization and are often subject matter experts from the various functional departments that are involved in the operation or maintenance of the process under study. Team members are expected to attend all team meetings, contribute to the work process, and complete assignments given to them by the project leader between meetings. Often, the subject matter experts (SMEs) are of greatest value assisting the team:

- When identifying key aspects of the problem and evaluating the appropriate goal for the project (define phase)
- During the process flow diagramming activity by contributing their expertise (measure phase)

- Collecting data about the parts of the process that they are most familiar with (measure and analyze phases)
- Identifying possible causes of the problem (measure phase)
- Identifying possible failure modes and ranking their severity, occurrence, and detection during completion of the PFMEA (measure phase)
- Developing possible solutions to the proven causes (improve phase)
- Identifying control subjects for ongoing measurements of the product and process (control phase)

Process Owners

Process owners are usually at the high supervisory or managerial level of the organization and are directly responsible for the successful creation of the product (goods, services, or information). They are typically not core team members, but may be called upon to assist the team with specific tasks as needed. Some of the most important needs for support from process owners occur during the Improve and Control phases when the team is

- Defining possible solutions to the proven causes of the problem
- Planning for dealing with cultural resistance
- Conducting pilot evaluations of possible solutions
- Implementing the selected improvements
- Designing the control plan and applying it to the everyday maintenance of the process performance
- Disbanding the teams after project completion and turning full responsibility back to the operating forces

Lean Six Sigma Deployment Roadmap

Any deployment of quality improvement methods, whether it is called Lean Six Sigma, Performance Excellence, Lean or by some other nomenclature, requires a methodical approach to be successful. A phased approach that starts small and then expands has been shown to be the most effective. See Fig. 15.5 for a description of such an approach using the Juran Transformation Roadmap.

Decide

During the decide phase of deployment, upper management is becoming familiar with the Six Sigma methodology (or whatever methodology is being considered) and evaluating how well the approach fits with their organization's strategies and goals, particularly those related to performance excellence. A decision must be made whether Six Sigma, or some other approach, best fits the organization's needs.

The upper managers must then decide what roadmap to follow. The one recommended here is an option, but variations on this could work as well. The important thing is that the managers have a roadmap to follow so that the deployment will be done in a methodical way.

A decision must be made at this point whom to select as a training partner. It would be quite rare for an organization to have qualified internal resources to train, consult with, and mentor the resources being developed during the deployment, so it is almost always

Decide	Prepare	Launch	Expand	Sustain
Learn about Lean and Six Sigma. Assess organization status-identify strategies, goals, and projects. Determine if Lean Six Sigma is to become part of the strategy. Decide on a "roadmap" to follow. Select external training/consulting partner. Attend an executive briefing. Discuss other initiatives and their impact on resources.	Organize the steering team to manage the process. Nominate and appoint champions. Nominate and select pilot projects. Train the Champions and Black Belts. Decide on the use of Green Belts and train them. Develop first wave plan. Select projects for first wave.	Support and mentor black and Green Belts. Support pilot project teams. Allow time to work on projects. Develop a project nomination process. Develop on-going cost of poor quality metric. Integrate participation reward and recognition. Establish assessments/measurements for on-going project selection.	Support infrastructure and review progress. Support expansion of all types and number of teams to other business units. Mandate improvement to all levels: other Belts. Begin product development and design teams. Create key macro-business process teams. Identify benchmarking opportunities.	Integrate process measures and move toward process owners. Fully Integrate Lean Six Sigma goals into next year's business plan. Deploy Lean Six Sigma to all business units. Enable employee participation with training and resources. Act on audits of business systems to drive new projects. Continue to assess culture and act on gaps. Sustain breakthrough performance.

FIGURE 15.5 The Juran transformation roadmap. (Juran Institute, Inc.)

necessary to contract outside resources. It's important to select a partner that fits well with the organization's culture, business style, and desires for implementation flexibility. The first training delivered by the partner selected should be an executive briefing, which is attended by the entire upper management team. For more on the importance of leading change from the highest levels of the organization.

Finally and very importantly, management must decide what they will stop doing. Resources in any organization are finite, and laying a Six Sigma deployment over lots of other projects that are underway is a recipe for failure. The current initiatives should be evaluated and prioritized, and only the vital few continued. For the organization to successfully weave Six Sigma into its culture, it must become the method of choice for creating breakthrough improvement. Launching Six Sigma projects that mirror others already underway will lead to confusion over ownership of the problem and, likely, failure to effectively solve it.

Consider the following case in point: A Juran client that has undertaken a Lean Six Sigma (LSS) deployment failed to do this "stop doing" exercise as fully as they should have. As a result, a couple of the first-wave projects devolved into turf battles over whose solution to or analysis of the problem was best. Since the other "non–Six Sigma" projects were initiated by the process owners before the LSS deployment, the Six Sigma teams ultimately had to abandon their projects for lack of implementation support and be assigned another project to complete their training requirements. These events led to tarnishing the reputation of the LSS deployment and reduced its chances for ultimate success. One key factor leading to this outcome was lack of involvement and buy-in at the highest levels of the organization.

Prepare

During this second phase of the deployment, upper management begins to define the support infrastructure. The steering team or teams are established. A team should be designated at the corporate level to oversee the wide deployment, and ultimately at divisional and even unit levels to oversee the deployment regionally and locally. For a graphic depiction of this arrangement, see Fig. 15.6.

In addition to the steering teams being developed, this is the time to select and train the initial group of project Champions. These people then can participate with the steering team in nominating the initial projects to be undertaken (see the discussion on nominating projects in the "Select the Problem" section later in the chapter).

The next step in this phase is very important: training the initial wave of Belts. There are a number of ways to approach this, including

- Only a Green Belt workshop at this phase and then Black Belt training during the expand phase.
- A combined Green Belt and Black Belt workshop with the whole class attending 10 days of training over a 4-week period and the Black Belt participants attending the full 20 days over that period. See Fig. 15.7 for a sample schedule for this method.
- Two workshops: a 10-day Green Belt and a 20-day Black Belt. This method is preferable if there are sufficient candidates to fill both workshops.

In any case, there should be 4 to 5 weeks between training weeks for the students to complete project work. By using this approach, within 1 month or 2 months of the completion of training, the first projects should be complete.

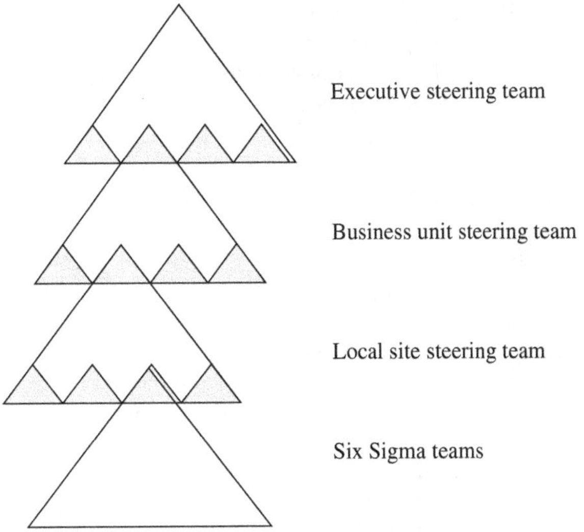

Figure 15.6 Linking steering teams together. (Juran Institute, Inc.)

	Week 1 define and team skills	Week 2 measure	Week 3 analyze	Week 4 improve/ control
Day 1	GB/BB	GB/BB	GB/BB	GB/BB
Day 2	GB/BB	GB/BB	GB/BB	GB/BB
Day 3	BB Only	GB/BB	GB/BB	BB Only
Day 4	BB Only	BB Only	BB Only	BB Only
Day 5	BB Only	BB Only	BB Only	BB Only

FIGURE 15.7 Combined GB/BB training schedule. (Juran Institute, Inc.)

Launch

Due to the practicum period between weeks of training noted above, the launch phase overlaps the prepare phase in some aspects. The primary effort during this phase is the execution of the first wave of projects, including mentoring of the Green Belts and Black Belts by the designated coaches. Those coaches should ideally come from the designated training partner identified during the decide phase.

Also during this phase, the ongoing project selection method should be institutionalized and the ongoing COPQ metric should be developed as a data source to help with future project selections.

A reward and recognition program should be established at this time as well. Monetary rewards, sometimes a share of the savings from a project, are often but not universally used. Intangible rewards, such as desirable career pathing, recognition, and pride in a job well done, can be effective instead of or in addition to any monetary rewards.

This is the time the organization should also decide how the improvement projects would be tracked and measured. There are a number of commercially available products, such as Power Steering, Minitab's Quality Companion, and i-nexus, which are widely used to track Six Sigma deployments. Internally developed solutions based on Microsoft Access or SharePoint are also widely used. Effective applications include a project "hopper" or "pipeline" for nominating future projects to be considered by the steering teams for execution.

Expand

This phase includes what the name implies: expansion of the methodology to other divisions, additional (often deeper) levels of the organization, and additional project methodologies (e.g., value stream improvement, DFSS). Expansion to different types of processes that may not have been considered during the initial wave of projects will also begin during this phase. This would include key macro business processes, for example, order processing, strategic planning, and price setting (see Fig. 15.8).

Benchmarking other divisions, companies, and industries for best practices in Six Sigma implementation is also a key activity of this deployment phase. This is aimed at bringing your deployment to a world-class level.

During this phase, companies will often decide to begin developing their own internal resources to conduct the workshops and coach projects going forward. The same training partner the organization has employed thus far usually facilitates the development of Master Black Belts and Lean Masters during this phase.

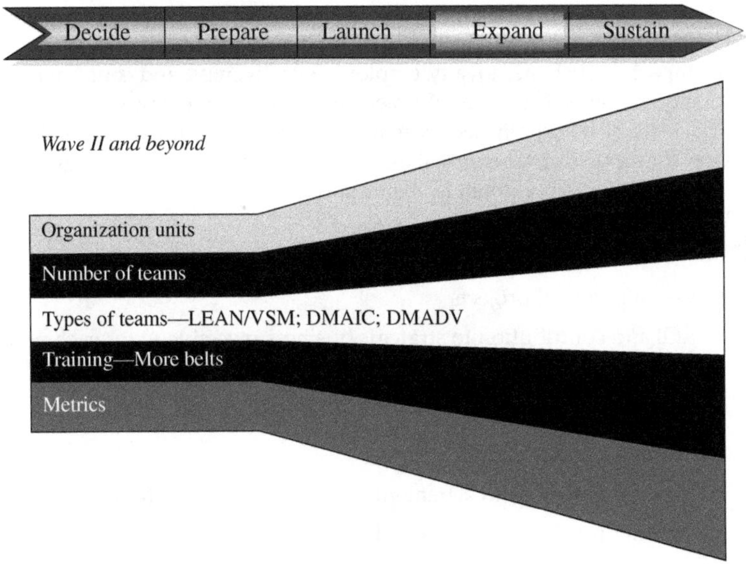

FIGURE 15.8 Areas of expansion. (Juran Institute, Inc.)

Sustain

The sustain phase is intended to solidify the results and methods implemented during the prior phases of deployment. This is the time when the deployment becomes ingrained in the corporate culture. The Six Sigma goals are integral to the organization's yearly strategy deployment and are widely deployed to all business units.

All employees should now have access to training in at least basic Six Sigma principles and tools and be empowered to improve quality in the workplace every day. Six Sigma should be a way of corporate life.

A key sustaining feature is the conducting of audits on a regular basis. Management should review the audit results and take action on any gaps identified. Often, these gaps will point to additional projects that should be undertaken. The audits may also point to gaps in adoption of the process by the corporate culture. The best way to address these remaining gaps is to up the level of involvement and extend it to all levels (if it hasn't already been) from the top of the organization to the bottom. Ultimately, there may be some who just refuse to get on board. In these cases, the ultimate decision may be that they should find employment elsewhere.

The Six Sigma DMAIC Steps

Experience with applying the five DMAIC steps shows that the team's DMAIC journey needs to be preceded by management selecting the project. The DMAIC journey with its five steps and some of the critical activities in each step are shown in Fig. 15.2. Here is a brief explanation of each step.

Select the Opportunity

In the select phase, potential projects are identified. Nominations can come from various sources, including customers, reports, and employees. To avoid suboptimization, management has to evaluate and select the projects. While evaluation criteria for project selection are many, the most frequent basis should be the COPQ at the organization or division level. Other criteria include impact on customer loyalty, employee effectiveness, and conformance with regulatory or other requirements. The project problem and goal statements are prepared and included in a team charter, which is confirmed by management. Management selects the most appropriate personnel for the project, assures that they are properly trained, and assigns the necessary priority. Project progress is monitored to ensure success.

Select: Deliverables

- List of potential projects
- ROI and contribution to strategic business objective(s) for each potential project
- List of potential projects
- Evaluation of projects
- Selected projects
- Project problem, goal statements, and a team charter for each project
- Formal project team(s) headed by Black Belt

Select: Questions to Be Answered

1. What customer-related issues confront us?
2. What mysterious, costly quality problems do we have that should be solved?
3. What are the likely benefits to be reaped by solving each of these problems?
4. Which of problems deserves to be tackled first, second, etc.?
5. What formal problem statement and goal statement should we assign to each project team?
6. Who should be the project team members and leader (Black Belt) for each project?

Define Phase

The define phase completes the project definition begun with the charter developed during selection. The team confirms the problem, goal, and scope of the project. The completed definition includes the following:

- Identify key customers related to the project
- Determine customer needs with respect to the project in the voice of the customer (VOC)
- Translate the VOC into CTQ requirement statements
- Define a high-level process flow to define the project limits

Define: Deliverables

- Confirmed project charter
- Voice of the customer
- CTQ statements
- A high-level flow, usually in the form of a supplier-input-process-output-customer (SIPOC) diagram

Define: Questions to Be Answered

1. Exactly what is the problem, in measurable terms?
2. What is the team's measurable goal?
3. What are the limits of the project? What is in and what is out of scope?
4. What resources are available—team members, time, finances—to accomplish the project?
5. Who are the customers related to this project?
6. What are their needs and how do we measure them in practical terms?

Measure Phase

The project team begins process characterization by measuring baseline performance (and problems) and documenting the process as follows:

- Understand and map the process in detail
- Measure baseline performance
- Map and measure the process creating the problem
- Plan for data collection
- Measure key product characteristics (outputs; Ys) and process parameters (inputs; Xs)
- Measure key customer requirements (CTQs)
- Measure potential failure modes
- Measure the capability of the measurement system
- Measure the short-term capability of the process

Map the Process

Focusing on the vital one (or few) outputs (Ys) identified by the Pareto analysis, graphically depict the process that creates it (them) by mapping the process with a flow diagram in order to understand the process anatomy.

Determine Baseline Performance

Measure the actual performance (outputs; Ys), such as costs of poor quality, number of defects, and cycle times of the process(es), which creates the problem to discover—by Pareto analysis—which vital outputs (Ys) make the greatest contribution to the problem.

Measure Potential Failure Modes

Referring to the analyzed process flow diagram for each process step, perform a failure mode and effect analysis (FMEA) by listing potential process defects (Ys) that could occur, their effects and their potential causes (Xs). (An additional source of ideas of possible Xs is the cause-effect diagram, which displays brainstormed possible causes for a given effect.) In addition, rate the severity of each effect, the likelihood of its occurrence, and the likelihood of its being detected should it occur. Upon completing the analysis, you will be able to identify those potential process failures that have the most risk associated with them. These results are used to further focus the project on those variables most in need of improvement.

Plan Data Collection for Short-Term Capability Study

- In preparation for determining the capability of the measurement system (which measures the Xs and Ys upon which the project team has focused) and the short-term capability of the process, create a sampling and data collection plan.
- In preparation for determining the short-term capability of the process, determine the capability of the measurement system to provide consistently accurate and precise data upon which the project team can depend to "tell the truth" about the process.
- If the measurement system is found to be not capable, take corrective action to make it so.
- If the measurement system is found to be capable, proceed with the next step—determining if the process is in statistical control with respect to given variables (Ys).

Measure the Short-Term Capability of the Process

- In preparation for measuring short-term capability of the process to meet given specifications (Ys), ascertain whether the process is in statistical control with respect to the given output (Y) of interest. A good way to measure process stability is to use a control chart to plot the process data and discover any indications of instability.
- If the process is not in statistical control—that is, if control charts detect special causes of variation in the process—take action to remove the special causes of variation before proceeding with the process baseline performance measurement.
- If the process is in statistical control—that is, the control charts do not detect special causes of variation in the process—perform a short-term capability study to provide baseline data of the ability of the process to consistently produce a given output (Y).

Confirm or Modify the Goal

In light of discoveries made during measurement of the current process performance, determine if the problem and goal statements are still appropriate for this project.

- Evaluate the project's problem statement and goal statement.
  - Does the problem statement and the goal statement meet the criteria of an effective problem and goal statement with clearly defined boundaries?
  - Are the same variables and units of measure found in the problem statement also found in the goal statement?
  - Can the project be handled by a single team?
  - Does it avoid unnecessary constraints but still specify clearly any necessary global constraints, such as organizational strategy?
  - Are there any points that need clarification or modification?
  - Are the team members representative of departments, divisions, or work units affected by the project? The detailed process flow diagram, particularly if it is constructed in "swim lane" fashion, can help with this.
- Verify that the problem truly exists. If the problem has not been accurately measured, the team must do so at this point.

- Validate project goal(s). Verify that the basis for the project goal(s) is (are) one or more of the following:
  - Technology
  - Market
  - Benchmarking
  - History
- Modify the problem statement and goal statement if either does not meet the criteria above.
- Obtain confirmation from the leadership team, Champion, Black Belt, or quality council on any necessary changes to the project goal or to team membership.
- Create a glossary (list of operational definitions) for your project that will serve as a "dictionary" for important terms relating to your project. Select a team member to act as glossary chief with the responsibility of maintaining the project glossary.

List Theories of Root Cause Based on the Process Flows and Measures

The team needs to develop a comprehensive and creative list of theories of root cause. A root cause is a factor that affects the outcome, would eliminate or reduce the problem if it were removed or mitigated. Tools typically used include Cause-Effect (fish bone or Ishikawa) diagrams, FMEA, and fault tree analysis.

Measure: Deliverables

- Baseline performance metrics describing outputs (Ys)
- Process flow diagram; key process input variables; key process output variables; cause-effect diagram; potential failure mode and effect analysis (FMEA) (to get clues to possible causes [Xs] of the defective outputs [Ys])
- Data collection plan, including sampling plan
- Gage reproducibility and repeatability or attribute measurement system analysis (to measure the capability of the measurement system itself)
- Capability measurement in terms of defect rates, capability indexes, and/or Sigma levels
- Confirmed or modified project goal
- Prioritized list of theories of cause based on cause-effect analysis, FMEA, or similar tools

Measure: Questions to Be Answered

1. How well is the current process performing with respect to the specific Ys (outputs) identified to Pareto analyses?
2. What data do we need to obtain in order to assess the capability of (a) the measurement system(s) and (b) the production process(es)?
3. What is the capability of the measurement system(s)?
4. Is the process in statistical control?

5. What is the capability of the process(es)?
6. Does the project goal need to be modified?
7. What are all the possible root causes for the problem?

Analyze Phase

In the analyze phase, the project team analyzes past and current performance data. Key information questions are answered through this analysis. Hypotheses on possible cause-effect relationships are developed and tested. Appropriate statistical tools and techniques are used: histograms, box plots, other exploratory graphical analysis, correlation and regression, hypothesis testing, contingency tables, analysis of variance (ANOVA), and other graphical and statistical tests may be used. In this way, the team confirms the determinants of process performance (i.e., the key or "vital few" inputs that affect response variable[s] of interest are identified). It is possible that the team may not have to carry out designed experiments (DOEs) in the next (Improve) phase if the exact cause-effect relationships can be established by analyzing past and current performance data.

Procedure to analyze response variables (outputs, Ys) and input variables (Xs):

- Perform graphical analysis using tools such as histograms, box plots, and Pareto analysis.
- Visually narrow the list of important categorically discrete input variables (Xs).
- Learn the effects of categorically discrete inputs (Xs) on variable outputs (Ys) and display the effects graphically.
- Perform correlation and regression to
  - Narrow the list of important continuous input variables (Xs) specifically to learn the "strength of association" between a specific variable input (Xs) and a specific variable output (Ys).
- Calculate confidence intervals to
  - Learn the range of values that, with a given probability, include the true value of our estimated population's parameter, which has been calculated from a sample (e.g., the population's center and/or spread).
  - Analyze relationships between specific Ys and Xs, to prove cause-effect relationships.
  - Confirm the vital few determinants (Xs) of process performance (Ys).
- Perform hypothesis testing using continuous variables data to
  - Answer the question, Is our population actual standard deviation the same as or different from its target standard deviation? Perform 1 variance test.
  - Answer the question, Is our population actual mean the same as or different from its target mean? Perform 1-sample *t*-tests.
  - Answer the questions, Is our population mean the same or different after a given treatment as it was before the treatment? or Is the average response at level 1 of the X factor the same or different as it is at level 2 of that factor? Perform 2 sample *t*-tests, or if there is a natural pairing of the response variable, paired *t*-tests.
  - Answer the question, Are several (>2) means the same or different? Perform analysis of variance.

Note: The above tests are referred to as parametric tests because they assume normally distributed response data and, in the case of ANOVA, equality of variances across all levels of the factor. For a discussion of nonparametric (also referred to as "distribution free") tests to use when assumptions of normality and or equality of variances are violated.

Perform hypothesis testing using attribute data to

- Answer the question, Is the proportion of some factor (e.g., defectives) in our sample the same or different from the target proportion? Perform a Minitab test and calculation of confidence interval for one proportion.
- Answer the question, Is proportion 1 the same or different from proportion 2? Perform the binomial proportions test and calculation of confidence interval for two proportions.
- Answer the question, Is a given output (Y) independent of or dependent on a particular input (X)? (This involves testing the theory that a given X is an important causal factor that should be included in our list of vital few Xs.) Perform a chi-squared test of independence (also called a contingency table).

Analyze: Deliverables

- Histograms, box plots, scatter diagrams, Pareto analysis, correlation and regression analyses (to analyze relationships between response variables [Ys] and potential causes [Xs])
- Results of hypothesis testing (to establish relationships between response variables [Ys] and input variables [Xs])
- List of vital few process inputs (Xs) that are proven root causes of the observed problem

Analyze: Questions to Be Answered

1. What patterns, if any, are demonstrated by current process outputs (Ys) of interest to the project team?
   - Analyze response variables (outputs; Ys).
   - Analyze input variables (Xs).
   - Analyze relationships between specific Ys and Xs, identifying cause-effect relationships.
2. What are the key determinants of process performance (vital few Xs)?
3. What process inputs (Xs) seem to determine each of the outputs (Ys)?
4. What are the vital few Xs on which the project team should focus?

Improve Phase

In the improve phase, the project team seeks to quantify the cause-effect relationship (mathematical relationship between input variables and the response variable of interest) so that process performance can be predicted, improved, and optimized. The team may utilize DOEs if applicable to the particular project. Screening experiments (fractional factorial designs) are used to identify the critical or "vital few" causes or determinants. A mathematical model of process performance is then established using 2k factorial experiments. If necessary, full factorial experiments are carried out. The operational range of input or process parameter settings is then determined. The team can further fine-tune or optimize process performance by using such techniques as response surface methods (RSM) and evolutionary operation (EVOP). Procedures to define, design, and implement improvements include

1. Plan designed experiments
2. Conduct screening experiments to identify the critical, vital few process determinants (Xs)
3. Conduct designed experiments to establish a mathematic model of process performance
4. Optimize process performance
5. Evaluate alternative improvements
6. Design the improvement

Plan Designed Experiments

- Learn about DOEs in preparation for planning and carrying out experiments to improve the "problem" process.
- Design in detail the experiments required by the project.

Conduct Fractional Factorial Screening Experiments

- Perform fractional factorial screening experiments to reduce even further the list of input variables to the vital few that strongly contribute to the outputs of interest. (A relatively large number of factors [Xs] are examined at only two levels in a relatively small number of runs.)

Conduct Further Experiments, If Necessary, to Develop Mathematical Model and Optimize Performance

- Perform 2k factorial experiments. Multiple factors (Xs, identified by screening experiments) are examined at only two levels to obtain information economically with relatively few experimental runs. Constructing equations that predict the effect on output Y of a given causal factor X discovers precise mathematical relationships between Xs and Ys. In addition, not only are the critical factors (X) identified, but also the level at which each factor performs the best and any significant interactions among the factors.
- If necessary, perform full factorial experiments. More information than is provided by 2k factorial experiments may be required. A full factorial experiment produces the same type of information as a 2k factorial does, but does so by examining multiple factors (Xs) at multiple levels.
- If necessary, and in addition, utilize RSM and/or EVOPs techniques to further assist in determining optimal process parameters.
- Using results of experiments derive mathematical models of the process and establish optimal settings for process parameters (Xs) to achieve desired (Ys).

Evaluate Alternatives and Choose Optimal Improvements

- Identify a broad range of possible improvements.
- Agree on criteria against which to evaluate the improvements and on the relative weight each criterion will have. The following criteria are commonly used:
  - Total cost
  - Impact on the problem
  - Benefit-cost relationship
  - Cultural impact or resistance to change
  - Implementation time
  - Risk
  - Health, safety, and the environment
- Evaluate the improvements using agreed-upon criteria
- Agree on the most suitable improvements

Design the Improvements

- Evaluate the improvements against the project goal.
- Verify that it will meet project goals.
- Identify the following customers:
  - Those who will create part of the improvements
  - Those who will operate the revised process
  - Those served by the improvements
- Determine customer needs with respect to the improvements.
- Determine the following required resources: people, money, time, and materials.
- Specify the procedures and other changes required.
- Assess human resource requirements, especially training.
- Verify that the design of the improvement meets customer needs.
- Plan to deal with any cultural resistance to change.

Improve: Deliverables

- Plan for designed experiments
- Reduced list of vital few inputs (Xs)
- Mathematical prediction model(s)
- Established process parameter settings
- Designed improvements
- Implementation plan
- Plans to deal with cultural resistance

Improve: Questions to Be Answered

1. What specific experiments should be conducted to arrive ultimately at the discovery of what the optional process parameter settings should be?

2. What are the vital few inputs (Xs, narrowed down still further by experimentation) that have the greatest impact on the outputs (Ys) of interest?
3. What is the mathematical model that describes and predicts relationships between specific Xs and Ys?
4. What are the ideal (optimal) process parameter settings for the process to produce output(s) at Six Sigma levels?
5. Have improvements been considered and selected that will address each of the vital few Xs proven during the analyze phase?
6. Has expected cultural resistance to change been evaluated and plans made to overcome it?
7. Has a pilot plan been developed and executed and the solutions appropriately adjusted based on the results?
8. Have all solutions been fully implemented along with required training, procedural changes, and revisions to tools and processes?

Control Phase

The project team designs and documents the necessary controls to ensure that gains from the improvement effort can be sustained once the changes are implemented. Sound quality principles and techniques are used, including the concepts of self-control and dominance, the feedback loop, mistake proofing, and statistical process control. Process documentations are updated (e.g., the failure mode and effects analysis), and process control plans are developed. Standard operating procedures (SOP) and work instructions are revised accordingly. The measurement system is validated, and the improved process capability is established. Implementation is monitored, and process performance is audited over a period to ensure that the gains are held. The project team reports the goal accomplished to management, and upon approval, turns the process totally over to the operating forces and disbands.

The activities required to complete the control step include

1. Design controls and document the improved process
2. Design for culture
3. Validate the measurement system
4. Establish the process capability
5. Implement and monitor

Design Controls and Document Improved Process

- Update FMEA to ensure that no necessary controls have been overlooked.
- Mistake-proof the improvement(s), if possible.
  - Identify the kind(s) of tactic(s) that can be incorporated into the improvements to make it mistake proof. Some options include
    - Designing systems to reduce the likelihood of error
    - Using technology rather than human sensing
    - Using active rather than passive checking
    - Keeping feedback loops as short as possible
    - Designing and incorporating the specific steps to mistake-proof as part of the improvements

- Design process quality controls to ensure that your improved levels of inputs (Xs) and outputs (Ys) are achieved continuously. Place all persons who will have roles in your improved process into a state of self-control to ensure that they have all the means necessary to be continuously successful.
- Provide the means to measure the results of the new process
  - Control subjects
  - Output measures (Ys)
  - Input measures and process variables (Xs)
  - Establish the control standard for each control subject
  - Base each control standard on the actual performance of the new process
- Determine how actual performance will be compared to the standard.
  - Statistical process control
- Design actions to regulate performance if it does not meet the standard. Use a control spreadsheet to develop an action plan for each control subject.
- Establish self-control for individuals so
  - They know exactly what is expected (product standards and process standards).
  - They know their actual performance (timely feedback).
  - They are able to regulate the process because they have
    - A capable process.
    - The necessary materials, tools, skills, and knowledge.
    - The authority to adjust the process.

Design for Culture to Minimize or Overcome Resistance

- Identify the likely sources of resistance (barriers) and supports (aids). Resistance typically arises because of
  - Fear of the unknown
  - Unwillingness to change customary routines
  - The need to acquire new skills
  - Unwillingness to adopt a remedy "not invented here"
  - Failure to recognize that a problem exists
  - Failure of previous solutions
  - Expense
- Rate the barriers and aids according to their perceived strengths
- Identify the countermeasures needed to overcome the barriers. Consider
  - Providing participation
  - Providing enough time
  - Keeping proposals free of excess baggage
  - Treating employees with dignity
  - Reversing positions to better understand the impact on the culture
  - Dealing with resistance seriously and directly
- Install statistical process control (SPC) where necessary to ensure that your process remains stable and predictable, and runs in the most economic manner.
- Consider introducing 6s standards to make the workplace function smoothly with maximum value-added activity and minimum non-value-added activity.

Validate Measurement System

Utilize commercially available software such as Minitab to evaluate measurement system capability (as in the measure phase) to ensure that the measurements utilized to evaluate control subjects can be depended on to tell the truth.

Establish Process Capability

- Prove the effectiveness of the new, improved process to ensure that the new controls work and to discover if your original problem has improved, and ensure that no new problems have inadvertently been created by your improvement(s).
- Decide how the improvements will be tested
  - Agree on the type of test(s)
  - Decide when, how long, and who will conduct the test(s)
  - Prepare a test plan for each improvement
  - Identify limitations of the test(s)
    - Develop an approach to deal with limitations.
    - Conduct the test.
    - Measure results.
    - Adjust the improvements if results are not satisfactory.
    - Retest, measure, and adjust until satisfied that the improved process will work under operating conditions.
- Utilizing control charts, ensure that the new process is in statistical control with respect to each individual control subject. If not, improve the process further until it is.
- When, and only when, the process is in statistical control, utilize Capability Analysis—as in the measure phase—to determine process capability for each individual control subject.

Implement the Controls and Monitor

- Transfer to the operating forces all the updated control plans, etc., and train the people involved in the process in the new procedure.
- Develop a plan for transferring the control plan to the operating forces. The plan for transferring should indicate:
  - How, when, and where the improvements will be implemented?
  - Why the changes are necessary and what they will achieve?
  - The detailed steps to be followed in the implementation.
- Involve those affected by the change in the planning and implementation.
- Coordinate changes with the leadership team, Black Belt, Champion, executive council, and the affected managers.
- Ensure preparations are completed before implementation, including
  - Written procedures
  - Training
  - Equipment, materials, and supplies
  - Staffing changes
  - Changes in assignments and responsibilities
  - Monitoring the results

- Periodically audit the process, and also the new controls, to ensure that the gains are being held
- Integrate controls with a balanced scorecard
- Develop systems for reporting results
- When developing systems for reporting results, determine
- What measures will be reported?
- How frequently?
- To whom (should be a level of management prepared to monitor progress and respond if gains are not held)?
- Document the controls

When documenting the controls, indicate

- The control standard
- Measurements of the process
- Feedback loop responsibilities (who does what if controls are defective)
- After a suitable period, transfer the audit function to the operating forces and disband the team (with appropriate celebrations and recognition)

Control: Deliverables

- Updated FMEA, process control plans, and standard operating procedures
- Validated capable measurement system(s)
- Production process in statistical control and able to get as close to Six Sigma levels as is optimally achievable, at a minimum accomplishing the project goal
- Updated project documentation, final project reports, and periodic audits to monitor success and hold the gains

Control: Questions to Be Answered

- What should be the plan to ensure the process remains in statistical control and produces defects only at or near Six Sigma levels?
- Is our measurement system capable of providing accurate and precise data with which to manage the process?
- Is our new process capable of meeting the established process performance goal?
- How do we ensure that all people who have a role in the process are in a state of self-control (have all the means to be successful on the job)?
- What standard procedures should be in place, and followed, to hold the gains?

Training and Certification of Belts

The introduction of Six Sigma in the past decade led to a surge in the certification of Belts. This was largely due to a lesson learned from the Total Quality Management (TQM) era. During TQM, many so-called experts were trained in the "methods of TQM." Unfortunately, few were trained in the tools to collect and analyze data. As a result, numerous organizations did not benefit from the TQM program.

Motorola introduced a core curriculum that all Six Sigma practitioners needed to learn. That evolved into a certification program that went beyond the borders of Motorola. As a result, there are many "certifiers" that will provide a certification as a Master Black Belt, Black Belt, Green Belt, and so on. Most certifications state that the person certified is an "expert" in the skills of Six Sigma or Lean or both. Certification did lead to improved performance, but also to some weak experts due to no oversight of the certifiers, many of which were consulting companies or universities not well versed in the methods or tools of Six Sigma and Lean.

The American Society for Quality (ASQ) for many years offered certification for quality technicians, quality auditors, quality engineers, and quality managers. As the Six Sigma movement grew, the ASQ and its affiliates around the world prepare a minimum body of knowledge for various belt levels. The ASQ provides a widely-accepted standard that any practitioner should at a minimum master. Certification must be based on legitimacy to be effective.

ASQ's Certified Quality Engineer (CQE) program is for people who want to understand the principles of product and service quality evaluation and control (ASQ, 2009). For a detailed list of the CQE body of knowledge, the reader is referred to the certification requirements for Certified Quality Engineer at www.asq.org.

ASQ also offers a certification for quality officers at the quality management level, called Certified Manager of Quality/Organizational Excellence. ASQ views the Certified Manager of Quality/Organizational Excellence as "a professional who leads and champions process-improvement initiatives—everywhere from small businesses to multinational corporations—that can have regional or global focus in a variety of service and industrial settings. A Certified Manager of Quality/Organizational Excellence facilitates and leads team efforts to establish and monitor customer/supplier relations, supports strategic planning and deployment initiatives, and helps develop measurement systems to determine organizational improvement. The Certified Manager of Quality/Organizational Excellence should be able to motivate and evaluate staff, manage projects and human resources, analyze financial situations, determine and evaluate risk, and employ knowledge management tools and techniques in resolving organizational challenges" (ASQ, 2009).

Note: No matter what organization you use to certify your experts, here are some lessons learned about certification:

- One project is not enough to make someone an expert
- Passing a written test that is not proctored is no guarantee the person who is supposed to be taking the test is actually taking it
- If you get someone in your organization to sign off on the success of the Belt project, you need independent evidence that the person is knowledgeable about the methods of Six Sigma
- Select a reputable certifying body

References

de Ansorena, R. (2009). "How Telefonica Makes Its Management Connections." *European Quality*, Vol. 8, No. 6, pp. 4–10.

Welch, J. W. (2009). *Winning: The Ultimate Business How-To Book*. Kindle edition. HarperCollins e-books.

CHAPTER 16

Root Cause Analysis Methods

Joseph A. De Feo and Kevin Caldwell

High Points of This Chapter

1. Juran's RCCA approach is an effective method for identifying and addressing the root cause of sporadic problems in products and processes. It follows four basic steps: identify a problem, diagnose the cause, remedy the cause, and hold the gains.
2. Root cause analysis is one of the most critical skills all employees need to understand. The heart of this is the "cause and effect" relationships that occur in all organizations.
3. Plan-do-study-act (PDSA) and plan-do-check-act (PDCA) are similar and useful methods for root cause analysis, planning, and executing tests of potentially beneficial changes. They are used in services and health care organizations.
4. "Just do it" can be a useful approach when the need to solve a problem is urgent, the penalties for risks of failed "solutions" are low, and the rewards of effective (experience-based) solutions are high.

Introduction

The heart of all organizational problem solving is to understand enough of the causes of the problem and finding an acceptable solution to it. Finding the "root cause or causes" of an organizational problem is the single most important determinant of success or failure of any problem-solving method. This is true of all types of problems: chronic problems that have been resistant to solution, day-to-day sporadic problems that occur infrequently but have a tendency to recur, problems that involve identifying and eliminating waste—all require effective root cause analysis and identification to reduce the risks and resistance associated with changing a process.

Why Do We Need to Know about Root Causes of Problems?

Though effective root cause analysis is at the heart of all problem-solving methods, such as Six Sigma, Lean, RCCA, and Plan, Do, Study/Check Act, its purpose is to solve problems that occur due to special or assignable causes. A special cause is one that occurs during daily operations. It is sporadic because it happens periodically, but it can cause havoc on the organization if it persists. These problems differ from the larger, chronic problems that quality improvement programs such as Six Sigma focus on. For more on root cause analysis used in solving the other types of problems noted earlier, see Chap. 15, Six Sigma: Breakthrough to in-Process Effectiveness. A comparison of the different methods to discover root causes and their uses is shown in Table 16.1.

Method	Purpose	Risk	Benefits	Level of Difficulty
Six Sigma DMAIC	Solve large, chronic, multifunctional problems	Low	High ROI (25:1)	High: Large scope problems require difficult diagnosis and expert skills
Juran's Breakthrough Model	Solve large, chronic, multifunctional problems	Low	High ROI (25:1)	High: Large scope problems require difficult diagnosis and expert skills
RCCA	Solve sporadic day-to-day problems	Low	Moderate ROI (5:1)	Low: Sporadic problems require finding out what changed; skills easy to gain by all staff
PDCA	Solve sporadic day-to-day problems	Low	Moderate ROI (5:1)	Low: Sporadic problems require finding out what changed; skills easy to gain by all staff
Lean Problem-Solving	Solve sporadic day-to-day problems	Med	Moderate ROI (1:1)	Low: Purpose is to identify waste and its causes, which tend to be well understood
PDSA	Solve sporadic day-to-day problems	Med	Moderate ROI (1:1)	Easy: Many services do not use tools to analyze data; rather, they move from symptom to solution
Just Do It	Make daily decisions based on what is already known	High	Moderate ROI (0:0)	Easy: Since this is not recommended, it is easy to do; there are no methods other than instinct

TABLE 16.1 Purpose of Root Cause Corrective Action versus Other Methods

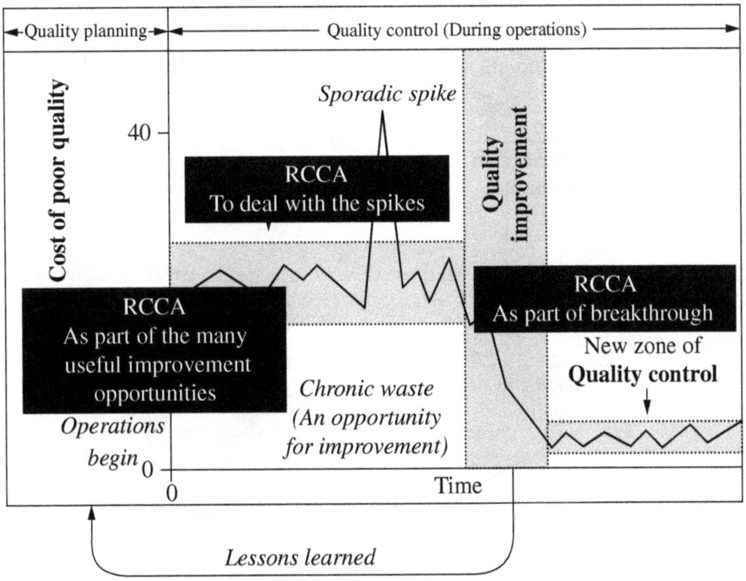

FIGURE 16.1 RCCA and the Juran Trilogy.

When to Apply Root Cause Analysis

The Juran Trilogy has been discussed in detail in Chap. 1, Universal Principles of Quality Management, but the question remains: At what stage of the trilogy does basic RCCA apply? To answer this question, refer to Fig. 16.1.

As seen in the figure, the application of the basic RCCA process is a control activity: restoration of the process performance to a previously acceptable level. Although control also has elements of the broad sense of continuous improvement, for the purposes of the present discussion, we will apply the definitions of the trilogy strictly; that is, control is restoring process performance to a previously acceptable level. Improvement, on the other hand, is defined as changing the very nature of the process, creating breakthrough and moving to a new and better level of performance with reduced waste and cost of poor quality. The former primarily deals with special causes, the latter with common causes of variation.

Juran's RCCA

Juran's four-step process for RCCA is an outgrowth of the work of Dr. J. M. Juran in which he described the universal process for quality improvement, shown in Fig. 16.2. This universal, as Dr. Juran stated in earlier chapters is the third universal of "control—the process for preventing adverse change." To ensure that all processes are in a state of control requires three basic elements:

1. The means to know the actual performance of the process
2. The ability to compare the actual performance to the targets or quality goals
3. The means to act on the difference to maintain control

The third step requires a means and a method to determine what correct action should be taken. There have been many versions to act on the difference for centuries. Walter Shewhart

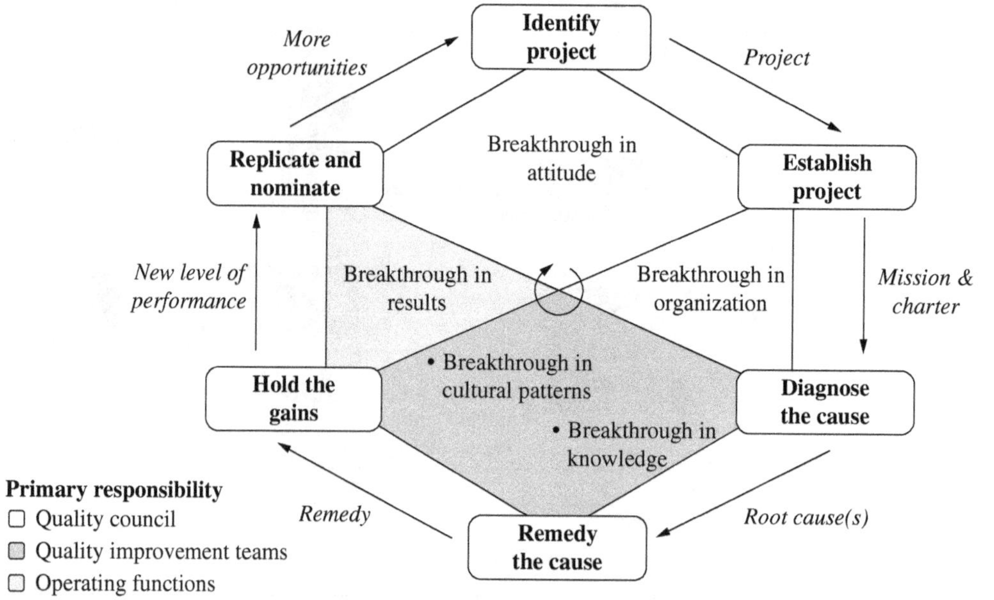

Figure 16.2 The six major steps of problem solving.

coined the term PDCA (Plan, Do, Check, Act) as a means to set up the control functions. Practitioners of PDCA still need to know how to perform the action. While there are many tools to aid in root cause analysis, a simple method is needed to solve daily, sporadic, small-scope problems.

The Juran RCCA method described here is a simplification of the universal process for improvement described by Dr. Juran, and consists of four steps:

1. *Define.* Identify a change in performance.
2. *Analyze.* Diagnose the cause.
3. *Improve.* Remedy the cause.
4. *Control.* Hold the gains.

"Quality control can be defined as the maintenance or restoration of the operating status quo as measured by [meeting] the acceptable level of defects and provision of customer needs" (Monroe, 2009). The mechanism of controlling quality is depicted in Fig. 16.3. The troubleshooting portion of the control feedback loop is where RCCA is needed. When a measurement of a control subject is outside the established standard of acceptability, some means for identifying the cause is needed. Once the root cause or causes are identified, a remedy must be put in place that will eliminate them. After the cause is eliminated, the control feedback loop continues to monitor the process so the cause and problem do not recur.

The four-step RCCA approach described above has several sub steps that must be undertaken to effectively diagnose and remedy the cause:

1. Define the problem
   - Identify frequency of the problem: sporadic or chronic (if the latter, apply breakthrough methods)

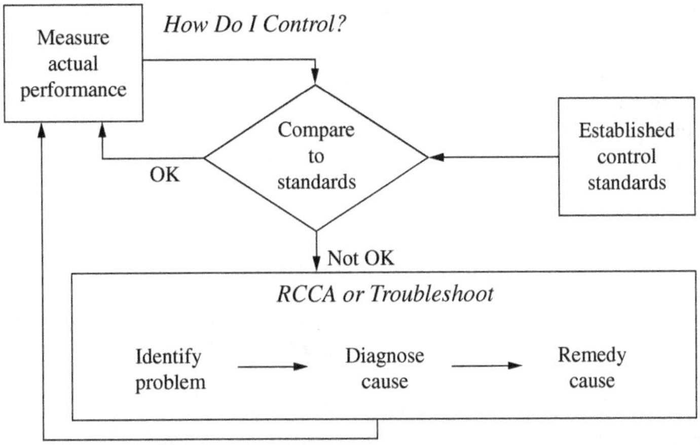

FIGURE 16.3 The control feedback loop.

- Establish responsibility to solve it, if it is not already established in a control plan
- Prepare a problem statement

2. Analyze and diagnose the cause
   - Analyze symptoms
   - Formulate theories
   - Test theories
   - Identify root cause(s)
3. Improve and remedy the cause
   - Design and implement the remedy
4. Control to hold the gains
   - Adjust controls

Each of the steps and sub steps of Juran's RCCA approach is discussed in more detail in the following sections.

The Medical Analogy

The Juran RCCA approach is analogous to the approach a physician takes in treating an ill patient. First, the doctor will want to understand what is wrong: What's the problem? Without a clear understanding of the problem, it will be impossible to solve.

Next, the doctor will want to know more about the outward evidence that the problem exists: the symptoms. He might take the patient's temperature, ask what kind of discomfort the patient is experiencing, look into the patient's throat and ears, and so on.

Based on the observed symptoms, the doctor will formulate tentative diagnoses—theories about what could be causing the patient's illness. At this point the doctor is still unsure what the true cause of the illness is, so he will order tests to determine which of his tentative diagnoses is true. Perhaps blood will be drawn from the patient for analysis; perhaps the patient will be given an MRI exam or other diagnostic tests.

Once the data about the possible causes of the illness have been gathered, the doctor is ready to settle on a final diagnosis based on the facts. Now, hopefully, the true root cause of the patient's illness is known and the doctor can apply an appropriate remedy. Perhaps the patient will be given medication, prescribed physical therapy, recommended to make certain lifestyle changes—whatever the appropriate remedy is to alleviate the proven cause of the illness.

Finally, the doctor might say, "Come back and see me in two weeks." This is the activity of holding the gains intended to ensure that the patient is continuing with the prescribed regimen and the remedy is effective: the patient is getting better.

Elements of Effective RCCA

These are the necessary elements for effective root cause corrective action:

- *A problem.* A problem is outward evidence that something is wrong and warrants a solution, for example, a visible performance deficiency in the output of an important design, manufacturing, service, or business processes. Time and resources are needed to analyze and solve problems.
- *Data and information.* We cannot solve the problem until we have the hard facts that prove what the root cause is. Without data, we are merely guessing at the causes of the problem, and our efforts to solve it will be hampered by our lack of knowledge. More importantly, we will create doubt and greater risk will be introduced into our system.
- *Tools.* When a problem arises, there are many questions that need answers. Those answers will come from data found within our processes. At times, we are often faced with a great deal of data, but little information or facts. We can use tools to help us organize and understand the data. They are invaluable aids to effective root cause analysis.
- *Structure.* A logical and structured approach is needed to guide the RCCA process. This structure becomes the "guide or boss," not the people trying to solve the problem. At a minimum, this structure needs to use and involve multiple functions to discover root causes. This structure will allow us to "torture the data until it confesses." Data contains information. We need a means to extract it.

An almost unlimited number of tools are available to the problem-solving team, but those most often used for basic RCCA are:

- *Affinity method and brainstorming.* Brainstorming is a quality tool intended to stimulate creativity. It is useful because it helps the team consider a full range of theories about possible causes. The affinity process helps organize those theories.
- *Cause and effect diagrams.* An effective way to organize and display the various theories of potential causes of a problem.
- *Data collection.* These methods are used to gather information about a quality problem. A typical data collection form is arranged for easy use and includes clear instructions.
- *Failure mode and effects analysis (FMEA).* This is a structured methodology for identifying potential failure modes in a process or design and assessing the risk associated with the failures. It helps identify the most likely possible causes and helps design a more robust remedy.
- *Graphs and charts.* These are pictorial representations of quantitative data. They can summarize large amounts of information in a small area and communicate complex situations concisely and clearly.
- *Histograms.* These are graphic summaries of variation in a set of data. The pictorial nature of the histogram enables us to see patterns that are difficult to see in a simple table of numbers.

- *Box plots.* Like histograms, box plots provide a graphic summary of the pattern of variation in a set of data. The box plot is especially useful when working with small sets of data or when comparing many different distributions.
- *Juran's Pareto analysis.* This is a ranked comparison of factors related to a quality problem. It helps identify and focus on the vital few factors.
- *Process control plans.* These plans summarize the plan of action for a process out of control. Their purpose is to document the actions necessary to bring the process back into control and assist the process owners in holding the gains achieved by the problem solving.
- *Scatter diagram.* This is a graphic presentation of the relationship between two variables. In root cause corrective action, scatter diagrams are usually used to explore cause-effect relationships in the diagnostic journey.
- *Stratification.* It is the separation of data into categories. Its most frequent use is during the diagnostic journey to identify which categories contribute the most to the problem being solved.

Phase 1: Define and Identify the Problem

It has been said that a problem well defined is half solved. The clear identification and definition of problems to be addressed in the RCCA project is an early key to success. In practice, a well-constructed control plan with effective feedback loops will identify problems to be addressed by RCCA nearly in real time. For further discussion of control activities, see the section "Phase 4: Control to Hold the Gains."

Select the Problem

Select the problem to be addressed. Once the data and information about potential problems to address have been gathered, tools must be applied to select the most important problems to address. Data collection and Juran's Pareto analysis are most often used to identify the vital few problems to address.

Once the problem for action has been selected, the nature of the problem must be stated clearly and concisely. A good problem statement should have the following characteristics, summarized by the acronym MOMS:

- *Measureable.* The problem must be stated in terms that can be measured, either by using an existing measurement system or creating a new one. Although the problem may not have been measured to date, the problem-solving team must be able to conceptualize how it could be measured in quantifiable terms.
- *Observable.* The problem must be seen and evidenced by its symptoms. Symptoms are the outward evidence that the problem exists.
- *Manageable.* The problem statement must be narrow enough in scope that the team can solve it with a reasonable application of resources over a reasonable period of time. "Boil the ocean" projects should be avoided.
- *Specific.* The problem statement should focus on specific products, services, or information; specific parts of the organization; or specific aspects of a larger problem.

In addition to the MOMS guidelines, problem statements should never include implications of a cause, blame for the problem occurring, or suggested solutions.

Is the Problem Sporadic or Chronic in Nature?

This is a fork in the road. If the problem involves a process that has gone out of control—for example, a fire has erupted and is burning—apply the RCCA process and tools discussed in this chapter to restore process control. If the problem is one that has been around for a while (chronic) and plaguing the operation with higher-than-tolerable COPQ, consider using the more sophisticated breakthrough improvement methods described in Chap. 14, Lean Techniques and the Shingo Prize and Chap. 15, Six Sigma Breakthrough to in-Process Effectiveness.

Tools most often used at this step are data collection and Juran's Pareto analysis.

Prepare a Goal Statement

Typically, the goal statement for an RCCA project is simple: eliminate the root cause or causes of the problem and restore control. In some cases, complete elimination may not be possible or practical; then the goal should be to reduce the impact of the causes so that the undesirable effects are minimized. In this case, the goal may be stated in terms of a percentage improvement, reduction in defect levels, etc.

Phase 2: Analyze and Diagnose the Cause

Analyze Symptoms

Analysis of symptoms is an important step in finding the root cause of the problem because this activity enables us to understand the current situation. How often is the problem occurring? How severe is it? What types of failures contribute most to the problem being analyzed? At what point in the process is the failure most often observed? These types of questions about the current situation must be answered to help us better understand where the root cause(s) may lie.

Think of it this way: If you were asked what route to take if driving to Cleveland, what would your response be? Typically, people will respond from their own frame of reference: "Well, go to route 224 east, and then take I-77 north ..." This misses an important point, however; the question didn't specify a starting point. Depending where one starts their journey to Cleveland, the route to get there will be entirely different. The point is that unless you know where you are starting from (the symptoms of the problem), it is difficult to map a route to where you want to be (achievement of the goal). This is why it is so important to do a thorough analysis of symptoms as a first step in diagnosing the cause. This analysis will be of great help when the team gets to the point of brainstorming possible causes and will result in a more thorough list of possible causes than would otherwise be achieved.

Tools that are often used at this step are data collection, process flow diagrams, Juran's Pareto analysis, and stratification.

Formulate Theories

A theory is simply an unproven statement of the cause of a certain condition. A student receiving a poor grade on an exam may tell his or her parents that the cause is that the teacher included material on the exam that was not discussed in class, but the parents may consider this only a theory. The parents may consider a number of other theories as well, such as the student did not read required chapters that explained the material or the student did not attend class every day. In the same way, when determining the cause of a quality problem, there must be speculation about its many possible causes. Jumping to conclusions before considering many theories and proving which one is correct could mean wasting time and resources on an inappropriate solution.

The formulation of theories follows a thought process moving from creative to empirical, divergent to convergent. Beginning with brainstorming, the team and any subject matter experts will attempt to identify as many causes as possible. Next, the team will organize these brainstormed theories into logical groups, probably using the affinity process. Finally, the group will begin to hone in on the most likely root causes by using cause and effect diagramming, FMEA, and possibly other prioritization tools.

These most likely theories of causes are the input for the next step of the diagnostic journey.

Prove Theories to Identify Root Cause(s)

"Before beginning to test theories, the team should be very clear on exactly which theories are being tested. A copy of the cause-effect diagram is an excellent guide for the team at this point. Diagram the theories that will be tested with a particular set of data. If the data demonstrate that a theory is not important, that theory can be crossed off as a possible cause. The cause-effect diagram also helps identify related theories that can be tested together. When the theories to be tested are stated clearly and precisely as they are understood, it is time to plan for collecting data to test them" (Juran Institute, Inc., 2008).

Assessing data that have been collected to answer questions regarding the truth or falsity of a given theory tests theories. The theory is assumed to be false unless the data indicate it to be otherwise. Once the data have been collected, appropriate analysis tools must be applied to convert the data into information. Information then becomes the answer to the question. This process is sometimes referred to as "torturing the data until it confesses."

The project team should recognize that rarely does the answer to one question constitute the end of the exploration. Testing of theories is typically an iterative process. The answer to one question leads to another question, and another, and another. Each time an answer is discovered, the team should ask again, why? Why does the analysis look the way it does? Why is the upper level (not root) cause we have proven occurring? When the "why" questions reach a level that has no more answers or goes beyond a level of cause that can be controlled, the team has arrived at the (operationally defined) root cause.

As an example, take the case of a problem the National Park Service experienced several years ago concerning the Jefferson Memorial*. The stone in the monument was crumbling due to frequent washing to remove bird droppings. The initial (mistaken) approach the Park Service took was to reduce by half the number of times the stone was cleaned. This saved some money and reduced the magnitude of the stone erosion, but it's easy to see how the "solution" led to other problems. People visiting the monument were dissatisfied with the unclean conditions.

So the Park Service undertook a more thorough analysis to find the root cause of the problem. They first asked, "Why are there so many bird droppings?" Of course, they considered several theories to answer the question. Perhaps the birds were attracted to food dropped by visitors. Perhaps they were attracted to the good roosting places in the structure. Perhaps there was an abundant natural food supply. Could they immediately determine which of these theories was true? Of course not, it was necessary that they visit the place where the problem was taking place (what the Japanese call the "gemba"), collect data about the possible causes, and identify the true cause of the proliferation of birds in the monument. It turned out that the third theory was true; hundreds of fat spiders were providing an ample food source for the birds. But was the investigation complete? No, it was not because the investigators had not yet reached the root cause of the problem.

---

*Paraphrased from Juran Institute, Inc. "The Quality Minutes: The Jefferson Memorial."

The next question to be answered was, "Why are there so many spiders?" A number of theories could have been forwarded about this question too:

- The crevices in the monument provide a good place to spin webs.
- There are insects there that provide food for the spiders.
- The spiders are attracted to and hide in the shadows inside the monument.

Further data-gathering proved that the second theory was true. Inside the Jefferson Memorial were thousands of tiny midges (a small flying insect that spiders eat). The investigators were nearing the root cause of the problem, but were not there yet.

"Why are there so many midges?" they asked. Possible answers included:

- Midges were attracted to a food supply inside the monument.
- The Jefferson Memorial, like many others in Washington, D.C., is near a body of water (the Potomac River), and the midges lay their eggs in the water.
- The midges are attracted to the lights that illuminate the memorial at night.

The second theory actually did explain why so many midges were in the vicinity of the monument, but not why they were on and inside it. Investigation revealed that the midges came out at sunset each evening in a "mating frenzy" at just the time that the lights were turned on. They were attracted to the illumination of the monument and took up residence where the spiders could feast on them. Now the investigators had found the true root cause of the problem: illuminating the monument each night at dusk. They delayed the lighting by 1 hour (the remedy), the midge population was dramatically reduced, and the food chain was broken. Now the Park Service could substantially reduce the washings and, therefore, the crumbling of the stone (the original problem). This application of the remedy to the true root cause resulted in many multiples of savings compared to the original solution of just reducing the washings. The solution was also one that could be replicated to other D.C. monuments to reap additional savings (The Juran Quality Minute: Jefferson Memorial).

One may ask, "How will I know when to stop asking 'Why?'" In other words, when have the investigators drilled down deeply enough to conclude they are at the level of the root cause?

There are two questions that will help you decide whether you have found the root cause:

1. Do the data suggest any other possible causes? After each data collection and analysis, it is usually possible to discard some theories and place more confidence in others. Theorizing is not a one-time activity, however. Each data display—the Pareto diagram, histogram, scatter diagram, or other chart—should always be examined by asking whether it suggests additional theories. If you have competing plausible theories that are consistent with the new data and cannot be discarded based on other data, then you have not arrived yet at the root cause.
2. Is the proposed root cause controllable in some way? Some causes are beyond our ability to control, like the weather. Turning up the heat or running a humidifier can control the effects of the weather, but the weather cannot be controlled directly. So no useful purpose is served by testing theories about why the weather is cold.

Tools most often used during the steps of formulating and testing theories are data collection, flow diagrams, graphs and charts, histograms, Juran's Pareto analysis, scatter diagrams, and stratification.

These steps of formulating and testing theories complete the diagnosis of the problem's root cause. Some may ask, why should I go to all that trouble just to find the root cause of the problem? Why is it important? Denise Robitaille (2009), an ASQ fellow and leading expert in root cause analysis provides useful answers in an article entitled "Four Things You Should Get from Root Cause Analysis." Emphasis on effective root cause analysis has gotten increased attention in several sectors. Registrars, for example, are requiring more substantial evidence of root cause analysis as part of responses to their requests for corrective action. All of this is good news. Except, my personal experience is that although people understand that they're required to do root cause analysis, they don't comprehend three issues:

1. What is root cause analysis?
2. How to conduct effective root cause analysis?
3. What the results of root cause analysis should yield?

Let's start by reviewing what root cause analysis is. It's an in-depth investigation into the cause of an identified problem. It asks why something happened. It should also investigate how something could have gone wrong, which will help to identify contributing factors and interim breakdowns.

There are two important things to remember at the outset. Root cause analysis is focused on cause, and the ultimate intent is to use the information to develop a corrective action plan. This perception is relevant to the next two issues people need to know.

People don't know how to do root cause analysis. They still treat it like it's a haphazard activity. Organizations fail to train individuals in good investigative techniques. They perpetuate a culture of blame: "Let's find out who screwed up." They simply don't treat root cause analysis like a controlled process.

Apart from the five whys there are many other tools that can be used. There are flowcharts, brainstorming, fish bone diagrams, Pareto charts, and design of experiment—just to name a few. Several tools should be used in concert to achieve the most productive results. For example, use brainstorming or the five whys to conjecture what could have gone wrong, then organize the results in a fish bone diagram that will direct you to the areas where you'll find the evidence you need to objectively conclude what the root cause of the problem really is. Organizations have to stop assigning people to do root cause analysis without giving them the necessary training and tools.

Finally, individuals need to understand what the expected outcome of this process is. It's great to say that we're going to conduct root cause analysis, but do people have any idea what they're supposed to do when they figure out the cause?

You should be able to get four things from root cause analysis:

1. Uncover the root cause or causes of the problem. Finding the root cause is the primary output of this process.
2. Identify weaknesses or other contributing factors, which, in and of themselves, are not necessarily nonconformance. They may be the outcome of shortsighted decisions to curtail activities so that efficiency or cost savings is perceived. You may have, for example, decided to wait until the first point of use to test components. The time-savings experienced at the receiving process could result in costly delays and scheduling snafus that dwarf any savings that had been anticipated. It wasn't a bad idea at the time, but it may have contributed to late deliveries.
3. Better understand the process surrounding the problem, as well as supporting processes. If you don't, you haven't done a thorough root cause analysis. Without that

heightened comprehension of the process, you can't understand interrelations, interdependencies, or other factors that are reliant on the outcome of seemingly unrelated processes. This takes us to the final outcome.

4. Create an architecture into which you can build your corrective action plan. Corrective action isn't just one activity. It needs to be a plan, reflective of all aspects of the problem. If you've done a good root cause analysis, you'll have identified not only the root cause, but the many different factors that need to be addressed to ensure that the problem doesn't recur, that you don't inadvertently create a new problem, and that your organization experiences some benefit from the action taken.

Your root cause analysis will let you see what processes may need to be modified, what documents and forms will have to be revised, who will require training, and a myriad of other considerations that go into a typical project plan.

Without root cause analysis, effective corrective action is impossible. Without corrective action, root cause analysis is a waste of time.

Phase 3: Improve or Remedy the Cause

Now that the project team has discovered the root cause(s) of the problem, the task is to restore control to the process. Applying appropriate remedies that will directly affect the cause and eliminate it, or at least drastically reduce its undesirable effects does this.

Evaluate Alternative Solutions

Like the formulate theories step, this step moves from creative to empirical, divergent to convergent thinking. Beginning with brainstorming, the team, subject matter experts, and process owners will attempt to identify as many alternatives for solutions as possible. Creativity is essential at this point, as often, solutions must be quite novel to fully address the root cause. Next, the team will evaluate these brainstormed potential solutions to determine which solution or combination of solutions will best address and eliminate the cause(s).

The team may construct flow diagrams of possible solution implementations to visualize which will act most effectively. They may also use a criteria-based selection matrix to assist their decision-making process and help them arrive at the best solutions (Fig. 16.4). The solution selection matrix can help the team optimize the ultimate solution by combining the best potential solutions from the matrix.

Tools most often used in this step are brainstorming, data collection, selection matrices, and flow diagrams.

Design and Implement the Remedy

Once the team selects a remedy, it designs the remedy by performing four tasks:

1. Ensure that the remedy achieves the project goals. Review project goals to verify that the remedy will achieve the desired results and that all involved are in agreement on this point. This is a final check before moving ahead.
2. Determine the required resources. Make every effort to determine, as accurately as possible; what resources are required to implement the proposed remedy. These resources include people, money, time, and materials.
3. Specify the procedures and other changes required. Before implementing the remedy, describe explicitly what procedures will be required to adopt the proposed remedy. Any changes that need to be made to existing organizational policies, procedures,

Solution Selection Matrix Rank possible solutions 1–10. 10 = fully meets criteria							Updated: 10/12/09
		Possible solutions					
Criteria	Weight	A	B	C	D	E	F
Low cost	3	9	8	10	7	9	7
High effectiveness	2	8	10	9	10	9	9
Low risk	2	8	8	7	9	9	7
Low resistance	1	9	8	5	8	9	10
Minimal process disruption	2	6	7	7	6	8	7
	Total score	80	82	81	79	88	77

Figure 16.4 Solution selection matrix.

systems, work patterns, reporting relationships, and other critical operations must also be described. Any surprises down the line may sabotage the remedy.

4. Assess human resource requirements. The success of any remedy depends on the people who will implement the required changes. Often, it will be necessary to train or retrain staff. Explore fully all training requirements, as well as the training resources needed.

Once these tasks have been performed, a flow diagram can be created to help specify the new procedures clearly.

As the team is designing the remedy, they should take into account the need to mistake-proof the remedy. They should consider and develop a variety of techniques to avoid, prevent, or reduce inadvertent errors that may occur even with the improved process.

The final action of this step is to implement the remedy. Depending on the complexity of the problem being addressed and the solutions to be implemented, a formal implementation plan may be needed. At a minimum, procedures, process standards, or work instructions will need to be modified to institutionalize the change.

Phase 4: Control to Hold the Gains

This phase is the most important one in the RCCA process for ensuring that the problem does not recur or, if it does, that the recurrence is recognized and remedied quickly. If a recurrence is recognized, it should be an indication to the project team that their job is not finished—they've missed a root cause during the course of their problem solving or designed and implemented an ineffective remedy.

If the problem solving has been done methodically, as described here, and a broad range of possible causes and remedies were considered during the formulate theories and evaluate alternatives steps, the remedy should be robust and the cause and problem should not recur. The controls put in place to hold the gains will indicate whether this is so.

Redesign Controls

The primary activity in designing controls is the development of a control plan. Hopefully, an effective control plan for the process in question is already in place and will only require modification to add control subjects related to the problem's solution.

The first step in building an effective control plan is selecting appropriate control subjects. Control subjects are those features of the product or process that will be measured to determine whether the process is remaining in control. Each control subject's performance is monitored using the feedback loop described in Fig. 16.3. A control plan matrix is used to keep track of the function of the feedback loop and to plan for action if the process or product does not meet standards. An important purpose of the process control matrix is to alert the process operator when the process is out of control and what to do to get it back under control.

In this matrix (Fig. 16.5), the horizontal rows describe the control elements for each subject. The vertical column headings indicate each element of the control activity:

- *Control subject.* Those features of the product or process that will be measured to determine whether the process is remaining in control.
- *Subject goal or standard.* The acceptable limits of performance for the product or process. Often, these are control limits on an SPC chart and are the primary basis for determining if the process is stable or out of control.
- *Unit of measure.* How will the measurement be stated? Inches? Millimeters? Percent defective?
- *Sensor.* What device, person, or combination of the two will be used to obtain the measurement?
- *Frequency of measurement.* How often will the control subject be measured (e.g., hourly, daily, weekly, etc.)?
- *Sample size.* How many measurements will be taken at the stated frequency?
- *Where are measurements recorded* (logbook, chart, database, etc.)?
- *Measured by whom.* Who is responsible for applying the sensor to the control subject and obtaining and recording the measurements?
- *Criteria for taking action.* This generally includes whatever process performance is outside the subject goal or standard. This variation is usually due to special causes and would prompt the troubleshooting part of the feedback loop.
- *What actions to take.* Knowing the cause of the out-of-control condition helps the assigned person take the appropriate action to bring the process back into conformance with the subject goal.
- *Who decides.* Who will make the call on the action to be taken?
- *Who acts.* Specific action(s) to be taken by the actor on the control subject to bring the process back into conformance with the subject goal.
- *Where action recorded.* Identifies where the actions taken to resolve the issue will be recorded. This recording is useful for analysis of similar problems in the future.

Implement Controls

Once a suitable control plan has been designed, implementation is a matter of training process owners and operators in its use. If SPC is a part of the plan, specific training on the proper use, interpretation of, and appropriate response to control charts must be included. The process owners also become the owners of the control plan, so their involvement in its implementation is essential.

Audit Controls

For a short time after the controls are in place, the project team, in conjunction with the process owners and operators, should monitor their effectiveness. This will provide the opportunity to recognize any ineffective elements of the plan and modify accordingly.

Process control plan for:						Date:		Revision level:		Approved by:		
Control subject	Subject goal (standard)	Unit of measure	Sensor	Frequency of measurement	Sample size	Where measurement recorded	Measured by whom	Criteria for taking action	What actions to take	Who decides	Who acts	Where action recorded

Figure 16.5 Control plan matrix.

By following the above four-phase approach to RCCA, project teams should consistently identify the root cause(s) and apply appropriate remedies in a relatively short time. During the time that it takes to identify and alleviate the causes, an interim action may be needed to ensure that defective products, services, or information do not reach the customer. These actions are sometimes referred to as containment. They should be designed to be effective and temporary until the root cause of the problem can be determined and alleviated.

Plan-Do-Study-Act

Plan-do-study-act (PDSA) is another problem-solving approach many use to find and address root causes of problems. The method was originally proposed by Dr. Walter Shewhart (as PDCA, plan-do-check-act) in his book *Economic Control of Quality of Manufactured Product* (1931) and later espoused by W. Edwards Deming. Deming referred to the method as the Shewhart cycle, but many, particularly after Deming achieved fame, refer to it as the Deming cycle.

The method differs from the root cause analysis method described previously in that it is primarily a guide for identifying root causes through experimentation. This implies that the analysis of symptoms and theorizing of causes are done before the cycle actually starts, and then iterative experiments are performed to drill down to the root causes of the problem being addressed.

The PDSA method is particularly popular in health care organizations, probably due to its promotion by the Institute for Healthcare Improvement (IHI) as a method for finding causes and stimulating improvement.

The work done prior to the actual PDSA cycle starts by "setting aims," which is analogous to the establishment of the goal in the Juran RCCA process. The piece of stating the problem to be solved, however, seems to be absent, so one might wonder how the activity of the team becomes focused. Then the team gathers knowledge about the process they are attempting to improve upon so they can come up with good ideas for changes to the process.

> … [T]he more complete the appropriate knowledge, the better the improvements will be when the knowledge is applied to making changes. Any approach to improvement, therefore, must be based on building and applying knowledge. This view leads to a set of fundamental questions, the answers to which form the basis of improvement:
>
> - What are we trying to accomplish?
> - How will we know that a change is an improvement?
> - What changes can we make that will result in improvement? (Langley et al., 1996)

In contrast to Juran's RCCA, the PDSA approach seeks to identify changes that might improve the process or outcomes of it, then implements those changes to see if they are effective in producing an improvement. The PDSA cycle is the method applied to this trial of changes. In a manner of thinking, PDSA seeks to confirm or refute ideas of problem causes by trial and error of solutions.

"These questions [above] provide a framework for a 'trial and learning' approach. The word 'trial' suggests that a change is going to be tested. The term 'learning' implies that criteria have been identified that will be used to study and learn from the trial" (Langley et al., 1996).

The PDSA approach follows these phases and steps:

1. Plan
   - Define the change to be tested
   - Design the experiment to test the change

2. Do
   - Carry out the experimental plan
   - Collect data about the effectiveness of the change
3. Study
   - Analyze the data from the experiment
   - Summarize what was learned
4. Act
   - Determine what permanent changes are to be implemented
   - Determine what additional changes need to be tested

Clearly, this approach has some advantages:

- It can yield results quickly if the experimenters are good at selecting solutions that will yield true improvement.
- It follows an experimental approach, which can yield a great deal of useful knowledge.
- It is widely accepted, particularly within health care and other organizations that typically rely on experimentation to determine beneficial changes (e.g., development of medications).

One might also note some disadvantages:

- Results can be slow to come if the experimenters are not good at selecting solutions that will yield true improvement.
- Changes that do not succeed may not yield a lot of useful information.
- Experimentation, unless it is done in a laboratory setting, can be disruptive to the process and can be resource-intensive.
- Experimentation can be costly in many cases.

Based on these pros and cons, the project team should choose the methodology that best fits their work style and organization's needs.

Just Do Its

As the name implies, "just do it" (JDIs) do not really include an analysis of the root cause of the problem because that root cause is usually readily apparent in what is sometimes referred to as a "blinding flash of the obvious." So analyzing the root cause in this case is done entirely by observation.

A number of years ago, consultants transitioned from the old way of teaching using overhead projectors to the new computerized method: constructing the materials to be taught in a presentation graphics program and projecting them using a liquid crystal display (LCD) projector. As the transition from the old way to the new way progressed, fewer and fewer meeting rooms had overhead projectors available, and more and more had LCD projectors. A problem arose for some training providers: If the trainer arrived at the training room prepared to show slides on an overhead projector and none were available, the training had to be either postponed or done in a less-than-desirable fashion, reading from and referring to printed materials only. What was the obvious cause of this problem? The consulting organization had not provided the consultant with the proper tools (either a laptop or some digital media that could be used on the training room PC) to do the job in the new environment. The JDI in this case, of course, was to provide the trainer with the needed tools.

Another situation where the JDI approach may be appropriate is when the need for a solution is urgent and delaying can have serious repercussions.

Such an example of an urgent need for a solution occurred in London in 1854. There had been a terrible outbreak of cholera, which ultimately claimed more than 500 lives in a period of ten days. Dr. John Snow came to the rescue. After analyzing the pattern of occurrence of the deaths using a concentration diagram, Snow recognized that most of the deaths were grouped around the Broad Street pump. Even though he did not recognize the root cause was bacteria in the water, Snow went directly to a solution and had the handle removed from the pump. Within days the cholera outbreak was over (The Juran Quality Minute: London Cholera Epidemic)†.

To implement JDIs without a thorough analysis and discovery of the root cause of the problem, three factors must be present:

1. The need for change must be urgent. Don't use the JDI approach just because it is quick and easy.
2. The change must carry a low cost of failure. What if you're wrong? The price to pay for making the change must be low, preferably zero. Dr. Snow had little if anything to lose by removing the pump handle. The worst that would happen is people would have to travel farther to get their water.
3. The change must have a significant potential reward. The decision here is, "Well, what if I'm right? Things will be a lot better if the change is effective."

JDIs used at the appropriate times and in the right situations can be a beneficial and effective method of attaining some quick wins.

References

The list and descriptions of tools in this section are edited from the pertinent sections of Juran Institute, Inc. (2008) "Root Cause, Corrective Action, Version 1.2," Southbury, CT.

Langley, G. G., Nolan, K. M., Norman, C. L., Provost, L. P., and Nolan, T. W. (1996). *The Improvement Guide: A Practical Approach to Enhancing Organizational Performance*, Jossey-Bass, New York, pp. 3–4.

Monroe, D. (2009). "Process Variables Controlled," *Quality Magazine*, November.

Nolan, K. M., Norman, C. L., Provost, L. P., Nolan, T. W., Langley, G. J. (1996). *The Improvement Guide: A Practical Approach to Enhancing Organizational Performance*, 1st ed. Jossey-Bass Publishers, San Francisco, California.

Robitaille, D. (2009). "Four Things You Should Get from Root Cause Analysis," *Quality Digest The Quality Insider*, November 10 (used by permission).

Shewhart, W. A. (1980). *Economic Control of Quality of Manufactured Product*, Martino Fine Books, Originally published in 1931 by D. Van Nostrand Company, Republished in 1980 as a 50th Anniversary Commemorative Reissue by ASQ Quality Press, Milwaukee, Wisconsin.

†Paraphrased from Juran Institute, Inc. "The Quality Minutes: London Cholera Epidemic."

CHAPTER 17

Continuous Innovation Using Design for Six Sigma

Joseph A. De Feo and John Early

High Points of This Chapter

1. Innovation is key to the survival of all organizations. Innovation, like continuous improvement, is the result of a systematic approach, not a haphazard one.
2. Continuous innovation (CI) is different from product development. Continuous innovation must happen in all areas of an organization, from creating products, services, or processes used to meet internal and external customer needs to designing new facilities' or office environments.
3. There have been many improvements in the methods used to design and develop products and services in the past decade. Design for Manufacturing, Design for Assembly, Design for Lean, Design for Environment, and Six Sigma all have become models to meet critical to quality customer needs—and lead to innovative products.

4. Continuous innovation using the steps of Design for Six Sigma or DMADV, as it is often referred to, is similar to the Juran Quality by Design model (see Chap. 4, Quality Planning and Design of New Goods and Services) and has become the basis for what we call "continuous innovation of goods, services, and processes."
5. Creating the habit of innovation requires that management create an infrastructure similar to that of continuous improvement. Set goals, select projects, and educate teams to create innovative goods and services—project by project.
6. Continuous innovation using Design for Six Sigma consists of carrying out five steps:
    1. *Define* the goals and objectives for the new good, service, or process.
    2. *Measure* and discover hidden customer needs.
    3. *Analyze* the customer needs and determine the innovative features that will meet those needs.
    4. *Design* by combining the features, thereby creating new products, services, or processes that incorporate the features.
    5. *Verify* that the new innovation meets the customers' and organization's needs.

Continuous Innovation and the Juran Trilogy

We have previously explored the Juran Trilogy as it relates to quality planning. Designing for customer needs always leads to higher-quality products and services as well as innovative outcomes because an effective design process uncovers hidden customer needs. This discovery and the subsequent solving of the problems that kept customer needs hidden lead to innovation (Fig. 17.1). This chapter addresses the use of the define, measure, analyze,

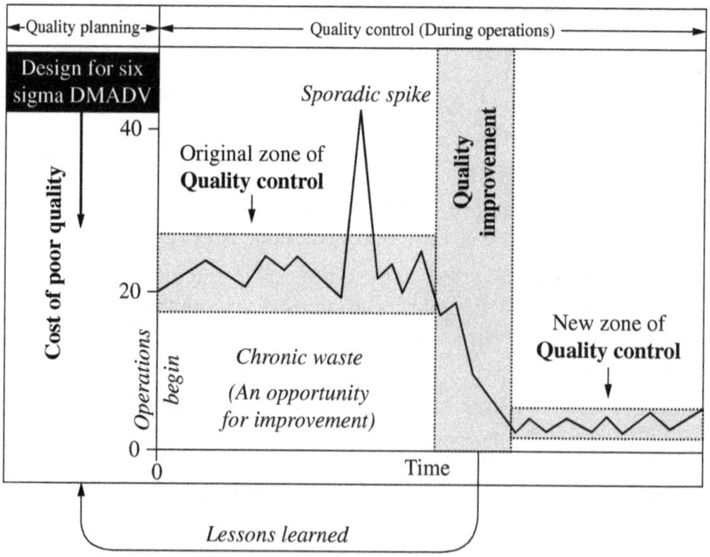

Figure 17.1 Design for Six Sigma and the Juran Trilogy. (Juran Institute, Inc., Southbury, CT.)

design, verify (DMADV) steps above and tools for creating continuous innovation (CI). Adapting the most effective models such as the Quality by Design used by the FDA and Design for Six Sigma (DFSS) model used by many such as GE, Samsung, and Microsoft, organizations can create the *habit of innovation,* which is similar to creating the *habit of improvement*. Deploying a CI program will ensure organization adaptability and sustainability in meeting societal and business needs.

CI using the Design for Six Sigma model and tools, which arose out of GE Medical's adaptation of the Juran quality planning model described in Chap. 4, Quality Planning and Design of New Goods and Services is a powerful engine available for those who want to truly plan quality into their products, typically goods, rather than services or processes.

Juran referred to the quality planning design steps (Fig. 17.2) as a framework for planning (designing) new products and services (or revisions). These steps apply to both the manufacturing and service sectors and to products for both external and internal customers.

Planning an effective solution for an improvement project (see Chap.5, Quality Improvement and Breakthrough Performance) may require one or more steps of this quality planning process. Early and Colleti (1999) and Juran (1988) provide extensive discussions of the steps. These quality planning steps must be incorporated with the technological tools for the product being developed. Designing an automobile requires automotive engineering disciplines; designing a path for treating diabetes requires medical disciplines. But both need the tools of quality planning to ensure that customer needs are met.

The road map is presented in greater detail in Fig. 2.4 (in Chap. 2, Developing an Excellence Culture). It is useful, however, to present an overview now to explain briefly the steps (Early and Colleti 1999).

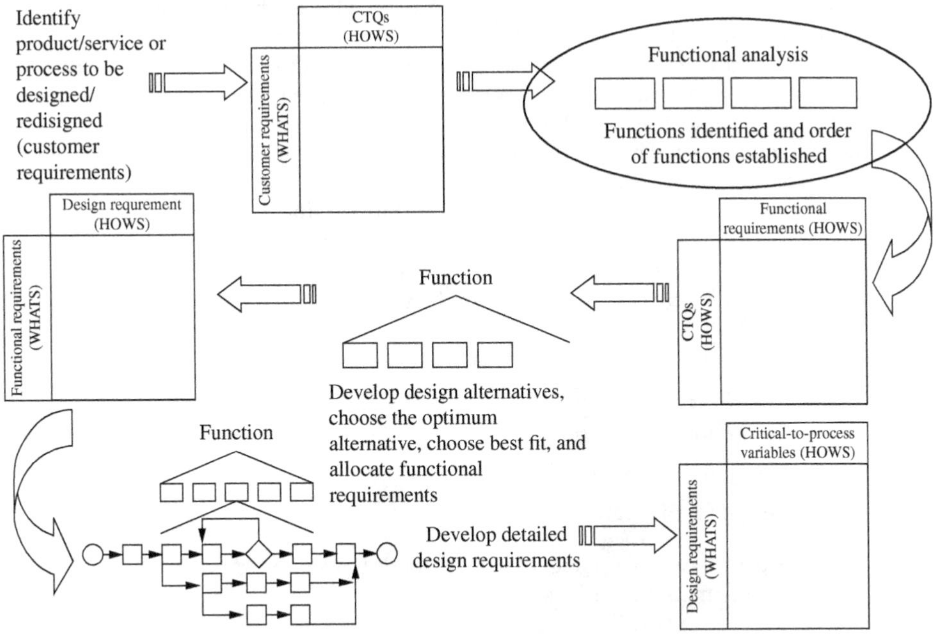

Figure 17.2 How to design matrices. (Juran Institute, Inc. Used by permission.)

New designs or innovations happen when one discovers hidden customer needs. Some examples include the following:

Abrasive cloth:	Lower internal cost of polishing parts due to better durability of cloth
Automobile:	Less effort in closing door; better "sound" when door closes
Dishwasher:	Greater durability because heavier parts make up the appliance
Electronics:	Simplicity all-in-one device, e.g., iPhone, iPod,
Software:	Understandable owner's manual
Fibers:	Lower number of breaks in processing fibers
Tire valve:	Higher productivity when tire manufacturer uses valve in a vulcanizing operation
Photographic film:	Fewer process adjustments when processing film due to lower variability
Commodity product:	Delivery of orders within 24 hours rather than the 48-hour standard requirement
Home mortgage application:	Decision in shorter time than that of competition

Traditionally, the main activities to capitalize on these insights were executed *sequentially*. For example, the planning department studied customer desires and then presented the results to design; design performed its tasks and handed the results to engineering; engineering created the detailed specifications; and the results were then given to manufacturing. Unfortunately, the sequential approach results in a minimum of communication between the departments as the planning proceeds—each department hands its output "over the wall" to the next department. This lack of communication often leads to problems for the next internal customer department. To prevent this from occurring, activities are organized as a team from the beginning of the project. Thus, for example, manufacturing works *simultaneously* with design and engineering before the detailed specifications are finalized. This approach allows the team to address production issues during the preparation of the specifications.

Creating new products and services contributes to the vitality of an organization. Many organizations have adopted numerous methods to improve the salability of their designs. From the 1980s to the present, there were a number of newly adopted methods based on Juran's Quality by Design to improve product salability. Many continue to pay dividends:

- Design and development phase gates
- Concurrent or Simultaneous Engineering
- Design for Manufacture
- Design for Assembly
- Design for Six Sigma

In this past decade, a number of new methods have popped up. Most recently there have been promising methods such as:

- Design for Environment
- Lean Design
- Sustainable Design

Today, Design for Six Sigma is a systematic methodology to provide the means to attain new services and innovative designs. The steps for designing new products and services that lead to innovation are as follows:

- Discover the customers and their needs
- Gather and research information, and observe the behaviors of these customers
- Generate and then design solutions to meet their needs
- Design the solution and validate that the needs are met
- Transfer the design to operations

Along the way, these steps force people to "think outside the box." They force people to gain new information in a structured and organized way, arriving sometimes at revolutionary means to create new services.

DFSS Works for Goods, Services, and Transactional Processes

The Design for Six Sigma (DFSS) model has been used within new product introduction (NPI) processes for a wide variety of physical goods including electronics, chemicals, sophisticated industrial equipment, transportation equipment, and a plethora of consumer goods. It has also been used successfully to develop high-quality new services in insurance, health care, banking, and public service.

In the design phase of DFSS, a multifunctional team develops both the detailed product design down to the full engineering drawings and the process design for delivering the product, including all equipment, work instructions, work cell organization, etc. The difference between product design and process design is fairly clear when physical goods are produced. It is sometimes less clear for services where the two are intertwined.

Making and acting on the distinction between the design of the service and the design of the process that delivers that service has proved to be very helpful. The *service design* is the flow of activity as experienced by the customer. The service *process design* is the flow of activity required to make the customer experience possible.

For example, the service for paying a customer's insurance claim will have features related to timeliness, ease of use, responsiveness, and transparency. These are what the customer sees, feels, hears, and touches. To deliver that seamless flow of activity to the customer, the production process will include features related to data processing, information access, payment procedures and policies, and interpersonal skills of individuals interacting with the customer during the process. The behind-the-scenes production process is largely invisible to the customer. In fact, when these invisible production processes become visible to the customer, it is usually because they have broken down and failed to deliver the seamless service as designed.

Experience shows that it is a useful division of the work to first design the customer-experienced service and then design the process that makes it possible. Teams that try to design both the service and the process as a single step usually subordinate the customer experience to the exigencies of operations.

An Example of Designing for Services

In an example from the service sector, the quality planning process was applied to re-planning the process of acquiring corporate and commercial credit customers for a major affiliate of a large banking corporation. Here is a summary of the steps in the quality planning process.

1. *Establish the project.* A goal of $43 million of sales revenue from credit customers was set for the year.

2. *Identify the customers.* This step identified 10 internal customer departments and 14 external customer organizations.
3. *Discover customers' needs.* Internal customers had 27 needs; external customers had 34 needs.
4. *Develop the product.* The product had nine product features to meet customers' needs.
5. *Develop the process.* To produce the product features, 13 processes were developed.
6. *Develop process controls and transfer to operations.* Checks and controls were defined for the processes, and the plans were placed in operation.

The revised process achieved the goal on revenue. Also, the cost of acquiring the customers was only one-quarter of the average of other affiliates in the bank. Quality planning generates a large amount of information that must be organized and analyzed systematically. The alignment and linkages of this information are essential for effective quality planning for a product. A useful tool is the quality planning spreadsheet or matrix (basically, a table). Figure 17.2 shows four spreadsheets corresponding to steps in the quality planning process. Note how the spreadsheets interact and build on one another; they cover both quality planning for the product and quality planning for the process that creates the product. The approach is often called *quality function deployment* (QFD). Thus QFD is a technique for documenting the logic of translating customer needs into product and process characteristics. The use of spreadsheets in the quality planning process unfolds later in this chapter.

These six quality planning steps apply to new or modified products (goods or services) or process in any industry. In the service sector the "product" could be a credit card approval, a mortgage approval, a response system for call centers, or hospital care. Also the product may be a service provided to internal customers. Endres (2000) describes the application of the six quality planning steps at the Aid Association for Lutherans insurance company and the Stanford University Hospital.

CI Requires Understanding Customer Needs and Solving Their Problems

Designing innovative and superior quality services and products requires gaining a clear understanding of the customers' needs and translating those needs into services aimed at meeting them. This information is the driver of most innovation, yet most do not recognize it as such.

Innovation has everything to do with creating something new. In competitive business situations, success often comes to the best innovators. Many organizations have design and development functions that create annual plans to develop new models and new services. Sometimes these functions design the good or service internally to the organization and then look for customers to sell it to. Other innovation comes from solving societal problems. And still other organizations look for customer problems to solve; as a result they create something new, something innovative. It is the latter that we have found to be the most economical and therefore provides the greatest return on its investment.

To create continuous innovation, an organization must design to meet customers' unmet (often hidden) needs. To do this one must:

- Capture the voice of the customers—the potential new customers or existing ones
- Discover hidden customers and needs
- Design solutions to meet those needs
- Use a systematic approach to ensure innovation happens—continuously

- Have tools to capture the information and use it to ensure that the good or service is produced efficiently
- Use multifunctional staff to carry out the systematic process to ensure the good or service can be produced as planned

One can learn about innovation, which means "making something new," by studying innovations and innovative methods from the past.

Polaroid Camera

The conventional photographic process involves exposing light-sensitive material, which in turn must be developed, fixed, and printed and the print developed and fixed, a procedure that can take hours (or days if the processing facility is far from the place where the photograph was taken). In 1947, a remarkable new system of developing and taking pictures was introduced by U.S. physicist Edwin Herbert Land (1909–1991). Land had left Harvard after his freshman year to conduct his own research on the polarization of light. Two years later, he invented a sheet polarization filter that could be used on camera lenses to eliminate reflection and glare. In 1937, Land founded the Polaroid Corporation to manufacture and market his filters, lamps, window shades, and sunglasses. In February 1947, he introduced Polaroid instant film for use in his own Polaroid Land Camera. The Land Camera (U.S. Patent 2,543,181) was first offered for sale on November 26, 1948. Polaroid film processes chemicals in a flat, hermetically sealed compartment attached to the photosensitive paper. A pair of pressure rollers spreads the chemicals uniformly across the paper when exposed, and the completed print is ready a minute later. In 1963, Polaroid introduced Polacolor, a full-color film that could be processed in less than a minute.

Life Savers Candy

In 1912, when candy maker Clarence Crane first marketed Crane's Peppermint Life Savers, life preservers were just beginning to be used on ships—the round kind with a hole in the center for tossing to a passenger who had fallen overboard. But that is not the whole story. Crane had been basically a chocolate maker. Chocolates were hard to sell in summer, however, so he decided to try to make a mint that would boost his summertime sales. At that time most of the mints available came from Europe, and they were square. Crane was buying bottles of flavoring in a drugstore one day when he noticed the druggist using a pill-making machine. It was operated by hand and made round, flat pills. Crane had his idea. The pill-making machines worked fine for his mints, and he was even able to add the life preserver touch by punching a tiny hole in the middle. In 1913, Crane sold the rights to his Life Savers candy to Edward Noble for only $2900. Noble then sold Life Savers in many flavors, including the original peppermint. Clarence Crane may have regretted that decision to sell, for Life Savers earned the new manufacturer many millions of dollars.

iPod

The iPod originated with a business idea dreamed up by Tony Fadell, an independent inventor. Fadell's idea was to take an MP3 player, build a Napster music sales service to complement it, and build a company around it. It resulted in Apple creating the iPod.

Segway

This new means of transportation meant reimagining virtually every piece of conventional wisdom about the last century of transportation, from how it moves, to the fuel it uses, to how you control it. The result is electric transportation that doesn't look, feel, or move like anything that has come before. And of all the conventional wisdom we've left in pieces behind us, none has been shattered more fully than the belief that we must choose between "more" and "less." In 2001, Dean Kamen announced the arrival of the first self-balancing, zero emissions personal transportation vehicle: the Segway Personal Transporter. Founded on the vision to develop highly efficient, zero emissions transportation solutions using dynamic stabilization technology, Segway focused its research and development on creating devices that took up a minimal amount of space, were extremely maneuverable, and could operate on pedestrian sidewalks and pathways. Today, Segway continues to develop safe, unique transportation solutions that address urban congestion and pollution.

Two Types of Innovation

There are two basic types of innovation. The first, type I, does happen, but rarely. Type I is something completely new. And new things under the sun do not occur as often as we think they do. The first automobile and internal combustion engine were certainly new innovations, but even they built on the wheel, cart, and other existing technologies.

Things such as nuclear power, radio, phones, electricity in the home, and manned flight are certainly good examples of something that was pretty close to new under the sun. All the great, really new innovations can often be traced back to a genius, a lucky accident, or both.

We know the names of many of the geniuses—Fermi, Wright, Edison, Benz, and Ford. However, this is not an endless list, and while lucky accidents are good, they are too chancy. Type II innovation presents a better way.

Type II innovation is much more common than type I. This second type of innovation can be reduced to three general approaches:

1. Making something that already exists larger
2. Making something that already exists smaller
3. Combining one thing that exists with something else that exists

The simplicity of type II is profound. It can create dramatic breakthroughs and change the way we live. Most of what we see and consider to be great innovations were derived from the three methods of type II innovations listed.

For example, the mobile phone or PDA in your pocket was once a fair-sized wooden box on the wall. The phone has been made smaller from the original wall model hardwired to the outside world. The phone has also been "combined" with a radio, calculator, computer, TV, and music player. The flat-screen television evolved from a device that was once considered a piece of furniture and that took up more room than an easy chair. Over time, the TV's depth and height have been "made smaller," and its width has been "made larger." Add the appropriate technology, and you have your flat-screen display.

An example is Web-based learning. Web-based learning came about when transparencies were replaced by electronic slides such as PowerPoint. This led to improved quality of presentation graphics, then added animation, placed on the Internet, with voice-over IP, and video, thus delivering Web-based learning.

The "bigger/smaller/combination" approach sounds simple when you look backward. But the trick is doing it in the present, as an innovation for the future. However, it is still much easier than becoming a genius.

The good news is you can get better at type II innovation. As good as we are today, we can also get better with practice.

The next time you are in a serious brainstorming meeting and need an innovation for a new product, service, marketing strategy, or similar task, put up three new header columns, and attack them one at a time.

A header is the place where you will hang your ideas. The three headers are, of course, "make it bigger," "make it smaller," and "combine it with." The "it" is whatever product or service or whatever you are working on. Have fun with it. Remember not to critique or scrub the ideas until after the generation of ideas is done. Most people are surprisingly good at type II innovation. Morph some of the wild ideas into something that is doable. The great innovator Henry Ford said, "If you think you can or can't, you're right."

Innovators are not born that way. If you have your heart set on being the next Thomas Edison, you are probably going a bit too far. But whatever your innovation quotient is now, you can make it better with practice and by using a methodology that causes innovation to happen.

For instance, how many times do we hear "Think outside the box"? That's all well and good, but what box? Few of us recognize that the box is in fact ourselves. Learning to temporarily let go, be foolish for a moment, and be comfortable with ambiguity is necessary for innovation.

Getting beyond our "boxed" selves is a skill that can be learned and improved with technique, practice, and courage. For example, imagining oneself as someone else and seeing everything through his or her eyes can be a great technique.

Arriving at this level of letting go will require a systematic methodology. Many methods have been used in developing simpler and better products. These design processes incorporate early involvement teams.

The teams are composed of a broad spectrum of employees, customers, and suppliers who work together through a systematic process of looking and thinking outside the box to solve problems. The results are significant, and new products can be discovered.

This concept of push innovations (e.g., toys and foods) is a short-term exercise that continues to flood the market with new products. Some are good and last a long time; many are short-lived. If you are trying to innovate to solve a customer or societal problem, the outcome of a purposeful design process often leads to products that benefit society for many years. Drug development is a good example. Aspirin has been around for more than 100 years. New drugs that reduce cholesterol will also be here for decades.

Why do some products last so long and others do not? This answer lies in the methods used to design or create the innovation. Innovation requires a systematic process and set of tools to create customer-focused, need-driven designs.

Designing world-class services and products requires gaining a clear understanding of the customers' needs and translating those needs into services aimed at meeting them. The process goes on to design and optimize the features and then develop and execute the new designs. This process is sometimes referred to as the service development process, the design process, or the DFSS process.

Random, innovative ideas, no matter how clever, will not deliver economic success unless they meet a customer need better than the current method or fulfill a previously unknown or unmet need. The talented design people we have working for our organizations give us excellent designs when we specify who wants it and what it is that they want—the "they" being the customers who make up a market segment.

The problem with most failed new products and services is not poor design. The problem is that the product or service did not have customers waiting and ready for the things that were actually produced. The question is whether there is a way to reliably get around this problem of good design. There are also innovations that are replaced or that evolve quickly. Foods based on fad diets and toys based on television shows come and go. Other innovations, such as the computer, stay for generations. Why do some innovative products and services splash onto the scene and evaporate while others last? The answer often lies in the reason for wanting to create them in the first place.

DFSS was developed to precisely fill this methodological void. DFSS is a rich concept with a well-developed core methodology. The process entails a five-phase service or product development method, and the phases are as follows:

Define

In the define phase, top management has to look critically at the business. It would help to revisit the organization's strategic plan. (If you do not have an up-to-date strategic plan, you should get one.) Management provides the design team with specific guidance on the need for the new service or product; management should not, however, design the product. It is okay to provide a high-level concept, but leave the design to the designers.

Measure

The measure phase is all about discovering and exploring customers and their needs—especially any unmet needs. This is the heart of DFSS. How do you ask a target audience for what they want in a service or product that does not exist? You cannot, at least not directly. It is best to focus on needs. Again, let the designers design the product, not the customer.

The team then transforms the customers' needs into something more technical. We will call these critical-to-quality characteristics (CTQs). In the CTQs, we transform the needs as articulated by the customer into words and phrases we can measure. The CTQs become the targets for the designers. This step makes it possible to design a product or service that will interest a target group of customers. (Recall that this was the failing of most unsuccessful products or services.)

Analyze

In the analyze phase, the designers try several concept designs with potential to meet the CTQs developed in the measure phase. The concepts are now traceable to one or more CTQs, which in turn are traceable to one or more customer needs. The team develops and matches functional requirements of the concept design to the CTQs. The analyze phase is the exciting part for most designers, but the foundation was laid during the define and measure phases.

Design

The detail design follows. In the design phase, we take the winning concept design and fill in all the details. When inevitable choices and tradeoffs must be made, we have ready-made selection criteria: the CTQs. The CTQs are like having the customer beside us at every decision point. We will develop and match the functional requirements from analysis to the design requirements of the detail design.

Verify

When the team is satisfied with the details of the design, they are ready to verify meeting the business needs given to them by management in the define phase and the customers' needs provided during the measure phase. Complete planning for procurement, production, delivery, advertising, warranty, and other items is also completed during the verify phase.

Innovation can be enhanced. Most innovation will flourish if organizations can develop their own creative talents. Type II innovation is the key—encouraging all employees to think in terms of making something bigger, smaller, or combined with something else. DFSS then helps us to identify customers, learn their needs, and deliver products or services that meet those needs. Innovation cannot be commanded. But innovation can certainly be encouraged and managed to achieve an organization's goals by assigning teams to solve customer problems, by creating new goods and services to solve them.

Evolution of Design and Innovation Methods

Quality by Design

Quality by Design was a concept first outlined by Dr. Juran in various publications, most notably *Juran on Quality by Design*, by Dr. Juran and the Juran Institute. It stated that quality must be planned into products, and that most quality crises and problems relate to the way in which quality was planned in the first place. While Quality by Design principles have been used to advance product and process quality in every industry, and particularly the automotive industry, they have most recently been adopted by the U.S. Food and Drug Administration (FDA) as a vehicle for the transformation of how drugs are discovered, developed, and commercially manufactured. The FDA defines Quality by Design as the level of effectiveness of the design function in determining a product's operational requirements (and their incorporation into design requirements) that can be converted into a finished product in a production process. Today Quality by Design has evolved into numerous other methods. Here are some of the most popular:

Concurrent Engineering

Concurrent Engineering was a popular new product development process in which all individuals responsible for development and production were involved at the earliest stages of product design. Some 70 to 80 percent of a product's cost is locked in at these early stages of development, when the product's configuration is determined and choices are made for the manufacturing processes and materials from which the product will be made. If a product is to end up cost-competitive, it is absolutely essential that cost be a consideration when these decisions are made.

One of the earliest forms of Design for Quality was the Design for Manufacturing and Assembly (DFMA) from University of Massachusetts Profs. Boothroyd and Dewhurst. They created a methodology and later software technology that help guide design teams through this critical stage of product development with cost information, even before prototype design models are created.

Design for Manufacture

Design for Manufacture (DFM) is a systematic approach that allows engineers to anticipate manufacturing costs early in the design process, even when only rough geometries are available on the product being developed. Given the large number of process technologies and materials available, few design engineers have detailed knowledge of all the major shape-forming processes. Consequently, engineers tend to design for manufacturing processes with which they are familiar. DFM methodology encourages individual engineers and concurrent development teams to investigate additional processes and materials and to develop designs that may be more economical to produce. With more information about viable processes and materials, users can quantify manufacturing costs for competing design alternatives and decide which design is best.

DFM provides guidance in the selection of materials and processes and generates piece part and tooling cost estimates at any stage of product design. DFM is a critical component of the DFMA process that provides manufacturing knowledge into the cost reduction analysis of Design for Assembly.

Design for Assembly

Design for Assembly (DFA) is a methodology for evaluating part designs and the overall design of an assembly. It is a quantifiable way to identify unnecessary parts in an assembly and to determine assembly times and costs. Using DFA software, product engineers assess the cost contribution of each part and then simplify the product concept through part reduction strategies. These strategies involve incorporating as many features into one part as is economically feasible. The outcome of a DFA-based design is a more elegant product with fewer parts that is both functionally efficient and easy to assemble. The larger benefits of a DFA-based design are reduced part costs, improved quality and reliability, and shorter development cycles.

Design for Environment

Meeting the needs of an increasingly eco-conscious marketplace, DFMA allows product designers to conduct an environmental assessment during the concept stage of design, where they can evaluate the impact of material selection as well as account for the end-of-life status of their product.

The analysis prompts designers to select, from the DFMA database, the materials they prefer to use or avoid, then reveals the proportions (by weight) of those materials in the product. It also estimates and designates the proportions of product that go to different end-of-life destinations, including reuse, recycling, landfill and incineration. These measures help manufacturers meet such requirements as the European Union's Restriction of Hazardous Substances (RoHS) regulations.

Sustainable Design

Sustainable Design (also called Environmental Design, Environmentally Sustainable Design, Environmentally Conscious Design, etc.) is a method of designing physical goods that comply with the principles of economic, social, and ecological sustainability. The intention of Sustainable Design is to prevent negative environmental impact by identifying potential impacts and applying creative or best practices to prevent or mitigate them. Manifestations of sustainable designs require no nonrenewable resources, impact the environment minimally, and relate people with the natural environment.

Design for Six Sigma

The evolution of many lessons learned has led to the development of DFSS. It is focused on creating new or modified designs that are capable of significantly higher levels of performance (approaching Six Sigma). The define, measure, analyze, design, verify (DMADV) sequence is a design methodology applicable to developing new or revised products, services, and processes. Although DFSS implies to design to the lowest level of defects possible, Six Sigma, it is more than that. The steps in DFSS enable one to understand the customers and their needs. DFSS actually focuses on both sides of quality: the right features and the fewest failures.

Design for Six Sigma—DMADV Steps

Table 17.1 summarizes the main activities within each of the DMADV steps. These are discussed in more detail in this section. Experience with applying the five DMADV steps has led us to believe that it is useful to define a step to select the project before the team actually begins its DMADV journey.

Select the Opportunity

The select phase in DFSS is more strategic than for quality improvement or DMAIC projects (see Chap. 15, Six Sigma: Breakthrough to in-Process Effectiveness). A target for

Define	Measure	Analyze	Design	Verify
Agree to opportunity	Identify customers	Develop alternative designs	Develop detailed designs	Execute manufacturing/ operations verification
Agree to goals	Discover customer needs	Complete functional analysis	Integrate designs	Execute pilot and ramp-up
Agree to scope	Translate needs into CTQs	Select best-fit design	Model predictions of performance	Execute control plan
Establish project plan	Establish design scorecard	Specify functional requirements	Optimize design parameters	Finalize design scorecard
Assign resources		Specify subsystem functional requirements Complete high-level design review Validate with customer Update design scorecard	Develop statistical tolerances Specify process features and detailed operations Design complete control plan Complete design verification test Validate with customer Complete design review Update design scorecard	Transition to operational owners and validate

Table 17.1 Major Activities in Phases of DFSS

a new product or capability is identified as part of the strategic and annual business planning processes. When a major opportunity is identified, leadership will determine that it is best served with a new design or redesign of something that exists. Typically this means that a new or emerging market has been targeted; it may also mean that customer needs in an existing market are shifting, or that competition has shifted, and a new approach is required.

This type of project selection is different from a DMAIC project in which specific deficiencies or wastes are targeted for an existing product or process. Rarely is an existing product or process so broken that the initial analysis in DMAIC leads to the conclusion that a total redesign is required. A major health insurer reached that conclusion with respect to payment of claims. Instead of multiple improvement projects, it redesigned the entire claims payment service so as to raise customer satisfaction from 75 to 93 percent, improve timeliness by a factor of 10, and reduce costs by more than one-half.

The project opportunity and goal statements are prepared and included in a team charter, which is confirmed by management. Unlike the rather simple and direct goal statements for a DMAIC project, the DMADV goal statement may, in fact, be multiple statements about the market to be served by the new product and the economic returns to be achieved, such as market penetration, growth, and profitability. Management selects the most appropriate team of personnel for the project, ensures that they are properly trained, and assigns the necessary priority. Project progress is monitored to ensure success.

Select: Deliverables

- Make a list of potential projects.
- Calculate the return on investment and contribution to strategic business objective(s) for each potential project.
- Identify potential projects.
- Evaluate projects and select a project.
- Prepare project opportunity statement and a team charter.
- Select and launch team.
- Formal project team leader should be a qualified practitioner or Black Belt.

Select: Questions to Be Answered

1. What new market opportunities do we have?
2. What new emerging customers or customer needs can we go after?
3. What are the likely benefits to be reaped by gaining or increasing that business?
4. Which of our list of opportunities deserves to be tackled first, second, etc.?
5. What formal opportunity statement and goal statement should we assign to each project team?
6. Who should be the project team members and leader (Black Belt) for each project?

Define Phase

A project begins with the define phase when it is officially launched by the management team. It may be necessary for the management team or Champion to work closely with the project design team to refine the design opportunity. This refinement will lead to an accurate scope of the project and will ensure a common understanding of the objectives and deliverables. Experience has shown that projects that fail to deliver the expected results frequently get off track at the start, when the project is being defined.

A key task in the define phase is to create the initial business case that validates the selection rationale and establishes the business justification through reduced product cost, increased sales, or entirely new market opportunities. The initial business casework is conducted under the auspices of the management team, and then it is validated and updated continuously by the design team through the subsequent phases of the design project. The management team selects a black belt to lead the design project. The Champion, who is the management sponsor with vested interest in the success of the design, in conjunction with the Black Belt, is responsible for selecting a cross-functional team that will conduct all the activities to complete the design and carry it into production.

Define: Deliverables

- Initial business case is developed.
- Design strategy and project are established; leaders and team are selected.
- Project charter is drafted, including project opportunity statement and design objectives.
- Team is launched and a list of customers defined: market customers, nonmarket customers—users, regulators, stakeholders etc.—and internal customers.

Define: Questions to Be Answered

1. What are the design goals or objectives of the project?
2. What are the specific goals of the project team?
3. What is the business case that justifies the project?
4. What charter will the team members receive from management empowering them to carry out the project?
5. What will be the project plan?
6. How will the project be managed?
7. Who will be the customers of this project?

Measure Phase

The measure phase in the DMADV sequence is mainly concerned with identifying the key customers, determining what their critical needs are, and developing measurable critical quality (CTQ) requirements necessary for a successfully designed product. An initial assessment of our markets and customer segmentation by various factors is required to identify the key customers. This assessment is often completed by the marketing organization and is then reviewed and verified by the design team. However, it is the design team's responsibility to complete the customer needs analysis and compile the results into a prioritized tabulation of customer needs. The design team transforms the critical customer needs into measurable terms from a design perspective. These translated needs become the measurable CTQs that must be satisfied by the design solution. Competitive benchmarking and creative internal development are two additional sources to generate CTQs. These methods probe into design requirements that are not generally addressed or possibly even known by the customer. The result is a set of CTQs stated in specific technical requirements for design in the voice of the organization that become the measurable goals (specifications) for product performance and ultimate success.

The project team may use several means to set the goals for each CTQ. Some tools include competitive benchmarking, competitive analysis, value analysis, criticality analysis, and stretch objectives for current performance. The result is a combination of customers' stated requirements, and requirements that may not be generally addressed or known by the customer. The measure phase ends with the assessment of the current baseline performance against the enumerated CTQs and performance of risk assessments. To establish these baselines, typical process capability methods and tools are utilized. These include the following:

- Establish the ability of the measurement system to collect accurate data using measurement system analysis (MSA)
- Measure the stability of the current or surrogate process(es) using statistical process control techniques

- Calculate the capability and sigma level of the current or surrogate process(es)
- Evaluate risk by using tools such as design failure mode effects analysis (DFMEA) and process failure mode effects analysis (PFMEA)

Another tool employed by some design project teams is the set of quality function deployment (QFD) matrices (see Fig. 17.2). Each matrix lists vertically some objectives to be fulfilled (the "what") and then horizontally the means to fulfill the objectives (the "how"). Within the body of the matrix are indicators for how well each objective is met by the respective means. For example, the first matrix displays how well each of the customer needs is addressed by the specific CTQs. As a group, the matrices are tied together, with the means (how) of one matrix becoming the objectives (what) of the next. In this way the customer needs are tied seamlessly to the CTQs, to the functional requirements, to the design requirements, finally to the process requirements, and ultimately to the control requirements. In this way nothing critical is lost and no extraneous matters are introduced.

The QFD matrix (or simpler version) is meant to highlight the strengths and weaknesses that currently exist. In particular, the weaknesses represent gaps that the design team must shrink or overcome. The demand on the team then is to provide innovative solutions that will economically satisfy customer needs. Keeping this matrix up to date provides a running gap analysis for the team.

Discover Customer Needs

- Plan to collect customer needs from internal customers and external customers.
- Collect list of customers' needs in their language.
- Discover and prioritize customer needs in terms of the customer-perceived benefit.

Translate and Prioritize Customer Needs

- Translate needs and benefits from the voice of the customer (VOC) into voice of the producer as CTQ requirements.
- Establish measurement for all prioritized CTQs, including units of measure, sensor, and validation.
- Establish targets and upper and lower specification limits for all CTQs.
- Establish target permissible defect rate (DPMO, Sigma) for each CTQ.

Establish Baseline and Design Scorecard

Once the prioritized list of CTQs is produced, the design team proceeds to determine the baseline performance of relevant existing product and production process. The current baseline performance is determined in terms of multiple components:

- Measurement systems analysis
- Product capability
- Production process capability
- Risk assessment by using tools such as product FMEA
- Competitive performance

Finally, a design scorecard is created that tracks the design evolution toward a Six Sigma product performance. This tool is used in the attempt to predict what the final product performance and defect levels will be after integration of all the design elements. The design scorecard is updated throughout the project to ensure that objectives are met.

Measure: Deliverables

In summary, the key deliverables that are required to complete the measure phase are

- A prioritized list of customer needs
- A prioritized list of CTQs
- A current baseline performance
- A design scorecard

Measure: Questions to Be Answered

1. What customer needs must the new product meet?
2. What are the critical product and process requirements that will enable the customer needs to be met?
3. How capable is our current product and production process of meeting these requirements?
4. How capable must any new product and production process be to meet these requirements?

Analyze Phase

The main purpose of the analyze phase is to select a high-level design and develop the design requirements that will be the targets for performance of the detailed design. This is sometimes referred to as system-level design versus the subsystem or component design levels.

The design team develops several high-level alternatives that represent different functional solutions to the collective CTQ requirements. A set of evaluation criteria is then developed, against which the design alternatives will be analyzed. The final configuration selected may be a combination of two or more alternatives. As more design information is developed during the course of the project, the design may be revisited and refined.

In developing the high-level design, the team establishes the system's functional architecture. The flow of signals, flow of information, and mechanical linkages indicate the relationship among the subsystems for each design alternative. Hierarchical function diagrams, functional block diagrams, function trees, and signal flow diagrams are commonly used to illustrate these interrelationships. Where possible, models are developed and simulations run to evaluate the overall system functionality.

The requirements for each subsystem are expressed in terms of their functionality and interfaces. The functionality may be expressed as the system transfer function, which would represent the desired behavior of the system or subsystem. Interfaces are described in terms of the input and output requirements and the controls (feedback, feed-forward, automatic controls). These specifications will be provided to the detail design teams in the design phase.

In the analyze phase, DMADV analysis tools enable the design team to assess the performance of each design alternative and to test the differences in performance of the competing design alternatives. The results of these tests lead to the selection of the best-fit design, which is then the basis to move into the next phase, detailed design. These analyses are accomplished using graphical and statistical tools including

- Competitive analysis
- Value analysis
- Criticality analysis
- Fault-tree analysis
- Risk analysis
- Capability analysis
- High-level design matrices from QFD
- TRIZ (teoriya resheniya izobretatelskikh zadach; Russian, literally "theory of the resolution of invention-related tasks")
- Updated design scorecard

One of the significant advances affecting this process is the availability of several statistical analysis tools. These software applications, running on desktops or laptop computers, speed up the number crunching required to perform the preceding analysis. This availability has also made it necessary for individuals who would not normally use these tools to be trained in the use and interpretation of the results.

Analyze: Deliverables

Develop a high-level product or service and process design and detail design requirements.

- Design alternatives
- Functional analysis
- Best alternative selected
- Best-fit analysis
- High-level quantitative design elements
- High-level resource requirements and operating ranges
- High-level design capability analysis and prediction
- Detail design requirements for subsystems/modules
- Key sourcing decisions
- Initial product introduction resources and plans
- Updated design scorecard
- QFD design matrices

Analyze: Questions to Be Answered

1. What design alternatives could be employed in the new product or process service?
2. Which is the "best" alternative?

3. What are the requirements for the detailed design?
4. Has customer feedback been obtained?
5. Does the high-level design pass a business and technical design review?
6. Has the design been validated with customers?

Design Phase

The design phase builds upon the high-level design requirements to deliver a detailed optimized functional design that meets operational manufacturing and service requirements. Detail designs are carried out on the subsystems and eventually integrated into the complete functional system (product). DMADV tools focus on optimizing the detail-level design parameters.

In particular, designed experiments and/or simulations serve several purposes. One purpose is to determine the best set of features (optimum configuration) to employ. Another purpose can be to obtain a mathematical prediction equation that can be used in subsequent modeling and simulations. Experiments are typically designed at differing levels of complexity, from minimal-run screening experiments to multilevel replicated design. Screening experiments typically try to establish which factors influence the system, providing somewhat limited results for modeling. More detailed experiments, including response surface and mixture designs, are conducted to determine system performance more accurately and produce a mathematical equation suitable for prediction and modeling applications. More complex products will often require nonlinear response surface models as well as mixture and multiple-response models.

During the design phase, the design team is also concerned about the processes that must be developed to provide the service or build the product. During the measure phase, the team examines the current capability of the business to deliver the product or service at the expected quality levels (approaching Six Sigma). During the design phase, the team continually updates the design scorecard with the results of designed experiments, benchmarking results, process capability studies, and other studies to track the design performance against the established goals, continuing the gap analysis that runs throughout the project. The product design is also reevaluated against the manufacturing or operational capability. Product designs may be revised as needed to ensure reliable, capable manufacturing and operations.

Part of the design for operations includes the validation of tolerances for each parameter. Designed experiments can contribute to developing these tolerances, and statistical tolerancing can also validate them.

To conclude the design phase requires the goals of the design for performance to be verified through testing of prototype, preproduction models, or initial pilot samples or pilot runs. The design team documents the set of tests, experiments, simulations, and pilot builds required to verify the product/service performance in a design verification test (DVT) plan. Upon completion of the several iterations that occur during the DVT and pilot runs, the design is solidified and the results of testing are summarized. A design review meeting marks the conclusion of the design phase, when the results of the DVT are reviewed. The design scorecard is updated, and each area of the development plan (quality plan, procurement plan, manufacturing plan, etc.) is adjusted as necessary.

Design: Deliverables

- Optimized design parameters (elements)—nominal values that are most robust
- Prediction models

- Optimal tolerances and design settings
- Detailed functional design
- Detailed designs and design drawings
- Detailed design for operations/manufacturing
- Standard operating procedures, standard work, and work instructions
- Reliability/lifetime analysis results
- Design verification test results
- Updated design scorecard

Design: Questions to Be Answered

1. What detailed product design parameters minimize variation in product performance?
2. What tolerances both are practical and ensure performance?
3. How do we ensure optimum product reliability?
4. How do we ensure simplicity and ease of manufacture or operations?
5. What detailed process parameters consistently and predictably minimize production process variation around target values?

Verify Phase

The purpose of the verify phase in the DMADV sequence is to ensure that the new design can be manufactured or service delivered and field supported within the required quality, reliability, and cost parameters. Following DVT, a ramp-up to full-scale production is accomplished via the manufacturing verification test (MVT) or operations verification test (OVT). The objective of this series of tests is to uncover any potential production or support issues or problems. The operations process is typically exercised through one or more pilot runs. During these runs, appropriate process evaluations occur, such as capability analyses and measurement systems analyses. Process controls are verified and adjustments are made to the appropriate standard operating procedures, inspection procedures, process sheets, and other process documentation. These formal documents are handed off to downstream process owners (e.g., manufacturing, logistics, and service). They should outline the required controls and tolerance limits that should be adhered to and maintained by manufacturing and service. These documents come under the stewardship of the company's internal quality systems. One of the considerations of the design team is to ensure that the project documentation will conform to the internal requirements of the quality system.

The design team should ensure that appropriate testing in a service and field support environment is accomplished to uncover potential lifetime or serviceability issues. These tests will vary greatly, depending on the product and industry. These tests may be lengthy and possibly not conclude before production launch. The risks associated with not having completed all tests depend on the effectiveness of earlier testing and the progress of final MVT/OVT tests that are underway. A final design scorecard should be completed, and all key findings should be recorded and archived for future reference. The team should complete a final report that includes a look back at the execution of the project. Identifying and discussing the positive and not-so-positive events and issues will help the team learn from any mistakes made and provide the basis for continuing improvement of the DFSS sequence.

Verify: Deliverables

- Verify product/process performance against project targets
- Pilot build is complete
- Pilot tests are completed and results are analyzed
- All operational and control documentation, procedures, controls, and training are complete
- Scale-up decision(s) are made
- Full-scale processes are built and implemented
- Business results are determined/analyzed
- Processes are transitioned to owners
- DFSS project is closed

Verify: Questions to Be Answered

1. Is the product or process meeting the specifications and requirements?
2. Is the production process "owned" by the business?

Examples of Continuous Innovation Process Using Design for Six Sigma

Example 1: A Design for Six Sigma (DMADV) Project*

Project Background

The current process to look up, retrieve, and interpret product engineering information such as component specification drawings and product structures has been in place since 1998. This system is complex and expensive to maintain. From the beginning, this process has had many shortcomings from the point of view of the primary users—the manufacturing plants. These shortcomings cost the company money in lost productivity and high system maintenance costs.

DMADV Process Implementation

With the long history of complaints and a limited customer base, areas of improvement were not difficult to determine. To provide focus for our team, a survey was developed and analyzed to prioritize customer groups and customer needs as well as their performance expectations for the new system. The needs became the customer CTQ items.

We worked with our customers to determine baseline capability against four criteria:

1. Accuracy of information
2. Fast retrieval of information
3. Easy retrieval of information
4. Easy-to-interpret information

*Adapted from the final report of a Six Sigma design project led by Dave Kinsel at a Juran Institute client; with acknowledgment of thanks.

From this list we constructed a quality function deployment flow-down matrix to convert the CTQs to product feature alternatives that support customer needs. The current process was mapped at high and then more detailed levels to identify areas of improvement.

A high-level design was prepared, and high-level capability was estimated. Next a more detailed design was developed, simulated, documented, and verified.

Results

- Accuracy level unchanged (Six Sigma capable)
- A 451 percent improvement in average print access/printout time (from 1.5 to 6 sigma)
- 100 percent improvement in virtual viewing/inquiry capability
- 300 percent improvement in drawing line weight differentiation
- Final expected savings: not insignificant

Project Details and Selected Slides

Problem statement: Plant quality and customer service/technical support personnel find that our current system to find and view product component and assembly information is cumbersome to access, interpret, and maintain.

Project definition: The purpose is to provide faster access to product engineering information in a consolidated format using a single user-friendly interface.

Mission statement: The project team will develop a user interface and training system to provide faster single-point access for plant quality managers, engineering, customer service, and technical support to product structures and related component and assembly specifications by July 2004.

The slides shown in Figs. 17.3 through 17.20 highlight the project for each phase: define, measure, analyze, design, and verify.

This example depicts well how the DFSS process takes place and can be used by the practitioner as a guide for his or her own projects.

Example 2: A Design for Six Sigma (DMADV) Project

This second project is an example of DFSS applied to a new product development and how that application can result in a more successful product being brought to market because it better meets customer needs. Due to the sensitive competitive nature of such a project, the example has intentionally been made generic for presentation here.

Project Background

The project was chartered to design a new, more competitive consumer medical device. The following sections detail the project background, important business considerations, and the customer characteristics.

Development Goals

Provide an improved consumer device that optimally meets feature and benefit requirements of the product line.

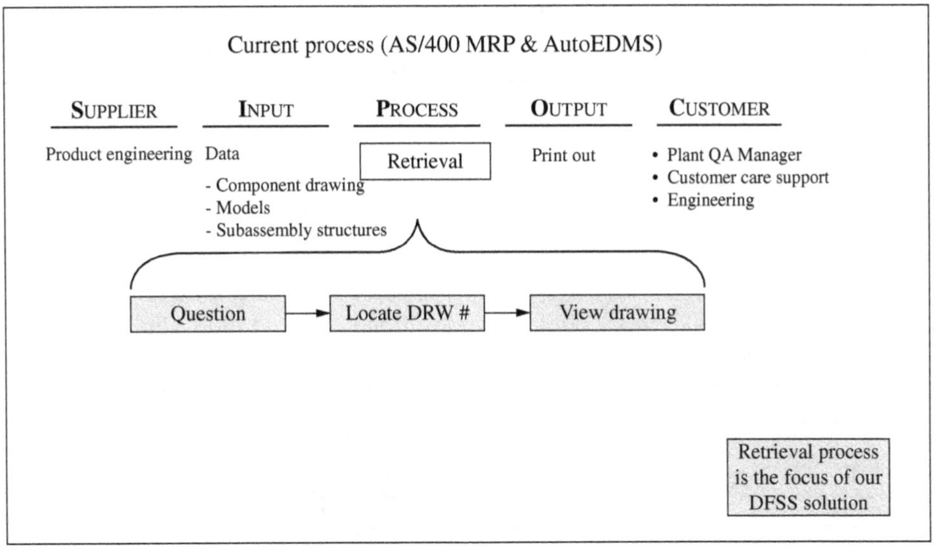

Figure 17.3 SIPOC (high-level process map).

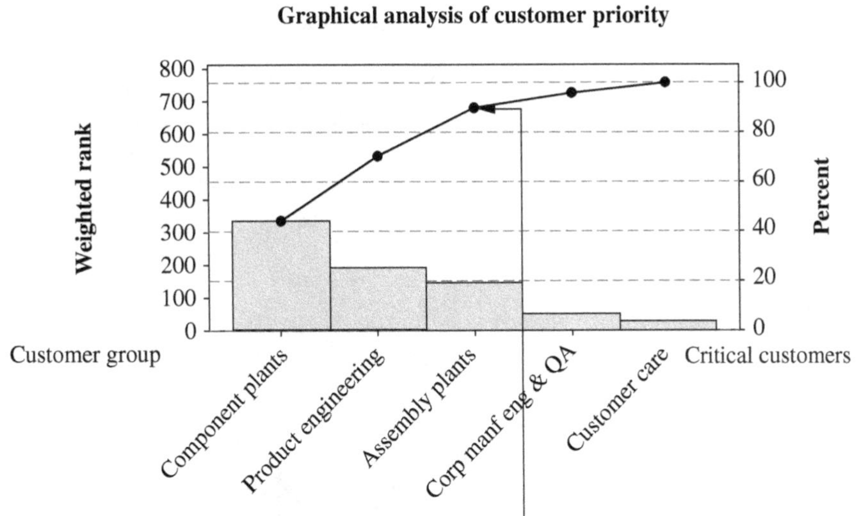

Count	34.1	192	144	50	29
Percent	45.1	25.4	19.0	6.6	3.8
Cum %	45.1	70.5	89.6	96.2	100.0

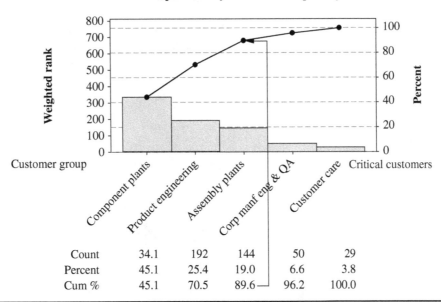

Figure 17.4 Pareto of customer prioritization.

Prioritized customers (based on # users in group & frequency of need)	Customer weighting	Customer needs (based on survey results)	Accuracy of information	Speed to retrieve	Ease of retrevial	Format
		Need weighting	**592**	**343**	**275**	**260**
		Association table customer Wt × Need Wt.				
Component plant	341		201,872	116,963	93,775	88,660
Product engineering	192		113,664	65,856	52,800	49,920
Assembly plant	144		85,248	49,392	39,600	37,440
		Totals:	400,784	232,211	186,175	176,020

Figure 17.5 Flow-down customer versus customer needs.

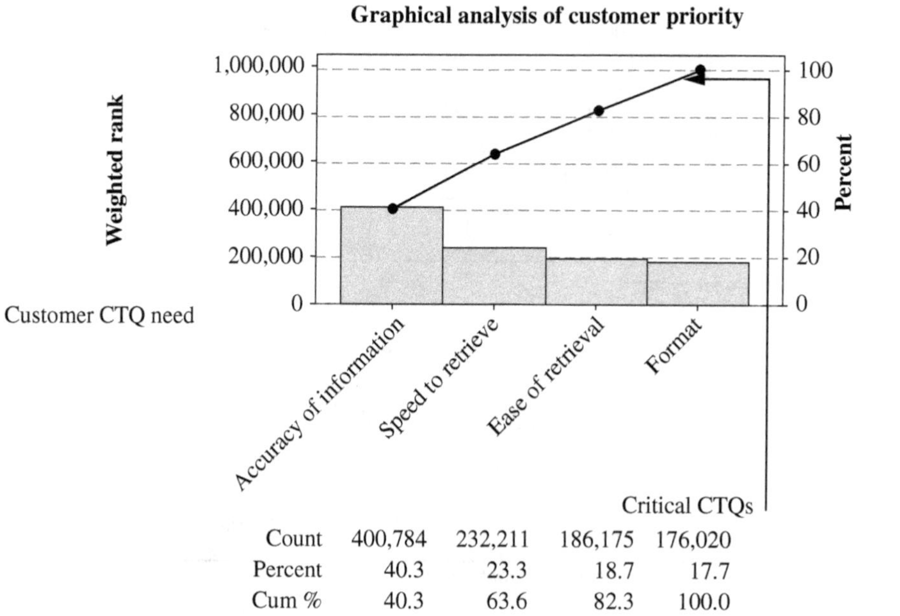

Figure 17.6 The vital few CTQs.

Need/ Expectation	Priority	Characteristic	Measure/ Sensor	Target	Upper Specification Limit	Allowable Defect Rate
Accuracy of information	400,784	Drawing represents part number correctly	Match: Y or N/ visual	Y	Must match	3.4 DPMO
Fast retrieval of information	232,211	Time to find a component drawing and print	Time/ Stopwatch	1.7 min	+1.6 min	10,700 DPMO
Easy retrieval of information	186,175	Number of user inputs to locate a drawing	Number of inputs/visual	10	+3	3.4 DPMO
Information is easy to interpret (format)	176,020	Different line weights are apparent on drawing	Number of multiple line weights/ visual	3	+1	3.4 DPMO

Figure 17.7 Translation of customer needs into measurable CTQs.

Baseline CTQ Capability Analysis

Drawing Access Spec (combined data from Dim and Non-Dim Lookup)

Normality Test

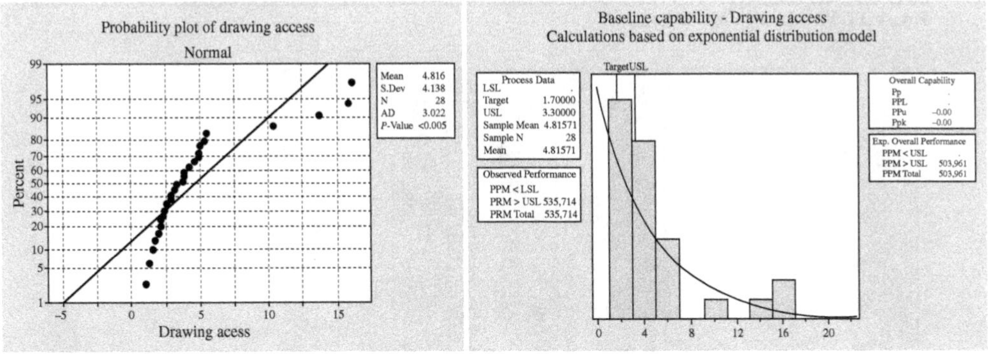

Alpha: 0.05

Ho: Data are normal

Ha: Data are not normal

P-value: <0.005, therefore reject Ho.

Data are not normal

Conclusion:

Use 1 sample Wilcoxon for data statistical analysis

Conclusion:

Baseline capability based on 500,000 DPMO is 1.4 sigma

Figure 17.8 Baseline CTQ.

Baseline CTQ Capability Analysis
Drawing Access Spec (combined data from Dim and Non-Dim Lookup)

Customer expectations are 1.70 minutes or faster to access and print a drawing, from CTQ survey. Is the current system running at the customers' expectations? This is the primary CTQ on our design scorecard.
See Appendix A for survey data analysis.

Ho = Sample median is equal to 1.70 minutes
Ha = Sample median is not equal to 1.70 minutes
Alpha = 0.05

Wilcoxon Signed Rank Test: Drawing Access

```
Test of median = 1.700 versus median not = 1.700

                      N
                     for   Wilcoxon           Estimated
                N   Test  Statistic      P      Median
Drawing Access 28    28       393.5  0.000       3.750
```

P-value = .000 < 0.05
Therefore reject Ho
Ha = Sample median is not equal to 1.70 minutes

Ho = Sample median is equal to 1.70 minutes
Ha = Sample median is greater than 1.70 minutes
Alpha = 0.05

Wilcoxon Signed Rank Test: Drawing Access

```
Test of median = 1.700 versus median not > 1.700

                      N
                     for   Wilcoxon           Estimated
                N   Test  Statistic      P      Median
Drawing Access 28    28       393.5  0.000       3.750
```

P-value = .000 < 0.05
Therefore reject Ho
Ha = Sample median is greater than 1.70 minutes

Conclusion: The current system is not meeting customer expectations

FIGURE 17.9 Baseline CTQ.

CTQs					
Description	Spec/Target LSL	Spec/Target USL	Current capability	High-level Capability	Feature Capability (From verification testing)
Information is accurate	0 Errors (6 Sigma)		0 Errors (6 Sigma)		
Fast retrieval	0 sec	3.3 min	4.82 min (1.4 Sigma)		
Easy retrieval (mimimal inputs)	8	12	17 (0 Sigma)		
Easy to Interpret format (number) of line weights)	2	4	1 (0 Sigma)		

FIGURE 17.10 Design scorecard.

CTQs	Functions: Retrieve specs by part number	Drawings retrievable by latest revision	Multiple line weights on drawing high print quality	Readily accessible	Attribute information on drawing (species/color etc.)	3D representation of cabinet	Capable of storing legacy data/drawings	
Easy to interpret (format)	□ 2	⊖ 1	■ 3	⊖ 1	■ 3	□ 2	⊖ 1	13
Fast retrieval	■ 3	⊖ 1	⊖ 1	■ 3	⊖ 1	⊖ 1	□ 2	12
Minimal number of inputs from user	■ 3	■ 3	⊖ 1	⊖ 1	⊖ 1	⊖ 1	⊖ 1	11
Information is accurate	■ 3	□ 2	⊖ 1	⊖ 1	⊖ 1	⊖ 1	⊖ 1	10
	11	7	6	6	6	5	5	

KEY	
■ 3	Strong relationship
□ 2	Moderate relationship
⊖ 1	Weak relationship

FIGURE 17.11 QFD flow down: CTQs versus functions.

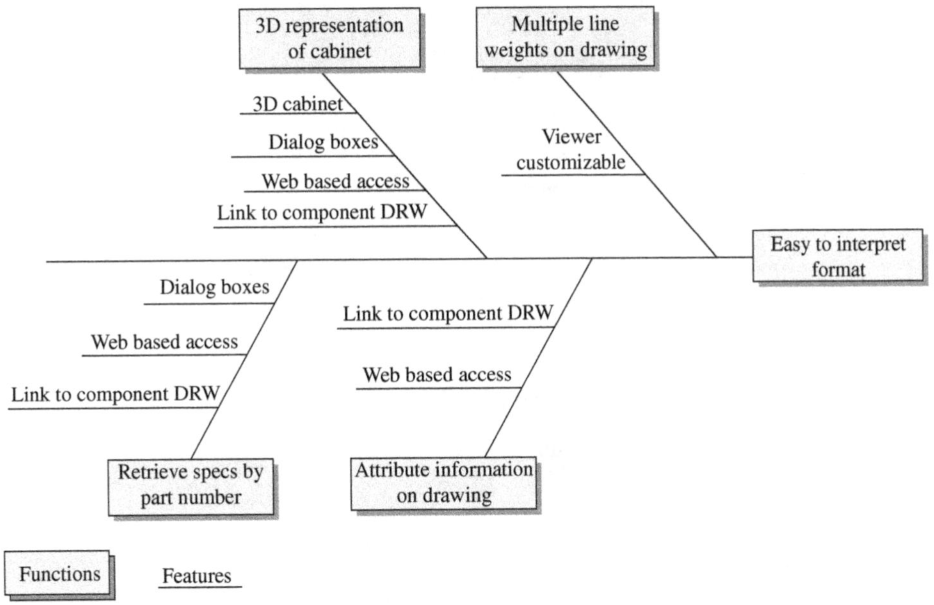

FIGURE 17.12 Easy to interpret function/feature diagram.

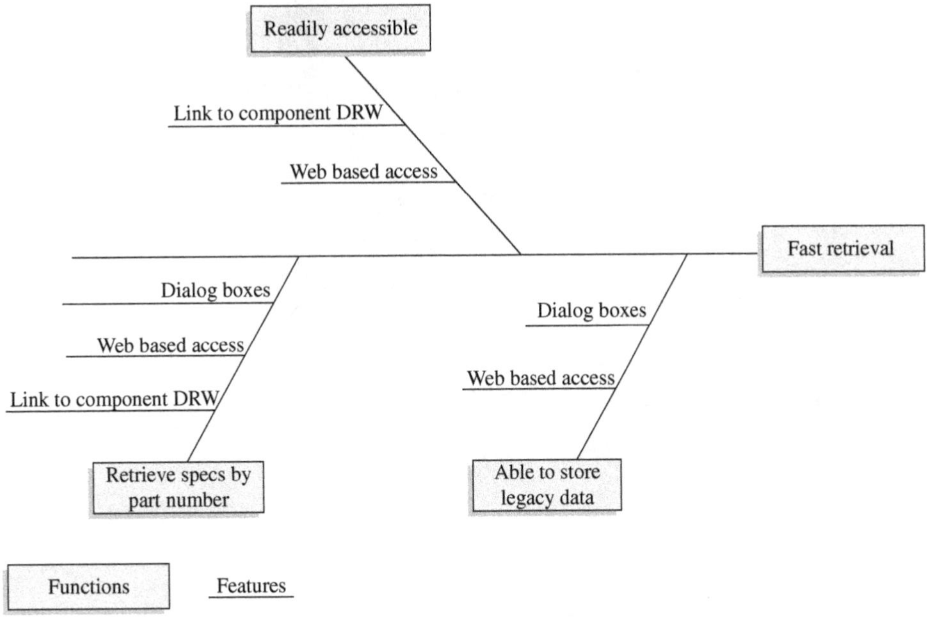

FIGURE 17.13 Function/feature diagram, fast retrieval.

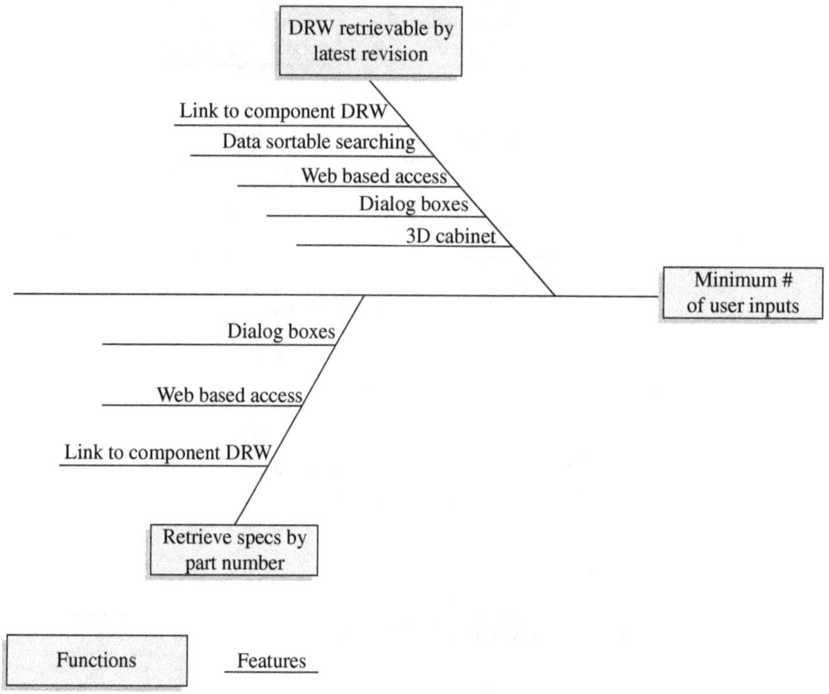

FIGURE 17.14 Function/feature diagram, minimum number of user inputs.

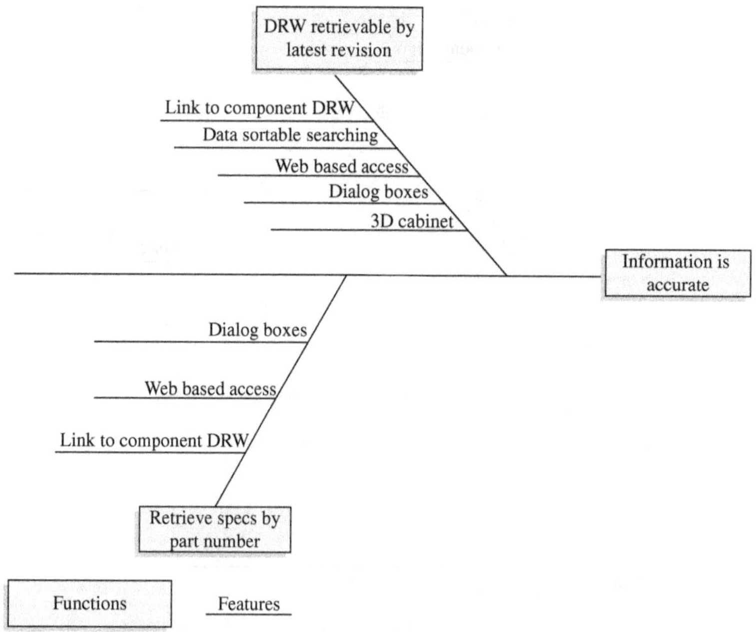

FIGURE 17.15 Function/feature diagram, information accurate.

Functions	Features: Link to component drawings	Web-based access	Dialog boxes	Virtual cab viewing	Data sortable searching	Viewer customization	
Drawings retrievable by latest revision	■ 3	□ 2	□ 2	□ 2	■ 3	⊖ 1	13
3D representation of cabinet	□ 2	□ 2	□ 2	■ 3	⊖ 1	⊖ 1	11
Readily accessible	■ 3	■ 3	⊖ 1	⊖ 1	⊖ 1	⊖ 1	10
Retrieve specs by part number	□ 2	□ 2	■ 3	⊖ 1	⊖ 1	⊖ 1	10
Attribute information on drawing (species/color etc.)	■ 3	□ 2	⊖ 1	⊖ 1	⊖ 1	⊖ 1	9
Capable of storing legacy data/drawings	⊖ 1	□ 2	□ 2	⊖ 1	⊖ 1	⊖ 1	8
Multiple line weights on drawing high print quality	⊖ 1	⊖ 1	⊖ 1	⊖ 1	⊖ 1	■ 3	8
	15	14	12	10	9	9	

KEY	
■ 3	Strong relationship
□ 2	Moderate relationship
⊖ 1	Weak relationship

Figure 17.16 QFD flowdown, functions versus features.

Features	Alternatives (X's): Use case-based training materials	JSP web page	Browser plug-in	Link from list	Java applet input box	Link from thumbnails	
Web based access	■ 3	■ 3	■ 3	■ 3	■ 3	■ 3	18
Virtual cab viewing	■ 3	□ 2	■ 3	■ 3	⊖ 1	■ 3	15
Link to component drawing	■ 3	■ 3	□ 2	■ 3	■ 3	⊖ 1	15
Dialog boxes	■ 3	■ 3	⊖ 1	⊖ 1	■ 3	⊖ 1	12
Data sortable searching	■ 3	■ 3	⊖ 1	⊖ 1	⊖ 1	⊖ 1	10
Viewer customization	■ 3	⊖ 1	■ 3	⊖ 1	⊖ 1	⊖ 1	10
	18	15	13	12	12	10	

KEY	
■ 3	Strong relationship
□ 2	Moderate relationship
⊖ 1	Weak relationship

Figure 17.17 QFD flowdown, features versus alternatives.

Verification CTQ Capability Analysis

Drawing Access Spec (combined data from Dim and non-Dim lookup)

Normality Test

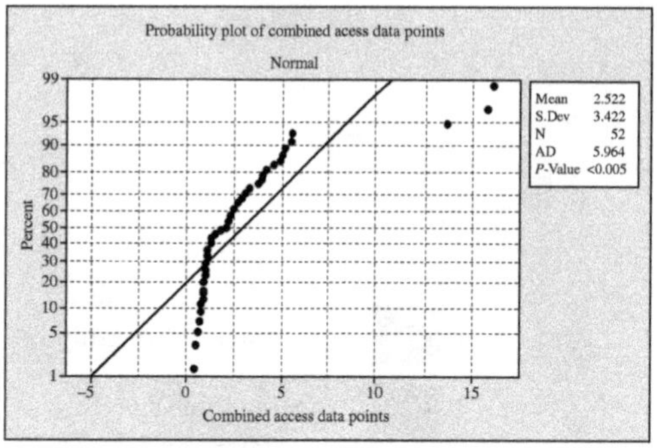

Alpha: 0.05
Ho: Data are normal
Ha: Data are not normal
P-value: <0.005, therefore reject Ho.
Data are not Normal. Perform Levenes test for equal variance
Conclusion:
Use Mann -Whitney for data statistical analysis

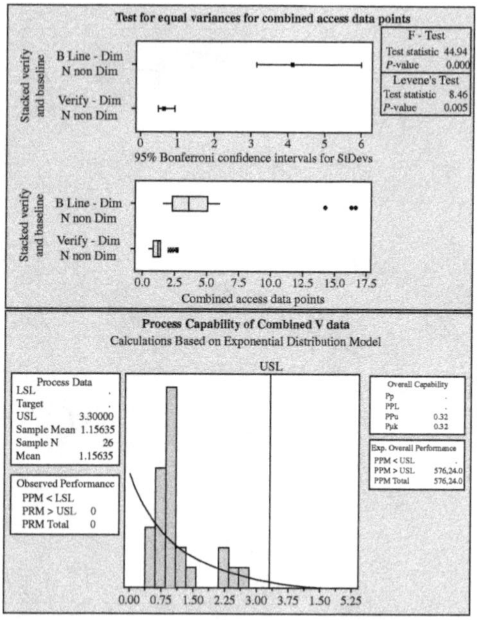

Conclusion:
Baseline capability based on 0 DPMO is 6 sigma

Figure 17.18 Verification CTQ capability analysis.

Verification CTQ Capability Analysis

Drawing Access Spec (combined data from Dim and non-Dim lookup)

Customer Expectations are 1.70 minutes or faster to access and print a drawing, from CTQ Survey. Is the new system running at the customers expectations? This is the primary CTQ on our design scorecard.

Ho = Sample median is equal to 1.70 minutes
Ha = Sample median is not equal to 1.70 minutes
Alpha = 0.05

Wilcoxon Signed Rank Test: Combined V data

```
Test of median = 1.700 versus median not = 1.700

                        N
                       for   Wilcoxon            Estimated
                   N  Test  Statistic       P       Median
Combined V data   26    26       53.5   0.002        1.033
```

P value = .002 < 0.05
Therefore reject Ho
Ha = Sample median is not equal to 1.70 minutes

Ho = Sample median is equal to 1.70 minutes
Ha = Sample median is less than 1.70 minutes
Alpha = 0.05

Wilcoxon Signed Rank Test: Combined V data

```
Test of median = 1.700 versus median < 1.700

                        N
                       for   Wilcoxon            Estimated
                   N  Test  Statistic       P       Median
Combined V data   26    26       53.5   0.001        1.033
```

P value = .001 < 0.05
Therefore reject Ho
Ha = Sample median is less than 1.70 minutes

Conclusion: DDL and the design of the user interface will meet customer expectations for drawing access time.

Figure 17.19 Verification CTQ capability analysis.

CTQs					
Description	Spec/Target		Current capability	High level capability	Feature capability (from verification testing)
	LSL	USL			
Information is accurate	0 Errors (6 Sigma)		0 Errors (6 Sigma)	0 Errors (6 Sigma)	0 Errors (6 Sigma)
Fast retrieval	0 sec	3.3 min	4.82 min (1.4 Sigma)	0.75 min (6 Sigma)	1.07 min (6 Sigma)
Easy retrieval (minimal inputs)	8	12	17 (0 Sigma)	10 (6 Sigma)	10 (6 Sigma)
Easy to interpret format (number of line weights)	2	4	1 (0 Sigma)	3 (6 Sigma)	3 (6 Sigma)

Figure 17.20 Design scorecard updated and verified.

Product Description

The product is a medical device for use by patients with specific conditions that lend themselves to use of self-monitoring systems.

Process(es) within Scope

- Industrial design
- Packaging configuration
- Device color, texture
- Device configuration
- Device ergonomics, ease of use
- Launch schedule

Market Strategy

The current market for this device mirrors the market for the higher-level devices it is used with. However, only 40 to 50 percent of our users of the higher-level device report using this device's current version. Among our competition this device is a much higher source of revenue, and this would imply that they can produce it at a lower cost.

Financial Strategy

Of today's similar devices 98 percent go into kits. Therefore, reducing cost is important.

Technology Strategy

No off-the-shelf original equipment manufacturer devices provide multiple capabilities. However, the project can leverage prior development efforts to implement enhanced capabilities in this design.

Product Strategy

The strategy is to provide a device that maximizes customer acceptance across all major higher-level device platforms.

Results Achieved and Champion Approval

Black Belt: Joe Black Belt
Project: Device
CTQs: Ease of Use, Performance, Cost

Champion:____________________
Approval:____________________
Date:____________________
Estimated Benefits: Save $0.13 per device, 6% increase in sales
Actual Benefits: Saved €0.23 ($0.335) per device, sales TBD

Define	Measure	Analyze (High-level design)	Design	Verify
Start Date: Jan 2010 *End Date: Jun 2010*	*Start Date: Jan 2010* *End Date: Sep 2010*	☑ *Start Date: Sep 2010* ☑ *End Date: Jan 2011*	☑ *Start Date: Jan 2011* ☑ *End Date: Jul 2011*	☑ *Start Date: Jan 2011* ☑ *End Date: Mar 2012*
☑ Design concept ☑ High-level process map ☑ Project charter ☑ Business case ☑ Formal project approval	☑ CTQs ☑ Performance objective(s): Project Y(s) ☑ Possible Xs ☑ Data collection plan ☑ Measurement system analysis ☑ Baseline product performance ☑ Baseline product performance	☑ Functional requirements ☑ QFD house of quality ☑ Design alternatives ☑ Selection criteria ☑ High-level design ☑ Design capability ☑ Transfer function	☑ Detail design requirements ☑ Design optimization ☑ Robust design ☑ Tolerancing ☑ Reliability study ☑ Cost analysis ☑ DVT	☑ MVT ☑ Process pilot test ☑ Sustainable process controls ☑ Validated: ☑ Control system ☑ Monitoring plan ☑ Response plan ☑ Standardize & translate ☑ Validated business case ☑ Formal champion approval

☐ Not complete
✓ Complete
❖ Not applicable

FIGURE 17.21 DFSS applied to medical device design.

Design Project Approach

Leverage existing device development, especially internal mechanism with reduced bounce (associated with pain).

Figure 17.21 shows the process the team followed, each step completed, and the associated timing. Also, you can see the actual results of the project as measured by the marketing and financial strategies above. The project in fact exceeded the cost reduction goal of $0.13 per device with the actual reduction of $0.335 per device.

It may also be noted from Fig. 17.21 that the total project took a little more than 2 years. Particularly for the application of DFSS to new product design, as in this case, it is common for DMADV projects to take a good bit longer than DMAIC projects.

References

Early, J. F., and Colleti, O. J. (1999). "The Quality Planning Process," *Juran's Quality Handbook*, 5th ed., McGraw-Hill, New York.

Endres, A. (2000). *Implementing Juran's Roadmap for Quality Leadership*, John Wiley and Sons, Inc., New York.

Juran, J. M. (1988). *Juran on Planning for Quality*, The Free Press, New York.

CHAPTER 18

Benchmarking: Best Practices for Market Leadership

Brad Wood

High Points of This Chapter

1. Benchmarking is a systematic and continuous process that facilitates the measurement and comparison of performance and the identification of leading practices that enable superior performance.
2. The main objective of benchmarking is to identify superior performance internal or external to your organization, determine the nature of and reasons for this performance, and determine if there are gaps in your organization to this performance.
3. Organizations may benchmark for many different reasons, but the strongest driver should be to improve organizational performance.
4. Benchmarking can be classified in many different ways, but the same principles and processes must be applied for benchmarking to be successful.
5. Benchmarking can provide vital input to the performance improvement process as depicted by the close interactions with the Juran Trilogy.
6. Benchmarking can have direct input to the strategic planning process by providing factual foundations to frame the vision, goals, and plans for world-class leadership.
7. A well-structured and systematic process such as the Juran 7-Step Benchmarking Process is essential to realize successful benchmarking.
8. Critical to success are a clear scope and objectives, good definitions, thorough validation, effective normalization, clear reporting, and a willingness to share information on leading practices.
9. It is essential that benchmarking always be conducted within a legal and ethical framework.
10. The resources required to conduct a benchmarking study and to act upon the findings must be provided by management if any real value is to be gained from benchmarking.

Benchmarking: What It Is and What It Is Not

Benchmarking has been in existence for a great many years. The concept of one individual observing how another performs a given task and then applying learning from that observation with the goal of adapting and improving how the task is executed is one of the fundamental ways in which human beings learn and develop. In the context of business, learning from one's competitors has also been in existence for as long as business has. However, the application of learning from leading practices to the business environment in a structured, methodical, and indeed legal and ethical way is relatively new. Xerox Corp. is most commonly credited with developing the modern form of benchmarking, and it is fair to say that the majority of today's benchmarking practices are built upon the approach developed by them in the 1970s.

Although their story has previously been well told in a multitude of management texts, it is still worthy of brief comment here to give context.

A combination of poor product quality, high overheads, and increasing competition from a growing number of Japanese organizations had left Xerox in a precarious position in the late 1970s. A visit to Japan provided the wake-up call that change was essential if they were to survive (Kearns and Nadler 1993). They put in place a series of benchmarking activities aimed at identifying the best-performing organizations in various aspects of their business and determining what it was that these organizations were doing that enabled a superior performance. Most famous is the benchmarking of logistics operations that they undertook with L.L. Bean (Camp 1989); with this, modern benchmarking was born.

Benchmarking has evolved to become an essential element of the business performance improvement tool kit and is now frequently used by many organizations in a wide range of different industries. But despite this, it remains one of the most widely misunderstood improvement tools. It means many different things to many different people, and all too frequently benchmarking projects fail to deliver on their promise of improvement or real results.

However, when executed correctly, benchmarking can provide a powerful focus for organizations, driving home the facts and convincing the organization of the need to embark upon improvement strategies. Benchmarking is a tool that enables the identification and ultimately the achievement of excellence, based upon the realities of the business environment rather than on internal standards and historical trends.

Benchmarking is not what we would term "industrial tourism" in which superficial industrial visits are undertaken, to perceived leading organizations, in the absence of any point of reference and do not assist in the improvement process. It is impossible to acquire detailed knowledge after only a quick glance or one short visit, and it is rare for such visits to result in an action plan that will lead to improvement. In the absence of prior benchmarking, it is also difficult to identify which organizations should be visited, and so there is a real risk that visits are made to organizations that are perceived as being the best or at least better, when the reality may be very different. However, there is a valuable role to be played by this type of site visit, when it is conducted following a structured benchmarking analysis and the organization being visited has been identified as a leading performer.

Benchmarking also should not be considered a personal performance appraisal tool. The focus should be on the organization and the individuals within it. Failure to adopt this philosophy will only lead to resistance and will undoubtedly add roadblocks to a successful benchmarking journey.

Nor should benchmarking be a momentary glimpse, but rather it should be considered a continuous process. Organizations must change performance rapidly to remain competitive in business environments today. This fast-paced tempo is further accelerated in sectors where benchmarking is commonplace, where businesses rapidly and continuously learn from one another. A prime example comes from the oil and gas industry where organizations have to respond to ever-increasing business, technological, and regulatory demands. The majority of the key players in this industry are participating in focused benchmarking consortia on an annual basis. It is also much more than a competitive analysis. Benchmarking goes further than examining the pricing and features of competitors' products or services. It considers not only the output but also the process by which the output was obtained. Benchmarking is also much more than market research. It considers the business practices in place that are enable the satisfaction of customer needs and thus realize superior business performance. It provides evidence-based input offering a powerful focus for management based on the facts that will convince the organization of the need to embark on improvement activities.

Participating in benchmarking should not be viewed as a stand-alone activity. To succeed, it must be part of a continuous improvement strategy, it must be conducted regularly, and it should be enveloped in the continuous improvement culture of an organization. Like any other project, it has to have the full support of senior management, the resources necessary to fulfill the objectives, and a robust project plan that is adhered to.

Finally, benchmarking should not be viewed as the answer in itself. It is a means to an end. An organization will not improve performance by benchmarking alone. It must act upon the findings of the benchmarking to improve. The output of benchmarking should provide input to decision-making or improvement action planning. This requires detailed consideration of the benchmarking analysis, formulation of learning points, and development of action plans in order to implement change and realize improvements.

So how can we define benchmarking? A scan of the literature will quickly reveal myriad definitions (Anand and Kodali 2008), each offering a slight variance on a common theme. Rather than repeat these here, we prefer to offer our own definition:

> *Benchmarking* is a systematic and continuous process that facilitates the measurement and comparison of performance and the identification of leading practices that enable superior performance.

This definition is deliberately generic so that it can encompass all types of benchmarking. In this context, measurement and comparison may be between organizations, business units, business functions, and business processes, products, or services. The benchmarking may be internal or external, between competitors, within the same industry, or cross-industry. Regardless of the category of benchmarking this definition still applies.

Objectives of Benchmarking

The objectives of benchmarking can be summarized as follows:

1. Determine superior performance levels.
2. Quantify any performance gaps.
3. Identify leading practices.
4. Evaluate reasons for superior performance.
5. Understand performance gaps in key business areas.
6. Share knowledge of working practices that enable superior performance.
7. Enable learning to build foundations for performance improvement.

When one is talking of superior performance, ultimately of course the aim should be for world-class performance. However, in reality it is often difficult to be able to ensure that the world's leading performers are participating in a given benchmarking exercise. Instead, benchmarking partners should be selected carefully to ensure the output would provide the required added value.

Once superior performance has been determined, the gap between this and the performance level of the benchmarker is quantified. The working practices enabling superior performance are identified and the enablers evaluated. This knowledge is then shared between benchmarkers to enable the learning to be taken away and implemented as part of a performance improvement program.

Thus benchmarking can be viewed as a three-phase process:

1. Phase one is a positioning analysis aimed at identifying gaps in performance.
2. Phase two is focused upon learning from those leading practices that enable superior performance.
3. Phase three relates to adapting and adopting the learning to transform to a leadership position.

Why Benchmark?

There are two good reasons for organizations to benchmark themselves. First, it will help them stay in business by offering opportunities to become better than other similar organizations, competitors. Second, it ensures that an organization is continually striving to improve its performance through learning. Benchmarking opens minds to new ideas from sources either within the same industry or from many other unrelated industries, identifying how those who have demonstrated performance leadership work.

Yet many organizations benchmark simply to be able to demonstrate to stakeholders, be they customers, shareholders, lenders, regulators, etc., that the organization is performing at an acceptable level. Of course this is a perfectly legitimate reason for benchmarking, although narrowing the focus in this way misses the real potential value of the technique.

Benchmarking also provides a very effective input to an organization's strategic planning processes by establishing credible goals and realistic targets based upon external references.

To really grasp the intent of benchmarking, an organization should be benchmarking not only to demonstrate good performance but also to identify ways in which it can change its practice to significantly improve its performance. Those organizations with a strong performance improvement culture will be benchmarking continuously as this provides them with objective evidence of where to focus improvement activities, how much they should be improving, and what changes to their working practices they might consider to realize improvements.

Classifying Benchmarking

There are many different ways to classify benchmarking (Table 18.1), and the literature is full of different classifications (Anand and Kodali 2008) that make it very confusing for someone new to the topic to really understand what benchmarking is and which approach is best for her or him. The fact of the matter is there is an underlying process that can be considered generic to almost all types of benchmarking. However, to provide some clarity on the differences in classification, we have considered benchmarking in terms of what is being benchmarked, who the benchmarking is going to involve, and how the benchmarking is to be conducted:

- Subject matter and scope (what)
- Internal and external, competitive and noncompetitive benchmarking (who)
- Data and information sources (how)

Classification Criteria		
Subject Matter (What)	**Participants (Who)**	**Data Sources (How)**
Functional benchmarking Process benchmarking Business unit or site (location) benchmarking Projects benchmarking Generic benchmarking Business excellence models	Internal benchmarking External benchmarking Competitive benchmarking Noncompetitive benchmarking (same industry and cross-industry)	Database benchmarking Survey benchmarking Self-assessment benchmarking One-to-one benchmarking Consortium benchmarking

(*Source:* Juran Institute, Inc. Copyright 1994. Used by permission.)

Table 18.1 Ways in Which Benchmarking Is Often Classified

Subject Matter and Scope (What)

Benchmarking is often categorized according to what is being benchmarked. Typical categories include:

- Functional benchmarking
- Process benchmarking
- Business unit or site (location) benchmarking
- Project benchmarking
- Generic benchmarking
- Business excellence models

Functional Benchmarking

Functional benchmarking describes the process whereby a specific business function forms the focus for the benchmarking. In the context of the organization, this may involve benchmarking several different business units or site locations. Typical examples of functional benchmarking include the analysis of the procurement, finance, internet technology (IT), safety, operations, or maintenance functions. The analysis focuses upon all aspects of the function rather than on the processes involved and the specific activities conducted.

Process Benchmarking

In process benchmarking, the focus of the study is upon a specific business process or a part thereof. Examples include product development, invoicing, order fulfillment, contractor management, and customer satisfaction management. Process benchmarking will often involve several functional groups and may also involve many different site locations. There is often a lot of overlap between what is termed *functional benchmarking* and *process benchmarking* (e.g., a benchmarking of the procurement process may look very similar to a benchmarking of the procurement function). Many business processes are not specific to any one industry and so can benefit from broadening participation in the analysis to organizations from a multitude of industries.

Business Unit or Site (Location) Benchmarking

Benchmarking individual business units or site locations against one another is often (but not always) seen in internal benchmarking studies within a single organization. The performance of each unit is analyzed and compared to that of other units. This analysis may incorporate all activities of each unit in their entirety or may be confined to selected functional groups or business processes. For example, Juran Global manages an annual benchmarking consortium comparing the performance of many of the world's oil and gas processing facilities. Each of the key business processes is included in the analysis, and participants come from a wide range of different organizations.

Project Benchmarking

Because projects vary widely in their nature, these studies are normally tailored for specific project types. For example, one may benchmark oil pipeline construction projects, software implementation projects, facility decommissioning projects, etc. Normally included are all the business processes pertaining to the project being analyzed, although the scope may often be limited to a subset of processes. For example, a construction project benchmarking may focus specifically upon contractor selection, procurement, and commissioning.

Generic Benchmarking

Generic benchmarking considers all business processes required to achieve a certain level of performance in a given area. The focus is upon the result and what is required to achieve it. For example, a hospital may undertake a generic benchmarking exercise to identify ways in which it can reduce treatment waiting times. In so doing it may benchmark across a number of different industries where customer waiting times are of paramount importance, for example, insurance claims processing, vessels clearance procedures for major waterways (e.g., Suez Canal), calamity response times for the different emergency services (police, fire, ambulance). Inevitably there is a lot of overlap between process benchmarking and generic benchmarking although in the latter there is often less emphasis on gap analysis and greater emphasis upon a detailed consideration of working practices.

Business Excellence Models

Business excellence models have been developed to provide a framework by which organizations can holistically measure and therefore improve their performance. The purpose of their design is such that they encompass all key aspects of an organization that drive performance. Two of the most well-known models are the United States Baldridge Award and the European Foundation for Quality Management (EFQM) Excellence Model. These models are similar to each other in many respects, and both identify a number of critical success criteria for realizing superior performance.

Although these models are designed to support a self-assessment process, they also lend themselves to providing an excellent framework for comparative benchmarking, although they are infrequently used for this purpose. Benchmarking in this way using such models is essentially a form of generic benchmarking whereby all elements required for excellence are considered. Furthermore, a requirement of both the Baldridge and EFQM models is that organizations be able to demonstrate benchmarking activity.

Internal and External, Competitive and Noncompetitive Benchmarking (Who)

Benchmarking studies are frequently classified by type of participant. Depending on the type of benchmarking being undertaken, it is not always possible to have any control over participant selection. But where this is possible, the selection of others to benchmark with is one of the first and often the most difficult tasks at the onset of any benchmarking study. Potential participants will be identified according to a range of criteria, the main one being the perceived performance level (where superior or world-class performance is the aim).

The four main types of benchmarking are

1. Internal benchmarking
2. External benchmarking
3. Competitive benchmarking
4. Noncompetitive benchmarking

Each of these has specific benefits and drawbacks that need to be considered when selecting the most suitable benchmarking approach. These include

- the similarity between the participants in terms of the subjects to be benchmarked,
- the level of control over the benchmarking process,

- the cost and time input required to conduct the benchmarking,
- the degree of openness that is possible and the level of confidentiality necessary, and
- the potential for learning and therefore performance improvement.

Internal Benchmarking

Internal benchmarking is the comparison of performance and practices of similar operations within the same organization. Depending upon the size of the organization and the nature of its business, this may or may not be feasible for the organization would need to have duplicate groups conducting the same activities. Should this be the case, internal benchmarking is often a popular first step as it allows organizations to prepare themselves for broader benchmarking activities within the safety of their own environment where they have full control over the process. This is likely also to be the least costly and time-consuming way to benchmark. But the potential for finding performance leaders is much smaller, and the opportunity for learning is usually more limited.

External Benchmarking

External benchmarking involves participants from different organizations. The opportunity for learning is normally greater than that achievable by internal benchmarking, but there is obviously a requirement to share information outside of the organization. This brings with it some potential restraints. There will almost certainly be limits on the data organizations are willing to share, especially if the other participants are competitors, and there will of course be a need for stricter confidentiality. External benchmarking is further categorized according to the nature of the participants who can be competitors: competitive benchmarking or noncompetitive benchmarking.

Competitive Benchmarking

By definition, the participants in a competitive benchmarking program are from the same industry, and the focus is normally upon industry-specific processes. For example, Juran has studied patient safety performance between different hospitals. This normally brings with it a high degree of sensitivity that needs to be carefully managed for a successful outcome to be realized; but when it is conducted properly, the results can be very valuable. Conversely, topics that are not directly related to the core business of the competing organizations are usually less sensitive to benchmark between competitors. But ironically these more generic topics are not normally those that an organization wishes to benchmark with its competitors as greater value is more frequently gained from benchmarking such topics cross-industry. It is likely there will be some subject areas that most organizations will not be willing to benchmark with competitors such as proprietary processes or products and innovations that provide competitive advantage.

Noncompetitive Benchmarking

Noncompetitive benchmarking is a form of external benchmarking in which the participants are not in direct competition with one another. They may be from within the same industry or cross-industry. For example, an organization operating a container port in the United States may benchmark with another in Europe. Although they are in the same industry, they are unlikely to be competitors as they are operating in different markets. Juran manages an annual global benchmarking consortium for gas pipelines. The participants are all in the same industry and primarily interested in benchmarking those processes specific to their industry. But they are not in direct competition with one another as they operate in totally different marketplaces, delineated by geographical region. This means that they are very willing to share knowledge and practices openly for mutual benefit without fear of giving anything away that may impact their competitiveness.

In cross-industry noncompetitive benchmarking, those subject areas that are not industry-specific are most commonly analyzed, and it is in this classification that most generic benchmarking studies sit. These tend to be support processes such as administration, human resources, R&D, finance, procurement, IT, and health, safety, and environmental (HSE). Cross-industry external benchmarking potentially offers the greatest opportunities for learning and performance improvement for a number of reasons. First, the pool of potential participants is much bigger. Second, participants thought to be superior performers in the subject area being benchmarked can be identified and invited to participate. Third, the willingness to share knowledge will be greatest where there is no fear of competitive sensitivity.

Data and Information Sources (How)

Benchmarking can also be classified according to the source of the data used in the comparative analysis. Such classification can be made in many ways, but the list below addresses the main categories:

- Database benchmarking
- Survey benchmarking
- Self-assessment benchmarking
- One-to-one benchmarking
- Consortium benchmarking

Database Benchmarking

In this type of benchmarking, data from a participant are compared to an existing database containing performance data. An analysis is performed and the results are provided to the participant. Benchmarking in this way normally requires a third party to administer the database and produce the analyses. The development of the Internet in recent years has led to the growth in this type of benchmarking as it can be easily administered online. The participating organization can submit its data via an online questionnaire and receive a report of the analyses online, usually in a very short time. A quick search of the Internet will reveal the large number and wide range of online benchmarking databases available.

This type of benchmarking is also sometimes offered by consultancy organizations that have accumulated performance data pertaining to specific activities. For example, Juran has been benchmarking in the oil and gas industry since 1995 and during this time has developed a comprehensive database of performance figures relating to this industry. And because the data have been well defined and thoroughly validated during the collection process it is extremely reliable. It is therefore an excellent data source against which oil and gas organizations can be benchmarked.

Many organizations start out on their benchmarking journey by purchasing data from proprietors of such databases. Although this type of benchmarking can be very useful in providing fast feedback on performance, it can have drawbacks. The participant has no control over the content of the analysis and has to accept the metrics that are used to determine performance. Often the source of the data is not disclosed; thus it may be difficult for the benchmarker to assure himself or herself of its relevance. The metrics used may not be clearly defined and may not be validated effectively, resulting in poor data quality and flawed analyses. Care should therefore be taken when entering into this type of benchmarking, and the participant should realize the potential shortcomings. For best results, only a bona fide consultant with a sterling reputation and a good track record should be sought.

Survey Benchmarking

This term is used to describe benchmarking exercises conducted via the completion of a survey or a review process. Typically, a survey document is sent to participating organizations to be completed and returned. Sometimes the survey documents are sent to organizations without their prior agreement to participate, in the hope that they will complete the survey and return it. Of course, this approach is nearly always less successful, with a relatively poor return rate.

The survey may be organized by a third-party consultant or by one of the participating organizations; although in the latter case there will be greater restrictions on what data can be shared directly between participants to ensure compliance with antitrust legislation. Sometimes there may be a fee involved for organizations to participate, and sometimes a single organization or even a consultancy may sponsor the entire exercise, in which case the output for the other participants is often less sophisticated.

The potential drawbacks of this approach are similar to those encountered with database benchmarking in that each organization has minimal control over the benchmarking process, the metrics may not be defined adequately, and the validation of submitted data may be limited. Nonetheless, this type of approach can provide a useful albeit limited comparative analysis with limited effort required on the part of each participating organization.

The survey process can be extended to include a review element, whereby the benchmarking coordinator (normally a third-party consultant) will visit each of the participating organizations as part of the survey process. This allows the consultant to delve more deeply into specific areas to gain richer data (often qualitative data) that can better inform the learning process of the participants, particularly in the area of assessing working practices that underpin superior performance.

Self-Assessment Benchmarking

As previously discussed, self-assessment is an integral part of many performance excellence models. These self-assessments can be used for generic benchmarking between organizations across all industries. These models provide an excellent framework for comparative benchmarking whereby all elements required for excellence are considered. The analysis will often focus on not just quantitative data analysis but also a qualitative view of working practices. However, there is an inherent weakness in the process associated with the subjective nature of self-assessment. Sometimes third parties (consultants) are employed to oversee the process to introduce some level of objectivity or even to conduct the assessments.

One-to-One Benchmarking

This type of benchmarking is probably most commonly reported in the literature, but as we pointed out, benchmarking is not industrial tourism, whereby relatively superficial site visits are conducted between organizations to explore performance. Such exchanges rarely bring fruitful insights, and there is always the uncertainty of whether the organization being visited is really a superior performer.

However, if having conducted a benchmarking study, an organization is identified as delivering superior performance levels in the subject area of interest, then a one-to-one benchmarking can investigate specific areas in much greater depth and deliver rich information pertaining to the drivers of superior performance. This approach is common in consortium benchmarking where, having received the benchmarking analysis, two organizations will agree to benchmark further one to one, to obtain more detailed understanding in specific working practices. A good example of this comes from a cross-industry procurement benchmarking consortium managed by Juran. Following participation in the study, two of the participants agreed to undertake a one-to-one benchmarking. One organization was particularly strong in contract tendering and contractor selection, while the other demonstrated superiority in strategic procurement. Cooperating in this way led to a greater understanding of leading working practices in each of these areas and improved performance for both parties.

Consortium Benchmarking

Without doubt, this form of benchmarking has the greatest potential to deliver improved performance for its participants. A consortium is formed between participants, usually (but not always) supported by a third-party facilitator. They agree on the participants to be invited; the subjects to be benchmarked; the methodology to be followed; the metrics (and their definitions) to be used; the validation criteria; the nature of the analysis, reporting, and deliverables; and the time scales to be adhered to. Thus the participants have a very high level of control over the entire process, and the outcomes from the process are normally reliable data, thorough analysis, and valuable results. This approach does require a great effort on behalf of each of the participants to achieve the desired outcome. It is therefore more time consuming and often more costly to undertake, but the added value is normally far in excess of that achieved through other benchmarking approaches.

We have demonstrated that benchmarking can be classified in many different ways according to the subject matter of the analysis, the nature of the benchmarking participants, data sources, and the methodologies employed. However, differentiating in this way is largely academic, and although these different approaches have their inherent pros and cons and some are clearly more effective than others, they all should have the same ultimate objective—to provide learning on how to improve business performance.

Benchmarking and Performance Improvement

The Juran Trilogy provides a model for effectively managing for quality to achieve performance breakthroughs. The ways in which benchmarking can provide vital input to the performance improvement process can be demonstrated by examining its interaction with the Juran Trilogy (Fig. 18.1).

Benchmarking findings often include quantified gap analyses in key performance areas of the business unit or process being assessed. Management will often use the Pareto principle, often referred to as the *80/20 rule*, in order to focus the right resources on the right improvement opportunities. Resources will be allocated to the "vital few" areas for improvement (typically with a high return on investment, or quick payback), as opposed to the "trivial many" areas with less improvement potential.

Benchmarking and Designing New Products

The main objective of designing for quality is to prepare organizations so that they are able to meet their performance goals. In so doing, they must do the following:

1. Identify customers.
2. Determine customer needs.
3. Develop product (service) features required to meet customer needs.
4. Establish quality goals that meet customer needs.

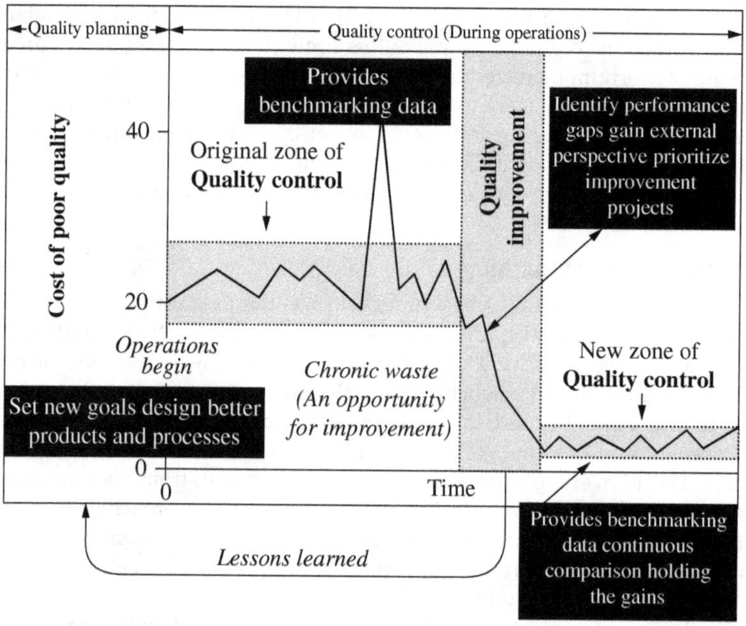

FIGURE 18.1 Benchmarking and the Juran Trilogy. (Juran Institute, Inc. Copyright 1994. Used by permission.)

5. Develop processes that deliver the product (service) features.
6. Prove that the organization's processes can meet the goals.

Benchmarking can contribute input to this process by providing the vehicle whereby organizations can learn from best practices and incorporate that learning into designing new and improved business processes. It also enables performance planning goals to be established based upon the reality of what is achievable by other benchmarked organizations. The benchmarking and product or service features identified may reveal hidden customer needs.

Benchmarking and Quality Control

Quality control activities are in place to ensure that an organization is in a position to meet its performance goals in a controlled and sustainable way. The key activities in quality control include

- Identify what needs to be controlled
- Measure what needs to be controlled
- Establish performance standards
- Interpret differences (i.e., actual performance versus the standard)
- Take action where differences arise

The relationship between benchmarking and quality control is two-way. The output from quality control provides data to be analyzed in the benchmarking process. This enables continuous comparisons to be made between the organization and its benchmarking

Process	Objective	Relationship with Benchmarking
Planning	Plan and prepare to meet performance goals.	• Learn form best practices • Set achievable goals • Design better processes
Control	Ensure performance goals are met.	• Provide data for benchmarking • Make continuous comparisons • Hold gains made
Improvement	Improve performance to significantly superior levels.	• Gain an external perspective • Identify performance gaps • Prioritize improvement projects

(*Source:* Juran Institute, Inc. Copyright 1994. Used by permission.)

TABLE 18.2 Benchmarking and the Juran Trilogy®

partners, who in turn help to ensure that any gains made through performance improvement can be maintained in the longer term. Benchmarking allows organizations to challenge the control standards with vigor. Improved process performance results in increased process capabilities, reduced process variation, fewer defects, and tighter controls.

Benchmarking and Breakthrough Improvement

The main objectives of the improvement process can be summarized as follows:

- Demonstrate the need for improvement.
- Identify specific projects for improvement.
- Diagnose problems to find root causes.
- Provide remedies.
- Prove that the remedies are effective in delivering a breakthrough in performance.
- Provide new controls to hold the gains made.

Benchmarking can support this process by providing an external perspective of what levels of performance are achievable and what working practices are required to achieve these levels of performance. It enables organizations to measure the gaps in their performance compared to the superior performance of other benchmarkers. This in turn allows organizations to identify where their performance is weakest and where it is strongest, thereby offering useful input into the prioritization of potential performance improvement projects. Table 18.2 summarizes how benchmarking relates to the Juran Trilogy.

Benchmarking and Strategic Planning

An organization's goals all too often fall short of stakeholder expectations. A primary contributor to this failure is that goal setting is based upon past trends and current practices in many organizations. Organizations are often inward looking, and the external perspective is frequently overlooked. Customer expectations are driven by the standards of the best providers in the industry and by their experiences with superior providers in other industries.

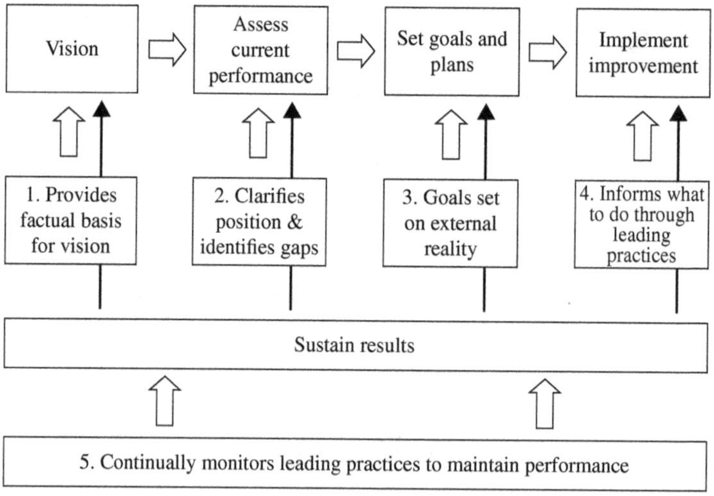

FIGURE 18.2 Benchmarking and strategic planning. (Juran Institute, Inc. Copyright 1994. Used by permission.)

Benchmarking can capture these external references to provide the basis for comparative analysis and learn from leading practices. Thus benchmarking can have a direct input to the strategic planning process. Five ways in which benchmarking can help enable sustainable results by shaping an organization's strategic direction are shown in Fig. 18.2.

The essentials are a well-defined process, a clear understanding of the scope of what is to be benchmarked and why, and a systematic approach that is thoroughly planned. If conducted correctly, benchmarking can contribute to the effectiveness of strategic performance management by providing information on what the best-practice organizations are already achieving. Through the establishment of evidence-based leading practices, benchmarking can provide the factual foundations to enable an organization to frame its vision, goals, and plans to realize world-class leadership.

Benchmarking and Vision Development

Depicted in Fig. 18.2 is a typical strategic planning process for realizing sustainable performance improvement that begins with an organization's vision for the future. The vision will always be influenced to some extent by the organization's business environment and what others have been able to achieve. Benchmarking supplies detailed analyses of that environment and provides a factual basis of what it is to be world class, thereby helping to bring the organization's vision into focus.

Assessing current performance and measuring the gap between this and the vision are critical to an organization's long-term sustainability. Many sources for measuring an organization's current performance exist, including market research, competitor analysis, and of course benchmarking. Benchmarking will clearly define an organization's current performance, clarify its position in relation to both the external business environment and the vision, and identify the performance gaps. This enables the organization to make adjustments to the strategy to close the gap between reality and the future vision.

Benchmarking for Goal Setting and Planning

Long-term plans or key strategies derived from the vision will comprise strategic goals addressing all aspects of the organization's performance including business process performance, product or service performance, customer satisfaction, the cost of poor quality, and the organization's competitive performance. By necessity, these strategic goals will be constantly evolving. Benchmarking analyses enable an organization to set these goals on the external reality and ensure it focuses on closing the gaps between actual and envisioned performance.

The findings from benchmarking enable organizations to understand exactly how much improvement is required for attainment of superior performance. Frequent and regular benchmarking supports the establishment of specific and measurable short-term plans, based upon reality rather than historical performance, resulting in step-by-step improvements in performance over time (Fig. 18.3). The objective is for the organization to overtake the performance leaders, turning a performance gap into superior performance leadership.

An implementation process is required to convert the long- and short-term plans into operational plans. This requires organizations to determine exactly how their specific strategic goals are to be met, identify the actions required to enable this, determine who has the responsibility for carrying out the actions, calculate and allocate the resources required, and plan, schedule, and control the implementation. The output from benchmarking once again provides the external perspective and feeds into this process by offering information relating to the leading practices that have been identified.

Organizations should review their performance on a regular basis to determine progress against the goals set and to measure the gap between the current state and the vision. Benchmarking is the perfect vehicle to support this review process by providing objective evidence of current performance, determining the gaps in performance levels being achieved by other organizations, identifying leading practices, and offering the opportunity to learn from the leading performers.

Thus it is clear that benchmarking is a powerful tool that can contribute significantly to an organization's ability to effectively and strategically manage its performance. It forces organizations and their managers to consider the broader perspective, to look outside of their comfort zones, to learn from those identified as excellent performers; and it fuels the drive for change. By revealing what the best-performing organizations are already achieving and by establishing a factual base for leading practice, benchmarking enables organizations to manage their performance to achieve sustainable world-class leadership.

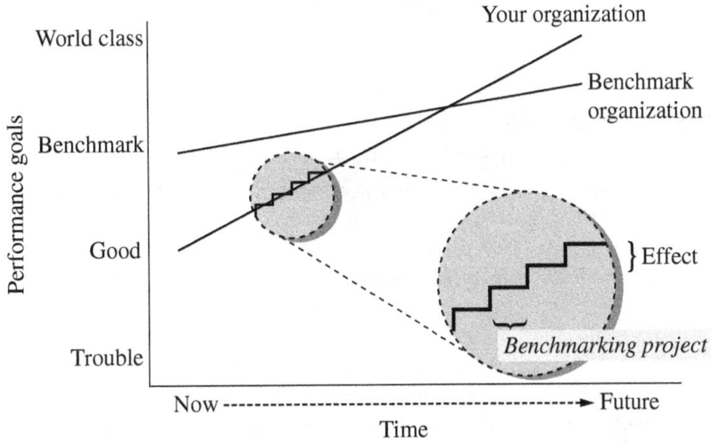

FIGURE 18.3 Benchmarking over time. (Juran Institute, Inc. Copyright 1994. Used by permission.)

The Benchmarking Process

Critical to the success of any benchmarking program are a number of key factors:

1. Determine objectives and scope the study.
2. Identify and define all metrics.
3. Agree on a schedule and stick to it.
4. Ensure resources are available to support the benchmarking.
5. Provide support to participants throughout the process.
6. Validate all data.
7. Normalize the data.
8. Clearly and effectively report the findings.
9. Enable sharing of leading practices.

Irrespective of the type of benchmarking undertaken, it is essential that a well-structured and systematic process be followed to realize these critical success factors. There are many such benchmarking processes described in the literature (Anand and Kodali 2008), but the pioneering model from which most others have been formulated is that used by Xerox and described by Camp (1989). Camp's 10-step benchmarking process was also described in the fifth edition of this handbook (Camp and DeToro 1999). Since that time Juran has published its own Benchmarking Process, which was developed over a period of many years and has formed the basis of a multitude of annual benchmarking consortia since 1995. Although described here in terms of external consortium benchmarking, the process is generic and equally applicable in principle to all types of benchmarking.

The Juran 7-Step Benchmarking Cycle depicted in Fig. 18.4 is divided into three phases. Phase 1 is a positioning analysis providing the benchmarker with a comprehensive study of the relative performance of all the benchmarking participants and a thorough consideration of the performance gaps to the top-performing superior or "best in class" organizations. The focus of phase 2 is upon learning from the phase 1 findings, and fully understanding the performance gaps, whereas phase 3 is concerned with the transformation of performance to a leading position through adopting and adapting leading practices, and developing improvement programs to implement changes required. Each of the steps in the process is described below.

Benchmarking Planning

The first step is to recognize the need for benchmarking, to clearly understand what is to be benchmarked and why, to determine the benchmarking methodology that is going to deliver the analysis required, and to identify who is to be benchmarked. A benchmarking project is no different from any other project. To succeed, thorough preparation and planning at the outset is essential. Often a business case will need to be made to justify the need for the benchmarking project.

Critical at this stage of designing the benchmarking is to clearly define the scope of the benchmarking envelope, what is to be benchmarked and what is to be excluded. The business processes to be compared, the information to be collected and the metrics to be used can then be agreed upon. These too must be clearly and unambiguously defined to ensure comparability of data collected. Finally, the most appropriate vehicle for data collection must be determined.

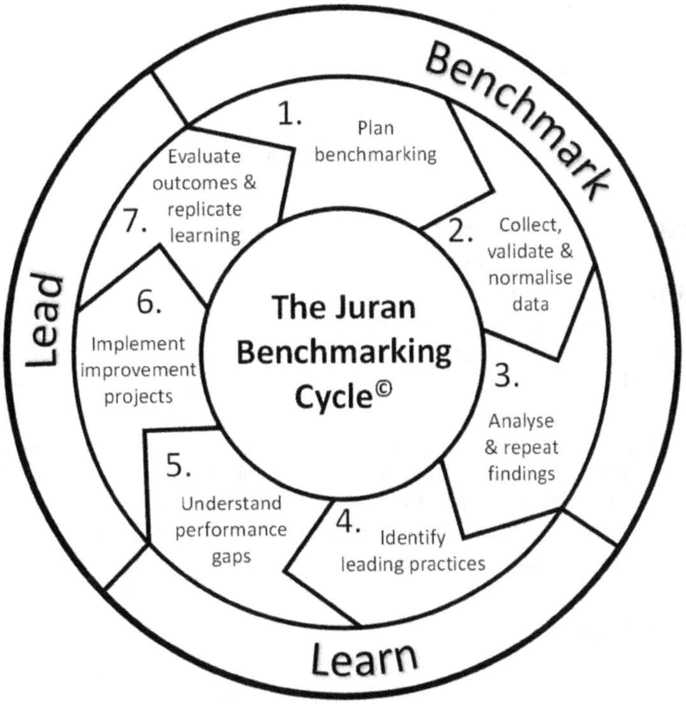

FIGURE 18.4 The Juran Benchmarking Cycle. (Juran Institute, Inc. Copyright 2016. Used by permission.)

Once the benchmarking topic has been well defined, the participants with whom the benchmarking will be conducted must be determined. As mentioned earlier, ideally those organizations that are known to be superior performers will be identified as participants in the benchmarking. However, the participants will be dependent upon the type of benchmarking being conducted as well as the way in which the participants are selected; but of course the ultimate aim is to benchmark with the recognized performance leaders.

During this initial planning step participants will also aim to

- Identify and agree on the key performance indicators (KPIs) to be used to assess performance.
- Create a metrics model that clearly demonstrates the interrelationships among the metrics in use.
- Develop clear and unambiguous definitions for all the metrics used.
- Produce a vehicle through which participants can collect data and information, and conduct some initial validation prior to submission.
- Agree on the project time schedule, milestones, and deadlines.

Data Collection, Validation, and Normalization

Once the precise KPIs and associated definitions are identified, a method for collecting data from each participant must be developed. Commonly a data collection document is produced and issued to all participants to enable them to collect and submit their data. Data submissions are increasingly conducted online via secure Web portals. Proprietary spreadsheets are also frequently used because they are widely available (all participants are likely to have access to them); they are easy to use, have very powerful calculating capability, and can be tailored to provide automated functionality for validation and calculation. The data collection document must be designed to be easy and quick to populate by the user, to provide a suite of validation checks to maximize data and information quality and minimize errors.

Participant Support during the Benchmarking Process

To facilitate the data submission and validation process, it is a good idea to operate a help desk that is available during the entire project duration. This is often provided when third-party consultants are facilitating the benchmarking. The desk can provide professional advice on how to fill the data collection document and answer questions related to specific program-related matters (e.g., interpretation of definitions used). The objective is to provide a swift response to participants so as not to delay them in the data collection process. Of course, by providing clear and thorough guidance notes and a well-structured data collection document, the need for participants to seek help will be minimized. Nonetheless, a help desk can be an extremely valuable and an essential source of benchmarking support, especially for newly established benchmarking programs. When required, a list of frequently asked questions (FAQs) can also be developed and forwarded to all participants.

Data Validation

Use of valid data is key to the success of any benchmarking program, where the adage "Garbage in, garbage out" has never been more appropriate. Incorrect or inaccurate data can easily result in misguided conclusions and inappropriate actions and can lead to the failure of any improvement program. Furthermore, endless rounds of clarification will lead to frustration by the participants and can delay the benchmarking process. Thus a high degree of emphasis should be placed on data validation. In Juran's benchmarking programs they adopt a two-phase approach, with the initial application of a suite of automated checks followed by a number of manual checks.

Automated checks are an integral part of the data collection document which is designed in such a way that it is easy to populate and has a number of built-in validation checks, thereby maximizing data quality and minimizing errors. The built-in automated error checks aim to prevent the input of spurious data and enable users to conduct their own first-pass manual check of the data prior to submission. Thorough initial checking by the users themselves significantly reduces the time and effort required for subsequent validation.

Once the data are submitted, the facilitator should conduct a number of manual checks according to a rigorous data validation process. Juran typically employs a three-step process:

1. Data completeness
2. Data integrity
3. Data consistency

An experienced individual who understands not only the benchmarking process but also the nature of the data being submitted and the interrelationship between different data points should carry out these checks. All data are first checked to ensure that they are complete. Comparing different interrelated data to ensure that the expected relationship between these data is observed should then carry out a check of the integrity of the data. Finally a range of intelligent triangulation checks can be conducted to further ensure consistency between data provided and any available historical data sets. Any anomalies should be raised one on one with the relevant participant to ensure that corrected data are provided.

Where either there are a large number of apparent errors or participants are experiencing difficulty in obtaining the data required, a data clinic may be held, attended by participants and designed to clarify any confusions relating to the data required.

Data Normalization

The single biggest problem in any benchmarking exercise is how to compare benchmarked subjects on a like-for-like basis (i.e., how to compare apples with pears). In some circumstances the benchmarkers will be similar enough to enable direct comparisons of performance between them. However, more typically the subjects being benchmarked will all be different from one another, be they organizations as a whole, business units, different sites, different functional groups, business processes, or products. No two subjects will be identical, although the extent of difference between them will vary considerably depending upon what and who is being benchmarked. Thus to be able to compare differences in performance levels requires some intervention. Some form of data normalization is usually required to enable like comparisons to be made between what may be very different subjects. Without it, direct comparisons of performance are normally impossible and may lead to misinformed conclusions. Normalization can be made on the basis of a wide range of factors including scope, scale, contractual arrangements, regulatory requirements, and geographical and political differences.

One solution is to organize benchmarkers into categories or peer groups with other benchmarkers or data sets with similar characteristics. The key is to be able to identify the factors that are driving the performance and then develop a method by which these drivers can be considered when comparing performance metrics. In its simplest form this may involve stratifying the data according to underlying criteria. For example, if a health authority wishes to compare death rates of people in different regions, it may stratify these according to gender or age. Another example comes from the chemical industry. A series of chemical organizations may decide to benchmark their performance in the field of managing for the environment, and in so doing they may wish to compare emission levels for a variety of polluting gases (e.g., oxides of nitrogen and sulfur, carbon dioxide and methane). These data could be stratified according to their harmful impact to the environment through the use of a standardized measurement such as the environmental impact unit (EIU). A further example might be those organizations comparing the efficiency of their R&D activities using a KPI that measures the percentage of their sales attributable to new products or services (e.g., products that have been on the market for less than 2 years).

But of course, even within these groups there may be differences between the benchmarking subjects. To conduct a valid comparison of performance, these differences in characteristics need to be taken into consideration in the analysis. The most effective way of doing this is through normalization of the performance data.

Normalization is essentially the process of converting metrics into a form that enables their comparison on a like-for-like basis, accounting for all (or as much as possible) the variation between the benchmarking subjects. The normalizing factor used should truly be the driver for the performance being benchmarked. For example, in benchmarking the operating costs of the invoicing function of an organization, perhaps a suitable normalizing factor is the number of invoices raised. For example, the raised costs could be compared on a per invoice raised basis. However, perhaps some invoices are more complicated to produce than others (e.g., they may contain more line items or be for a higher total value that requires more checks before the invoice is raised), so this way of normalizing may not be appropriate after all.

The most common way of doing this is by looking at performance per unit or per hour. For example, if we are measuring the cost of manufacturing a motorcar, we might compare the cost per vehicle produced; or if we are looking at the time taken to treat a hospital patient with a given ailment, we might consider the number of patients examined per hour.

In some cases, a simple measurement per unit is not sufficient to accommodate the variation observed between benchmarking subjects, and a more sophisticated approach has to be developed. In such cases the use of weighting factors that represent the variation of the different benchmarking subjects is often a very effective means of normalization. Weighting factors may be developed in relation to costs, time, and efficacy. An example of a highly effective weighting factor is the Juran Complexity Factor (JCF). The JCF was developed to enable like-for-like comparisons to be made between oil and gas production facilities of very different size and design. The normalizing factor takes into consideration the equipment present in the facility and the time it takes to operate and maintain this equipment under normal conditions. The JCF is then used to normalize all cost performance between facilities in the benchmarking. This enables organizations to directly benchmark their facilities with those of other organizations even though they may be very different in design and size.

The efficacy of any normalization method should be fully tested before it is implemented. As mentioned, for a normalizing factor to be effective, it must be representative of the driving force for the performance subjects being benchmarked. Thus there must be a good relationship between the performance metric and the normalizing factor. A good way of testing this is to examine the correlation between the normalizing factor and the performance metric being normalized. There should be a strong direct relationship between the two. For example, an increase in the normalizing factor should lead to an increase in the metric being benchmarked (and vice versa), although this relationship may or may not be linear.

Analysis and Reporting of Benchmarking Findings

The aim of the analysis is to determine the findings from the data collected in the benchmarking in conjunction, where appropriate, with other pertinent data and information from a number of different sources including the public domain, the participants themselves, and any previous editions of the benchmarking study. The level of analysis will be dependent upon the scope and objectives agreed upon at the commencement of the benchmarking.

It is essential that the analysis be impartial and totally objective. It must also be aligned to the benchmarking objectives; and to be of value, it must indicate the benchmarker's strengths and weaknesses, determine, and where possible quantify, the gaps to the leading performers, and identify as far as possible the reasons for these gaps. It is important that the metrics be considered collectively and not in isolation as the results from one metric may help to explain those of another. The strategies and working practices of each of the participants should also be explored and used to determine how they may influence performance.

The performance data and any normalization data streams are analyzed to compare participant performance and determine performance gaps. It is also important to consider the level of statistical testing of the data to ensure that comparisons being made are statistically significant and the conclusions drawn thereafter are valid.

Quantitative analyses are typically made in relation to the top quartile (i.e., the boundary to the 25th percentile), the best in class (i.e., the single leading performer), or the average (mean) of the benchmarking population. There are pros and cons to comparisons of each of these criteria. Analysis of the gap to the best in class is probably the most common and on the face of it seems the most obvious; after all, the objective is to close the gap to the leading performer. However, making comparisons to the data of a single benchmarker at a single point in time will always carry a risk that there is error in the data value (although the validation process should minimize this error) or that the performance level reported is not sustainable in the longer term and is therefore not realistic. In contrast, comparison to the top quartile and in particular to the average is more stable and reliable, as these comprise data from more than one participant.

Reasons for apparent differences in performance should be considered during the analyses. With multinational or global benchmarking studies, it is important to consider the impact that may be attributed to differences in geographical location. For example, when one is analyzing costs, it is clear that cost levels (e.g., salaries) in the West cannot be easily compared to the East, Russia, Africa, or Latin America. In addition, fluctuations in exchange rates between currencies can have a dramatic effect. Likewise different tax regimes, regulatory requirements, political policies, and cultural differences can all significantly influence performance.

Report Development

Once the analysis is complete, it must be reported to the benchmarking participants. The content of the report and the medium used for reporting were agreed to at the outset of the benchmarking exercise and in part are determined by the type of benchmarking being undertaken.

Reports may be delivered online, electronically or in paper hard-copy format. Whatever the medium selected, the report must present the benchmarking findings in a clear, concise, and easily understood form. Optimum use of color, diagrams, pictures, and charts should be made to facilitate communication of the findings. Charts and tables should be annotated to provide guidance to the reader. The analysis should be reported in full together with recommendations for the focus of performance improvement efforts required to close gaps.

One very important point that must be addressed is the level of data anonymization that will be employed in the report. There is always a tradeoff between confidentiality and learning opportunity. The higher the level of confidentiality, the lower the potential for learning. If the identity of the superior performers cannot be revealed, then the opportunity to learn from them is minimized. However, compliance with antitrust legislation is always a primary requirement. Thus the report must always be in line with any confidentiality agreement made between the benchmarkers and must meet legal requirements. But to maximize the learning potential, the degree of openness should also be maximized.

Unfortunately, many benchmarking exercises will stop at this point. But to maximize the value gained from benchmarking, organizations must go further to try to understand the practices that enable the leaders to attain their superior performance levels. This is the purpose of phase 2 of the Juran Benchmarking Cycle.

Identifying Leading Practices

A benchmarking program must go well beyond the comparison of performance data. The transfer of knowledge from the leading practitioners to the other benchmarkers is critical. This is essential to maximize the effectiveness of knowledge transfer, which leads to highly successful change/process improvement programs. This can be achieved in a number of ways:

- Internal forums
- One-to-one benchmarking
- Leading practice sharing forums

Internal Forums

Organizations participating in the benchmarking should fully review the benchmarking report and consider the findings in detail. Thereafter many organizations find it beneficial to organize an internal forum attended by all parties within the organization affected by the benchmarking to discuss the findings openly and determine first actions required to begin the process of gap closure and performance improvement. In cases where an organization may have numerous participants in the benchmarking (e.g., an organization may benchmark several different business units simultaneously), these internal forums can be an excellent platform for sharing knowledge among those business units. Juran's oil and gas experts have had the opportunity to attend and facilitate a series of internal knowledge exchange forums in the oil and gas industry (e.g., Pemex Mexico in 2004 and Qatar Petroleum and Saudi Aramco in 2008), where up to 500 members of staff from different departments within one organization gathered with the sole aim of sharing knowledge and leading practices related to vital topics and processes.

One-to-One Benchmarking

It is common for organizations to benchmark one to one following participation in a group benchmarking study. Having identified superior performers in various areas of the benchmarking, organizations collaborate on a one-to-one basis to explore specific issues in greater detail perhaps by on-site visits or further data exchange and analyses, to maximize the learning outcomes.

Leading Practice Sharing Forums

This involves the sharing of leading practices between top performing organizations to the mutual benefit of all benchmarkers. Of course, when one is benchmarking with true competitors, the options for this may be limited and alternative approaches may be required to establish learning.

Once findings have been reported to all participating organizations, a leading practice sharing forum can be organized, attended by all participating benchmarking organizations. The leading practitioners in each of the elements of the benchmarking model are asked to make presentations to the closed forum. Prior to the forum, all participants would be invited to submit any questions they might have for the leading practitioners, which they would like to be addressed by the leading practitioners in their presentations.

The objective of this forum is the identification of master class opportunities and the transfer of knowledge from the leading practitioners to the other benchmarking partners. Leading practitioners will present to their peers the "whys and the hows" of their practices. The intention is that audience members will learn from these presentations, which will subsequently help them to formulate their own improvement programs. Participants should leave the leading practice forums in a position to develop initial action plans to take the first steps towards the implementation of improvement programs.

Understanding Performance Gaps

Once the benchmarking findings have been reviewed and learning points ascertained, each benchmarker needs to fully understand its performance gaps. The output of a benchmarking exercise should become input for action planning. To do this, each organization needs to develop and communicate an action plan for changes required to realize improvements. Here the learning from the benchmarking will feed into the organization's strategic plan and be implemented using its performance improvement processes. But how can organizations translate benchmarking findings into action plans that will lead to performance improvement?

A typical output from a benchmarking exercise will include a series of performance gaps between a participant and the leading practitioners in key business processes. Often organizations react with disbelief and denial when they are confronted with the performance gaps translated into monetary terms: Comments such as "It is impossible to save this much, these numbers cannot be right!" are commonplace. It is of paramount importance, for both the credibility of the benchmarking findings and the subsequent level of managerial buy-in, that one additional internal journey be embarked upon prior to moving to action planning. The organization needs to truly understand the performance gaps identified. Therefore it must eliminate any distorting elements from the gaps presented. To render the performance gaps actionable, the organization needs to break them down into controllable and noncontrollable gaps. Noncontrollable gaps are those relating to aspects of an organization's activities that are not under the direct control of that organization at the time. For instance, these could include start-up costs, one-off expenditures, extraordinary incidentals, regulations that one has to comply with, site-specific operational issues (e.g., climate, geography, topography), and geopolitical and safety-related costs.

An actionable performance gap should be free from noncontrollable elements (see Fig. 18.5). This will allow management to do the following:

1. Assess a performance gap that they can relate to and therefore buy into.
2. Prioritize improvement areas and distinguish the "vital few" versus the "useful many" opportunities for improvement (Juran and Godfrey 1995).
3. Allocate resources to fix the problems and bridge the gaps including an accountable project manager, budget, time frame, and targets.
4. Put controls in place by embedding the requirement to action the improvements into managerial and individual employee target setting, compensation schemes, and business planning.

Thus, benchmarking findings will have been embedded into performance improvement action plans and integrated into routine business cycles, helping to ensure that resources are focused, individuals have bought into the process, and goals are achievable.

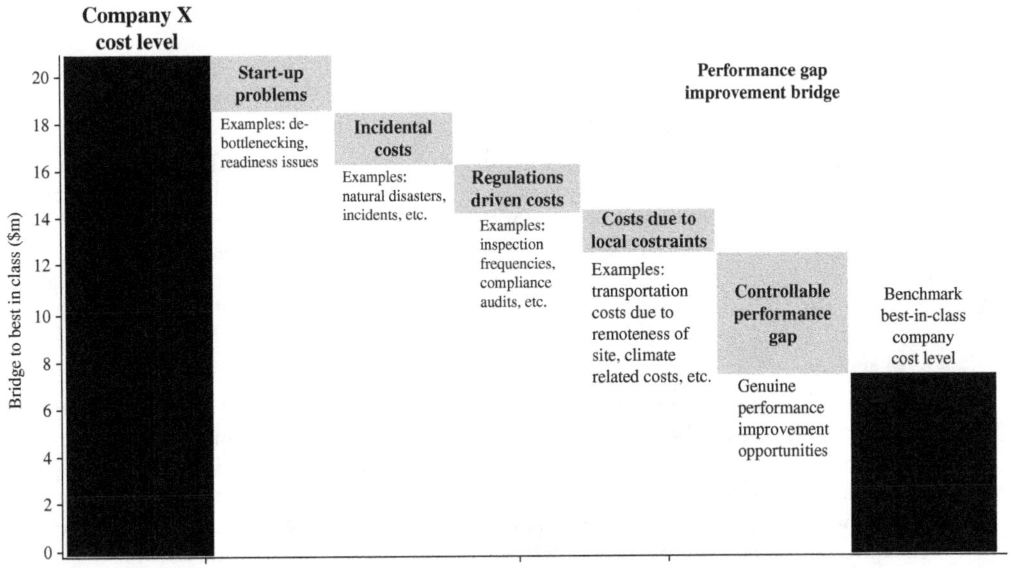

FIGURE 18.5 Performance bridge analysis from "gross to net performance gap." (Juran Institute, Inc. Copyright 1994. Used by permission.)

Implementing Improvement Projects

As mentioned earlier, the output from the benchmarking should form one of many inputs to an organization's performance improvement activities. Following the initial action planning described above, organizations need to develop these further into projects to realize improvements and close performance gaps. The many methods that can be utilized for this will be dependent upon the mature of the problems identified and are described in detail elsewhere in this text.

Evaluating Outcomes and Replicating Learning

The learning gained and the improvements to performance realized must be fully evaluated and, where appropriate, embedded to ensure all gains are rolled out throughout the organization and are sustained over time. Benchmarking may take place at the corporate, operational, or functional level within an organization, and it is important that each of these levels be linked via a cascading series of goals, interlinked to ensure systematic progress toward attaining the vision. As improvement opportunities arise, they should be embedded into and replicated throughout participant organizations.

Finally, because benchmarking should be a continuous process, that process itself should be evaluated, and where necessary, the objectives and scope refined to meet the evolving needs of the benchmarkers before repeating.

A New Benchmarking Standard

Juran has been instrumental in the development of a new standard for conducting benchmarking. Publically Available Specification PAS 7070:2016 (BSI 2016) represents the first reference methodology for benchmarking that is not proprietary in nature. It has been produced with the objective of providing a specification for the benchmarking process that can be used by small and large organizations alike to ensure a robust and uniformly applicable process for benchmarking which have objectively verifiable outcomes.

Legal and Ethical Aspects of Benchmarking

The legality of benchmarking is governed by competition (antitrust) law and intellectual property law, and all benchmarkers must be aware of the legal and ethical implications of their benchmarking activity. While the ethos of benchmarking is the sharing of knowledge and information to the mutual benefit of all participants, organizations must not lose sight of the potential value of their corporate knowledge and therefore the necessity to adequately control its use.

Legal Issues

During the benchmarking process, knowledge and information are often shared among the participants. Should this be subsequently applied by one of the benchmarking participants as part of the performance improvement process, there arises the possibility of infringing the intellectual property rights of the benchmarker who offered the information. Often organizations fail to recognize the risks of such infringements, albeit inadvertent, and the potential for conflict between benchmarking and intellectual property law (Boulter 2003).

Care has to be taken when exchanging information in a benchmarking study. This is particularly the case if the benchmarking is between competitors and if price information and business secrets are being benchmarked. Of course, the vast majority of benchmarking activities steer clear of both these topics; nonetheless they may still be considered to constitute anticompetitive behavior.

Although competition laws as depicted by antitrust laws in the United States and the Treaty of Rome (Article 81) in the European Union are well documented, their applicability to benchmarking is a gray area. They are primarily aimed at prohibiting cartel formation and price fixing. Other countries also have legal requirements pertaining to competition, and so the legal situation is further complicated where benchmarking exercises span a number of different countries. As a result, any organization considering entering into a benchmarking exercise is strongly urged to seek legal advice before doing so.

The Benchmarking Code of Conduct

The Benchmarking Code of Conduct was first developed by the International Benchmarking Clearinghouse, a service of the American Productivity and Quality Center (APQC) in 1992 (see www.apqc.org). In 1996 a European version of the code was developed (see www.efqm.org), based upon the American version, to comply with European competition law. Neither document is legally binding, but they do lay down the principles for ethical and legal benchmarking. The main principles address legality, confidentiality, and information exchange, and all benchmarking programs should ensure that participants comply with these.

Confidentiality

Essential in all benchmarking studies is the requirement for some degree of confidentiality. The strictness of the level of confidentiality will be dependent upon the sensitivity of the subjects being benchmarked, the requirement to comply with competition law, and the degree of willingness by the participants to share data and information openly. It is clear that great care must be exercised when benchmarking prices. In many cases costs are considered an indicator of prices, and therefore strict confidentiality is normally also expected when comparing costs.

The degree of confidentiality exercised in benchmarking studies can vary enormously. At one end of the scale participants are totally unaware of whom they are benchmarking with as the identities of the other participants are withheld and only the third-party facilitator is aware of who each participant is. Unfortunately a major drawback of such a strict level of confidentiality is that the learning potential is greatly reduced. If participants do not know the identity of the better performers in a benchmarking process, how can they possibly learn from the findings? The whole object of the benchmarking is lost, and the study becomes nothing more than a league table.

Therefore a more pragmatic approach is preferred whereby sensitive data (e.g., costs) can be anonymized whereas other less sensitive data can be shared more openly. And with the skillful support of a third-party facilitator the participants can still maximize the learning potential from the study.

Irrespective of the level of confidentiality and anonymity decided upon, it is essential that all parties in a benchmarking study, including the facilitator (be they a consultant or a participating organization), sign a confidentiality agreement. This agreement will be legally binding and will spell out how the data, information, and findings of the study will be shared, used, and disseminated by all parties.

Managing for Effective Benchmarking

For any benchmarking initiative to succeed, it must be managed effectively. Each organization participating in a benchmarking exercise will normally establish a benchmarking team comprising individuals from a range of disciplines who will manage the benchmarking activities from outset to completion. The size of this team will be dependent upon the size of the organization and the scope and scale of the benchmarking. A project owner should be established who will lead the team and act as a focal point both within the organization and with the other participants and the facilitator. The owner will be responsible for briefing the team on the project findings and ensuring that all the necessary resources and support is forthcoming from senior management. The team will be responsible for delivering the benchmarking project, setting internal targets and ensuring they are met.

While senior managers are unlikely to be involved directly in conducting the benchmarking, they play a key role in ensuring it is executed successfully. Key roles of senior management are to do the following:

- Set benchmarking goals.
- Integrate benchmarking into the organization's strategic plan.
- Act as a role model.
- Establish the environment for change.
- Create the infrastructure for benchmarking.
- Monitor progress.

Set Benchmarking Goals

The intent to benchmark must be established and clearly communicated within the organization, identifying the reasons for and objectives of benchmarking. A benchmarking policy should be developed and documented.

Integrate Benchmarking into the Organization's Strategic Plan

Benchmarking must form a fundamental element of the organization's business plan. The benchmarking direction must be set and the findings communicated throughout the organization. The findings must also be incorporated into the organizational goals and driven down through the organization.

Act as a Role Model

Senior managers must openly demonstrate their commitment to the benchmarking effort even though they are unlikely to be involved day to day in the benchmarking exercise. They must remove any roadblocks that may derail the benchmarking team, and they must also commit the resources required to conduct the benchmarking and fully realize the benefits from it. This is likely to mean both a financial commitment and a time commitment, releasing individuals from the regular positions to perform their role in the benchmarking team.

Establish the Environment for Change

Managers must demonstrate a willingness to accept the findings of the benchmarking and to act upon them, creating a change environment to realize the potential improvements in performance.

Create the Infrastructure for Benchmarking

Where necessary, training must be provided to those involved in the process as well as reward and recognition for benchmarking team members and those members of the organization whose efforts have led to superior performance as determined by the benchmarking. However, it is essential that any poor performance identified by the benchmarking not be punished. As we mentioned, benchmarking should not be used as a personnel appraisal tool!

Monitor Progress

There must be a commitment from the outset to provide the resources required for successful benchmarking. Where necessary, training must be provided to those involved in the process.

Case Study

This case study describes the experiences of an oil terminal that was able to realize significant improvements in performance driven by its participation in a benchmarking consortium.

In 2005, the terminal benchmarked its performance for the first time. Prior to benchmarking, the terminal believed that it would be among the highest performers. However, the findings of the study, which examined 2004 performance, indicated that its performance was fourth quartile in many areas including operations and maintenance expenditure and workers' time, and their total expenditure was nearly twice as high as the average for the benchmarking group. Even more staggering was the ninefold gap to the best-performing terminal. The biggest performance gap lay in maintenance expenditure and workers' time that were both the highest of the study. Of particular note were very high third-party (contractors and services) costs and workers' time.

The terminal managers' first reaction was one of denial, but after comprehensively reviewing the findings and recovering from the shock of the reality of their relative performance, they established a performance improvement team led by a newly appointed performance improvement manager. An action plan was developed that focused upon the weaknesses identified in the benchmarking exercise. The first step they took was to differentiate the costs they felt they could control from those they could not. For example, there were some costs that management agreed could not be reduced, at least in the initial term, due to the remote location of the facility. These costs were therefore parked for the time being, and focus was placed on those areas that were controllable and represented the biggest performance gaps.

The team conducted an organizational design review and developed a restructuring that implemented clear accountabilities for key areas including operations, maintenance, integrity management, and projects. They focused upon their main weakness which was maintenance. A series of step change programs were initiated to realize rapid improvements in performance. Changes made were characterized by the following:

- A move from inefficient hierarchies to more efficient self-directing work teams
- A change in shift working patterns with a move from 8- to 12-hour shifts that resulted in a 30 percent increase in efficiency
- Improved control of third-party service providers and contractors resulting in a 30 percent reduction in contractor spending
- Reengineering of work planning and execution processes
- Optimized maintenance work frequencies through the adoption of a risk-based approach to maintenance

The terminal was benchmarking its performance on an annual basis and by 2007 began to see the fruits of its performance improvement initiative. The 2007 benchmarking program, which examined terminal performance in 2006, revealed some marked improvements. The total expenditure and workers' time were now third quartile with total expenditure now reduced to 25 percent higher than the group average and the gap to the lowest-cost terminal reduced to a factor of 3.5. Overall, the terminal's total storage and loading costs were reduced by some 27 percent during this period.

Greatest improvements were made in the areas of operations and maintenance. The terminal was now a first-quartile performer in terms of operations expenditure and workers' time. Between 2005 and 2007, the terminal's maintenance expenditure had shifted from fourth to third quartile and maintenance workers' time from fourth to second quartile. This shift represented a very significant 58 percent reduction in costs and a 50 percent reduction in workers' time for maintenance.

A number of conclusions can be drawn from our benchmarking experience and in particular the findings of this case study.

First and most importantly, it is possible to compare the performance of different terminals on a like-for-like basis. However, to do this, it is critical that the right normalization method be used and that the efficacy of the method be checked to ensure a healthy correlation with the subjects being measured.

Second, benchmarking is a powerful improvement tool. Furthermore, it can provide a wake-up call for complacency, as in the terminal described here; your performance may not be quite as good as you think when compared to that of your peers. However, it is also important to understand that to realize any improvement in performance, it is essential to act upon the findings of a benchmarking exercise.

Third, it is clear that if a structured and well-researched approach to performance is adopted, then significant gains are achievable.

Finally, it is important to understand the importance of benchmarking on a continuous basis for two main reasons. In so doing, a terminal will be able to trend its own performance over time and therefore to determine the effectiveness of its performance improvement activities. Also your peers will be striving to improve their performance, and so the benchmark is constantly moving. Keeping abreast of these improvements and staying in touch with the leaders will require a continuous approach.

References

Anand, G., and Kodali, R. (2008). "Benchmarking the Benchmarking Models." *Benchmarking: An International Journal*, vol. 15, no. 3, pp. 257–291.

Benchmarking Process Management—Specification. (2016). Publically Available Specification PAS 7070:2016. *BSI*.

Boulter, L. (2003). "Legal Issues in Benchmarking." *Benchmarking: An International Journal*, vol. 10, no. 6, pp. 528–537.

Camp, R. C. (1989). *Benchmarking: The Search for Industry Best Practices That Lead to Superior Performance*. ASQC Quality Press, Milwaukee, WI.

Camp, R. C., and DeToro, I. J. (1999). "Benchmarking." In: Juran, J. M., Godfrey, A. B., Hoogstoel, R. E., and Schilling, E. G. (eds.), *Juran's Quality Handbook*. 5th ed. McGraw-Hill, New York.

Juran, J. M., and Godfrey, A. B. (1995). *Managerial Breakthrough*. Barnes & Noble, New York.

Kearns, D. T., and Nadler, D. A. (1993). *Prophets in the Dark—How Xerox Reinvented Itself and Beat Back the Japanese*. Harper Business, New York.

CHAPTER 19

Graphical Tools to Improve Process Performance

Brian A. Stockhoff

High Points of This Chapter

1. Obtaining accurate, reliable, and relevant information happens when asking the right questions. Ask the right question, you will get the right data. But what do you do with it, how much do you need, what tool should you use. These are the questions this chapter answers.
2. The core tools are used and integrated within the structure of design, control, and improvement methods; the list here is a useful starting place for managers and teams to master these tools to be well prepared for many of the organizational problems they are likely to face.
3. There are many tools useful to manage an organization and available for process improvement, design and control. Tools for improvement require the testing of theories and finding root causes.

4. Design tools require the collection of opinions, specifications, and then determine means to develop new services or products that are reliable.
5. Control requires the use of statistical tools to help distinguish from common and special causes of variation, to reduce risk, thereby facilitating appropriate intervention.
6. Mastering the top core tools can lead to a great improvement in the ability of an organization to attain the needed data and hence information to be successful. Other tools are less popular and used for special cases.
7. This chapter will cover the core tools from A to Z.

Introduction

Dr. Ishikawa, distinguished quality guru from Japan, wrote the classic *Guide to Quality Control* (1972). It is generally credited as the first training manual of problem-solving tools specifically presented for use in quality improvement. In the first publication of this book it was used as a training reference for factory workers who were members of quality control (QC) circles. Quality Circles were a group of employees that work together to improve the performance of their work area.

In this handbook, we expand on the "seven quality tools" of Ishikawa's work to include tools that find utility more broadly across design, control, and improvement methods. There remain many other useful tools; this list is not exhaustive, nor could it or any list be so. However, the list here is a useful starting place for managers and teams to master these tools to be well prepared for many of the organizational problems they are likely to face.

The core tools are used and integrated within the structure of design, control, and improvement methods. For each method, such as Lean or Six Sigma (both improvement methods) use tools to complete each step in the method. In Fig. 19.1, we provide an example of how each tool is used in an application map (in matrix form), with each column corresponding to a tool, and each row corresponding to a process step, with improvement expanded in detail. At each intersection is a symbol indicating the frequency of use of that tool at that process step (frequent, infrequent, and very rarely).

The process map is a valuable guide to problem-solving teams in these ways:

1. It reminds the team that there is a structured order to the problem-solving process and helps keep the team on track.
2. At a given step, if the team is at a loss what to do next, one of the frequently used tools may suggest the next action.
3. At a given step, using a tool indicated as rarely used is a signal to the team to reconsider its course of action. A convenient example is the use of brainstorming (which is an effective way to develop a list of theories, ideas, and opinions of group members) to test theories (which always requires data, not the opinions of the team members). After an introduction to each tool (what is it, and why do we use it?), steps are provided for tool creation and use. An example then is provided to demonstrate its use, enabling readers to better understand the application and potential relevance to their own situation. A more thorough treatment is available from a number of texts, including the course notes for *Quality Improvement Tools* or from *Modern Methods for Quality Control and Improvement* by Wadsworth et al. (1986).

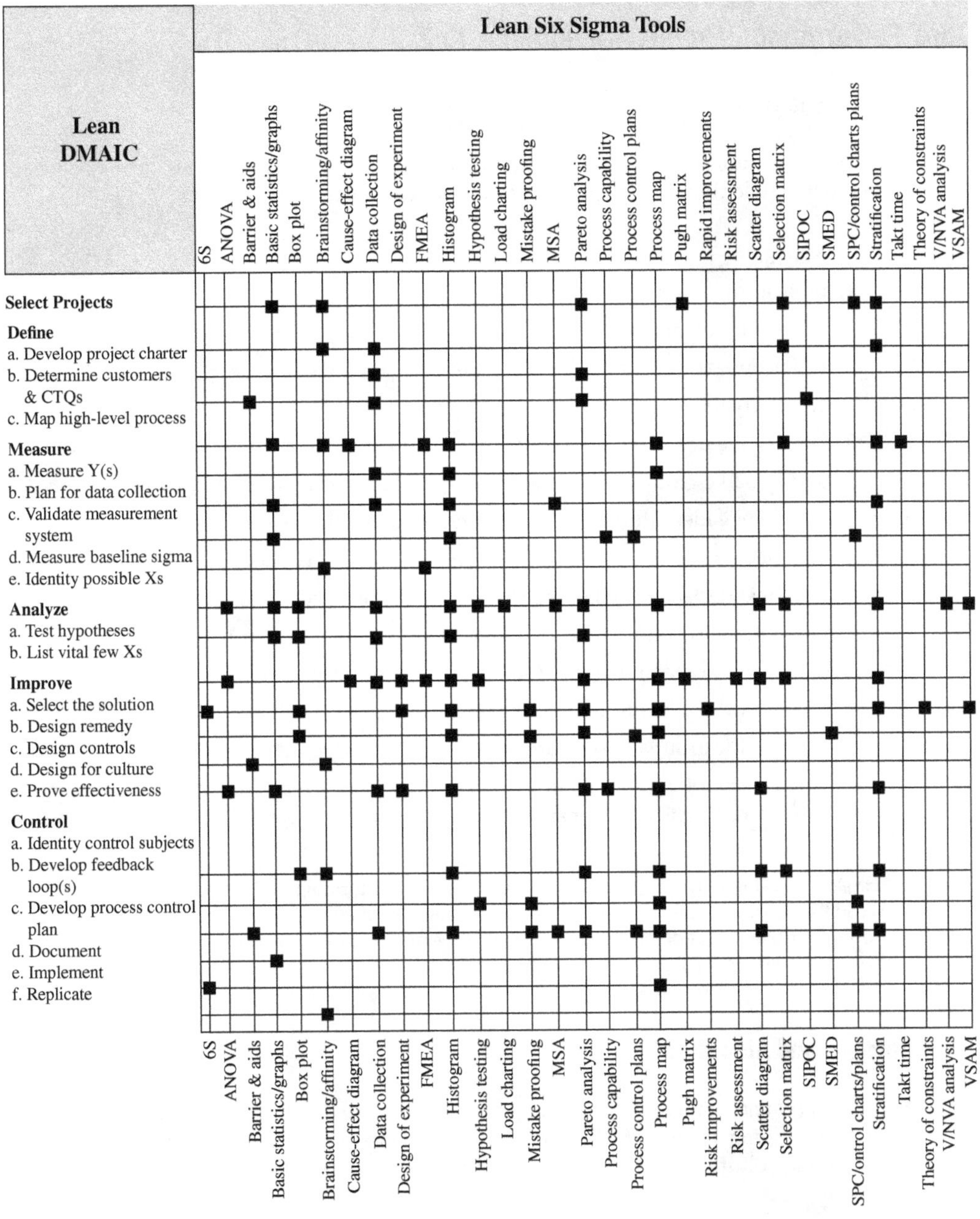

FIGURE 19.1 Applications for Quality Improvement Tools.

Core Performance Excellence Tools: A to Z

Affinity Diagram

Purpose The affinity process takes many items and sorts them into meaningful groups. It is used when soliciting variable information from customers or employees.

Steps to Create

1. Brainstorm ideas:
   a. Set a time limit.
   b. Record each idea on adhesive notes or 3 × 5 cards.
   c. Clarify the ideas and eliminate duplicates.
2. Display the unsorted ideas on a table or stick them to a wall.
3. Sort into like groups; do this without speaking, based on individual perception.
   a. Arrange ideas into meaningful categories of "like issues."
   b. If one person does not like the placement of an idea, he/she can move it.
   c. If one idea seems to belong in more than one place, make a duplicate card.
   d. Continue sorting until consensus is reached; aim for 5 to 10 groups.
   e. Consider breaking large groups into smaller ones.
4. Create a title or heading for each category.
5. Transfer the groups into an organized affinity diagram.
6. Discuss groupings and understand how they relate to each other; if necessary, move items to complete a consensus affinity diagram.

Example A customer focus group identified through brainstorming various positive attributes they would like to see in a child's toy. Each idea was written on an adhesive note and placed onto a board. Through sorting, it was determined there were three distinct groups of attributes; these groups were given brief, descriptive names as shown in Fig. 19.2.

Barriers and Aids Chart

Purpose A barriers and aids chart is a graphical means to define obstacles to improvement and corresponding means of alleviating these. It helps identify and overcome technical problems and cultural resistance that slow or prevent quality improvement.

Steps to Create The major steps are to identify all likely sources of resistance (barriers) and support (aids), rate the barriers and known aids according to their perceived strengths, and identify countermeasures needed to overcome barriers. More specifically,

1. Place a clear description of the objective (remedy) at the far right of the surface being used. Draw a heavy arrow pointing to the objective.

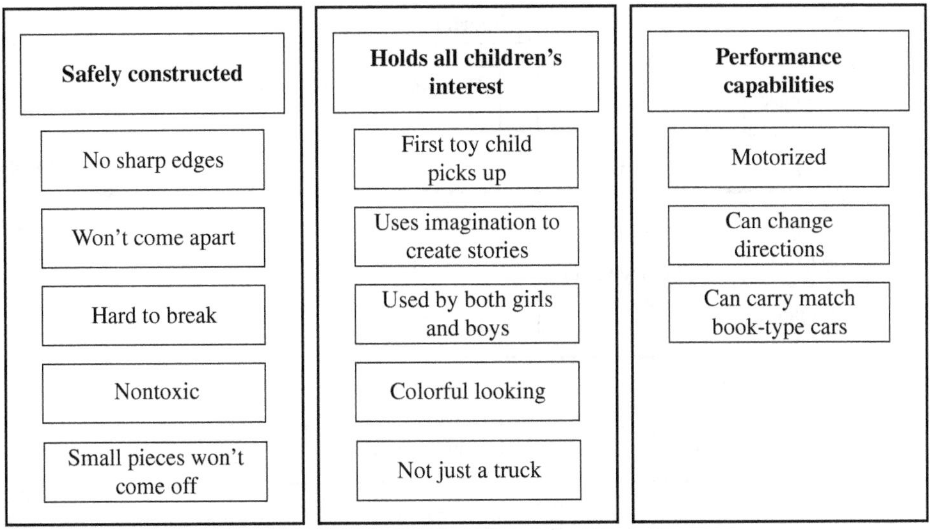

FIGURE 19.2 Affinity diagram.

2. Brainstorm a list of potential barriers.
3. Select those vital few barriers that should be overcome, and place them above the heavy arrow, labeled as barriers, with smaller arrows pointing down.
4. Brainstorm a list of existing aids for overcoming the selected barriers.
5. Select those aids which will help overcome the barriers, and place each one opposite the barrier(s) it will help overcome. Label them as aids.
6. Identify any barriers that will not have adequate aids.
7. Draw a horizontal line below the aids with a "countermeasures" label.
8. Design the countermeasure for barriers without adequate aids, placing them opposite the barriers.
9. Review the chart for missing vital-few barriers, the effectiveness of aids, and the effectiveness of countermeasures.

Example New Procedure for Engineering Studies. An improvement team developed a new procedure for completing engineering studies. When planning for cultural resistance to the remedy, the team came up with a list of barriers and aids most likely to influence its success. It also developed some countermeasures to overcome some of the barriers. The major barriers, aids, and countermeasures are shown in Fig. 19.3. Note that while this chart is relatively simple, there often will be dozens of concerns to consider. The major barriers are placed at the top of the chart, the desired result at the right—much as in a cause-effect diagram.

The team brainstormed a list of existing aids that might help overcome the major barriers. Next they reviewed the brainstormed list of aids and selected those which they believed might overcome the effects of some of the major barriers. Each aid is positioned underneath the barrier it is expected to overcome.

Finally, some barriers were not overcome adequately by existing aids, so the team developed specific countermeasures that it would implement. Those are placed below the aids.

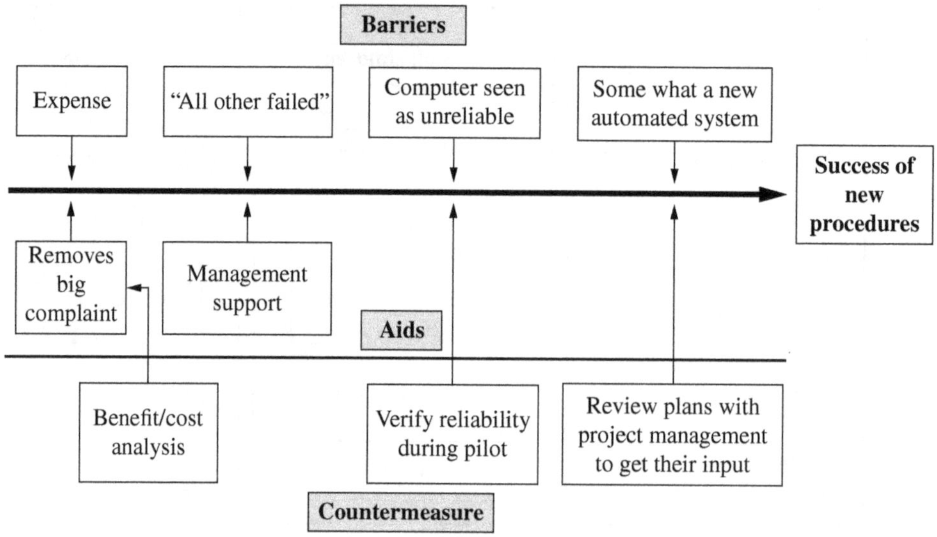

Figure 19.3 Barrier and aids chart.

Basic Statistics

Purpose Statistics are necessary to analyze and interpret data collected on a problem (See Fig. 19.4). Descriptive statistics help characterize problems, and provide a starting point for more advanced statistical methods such as hypothesis testing.

- Measures of central tendency:
  - Mean—the average value of a list of numbers
  - Median—the middle value in a sequential list of numbers
  - Mode—the value that occurs most often in a list of numbers
- Measures of dispersion
- Dispersion or distribution refers to the scattering of data around the central tendency. Its measures are:
  - Range—the difference between the maximum and the minimum values
  - Variance—average squared deviation from each data point from the mean
  - Standard deviation—square root of the variance
- Types of data
- Types of data direct the type of analysis to be done:
  - Continuous data can be measured to an infinite level, e.g., time, temperature, thickness
  - Categorical data falls into categories:
    i. Ordinal—can be arranged into some natural order, e.g., short, medium, tall
    ii. Nominal—cannot be arranged into any natural order, e.g., colors, departments
- Sampling

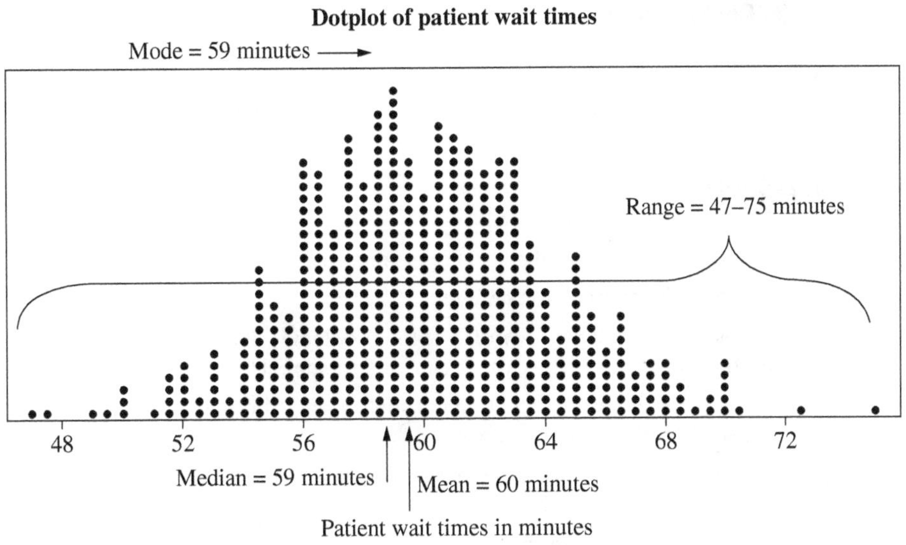

FIGURE 19.4 Central tendency and dispersion.

At times, it may be necessary to collect data on a sample of the population rather than use data from the entire population. The purpose of sampling is to draw conclusions about the population using the sample. This is known as statistical inference.

Key considerations for sampling are

- Sampling scheme: random, stratified
- Precision required (+/−?)
- Amount of characteristic's variation
- Confidence level (e.g., 95 percent)
- Sample size

Qualities of a good sample include

- Free from bias—bias is the presence or influence of any factor that causes the population or process being sampled to appear different from what it actually is.
- Representative—the data collected should accurately reflect a population or process. Representative sampling helps avoid biases specific to segments under investigation.
- Random—in a random sample, data is collected in no predetermined order, and each element has an equal chance of being selected for measurement. Random sampling helps avoid biases specific to the time and order of data collection, operator, or data collector.

Example A project team in a clinical setting was curious to know what the wait times of their patients looked like. They decided to plot the last 500 patients' times on a graph. The *X* axis represents the total wait times while the *Y* axis represents the total count of patients. The team concluded that most of the data centered around 60 minutes; however, some patients waited as little as 47 minutes while others as long as 75 minutes.

Benefit/Cost Analysis

Purpose A benefit/cost analysis characterizes the pros and cons of a solution. It is particularly useful when trying to make a business case for a quality improvement, or to decide among several alternatives.

Steps to Create

1. Estimate the one-time costs.
2. Estimate the additional annual operating costs.
3. Estimate the annual cost savings.
4. If possible, calculate how much of the problem is likely to be eliminated because of each proposed remedy. Otherwise, rank the impact of alternatives.
5. Assess the impact on customer satisfaction. Alternatives that reduce customer satisfaction should be discarded.
6. Calculate the net annual operating costs. A negative number means that net savings are expected.
7. Calculate annual costs of one-time costs.
8. Calculate total annual costs as the algebraic sum of net annual costs and the annual costs of one-time costs.
9. Review data and rank the alternatives.

Example Computerized versus Manual Solution. After identifying root causes of a quality problem, a team decided that both computerized and manual solutions were feasible and met customer criteria. To compare the alternatives on a financial basis, they applied a cost/benefit analysis, as shown in Fig. 19.5. Using this information, the team ranked the two remedies. The better rank of the computerized solution (closer to one) suggests this would be the better of the two alternatives.

Remedy alternative	Manual	Computerized
One-time costs	$7,500	$134,000
Annual cost of one-time costs	$1,500	$26,800
Additional annual operating costs	0	$17,000
Annual cost savings	$1,462,200	$1,562,200
Net annual operating costs (savings)	($1,462,200)	($1,545,000)
Total annual costs (savings)	($1,460,500)	($1,518,200)
Problem impact	70%	75%
Customer satisfaction impact	Low	Low
Benefit/cost assessment rank team average	**1.7**	**1.3**

FIGURE 19.5 Benefit/cost analysis.

Box Plot

Purpose This is a graphic, five-number summary of variation in a data set. The data are summarized by: the smallest value, second quartile, median, third quartile, and largest value. The box plot can be used to display the variation in a small sample of data or for comparing the variation among groups.

Steps to Create

1. Collect the raw data and convert it to an ordered data set by arranging the values from the lowest to the highest.
2. Decide on the type of box plot you wish to construct.
3. Calculate the appropriate summaries.

$$\text{Depth of the median} = d(\text{M}) = (n+1)/2$$

$$\text{Depth of the first quartile} = d(\text{Q1}) = (n+2)/4$$

$$\text{Depth of the third quartile} = d(\text{Q3}) = (3n+2)/4$$

$$\text{Upper adjacent} = \text{the largest observation that is less then } \{\text{Third quartile} - (1.5 \times \text{IQR})\}$$

$$\text{Lower adjacent} = \text{the smallest observation that is greater than } \{\text{First quartile} - (1.5 \times \text{IQR})\}$$

4. Draw and label the horizontal axis.
5. Draw and label the vertical axis.

Example Due to complaints that photocopiers were breaking down and failing to work, a study was conducted to see which contractor responded to the maintenance calls the quickest. The faster the contractor could respond, the faster the machines were up and running. For each contractor, the team gathered data from the last 10 calls and graphed the information to produce the box plots shown below Contractor B was not only quicker in responding to the calls, but the variation in response times were smaller showing that Contractor B was also more consistent than A (Fig. 19.6).

Brainstorming

Purpose A group technique for generating constructive and creative ideas from all participants. Use of this tool should provide new ideas, or new applications and novel use of existing ideas. The technique is outlined in Fig. 19.7.

Steps to Create

1. Phrase the statement to brainstorm.
2. Prepare for brainstorming.
   a. Communicate the statement ahead of time.
   b. Provide appropriate surfaces for contributions.

3. Introduce the session.
    a. Review the conceptual rules.
        (1) No criticism or evaluation of any kind.
        (2) Be unconventional.
        (3) Aim for quantity of ideas in a short time.
        (4) "Hitchhike" on others' ideas.

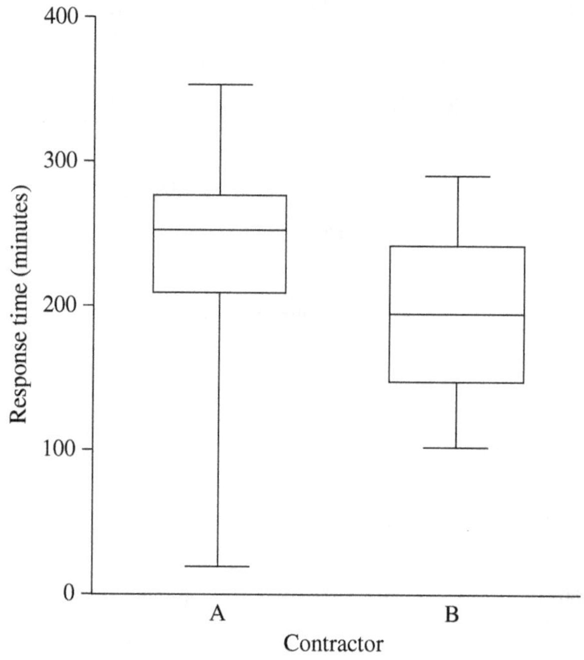

Figure 19.6 Basic box plot.

- Good ideas are not praised or endorsed. All judgment is suspended initially in preference to generating ideas.
- Thinking must be unconventional, imaginative, or even outrageous. Self-criticism and self-judgment are suspended.
- To discourage analytical or critical thinking, team members are instructed to aim for a large number of new ideas in the shortest possible time.
- Team members should "hitchhike" on other ideas, by expanding them, modifying them, or producing new ones by association.

Figure 19.7 Brainstorming.

   b. Review the practical rules.
      (1) Make contributions in turn.
      (2) Only one idea per turn.
      (3) You may pass.
      (4) Do not provide explanations.
4. Warm up.
5. Brainstorm.
   a. Write issue where it will be visible to all.
   b. Have another person write all contributions where visible.
   c. Stop before fatigue sets in.
6. Process ideas.

Example Prior to meeting, members of a focus group were provided the following statement: "What are positive attributes you would like to see in a child's toy?" Upon meeting, the team used adhesive notes to record and post ideas on a wall, making one contribution per turn. After ideas were all recorded, they were processed to clarify and eliminate duplicates. Refer to the affinity diagram section above for additional processing of ideas.

Cause-Effect Diagram

Purpose Developed by Kaoru Ishikawa, this tool is frequently called the Ishikawa diagram in his honor. Its purpose is to organize and display the interrelationships of various theories of root cause of a problem. By focusing attention on the possible causes of a specific problem in a structured, systematic way, the diagram enables a problem-solving team to clarify its thinking about those potential causes, and enables the team to work more productively toward discovering the true root cause or causes.

Steps to Create

1. Define clearly the effect (the *Y*) for which the cause must be identified.
2. Place the effect or symptom being explained at the right, enclosed in a box. Draw the central spine as a thick line pointing to it.
3. Use brainstorming or a rational step-by-step approach to identify the possible causes (the Xs).
4. Each of the major areas of potential (not less than 2 and normally not more than 5) should be placed in a box and connected to the central spine by a line at an angle of about 70°.
5. Add potential for each main area, placing them on the horizontal lines.
6. Add subsidiary causes for each cause already entered.
7. Continue adding possible causes until each branch reaches a potential root cause.
8. Check the logical validity of each causal chain. It should read "negative," that is, a flat tire caused the car to swerve, a nail caused the tire to go flat, a person left a nail on the driveway.
9. Check for completeness.

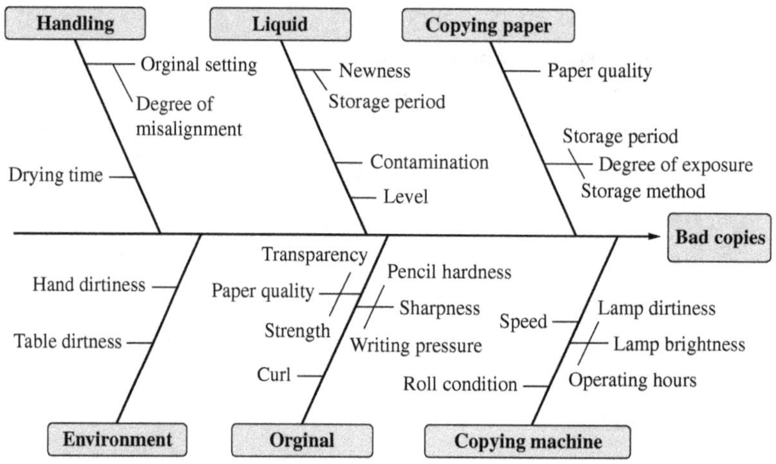

Figure 19.8 Cause-effect diagram.

Example A team was tasked with identifying causes of bad photocopies (the effect). The team considered possible causes related to the 5Ms (man, machine, materials, methods, measurement) and initially structured their fishbone diagram using these categories (they also briefly considered the 5Ps of plant, product, people, policies, procedures). However, after further brainstorming, they realized their possible causes fell into categories of handling, liquid, copying paper, environment, original and copying machine. The team restructured their original diagram into that shown in Fig. 19.8.

Check Sheets

Purpose Used to collect and analyze data, a check sheet is a type of graph or chart that is formatted to allow immediate conclusions to be drawn regarding the data, including patterns and trends (See Fig. 19.9).

Steps to Create

1. Title the top of the sheet with the name of the item or process that is being analyzed.
2. Determine if an immediate analysis can be completed or if observation of the subject is necessary to gather the date. If so, record the time period.
3. Provide any additional information about the analysis that will be helpful for others to review, that is, who is conducting the analysis, the date and time of the analysis, where and why the analysis is being done.
4. List in a column on the right-hand side of the sheet what subcategories or items are being tallied. If more than one part is being analyzed, create separate sections with individual headers.
5. Record the information in bundles of up to 5 tallies per bundle, as shown in the example.

Example Customer service for a large television manufacturer noticed higher than usual complaints about failing parts on three types of televisions. The repair department began a quick assessment of the televisions returned for repair. They used a check sheet to determine which parts needed

COMPONENTS REPLACED BY LAB

Enter a mark for each component replaced. Mark like the following: / // /// //// 𝍸

Time Period: 22 Feb to 27 Feb 1988
Repair Technician: Bob

TV SET MODEL 1013

Integrated circuits	𝍸
Capacitors	𝍸 𝍸 𝍸 𝍸 𝍸 //
Resistors	//
Transformers	////
Commands	
CRT	/

TV SET MODEL 1017

Integrated circuits	///
Capacitors	𝍸 𝍸 𝍸 𝍸 𝍸 //
Resistors	/
Transformers	//
Commands	𝍸 𝍸 𝍸 ///
CRT	/

TV SET MODEL 1019

Integrated circuits	/
Capacitors	𝍸 𝍸 𝍸 𝍸 ///
Resistors	/
Transformers	//
Commands	
CRT	/

FIGURE 19.9 Components replaced by lab.

replacing and in what quantity. The following check sheet was created for television set models 1013, 1017, and 1019.

Control Plan

Purpose A control plan is used to hold the gain obtained through a quality improvement project. The plan specifies the control subjects to be monitored, measurement of these, and the actions to take based on explicit criteria (See Fig. 19.10).

Steps to Create

1. Identify variables that affect the remedy and the customer directly or indirectly.
2. Establish the standard that will trigger action. The best standard is a control limit from a control chart because it is achievable.

Process control plan for: improvement team 3				Date: 11/22/2009				Revision level: 2.1		Approved by: Champion	
Control subject	Subject goal	Unit of measure	Sensor	Frequency of measurement	Sample size	Recording of measurement/ tool used	Measured by whom	Criteria for taking action (i.e., when to take action)	What actions to take	Who decides	Who acts	Record of action taken
Spray delivery capacity	10 gallons per minute	Gallons per minute	Water meter	During start-up of every job	Each job	Gauge	Foreman	>11 gallons and <9 gallons	Reduce flow speed Increase flow speed	On-site worker assigned to "decision" role for project	Specialist II	None
Crew size	One person per 100,000 sq. t. of yard	Number of workers per 10,000 sq. feet	Foreman	During start of daily run for each job and yard	Each job	Visual count	Foreman	> = 12 and < = 3 workers call office	See office manager	One-site foreman	Specialist II	None
Schedule forecast no PC to determine to/from	Forecast times always within 10% of actual	Number of workers	Foreman	Every job	Each job	Visual count	Foreman	Actual vs. estimated # of workers varies by 7	Adjust program such that variance –3	On-site foreman	Supervisor	TBD
Schedule forecast on PC to determine work need	Forecast times always within 10% of actual	Location of job	Foreman	Every job	Each job	Visual count	Foreman	Actual vs. estimated # of workers varies by 7	Adjust program such that variance –3	On-site foreman	Supervisor	TBD

FIGURE 19.10 Process control plan.

3. Establish how each control variable will be measured. Enter this information on the same line as the control variable under the column "How measured." Make similar entries for each of the following steps.
4. Establish where and when the measurements will be made and how they will be recorded, including the type of control chart.
5. Decide who will analyze the ongoing measurement—that is, who will determine that the process is out-of-control.
6. Decide who will act to diagnose and eliminate the assignable cause for the out-of-control condition.
7. Decide what steps can be taken to bring the process back into control. While it may not be possible to foresee all problems, identifying specific actions ahead of time will make control much more effective.
8. Review the matrix to verify that
   a. All critical control variables have been identified.
   b. The control plan will bring the process back into control quickly.
   c. The control plan makes maximum use of self-control.

Example Improvement Team 3 completed their analysis of the process and concluded that 4 main subjects would be monitored to sustain the overall improvements. If the process becomes unstable and the "criteria for taking action" is met, management and team members know exactly who is responsible and properly trained to address the out of control situation. A detailed description of the action taken to bring the process back into control is later documented on the plan.

Customer Needs Spreadsheet

Purpose To help analyze and prioritize customer needs. Often times, information that is collected from customers is too broad, too vague, and too voluminous to be used directly in designing a product. Prioritizing the customers and their needs ensures that a team focuses on what is most important to a design, and ensures the budgeted resources are allocated accordingly (See Fig. 19.11).

Steps to Create

1. Create a multicolumn spreadsheet.
2. Label the first column, "Customers." List, in priority order, the vital few customers. Include groups of "useful many" customers that, collectively, can be considered as vital few.
3. Label the top row, "Customer Needs," and list all discovered needs in the columns below. Enter one need for each column.
4. Correlate the relationships between customers and needs:
   a. Create a legend to define the relationship.
   b. Base the relationship on solid evidence.
   c. More than one customer can be addressed by the same need.
   d. Enter the appropriate value where needs and customers intersect.
   e. Review the spreadsheet and add any additional customers or needs that have been left off the list.

Customers	Customer needs							
	Attractive	Informative and well-written articles	Catchy cover lines	Stable circulation	It sells	Enough complete	Material complete	No last minute changes
Readers	●	●	○					
Advertisers	●	○	●	●	●			
Printers						●	●	●
Typesetters						●	●	●
Color separators						●	●	●
Newsstand	●	○	●	○	●			

Legend
● Very strong relationship
○ Strong relationship
△ Weak relationship

Figure 19.11 Customer needs spreadsheet.

5. Go back and summarize the data you have collected.
6. Analyze each need in terms of:
   a. The strength of the relationship between needs and customers.
   b. The customer's importance.
7. Determine criteria and prioritize the needs from most critical to least critical.

Example An analysis was conducted for the design and production of a new magazine. A list of customers and their needs were determined from previous analysis and placed in a customer needs spreadsheet. The left column lists, in priority order, all the external and internal customers. The column headings are the various needs that have been discovered. Upon completion, the team determined that they would need to focus on two main areas of the magazine in order to ensure the most optimal results. First, they would need make the magazine attractive to their readers and create catchy cover lines. Second, they would need to ensure that the content is complete and free of errors so that production of the magazine could begin on schedule.

Failure Mode and Effects Analysis

Purpose A failure mode and effect analysis (FMEA) helps identify possible ways in which failures can occur, and the effects of these failures. The tool also helps prioritize these based on risk, and track subsequent actions to reduce risk. Many types exist, but the most common are design (or product) and process FMEA (See Fig. 19.12).

Steps to Create

1. Create a nine column spreadsheet.
2. Create an assigned-value table.

3. In column one, list all possible modes of failure. Each item should have a separate line.
4. In the next column, identify all possible causes of failure for each mode.
5. In column three, determine the affect each failure will have on the customer, the overall product, other components and the entire system.

   Note: For steps 6, 7, 8 use values established in assigned values table.
6. Evaluate the frequency of occurrence. Enter the appropriate integer in column four.
7. Evaluate the degree of severity of each failure effect. Record appropriate value in column five.
8. Evaluate the chance of detection for each cause of failure. Place this number in column six.
9. Calculate the risk priority factor by multiplying columns 4, 5, and 6. Put the product in column seven.
10. Design action/remedy for only those "vital few" causes with the highest risk factors. Reduce the level of failure to a rate that is acceptable.
11. Validate each action/remedy.

Example A bank established an improvement team to improve its services related to new checking accounts. The team looked at a variety of different components of this service, including the printing of new checks. Because failures in this component could directly and indirectly affect customers, an FMEA approach was taken to identify and characterize potential failures.

Product: New checking account
Component: Printing new checks

1	2	3	4	5	6	7	8	9
Mode of failure	**Cause of failure**	**Effect of failure**	**Frequency of occurrence (1–10)**	**Degree of severity (1–10)**	**Chance of detection (1–10)**	**Risk priority (1–1000) (4) × (5) × (6)**	**Design action**	**Design validation**
Checks being printed incorrectly	Incorrect information on application form	Checks have to be re-issued	4	6	8	192	Clerk reviews information with customer	Clerk initials form after review
	Data entry error	Ditto	8	6	5	240	Review step in software	Run software
	Information entered in the wrong field on application field	Ditto	5	6	2	60		

Note following assigned values

Column/value	1	2	3	4	5	6	7	8	9	10
4. Frequency (errors per 10,000 customers)	<2	4	8	10	15	20	25	30	35	<35
5. Severity for customer	Trivial				Cause complaint			Major time or $		Loss of customer
6. Detection	Certain				Possible					None

Figure 19.12 Failure mode and effects analysis.

Flow Diagram/Process Map

Purpose A graphic representation of the sequence of steps needed to produce some output. The output may be a physical product, a service, information, or a combination of the three. The symbols of a flow diagram are specific to function and are explained in Fig. 19.13.

Steps to Create

1. Discuss how you intend to use the flow diagram or process map.
2. Decide on the desired outcome of the session.
3. Define the boundaries of the process. Show the first and last steps, using appropriate flow diagram symbols.
4. Document each step in sequence, starting with the first (or last) step. Lay out the flow consistently from the top to bottom or left to right.
5. When you encounter a decision or branch point, chose one branch and continue.

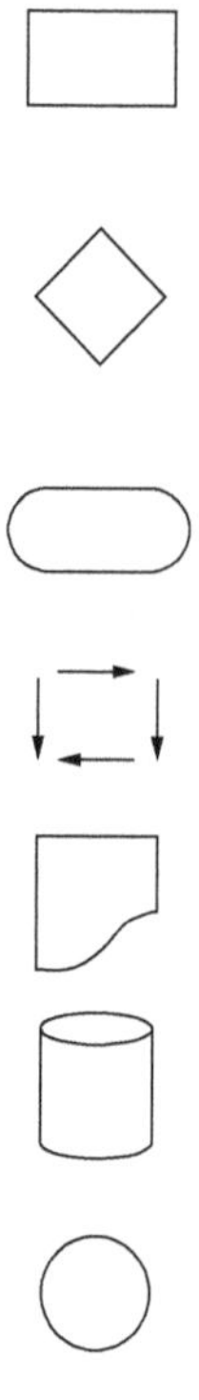

The *activity symbol* is a rectangle that indicates a single step in the process. A brief description of the activity is shown inside the rectangle.

The *decision symbol* is a diamond that designates a decision or branch point in the process. The description of the decision or branch is written inside the symbol, usually in the form of a question. The answer to the question determines the path that will be taken out of the decision symbol. Each path is labeled to correspond to an answer.

The *terminal symbol* is a rounded rectangle that identifies the beginning or the end of a process. "Start" or "End" is shown inside the symbol.

Flow lines are used to represent the progression of steps in the sequence. The arrowhead on the flow line indicates the direction of the process flow.

The *document symbol* represents written information pertinent to the process. The title or description of the document is shown inside the symbol.

The *database symbol* represents electronically stored information pertinent to the process. The title or description of the data base is shown inside the symbol.

The *connector* is a circle used to indicate a continuation of the flow diagram. A letter or number is shown inside the circle. This same letter or number is used in a connector symbol on the continued flow diagram to indicate how the processes are connected.

These basic symbols are arranged to show the actual sequence of steps in a process, running consistently from top to bottom or left to right on the page. As we have seen, decision diamonds can lead us to branch back and repeat an earlier step.

FIGURE 19.13 Symbols used in flow diagramming.

6. If you encounter an unfamiliar segment, make a note and continue.
7. Repeat steps 4, 5, and 6 until you reach the last (or first) step in the process.
8. Go back and flow diagram the other branches from the decision symbols.
9. Review the completed chart to see if you have missed any decision points or special cases.
10. Fill in unfamiliar segments and verify accuracy.
    a. Observe process
    b. Interview knowledgeable people
11. Analyze the flow diagram.

Example A team tasked with improving the process for distribution of technical manuals realized they needed a better understanding of the current process steps and boundaries. Drawing upon the expertise of people directly involved in the process, they developed the flow diagram shown in Fig. 19.14. This baseline process map also served later to help the team identify potential problems contributing to errors and delays in distribution.

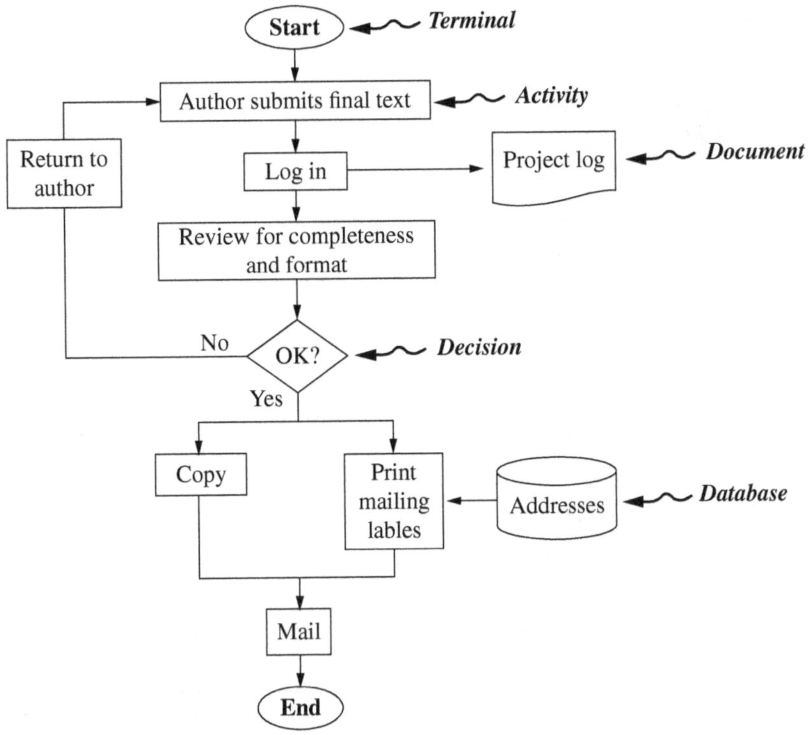

FIGURE 19.14 Flow diagram.

Graphs and Charts

Purpose A broad class of tools used to summarize quantitative data in pictorial representations. Three types of graphs and charts that prove especially useful in quality improvement are line graphs, bar graphs, and pie charts. A line graph connects points which represent pairs of numeric data, to display the relationship between two continuous numerical variables (e.g., cost and time). A bar graph also portrays the relationship between pairs of variables, but only one of the variables need be numeric. A pie chart shows the proportions of the various classes of a phenomenon being studied that make up the whole. See Fig. 19.15 for examples of a line graph, bar graph, and a pie chart.

Steps to Create

Line Graphs

1. Determine the range of the vertical axis and the size of each increment. Label the vertical axis.
2. Do the same for the horizontal axis.
3. Draw axes and if needed, a grid.
4. Plot each data point.
5. Connect the points with a line.
6. Label and title the graph.

Bar Graphs

1. Determine the range of the vertical axis and the size of each increment. Label the vertical axis.
2. Choose a simple, grouped or stacked bar graph.
3. Determine the number of bars. Draw the horizontal axis. Label the horizontal axis.
4. Determine the order of the bars.
5. Draw the bars.
6. Label and title the graph.

Pie Charts

1. Determine the percentage for each category.
2. Convert the percentage values into degrees.
3. Draw and circle with a compass and mark the segments of the pie chart with a protractor.
4. Label the segments and title the chart.

When creating graphs and charts, keep the following in mind:

Graphic integrity: A graph must not lie. It should be constructed so that the viewer is not misled. Rather than relying solely on the graphics, look at the written data to ensure that the true information is conveyed by the graph.

Consistent scale: Numeric scales must maintain regular intervals. Different graphs that might be compared to each other should all be drawn to the same scale.

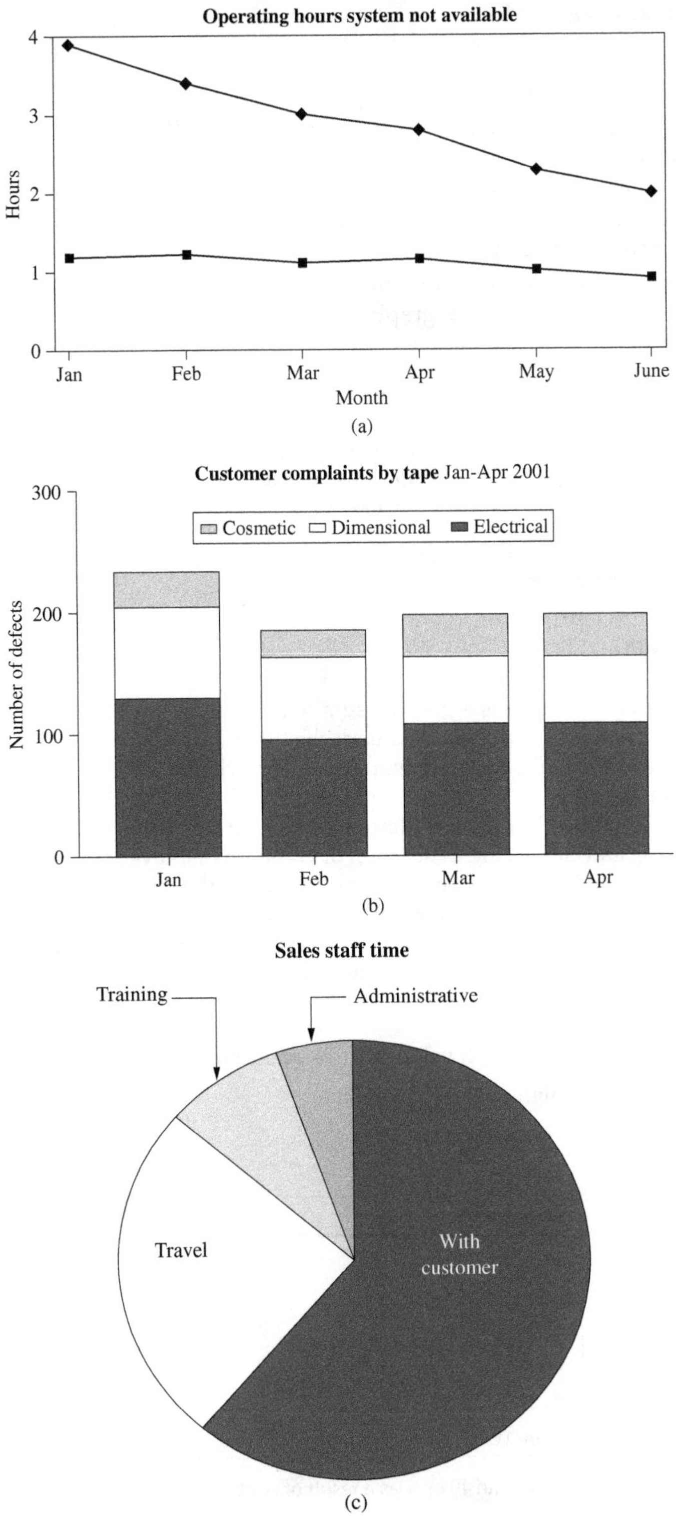

FIGURE 19.15 (*a*) Line graph; (*b*) bar graph; (*c*) pie chart.

Ease of reading: How well a graph is understood and remembered depends on how easy it is to read. Use labels to improve clarity. Place labels close to the object they identify.

Consistency of symbols: When two or more graphs are to be compared, it is important to maintain consistency along many dimensions, in order to minimize confusion in interpreting the graphs.

Simplicity: Do not obscure information with unnecessary decoration. Before adding text or decoration to a graph, ask, "What additional value or information am I adding?"

Example

Line Graph The hours a computer operating system was not available is plotted over time for two different computing center bar graph. Customer complaints stratified by type (cosmetic, dimensional, electrical) are shown across different months.

Pie Chart The proportion of time sales staff spent in different activities is shown as different sized slices of the overall pie.

Histogram

Purpose A histogram is a graphic summary of variation in a set of data. Four concepts related to variation in a set of data underlie the usefulness of the histogram: (1) values in a set of data almost always show variation, (2) variation displays a pattern, (3) patterns of variation are difficult to see in simple numerical tables, and (4) patterns of variation are easier to see when the data are summarized pictorially in a histogram. Analysis consists of identifying and classifying the pattern of variation displayed by the histogram (such as the shape, the location of the center, or the spread of the data from the center) then relating what is known about the characteristic pattern to the physical conditions under which the data were created to explain what in those conditions might have given rise to the pattern. Figure 19.16 illustrates some common patterns.

Steps to Create

1. Determine the high value, the low value and the range.

   Range = (high value) – (low value)

2. Decide on the number of cells.

Data Points	Number of Cells*
20–50	6
51–100	7
101–200	8
201–500	9
501–1000	10
Over 1000	11–20

*Less than 40 only as a result of stratification.

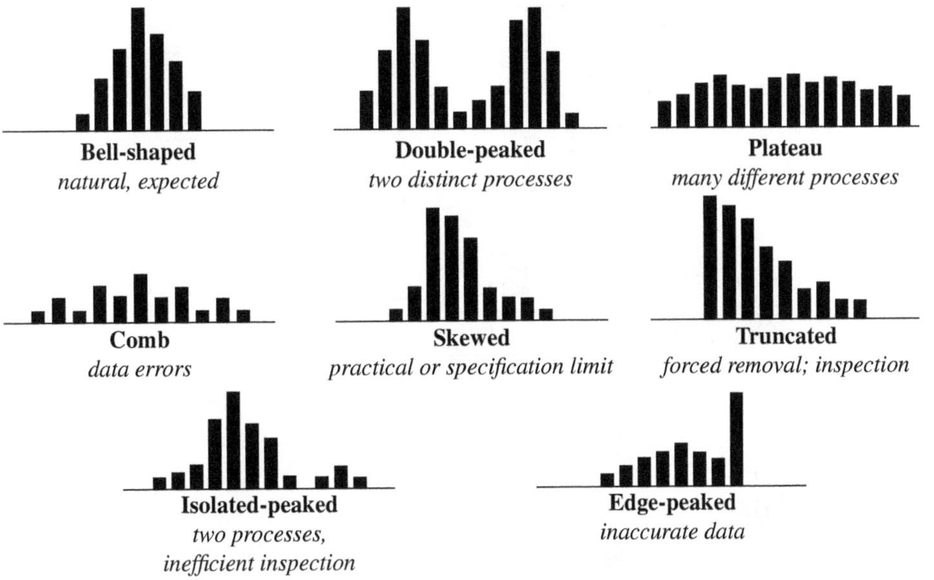

FIGURE 19.16 Histograms.

3. Calculate the approximate cell width.

$$\text{Approximate cell width} = (\text{range})/(\text{number of cells})$$

4. Round the cell width to a convenient number.
5. Construct the cells by listing the cell boundaries.
6. Tally the number of data points in each cell.
7. Draw and label the horizontal axis.
8. Draw and label the vertical axis.
9. Draw bars to represent the number of data points in each cell.
10. Title chart and indicate total observations.
11. Identify and classify the pattern of variation.
12. Develop an explanation for the pattern.

Example The data in Fig. 19.17*a* show days elapsed between an interdepartmental request for an interview and the actual interview. In Fig. 19.17*b*, the histogram helped a team recognize the unacceptable range of time elapsed from request to interview. It also provided the team with a vivid demonstration of a human-created phenomenon—the rush at day 15 to get as many requests completed within the 15-day goal. The histogram directed the team's attention toward steps to reduce the duration (and with it the spread) of the process.

14	15	18	19	13
12	18	22	14	18
15	17	19	15	17
18	20	10	15	15
20	14	17	20	21
15	24	15	18	14
23	13	23	21	20
18	21	18	15	15

(a)

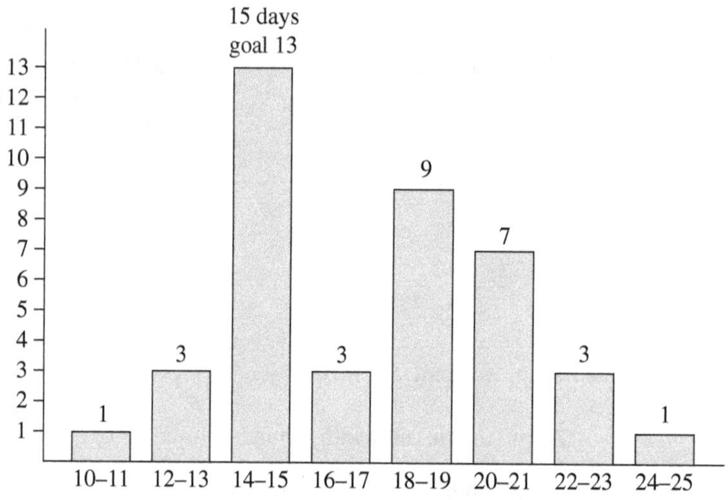

(b)

Figure 19.17 (*a*) Data showing elapsed time; (*b*) elapsed working days.

Pareto Analysis

Purpose This is a tool used to establish priorities, dividing contributing effects into the "vital few" and "useful many." (See Fig. 19.18). A Pareto diagram includes three basic elements: (1) the contributors to the total effect, ranked by the magnitude of contribution; (2) the magnitude of the contribution of each expressed numerically; and (3) the cumulative-percent-of-total effect of the ranked contributors.

Steps to Create Pareto charts are not as commonly present in software as similar types of graphical analysis tools, and the steps for manual creation assist in understanding the different elements.

1. Total the data of each contributor and sum these to determine the grand total.
2. Reorder the contributors from the largest to the smallest.

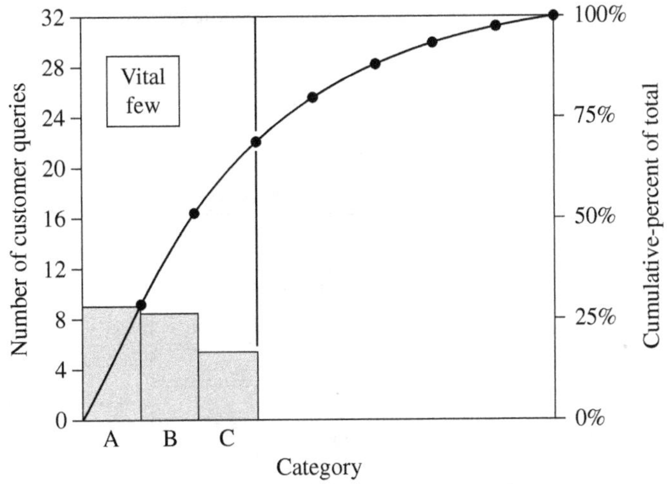

FIGURE 19.18 Pareto analysis.

3. Determine the cumulative-percent of total for each contributor.
4. Draw and label the left vertical axis from 0 to the grand total or just beyond.
5. Draw and label the horizontal axis. List the contributors from largest to smallest, going from left to right.
6. Draw and label the right vertical axis from 0 to 100 percent. Line up the 100 percent with the grand total on the left axis.
7. Draw bars to represent the magnitude of each contributor's effect.
8. Draw a line graph to represent the cumulative percent-of-total.
9. Analyze the diagram. Look for a break point on the cumulative-percent graph.
10. Title the chart; label the "vital few" and the "useful many."

Example To assist in the decision process of selecting new improvement projects, a Pareto chart was created to better understand the types of questions received from customers. Based on this information, it was decided to charter a team to address queries falling into categories A, B, and C.

Planning Matrix and Tree Diagram

Purpose A tree diagram is a graphical method for identifying all the parts that are needed to create some final objective. The planning matrix is an extension of the tree diagram. (See Fig. 19.19). It shows all the factors, components, and tasks required to achieve the final objectives. It specifies who will complete each step and when. It may also specify for each step who will help, what the budget will be, who the team contact is for work done outside the team, the status of the task, etc.

Steps to Create

1. Use a tree diagram to identify all the tasks needed to complete a specific piece of work.
2. List each task on an adhesive note and post on the wall or flip chart in a vertical column.

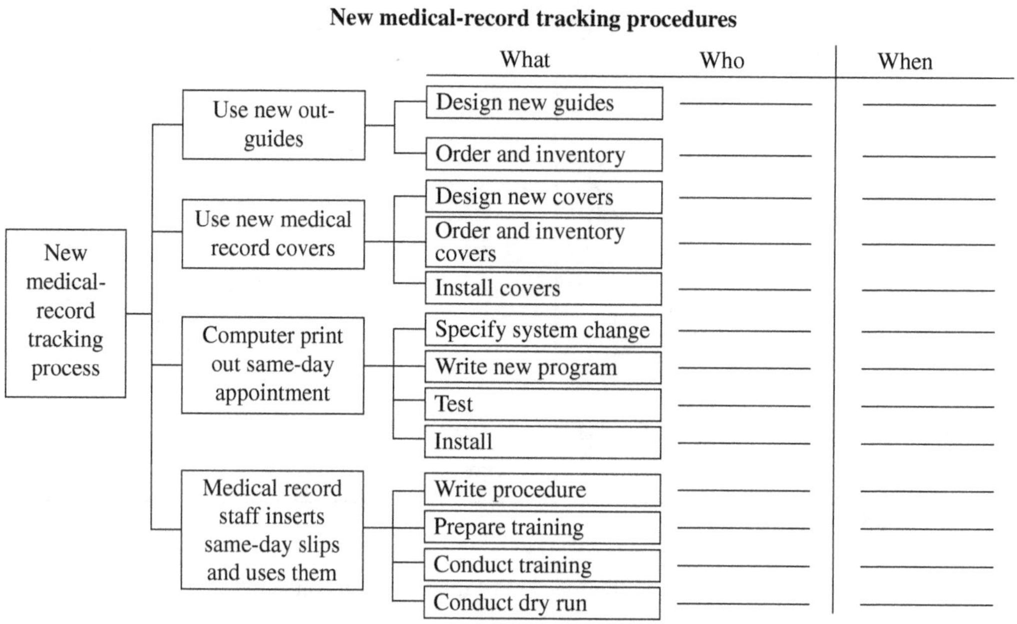

Figure 19.19 Planning matrix and tree diagram.

3. Label other columns with "who" and "when."
4. Work through the task one by one. Take the following steps:
   a. Discuss and identify the most appropriate person or group of persons to do the work.
   b. Agree on the necessary completion date.
5. Agree on how the team will monitor progress on the plan. Possibilities include the following:
   a. An agenda item at each meeting, obtain a brief report on active tasks.
   b. Have the team leader (or a designated member) check with each responsible person before the meeting and enter on the agenda only those tasks that require discussion.
6. Transfer the matrix to standard paper and include it with future team minutes and agendas.

Example An improvement team determined that a new medical record tracking process was needed to reduce lost and missing paper records. To help plan for the creation of this new process, they broke down the process into four major parts, and subsequently broke these down further into the elements necessary and sufficient to complete the major step (answering the question "What needs to be done?"). Using this tree diagram, the team proceeded to develop it further into a planning matrix by the addition of "who" and "when" information.

Pugh Matrix

Purpose This is a useful tool for comparing several alternative concepts against preestablished criteria. It allows you to

- Compare alternative solutions against project CTQs (critical to quality characteristics or requirements).
- Create strong alternative solutions from weaker ones.
- Arrive at an optimum solution that may be a hybrid or variant of other solutions.

Steps to Create

Enter the Criteria Customer CTQs must be included. Business criteria such as time to market, complexity, ability to patent, etc., can also be added:

1. Weight the criteria in terms of importance.
2. Rate each alternative as better (+), worse (–), or same (s) at achieving criteria as compared to datum alternative.
3. Compare the number of positives, negatives, and sames between alternatives. Can you create a new alternative that leverages the best of the initial alternatives?

Example An example template for the construction of a *Pugh matrix* is shown in Fig. 8.20.

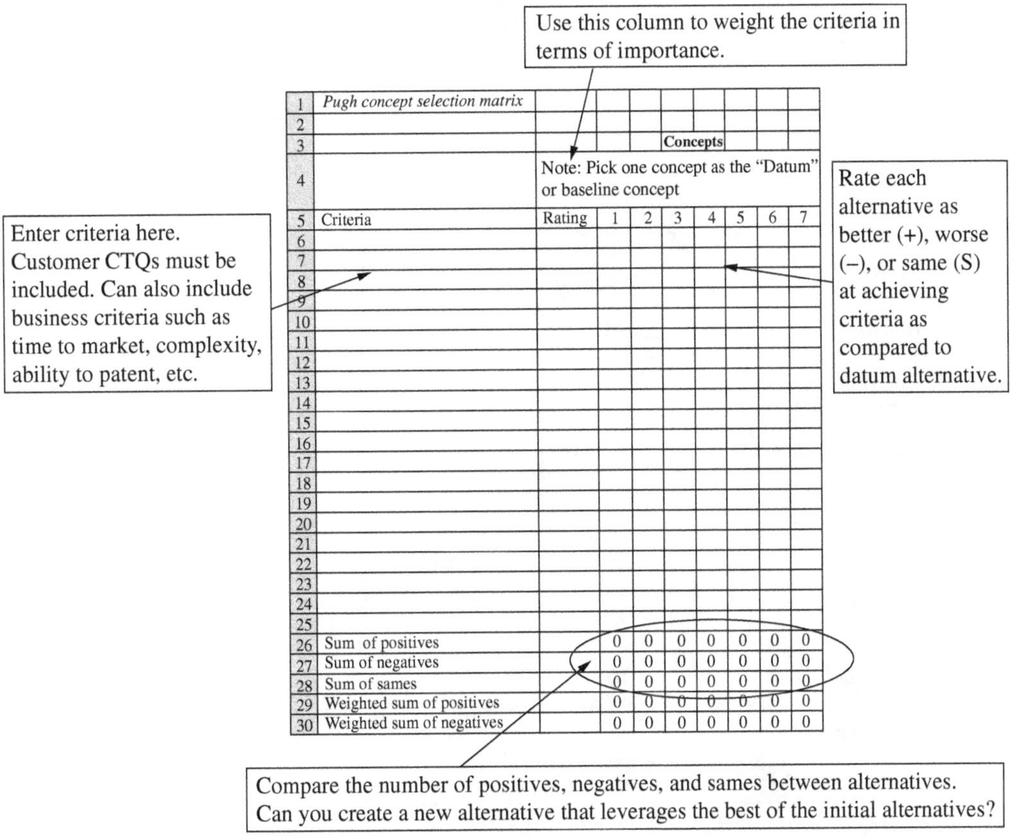

FIGURE 19.20 Pugh matrix.

Selection Matrix

Purpose This tool assists in reducing a set of alternatives based on explicit criteria. Selection matrices often are applied when deciding among different solutions that each may have positive and negative attributes, such that the "best" solution is not readily apparent. Unlike methods such as voting techniques that can introduce considerable subjectivity, selection matrices help introduce establish objectivity to the decision process.

Steps to Create

1. Agree on the criteria to be used to evaluate the alternatives.
2. Each team member allocates a total of 100 points among the criteria.
3. Calculate the average number of points allocated to each criterion.
4. Review and agree on the weights.
5. Assemble the list of alternatives to be evaluated.
6. Rate each alternative. Each team member rates each alternative according to how well it meets each criterion, using a scale of 1 (worst possible) to 5 (best possible).
7. Calculate each team member's average rating for each alternative.
8. Give each team member a table showing how each member rated each alternative.
9. Discuss the ratings and reach consensus on the next steps.

This matrix can be used to select projects and, as in this example, to select solutions.

Criterion	Weight	Alternative 1	Alternative 2	Alternative 3
Remedy Name		Overhaul and change speed	Replace equipment	Retain and change speed
Total cost	20	5	2	4
Impact on the problem	10	3	5	4
Benefit and cost relationship	30	4	3	5
Cultural impact and resistance to change	20	4	5	1
Implementation time	2	5	4	2
Uncertainty about effectiveness	6	4	5	3
Health and safety	10	4	5	5
Environment	2	3	3	3
Average rating		**4.10**	**3.74**	**3.68**

FIGURE 19.21 Selection matrix.

Example A team was evaluating possible alternatives to improve the performance of saws used to cut wood for high-end cabinetry. The team brainstormed criteria and then differentiated these through a weighting method in which the weights summed to 100. Each alternative then was scored as to how well it met each individual criterion (scored 1 to 5, with 5 a best score). An average score then was calculated for each alternative, with results as shown in Fig. 19.21. Using the results as a starting point for discussion, the team ultimately determined that Alternative 1 was the best option.

SIPOC

Purpose The acronym SIPOC stands for supplier, input, process, output, and customer. It is a high-level map showing a process's primary suppliers, the inputs received from them, and the process that adds value to those inputs. That process produces an output that is intended to meet or exceed customer requirements. The SIPOC typically is used at the early stages of a project to help characterize a process, and to identify appropriate team members. This model is applicable to both product and service processes.

Steps to Create

1. Define the process, name it and define the start and stop points.
2. Identify suppliers and the critical inputs the process receives from them.
3. Identify the customers of the process (those who receive the outputs) and the outputs of the process that respond to customer needs.
4. Identify the 5 to 8 major process steps that produce the output.
5. Validate the process map by working with the key functions that perform the major steps.

Example An improvement team was created to address the Order Receiving process. To help identify the high level steps and scope of their project, the team created the SIPOC shown in Fig. 19.22, beginning the process with receiving the order and following the process through to the time the product is scheduled for production.

Supplier	Input	Process	Output	Customer
Store location	Electronic order	Receive order	Queue-created file	Order sorter system
Order sorter system	Queue-created file	Hold order in queue	Purchase order TIF file	Order entry
Order entry	Purchase order TIF file	Enter order in system	Order ready for scheduling	Order checker
Order checker	Electronic order	Check order	Scheduling form	Scheduling
Scheduling	Scheduling form	Schedule production	Paper work	Production plant

FIGURE 19.22 SIPOC.

Statistical Process Control

Purpose The daily life of many employees involves operating a process within the intended boundaries; that is, to maintain it according to specifications established through quality planning and improvement. Historically, this relied heavily on inspection, with detection and elimination of nonconforming product after the fact. In contrast, the concept of control over a process entails predicting its performance, within certain limits. Rather than merely detecting non-conforming output ("inspecting quality into a product"), control is forward-looking and seeks incremental but continuous improvement through identification and elimination of special causes that create unpredictable variation (and potentially, but not necessarily, nonconformity to specification).

Statistical process control is the application of statistical methods to the measurement and analysis of variation in a process. A process is a collection of activities that converts inputs into outputs or results. Through use of control charts, statistical process control assists in detecting special (or assignable) causes of variation in both in-process parameters and end-of-process (product) parameters. The objective of a control chart is not to achieve a state of statistical control as an end in itself but, to reduce variation.

Before proceeding with the steps to create a control chart, further discussion is warranted regarding common and special cause variation in the context of process control. A statistical control chart compares process performance data to computed "statistical control limits," drawn as limit lines on the chart. The process performance data usually consist of groups of measurements (called rational subgroups) from the regular sequence of production while preserving the order of the data. A prime objective of a control chart is detecting special (or assignable) causes of variation in a process. Knowing the meaning of "special causes" and the distinction from common (random or chance) causes is essential to understanding the control chart concept.

Process variations have two kinds of causes: (1) common (random or chance), which are inherent in the process, and (2) special (or assignable), which cause excessive variation. See Table 19.1 for distinctions between common and special causes. Ideally, only common causes are present in a process because they represent a stable and predictable process that leads to minimum variation. A process that is operating without special causes of variation is said to be "in a state of statistical control." The control chart for such a process has all of the data points within the statistical control limits, and exhibits no discernible patterns.

The control chart distinguishes between common and special causes of variation through the choice of control limits. These are calculated by using the laws of probability so that highly improbable causes of variation are presumed to be due to special causes not to random causes. When the variation exceeds the statistical control limits, it is a signal that special causes have entered the process and the process should be investigated to identify these causes of excessive variation. Random variation within the control limits means that only common (random) causes are present; the amount of variation has stabilized, and minor process adjustments (tampering) should be avoided. Note that a control chart detects the presence of a special cause but does not find the cause—that task must be handled by subsequent investigation of the process.

Steps to Create Setting up a control chart requires the following steps:

1. Choosing the characteristic to be charted.
2. Giving high priority to characteristics that are currently running with a high defective rate. A Pareto analysis can establish priorities.
3. Identifying the process variables and conditions that contribute to the end-product characteristics, in order to define potential charting applications from raw materials through processing steps to final characteristics. For example, the pH, salt concentration, and temperature of a plating solution are process variables contributing to plating smoothness.

Random (Common) Causes	Assignable (Special) Causes
Description	
Consists of many individual causes.	Consists of one or just a few individual causes.
Any one random cause results in a minute amount of variation (but many random causes act together to yield a substantial total).	Any one assignable cause can result in a large amount of variation.
Examples are human variation in setting control dials; slight vibration in machines; slight variation in raw material.	Examples are operator blunder, a faulty setup, or a batch of defective raw material.
Interpretation	
Random variation cannot be eliminated from a process economically.	Assignable variation can be detected; action to eliminate the causes is usually economically justified.
An observation within the control limits of random variation means that the process should not be adjusted.	An observation beyond control limits means that the process should be investigated and corrected.
With only random variation, the process is sufficiently stable to use sampling procedures to predict the quality of total production or do process optimization studies.	With assignable variation present, the process is not sufficiently stable to use sampling procedures for prediction.

(*Source: Quality Planning & Analysis,* Juran Institute, Inc., Copyright 2007. Used by permission.)

TABLE 19.1 Distinction between Random and Assignable Causes of Variation

4. Verifying that the measurement process has sufficient accuracy and precision to provide data that does not obscure variation in the manufacturing or service process. The observed variation in a process reflects the variation in the manufacturing process and also the combined variation in the manufacturing and measurement processes. Anthis et al. (1991) describe how the measurement process was a roadblock to improvement by hiding important clues to the sources of variation in a manufacturing process. Dechert et al. (2000) explain how large measurement variation can be controlled and result in effective statistical process control methods.
5. Determining the earliest point in the production process at which testing can be done to get information on assignable causes so that the chart serves as an effective early-warning device to prevent defectives.
6. Choosing the type of control chart. Table 19.2 compares three basic control charts. Schilling (1990) provides additional guidance in choosing the type of control chart.
7. Deciding on the central line to be used as the basis of calculating the limits. The central line may be the average of past data, or it may be a desired average (i.e., a standard value). The limits are usually set at 3s, but other multiples may be chosen for different statistical risks.
8. Choosing the "rational subgroup." Each point on a control chart represents a subgroup (or sample) consisting of several units of product. For process control, rational subgroups should be chosen so that the units within a subgroup have the greatest chance of being alike and the units between subgroups have the greatest chance of being different.

Statistical Measure Plotted	Average X-Bar and Range *R*	Percentage Nonconforming (*p*)	Number of Nonconformities (*c*)
Type of data required	Variable data (measured values of a characteristic)	Attribute data (number of defective units of product)	Attribute data (number of defects per unit of product)
General field of application	Control of individual characteristics	Control of overall fraction defective of a process	Control of overall number of defects per unit
Significant advantages	Provides maximum use of information available from data Provides detailed information on process average and variation for control of individual dimensions	Data required are often already available from inspection records Easily understood by personnel Provides an overall picture of quality	Same advantages as *p* chart but also provides a measure of defectiveness
Significant disadvantages	Not understood unless training is provided; can cause confusion between control limits and tolerance limits Cannot be use with go/no go type of data	Does not provide detailed information for control of individual characteristics Does not recognize different degrees of defectiveness in units of product	Does not provide detailed information for control of individual characteristics
Sample size	Usually four or five	Use given inspection results or samples of 25, 50, or 100	Any convenient unit of product such as 100 feet of wire or one television set

(*Source: Quality Planning & Analysis,* Juran Institute, Inc., Copyright 2007. Used by permission.)

TABLE 19.2 Comparison of Some Control Charts

9. Providing a system for collecting the data. If the control chart is to serve as a day-to-day shop tool, it must be simple and convenient to use. Measurement must be simplified and kept free of error. Indicating instruments must be designed to give prompt, reliable readings. Better yet, instruments should be designed that can record as well as indicate. Recording of data can be simplified by skillful design of data or tally sheets. Working conditions are also a factor.
10. Calculating the control limits and providing specific instructions for the interpretation of the results and the actions that various production personnel are to take (see below). Control limit formulas for the three basic types of control charts are given in Table 19.3. These formulas are based on $\pm 3\sigma$ and use a central line equal to the average of the data used in calculating the control limits. Values of the A2, D3, and D4 factors used in the formulas are given in Table 19.4. Each year, *Quality Progress* magazine publishes a directory that includes software for calculating sample parameters and control limits and for plotting the data. The general rule of thumb is to collect 20 to 30 samples (rational subgroups) before attempting to establish control limits.
11. Plotting the data and interpreting the results.

Chart For	Central Line	Lower Limit	Upper Limit
Averages $\bar{X}$	$\bar{\bar{X}}$	$\bar{\bar{X}} - A_2\bar{R}$	$\bar{\bar{X}} + A_2\bar{R}$
Ranges R	$\bar{R}$	$D_3\bar{R}$	$D_4\bar{R}$
Proportion nonconforming p	$\bar{p}$	$\bar{p} - 3\sqrt{\frac{\bar{p}(1-\bar{p})}{n}}$	$\bar{p} + 3\sqrt{\frac{\bar{p}(1-\bar{p})}{n}}$
Number of nonconformities c	$\bar{c}$	$\bar{c} - 3\sqrt{\bar{c}}$	$\bar{c} + 3\sqrt{\bar{c}}$

(*Source: Quality Planning & Analysis* Juran Institute, Inc., Copyright 2007. Used by permission.)

TABLE 19.3 Control Chart Limits—Attaining a State of Control

Factors for X and R Control Charts;* Factors for Estimating s from R†				
Number of Observations in Sample	A_2	D_3	D_4	**Factor for Estimate from** $\bar{R}$: $d_2 = \bar{R}/s$
2	1.880	0	3.268	1.128
3	1.023	0	2.574	1.693
4	0.729	0	2.282	2.059
5	0.577	0	2.114	2.326
6	0.483	0	2.004	2.534
7	0.419	0.076	1.924	2.704
8	0.373	0.136	1.864	2.847
9	0.337	0.184	1.816	2.970
10	0.308	0.223	1.777	3.078
11	0.285	0.256	1.744	3.173
12	0.266	0.284	1.717	3.258
13	0.249	0.308	1.692	3.336
14	0.235	0.329	1.671	3.407
15	0.223	0.348	1.652	3.472

Upper control limit for $\bar{X} = UCL_{\bar{X}} = \bar{X} + A_2\bar{R}$
Lower control limit for $\bar{X} = LCL_{\bar{X}} = \bar{X} - A_2\bar{R}$

Upper control limit for $R = UCL_R = D_4\bar{R}$
Lower control limit for $R = LCL_R = D_3\bar{R}$

$s = \bar{R}/d^2$

TABLE 19.4 Factors for X and R Control Charts

Stage	Step	Method
Preparatory	State purpose of investigation Determine state of control Determine critical variables Determine candidates for control Choose appropriate type of chart Decide how to sample Choose subgroup size and frequency	Relate to quality system Attributes chart Fishbone Pareto Depends on data and purpose Rational subgroups Sensitivity desired
Initiation	Ensure cooperation Train user Analyze results	Team approach Log actions Look for patterns
Operational	Assess effectiveness Keep up interest Modify chart	Periodically check usage and relevance Change chart, involve users Keep frequency and nature of chart current with results
Phase-out	Eliminate chart after purpose is accomplished	Go to spot checks, periodic sample inspection, overall *p*, *c* charts

(*Source:* Schilling, 1990.)

TABLE 19.5 Life Cycle of Control Chart Applications

The control chart is a powerful statistical concept, but its use should be kept in perspective. The ultimate purpose of an operations process is to make product that is fit for use—not to make product that simply meets statistical control limits. Once the charts have served their purpose, many should be taken down and the effort shifted to other characteristics needing improvement. Schilling (1990) traces the life cycle of control chart applications (Table 19.5). A given application might employ several types of control charts. Note that, in the "phase out" stage, statistical control has been achieved, and some of the charts are replaced with spot checks.

Types of Control Charts The traditional "Shewhart" control charts (named for Dr. Walter A. Shewhart; see JQH5, Section 45 for a historical account of their development) are divided into two categories: variable charts (those using continuous, measurement data), and attribute charts (those using count data). Selection of the proper control chart type is shown in Fig. 19.23 (Control Chart Selection Roadmap) and the types are described further below.

Regardless of the specific chart type or statistic (e.g., average, range, standard deviation, proportion), control limits are established such that it would be very unlikely that the values would fall outside if the process were stable; usually this is set at plus or minus three standard deviations.

Examples of Control Charts for Variables Data

In these charts, the mean and either range or standard deviation are the typical statistics that are monitored. These statistics are monitored in a pair of charts. The averages chart plots the sample averages, specifically, the average of each rational subgroup (if the rational subgroup size is one, then an X-mR (also known as I-mR chart) of individuals is used instead). The range chart or standard deviation chart plots the range or standard deviation of the rational subgroups. The specific subtypes are X-bar and R chart. Also called *average and range* chart.

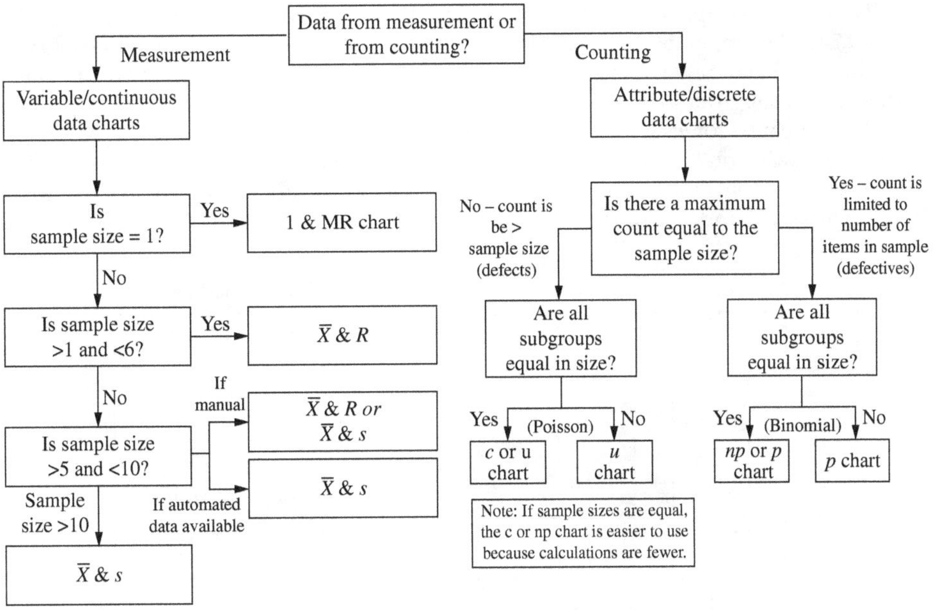

Note that "measurement" implies a continuous scale is used to assess a characteristic. "Counting" occurs when outcomes are tabulated as pass/fail, accept/reject, or some other categorization, even if the original measurement was continuous. Also, control limits depend on number of observations in each sample. When subgroup sizes vary, control limits will not be straight lines on any of the charts.

FIGURE 19.23 Control chart selection roadmap.

The "X-bar" refers to the average of a rational subgroup, and measures the central tendency of the response variable over time. R is the range (difference between the highest and lowest values in each subgroup), and the R chart measures the gain or loss in uniformity within a subgroup which represents the variability in the response variable over time. Note that, because specification limits apply to individual values rather than averages (averages inherently vary less than the component individual values), control limits cannot be compared to specification limits which should not be placed on a control chart for averages.

X-bar and s chart. The average and standard deviation chart is similar to the X-bar and R chart, but the standard deviation (instead of the range) is used in the s chart. Although an s chart is statistically more efficient than the range for subgroup sizes greater than 2, a range chart is easier to compute and understand, and so traditionally is used for subgroup sizes smaller than about 10.

X-mR chart. Also known as an I-mR (individuals and moving range) chart, this charts individual measures and a moving range. It is used when the rational subgroup size = 1 (such that there are not multiple measures from which to obtain an average).

Z-mR chart. This is similar to the X-mR chart, except that the individual values are standardized through a Z transformation. This is useful for short runs in which there are fewer than the recommended 20 to 30 needed to establish one of the preceding charts.

Individuals Chart Also called a run chart, this is an alternative to the X-bar and R chart, and simply is a plot of individual values against time. In the simplest case, specification limits are added to the chart; in other cases, $\pm 3\sigma$ limits of individual values are added. A chart of individual values is not as sensitive as the X-bar chart, however.

By way of example of variable control charting, refer to the X-bar and R charting in Fig. 19.24. The upper part of the figure displays the individual observations for two machines, N-5 and N-7. For each machine, the data consist of 10 samples (with six units in each rational subgroup) plotted in time order of production. The lower portion shows the X-bar and R charts for each machine. For machine N-5, all points fall within the control limits, so that (based on this rule), the process appears to be free of assignable causes of variation, and is "in control." However, machine N-7 has both within-sample variation (seen in the range chart) and between-sample variation (seen in the chart of sample averages).

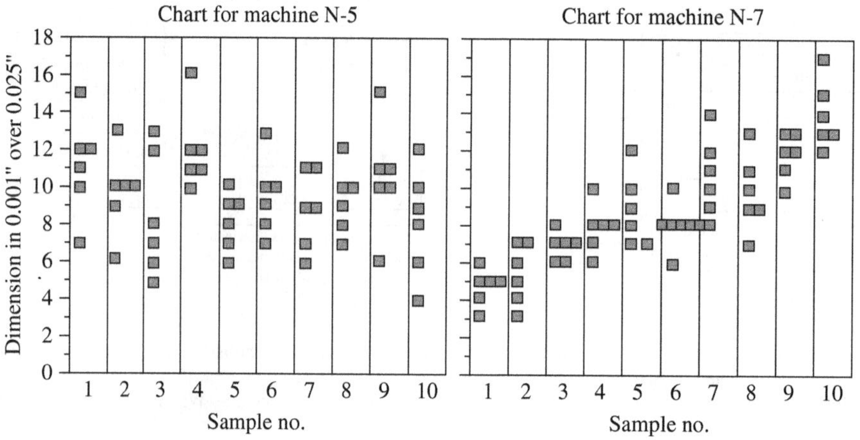

For machine N-5:

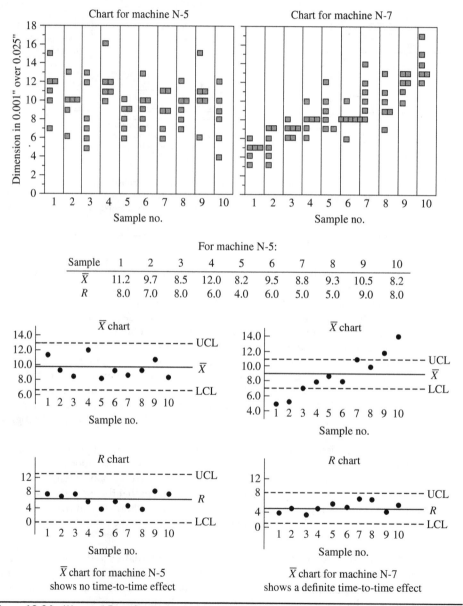

Sample	1	2	3	4	5	6	7	8	9	10
$\overline{X}$	11.2	9.7	8.5	12.0	8.2	9.5	8.8	9.3	10.5	8.2
R	8.0	7.0	8.0	6.0	4.0	6.0	5.0	5.0	9.0	8.0

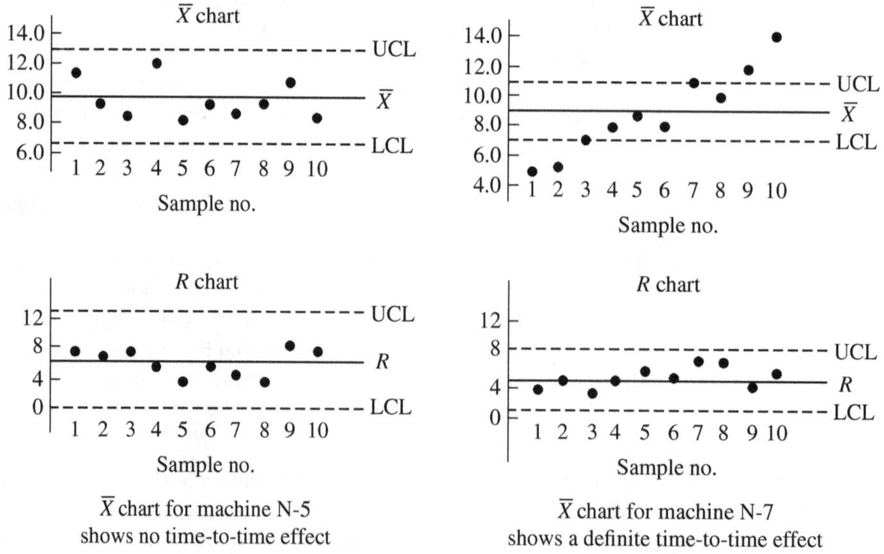

FIGURE 19.24 X-bar and R confirm suggested machine.

The X-bar chart indicates some factor (special cause) such as tool wear is present that results in larger values of the characteristic with the passing of time (note the importance of preserving the order of measurements when collecting data).

Interpreting Variables Charts Place the charts for X-bar and R (or s) one above the other so the average and range for any one subgroup are on the same vertical line. Observe whether either or both indicate lack of control for that subgroup. Usually, the R (or s) chart is interpreted first because the range or standard deviation is used in calculating limits for the X-bar chart.

R's outside control limits are evidence that the uniformity of the process has changed. Typical causes are a change in personnel, increased variability of material, or excessive wear in the process machinery. If the R or s chart exhibits special cause variation, then the within-subgroup variation will contain both common and special cause variation, and its use in calculating control limits for the X-bar chart will result in excessively large control limits (reducing its ability to detect out-of-control conditions). A single out-of-control R can be caused by a shift in the process that occurred while the subgroup was being taken.

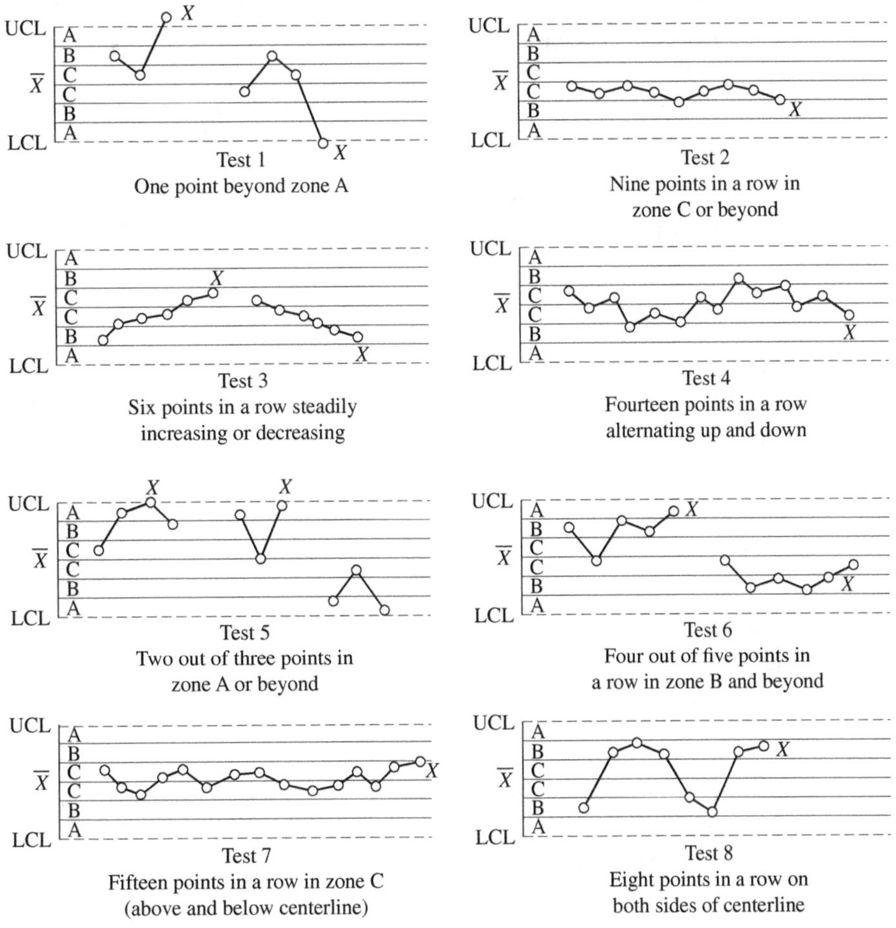

Figure 19.25 Illustrations of tests.

X-bars outside the control limits are evidence of a general change affecting all pieces after the first out-of-limits subgroup. The log kept during data collection, the operation of the process, and the worker's experience should be studied to discover a variable that could have caused the out-of-control subgroups. Typical causes are a change in material, personnel, machine setting, tool wear, temperature, or vibration.

Look for unusual patterns and nonrandomness. Nelson (1984, 1985) provides eight tests to detect such patterns on control charts using 3σ control limits (Fig. 19.25). Each of the zones shown is 1σ wide. (Note that test 2 in Fig. 19.25 requires nine points in a row; other authors suggest seven or eight points in a row; see Nelson, 1985, for elaboration.)

Ott and Schilling (1990) provide a definitive text on analysis after the initial control charts by presenting an extensive collection of cases with innovative statistical analysis clearly described.

Examples of Control Charts for Attribute Data

Whereas control charts for variables data require numerical measurements (e.g., line width from a photoresist process), control charts for attribute data require only a count of observations of a characteristic (e.g., the number of nonconforming items in a sample). These also are called categorical data, because units are classified into groups such as pass and fail.

p chart. Also called a proportions chart, it tracks the proportion or percentage of nonconforming units (percentage defective) in each sample over time.

np chart. This chart is used to track the number of nonconforming (defective) units in each sample over time. An np chart should only be used when the number of units sampled is constant (or nearly so).

c chart. Used to track the number of nonconformities (i.e., defects, rather than defective units as in the p chart).

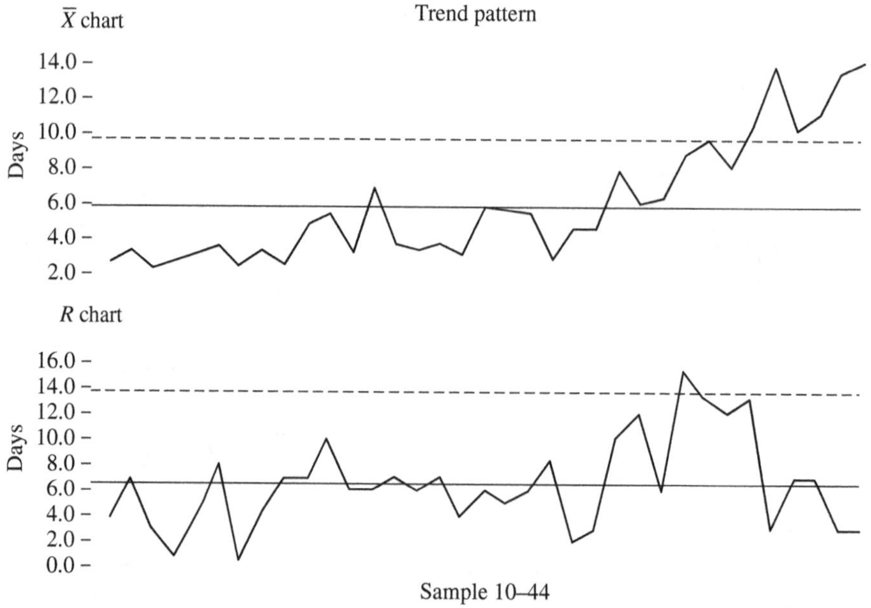

FIGURE 19.26 Average and range control charts.

u chart. A variation of the c chart, and analogous to the np chart, this chart tracks the number of nonconformities (defects) per unit in a sample of n units. As with the np chart, the number of units should be approximately constant.

As an example of attribute control charting, the fraction nonconforming (p) chart can be illustrated with data on magnets used in electrical relays. For each of 19 weeks, the number of magnets inspected and the number of nonconforming magnets were recorded. The total number of magnets tested was 14,091. The total number nonconforming was 1030, or 7.3 percent. The resulting control chart (calculating control limits based on average sample size of 741.6) is shown in Fig. 19.26. Note that several points fall beyond the control limits, suggesting some special cause(s) at work. In the case of the unusually low point for the last sample, it may be useful to identify and reinforce any special cause of the exceptionally good quality. The same rules as described above in Fig. 19.25 also apply to attribute charts.

Stratification

Purpose Stratification is the separation of data into categories. (See Fig. 19.27 for an example of stratification of data.) The most frequent use is during problem analysis to identify which categories contribute to the problem being solved. However, it can be applied when identifying projects, analyzing symptoms, testing hypotheses, and developing solutions. Stratification helps answer questions as to the frequency of defects, factors that may be contributing to a quality problem, and the degree to which results may differ across groups (strata).

Steps to Create

1. Select the stratification variables. If new data are to be collected, be certain that all potential stratification variables are collected as identifiers.
2. Establish categories that are to be used for each stratification variable. The categories may be either discrete values or ranges of values.
3. Sort observations into the categories of one of the stratification variables. Each category will have a list of the observations that belong to it.

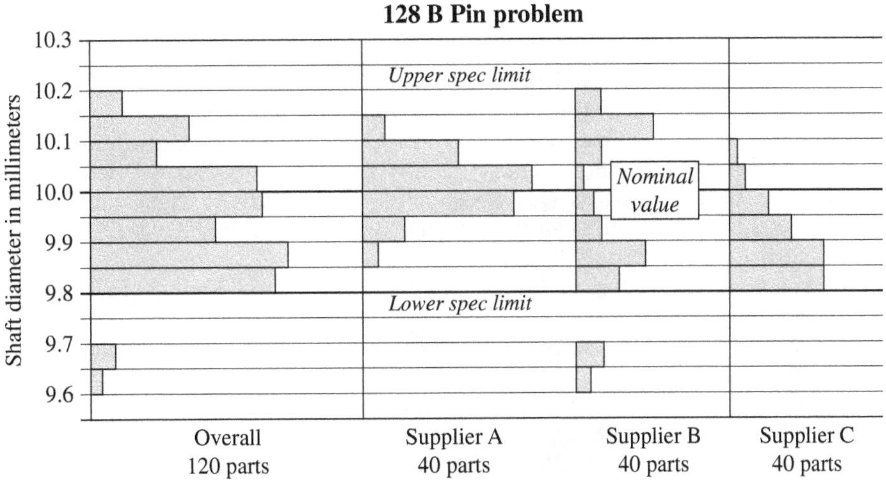

Figure 19.27 Stratification.

4. Calculate the phenomenon being measured for each category. These calculations can be a count of the number of observations in the category, an average value for those observations or a display (like a histogram) for each category.
5. Display the results. Bar graphs are usually the most effective.
6. Prepare and display the results for other stratification variables. Repeat steps 2 through 5. Do second-stage stratification as appropriate.
7. Plan for addition confirmation.

Example A manufacturer of mechanical equipment had recently received a rash of complaints about pins (Stock Number 128B) coming loose from press-fit sockets. The sockets were produced internally by the manufacturer under good statistical process control. The steel pins that fit into the sockets were purchased from three different suppliers.

The quality improvement team looking into the complaints measured the diameter of 120 pins from inventory, 40 from each of the three suppliers. The nominal value for the pin diameter was 10 mm. The upper specification limit was 10.2 mm, and the lower limit, 9.8 mm.

To get a better understanding of the data, the team produced a histogram of all 120 parts. The histogram showed the pin diameter measurements to have a broad, multipeaked distribution, with most of the data between the lower specification limit and the nominal value. Because most of the pins were smaller than nominal, there was indeed a good chance of a loose fit.

This summary histogram, however, did not tell the team much about what the cause of the problem was. So the team decided to stratify the data by supplier and to plot new histograms.

On the basis of the histograms on the previous page, the team drew the following conclusions:

- *Supplier A* has good controls on its process. Most of the product is close to the nominal value, and because the inherent variability in the process is smaller than the width of the specification limits, there is little chance of producing a part outside the limits.
- *Supplier B* appears to be running two distinct processes, neither of which has been set up to produce pins with diameters close to the nominal. The shape of the distribution for supplier B looks like the sum of two distributions similar to that of supplier A, one of which has been shifted up a bit, the other shifted down.
- *Supplier C* has a process that is highly variable and not set up to produce pins at the nominal value. The abruptly ended (or truncated) nature of the distribution suggests that the supplier is using inspection to screen out off-spec pins.

References

Anthis, D. L., Hart, R. F., and Stanula, R. J. (1991). "The Measurement Process: Roadblock to Product Improvement?" *Quality Engineering*, 3(4), pp. 461–470.

Dechert, J., Case, K. E., and Kautiainen, T. L. (2000). Statistical process control in the presence of large measurement variation. *Quality Engineering*, 12(3), pp. 417–423.

Ishikawa, K. (1972). *Guide to Quality Control*. Asian Productivity Organization, Tokyo, Japan.

Nelson, L. S. (1984). "The Stewhart Control Chart-Tests for Special Causes." Journal of Quality Technology, 16(4), pp. 237–239.

Nelson, L. S. (1985). "Interpreting Stewhart Control Charts." Journal of Quality Technology, 17(2), pp. 114–116.

Ott, E. R., and Schilling, E. G. (1990). "Elements of Process Control." Quality Engineering, 2(2), p. 132.

Schilling, E. G. (1990). Elements of Process Control, Quality Engineering, 2(2), p. 132.

Wadsworth, H. M., Stephens, K. S., Godfrey, A. B. (1986). *Modern Methods for Quality Control and Improvement*, John Wiley & Son, Inc., New York.

CHAPTER 20

Accurate and Reliable Data and Measurement Systems

John Early

High Points of This Chapter

1. Obtaining accurate, reliable, and relevant information happens when asking the right questions. Ask the right question, and you will get the right data.
2. Ten principles for effective measurement can help to develop accurate and reliable measures of performance.
3. When planning for data collection, the key issue is not how to collect data; rather, the key issue is how to generate useful information. This is accomplished by a thorough understanding of the measurement system and comparative advantages among data collection and analysis choices and beginning with the end in mind.

4. There are many tools useful to manage an organization that provides process improvement, design, and control. Tools for improvement require the testing of theories and finding root causes. Design tools require the collection of opinions and specifications, and then determine means to develop new services or products that are reliable. Control requires the use of statistical tools to help distinguish between common and special causes of variation to reduce risk, thereby facilitating appropriate intervention.
5. Chapter 19, Graphical Tools to Improve Process Performance, covered some of the basic tools for planning, improving, and controlling quality. This chapter provides some additional tools for more detailed and complex analysis.

Measurement and Superior Results

In other chapters of this handbook, the reader will have seen many times the need for accurate, reliable, and relevant data in order to make the necessary decisions to achieve superior results. First and foremost, information must be directly *relevant* to the question being asked. If we wish to retain loyal customers, we require data on the actual loyalty and spending of the set of existing and potential customers. We will also need data that demonstrate the causative factors for customers' behavior. Next, the data must be *accurate*. As we will see, accuracy has two components: freedom from bias and sampling error (the uncertainty associated with using a sample of the whole) that is small enough to support the decision we must make. If we wish to improve the clinical status of the asthmatic population in a health plan, for example, we need to demonstrate that our measurement of that status is free from significant bias and that the samples we use to estimate the status are large enough so that our uncertainty arising from the samples size is small compared to the improvement we wish to achieve. Finally, the data must be *reliable*. Reliability encompasses both accuracy and relevance, but goes a step further, ensuring that the measurements will continue to be accurate and relevant on an ongoing basis within the operating and business environment so that we can continue to rely on it for decision-making. For example, if we have established that a process for manufacturing a complex electronic connector is capable of meeting customer requirements and that key variables of temperature, pressure, and speed must be controlled within proven limits, then our business success depends on those measures of temperature, pressure, and speed continuing to be both accurate and relevant in order to yield defect-free connectors.

Measurement and Analysis and the Juran Trilogy®

The Juran Trilogy® of Quality Planning, Quality Improvement, and Quality Control each rely on a foundation of accurate, reliable, and relevant information. While each is a distinct managerial process, they share a common need for information, apply many of the same tools, and often use the same data to ensure their ends. However, each has unique information requirements that form the basis for our pursuit of measurement and analysis for quality. The following sections list some of the key questions for each phase of the Juran Trilogy.

Quality Planning Measurement Questions

- What level of product market performance is required to meet strategic objectives?
- How does our product perform vis-à-vis the competition?
- How does our product perform with respect to customer expectations?

- What is the magnitude of the customer demand for the product (good or service) or process, and how much are they willing to pay?
- Who are the customers for the product (good or service) or process?
- How important is each customer?
- What are the needs/benefits for each significant customer?
- What is the relative importance of each of these needs and benefits?
- What is the impact of each product feature on each customer need?
- What are the mathematical tradeoffs among the various product features?
- What is the impact of each process feature/parameter on delivery of the product feature?
- What is the capability of the process to deliver the product?
- What are the optimal tolerances for the target of each product and process feature?
- How much of the strategic objective for the planning project was achieved?

Quality Improvement Measurement Questions

- What are the most important deficiencies driving customer disloyalty?
- What are the largest detailed categories of costs of poor quality?
- How much improvement in cost and customer loyalty is needed to meet strategic objectives?
- What are the major contributors to the identified problem?
- How much does each theory of cause contribute to the overall problem?
- Which theories are proven as root causes?
- How much improvement will the proposed remedy create?
- How much improvement did the project finally achieve?

Quality Control Measurement Questions

- What variables have the largest impact on the variability of the process?
- What is the normal random variation for the control variables?
- Is a variable exhibiting a sporadic spike in its variation due to an assignable cause?

Ten Principles of Effective Measurement

Quality measurement is central to quality control and improvement: "What gets measured, gets done." Before embarking on the details for good measurement and analysis, we need to consider the following principles that can help to develop effective measurements for quality:

1. Define the purpose and use that will be made of the measurement. An example of particular importance is the application of measurements in quality improvement. Final measurements must be supplemented with intermediate measurements for diagnosis.
2. Emphasize customer-related measurements; be sure to include both external and internal customers.

3. Focus on measurements that are useful—not just easy to collect. When quantification is too difficult, surrogate measures can at least provide a partial understanding of an output.
4. Provide for participation from all levels in both the planning and implementation of measurements. Measurements that are not used will eventually be ignored.
5. Provide for making measurements as close in time as possible to the activities they affect. This timing facilitates diagnosis and decision making.
6. Provide not only concurrent indicators but also leading and lagging indicators. Current and historical measurements are necessary, but leading indicators help to look into the future.
7. Define in advance plans for data collection and storage, analysis, and presentation of measurements. Plans are incomplete unless the expected use of the measurements is carefully examined.
8. Seek simplicity in data recording, analysis, and presentation. Simple check sheets, coding of data, and automatic gauging are useful. Graphical presentations can be especially effective.
9. Provide for periodic evaluations of the accuracy, integrity, and usefulness of measurements. Usefulness includes relevance, comprehensiveness, level of detail, readability, and interpretability.
10. Realize that measurements alone cannot improve products and processes.

Measurements must be supplemented with the resources and training to enable people to achieve improvement.

Planning for Measurement and Data Collection

"Begin with the end in mind" is an appropriate maxim when starting any effort that requires collection of data. The "end" in this case is obtaining the information needed to effectively and efficiently plan, control, or improve. Before launching into a discourse on statistical analysis, we first will consider the need to plan for data collection. Part of this planning process is to consider the source of a set of data that we desire to analyze to solve a problem; the most common sources are historical data, newly collected operational data, and data from planned experimentation. These sources each have their advantages and drawbacks. Regardless of the source, all data need careful review before proceeding with an analysis and communication of the information gained from it.

Planning for Collection and Analysis of Data

In collecting and analyzing data, quality teams are seeking the answer to questions such as, *How often does the problem occur?* or, *What is causing the problem?* In other words, they are seeking information. However, although good information always is based on data (the facts), simply collecting data does not necessarily ensure that useful information has been obtained. The key issue, then, is not, *How do we collect data?* Rather, the key issue is, *How do we generate useful information?* Although most organizations have vast stores of data about their operations, frequently, the data needed to provide truly useful information do not exist. The all-too-common practice at many organizations is to go "data diving," looking at much of or all of the data available to learn whatever they can about the process. While this practice can yield some useful information, it is inherently wasteful and can add time to the execution of a project. The process of planning the data collection with the end in mind that is described here is far more efficient and effective.

Information generation begins and ends with questions. To generate information, we need to

- Formulate precisely the question we are trying to answer.
- Collect data relevant to that question.
- Analyze the data to determine the factual answer to the question.
- Present the data in a way that clearly communicates the answer to the question.

Learning to "ask the right questions" is the key skill in effective data collection. Accurate, precise data, collected through an elaborately designed statistical sampling plan, is useless if it is not relevant to answering a question that someone cares about.

Notice in Fig. 20.1 how this planning process "works backwards" through the model. We start by defining the question. Then, rather than diving into the details of data collection, we consider how we might communicate the answer to the question and what types of analysis we will need to perform. This helps us define our data needs and clarifies which characteristics are most important in the data. With this understanding as a foundation, we can deal more coherently with the where, who, how, and what of data collection.

To generate useful information, planning for good data collection, analysis, and communication proceeds through the following steps and associated considerations:

1. Establish data collection objectives and formulate the question in a specific statement:

    - What is your goal for collecting data?
    - What process or product will you monitor to collect the data?
    - What is the "theory" you are trying to test?
    - What is the question you are attempting to answer?

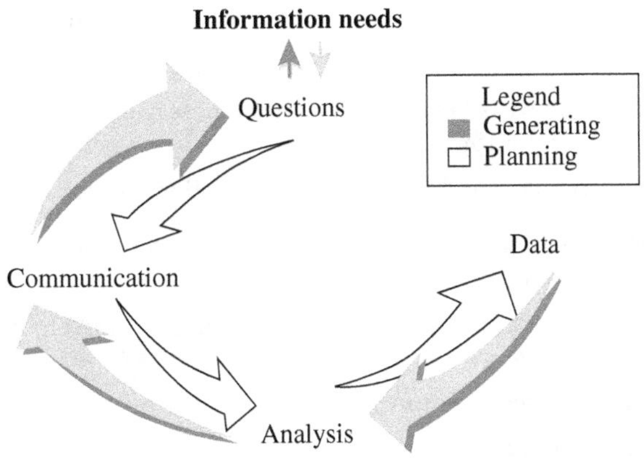

Figure 20.1 Planning for data collection.

2. Decide what to measure, with consideration as to how the data will be communicated and analyzed:
    - What data do you need?
    - What type of measure is it? Time and physical measures such as length, mass, volume, and temperature are common; other measures include rankings (e.g., low-medium-high), ratios (e.g., speed), and indexes (e.g., case-mix adjusted hospital length of stay, refractive index). See the section "Types of Measures" for a discussion of scales of measurement.
    - What type of data is it? Variables data (readings on a scale of measurement) may be more expensive than attributes data (go or no-go data), but the information is much more useful.
    - What is the operational definition of each measure? An operational definition is a detailed description of a process, activity, or project term written to ensure common understanding among the members of a group.
    - How will the data be communicated and analyzed?
    - Are past data available that are applicable (however, bear in mind the hazards of historical data sets, discussed below)?
3. Decide how to measure a population or sample:
    - What measurement tool will you use? Calipers, Likert scale survey?
    - What is your sampling strategy? Simple random sampling? Stratified random sampling?
    - How much data will be collected? Calculate sample size considering the desired precision of the result, statistical risk, variability of the data, measurement error, economic factors, etc.
    - What is the measurement method?
4. Collect the data with a minimum of bias.
5. Define comprehensive data collection points:
    - Where in the process can we get appropriate data?
6. Select and train unbiased collectors:
    - Understand data collectors and their environment.
    - Who in the process can give us these data?
    - How can we collect these data from these sources with minimum effort and least chance for error?
7. Design, prepare, and then test data collection methods, forms, and instructions:
    - What additional information should be captured for future analysis, reference, or traceability?
    - Conduct a measurement systems analysis (MSA) to confirm that the measures are accurate.
8. Audit the collection process and validate the results.
9. Screen the data.
10. Analyze the data.

11. Evaluate assumptions for determining the sample size and analyzing the data. Take corrective steps (including additional observations) if required.
12. Apply graphical and statistical techniques to evaluate the original problem.
13. Determine if further data and analysis are needed.
14. Consider a sensitivity analysis (e.g., by varying key sample estimates and other factors in the analysis and noting the effect on final conclusions).
15. Review the conclusions of the data analysis to determine if the original technical problem has been evaluated or if it has been changed to fit the statistical methods.
16. Present the results:
    - Write a report, including an executive summary.
    - State the conclusions in meaningful form by emphasizing results in terms of the original problem rather than the statistical indexes used in the analysis.
    - Present the results in graphic form where appropriate. Use simple statistical methods in the body of the report, and place complicated analyses in an appendix.
17. Determine if the conclusions of the specific problem apply to other problems or if the data and calculations could be a useful input to other problems.

Types of Measures

In planning for data collection, one needs to be clear about the characteristics of the data being collected and the implications of those characteristics for the questions to be answered. Two classifications of data types are useful here: the mathematical distinctions and the substantive quality questions answered.

The mathematical distinctions are typically known as measurement scales and are part of a system of measurement. The most useful scale is the *ratio scale* in which we record the actual amounts of a parameter such as weight. Ratio scales are also referred to as *continuous variables* data. An *interval scale* records ordered numbers but lacks an arithmetic origin such as zero—clock time is an example.

An *ordinal scale* records information in ranked categories—an example is customer preference for the flavor of various soft drinks. An unusual example of a measurement scale is the Wong-Baker FACES pain rating scale used widely in hospitals for children to communicate the intensity of pain felt to nurses (Wong and Baker 1998). The scale shows six faces to which a child can point, ranging from a very happy face (to indicate no hurt) to a very sad face (hurts most).

Finally, the *nominal scale* classifies objects into categories without an ordering or origin point—for example, the classification good or no-good, individual gender, color, the production shift, product, or geographic location.

Ordinal and nominal scales constitute a type of data referred to as *discrete* or *categorical* data.

The type of measurement scale determines the statistical analysis that can be applied to the data. In this regard, the ratio scale is the most powerful scale. For elaboration, see Emory and Cooper (1991).

For quality purposes, there are five general classes of quality measures:

1. Defects (deficiencies, failures)
2. Costs of poor quality

3. Product and process features
4. Customer needs
5. Customer behavior

Units of measure for product deficiencies usually take the form of a fraction:

$$\frac{\text{Number of occurrences}}{\text{Opportunity for occurrences}}$$

The numerator may be in such terms as number of defects produced, number of field failures, or cost of warranty charges. The denominator may be in such terms as number of units produced, dollar volume of sales, number of units in service, or length of time in service. The deficiencies are determined by comparing the product delivered to its specification. In physical products, those specifications are in terms of physical dimensions, electrical or physical properties, or performance characteristics. In service products, the most common specification is in terms of timeliness. Other specifications usually relate to the actual performance versus the rules or specifications for the service—see the discussion of service features below.

Costs of poor quality are usually denominated in the currency of organization, but may also be expressed as fractions of sales, total costs, or gross margin.

Units of measure for product features are more difficult to create. The number and variety of these features may be large. Sometimes inventing a new unit of measure is a fascinating technical challenge. In one example, a manufacturer of a newly developed polystyrene product had to invent a unit of measure and a sensor to evaluate an important product feature. It was then possible to measure that feature of both the product and of competitors' products before releasing the product for manufacture. In another case, the process of harvesting peas in the field required a unit of measure for tenderness and the invention of a "tenderometer" gauge. A numerical scale was created, and measurements were taken in the field to determine when the peas were ready for harvesting.

Timeliness of execution is typically one important feature for a service product. Generally, the content of the service will also have certain performance features. A repair service will have features on the effectiveness and reliability of the repair.

Financial services will measure such features as the eligibility of the customer to receive a service or a specific return or interest rate. They also have specifications for calculating returns, payments, and value. These rules yield results that can be measured. The rules are extensively applied through automated decision engines, but the accuracy of these automated methods need to be validated, and there is often a human element in the setup and execution as well.

Health care has both process and outcome quality measures. The processes describe the application of established standards of care for a given set of symptoms and signs. The outcomes measure the success of the treatment in restoring heath, avoiding further adverse episodes, and the safety of the patient from adverse events within the care setting, such as medication errors, falls, or procedural error.

Insurance pays claims according to the coverage of the policy and the nature of the insurable event. The insurance policy incorporates these rules for reimbursement for loss. "Claim engines" do most of the calculations, but require human specification and input. The accuracy of these payments can be expressed in monetary terms as well as in percent-defective terms.

Often a number of important product features exist. To develop an overall unit of measure, we can identify the important product features and then define the relative importance of each feature. In subsequent measurement, each feature receives a score. The overall measure is calculated as the weighted average of the scores for all features. This approach is illustrated in Table 20.1. In using such an approach for periodic or continuous measurement, some cautions should be cited (Early 1989). First, the relative importance of each feature is not precise and

Attribute	Relative Importance %	Company X Rating	Company X Weighted Rating	Company A Rating	Company A Weighted Rating	Company B Rating	Company B Weighted Rating
Safety	28	6	168	5	140	4.5	126
Performance	20	6	120	7	140	6.5	130
Quality	20	6	120	7	140	4	80
Field service	12	4	48	8	96	5	60
Ease of use	8	4	32	6	48	5	40
Company image	8	8	64	4	32	4	32
Plant service	4	7.5	30	7.5	30	5	20
Total			582		626		488

Source: *Quality Planning and Analysis*, Copyright 2007. Used by permission.

TABLE 20.1 Multiattribute Study

may change greatly over time. Second, improvement in certain features can result in an improved overall measure but can hide deterioration in one feature that has great importance.

The Sensor

The sensor is the means used to make the actual measurement. Most sensors are designed to provide information in terms of units of measure. For operational control subjects, the sensors are usually technological instruments or human beings employed as instruments (e.g., inspectors, auditors); for managerial and service subjects, the sensors are often data systems. Choosing the sensor includes defining how the measurements will be made—how, when, and who will make the measurements—and the criteria for taking action.

Clearly, sensors must be economical and easy to use. In addition, because sensors provide data that can lead to critical decisions on products and processes, sensors must be both accurate and precise, as discussed in the section "Measurement System Analysis."

Historical Data, Operational Data, and Experimental Data

Historical data are data that we already have and that may seem relevant to a question or problem at hand. Data often are saved during the production process, for example. If a satisfactory process goes out of control after some years of operation, it frequently is suggested that it would save both time and expense to analyze the historical data statistically rather than collect new data or perform a planned experiment to obtain new data that could lead to process correction. Thus, we have available data that may consist of measurements Y (such as a process yield, e.g., the strength of a material produced) and associated process variables $x1, x2, \ldots, xk$ (such as $x1$ = pressure and $x2$ = acid concentration, with $k = 2$). If such data do not exist, we might set up a data collection scheme to collect new operational data.

Historical or new operational data can both be invaluable for the following reasons:

- It is less time consuming and expensive to collect. Especially when multiple theories are at issue, even very lean eighth-fraction screening experimental designs can be prohibitive for some processes if they look at seven or more factors.
- For some types of operations that have a significant human performance component, the mere act of collecting new data, not to mention conducting experiments, can have unintended consequences on human behavior and, hence, the process—the famous Hawthorne effect.

- For out-of-control situations in previously stable processes, the information question is "what changed?" which usually is a specific unique occurrence of an assignable cause that does not require significant experimentation.
- Substantial chronic random variation typically has root causes that are at least identifiable, and often quantifiable, from operational data.
- Although caveats need to be observed, operational data can be helpful in developing and testing the theories that will be used ultimately in an experiment.
- When dealing with either operational data or experimental data, the same pitfalls apply if one fails to test all the possible causes or extends results beyond the actual measured operating range.

Nevertheless, historical and operational data have potential drawbacks that include

- The x's may be highly correlated with each other in practice; hence, it may not be possible to separate the effects among them.
- The x's may cover a very small part of the possible operating range, so small that any indications of changes in Y attributable to changes in the x's may be overwhelmed by the size of the variability of the process.
- Other variables that affect the output of the process (e.g., time of day, atmospheric conditions, operator running the process, etc.) may not have been held constant and may in fact be the real causes of changes observed in the process.

In such cases, experimental data may be superior. Experiments are run at each of a number of combinations of settings that are selected in advance by statistical design criteria for each variable $x1, \ldots, xk$.

Measurement System Analysis

Control of a process, design of a new product, and elimination of chronic random variation all require accurate measurement of both the desired results and the contributing factors. A good measurement system that provides this critical information should have the following attributes:

- *Minimal bias*. Bias is the difference between the average measured value and a reference value. A reference value, in turn, is an agreed-upon standard, such as a traceable national standard. The reference standard is used to calibrate a measurement system, thereby bringing the reported measure in line with the accepted, known value. Bias sometimes is referred to as "accuracy." However, because accuracy has several meanings in the literature, "bias" is the recommended term in the present context.
- *Repeatability*. Repeatability is the variation in measurements obtained with one measurement instrument when used several times by an appraiser while measuring the identical characteristic on the same part.
- *Reproducibility*. Reproducibility is the variation in the average of the measurements made by different appraisers using the same measuring instrument when measuring the identical characteristic on the same part.
- *Stability*. Stability (or drift) is the total variation in the measurements obtained with a measurement system on the same master or parts when measuring a single characteristic over an extended period. A measurement system is stable if the same results are obtained at different points in time.
- *Linearity*. Linearity is the difference in bias values at different points along the expected operating range of a measurement instrument.

- *Precision*. Repeatability, reproducibility, and stability tend to be random, and the three together are often referred to as "precision." Refer to Fig. 20.2 for a graphic depiction of the difference between bias and precision.

These five sources of measurement variation are illustrated in Fig. 20.3 and are generally consistent with the definitions provided by the AIAG *Measurement Systems Analysis Reference Manual* (Automotive Industry Action Group 2003).

Any statement of bias and precision must be preceded by three conditions:

1. Definition of the test method. This definition includes the step-by-step procedure, equipment to be used, preparation of test specimens, test conditions, etc.
2. Definition of the system of causes of variability, such as material, analysts, apparatus, laboratories, days, etc. American Society for Testing and Materials (ASTM) recommends that modifiers of the word "precision" be used to clarify the scope of the precision measure. Examples of such modifiers are single-operator, single-analyst, single-laboratory-operator-material-day, and multilaboratory.
3. Existence of a statistically controlled measurement process. The measurement process must have stability for the statements on bias and precision to be valid. This stability can be verified by a control chart.

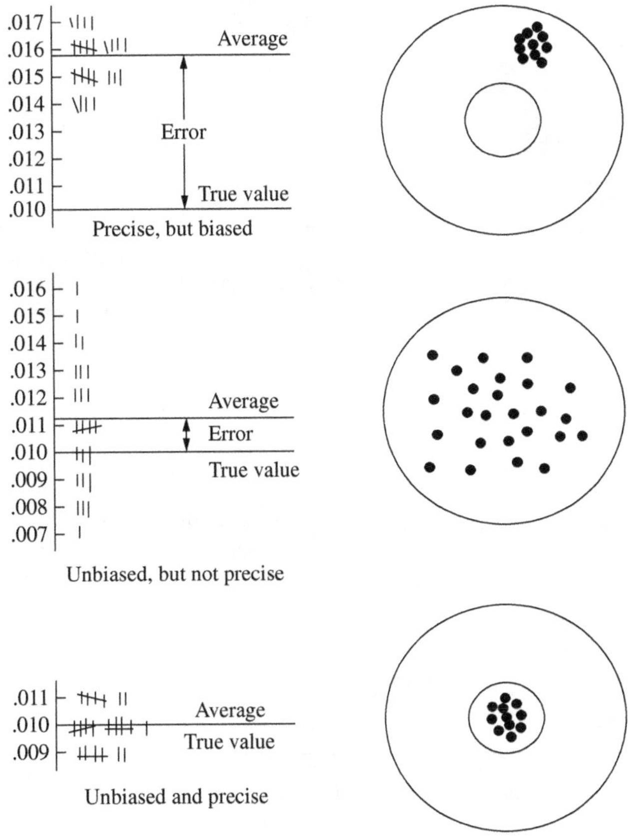

FIGURE 20.2 Bias and precision. (*Quality Planning and Analysis, Copyright 2007. Used by permission.*)

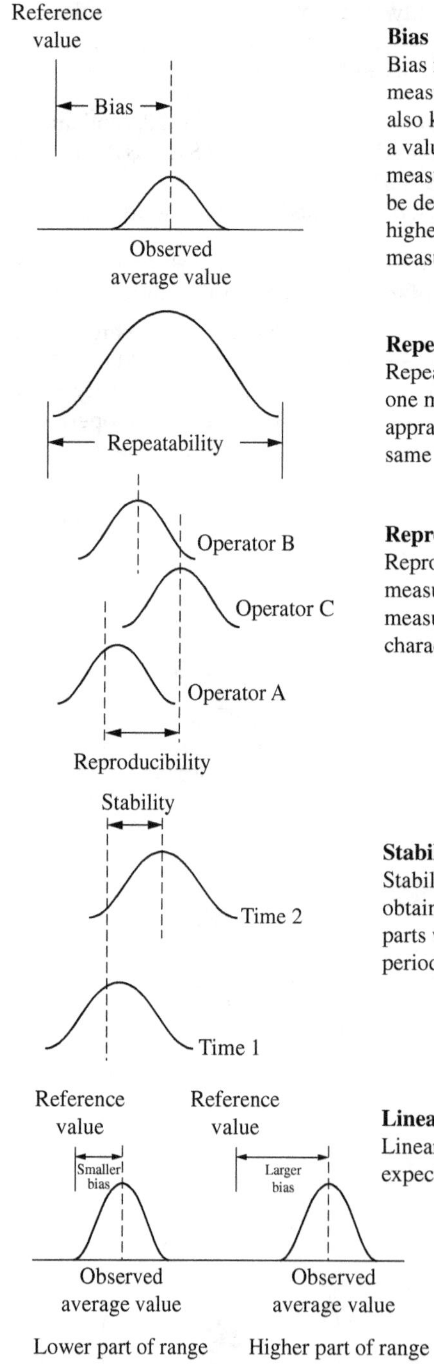

Bias

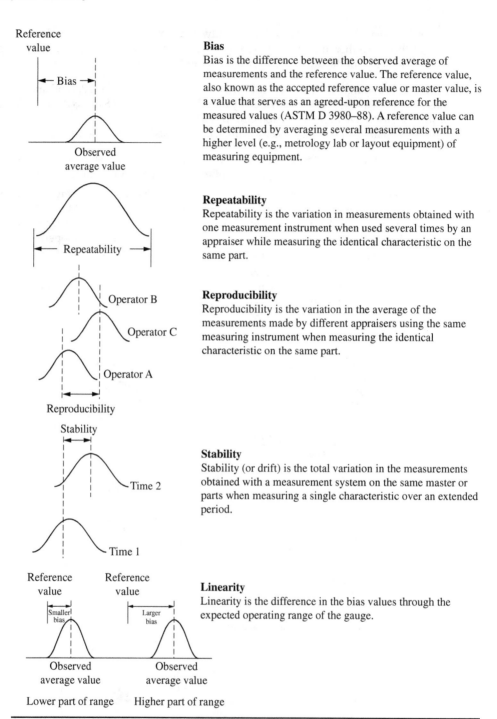

Bias is the difference between the observed average of measurements and the reference value. The reference value, also known as the accepted reference value or master value, is a value that serves as an agreed-upon reference for the measured values (ASTM D 3980–88). A reference value can be determined by averaging several measurements with a higher level (e.g., metrology lab or layout equipment) of measuring equipment.

Repeatability

Repeatability is the variation in measurements obtained with one measurement instrument when used several times by an appraiser while measuring the identical characteristic on the same part.

Reproducibility

Reproducibility is the variation in the average of the measurements made by different appraisers using the same measuring instrument when measuring the identical characteristic on the same part.

Stability

Stability (or drift) is the total variation in the measurements obtained with a measurement system on the same master or parts when measuring a single characteristic over an extended period.

Linearity

Linearity is the difference in the bias values through the expected operating range of the gauge.

FIGURE 20.3 Five sources of measurement variation. *(Reprinted with permission from the MSA Manual DaimlerChrysler, Ford, General Motors Supplier Quality Requirements Task Force.)*

Effect of Measurement Error on Acceptance Decisions

Error of measurement can cause incorrect decisions on (1) individual units of product and (2) lots submitted to sampling plans. In one example of measuring the softening point of a material, the standard deviation of the test precision is 2 degrees, yielding two standard deviations of ±4 degrees. The specification limits on the material are ±3 degrees. Imagine the incorrect decisions that are made under these conditions.

Two types of errors can occur in the classification of a product: (1) a nonconforming unit can be accepted (the consumer's risk) and (2) a conforming unit can be rejected (the producer's risk). In a classic paper, Eagle (1954) showed the effect of precision on each of these errors.

The probability of accepting a nonconforming unit as a function of measurement error (called test error, σ_{TE}, by Eagle) is shown in Fig. 20.4. The abscissa expresses the test error as the standard deviation divided by the plus-or-minus value of the specification range (assumed

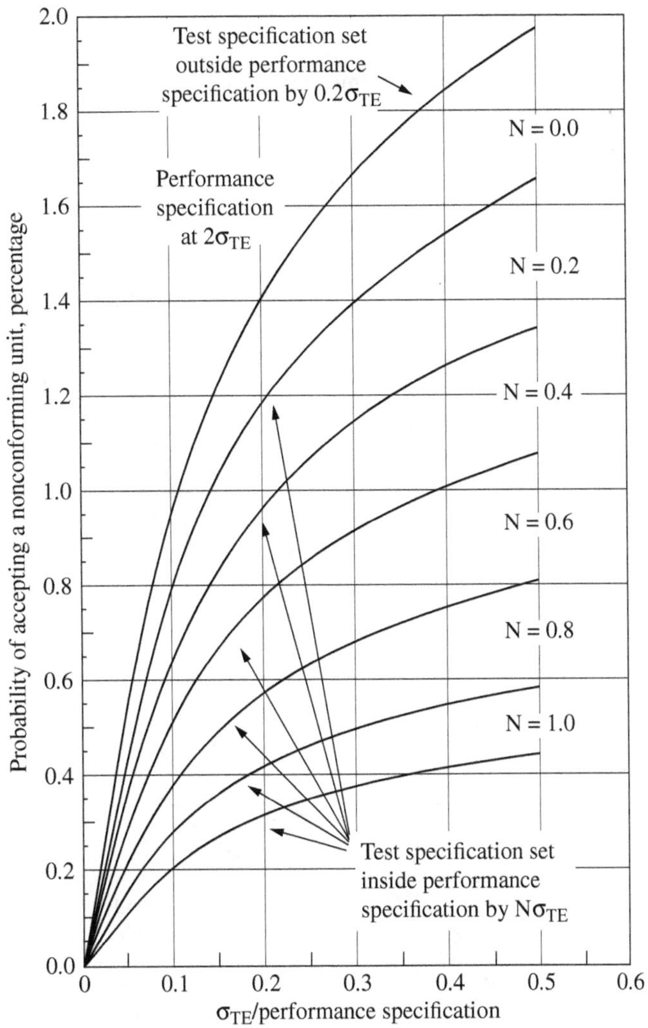

FIGURE 20.4 Probability of accepting a nonconforming unit.

equal to two standard deviations of the product). For example, if the measurement error is one-half of the tolerance range, the probability is about 1.65 percent that a nonconforming unit will be read as conforming (due to the measurement error) and therefore will be accepted.

Figure 20.5 shows the percentage of conforming units that will be rejected as a function of the measurement error. For example, if the measurement error is one-half of the plus-or-minus tolerance range, about 14 percent of the units that are within specifications will be rejected because the measurement error will show that these conforming units are outside specification.

The test specification can be adjusted with respect to the performance specification (see Figs. 20.4 and 20.5). Moving the test specification inside the performance specification reduces the probability of accepting a nonconforming product but increases the probability of rejecting a conforming product. The reverse occurs if the test specification is moved outside the performance specification. Both risks can be reduced by increasing the precision of the test (i.e., by reducing the value of σ_{TE}).

Hoag et al. (1975) studied the effect of inspector errors on type I (α) and type II (β) risks of sampling plans (see the section "Hypothesis Testing" for definitions of type I and II risks). For a single sampling plan and an 80 percent probability of the inspector detecting a defect, the real value of β is two to three times that specified, and the real value of α is about one-fourth to one-half of that specified.

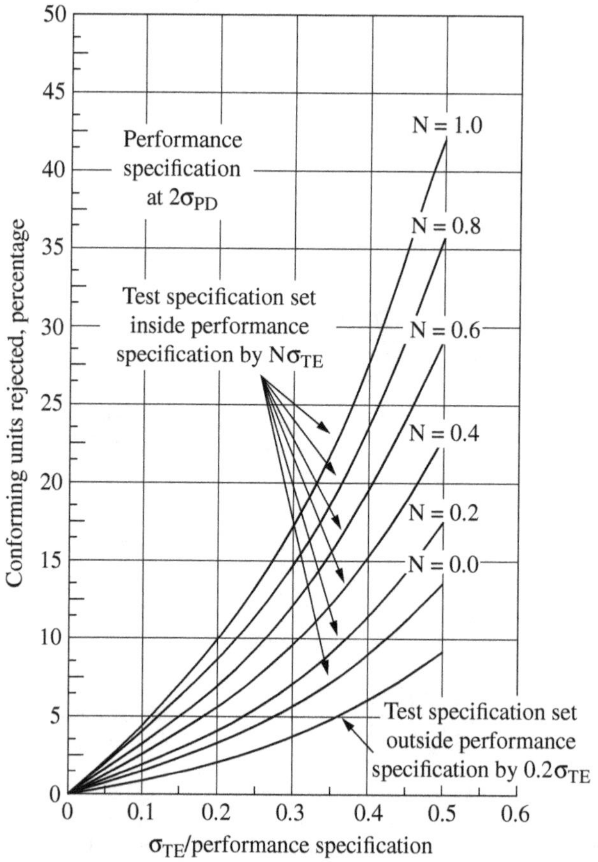

FIGURE 20.5 Conforming units rejected (percentage).

Case et al. (1975) investigated the effect of inspection error on the average outgoing quality (AOQ) of an attribute sampling procedure. They concluded that the AOQ values change and significant changes can occur in the shape of the AOQ curve.

All these investigations concluded that measurement error can be a serious problem.

Components of Variation

In drawing conclusions about measurement error, it is worthwhile to study the causes of variation in observed values. The relationship is

$$\sigma \text{ observed} = \sqrt{\sigma^2_{\text{cause A}} + \sigma^2_{\text{cause B}} + \cdots + \sigma^2_{\text{cause N}}}$$

The formula assumes that the causes act independently.

It is valuable to find the numerical values of the components of observed variation because the knowledge may suggest where effort should be concentrated to reduce variation in the product. A separation of the observed variation into product variation plus other causes of variation may indicate important factors other than the manufacturing process. Thus, if it is found that the measurement error is a large percentage of the total variation, this finding must be analyzed before proceeding with a quality improvement program. Finding the components (e.g., instrument, operator) of this error may help to reduce the measurement error, which in turn may completely eliminate a problem.

Observations from an instrument used to measure a series of different units of product can be viewed as a composite of (1) the variation due to the measuring method and (2) the variation in the product itself. This value can be expressed as

$$\sigma_O = \sqrt{\sigma_P^2 + \sigma_E^2}$$

where $\sigma_O = \sigma$ of the observed data
$\sigma_P = \sigma$ of the product
$\sigma_E = \sigma$ of the measuring method

Solving for σ_P yields $\sigma_P = \sqrt{\sigma_O^2 + \sigma_E^2}$.

The components of measurement error often focus on repeatability and reproducibility (R&R). Repeatability primarily concerns variation due to measurement gauges and equipment; reproducibility concerns variation due to human "appraisers" who use the gauges and equipment. Studies to estimate these components are often called "gauge R&R" studies.

A gauge R&R study can provide separate numerical estimates of repeatability and reproducibility. Two methods are usually used to analyze the measurement data. Each method requires a number of appraisers, a number of parts, and repeat trials of appraisers measuring different parts. For example, an R&R study might use three appraisers, ten parts, and two trials.

One method analyzes averages and ranges of the measurement study data. This method requires minimum statistical background and does not require a computer. The second method is the analysis of variance, ANOVA (see the section "Statistical Tools for Improvement"). Compared to the first method, ANOVA requires a higher level of statistical knowledge to interpret the results, but can evaluate the data for possible interaction between appraisers and parts. ANOVA is best done on a computer using Minitab or other software. Overall, the ANOVA method is preferred to analyzing the averages and ranges. Detailed illustrations of each method are provided in the Automotive Industry Action Group booklet *Measurement Systems Analysis* (1995). Also, see Tsai (1988) for an example using ANOVA and considering both no interaction and interaction of operators and parts. Burdick and Larsen (1997) provide methods for constructing confidence intervals on measures of variability in R&R studies.

When the total standard deviation of repeatability and reproducibility is determined from ANOVA, a judgment must then be made on the adequacy of the measurement process. A common practice is to calculate 5.15σ ($\pm 2.575\sigma$) as the total spread of the measurements that will include 99 percent of the measurements. If 5.15σ is equal to or less than 10 percent of the specification range for the quality characteristic, the measurement process is viewed as acceptable for that characteristic; if the result is greater than 10 percent, the measurement process is viewed as unacceptable. Engel and DeVries (1997) examine how the practice of comparing measurement error with the specification interval relates to making correct decisions in product testing.

Reducing and Controlling Errors of Measurement

Steps can be taken to reduce and control errors for all sources of measurement variation. The systematic errors that contribute to bias can sometimes be handled by applying a numerical correction to the measured data. If an instrument has a bias of 0.001, then, on average, it reads 0.001 too low. The data can be adjusted by adding 0.001 to each value of the data. Of course, it is preferable to adjust the instrument as part of a calibration program.

In a calibration program, the measurements made by an instrument are compared to a reference standard of known accuracy (a calibration program should include provisions for periodic audits). If the instrument is found to be out of calibration, an adjustment is made.

A calibration program can become complex for these reasons:

- The large number of measuring instruments
- The need for periodic calibration of many instruments
- The need for many reference standards
- The increased technological complexity of new instruments
- The variety of types of instruments (i.e., mechanical, electronic, chemical, etc.)

Precision in measurement can be improved through either or both of the following procedures:

- Discovering the causes of variation and remedying these causes. A useful step is to resolve the observed values into components of variation (see earlier). This process can lead to the discovery of inadequate training, perishable reagents, lack of sufficient detail in procedures, and other such problems. This fundamental approach also points to other causes for which the remedy is unknown or uneconomic (i.e., basic redesign of the test procedure).
- Using multiple measurements and statistical methodology to control the error in measurement. The use of multiple measurements is based on the following relationship:

$$\sigma_{\overline{X}} = \frac{\sigma}{\sqrt{n}}$$

As in all sampling schemes, halving the error in measurement requires quadrupling (not doubling) the number of measurements.

As the number of tests grows larger, a significant reduction in the error in measurement can be achieved only by taking a still larger number of additional tests. Thus, the cost of the additional tests versus the value of the slight improvement in measurement error becomes an issue. The alternatives of reducing the causes of variation must also be considered.

For an in-depth discussion of reducing other forms of measurement error, see Automotive Industry Action Group (2003) and Coleman et al. (2008).

A successful measurement system analysis (MSA) is critical not only for control but also for validating the measures used in quality planning and improvement, as illustrated by this Six Sigma improvement project (DMAIC) (courtesy of Steve Wittig and Chris Arquette at a Juran Institute client forum). It also illustrates the use and importance of attribute MSA studies for discrete variables.

Background

The paint line has a first run yield of 74 percent. This means that 26 percent of all frames need to be reworked at least once. Defects due to finish issues account for 15 percent, and material (wood) issues account for 11 percent. This project looks only at finish defects because this is readily within our control. Any rework is nonvalue-added and contributes to wasted paint/primer, labor, utilities, work in progress, capacity, and more hazardous waste. Our goal is to improve first-run yield to 90 percent for finish defects.

Summary of MSA Effort

The paint line is old and somewhat neglected. Our first MSA results were expectedly poor, and the appraisers were contributing to the defect rate by rejecting good frames. We improved this by continued training of the appraisers by quality control. We did two more MSAs with acceptable results. This will need to be an ongoing test/train routine. Figs. 20.6 to 20.8 are attribute MSA results, and Figs. 20.9 and 20.10 are results of a variable MSA.

Data Screening

As a practical matter, many data sets contain some instances of incorrectly transcribed values, values from points where an experiment goes awry for some reason (such as

Validate measurement system attribute data analysis—MSA 1

Sample #	Expert	Operator 1		Operator 2		Operator 3	
		Try 1	Try 2	Try 1	Try 2	Try 1	Try 2
1	Blister	Blister	Blister	Blister	Blister	Blister	Blister
2	Good	Light Ed.	Good	Good	Good	Good	Good
3	Drip	Drip	Drip	Dirt	Dirt	Drip	Drip
4	Dirt	Contam	Dirt	Dirt	Dirt	Dirt	Dirt
5	Contam	Contam	Contam	Good	Good	Over run	Over run
6	Blister	Blister	Blister	Blister	Blister	Dirt	Blister
7	Good	Good	Good	Good	Good	Good	Good
8	Dirt	Light Ed.	Contam	Good	Good	Dirt	Dirt
9	Good	Good	Drip	Dirt	Good	Drip	Drip
10	Good	Orange P.	Good	Good	Good	Good	Good
11	Dirt	Dirt	Good	Dirt	Dirt	Dirt	Dirt
12	Good	Contam	Light Ed.	Good	Good	Over run	Over run
13	Good	Light Ed.	Light Ed.	Good	Good	Light Ed.	Over run
14	Contam	Contam	Contam	Good	Good	Good	Good
15	Drip	Drip	Light Ed.	Dirt	Good	Drip	Drip
16	Light Ed.	Light Ed.	Light Ed.	Good	Good	Good	Good
17	Dirt	Contam	Contam	Good	Dirt	Dirt	Dirt
18	Dirt	Contam	Contam	Dirt	Dirt	Dirt	Dirt
19	Blister	Blister	Good	Good	Blister	Blister	Blister
20	Good	Good	Good	Good	Good	Orange P.	Orange P.

FIGURE 20.6 Baseline attribute MSA on appraisers' accept/reject decisions. (*Juran Institute, Inc.*)

Attribute data analysis—MSA 1 results

Within appraiser

Assessment agreement

Appraiser	# Inspected	# Matched	Percent (%)	95.0% CI
1	20	11	55.0	(31.5, 76.9)
2	20	16	80.0	(56.3, 94.3)
3	20	18	90.0	(68.3, 98.8)

Matched: Appraiser agrees with him/herself across trials.

Each appraiser vs. standard

Assessment agreement

Appraiser	# Inspected	# Matched	Percent (%)	95.0% CI
1	20	8	40.0	(19.1, 63.9)
2	20	11	55.0	(31.5, 76.9)
3	20	12	60.0	(36.1, 80.9)

Matched: Appraisers' assessment across trials agrees with standard.

Between appraisers

Assessment agreement

# Inspected	# Matched	Percent (%)	95.0% CI
20	2	10.0	(1.2, 31.7)

Matched: All appraisers' assessments agree with each other.

All appraisers vs. standard

Assessment agreement

# Inspected	# Matched	Percent (%)	95.0% CI
20	2	10.0	(1.2, 31.7)

Matched: All appraisers' assessments agree with standard.

Note: 38% were called bad that were good. 22% were called good that were bad. This potentially could yield an improvement in the defect rate by 16%.

Figure 20.7 Results of baseline attribute MSA. Results are not acceptable. (*Juran Institute, Inc.*)

Attribute data analysis—MSA 2 results

Within appraiser

Assessment agreement

Appraiser	# Inspected	# Matched	Percent (%)	95.0% CI
1	20	16	80.0	(56.3, 94.3)
2	20	19	95.0	(75.1, 99.9)
3	20	20	100.0	(86.1, 100.0)

Matched: Appraiser agrees with him/herself across trials.

Each appraiser vs. standard

Assessment agreement

Appraiser	# Inspected	# Matched	Percent (%)	95.0% CI
1	20	15	75.0	(50.9, 91.3)
2	20	19	95.0	(75.1, 99.9)
3	20	18	90.0	(68.3, 98.8)

Matched: Appraisers' assessment across trials agrees with standard.

Between appraisers

Assessment agreement

# Inspected	# Matched	Percent (%)	95.0% CI
20	13	65.0	(40.8, 84.6)

Matched: All appraisers' assessments agree with each other.

All appraisers vs. standard

Assessment agreement

# Inspected	# Matched	Percent (%)	95.0% CI
20	13	65.0	(40.8, 84.6)

Matched: All appraisers' assessments agree with standard.

Conclusion: MSA is greatly improved over first. Continue Q.C. training of inspectors to bring up agreement between all appraisers and standard.

Figure 20.8 Attribute MSA after improvement. (*Juran Institute, Inc.*)

Measurement system analysis Sheen Gage study

Gage R&R Source	**VarComp**	**%Contribution (of VarComp)**
Total Gage R&R	0.0519	0.69
Repeatability	0.0298	0.39
Reproducibility	0.0221	0.29
Operator	0.0028	0.04
Operator* measurement	0.0193	0.26
Part-to-part	7.4851	99.31
Total variation	7.5370	100.00

Source	**StdDev (SD)**	**Study Var (5.15*SD)**	**%Study Var (%SV)**	
Total Gage R&R	0.22776	1.1730	8.30	
Repeatability	0.17248	0.8883	6.28	
Reproducibility	0.14874	0.7660	5.42	
Operator	0.05294	0.2726	1.93	
Operator* measurement		0.13900	0.7159	5.06
Part-to-part	2.73590	14.0899	99.66	
Total variation	2.74536	14.1386	100.00	

Number of distinct categories = 17

Figure 20.9 Results of baseline variable data MSA on Sheen Gage results are acceptable. (*Juran Institute, Inc.*)

equipment malfunction), and the measurement system fails, or other factors lead to observational error. Procedures for finding these problems are called data screening and should be performed.

Data-Screening Methods

Numerous tests are available to detect outliers, that is, "observation(s) [or (a)subset of observations] which appear to be inconsistent with the remainder of that set of data" (Barnett and Lewis 1994). One of the most common methods of data screening is to classify observations as outliers if they are outside an interval of L multiples of the standard deviation about the mean. The number L is commonly taken to be 2.5, 3, or 4. The larger L is, the less likely it is that outliers will be detected, while the smaller L is, the more good observations one will wrongly detect as potential outliers. For example, because approximately 99.73 percent of a population lies within ±3 standard deviations from the mean, application of $L = 3$ will yield (100) (0.0027) = 0.27 percent of the observations being further than 3 standard deviations from the mean even if there are no true outliers in the data set (this assumes a normal distribution for the observations). As the data set being considered becomes larger, the more possible outliers one will identify, even if there are no problems with the data. For this reason, outliers identified in this way should be deleted from the analysis only if they can be traced to specific causes (such as recording errors, experimental errors, and the like). Otherwise, there is a substantial risk of eliminating data that are, in a sense, "trying to tell you something."

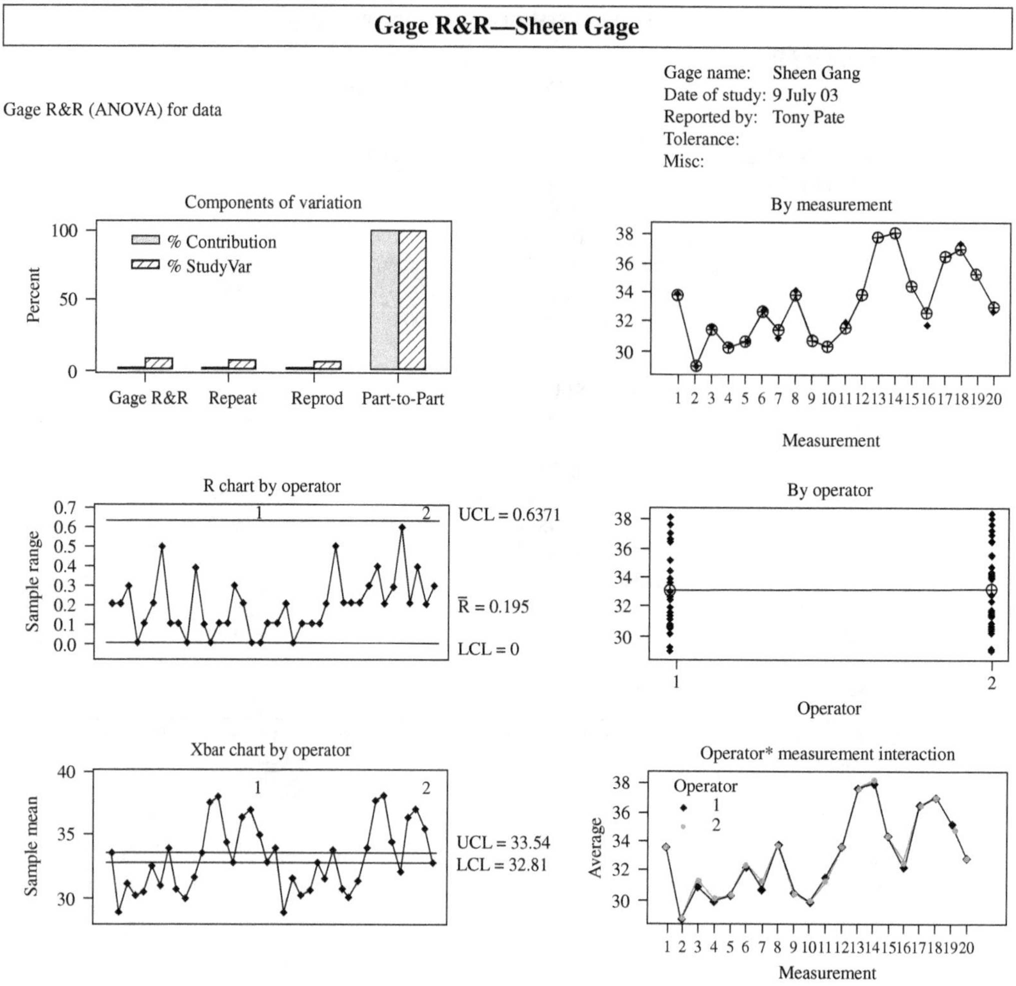

Figure 20.10 Gage R&R. (*Juran Institute, Inc.*)

Typically, one adjusts L based on the size of the data set to be screened; with n = 1000 points, $L = 3$ is reasonable; with $n = 100$, $L = 2.5$ can be used, and only (100) (0.0124) = 1.24 outliers will be expected to be found if the data have no problems. After bad data are deleted or replaced (this is desirable if the experiment can be rerun under comparable conditions to those specified in the experimental plan), the data should be screened again. With the "worst" points removed/corrected, less extreme cases may come to be identified as possible outliers, and should again be investigated.

Another commonly used method is to visualize the data in some way (e.g., to plot variables in box plots or scatter plots). Points visually distant from the others should be scrutinized and eliminated as outliers if (and only if) they are reasonably attributable to some specific cause unrelated to the question at hand. Regression analysis can be helpful as an additional step, using residuals (the differences between the observed and predicted values) to flag potential data points that are unusual (and may have undue leverage on the regression model). Regression is discussed later in this chapter.

Summarization of Data

A mantra for the data analyst is that the first three steps of any data analysis are to (1) plot the data, (2) plot the data, and (3) plot the data. Clearly important, many of the most practical methods of summarizing data are quite simple in concept. Depending on the goals of the data summarization, sometimes one method will provide a useful and complete summarization. More often, two or more methods will be needed to obtain the clarity of description that is desired. Several key methods are plots versus time order of data, frequency distributions and histograms, and sample characteristics such as measures of central tendency/location (mean, median, and mode) and measures of dispersion (range, standard deviation, and variance) displayed graphically.

Plots versus Time Order of Data

Plotting the output *Y* against the time order in which the data were obtained (essentially a scatter plot of *Y* versus time) can reveal several possible phenomena:

- A few observations are far from the others. They should be investigated as to their cause and, if erroneous, corrected or discarded.
- There are trends or cycles within a time period—a day, week, etc. This may represent such phenomena as warming of a machine, operator fatigue, seasonal demand, customer timing preferences, or similar time-related trends.
- Variability decreases or increases with time; this may be due to a learning curve or raw material characteristics, as when one lot of material is used up and the next lot has lesser or greater heterogeneity. It may also reflect changes in customer behavior for services.

While the preceding trends may be apparent even in a plot of the original observations *Y* versus time, they are often more easily spotted in plots of the residuals of the observations after a regression analysis (see the section "Correlation and Regression Analysis" later in this chapter) or using a control chart.

Histograms

A frequency distribution is a tabulation of data arranged according to size. Presenting data in this form clarifies the central tendency and the dispersion along the scale of measurement, as well as the relative frequency of occurrence of the various values (i.e., the shape of the distribution of data). The shape of the distribution may suggest some theories of cause for variation in the process, and reduce the likelihood of others. Histograms usually require at least 40 data points to provide useful insight.

Box Plots

These also display frequency distribution and indications with regard to central tendency and the dispersion along the scale of measurement. They provide less rich detail than histograms, but can be used with as few as eight data points, and facilitate the comparison of many distributions.

Sample Characteristics

Descriptive statistics such as the mean (average), median, mode, range, variance, and standard deviation provide numerical ways of summarizing data, and should be used in conjunction with graphical displays of the type discussed previously.

Analysis

The emphasis in this section is on statistical tools used in the analysis of data for quality improvement, control, and planning. Statistics, for our purposes, is the use of a small sample of data to infer properties of a larger population or universe in which we are interested. Statistics is grounded in probability. Probability is a measure that describes the chance that an event will occur. Based on appropriately collected data, statistics and probability are used to understand explicitly the accuracy of the information we have for managing quality and assess the risks of both acting and not acting on the basis of that data.

The following are some types of problems that can benefit from statistical analysis:

- Determining the usefulness of a limited number of test results in predicting the true value of a product characteristic
- Determining the number of tests required to provide adequate data for evaluation
- Comparing test data between two alternative designs
- Predicting the amount of product that will fall within specification limits
- Predicting system performance
- Controlling process quality by early detection of process changes
- Planning experiments to discover the factors that influence a characteristic of a product or process (i.e., exploratory experimentation)
- Determining the quantitative relationship between two or more variables

The Concept of Statistical Variation

Variety is the so-called "spice of life," and this is no less true when it comes to statistics. The concept of variation is that no two items are perfectly identical. Variation is a fact of nature and a bane of industrial life. For example, even "identical" twins vary slightly in height and weight at birth. The dimensions of an integrated chip vary from chip to chip; cans of tomato soup vary slightly from can to can; the time required to assign a seat at an airline check-in counter varies from passenger to passenger. To disregard the existence of variation (or to rationalize falsely that it is small) can lead to incorrect decisions on major problems. Statistics helps to analyze data properly and draw conclusions, taking into account the existence of variation.

Statistical variation—variation due to random causes—is much greater than most people think. Often, we decide what action to take based on the most recent data point, and we forget that the data point is part of a history of data.

In order to make decisions and improve processes, statistical variation must be taken into account. Variation can be visualized through the use of histograms, box plots, and similar tools. Frequently, such tools are sufficient to draw practical conclusions because differences in central tendency are large and variation is relatively small. However, statistical tools become necessary when the picture (quite literally) is less clear.

Building on the foundation of descriptive statistics, we start with an overview of the probability distributions that underlie many statistical tools and are used to model data and allow estimation of probabilities. Terms are defined as they are encountered, including further discussion of enumerative and analytical studies. Following an introduction to statistical inference and hypothesis testing, specific methods are discussed by way of example.

Probability Distributions

Before diving in, we should make a distinction between a sample and a population. A population is the totality of the phenomenon under study. A sample is a limited number of items taken from that population. Measurements are made on the smaller subset of items, and we can calculate a sample statistic (e.g., the mean). A sample statistic is a quantity computed from a sample to estimate a population parameter. Samples for statistics must be random. Simple random samples require that every element of the population have the same equal probability of selection for the sample. More complex sampling, such as stratified sampling, requires still requires that each element have a known, but not necessarily equal, chance of selection.

A probability distribution function is a mathematical formula that relates the values of the characteristic with their probability of occurrence in the population. The collection of these probabilities is called a probability distribution. The mean (μ) of a probability distribution often is called the expected value. Some distributions and their functions are summarized in Fig. 20.11. Distributions are of two types:

Distribution	Form	Probability function	
Normal	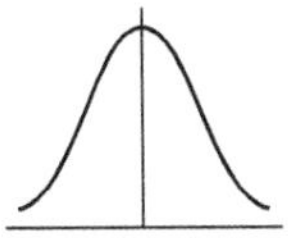	$y = \frac{1}{\sigma\sqrt{2\pi}} e^{-\frac{(x-\mu)^2}{2\sigma^2}}$ μ = Mean σ = Standard deviation	Applicable when there is a concentration of observations about the average and it is equally likely that observations will occur above and below the average. Variation in observations is usually the result of many small causes.
Exponential	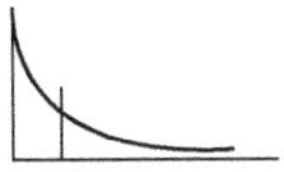	$y = \frac{1}{\mu} e^{-\frac{x}{\mu}}$	Applicable when it is likely that more observations will occur below the average than above.
Weibull	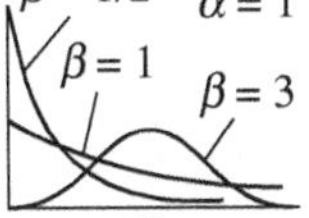	$y = \alpha\beta(X-\gamma)^{\beta-1}e^{-\alpha(X-\gamma)^{\alpha}}$ α = Scale parameter β = Shape parameter γ = Location parameter	Applicable in describing a wide variety of patterns in variation, including departures from the normal and exponential.
Poisson*	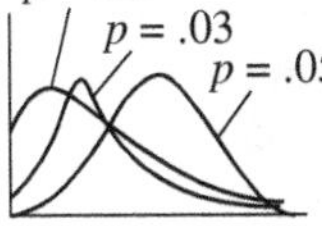	$y = \frac{(np)^r e^{-np}}{r!}$ n = Number of trials r = Number of occurrences p = Probability of occurrence	Same as binomial but particularly applicable when there are many opportunities for occurrence of an event but a low probability (less than .10) on each trial.
Binomial*	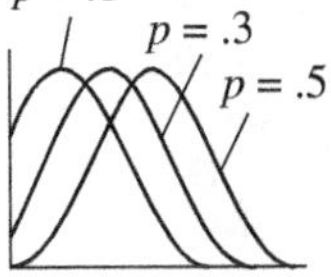	$y = \frac{n!}{r!(n-r)!} p^r q^{n-r}$ n = Number of trials r = Number of occurrences p = Probability of occurrence $q = 1 - p$	Applicable in defining the probability of r occurrences in n trials of an event that has constant probability of occurrence on each independent trial.

FIGURE 20.11 Summary of common probability distributions. (*Quality Planning and Analysis, Copyright 2007. Used by permission.*)

3.37	3.34	3.38	3.32	3.33	3.28	3.34	3.31	3.33	3.34
3.29	3.36	3.30	3.31	3.33	3.34	3.34	3.36	3.39	3.38
3.35	3.36	3.30	3.32	3.33	3.35	3.35	3.34	3.32	3.38
3.32	3.37	3.34	3.38	3.36	3.37	3.36	3.31	3.33	3.30
3.35	3.33	3.38	3.37	3.44	3.32	3.36	3.32	3.29	3.35
3.38	3.39	3.34	3.32	3.30	3.39	3.36	3.40	3.32	3.33
3.29	3.41	3.27	3.36	3.41	3.37	3.36	3.37	3.33	3.66
3.31	3.33	3.35	3.34	3.35	3.34	3.31	3.36	3.37	3.35
3.40	3.35	3.37	3.35	3.32	3.36	3.38	3.35	3.31	3.34
3.35	3.36	3.39	3.31	3.31	3.30	3.35	3.33	3.35	3.31

Source: Quality Planning and Analysis, Copyright 2007. Used by permission.

TABLE 20.2 Resistance of 100 Coils, Ω

Continuous (for "Variable" Data)

When the characteristic being measured can take on any value (subject to the fineness of the measuring process), its probability distribution is called a "continuous probability distribution." For example, the probability distribution of the resistance data in Table 20.2 is an example of a continuous probability distribution because the resistance could have any value, limited only by the fineness of the measuring instrument. Most continuous characteristics follow one of several common probability distributions: the normal distribution, the exponential distribution, or the Weibull distribution.

Discrete (for "Attribute" Data)

When the characteristic being measured can take on only certain specific values (e.g., integers 0, 1, 2, 3), its probability distribution is called a "discrete probability distribution." For example, the distribution of the number of defects r in a sample of five items is a discrete probability distribution because r can be only 0, 1, 2, 3, 4, or 5 (and not 1.25 or similar intermediate values). The common discrete distributions are the Poisson and binomial.

Statistical Inference

Statistical inference is the process of estimating, through sampling and application of statistical methods, certain characteristics of a population. In the world of quality, these estimates and statistical conclusions are used to draw practical conclusions, typically providing the practitioner confidence in taking subsequent action (or inaction) to improve a process.

Sampling Variation and Sampling Distributions

Suppose that a battery is to be evaluated to ensure that life requirements are met. A mean life of 30 hours is desired. Preliminary data indicate that the life follows a normal distribution and that the standard deviation is equal to 10 hours. A sample of four batteries is selected at random from the population and tested. If the mean of the four is close to 30 hours, it is concluded that the population of batteries meets the specification. Figure 20.12 plots the distribution of individual batteries from the population, assuming that the true mean of the population is exactly 30 hours.

If a sample of four is life-tested, the following lifetimes might result: 34, 28, 38, and 24, giving a mean of 31.0 hours. However, this random sample is selected from the many batteries

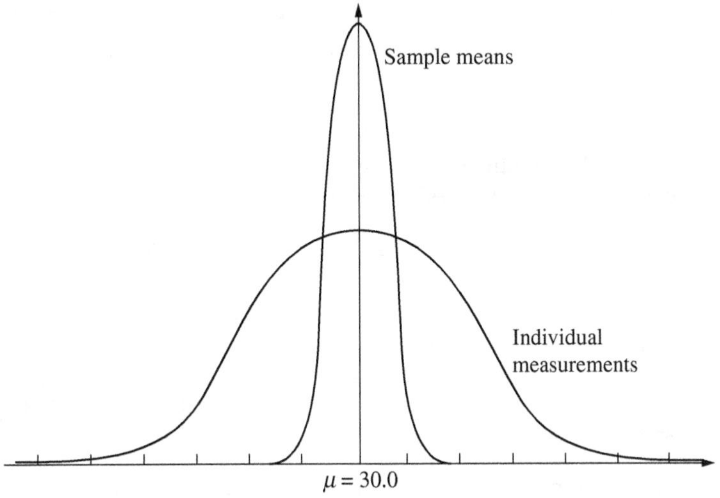

FIGURE 20.12 Distributions of individual measurements and sample means. (*Juran Institute, Inc., 1994.*)

made by the same process. Suppose that another sample of four is taken. The second sample of four is likely to be different from the first sample. Perhaps the results would be 40, 32, 18, and 29, giving a mean of 29.8 hours. If the process of drawing many samples (with four in each sample) is repeated over and over, different results would be obtained in most samples. The fact that samples drawn from the same process can yield different sample results illustrates the concept of sampling variation.

Returning to the problem of evaluating the battery, a dilemma exists. In the actual evaluation, let's assume only one sample of four can be drawn (e.g., because of time and cost limitations). Yet the experiment of drawing many samples indicates that samples vary. The question is, How reliable is the single sample of four that will be the basis of the decision? The final decision can be influenced by the "luck" of which sample is chosen. The key point is that the existence of sampling variation means that any one sample cannot always be relied upon to give an adequate decision. The statistical approach analyzes the results of the sample, taking into account the possible sampling variation that could occur.

Formulas have been developed to define the expected amount of sampling variation. In particular, the central limit theorem states that if $x_1, x_2, \ldots, x_n$ are outcomes of a sample of n independent observations of a random variable x, then the mean of the samples of n will approximately follow a normal distribution, with mean μ and standard deviation. When n is large ($n > 30$), the normal approximation is very close. For smaller samples, a modified Student-T distribution applies. The central limit theorem is very helpful to much practical statistical work. First, the variation of means is smaller than the variation of the underlying population, which makes conclusions easier. Second, because means are approximately normally distributed, we can apply the wide variety of techniques that rely on the assumption of normality.

Statistical Tools for Improvement

This concept of a sampling distribution is fundamental to the two major areas of statistical inference, estimation and tests of hypotheses, which are discussed next.

Statistical Estimation: Point Estimation and Confidence Intervals

Estimation is the process of analyzing a sample result to predict the corresponding value of the population parameter. In other words, the process is to estimate a desired population parameter by an appropriate measure calculated from the sample values. For example, the sample of four batteries previously mentioned had a mean life of 31.0 hours. If this is a representative sample from the process, what estimate can be made of the true average life of the entire population of batteries? The estimation statement has two parts:

1. The point estimate is a single value used to estimate the population parameter. For example, 31.0 hours is the point estimate of the average life of the population.
2. The confidence interval is a range of values that include (with a preassigned probability called a confidence level*) the true value of a population parameter. Confidence limits are the upper and lower boundaries of the confidence interval. Confidence limits should not be confused with other limits (e.g., control limits, statistical tolerance limits).

Table 20.3 summarizes confidence limit formulas for common parameters. The following example illustrates one of these formulas:

Problem Twenty-five specimens of brass have a mean hardness of 54.62 and an estimated standard deviation of 5.34. Determine the 95 percent confidence limits on the mean. The standard deviation of the population is unknown.

Solution Note that when the standard deviation is unknown and is estimated from the sample, the t distribution in Table 20.4 must be used. The t value for 95 percent confidence is found by entering the table at 0.975 and 25 – 1, or 24, degrees of freedom† and reading a t value of 2.064.

$$\text{Confidence limits} = \overline{X} = \pm t\frac{s}{\sqrt{n}}$$

$$= 54.62 \pm (2.064)\frac{5.34}{\sqrt{25}}$$

$$= 52.42 \text{ and } 56.82$$

There is 95 percent confidence that the true mean hardness of the brass is between 52.42 and 56.82.

Determination of Sample Size

The only way to obtain the true value of a population parameter such as the mean is to measure (with a perfect measurement system) each and every individual within the population. This is not realistic (and is unnecessary when statistics are properly applied), so samples are taken instead. But how large a sample should be taken? The answer depends on (1) the sampling risks

---

*A confidence level is the probability that an assertion about the value of a population parameter is correct. Confidence levels of 90, 95, or 99 percent are usually used in practice.

†A mathematical derivation of degrees of freedom is beyond the scope of this book, but the underlying concept can be stated. Degrees of freedom (DF) is the parameter involved when, for example, a sample standard deviation is used to estimate the true standard deviation of a universe. DF equals the number of measurements in the sample minus some number of constraints estimated from the data to compute the standard deviation. In this example, it was necessary to estimate only one constant (the population mean) to compute the standard deviation.
Therefore, DF = 25 – 1 = 24.

Mean of a normal population (standard deviation known)	$\bar{X} \pm Z_{\alpha/2}\frac{\sigma}{\sqrt{n}}$ where $\bar{X}$ = sample average Z = normal distribution coefficient σ = standard deviation of population n = sample size
Mean of a normal population (standard deviation unknown)	$\bar{X} \pm t_{\alpha/2}\frac{s}{\sqrt{n}}$ where t = distribution coefficient (with $n - 1$ degrees of freedom) s = estimated σ (s is the sample standard deviation)
Standard deviation of a normal population	Upper confidence limit $= s\sqrt{\frac{n-1}{x^2_{\alpha/2}}}$ Lower confidence limit $= s\sqrt{\frac{n-1}{x^2_{1-\alpha/2}}}$ where x^2 = chi-square distribution coefficient with $n - 1$ degrees of freedom $1 - \alpha$ = confidence level
Population fraction defective	See charts: *Ninety-five percent confidence belts for population proportion* and *Binomial Distribution* at the end of this chapter, pages 675–676.
Difference between the means of two normal populations (standard deviations σ_1 and σ_2 known)	$(\bar{X}_1 - \bar{X}_2) \pm Z_{\alpha/2}\sqrt{\frac{\alpha_1^2}{n_1} + \frac{\alpha_2^2}{n_2}}$
Difference between the means of two normal populations ($\sigma_1 = \sigma_2$ but unknown)	$(\bar{X}_1 - \bar{X}_2) \pm t_{\alpha/2}\sqrt{\frac{1}{n_1} + \frac{1}{n_2}} \times \sqrt{\frac{\Sigma(X - \bar{X}_1)^2 + \Sigma(X - \bar{X}_2)^2}{n_1 + n_2 - 2}}$
Mean time between failures based on an exponential population of time between failures	Upper confidence limit $= \frac{2rm}{X^2_{\alpha/2}}$ Lower confidence limit $= \frac{2rm}{X^2_{1-\alpha/2}}$ where r = number of occurrences in the sample (i.e., number of failures) m = sample mean time between failures DF = $2r$

Source: Quality Planning and Analysis, Copyright 2007. Used by permission.

TABLE 20.3 Summary of Confidence Limit Formulas ($1 - \alpha$) (Confidence Level)

Distribution of *t*

Value of *t* corresponding to certain selected probabilities (i.e., tail areas under the curve). To illustrate: the probability is .975 that a sample with 20 degrees of freedom would have $t = +2.086$ or smaller.

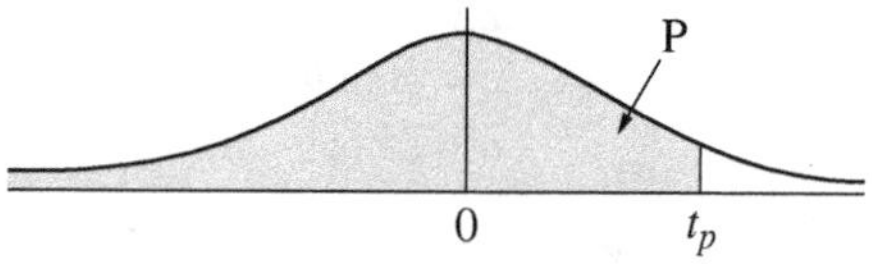

DF	$t_{.60}$	$t_{.70}$	$t_{.80}$	$t_{.90}$	$t_{.95}$	$t_{.975}$	$t_{.99}$	$t_{.995}$
1	0.325	0.727	1.376	3.078	6.314	12.706	31.821	63.657
2	0.289	0.617	1.061	1.886	2.920	4.303	6.965	9.925
3	0.277	0.584	0.978	1.638	2.353	3.182	4.541	5.841
4	0.271	0.569	0.941	1.533	2.132	2.776	3.747	4.604
5	0.267	0.559	0.920	1.476	2.015	2.571	3.365	4.032
6	0.265	0.553	0.906	1.440	1.943	2.447	3.143	3.707
7	0.263	0.549	0.896	1.415	1.895	2.365	2.998	3.499
8	0.262	0.546	0.889	1.397	1.860	2.306	2.896	3.355
9	0.261	0.543	0.883	1.383	1.833	2.262	2.821	3.250
10	0.260	0.542	0.879	1.372	1.812	2.228	2.764	3.169
11	0.260	0.540	0.876	1.363	1.796	2.201	2.718	3.106
12	0.259	0.539	0.873	1.356	1.782	2.179	2.681	3.055
13	0.259	0.538	0.870	1.350	1.771	2.160	2.650	3.012
14	0.258	0.537	0.868	1.345	1.761	2.145	2.624	2.977
15	0.258	0.536	0.866	1.341	1.753	2.131	2.602	2.947
16	0.258	0.535	0.865	1.337	1.746	2.120	2.583	2.921
17	0.257	0.534	0.863	1.333	1.740	2.110	2.567	2.898
18	0.257	0.534	0.862	1.330	1.734	2.101	2.552	2.878
19	0.257	0.533	0.861	1.328	1.729	2.093	2.539	2.861
20	0.257	0.533	0.860	1.325	1.725	2.086	2.528	2.845
21	0.257	0.532	0.859	1.323	1.721	2.080	2.518	2.831
22	0.256	0.532	0.858	1.321	1.717	2.074	2.508	2.819
23	0.256	0.532	0.858	1.319	1.714	2.069	2.500	2.807
24	0.256	0.531	0.857	1.318	1.711	2.064	2.492	2.797
25	0.256	0.531	0.856	1.316	1.708	2.060	2.485	2.787
26	0.256	0.531	0.856	1.315	1.706	2.056	2.479	2.779
27	0.256	0.531	0.855	1.314	1.703	2.052	2.473	2.771
28	0.256	0.530	0.855	1.313	1.701	2.048	2.467	2.763
29	0.256	0.530	0.854	1.311	1.699	2.045	2.462	2.756
30	0.256	0.530	0.854	1.310	1.697	2.042	2.457	2.750
40	0.255	0.529	0.851	1.303	1.684	2.021	2.423	2.704
60	0.254	0.527	0.848	1.296	1.671	2.000	2.390	2.660
120	0.254	0.526	0.845	1.289	1.658	1.980	2.358	2.617
∞	0.253	0.524	0.842	1.282	1.645	1.960	2.326	2.576

Source: Introduction to Statistical Analysis, Copyright 1969. Used by permission.

TABLE 20.4 Distribution of *t*

Null hypothesis (H_0): Statement of no change or no difference. This statement is assumed true until sufficient evidence is presented to reject it.
Alternative hypothesis (H_a): Statement of change or difference. This statement is considered true if H_0 is rejected.
Type I error: The error in rejecting H_0 when it is true or in saying there is a difference when there is no difference.
Alpha risk: The maximum risk or maximum probability of making a type I error. This probability is preset, based on how much risk the researcher is willing to take in committing a type I error (rejecting H_0 wrongly), and it is usually established at 5% (or .05). If the *p*-value is less than alpha, reject H_0.
Significance level: The risk of committing a type I error.
Type II error: The error in failing to reject H_0 when it is false or in saying there is no difference when there really is a difference.
Beta risk: The risk or probability of making a type II error or overlooking an effective treatment or solution to the problem.
Significant difference: The term used to describe the results of a statistical hypothesis test where a difference is too large to be reasonably attributed to chance.
p-value: The probability of obtaining different samples when there is really no difference in the population(s)—that is, the actual probability of committing a type I error. The *p*-value is the actual probability of incorrectly rejecting the null hypothesis (H_0) (i.e., the chance of rejecting the null when it is true). When the *p*-value is less than alpha, reject H_0. If the *p*-value is greater than alpha, fail to reject H_0.
Power: The ability of a statistical test to detect a real difference when there really is one, or the probability of being correct in rejecting H_0. Commonly used to determine if sample sizes are sufficient to detect a difference in treatments if one exists. Power = $(1 - \beta)$, or 1 minus the probability of making a type II error.

Source: Quality Planning and Analysis, Copyright 2007. Used by permission.

TABLE 20.5 Hypothesis Testing Definitions

desired (alpha and beta risk, discussed further below and defined in Table 20.5), (2) the size of the smallest true difference that is desired to be detected, and (3) the variation in the characteristic being measured.

For example, suppose it was important to detect that the mean life of the battery cited previously was 35.0 hours (recall that the intended value is 30.0 hours). Specifically, we want to be 80 percent certain of detecting this difference (this is the "power" of the test, and has a corresponding risk of $\beta = 0.2$; this means we are willing to take a 20 percent chance of failing to detect the five-hour difference when, in fact, it exists). Further, if the true mean was 30.0 hours, we want to have only a 5 percent risk of wrongly concluding it is not 30.0 hours (a risk of $\alpha = 0.05$). Then, using the following formula:

$$n = \left[\frac{(Z_{\alpha/2} + Z_{\beta})_{\sigma}}{\mu - \mu_{o}}\right]^2$$

we plug in our values to obtain

$$n = \left[\frac{(1.96 + 0.84)10}{35 - 30}\right]^2 = 31.4$$

The required sample size is 32 (Gryna et al. 2007, p. 605).

Note that sample size sometimes is constrained by cost or time limitations; in addition, rules of thumb exist to estimate sample size. However, these potentially lead to gross under- or oversampling, with wasted time and effort. The recommended approach is to use power and sample size calculators (available online and in statistical software; these readily apply formulas appropriate for different sampling situations) in order to enter data collection and hypothesis testing with full knowledge of the statistically appropriate sample size.

Hypothesis Testing

A hypothesis, as used here, is an assertion about a population. Typically, the hypothesis is stated as a pair of hypotheses as follows: the null hypothesis (H_0) and an alternative hypothesis, H_a. The null hypothesis, H_0, is a statement of no change or no difference—hence, the term "null." The alternative hypothesis is the statement of change or difference—that is, if we reject the null hypothesis, the alternative is true by default.

For example, to test the hypothesis that the mean life of a population of batteries equals 30 hours, we state:

$$H_0: \mu = 30.0 \text{ hours}$$

$$H_a: \mu \neq 30.0 \text{ hours}$$

A hypothesis test is a test of the validity of the assertion, and is carried out by analyzing a sample of data. Sample results must be carefully evaluated for two reasons. First, there are many other samples that, by chance alone, could be drawn from the population. Second, the numerical results in the sample actually selected can easily be compatible with several different hypotheses. These points are handled by recognizing the two types of sampling errors, already alluded to above.

The Two Types of Sampling Errors

In evaluating a hypothesis, two errors can be made

1. Reject the null hypothesis when it is true. This is called a type I error, or the level of significance. The maximum probability of a type I error is denoted by α.
2. Fail to reject the null hypothesis when it is false. This is called type II error, and the probability is denoted by β.

These errors are defined in terms of probability numbers and can be controlled to desired values. The results possible in testing a hypothesis are summarized in Table 20.6. Definitions are found in Table 20.5. For additional detail on sampling errors in the context of quality, see Gryna et al. (2007).

Suppose Decision of Analysis Is	Suppose the H_0 Is True	Suppose the H_0 Is False
Fail to reject H_0	Correct decision $p = 1 - \alpha$	Wrong decision $p = \beta$
Reject H_0	Wrong decision $p = \alpha$	Correct decision $p = 1 - \beta$

Source: Quality Planning and Analysis, Copyright 2007. Used by permission.

Table 20.6 Type I (α) Error and Type II (β) Error

Steps to Hypothesis Testing

As emphasized earlier, it is important to plan for data collection and analysis; an investigator ideally should arrive at the point of actual hypothesis testing with elements such as sample size already defined. Hypothesis testing often is an iterative process, however, and as mentioned above in the opening discussion of data collection, further data may be needed after initial collection, for example, to bolster sample sizes to obtain the desired power so that both type I and type II errors are defined in advance.

Generally, then, the steps to test a hypothesis are as follows:

1. State the practical problem.
2. State the null hypothesis and alternative hypothesis.
3. Choose a value for α (alpha). Common values are 0.01, 0.05, and 0.10.
4. Choose the test statistic for testing the hypothesis.
5. Determine the rejection region for the test (i.e., the range of values of the test statistic that results in a decision to reject the null hypothesis).
6. Obtain a sample of observations, compute the test statistic, and compare the value to the rejection region to decide whether to reject or fail to reject the hypothesis.
7. Draw the practical conclusion.

Common Tests of Hypotheses

No single means of organizing hypothesis tests can convey all the information that may be of interest to an investigator. Table 20.7 summarizes some common tests of hypotheses in terms of the formulas. Table 20.8 categorizes tests according to the question being asked and type of data. Figure 20.13 provides similar information but in the form of a roadmap to assist in deciding what hypothesis test(s) are appropriate. Readers may find that the combination of these presentations will provide the best understanding of what is a multifaceted topic.

The hypothesis testing procedure is illustrated through the following example:

1. State the practical problem. To investigate a problem with warping wood panels, it was proposed that warping was caused by differing moisture content in the layers of the laminated product before drying. The sample data shown in Table 20.9 were taken between layers 1-2 and 2-3. Is there a significant difference in the moisture content?
2. State the null hypothesis and alternative hypothesis:

$$H_o: \mu 1\text{-}2 = \mu 2\text{-}3$$

$$H_a: \mu 1\text{-}2 \neq \mu 2\text{-}3$$

3. Choose a value for α. In this example, a type I error (α) of 0.05 will be assumed.
4. Choose the test statistic for testing the hypothesis.

   Because we have two samples and desire to test for a difference in the means, a two-sample t-test is appropriate. (Note: A probability plot or test for normality will confirm the assumption of normality in the data. Also, an equal variance test concludes variances are approximately equal.)
5. Determine the rejection region for the test.

Hypothesis	Test Statistic and Distribution
H_0: $\mu = \mu_0$ (the mean of a normal population is equal to a specified value μ_0; s is known)	$Z = \frac{\bar{X} - \mu_0}{\sigma/\sqrt{n}}$ Standard normal distribution
H_0: $\mu = \mu_0$ (the mean of a normal population is equal to a specified value μ_0; σ is estimated by s)	$t = \frac{\bar{X} - \mu_0}{s/\sqrt{n}}$ t distribution with $n - 1$ degrees of freedom (DF)
H_0: $\mu_1 = \mu_2$ (the mean of population 1 is equal to the mean of population 2; assume that $\sigma_1 = \sigma_2$ and that both populations are normal)	$t = \frac{\bar{X}_1 - \bar{X}_2}{\sqrt{1/n_1 + 1/n_2}\sqrt{[(n_1-1)s_1^2(n_2-1)s_2^2]/(n_1+n_2-2)}}$ t distribution with DF = $n_1 + n_2 - 2$
H_0: $\sigma = \sigma_0$ (the standard deviation of a normal population is equal to a specified value σ_0)	$X^2 = \frac{(n-1)s^2}{\sigma_0^2}$ Chi-square distribution with DF = $n - 1$
H_0: $\sigma_1 = \sigma_2$ (the standard deviation of population 1 is equal to the standard deviation of population 2; assume that both populations are normal)	$F = \frac{s_1^2}{s_2^2}$ F distribution with $DF_1 = n_1 - 1$ and $DF_2 = n_2 - 1$
H_0: $\hat{p} = p_0$ (the fraction defective in a population is equal to a specified value p_0; assume that = sample proportion	$Z = \frac{\hat{p} - p_0}{\sqrt{p_0(1-p_0)/n}}$ Standard normal distribution
H_0: $p_1 = p_2$ (the fraction defective in population 1 is equal to the fraction defective in population 2; assume that n_1p_1 and n_2p_2 are each ≥5)	$Z = \frac{X_1/n_1 - X_2/n_2}{\sqrt{\hat{p}(1-\hat{p})(1/n_1 + 1/n_2)}}$ $\hat{p} = \frac{X_1 + X_2}{n_1 + n_2}$ Standard normal distribution
To test for independence in a J × K contingency table that cross-classifies the variable A and B H_0: A is independent of B H_a: A is dependent on B	$X^2 = \sum_{j=1}^{J}\sum_{k=1}^{K}\frac{(f_{jk} - e_{jk})^2}{e_{jk}}$ Chi-square distribution with DF = $(J - 1)(K - 1)$ Where f_{jk} = the observed frequency of data for category j of variable A and to category k of variable B e_{jk} = the expected frequency = $f_{jO}f_{Ok}/f_{OO}$ f_{jO} = frequency total for category j for variable A f_{Ok} = frequency total for category k of variable B f_{OO} = frequency total for $J \times K$ table

Source: Quality Planning and Analysis, Copyright 2007. Used by permission.

TABLE 20.7 Summary of Formulas on Tests of Hypotheses

Tests of hypotheses organized by the question being asked. All tests assume a categorical X in the Y = *f* (X) format. For example, X might be manufacturing plant, and there could be 1, 2, or more than two plants of interest in terms of output, Y. A continuous Y might be mean or standard deviation of daily units produced, a categorical Y might be proportion defective units produced in a single day.

Question: Is There a Difference in the Parameter	Number of Sample Groups	Continuous Y (Normal)		Categorical Y	
		Parameter of Interest	**Test**	**Parameter of Interest**	**Test**
Compared to a target?	1	μ σ	1-sample *t* Chi-square	Proportion	1-proportion test
Between two groups?	2	μ σ	2-sample *t* F-test	Proportion	2-proportion test
Among all groups?	≥2	μ σ	ANOVA* Bartlett's	Proportion	Chi-square test of Independence

*ANOVA assumes both equal variances and normality.

Source: Juran Institute, Inc. Used by permission.

Table 20.8 Hypothesis Testing Table

The critical value defining the rejection region is approximately 2.0 (see Table 20.4); if the absolute value of the calculated *t* is larger than the critical value, then we reject the null hypothesis.

6. Obtain a sample of observations, compute the test statistic, and compare the value to the rejection region to decide whether to reject or fail to reject the hypothesis. A box plot (remember to plot the data!) suggests that the moisture content in Layer 1-2 tends to be higher than in Layer 2-3. Minitab output (see Fig. 20.14) shows that the calculated *t* is 4.18, which is in the rejection region.

Because the calculated *t* is larger than the critical value, the associated *p*-value is < α, and we reject the null hypothesis, H_0.

N	Mean	StDev	SE Mean	
Layer 1-2	25	5.350	0.613	0.12
Layer 2-3	25	4.689	0.499	0.10

Difference = mμ (Layer 1-2) – mμ (Layer 2-3)

Estimate for difference: 0.660901

95 percent CI for difference: (0.343158, 0.978644)

T-test of difference = 0 (vs. not =): *t*-value = 4.18 *p*-value = 0.000 DF = 48

Both use pooled StDev = 0.5587.

1. Draw the practical conclusion. We conclude that the moisture content in Layer 1-2 is higher than the moisture content of Layer 2-3.

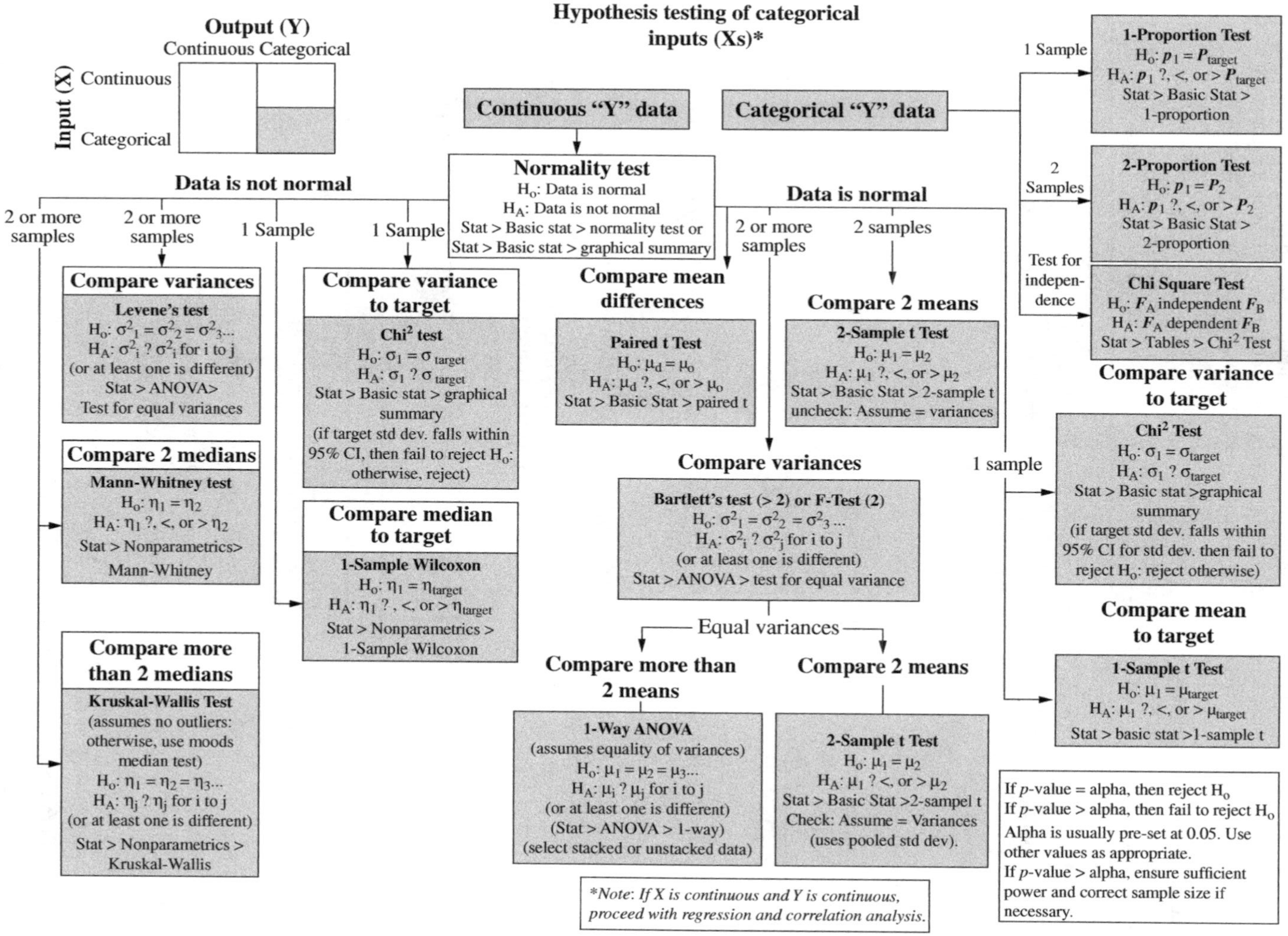

Figure 20.13 Hypothesis testing.

Layer 1-2		Layer 2-3	
4.43	4.40	3.74	5.14
6.01	5.99	4.30	5.19
5.87	5.72	5.27	4.16
4.64	5.25	4.94	5.18
3.50	5.83	4.89	4.78
5.24	5.44	4.34	5.42
5.34	6.15	5.30	4.05
5.99	5.14	4.55	3.92
5.75	5.72	5.17	4.07
5.48	5.00	5.09	4.54
5.64	5.01	4.74	4.23
5.15	5.42	4.96	5.07
5.64		4.21	

Source: Quality Planning and Analysis, Copyright 2007. Used by permission.

TABLE 20.9 Moisture Content

Nonparametric Hypothesis Tests, Data Transformation, and Bootstrapping

The preceding discussion has focused on "parametric" hypothesis tests (so-called because they rely on parameter estimation). Often, it is the case that one or more of the assumptions underlying the parametric tests are violated. In particular, practitioners frequently face skewed or otherwise nonnormal data, and application of parametric tests that assume bell-shaped data distribution may lead to erroneous conclusions and inappropriate action. Fortunately, options are available; these include nonparametric tests, data transformation, and bootstrapping.

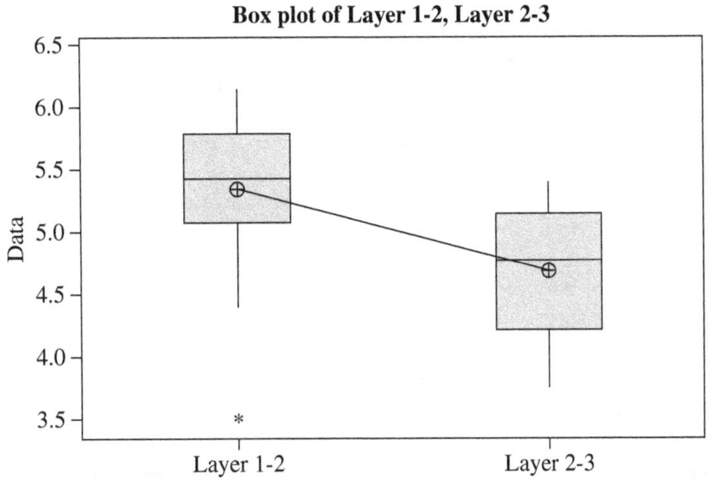

FIGURE 20.14 Box plot of Layer 1-2, Layer 2-3. (*Quality Planning and Analysis, Copyright 2007. Used by permission.*)

Nonparametric hypothesis tests avoid violating key assumptions by virtue of being "distribution-free"; that is, they are not strictly dependent on particular distributions (such as a normal distribution); however, nonparametric tests have their own set of assumptions of which investigators should be aware. In effect, these methods typically transform the original data into ranks, and hypothesis tests then are carried out on the ranked data. Although nonparametric methods are not nearly as well developed and frequently are statistically less powerful compared to parametric tests, they are available for basic one-, two-, and two or more sample tests (see the bottom of Table 20.7 and the left side of the roadmap in Fig. 20.13). See Sprent and Smeeton (2001) for more on traditional nonparametric methods. New methods continue to emerge, for example, wavelets and nonparametric Bayesian techniques; see Kvam and Vidakovic (2007).

Data transformation allows one to take data that violate some assumption of a parametric test and change them so that the assumption no longer is violated. For example, nonnormal data, or sample data with unequal variances can be changed to new numbers that are normal or have equal variances. Three common methods are

Power Functions

Traditionally, standard functions such as taking the square (x^2), square root ($x^{1/2}$), log (log10(x)), natural log (ln(x)), or inverse (x^{-1}) were used because they could easily be done with a calculator. Trial and error often is needed to find a function that appropriately transforms the data to meet the test assumptions.

Box-Cox Transformation

This method provides simultaneous testing of power functions to find an optimum value λ that minimizes the variance. Typically, one selects a power (value of λ) that is understandable and within a 95 percent confidence interval of the estimated λ (e.g., square: $\lambda = 2$; square root: $\lambda = 0.5$; natural log: $\lambda = 0$; inverse: $\lambda = -1$). The Box-Cox transformation does not work with negative numbers.

Johnson Transformation

This method selects an optimal function among three families of distributions (bounded, unbounded, lognormal). While effective in situations where Box-Cox does not work, the resulting transformation is not intuitive.

These methods are easy to apply (with software), and allow use of the more powerful parametric tests. However, the transformed data do not necessarily have intuitive meaning.

Bootstrapping is one of a broader class of computation-intensive resampling methods. Rather than assuming any particular distribution of a test statistic (such as normal), the distribution is determined empirically. More specifically, a statistic of interest (such as the mean) is repeatedly calculated from different samples drawn themselves, with replacements, from a sample. The distribution of these calculated statistics then is used as the basis for determining the probability of obtaining any particular value by chance. Itself a nonparametric approach, bootstrapping is a flexible method that gradually is gaining acceptance. For more information on the method and applications, see Davison and Hinkley (2006).

Correlation and Regression Analysis

Correlation and regression analysis help us understand relationships. More specifically, regression analysis is the modeling of the relationships between independent and dependent variables, while correlation analysis is a study of the strength of the linear relationships among variables. From a practical perspective, simple linear regression examines the distribution of one variable (the response, or dependent variable) as a function of one or more independent variables (the predictor, or independent variable) held at each of several levels. Note that the cause-and-effect relationship is stated explicitly, and it is this relationship that is tested to determine its statistical significance. In addition, regression analysis is used in forecasting and prediction based on the important independent variables, and in locating optimum operating conditions. In contrast, correlation typically looks at the joint variation of two variables that have not been manipulated by the experimenter, and there is no explicit cause-and-effect hypothesis.

For example, suppose that the life of a tool varies with the cutting speed of the tool and we want to predict life based on cutting speed. Thus, life is the dependent variable (Y) and cutting speed is the independent variable (X). Data are collected at four cutting speeds (Table 20.10).

Remembering to always plot the data, we note that a scatter plot (Fig. 20.15) suggests that life varies with cutting speed (specifically, life decreases with an increase in speed) and also varies in a linear manner (i.e., increases in speed result in a certain decrease in life that is the same over the range of the data). Note that the relationship is not perfect—the points scatter about the line.

Often, it is valuable to obtain a regression equation. In this case, we have a linear relationship in the general form provided by

$$Y = \beta_0 + \beta_1 X + \varepsilon$$

X	Y	X	Y	X	Y	X	Y
90	41	100	22	105	21	110	15
90	43	100	35	105	13	110	11
90	35	100	29	105	18	110	6
90	32	100	18	105	20	110	10

(X, in feet per minute versus tool life; Y, in minutes)

Source: Quality Planning and Analysis, Copyright 2007. Used by permission.

Table 20.10 Cutting Speed

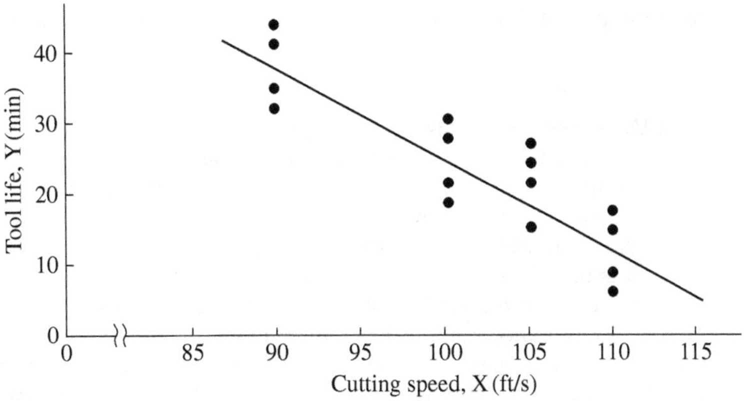

Figure 20.15 Tool life (Y) versus cutting speed (X). (*Quality Planning and Analysis, Copyright 2007.*) *Used by permission.*

where β_0 and β_1 are the unknown population intercept and slope, and ε is a random-error term that may be due to measurement errors and/or the effects of other independent variables. This model is estimated from sample data by the form

$$\hat{Y} = b_0 + b_1X$$

where $\hat{Y}$ is the predicted value of Y for a given value of X and b_0 and b_1 are the sample estimates of β_0 and β_1. Estimates usually are found by least-squares methods; formulas can be found in statistics books such as Kutner et al. (2004).

For this example, the resulting prediction equation is

$$\text{Tool life} = 106.90 - 1.3614\ (\text{cutting speed})$$

This equation can be used to predict tool life by plugging in values of cutting speed. Extreme caution should be used in making predictions outside the actual sample space (e.g., for cutting speeds above or below the tested maximum or minimum), however, as these are tenuous without confirmation by observation.

Although a prediction equation can be found mathematically, it should not be used without knowing how "good" it is. A number of criteria exist for judging the adequacy of the prediction equation. One common measure is R_2, the proportion of variation explained by the prediction equation. R_2, or the coefficient of determination, is the ratio of the variation due to the regression to the total variation. The higher R_2, the greater the probable utility of the prediction equation in estimating Y based on X.

Another measure of the degree of association between two variables is the simple linear correlation coefficient, r. This is the square root of the coefficient of determination, so that the values of r range from −1 to +1. A positive r is consistent with a positive relationship (an increase in one variable is associated with an increase in the other), whereas the opposite is true of a negative r (an increase in one variable is associated with a decrease in the other). Scatter plots are strongly recommended when interpreting correlations, especially as very different patterns can result in identical values of r. The significance level of r varies with sample size; statistical software is recommended to obtain exact significance levels.

The above discussion introduces simple linear correlation and regression—the direction and strength of a relationship between two variables, or prediction of a dependent variable, Y, from a single predictor variable, X. A natural extension of this is multiple regression that allows for two or more independent variables. For a discussion of how to estimate and examine a multiple regression prediction equation, see Kutner et al. (2004).

Analysis of Variance

Analysis of Variance (ANOVA) is an approach related to linear regression, falling into the class of what are called general linear models. However, unlike regression, the X is discrete rather than continuous (noting that general linear models actually can blend characteristics of both regression and ANOVA). In ANOVA, the total variation of all measurements around the overall mean is divided into sources of variation that are then analyzed for statistical significance. It is used in situations where the investigator is interested in comparing the means among two or more discrete groups. For example, an investigator may be interested in comparing performance among three different machine configurations. The ANOVA analysis detects a difference somewhere among the means (i.e., at least one mean is different from the others), and confidence intervals or follow-up tests such as pairwise comparisons can be applied to determine which mean (or means) is different. ANOVA is the basis for design of experiments, discussed next.

Design of Experiments

With origins in the pioneering work in agriculture of Sir Ronald A. Fisher, designed experiments have taken on an increasingly significant role in quality improvement in the business world. This section will first compare the classical and designed approaches to experimentation, thereby providing the reader with an understanding as to the limitations of traditional methods and the power of contemporary methods. Next, basic concepts and terminology will be introduced in the context of an example improvement problem, followed by an overview of different types of designs and the typical progression through a series of designed experiments. The section finishes with the related topic of Taguchi designs.

Contrast between the Classical and Contemporary Methods of Experimentation

The classical method of experimentation is to vary one factor at a time (sometimes called OFAT), holding everything else constant. By way of example, and to illustrate the need for designed experiments, consider the case of a certain fellow who decided he wanted to investigate the causes of intoxication. As the story goes, he drank some whiskey and water on Monday and became highly inebriated. The next day, he repeated the experiment holding all variables constant except one… he decided to replace the whiskey with vodka. As you may guess, the result was drunkenness. On the third day, he repeated the experiment for the last time. On this trial, he used bourbon in lieu of the whiskey and vodka. This time it took him two days just to be able to gather enough of his faculties to analyze the experimental results. After recovering, he concluded that water causes intoxication. Why? Because it was the common variable!

The contrast between this traditional method and the designed approach is striking. In particular, a designed approach permits the greatest information to be gained from the fewest data points (efficient experimentation), and allows the estimation of interaction effects among factors. Table 20.11 compares these two approaches in more detail for an experiment in which there are two factors (or variables) whose effects on a characteristic are being investigated (the same conclusions hold for an experiment with more than two factors).

Concepts and Terminology—An Example Designed Experiment

Suppose that three detergents (A, B, C) are to be compared for their ability to clean clothes in an automatic washing machine. The "whiteness" readings obtained by a special measuring procedure are the dependent, or response, variable. The independent variable under investigation (detergent) is a factor, and each variation of the factor is called a level; in this case, there are three levels. A treatment is a single level assigned to a single factor, detergent A. A treatment combination is the set of levels for all factors in a given experimental run. A factor may be qualitative (different detergents) or quantitative (water temperature). Finally, some experiments have a fixed-effects model (i.e., the levels investigated represent all levels of concern to the investigator—for example, three specific washing machines or brands). Other experiments have a random effects model, that is, the levels chosen are just a sample from a larger population (e.g., three operators of washing machines). A mixed-effects model has both fixed and random factors.

Figure 20.16 outlines six possible designs of experiments, starting with the classical design in (a). Here, all factors except detergent are held constant. Thus, nine tests are run, three with each detergent with the washing time, make of machine, water temperature, and all other factors held constant. One drawback of this design is that the conclusions about detergent brands apply only to the specific conditions of the experiment.

Design (b) recognizes a second factor at three levels (i.e., washing machines brands I, II, and III). However, in this design, it would not be known whether an observed difference was due to detergents or washing machine (they are said to be confounded).

Criteria	Classical	Modern
Basic procedure	Hold everything constant except the factor under investigation. Vary that factor and note the effect on the characteristic of concern. To investigate a second factor, conduct a separate experiment in the same manner.	Plan the experiment to evaluate both factors in one main experiment. Include in the design measurements to evaluate the effect of varying both factors simultaneously.
Experimental conditions	Care should be taken to have material, workers, and machine constant throughout the entire experiment.	Realizes difficulty of holding conditions reasonably constant throughout an entire experiment. Instead, experiment is divided into several groups or blocks of measurements. Within each block, conditions must be reasonably constant (except for deliberate variation to investigate a factor).
Experimental error	Recognized but not stated in quantitative terms.	Stated in quantitative terms.
Basis of evaluation	Effect due to a factor is evaluated with only a vague knowledge of the amount of experimental error.	Effect due to a factor is evaluated by comparing variation due to that factor with the quantitative measure of an experimental error.
Possible bias due to sequence of measurements	Often assumed that sequence has no effect.	Guarded against by randomization.
Effect of varying both factors simultaneously ("interaction")	Not adequately planned into experiment. Frequently assumed that the effect of varying factor 1 (when factor 2 is held constant at some value) would be the same for any value of factor 2.	Experiment can be planned to include an investigation for interaction between factors.
Validity of results	Misleading and erroneous if interaction exists and is not realized.	Even if interaction exists, a valid evaluation of the main factors can be made.
Number of measurements	For a given amount of useful and valid information, more measurements are needed than in the modern approach.	Fewer measurements needed for useful and valid information.
Definition of problem	Objective of experiment frequently not defined as necessary.	Designing the experiment requires defining the objective in detail (how large an effect do we want to determine, what numerical risks can be taken, etc.).
Application of conclusions	Sometimes disputed as applicable only to the controlled conditions under which the experiment was conducted.	Broad conditions can be planned in the experiment, thereby making conclusions applicable to a wider range of actual conditions.

Source: Quality Planning and Analysis, Copyright 2007. Used by permission.

Table 20.11 Comparison of Classical and Modern Methods of Experimentation

A	B	C
-	-	-
-	-	-
-	-	-

(a)

I	II	III
A	B	C
A	B	C
A	B	C

(b)

I	II	III
C	B	B
A	C	B
A	A	C

(c)

I	II	III
B	A	C
C	C	A
A	B	B

(d)

	I	II	III
1	C	A	B
2	B	C	A
3	A	B	C

(e)

I	II	III
ABC	ABC	ABC
1 ---	---	---
2 ---	---	---
3 ---	---	---

(f)

Figure 20.16 Some experimental designs. (*Quality Planning and Analysis, Copyright 2007. Used by permission.*)

In design (c), the nine tests are assigned completely at random, thus the name "completely randomized design." However, detergent A is not used with machine brand III, and detergent B is not used with machine brand I, thus complicating the conclusions.

Design (d) shows a randomized block design. Here each block is a machine brand, and the detergents are run in random order within each block. This design guards against any possible bias due to the order in which the detergents are used and has advantages in the subsequent data analysis and conclusions. First, a test of hypothesis can be run to compare detergents and a separate test of hypothesis run to compare machines; all nine observations are used in both tests. Second, the conclusions concerning detergents apply for the three machines and vice versa, thus providing conclusions over a wider range of conditions.

Now suppose that another factor such as water temperature is also to be studied, using the Latin square design shown in (e). Note that this design requires using each detergent only once with each machine and only once with each temperature. Thus, three factors can be evaluated (by three separate tests of hypothesis) with only nine observations. However, there is a danger. This design assumes no interaction among the factors. No interaction between detergent and machine means that the effect of changing from detergent A to B to C does not depend on which machine is used, and similarly for the other combinations of factors. The concept of interaction is shown in Fig. 20.17. There is no interaction among the detergents and the machines. But the detergents do interact with temperature. At high temperatures, C is the best performer. At low temperatures, A performs best.

Finally, the main factors and possible interactions could be investigated by the factorial design in design (f) in Fig. 20.16. Factorial means that at least one test is run for every combination of main factors, in this case $3 \times 3 \times 3$ or 27 combinations. Separate tests of hypothesis can be run to evaluate the main factors and also possible interactions. Again, all the observations contribute to each comparison. When there are many factors, a portion of the complete factorial (i.e., a "fractional factorial") is useful when experimental resources are limited (see its application in a sequential testing approach, below).

Most problems can be handled with one of the standard experimental designs or a series of these. Designs can be classified by the number of factors to be investigated, the structure of the experimental design, and the kind of information the experiment is intended to provide (Table 20.12). For a description of both the design and analysis of various design structures, see Box et al. (2005). Another excellent general reference is Myers et al. (2009) for a detailed look at response surface designs.

A sequential approach to experimentation often can be helpful. Briefly, a typical sequence of designed experiments will allow an experimenter to quickly and efficiently narrow down a large number of possible factors (or X's in the Y = f(X) terminology of Lean Six Sigma) to find out which are most important, and then refine the relationships to find optimal settings for each of the vital few factors. The steps might be as follows:

1. *Screening experiment*. In this stage, a fractional factorial design may be applied that does not allow interactions to be detected, but can ferret out which of many factors have the greatest main effect.
2. *Fractional factorial design*. The smaller number of factors identified in the screening experiment are tested to allow detection of interaction effects.

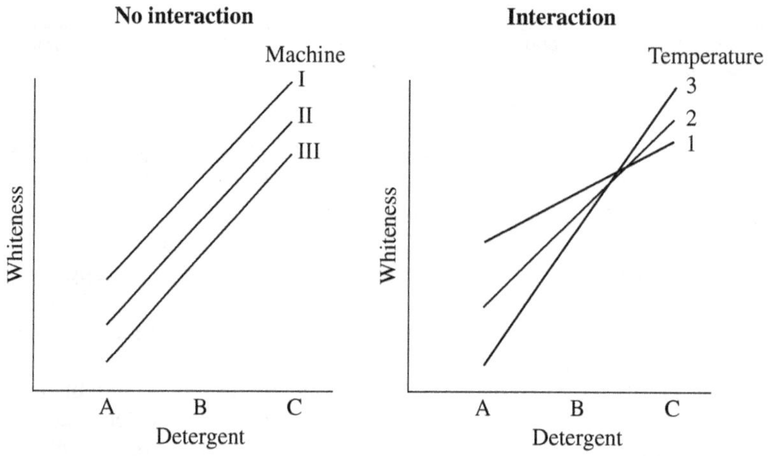

Figure 20.17 Interaction. (*Quality Planning and Analysis, Copyright 2007. Used by permission.*)

Design	Type of Application
Completely randomized	Appropriate when only one experimental factor is being investigated
Factorial	Appropriate when several factors are being investigated at two or more levels and interaction of factors may be significant
Blocked factorial	Appropriate when number of runs required for factorial is too large to be carried out under homogeneous conditions
Fractional factorial	Appropriate when many factors and levels exist and running all combinations is impractical
Randomized block	Appropriate when one factor is being investigated and experimental material or environment can be divided into blocks or homogeneous groups
Balanced incomplete block	Appropriate when all the treatments cannot be accommodated in a block
Partially balanced incomplete block	Appropriate if a balanced incomplete block requires a larger number of blocks than is practical
Latin square	Appropriate when one primary factor is under investigation and results may be affected by two other experimental variables or by two sources of nonhomogeneity. It is assumed that no interactions exist.
Youden square	Same as Latin square, but number of rows, columns, and treatments need not be the same
Nested	Appropriate when objective is to study relative variability instead of mean effect of sources of variation (e.g., variance of tests on the same sample and variance of different samples)
Response surface	Objective is to provide empirical maps (contour diagrams) illustrating how factors under the experimenter's control influence the response
Mixture designs	Use when constraints are inherent (e.g., the sum of components in a paint must add to 100%)

Source: Adapted from JQH5, Table 47.3.

TABLE 20.12 Classification of Designs

3. *Full factorial design*. A small number of factors (usually no more than five) are tested to allow all main effects and higher-order (e.g., three-way, four-way) interactions to be detected and accounted for. Such designs also can detect curvature that indicates a potential optimum.
4. *Response surface design*. By adding data points in particular ways (e.g., a composite design), an experimenter can build on earlier experiments to fully characterize nonlinear relationships and pinpoint optimal settings.
5. *Evolutionary Operation (EVOP)*. Once an improved process is in production mode, evolutionary operation techniques can be used to conduct many small experiments on production units over time. Although individual changes are small, the cumulative effect over time can be quite large, and exemplifies the power of continuous improvement. See Box and Draper (1969) for a classic text on this subject.

For a series of four papers on sequential experimentation, see Carter (1996). Emanuel and Palanisamy (2000) discuss sequential experimentation at two levels and a maximum of seven factors.

Taguchi Approach to Experimental Design

Professor Genichi Taguchi uses an approach to experimental design that has three purposes:

1. Design products and processes that perform consistently on target and are relatively insensitive ("robust") to factors that are difficult to control.
2. Design products that are relatively insensitive (robust) to component variation.
3. Minimize variation around a target value.

Thus, although cited in this "improvement tools" section because of its association with DOE, the approach is meant to provide valuable information for product design and development (see the section "Statistical Tools for Designing for Quality" in this chapter). Taguchi divides quality control into online control (e.g., diagnosing and adjusting a process during production) and offline control that encompasses the engineering design process and its three phases: systems design, parameter design, and tolerance design. For an extensive bibliography and a summary of some controversial aspects of the Taguchi approach, see Box and Draper (1969, pp. 47.58 and 47.59).

Many books are available that cover DOE for engineering and manufacturing applications. For readers in nonmanufacturing environments, Ledolter and Swersey (2007) may be of interest. Another text readers may find useful for not only classical but more contemporary techniques (e.g., Bayesian inference, kriging) is del Castillo (2007).

Discrete Event and Monte Carlo Simulation

Advances in user-friendly software make computer simulations increasingly accessible to quality practitioners that do not have a strong background in mathematics, programming, or modeling. Numerous types of simulation models exist, but two that may be of most interest to readers are discrete event and Monte Carlo simulations. These can be powerful methods for making process improvements; in particular, modeling provides a means of asking "what if?" questions and rapidly testing the effects of process changes and potential solutions in a safe, low-risk environment.

Discrete Event Simulation

Discrete event simulation (DES) attempts to mimic situations in which there are distinct, recognizable events and transactions. In a hospital, for example, arrival of patients at an emergency department and subsequent steps in patient care represent specific events that combine into a flow of transactions: arrival, registration, triage, nursing assessment, physician assessment, etc., through inpatient admission, discharge, or transfer. Discrete event simulation enables system components to be changed and tracks the resulting process flow over time to help understand the relationships among inputs, outputs, and process variables.

Typically, a process flow diagram (or process "map") that graphically displays the sequence and flow of activities forms the basis for a discrete event simulation. A discrete event simulation takes this basic flow diagram and adds inputs and process variables that govern the flow of transactions. Following on the hospital example, these include inputs (such as patient arrivals), human resources (e.g., number of nurses, physician schedules, overtime availability, skill levels, pay rates, etc.), equipment resources (e.g., types and number of beds, imaging equipment, etc.), rules for flow (the required sequence of steps, batching of inputs or outputs, priority rules, exceptions, decisions), resource acquisition (what resources are needed to complete an activity (e.g., one RN or one physician's assistant; two RNs; one RN and one physician, etc.), activity cycle times (work time, wait time), and similar details.

Once these details are built into the model, it "runs" by tracing the path of units (patients, in the hospital example) from arrival through to exit from the process. Patients are processed in accordance with the activities, rules, and constraints, and any relevant attributes (patient-specific characteristics) that may be assigned to them (e.g., acuity level, age, gender). The output consists of a multitude of descriptive statistics and measures that portray the collective behavior of the process as the various players interact and move through time.

Although every model is different and details vary, there are basic steps that should be a part of every simulation study. These steps and related questions are (adapted from Law and Kelton 2000):

1. State the problem and question(s) being asked. What is the business need for the simulation? What problem is to be fixed? What answers are being sought?
2. Prepare a plan for the simulation study. Who needs to be involved? What data are needed and how will data be collected? What alternative scenarios are to be tested? What are the milestones and timeline for completion?
3. Collect data. What is my current state? What are the data for alternative scenarios? Are there gaps in the data, and how will they be handled?
4. Build and validate a conceptual model. Given available data, what is the general structure of the model? What will be the inputs, process variables, and outputs? What statistical accumulators are needed, and where? If the model is built, will it provide the answers to the questions?
5. Build and validate an operational model. Are the model components necessary and sufficient? Does the model produce results consistent with the current state?
6. Design scenarios or experiments needed to answer the questions. What model parameters will be changed? Which are fixed? What combinations of factors need to be tested?
7. Run the scenarios or experiments to obtain the needed outputs. Are the results reproducible? Are additional scenarios or experiments suggested?
8. Analyze and interpret the data. What are the statistical results? Do the descriptive statistics and/or statistical tests indicate meaningful effects? What are the answers to the original questions? Are additional questions raised?

As emphasized at the beginning of this chapter, formulation of the question(s) being asked is a critical first step to the successful application of simulation modeling. Failure to have a clear understanding of what the model is being asked to do leads to poorly constructed models, models with insufficient inputs or process detail, or overly complicated models that take unnecessary time and effort to build and run. In addition, a clearly communicated business need will garner the stakeholder support needed to collect data, evaluate the model, and implement suggested changes.

Monte Carlo

Named after the famed gambling destination, this method seeks to account for uncertainty (variability) in inputs and carry this forward into probability distributions of outcomes. Essentially, instead of using single, fixed values in equations [such as $Y = f(X)$], distributions are used for the inputs (Xs), and samples repeatedly are drawn from the distributions, yielding a distribution of outputs (Y values) instead of a single value. For example, while the forecasted net return on a new product could simply be stated as an expected $10 million, it would be useful to know the probability of achieving this, or that the uncertainty in the forecast is such that there is a high probability of a negative return.

By way of illustration, assume we have three components, A, B, and C that are assembled end-to-end to create a final product. If the mean lengths are 5, 10, and 15 mm, then we can simply add these together to arrive at an expected mean combined total length of 5 mm + 10 mm + 15 mm = 30 mm. However, we know from the concept of statistical variation that there will be variation in the components. Assuming we sample populations of each component and find the respective distributions for each of A, B, and C, what can we expect the overall distribution of assembled product length to look like? By repeatedly taking a random sample from each distribution and adding the lengths, Monte Carlo simulation generates a distribution of the total length. Figure 20.18 shows the relative frequency distribution of the combined lengths of the three components from a Monte Carlo simulation with each of the three components having a standard deviation of 0.1 mm. The mean expected combined total length is almost exactly 30 mm, but the simulation shows the variation around this, with only 45% of assembled components expected to be within +/− 0.1 mm of the total mean value. This approach provides substantially more information than the single estimate of 30 mm.

Simulated DOE

As tools evolve, they are being combined in new ways. One example is the combination of Monte Carlo, discrete event simulation, and DOE. Briefly, this approach involves a discrete event simulation (DES) that uses probability distributions for the input and/or process variables (Monte Carlo), and the investigator changes these variables (as factors) following a structured, designed approach (DOE). While any results and conclusions should be treated

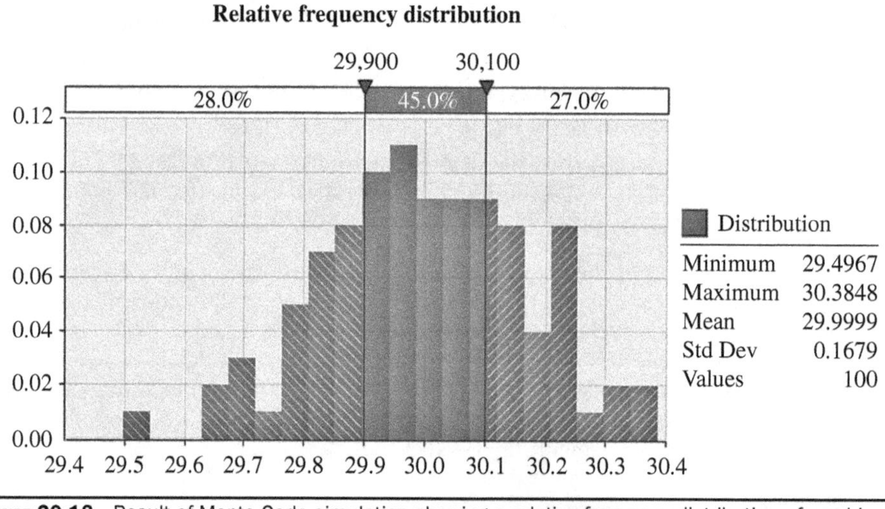

FIGURE 20.18 Result of Monte Carlo simulation showing a relative frequency distribution of combined total length of three components A, B, and C that individually have normal distributions of 5, 10, and 15 mm, respectively, each with a standard deviation of 0.1 mm. The mean expected combined total length is approximately 30 mm, but the simulation shows the variation around this, e.g., that only 45% of assembled components are expected to be within +/− 0.1 mm of this mean value.

as preliminary until verified by actual experimentation, this can be particularly useful in environments where real-life changes may be difficult or dangerous to make.

Additional Advanced Analysis Tools

For practitioners faced with more complex scenarios such as multiple variables (more than one y and/or x), nonlinear data, or categorical outputs, extensions of the general linear models and other alternatives are available. In particular are methods for multivariate analysis; this refers to statistical techniques that simultaneously analyze multiple measurements on subjects. Many techniques are extensions of the univariate (single-variable distributions) and bivariate (correlation, regression) methods dealt with above. Beyond the scope of this chapter, these include:

- *Multiple regression.* Applies when the investigator has a single, continuous dependent variable and multiple, continuous independent variables (Xs) of interest.
- *Nonlinear regression.* Useful when data cannot easily be treated by standard linear methods (note that curvilinear data do not necessarily require nonlinear methods).
- *Nonparametric linear regression.* Applies when the usual assumptions of regression are violated.
- *Multiple discriminant analysis.* Used in situations with a single, categorical (dichotomous or multichotomous) dependent variable (Y) and continuous independent variables (Xs).
- *Logistic regression.* Also known as logit analysis, this is a combination of multiple regression and multiple discriminant analysis in which one or more categorical or continuous independent variables (Xs) are used to predict a single, categorical dependent variable (Y). Odds ratios often are computed with this method.
- *Multivariate analysis of variance and covariance (MANOVA, MANCOVA).* Dependence techniques that extend ANOVA to allow more than one continuous, dependent variable (Y) and several categorical independent variables (Xs).
- *Principal component analysis (PCA) and common factor analysis.* These methods analyze interrelationships among a large number of variables and seek to condense the information into a smaller set of factors without loss of information.
- *Cluster analysis.* An interdependence technique that allows mutually exclusive subgroups to be identified based on similarities among the individuals. Unlike discriminant analysis, the groups are not predefined.
- *Canonical correlation analysis.* An extension of multiple regression that correlates simultaneously several continuous dependent variables (Ys) and several continuous independent variables (Xs).
- *Conjoint analysis.* Often used in marketing analyses, this method helps assess the relative importance of both attributes and levels of complex entities (e.g., products). It is useful when trade-offs exist when making comparisons.
- *Multidimensional scaling.* An interdependence method (also called perceptual mapping), this seeks to transform preferences or judgments of similarity into a representation by distance in multidimensional space.
- *Correspondence analysis.* Another interdependence technique; this accommodates the perceptual mapping of objects (such as products) onto a set of categorical attributes. This method allows both categorical data and nonlinear relationships.

Readers are encouraged to research any techniques that appear to fit their need; although complex, these are powerful means of getting useful information from data. Some useful references include:

Multivariate techniques:

Hair, J. F., Jr., Black, W. C., Babin, B. J., Anderson, R. E., and Tatham, R. L. (2006). *Multivariate Data Analysis*. Pearson Prentice-Hall, Upper Saddle River, NJ.

Affifi, A., Clark, V. A., and May, S. (2004). *Computer-Aided Multivariate Analysis* (4th ed.). Chapman and Hall/CRC Press, Boca Raton, FL.

Coleman, S, Greenfield, T., Stewardson, D., and Montgomery, D. C. (2008). *Statistical Practice in Business and Industry*. John Wiley & Sons, Hoboken, NJ. (see Chapter 13).

Hypothesis testing and DOE:

Box, G. E. P., Hunter, J. S., and Hunter, W. G. (2005). *Statistics for Experimenters: Design, Innovation and Discovery* (2nd ed.). Wiley-Interscience, Hoboken, NJ.

Logistic regression, Poisson regression, odds ratios:

Agresti, A. (1996). *An Introduction to Categorical Data Analysis*. John Wiley & Sons, New York.

Nonparametric:

Sprent, P., and Smeeton, N. C. (2001). *Applied Nonparametric Statistical Methods* (3rd ed.). Chapman and Hall/CRC Press, Boca Raton, FL.

Statistical Tools for Designing for Quality

Statistical tools for quality in the design and development process include techniques such as graphical summaries, probability distributions, confidence limits, tests of hypotheses, design of experiments, regression, and correlation analysis. These topics are covered in earlier sections of this chapter. To supplement these techniques, this section explains some statistical tools for reliability and availability, and tools for setting specification limits on product characteristics.

Failure Patterns for Complex Products

Methodology for quantifying reliability was first developed for complex products. Suppose that a piece of equipment is placed on test, is run until it fails, and the failure time is recorded. The equipment is repaired and again placed on test, and the time of the next failure is recorded. The procedure is repeated to accumulate the data shown in Table 20.13. The failure rate is calculated, for equal time intervals, as the number of failures per unit of time. When the failure rate is plotted against time, the result (Fig. 20.19) often follows a familiar pattern of failure known as the bathtub curve. Three periods are apparent that differ in the frequency of failure and in the failure causation pattern:

- *The infant mortality period*. This period is characterized by high failure rates that show up early in use (see the lower half of Fig. 20.18). Commonly, these failures are the result of blunders in design or manufacture, misuse, or misapplication. Once corrected, these failures usually do not occur again (e.g., an oil hole that is not drilled). Sometimes it is possible to "debug" the product by a simulated use test or by overstressing (in electronics this is known as burn-in). The weak units still fail, but the failure takes place in the test rig rather than in service. O'Connor (1995) explains the use of burn-in tests and environmental screening tests.
- *The constant-failure-rate period*. Here the failures result from the limitations inherent in the design, changes in the environment, and accidents caused by use or maintenance.

Time of Failure, Infant Mortality Period		Time of Failure, Constant Failure Rate Period		Time of Failure, Wear-Out Period	
1.0	7.2	28.1	60.2	100.8	125.8
1.2	7.9	28.2	63.7	102.6	126.6
1.3	8.3	29.0	64.6	103.2	127.7
2.0	8.7	29.9	65.3	104.0	128.4
2.4	9.2	30.6	66.2	104.3	129.2
2.9	9.8	32.4	70.1	105.0	129.5
3.0	10.2	33.0	71.0	105.8	129.9
3.1	10.4	35.3	75.1	106.5	
3.3	11.9	36.1	75.6	110.7	
3.5	13.8	40.1	78.4	112.6	
3.8	14.4	42.8	79.2	113.5	
4.3	15.6	43.7	84.1	114.8	
4.6	16.2	44.5	86.0	115.1	
4.7	17.0	50.4	87.9	117.4	
4.8	17.5	51.2	88.4	118.3	
5.2	19.2	52.0	89.9	119.7	
5.4		53.3	90.8	120.6	
5.9		54.2	91.1	121.0	
6.4		55.6	91.5	122.9	
6.8		56.4	92.1	123.3	
6.9		58.3	97.9	124.5	

Source: *Quality Planning and Analysis*, Copyright 2007. Used by permission.

TABLE 20.13 Failure History for a Unit

The accidents can be held down by good control of operating and maintenance procedures. However, a reduction in the failure rate requires basic redesign.

- *The wear-out period.* These failures are due to old age (e.g., a metal becomes brittle or insulation dries out). A reduction in failure rates requires preventive replacement of these dying components before they result in catastrophic failure.

The top portion of Fig. 20.19 shows the corresponding Weibull plot when $\alpha = 2.6$ was applied to the original data (Table 20.14). The values of the shape parameter, b, were approximately 0.5, 1.0, and 6.0, respectively. A shape parameter less than 1.0 indicates a decreasing failure rate, a value of 1.0 a constant failure rate, and a value greater than 1.0 an increasing failure rate.

The Distribution of Time Between Failures

Users desire low failure rates during the infant mortality period, and after this are concerned with the length of time that a product will perform without failure. Thus, for repairable products, the time between failures (TBF) is a critical characteristic. The variation in time between failures can be studied statistically. The corresponding characteristic for non-repairable products is usually called the time to failure.

When the failure rate is constant, the distribution of time between failures is distributed exponentially. Consider the 42 failure times in the constant failure rate portion of Table 20.13.

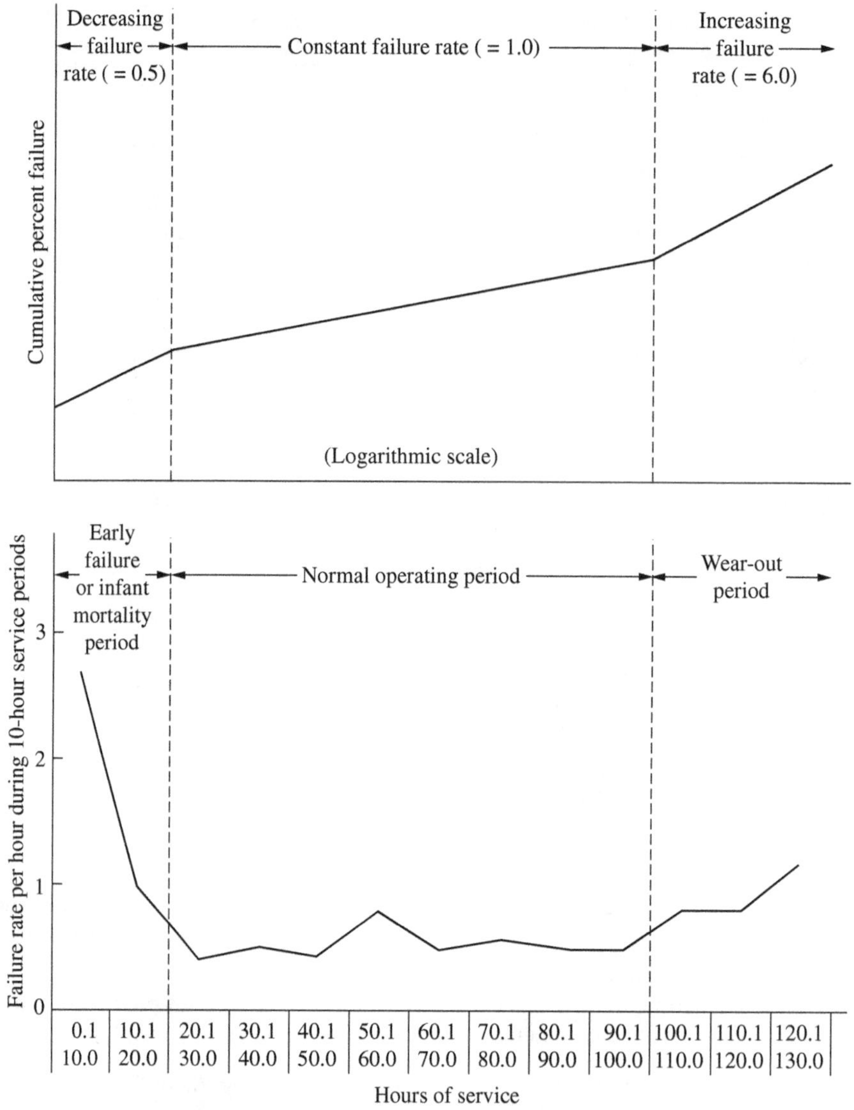

FIGURE 20.19 Failure rate versus time. (*Quality Planning and Analysis, Copyright 2007. Used by permission.*)

The time between failures for successive failures can be tallied, and the 41 resulting TBFs can be formed into the frequency distribution shown in Fig. 20.20a. The distribution is roughly exponential in shape, indicating that when the failure rate is constant, the distribution of time between failures (not mean time between failures) is exponential. This distribution is the basis of the exponential formula for reliability.

The Exponential Formula for Reliability

The distribution of TBF indicates the chance of failure-free operation for the specified time period. The chance of obtaining failure-free operation for a specified time period or longer

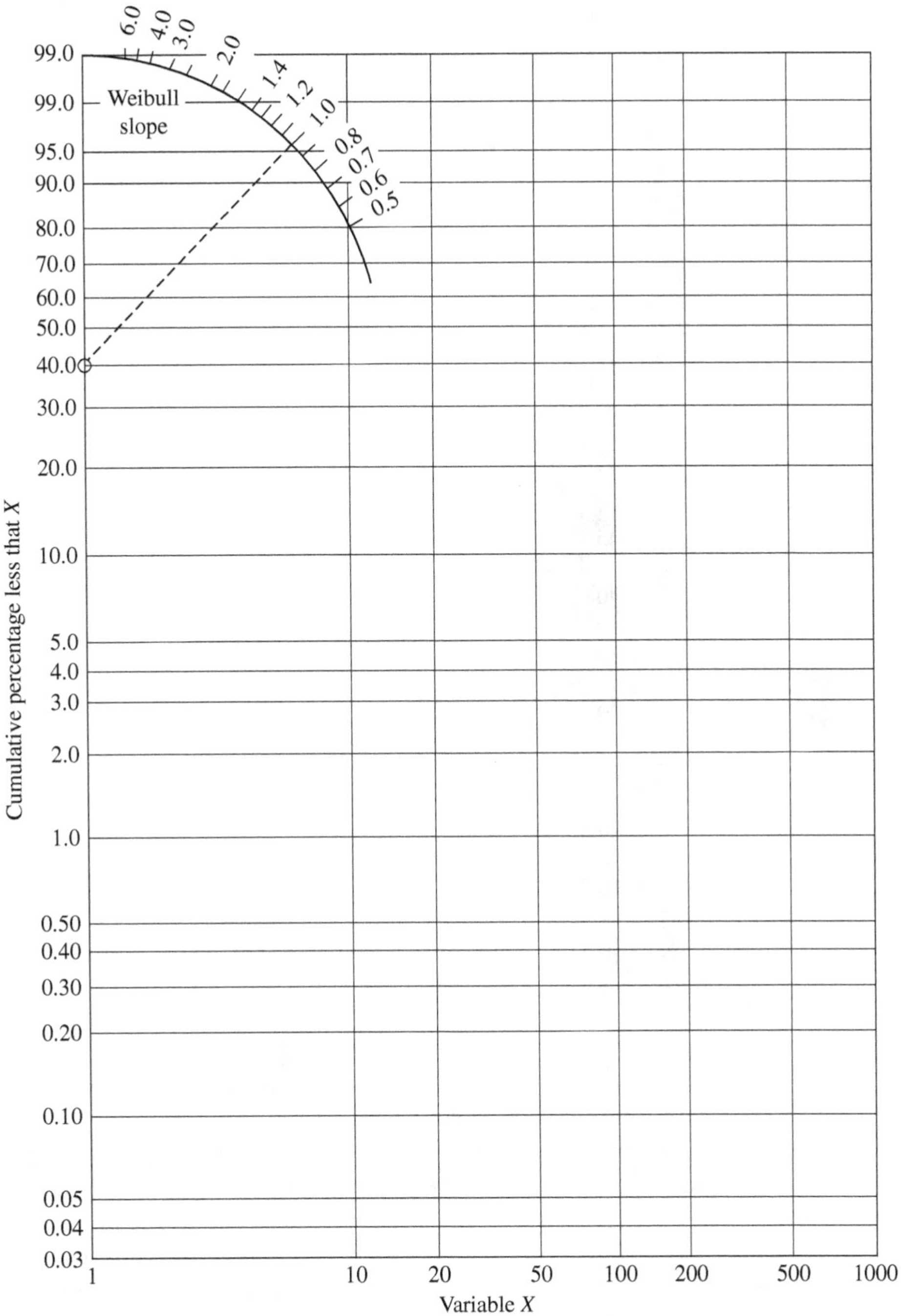

Source: Quality Planning and Analysis, Copyright 2007. Used by permission.

Table 20.14 Weibull Paper

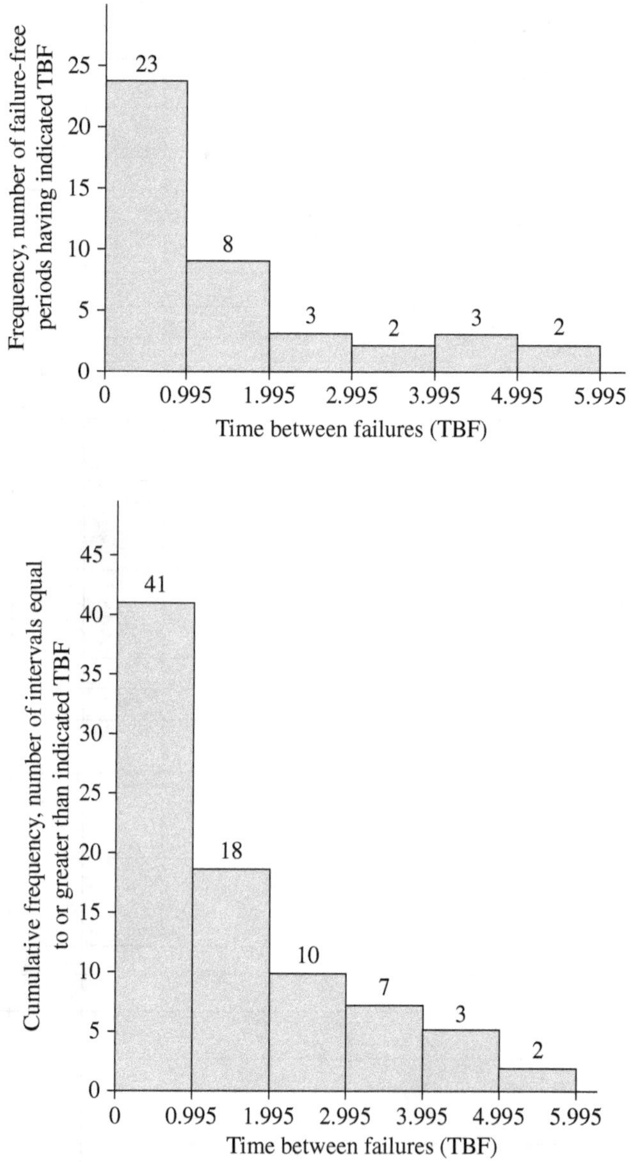

FIGURE 20.20 (a) Histogram of TBF. (*Quality Planning and Analysis, Copyright 2007. Used by permission.*) (b) Cumulative histogram of TBF. (*Quality Planning and Analysis, Copyright 2007. Used by permission.*)

can be shown by changing the TBF distribution to a distribution showing the number of intervals equal to or greater than a specified time length (Fig. 20.20b). If the frequencies are expressed as relative frequencies, they become estimates of the probability of survival. When the failure rate is constant, the probability of survival (or reliability) is

$$P_s = R = e^{-t/\mu} = e^{-t\lambda}$$

Where $P_s = R$ = probability of failure-free operation for a time period equal to or greater than t

$e = 2.718$

t = specified period of failure-free operation

μ = mean time between failures (the mean of TBF distribution)

λ = failure rate (the reciprocal of μ)

Note that this formula is simply the exponential probability distribution rewritten in terms of reliability.

Problem A washing machine requires 30 minutes to clean a load of clothes. The mean time between failures of the machine is 100 hours. Assuming a constant failure rate, what is the chance of the machine completing a cycle without failure?

Solution Applying the exponential formula, we obtain

$$R = e^{-t/\mu} = e^{-0.5/100} = 0.995$$

There is a 99.5 percent chance of completing a washing cycle.

How about the assumption of a constant failure rate? In practice, sufficient data usually are not available to evaluate the assumption. However, experience suggests that this assumption often is true, particularly when (1) infant mortality types of failures have been eliminated before delivery of the product to the user, and (2) the user replaces the product or specific components before the wear-out phase begins.

The Meaning of Mean Time Between Failures

Confusion surrounds the meaning of mean time between failures (MTBF). Further explanation is warranted:

- The MTBF is the mean (or average) time between successive failures of a product. This definition assumes that the product in question can be repaired and placed back into operation after each failure. For nonrepairable products, the term "mean time to failure" (MTTF) is used.
- If the failure rate is constant, the probability that a product will operate without failure for a time equal to or greater than it's MTBF is only 37 percent. This outcome is based on the exponential distribution (R is equal to 0.37 when t is equal to the MTBF). This result is contrary to the intuitive feeling that there is a 50-50 chance of exceeding an MTBF.
- MTBF is not the same as "operating life," "service life," or other indexes, which generally connote overhaul or replacement time.
- An increase in an MTBF does not result in a proportional increase in reliability (the probability of survival). If $t = 1$ hour, the following table shows the MTBF required to obtain various reliabilities.

MTBF	*R*
5	0.82
10	0.90
20	0.95
100	0.99

A fivefold increase in MTBF from 20 to 100 hours is necessary to increase the reliability by 4 percentage points compared with a doubling of the MTBF from 5 to 10 hours to get an 8 percentage point increase in reliability.

MTBF is a useful measure of reliability, but it is not correct for all applications.

The Relationship Between Part and System Reliability

It often is assumed that system reliability (i.e., the probability of survival, P_s) is the product of the individual reliabilities of the n parts within the system:

$$P_s = P_1 P_2 \ldots P_n$$

For example, if a communications system has four subsystems with reliabilities of 0.970, 0.989, 0.995, and 0.996, the system reliability is the product, or 0.951. The formula assumes that (1) the failure of any part causes failure of the system and (2) the reliabilities of the parts are independent of one another (i.e., the reliability of one part does not depend on the functioning of another part).

These assumptions are not always true, but in practice, the formula serves two purposes. First, it shows the effect of increased complexity of equipment on overall reliability. As the number of parts in a system increases, the system reliability decreases dramatically (see Fig. 20.21). Second, the formula often is a convenient approximation that can be refined as information on the interrelationships of the parts becomes available.

When it can be assumed that (1) the failure of any part causes system failure, (2) the parts are independent, and (3) each part follows an exponential distribution, then

$$P_s = e^{-t_1\lambda_1} e^{-t_2\lambda_2} \ldots e^{-t_n\lambda_n}$$

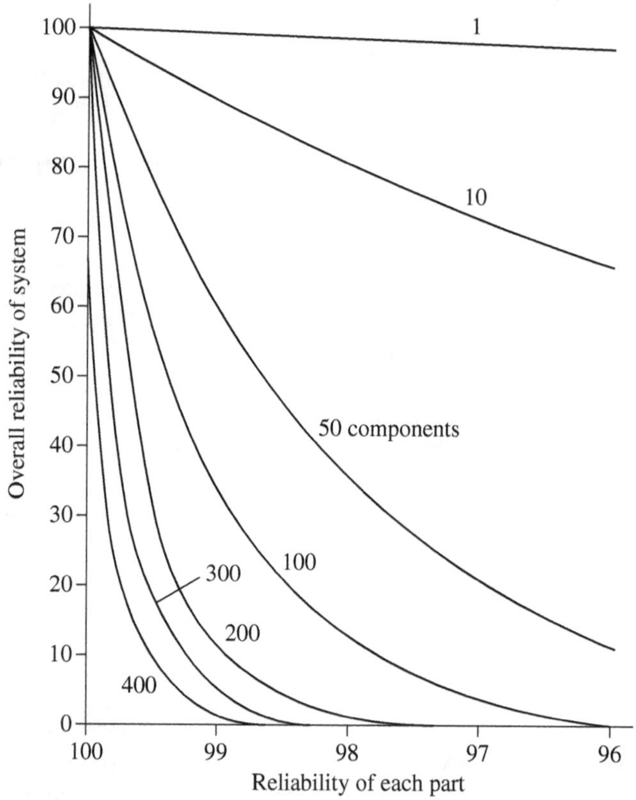

FIGURE 20.21 Relationship between part and system reliability. (*Quality Planning and Analysis, Copyright 2007. Used by permission.*)

Further, if t is the same for each part,

$$P_s = e^{-1\Sigma\lambda}$$

Thus, when the failure rate is constant (and therefore the exponential distribution can be applied), the reliability of a system can be predicted based on the addition of the part failure rates (see the section "Predicting Reliability During Design," next).

Sometimes designs are planned with redundancy so that the failure of one part will not cause system failure. Redundancy is an old (but still useful) design technique invented long before the advent of reliability prediction techniques. However, the designer can now predict the effect of redundancy on system reliability in quantitative terms.

Redundancy is the existence of more than one element for accomplishing a given task, where all elements must fail before there is an overall failure of the system. In parallel redundancy (one of several types of redundancy), two or more elements operate at the same time to accomplish the task, and any single element is capable of handling the job itself in case of failure of the other elements. When parallel redundancy is used, the overall reliability is calculated as follows:

$$P_s = 1 - (1 - \mathrm{P1})n$$

where P_s = reliability of the system
P1 = reliability of the individual elements in the redundancy
n = number of identical redundant elements

Problem Suppose that a unit has a reliability of 99.0 percent for a specified mission time. If two identical units are used in parallel redundancy, what overall reliability will be expected?

Solution Applying the formula above, we obtain

$$R = 1 - (1 - 0.99)(1 - 0.99) = 0.9999, \text{ or } 99.99 \text{ percent}$$

Predicting Reliability During Design

Reliability prediction methods continue to evolve, but include such standards as failure mode and effects analysis (FMEA) and testing. Ireson et al. (1996) provide an extensive discussion of reliability prediction, and should be consulted beyond the methods discussed in this handbook.

The following steps make up a reliability prediction method:

1. Define the product and its functional operation. The system, subsystems, and units must be precisely defined in terms of their functional configurations and boundaries. This precise definition is aided by preparation of a functional block diagram that shows the subsystems and lower-level products, their interrelationships, and the interfaces with other systems. Given a functional block diagram and a well-defined statement of the functional requirements of the product, the conditions that constitute failure or unsatisfactory performance can be defined.
2. Prepare a reliability block diagram. For systems in which there are redundancies or other special interrelationships among parts, a reliability block diagram is useful. This diagram is similar to a functional block diagram, but the reliability block diagram shows exactly what must function for successful operation of the system. The diagram shows redundancies and alternative modes of operation. The reliability block diagram is the foundation for developing the probability model for reliability. O'Connor (1995) provides further discussion.

3. Develop the probability model for predicting reliability. A simple model may add only failure rates; a complex model can account for redundancies and other conditions.
4. Collect information relevant to parts reliability. The data include information such as parts function, parts ratings, stresses, internal and external environments, and operating time. Many sources of failure-rate information state failure rates as a function of operating parameters. For example, failure rates for fixed ceramic capacitors are stated as a function of (1) expected operating temperature and (2) the ratio of the operating voltage to the rated voltage. Such data show the effect of derating (assigning a part to operate below its rated voltage) on reducing the failure rate.
5. Select parts reliability data. The required parts data consist of information on catastrophic failures and on tolerance variations with respect to time under known operating and environmental conditions. Acquiring these data is a major problem for the designer because there is no single reliability data bank comparable to handbooks such as those for physical properties of materials. Instead, the designer must build a data bank by securing reliability data from a variety of sources:
   - Field performance studies conducted under controlled conditions
   - Specified life tests
   - Data from parts manufacturers or industry associations
   - Customers' parts qualification and inspection tests
   - Government agency data banks such as the Government Industry Data Exchange Program (GIDEP) and the Reliability Information Analysis Center (RIAC)

Combine all of the above to obtain the numerical reliability prediction.

Ireson et al. (1996) and O'Connor (1995) are excellent references for reliability prediction. Included are the basic methods of prediction, repairable versus non-repairable systems, electronic and mechanical reliability, reliability testing, and software reliability. Box and Draper (1969) provides extensive discussion of reliability data analysis, including topics such as censored life data (not all test units have failed during the test) and accelerated-life test data analysis. Dodson (1999) explains how the use of computer spreadsheets can simplify reliability modeling using various statistical distributions.

Reliability prediction techniques based on component failure data to estimate system failure rates have generated controversy. Jones and Hayes (1999) present a comparison of predicted and observed performance for five prediction techniques using parts count analyses. The predictions differed greatly from observed field behavior and from each other. The standard ANSI/IEC/ASQC D60300-3-1-1997 (Dependability Management—Part 3: Application Guide—Section 1—Analysis Techniques for Dependability) compares five analysis techniques: FMEA/FMECA, fault tree analysis, reliability block diagram, Markov analysis, and parts count reliability prediction.

The reliability of a system evolves during design, development, testing, production, and field use. The concept of reliability growth assumes that the causes of product failures are discovered and action is taken to remove the causes, thus resulting in improved reliability of future units ("test, analyze, and fix"). Reliability growth models provide predictions of reliability due to such improvements. For elaboration, see O'Connor (1995). Also, ANSI/IEC/ASQC D601164-1997 (Reliability Growth—Statistical Test and Estimation Methods) and the related IEC 61164 Ed. 2.0 (2004) (Reliability growth—Statistical test and estimation methods) describe methods of estimating reliability growth.

Predicting Reliability Based on the Exponential Distribution

When the failure rate is constant and when study of a functional block diagram reveals that all parts must function for system success, then reliability is predicted to be the simple total of failure rates. An example of a subsystem prediction is shown in Table 20.15. The prediction for the subsystem is made by adding the failure rates of the parts; the MTBF is then calculated as the reciprocal of the failure rate.

For further discussion of reliability prediction, including an example for an electronic system, see Gryna et al. (2007).

Predicting Reliability Based on the Weibull Distribution

Prediction of overall reliability based on the simple addition of component failure rates is valid only if the failure rate is constant. When this assumption cannot be made, an alternative approach based on the Weibull distribution can be used.

1. Graphically, use the Weibull distribution to predict the reliability R for the time period specified. $R = 100 - \%$ failure. Do this for each component (Table 20.14).
2. Combine the component reliabilities using the product rule and/or redundancy formulas to predict system reliability.

Predictions of reliability using the exponential distribution or the Weibull distribution are based on reliability as a function of time. Next we consider reliability as a function of stress and strength.

Reliability as a Function of Applied Stress and Strength

Failures are not always a function of time. In some cases, a part will function indefinitely if its strength is greater than the stress applied to it. The terms "strength" and "stress" here are used in the broad sense of inherent capability and operating conditions applied to a part, respectively. For example, operating temperature is a critical parameter, and the maximum

Part Description	Quantity	Generic Failure Rate per Million Hours	Total Failure Rates per Million Hours
Heavy-duty ball bearing	6	14.4	86.4
Brake assembly	4	16.8	67.2
Cam	2	0.016	0.032
Pneumatic hose	1	29.28	29.28
Fixed displacement pump	1	1.464	1.464
Manifold	1	8.80	65.0
Guide pin	5	13.0	65.0
Control valve	1	15.20	15.20
Total assembly failure rate			273.376

MTBF = 1/0.000273376 = 3.657.9 hours

Source: Adapted from Ireson et al., p. 19.9. *Quality Planning and Analysis*, Copyright 2007. Used by permission.

Table 20.15 Example of Mechanical Parts and Subsystem Failure Rates

expected temperature is 145°F (63°C). Further, capability is indicated by a strength distribution having a mean of 172°F (78°C) and a standard deviation of 13°F (7°C) (Fig. 20.22). With knowledge of only the maximum temperatures, the safety margin is:

$$\frac{172-145}{13}=2.08$$

The safety margin says that the average strength is 2.08 standard deviations above the maximum expected temperature of 145°F (63°C). Table 20.16 can be used to calculate a reliability of 0.981 [the area beyond 145°F (63°C)].

This calculation illustrates the importance of variation in addition to the average value during design. Designers have always recognized the existence of variation by using a safety factor in design. However, the safety factor is often defined as the ratio of average strength to the worst stress expected.

Note that in Fig. 20.23, all designs have the same safety factor. Also note that the reliability (probability of a part having strength greater than the stress) varies considerably. Thus, the uncertainty often associated with this definition of safety factor is, in part, due to its failure to reflect the variation in both strength and stress. Such variation is partially reflected in a safety margin, defined as:

$$\frac{\text{Average strength} - \text{worst stress}}{\text{Standard deviation of strength}}$$

This recognizes the variation in strength but is conservative because it does not recognize a variation in stress.

Availability

Availability has been defined as the probability that a product, when used under given conditions, will perform satisfactorily when called upon. Availability considers the operating time of the product and the time required for repairs. Idle time, during which the product is not needed, is excluded.

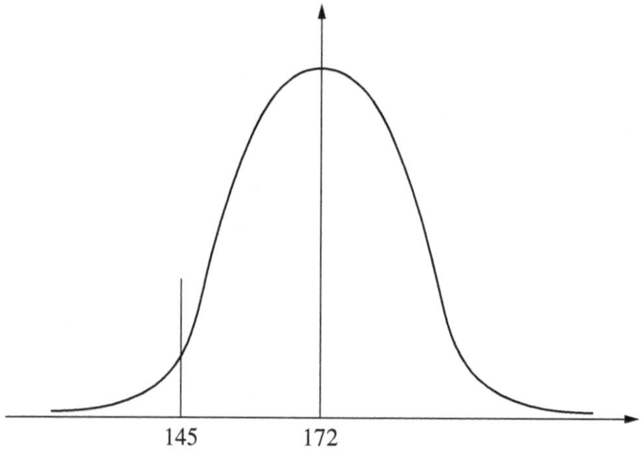

FIGURE 20.22 Distribution of strength.

TABLE A
Normal distribution

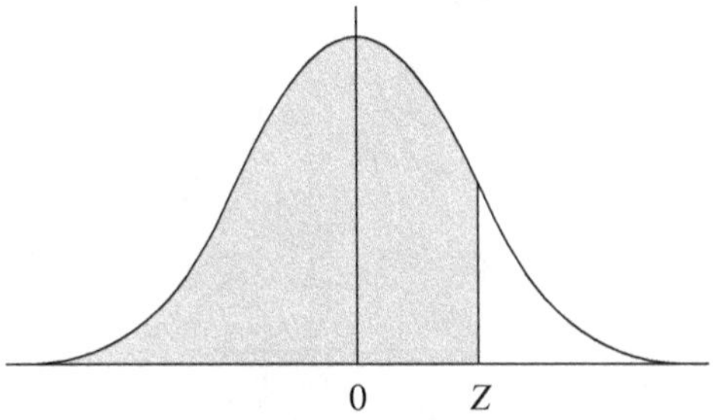

Proportion of total areas under the curve from $-\infty$ to $Z = \frac{X - \mu}{\sigma}$, To illustrate when $Z = 2$, the probability is .9773 of obtaining a value equal to or less then X.

Z	0.09	0.08	0.07	0.06	0.05	0.04	0.03	0.02	0.01	0.00
–3.0	.00100	.00104	.00107	.00111	.00114	.00118	.00122	.00126	.00131	.00135
–2.9	.0014	.0014	.0015	.0015	.0016	.0016	.0017	.0017	.0018	.0019
–2.8	.0019	.0020	.0021	.0021	.0022	.0023	.0023	.0024	.0025	.0026
–2.7	.0026	.0027	.0028	.0029	.0030	.0031	.0032	.0033	.0034	.0035
–2.6	.0036	.0037	.0038	.0039	.0040	.0041	.0043	.0044	.0045	.0047
–2.5	.0048	.0049	.0051	.0052	.0054	.0055	.0057	.0059	.0060	.0062
–2.4	.0064	.0066	.0068	.0069	.0071	.0073	.0075	.0078	.0080	.0082
–2.3	.0084	.0087	.0089	.0091	.0094	.0096	.0099	.0102	.0104	.0107
–2.2	.0110	.0113	.0116	.0119	.0122	.0125	.0129	.0132	.0136	.0139
–2.1	.0143	.0146	.0150	.0154	.0158	.0162	.0166	.0170	.0174	.0179
–2.0	.0183	.0188	.0192	.0197	.0202	.0207	.0212	.0217	.0222	.0228
–1.9	.0233	.0239	.0244	.0250	.0256	.0262	.0268	.0274	.0281	.0287
–1.8	.0294	.0301	.0307	.0314	.0322	.0329	.0336	.0344	.0351	.0359
–1.7	.0367	.0375	.0384	.0392	.0401	.0409	.0418	.0427	.0436	.0446
–1.6	.0455	.0465	.0475	.0485	.0495	.0505	.0516	.0526	.0537	.0548
–1.5	.0559	.0571	.0582	.0594	.0606	.0618	.0630	.0643	.0655	.0668
–1.4	.0681	.0694	.0708	.0721	.0735	.0749	.0764	.0778	.0793	.0808
–1.3	.0823	.0838	.0853	.0869	.0885	.0901	.0918	.0934	.0951	.0968
–1.2	.0985	.1003	.1020	.1038	.1057	.1075	.1093	.1112	.1131	.1151
–1.1	.1170	.1190	.1210	.1230	.1251	.1271	.1292	.1314	.1335	.1357

Source: Quality Planning and Analysis, Copyright 2007. Used by permission.

TABLE 20.16 Normal Distribution

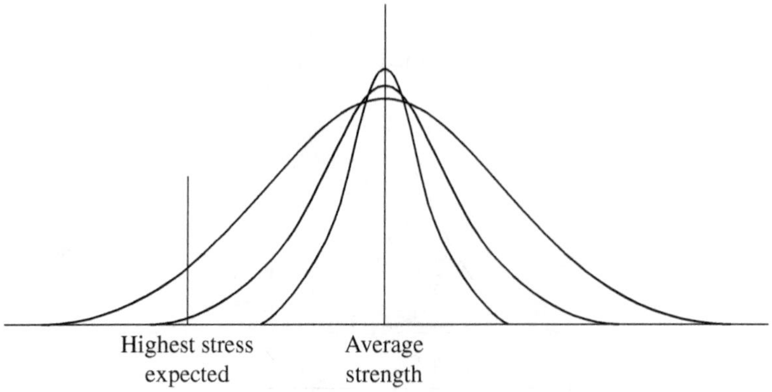

FIGURE 20.23 Variation and safety factor. (*Quality Planning and Analysis, Copyright 2007. Used by permission.*)

Availability is calculated as the ratio of operating time to operating time plus downtime. However, downtime can be viewed in two ways:

1. *Total downtime.* This period includes active repair (diagnosis and repair time), preventive maintenance time, and logistics time (time spent waiting for personnel, spare parts, etc.). When total downtime is used, the resulting ratio is called operational availability (A_0).
2. *Active repair time.* The resulting ratio is called intrinsic availability (A_i). Under certain conditions, availability can be calculated as:

$$A_0 = \frac{\text{MTBF}}{\text{MTBF} + \text{MDT}} \quad \text{and} \quad A_i = \frac{\text{MTBF}}{\text{MTBF} + \text{MTTR}}$$

where MTBF = mean time between failures
MDT = mean downtime
MTTR = mean time to repair

This is known as the steady-state formula for availability. The steady-state formula for availability has the virtue of simplicity. However, the formula is based on several assumptions that are not always met in the real world. The assumptions are

- The product is operating in the constant failure rate period of the overall life. Thus, the failure-time distribution is exponential.
- The downtime or repair-time distribution is exponential.
- Attempts to locate system failures do not change the overall system failure rate.
- No reliability growth occurs (such growth might be due to design improvements or through debugging of bad parts).
- Preventive maintenance is scheduled outside the time frame included in the availability calculation.

More precise formulas for calculating availability depend on operational conditions and statistical assumptions. These formulas are discussed by Ireson et al. (1996).

Setting Specification Limits

A major step in the development of physical products is the conversion of product features into dimensional, chemical, electrical, and other characteristics of the product. Thus, a heating system for an automobile will have many characteristics for the heater, air ducts, blower assembly, engine coolant, etc.

For each characteristic, the designer must specify (1) the desired average (or "nominal value") and (2) the specification limits (or "tolerance limits") above and below the nominal value that individual units of product must meet. These two elements relate to parameter design and tolerance design, as discussed in Gryna et al. (2007).

The specification limits should reflect the functional needs of the product, manufacturing variability, and economic consequences. These three aspects are addressed in the next three sections. For greater depth in the statistical treatment of specification limits, see Anand (1996).

Specification Limits and Functional Needs

Sometimes data can be developed to relate product performance to measurements of a critical component. For example, a thermostat may be required to turn on and shut off a power source at specified low and high temperature values, respectively. A number of thermostat elements are built and tested. The prime recorded data are (1) turn-on temperature, (2) shutoff temperature, and (3) physical characteristics of the thermostat elements. We can then prepare scatter diagrams (Fig. 20.24) and regression equations to help establish critical component tolerances on a scientific basis within the confidence limits for the numbers involved. Ideally, the sample size is sufficient, and the data come from a statistically controlled process—two conditions that are both rarely achieved. O'Connor (1995) explains how this approach can be related to the Taguchi approach to develop a more robust design.

Specification Limits and Manufacturing Variability

Generally, designers will not be provided with information on process capability. Their problem will be to obtain a sample of data from the process, calculate the limits that the process can meet, and compare these to the limits they were going to specify. If they do not

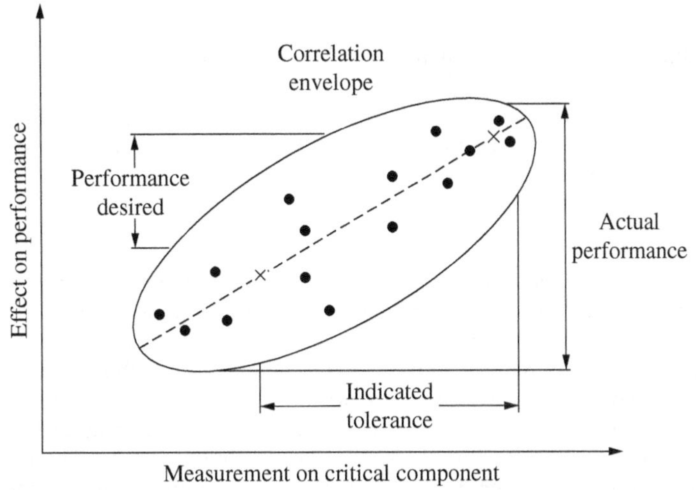

Figure 20.24 Approach to functional tolerancing. (*Quality Planning and Analysis, Copyright 2007. Used by permission.*)

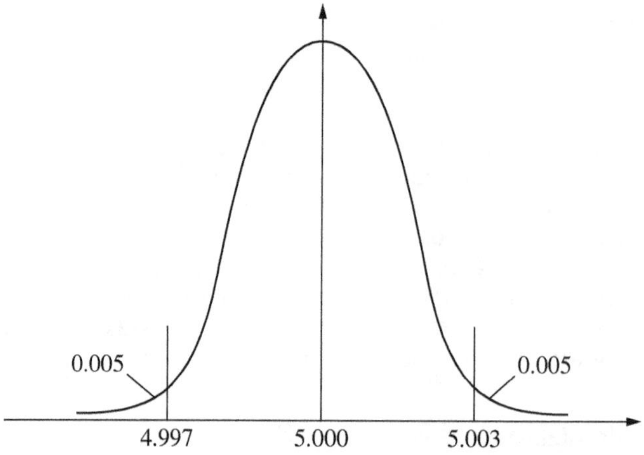

Figure 20.25 Distribution with 99 percent limits. (*Quality Planning and Analysis, Copyright 2007. Used by permission.*)

have any limits in mind, the capability limits calculated from process data provide a set of limits that are realistic from the viewpoint of producibility. These limits must then be evaluated against the functional needs of the product.

Statistically, the problem is to predict the limits of variation of individual items in the total population based on a sample of data. For example, suppose that a product characteristic is normally distributed with a population average of 5.000 in (12.7 cm) and a population standard deviation of 0.001 in (0.00254 cm). Limits can then be calculated to include any given percentage of the population. Figure 20.25 shows the location of the 99 percent limits. Table 20.16 indicates that 2.575 standard deviations will include 99 percent of the population. Thus, in this example, a realistic set of tolerance limits would be:

$$5.000 \pm 2.575(0.001) = \frac{5.003}{4.997}$$

Ninety-nine percent of the individual pieces in the population will have values between 4.997 and 5.003.

In practice, the average and standard deviation of the population are not known but must be estimated from a sample of product from the process. As a first approximation, tolerance limits are sometimes set at:

$$\overline{X} \pm 3s$$

Here, the average $\overline{X}$ and standard deviation s of the sample are used directly as estimates of the population values. If the true average and standard deviation of the population happen to be equal to those of the sample, and if the characteristic is normally distributed, then 99.73 percent of the pieces in the population will fall within the limits calculated. These limits are frequently called natural tolerance limits (limits that recognize the actual variation of the process and therefore are realistic). This approximation ignores the possible error in both the average and standard deviation as estimated from the sample.

Methodology has been developed for setting tolerance limits in a more precise manner. For example, formulas and tables are available for determining tolerance limits based on a normally distributed population. Table 20.17 provides factors for calculating tolerance limits

Tolerance Factors for Normal Distributions (Two Sided)										
P	γ = 0.75					γ = 0.90				
N	0.75	0.90	0.95	0.99	0.999	0.75	0.90	0.95	0.99	0.999
2	4.498	6.301	7.414	9.531	11.920	11.407	15.978	18.800	24.167	30.227
3	2.501	3.538	4.187	5.431	6.844	4.132	5.847	6.919	8.974	11.309
4	2.035	2.892	3.431	4.471	5.657	2.932	4.166	4.943	6.440	8.149
5	1.825	2.599	3.088	4.033	5.117	2.454	3.494	4.152	5.423	6.879
6	1.704	2.429	2.889	3.779	4.802	2.196	3.131	3.723	4.870	6.188
7	1.624	2.318	2.757	3.611	4.593	2.034	2.902	3.452	4.521	5.750
8	1.568	2.238	2.663	3.491	4.444	1.921	2.743	3.264	4.278	5.446
9	1.525	2.178	2.593	3.400	4.330	1.839	2.626	3.125	4.098	5.220
10	1.492	2.131	2.537	3.328	4.241	1.775	2.535	3.018	3.959	5.046
11	1.465	2.093	2.493	3.271	4.169	1.724	2.463	2.933	3.849	4.906
12	1.443	2.062	2.456	3.223	4.110	1.683	2.404	2.863	3.758	4.792
13	1.425	2.036	2.424	3.183	4.059	1.648	2.355	2.805	3.682	4.697
14	1.409	2.013	2.398	3.148	4.016	1.619	2.314	2.756	3.618	4.615
15	1.395	1.994	2.375	3.118	3.979	1.594	2.278	2.713	3.562	4.545
16	1.383	1.977	2.355	3.092	3.946	1.572	2.246	2.676	3.514	4.484
17	1.372	1.962	2.337	3.069	3.917	1.552	2.219	2.643	3.471	4.430
18	1.363	1.948	2.321	3.048	3.891	1.535	2.194	2.614	3.433	4.382
19	1.355	1.936	2.307	3.030	3.867	1.520	2.172	2.588	3.399	4.339
20	1.347	1.925	2.294	3.013	3.846	1.506	2.152	2.564	3.368	4.300
21	1.340	1.915	2.282	2.998	3.827	1.493	2.135	2.543	3.340	4.264
22	1.334	1.906	2.271	2.984	3.809	1.482	2.118	2.524	3.315	4.232
23	1.328	1.898	2.261	2.971	3.793	1.471	2.103	2.506	3.292	4.203
24	1.322	1.891	2.252	2.950	3.778	1.462	2.089	2.480	3.270	4.176
25	1.317	1.883	2.244	2.948	3.764	1.453	2.077	2.474	3.251	4.151
26	1.313	1.877	2.236	2.938	3.751	1.444	2.065	2.460	3.232	4.127
27	1.309	1.871	2.229	2.929	3.740	1.437	2.054	2.447	3.215	4.106
30	1.297	1.855	2.210	2.904	3.708	1.417	2.025	2.413	3.170	4.049
35	1.283	1.834	2.185	2.871	3.667	1.390	1.988	2.368	3.112	3.974
40	1.271	1.818	2.166	2.846	3.635	1.370	1.959	2.334	3.066	3.917
100	1.218	1.742	2.075	2.727	3.484	1.275	1.822	1.172	2.854	3.646
500	1.177	1.683	2.006	2.636	3.368	1.201	1.717	2.046	2.689	3.434
1000	1.169	1.671	1.992	2.617	3.344	1.185	1.695	2.019	2.654	3.390
∞	1.150	1.645	1.960	2.576	3.291	1.150	1.645	1.960	2.576	3.291

TABLE 20.17 Tolerance Factors for Normal Distributions

Tolerance Factors for Normal Distributions (Two Sided)									
$\gamma = 0.75$					$\gamma = 0.90$				
0.75	**0.90**	**0.95**	**0.99**	**0.999**	**0.75**	**0.90**	**0.95**	**0.99**	**0.999**
22.858	32.019	37.674	48.430	60.573	114.363	160.363	188.491	242.300	303.054
5.922	8.380	9.916	12.861	16.208	13.378	18.930	22.401	29.055	36.616
3.779	5.369	6.370	8.299	10.502	6.614	9.398	11.150	14.527	18.383
3.002	4.275	5.079	6.634	8.415	4.643	6.612	7.855	10.260	13.015
2.604	3.712	4.414	5.775	7.337	3.743	5.337	6.345	8.301	10.548
2.361	3.369	4.007	5.248	6.676	3.233	4.613	5.488	7.187	9.142
2.197	3.136	3.732	4.891	6.226	2.905	4.147	4.936	6.468	8.234
2.078	2.967	3.532	4.631	5.899	2.677	3.822	4.550	5.966	7.600
1.987	2.839	3.379	4.433	5.649	2.508	3.582	4.265	5.594	7.129
1.916	2.737	3.259	4.277	5.452	2.378	3.397	4.045	5.308	6.766
1.858	2.655	3.162	4.150	5.291	2.274	3.250	3.870	5.079	6.477
1.810	2.587	3.081	4.044	5.158	2.190	3.130	3.727	4.893	6.240
1.770	2.529	3.012	3.955	5.045	2.120	3.029	3.608	4.737	6.043
1.735	2.480	2.954	3.878	4.949	2.060	2.945	3.507	4.605	5.876
1.705	2.437	2.903	3.812	4.865	2.009	2.872	3.421	4.492	5.732
1.679	2.400	2.858	3.754	4.791	1.965	2.808	3.345	4.393	5.607
1.655	2.366	2.819	3.702	4.725	1.926	2.753	3.279	4.307	5.497
1.635	2.337	2.784	3.656	4.667	1.891	2.703	3.221	4.230	5.399
1.616	2.310	2.752	3.615	4.614	1.860	2.659	3.168	4.161	5.312
1.599	2.286	2.723	3.577	4.567	1.833	2.620	3.121	4.100	5.234
1.584	2.264	2.697	3.543	4.523	1.808	2.584	3.078	4.044	5.163
1.570	2.244	2.673	3.512	4.484	1.795	2.551	3.040	3.993	5.098
1.557	2.225	2.651	3.483	4.447	1.764	2.522	3.004	3.947	5.039
1.545	2.208	2.631	3.457	4.413	1.745	2.494	2.972	3.904	4.985
1.534	2.193	2.612	3.432	4.382	1.727	2.460	2.941	3.865	4.935
1.523	2.178	2.595	3.409	4.353	1.711	2.446	2.914	3.828	4.888
1.497	2.140	2.549	3.350	4.278	1.668	2.385	2.841	3.733	4.768
1.462	2.090	2.490	3.272	4.179	1.613	2.306	2.748	3.611	4.611
1.435	2.052	2.445	3.213	4.104	1.571	2.247	2.677	3.518	4.493
1.311	1.874	2.233	2.934	3.748	1.383	1.977	2.355	3.096	3.954
1.215	1.737	2.070	2.721	3.475	1.243	1.777	2.117	2.783	3.555
1.195	1.709	2.036	2.676	3.418	1.214	1.736	2.068	2.718	3.472
1.150	1.645	1.960	2.576	3.291	1.150	1.645	1.960	2.576	3.291

*Table H—"Tolerance factors for normal distributions" from *Selected Techniques of Statistical Analysis—OSRD* by C. Eisenhart, M. W. Hastay, and W. A. Wallis, Copyright 1947 by The McGraw-Hill Companies, Inc. Reprinted by permission of The McGraw-Hill Companies, Inc.

γ = confidence level

P = percentage of population within tolerance limits

N = number of values in sample

Source: Quality Planning and Analysis, Copyright 2007. Used by permission.

TABLE 20.17 Tolerance Factors for Normal Distributions (*Continued*)

that recognize the uncertainty in the sample mean and sample standard deviation. The tolerance limits are determined as:

$$\overline{X} \pm Ks$$

The factor K is a function of the confidence level desired, the percentage of the population to be included within the tolerance limits, and the number of data values in the sample.

For example, suppose that a sample of 10 resistors from a process yielded an average and standard deviation of 5.04 and 0.016, respectively. The tolerance limits are to include 99 percent of the population, and the tolerance statement is to have a confidence level of 95 percent. Referring to Table 20.17, the value of K is 4.433, and tolerance limits are then calculated as:

$$5.04 \pm 4.443(0.016) = \frac{5.11}{4.97}$$

We are 95 percent confident that at least 99 percent of the resistors in the population will have resistance between 4.97 and 5.11 Ω. Tolerance limits calculated in this manner are often called statistical tolerance limits. This approach is more rigorous than the 3s natural tolerance limits, but the two percentages in the statement are a mystery to those without a statistical background.

For products in some industries (e.g., electronics), the number of units outside of specification limits is stated in terms of parts per million (ppm). Thus, if limits are set at three standard deviations, 2700 ppm (100 to 99.73 percent) will fall outside the limits. For many applications (e.g., a personal computer with many logic gates), such a level is totally unacceptable. Table 20.18 shows the ppm for several standard deviations. These levels of ppm assume that the process average is constant at the nominal specification. A deviation from the nominal value will result in a higher ppm value. To allow for modest shifts in the process average, some manufacturers follow a guideline for setting specification limits at $\pm 6\sigma$.

Designers often must set tolerance limits with only a few measurements from the process (or more likely from the development tests conducted under laboratory conditions). In developing a paint formulation, for example, the following values of gloss were obtained: 76.5, 75.2, 77.5, 78.9, 76.1, 78.3, and 77.7. A group of chemists was asked where they would set a minimum specification limit. Their answer was 75.0—a reasonable answer for those without statistical knowledge. Figure 20.26 shows a plot of the data on normal probability paper. If the line is extrapolated to 75.0, the plot predicts that about 11 percent of the population will fall below 75.0, even though all of the sample data exceed 75.0. Of course, a larger

Number of Standard Deviations	Part per Million (ppm)
$\pm 3\sigma$	2700
$\pm 4\sigma$	63
$\pm 5\sigma$	0.57
$\pm 6\sigma$	0.002

*If the process is not centered and the mean shifts by up to 1.5σ, then $\pm 6\sigma$ will be 3.4 ppm.

Source: Quality Planning and Analysis, Copyright 2007. Used by permission.

Table 20.18 Standard Deviations and PPM (Centered Process)*

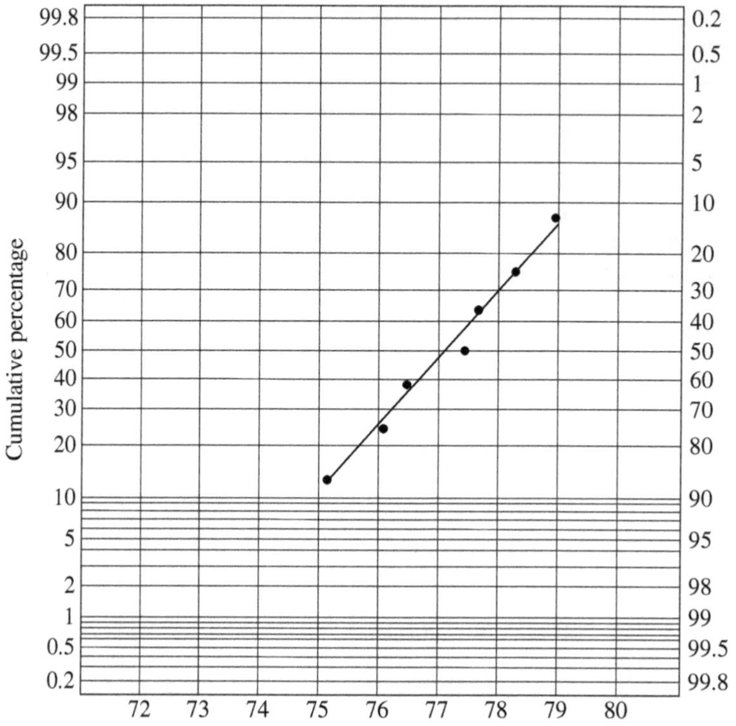

FIGURE 20.26 Probability plot of development data. (*Quality Planning and Analysis, Copyright 2007. Used by permission.*)

sample size is preferred and further statistical analyses could be made, but the plot provides a simple tool for evaluating a small sample of data.

All methods of setting tolerance limits based on process data assume that the sample of data represents a process that is sufficiently stable to be predictable. In practice, the assumption is often accepted without any formal evaluation. If sufficient data are available, the assumption should be checked with a control chart.

Statistical tolerance limits are sometimes confused with other limits used in engineering and statistics. Table 20.19 summarizes the distinctions among five types of limits (see also Box and Draper 1969, pp. 44.47–44.58).

Specifications Limits and Economic Consequences

In setting traditional specification limits around a nominal value, we assume that there is no monetary loss for product falling within specification limits. For product falling outside the specification limits, the loss is the cost of replacing the product.

Another viewpoint holds that any deviation from the nominal value causes a loss. Thus, there is an ideal (nominal) value that customers desire, and any deviation from this ideal results in customer dissatisfaction. This loss can be described by a loss function (Fig. 20.27).

Many formulas can predict loss as a function of deviation from the target. Taguchi proposes the use of a simple quadratic loss function:

$$L = k(X - T)^2$$

Name of Limit	Meaning
Tolerance	Set by the engineering design function to define the minimum and maximum values allowable for the product to work properly
Statistical tolerance	Calculated from process data to define the amount of variation that the process exhibits; these limits will contain a specified proportion of the total population
Prediction	Calculated from process data to define the limits which will contain all of *k* future observations
Confidence	Calculated from data to define an interval within which a population parameter lies
Control	Calculated from process data to define the limits of chance (random) variation around some central value

Source: Quality Planning and Analysis, Copyright 2007. Used by permission.

Table 20.19 Distinctions Among Limits

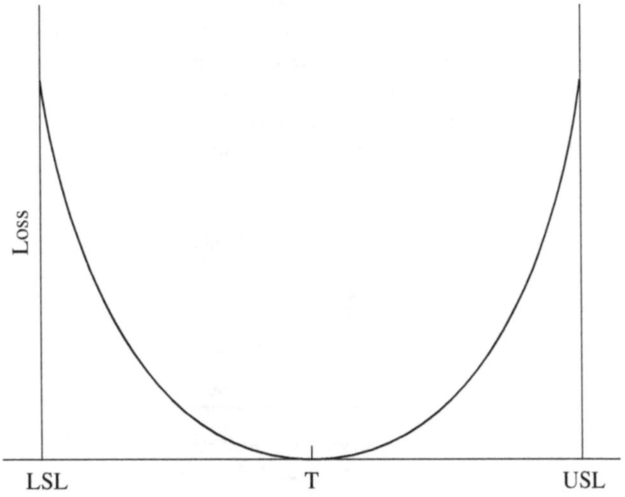

Figure 20.27 Loss function. (*Quality Planning and Analysis, Copyright 2007. Used by permission.*)

where L = loss in monetary terms
k = cost coefficient
X = value of quality characteristic
T = target value

Ross (1996) provides an example to illustrate how the loss function can help to determine specification limits. In automatic transmissions for trucks, shift points are designed to occur at a certain speed and throttle position. Suppose it costs the producer $100 to adjust a valve body under warranty when a customer complains of the shift point. Research indicates that the average customer would request an adjustment if the shift point is off from the nominal by 40 rpm transmission output speed on the first-to-second gear shift. The loss function is then:

$$\text{Loss} = k(X - T)^2$$
$$100 = k(40)^2$$
$$k = \$0.0625$$

This adjustment can be made at the factory at a lower cost, about $10. The loss function is now used to calculate the specification limits:

$$\$10 = 0.0625(X - T)^2$$
$$(X - T) = \pm 12.65 \text{ or } \pm 13 \text{ rpm}$$

The specification limits should be set at 13 rpm around the desired nominal value. If the transmission shift point is further than 13 rpm from the nominal, adjustment at the factory is less expensive than waiting for a customer complaint and making the adjustment under warranty in the field. Ross (1996) discusses how the loss function can be applied to set one-sided specification limits (e.g., a minimum value or a maximum value).

Specification Limits for Interacting Dimensions

Interacting dimensions mate or merge with other dimensions to create a final result. Consider the simple mechanical assembly shown in Fig. 20.28. The lengths of components A, B, and C are interacting dimensions because they determine the overall assembly length.

Suppose the components were manufactured to the specifications indicated in Fig. 20.28. A logical specification for the assembly length would be 3.500 + 0.0035, giving limits of 3.5035 and 3.4965. This logic may be verified from the two extreme assemblies shown in the following table:

Maximum	Minimum
1.001	0.999
0.5005	0.4995
2.002	1.998
3.5035	3.4965

The approach of adding component tolerances is mathematically correct, but is often too conservative. Suppose that about 1 percent of the pieces of component A are expected to be below the lower tolerance limit for component A and suppose the same for components B

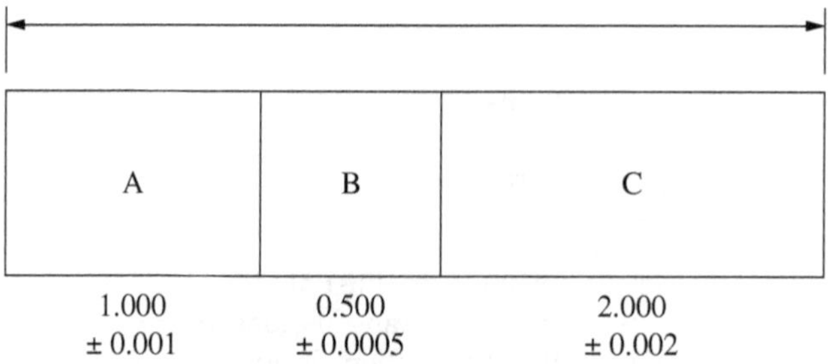

Figure 20.28 Mechanical assembly. (*Quality Planning and Analysis, Copyright 2007. Used by permission.*)

and C. If a component A is selected at random, there is, on average, 1 chance in 100 that it will be on the low side, and similarly for components B and C. The key point is this: If assemblies are made at random and if the components are manufactured independently, then the chance that an assembly will have all three components simultaneously below the lower tolerance limit is:

$$\frac{1}{100} \times \frac{1}{100} \times \frac{1}{100} = \frac{1}{1{,}000{,}000}$$

There is only about one chance in a million that all three components will be too small, resulting in a small assembly. Thus, setting component and assembly tolerances based on the simple addition formula is conservative in that it fails to recognize the extremely low probability of an assembly containing all low (or all high) components.

The statistical approach is based on the relationship between the variances of a number of independent causes and the variance of the dependent or overall result. This may be written as:

$$\sigma_{\text{result}} = \sqrt{\sigma^2_{\text{cause A}} + \sigma^2_{\text{cause B}} + \sigma^2_{\text{cause C}} + \cdots}$$

In terms of the assembly example, the formula is:

$$\sigma_{\text{assembly}} = \sqrt{\sigma^2_{\text{A}} + \sigma^2_{\text{B}} + \sigma^2_{\text{C}}}$$

Now suppose that for each component, the tolerance range is equal to three standard deviations (or any constant multiple of the standard deviation). Because σ is equal to T divided by 3, the variance relationship may be rewritten as:

$$\frac{T}{3} = \sqrt{\left(\frac{T_A}{3}\right)^2 + \left(\frac{T_B}{3}\right)^2 + \left(\frac{T_C}{3}\right)^2}$$

or

$$T_{\text{assembly}} = \sqrt{T_A^2 + T_B^2 + T_C^2}$$

Thus, the squares of tolerances are added to determine the square of the tolerance for the overall result. This formula compares to the simple addition of tolerances commonly used.

The effect of the statistical approach is dramatic. Listed below are two possible sets of component tolerances that will yield an assembly tolerance equal to 0.0035 when used with the previous formula:

Component	Alternative 1	Alternative 2
A	±0.002	±0.001
B	±0.002	±0.001
C	±0.002	±0.003

With alternative 1, the tolerance for component A has been doubled, the tolerance for component B has been quadrupled, and the tolerance for component C has been kept the same as the original component tolerance based on the simple addition approach. If alternative 2 is chosen, similar significant increases in the component tolerances may be achieved. This formula, then, may result in a larger component tolerance with no change in the manufacturing processes and no change in the assembly tolerance.

The risk of this approach is that an assembly may fall outside the assembly tolerance. However, this probability can be calculated by expressing the component tolerances as standard deviations, calculating the standard deviation of the result, and finding the area under the normal curve outside the assembly tolerance limits. For example, if each component tolerance is equal to 3*s*, then 99.73 percent of the assemblies will be within the assembly tolerance, that is, 0.27 percent, or about 3 assemblies in 1000 taken at random would fail to meet the assembly tolerance. The risk can be eliminated by changing components for the few assemblies that do not meet the assembly tolerance.

The tolerance formula is not restricted to outside dimensions of assemblies. Generalizing, the left side of the equation contains the dependent variable or physical result, and the right side of the equation contains the independent variables of physical causes. If the result is placed on the left and the causes on the right, the formula always has plus signs under the square root—even if the result is an internal dimension (such as the clearance between a shaft and hole). The causes of variation are additive wherever the physical result happens to fall.

The formula has been applied to a variety of mechanical and electronic products. The concept may be applied to several interacting variables in an engineering relationship. The nature of the relationship need not be additive (assembly example) or subtractive (shaft-and-hole example). The tolerance formula can be adapted to predict the variation of results that are the product and/or the division of several variables.

Assumptions of the Formula

The formula is based on several assumptions:

- The component dimensions are independent and each component to be assembled is chosen randomly. These assumptions are usually met in practice.
- Each component dimension should be normally distributed. Some departure from this assumption is permissible.
- The actual average for each component is equal to the nominal value stated in the specification. For the original assembly example, the actual averages for components A, B, and C must be 1.000, 0.500, and 2.000, respectively. Otherwise, the nominal value of 3.500 will not be achieved for the assembly and tolerance limits set at about 3.500 will not be realistic. Thus it is important to control the average value for interacting dimensions. Consequently, process control techniques are needed using variables measurement.

Use caution if any assumption is violated. Reasonable departures from the assumptions may still permit applying the concept of the formula. Notice that in the example, the formula resulted in the doubling of certain tolerances. This much of an increase may not even be necessary from the viewpoint of process capability.

Bender (1975) has studied these assumptions for some complex assemblies and concluded, based on a "combination of probability and experience," that a factor of 1.5 should be included to account for the assumptions:

$$T_{\text{result}} = 1.5\sqrt{T_A^2 + T_B^2 + T_C^2 + \cdots}$$

Graves (1997) suggests developing different factors for initial versus mature production, high versus low volume production, and mature versus developing technology and measurement processes.

Finally, variation simulation analysis is a technique that uses computer simulation to analyze tolerances. This technique can handle product characteristics with either normal or non-normal distributions. Dodson (1999) describes the use of simulation in the tolerance design of circuits; Gomer (1998) demonstrates simulation to analyze tolerances in engine design. For an overall text on reliability, see Meeker and Escobar (1998).

Statistical Tools for Control

In addition to the fundamental control charts, there are some special-purpose methods for control that are sometimes helpful.

PRE-Control

PRE-Control is a statistical technique for detecting process conditions and changes that may cause defects (rather than changes that are statistically significant). PRE-Control focuses on controlling conformance to specifications, rather than statistical control. PRE-Control starts a process centered between specification limits and detects shifts that might result in making some of the parts outside a specification limit. It requires no plotting and no computations, and it needs only three measurements to give control information. The technique uses the normal distribution curve to determine significant changes in either the aim or the spread of a production process that could result in increased production of defective work.

The relative simplicity of PRE-Control versus statistical control charts can have important advantages in many applications. The concept, however, has generated some controversy. For a comparison of PRE-Control versus other approaches and the most appropriate applications of PRE-Control, see Ledolter and Swersey (1997) and Steiner (1997). For a complete story, also see the references in both of these papers.

Short-Run Control Charts

Some processes are carried out in such short runs that the usual procedure of collecting 20 to 30 samples to establish a control chart is not feasible. Sometimes these short runs are caused by previously known assignable causes that take place at predetermined times (such as a frequent shift in production from one product to another, as may be the case in lean production systems). Hough and Pond (1995) discuss four ways to construct control charts in these situations:

1. Ignore the systematic variability, and plot on a single chart.
2. Stratify the data, and plot them on a single chart.
3. Use regression analysis to model the data, and plot the residuals on a chart.
4. Standardize the data, and plot the standardized data on a chart.

The last option has received the most consideration. It involves transforming the data via the Z-transformation

$$Z = \frac{X - \mu}{\sigma}$$

to remove any systematic changes in level and variability (thereby normalizing the data to a common baseline). This standardization of Shewhart charts has been discussed by Nelson (1989), Wheeler (1991), and Griffith (1996). Pyzdek (1993) also provides a good discussion of short and small runs.

Cumulative Sum Control Chart

The cumulative sum (CUMSUM or CUSUM) control chart is a chronological plot of the cumulative sum of deviations of a sample statistic (e.g., $\bar{X}$, p, number of nonconformities) from a reference value (e.g., the nominal or target specification). By definition, the CUMSUM chart focuses on a target value rather than on the actual average of process data. Each point plotted contains information from all observations (i.e., a cumulative sum). CUMSUM charts are particularly useful in detecting small shifts in the process average (say, 0.5σ to 2.0σ). The chart shown in Fig. 20.29 is one way of constructing CUMSUM charts. The method is as follows:

1. Compute the control statistic (x-bar for the example in Fig. 20.29).
2. Determine the target value T (10 in Fig. 20.29).
3. Compute the standard deviation s (1.96 in Fig. 20.29).
4. Draw a reference line at zero and upper and lower control limits (UCL and LCL respectively) at ±4s.
5. Compute the upper cumulative sum C_U for each sample point k as follows:

$$C_{U,k} = \text{Maximum}\left\{0, \sum_{i=1}^{k}[\bar{x}_i - (T + s/2)]\right\}$$

6. Compute the lower cumulative sum C_L for each sample point k as follows:

$$C_{L,k} = \text{Minimum}\left\{0, \sum_{i=1}^{k}[\bar{x}_i - (T - s/2)]\right\}$$

7. Plot C_U and C_L as two separate lines.
8. When C_U exceeds the UCL, then an upward shift has occurred. When C_L drops below LCL, then a downward shift has occurred.

Moving Average Control Charts

Another special chart is the moving average chart. This chart is a chronological plot of the moving average, which is calculated as the average value updated by dropping the oldest

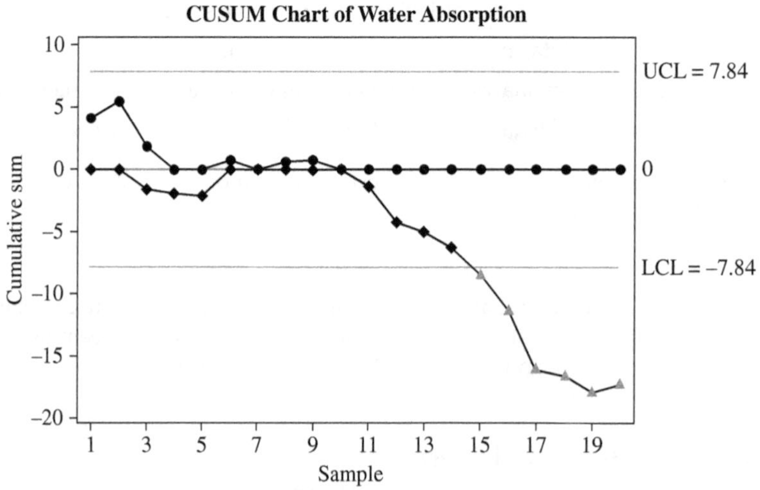

FIGURE 20.29 Cumulative sum control chart. (*Juran Institute, Inc., Copyright 1994. Used by permission.*)

individual measurement and adding the newest individual measurement. Thus, a new average is calculated with each individual measurement. A further refinement is the exponentially weighted moving average (EWMA) chart. In the EWMA chart, the observations are weighted, and the highest weight is given to the most recent data. Moving average charts are effective in detecting small shifts, highlighting trends, and using data in processes in which it takes a long time to produce a single item.

Box-Jenkins Manual Adjustment Chart

Still another chart is the Box-Jenkins manual adjustment chart. The average and range, CUMSUM, and EWMA charts for variables focus on monitoring a process and reducing variability due to special causes of variation identified by the charts. Box-Jenkins charts have a different objective: to analyze process data to regulate the process after each observation and thereby minimize process variation. For elaboration on this advanced technique, see Box and Luceño (1997).

Multivariate Control Charts

Finally, we consider the concept of multivariate control charts. When there are two or more quality characteristics on a unit of product, these could be monitored independently with separate control charts. Then the probability that a sample average on either control chart exceeds three sigma limits is 0.0027. But the joint probability that both variables exceed their control limits simultaneously when they are both in control is (0.0027)(0.0027) or 0.00000729, which is much smaller than 0.0027. The situation becomes more distorted as the number of characteristics increases. For this and other reasons, monitoring several characteristics independently can be misleading. Multivariate control charts and statistics (e.g., Hotelling's T^2 charts, multivariate EWMA) address this issue. See Montgomery (2000, Section 8.4) for a highly useful discussion.

Process Capability

In planning the quality aspects of operations, nothing is more important than advance assurance that the processes will meet the specifications. In recent decades, a concept of process capability has emerged to provide a quantified prediction of process adequacy. This ability to predict quantitatively has resulted in widespread adoption of the concept as a major element of quality planning. Process capability is the measured, inherent variation of the product turned out by a process.

Basic Definitions

Each key word in this definition must itself be clearly defined because the concept of capability has an enormous extent of application, and nonscientific terms are inadequate for communication within the industrial community.

- Process refers to some unique combination of machine, tools, methods, materials, and people engaged in production. It is often feasible and illuminating to separate and quantify the effect of the variables entering this combination.
- Capability refers to an ability, based on tested performance, to achieve measurable results.
- Measured capability refers to the fact that process capability is quantified from data that, in turn, are the results of measurement of work performed by the process.

- Inherent capability refers to the product uniformity resulting from a process that is in a state of statistical control (i.e., in the absence of time-to-time "drift" or other assignable causes of variation). "Instantaneous reproducibility" is a synonym for inherent capability.
- The product is measured because product variation is the end result.

Uses of Process Capability Information

Process capability information serves multiple purposes:

- Predicting the extent of variability that processes will exhibit. Such capability information, when provided to designers, provides important information in setting realistic specification limits.
- Choosing from among competing processes that are most appropriate to meet the tolerances.
- Planning the interrelationship of sequential processes. For example, one process may distort the precision achieved by a predecessor process, as in hardening of gear teeth. Quantifying the respective process capabilities often points the way to a solution.
- Providing a quantified basis for establishing a schedule of periodic process control checks and readjustments.
- Assigning machines to classes of work for which they are best suited.
- Testing theories of causes of defects during quality improvement programs.
- Serving as a basis for specifying the quality performance requirements for purchased machines.

These purposes account for the growing use of the process capability concept.

Planning for a Process Capability Study

Capability studies are conducted for various reasons, for example, to respond to a customer request for a capability index number or to evaluate and improve product quality. Prior to data collection, clarify the purpose for making the study and the steps needed to ensure that it is achieved.

In some cases, the capability study will focus on determining a histogram and capability index for a relatively simple process. Here the planning should ensure that process conditions (e.g., temperature, pressure) are completely defined and recorded. All other inputs must clearly be representative (i.e., specific equipment, material, and, of course, personnel).

For more complex processes or when defect levels of 1 to 10 parts per million are desired, the following steps are recommended:

1. Develop a process description, including inputs, process steps, and output quality characteristics. This description can range from simply identifying the equipment to developing a mathematical equation that shows the effect of each process variable on the quality characteristics.
2. Define the process conditions for each process variable. In a simple case, this step involves stating the settings for temperature and pressure. But for some processes, it means determining the optimum value or aim of each process variable. The statistical design of experiments provides the methodology. Also, determine the operating ranges of the process variables around the optimum because the range will affect the variability of the product results.

3. Make sure that each quality characteristic has at least one process variable that can be used to adjust it.
4. Decide whether measurement error is significant. This can be determined from a separate error of measurement study. In some cases, the error of measurement can be evaluated as part of the overall study.
5. Decide whether the capability study will focus only on variability or will also include mistakes or errors that cause quality problems.
6. Plan for the use of control charts to evaluate the stability of the process.
7. Prepare a data collection plan, including adequate sample size that documents results on quality characteristics along with the process conditions (e.g., values of all process variables) and preserves information on the order of measurements so that trends can be evaluated.
8. Plan which methods will be used to analyze data from the study to ensure that before starting the study, all necessary data for the analysis will be available. The analyses should include process capability calculations on variability and also analysis of attribute or categorical data on mistakes and analysis of data from statistically designed experiments built into the study.
9. Be prepared to spend time investigating interim results before process capability calculations can be made. These investigations can include analysis of optimum values and ranges of process variables, out-of-control points on control charts, or other unusual results. The investigations then lead to the ultimate objective, that is, improvement of the process.

Note that these steps focus on improvement rather than just on determining a capability index.

Standardized Process Capability Formula

The most widely adopted formula for process capability is:

$$\text{Process capability} = \pm 3\sigma \text{ (a total of } 6\sigma\text{)}$$

where σ is the standard deviation of the process under a state of statistical control (i.e., under no drift and no sudden changes). If the process is centered at the nominal specification and follows a normal probability distribution, 99.73 percent of production will fall within 3σ of the nominal specification.

Relationship to Product Specifications

A major reason for quantifying process capability is to compute the ability of the process to hold product specifications. For processes that are in a state of statistical control, a comparison of the variation of $6s$ to the specification limits permits ready calculation of percentage defective by conventional statistical theory.

Planners try to select processes with the $6s$ process capability well within the specification width. A measure of this relationship is the capability ratio:

$$C_p = \text{capability ratio} = \frac{\text{specification range}}{\text{process capability}} = \frac{\text{USL} - \text{LSL}}{6s}$$

where USL is the upper specification limit and LSL is the lower specification limit. Note that $6s$ is used as an estimate of 6σ.

Some companies define the ratio as the reciprocal. Some industries now express defect rates in terms of parts per million. A defect rate of one part per million requires a capability ratio (specification range over process capability) of about 1.63.

Figure 20.30 shows four of many possible relations between process variability and specification limits and the likely courses of action for each. Note that in all of these cases, the average of the process is at the midpoint between the specification limits.

Table 20.20 shows selected capability ratios and the corresponding level of defects, assuming that the process average is midway between the specification limits. A process that is just meeting specification limits (specification range $\pm 3\sigma$) has a C_p of 1.0. The criticality of many applications and the reality that the process average will not remain at the midpoint of the specification range suggest that C_p should be at least 1.33. Note that a process operating at $C_p = 2.0$ over the short term (and centered midway between the specification limits) will correspond to a process sigma capability measure of $3C_p$, or 6 sigma (allowing for a 1.5s

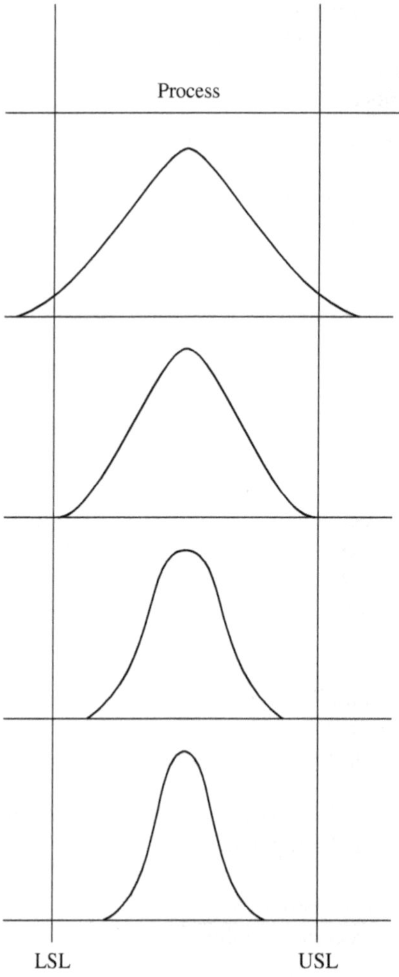

C_p	Total amount outside limits	Typical actions to be taken
<1.0	≥5.0%	Heavy process control, sorting, rework, etc.
1.0	0.3%	Heavy process control, inspection
1.33	64 ppm	Reduced inspection, selected use of control charts
1.63	1 ppm	Spot checking, selected use of control charts

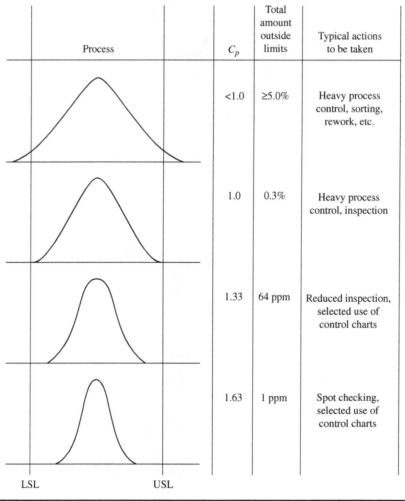

FIGURE 20.30 Four examples of process variability. (*Quality Planning and Analysis, Copyright 2007. Used by permission.*)

Process Capability Index (C_p)	Total Product Outside Two-Sided Specification Limits*
0.5	13.36%
0.67	4.55%
1.00	0.3%
1.33	64 ppm
1.63	1 ppm
2.00	0

*Assuming that the process is centered midway between the specification limits.

Source: Quality Planning and Analysis, Copyright 2007. Used by permission.

TABLE 20.20 Process Capability Index (C_p) and Product Outside Specification Limits

shift over the long term. This corresponds to $6s - 1.5s = 4.5s$, which is expected to produce 3.4 ppm outside of the two-sided specification limits over the long term).

Note that the C_p index measures whether the process variability can fit within the specification range. It does not indicate whether the process is actually running within the specification because the index does not include a measure of the process average (this issue is addressed by another measure, C_{pk}).

Three capability indexes commonly in use are shown in Table 20.21. Of these, the simplest is C_p. The higher the value of any indexes, the lower the amount of product outside the specification limits.

Pignatiello and Ramberg (1993) provide an excellent discussion of various capability indexes. Bothe (1997) provides a comprehensive reference book that includes extensive discussion of the mathematical aspects. These references explain how to calculate confidence bounds for various process capability indexes.

The C_{pk} Capability Index

Process capability, as measured by C_{pk}, refers to the variation in a process about the average value. This concept is illustrated in Fig. 20.31. The two processes have equal capabilities (C_p) because 6σ is the same for each distribution, as indicated by the widths of the distribution curves. The process aimed at μ_2 is producing defectives because the aim is off center, not because of the inherent variation about the aim (i.e., the capability).

Process Capability	Process Performance
$C_p = \dfrac{USL - LSL}{6\sigma}$	$P_p = \dfrac{USL - LSL}{6s}$
$C_{pk} = \min\left[\dfrac{USL - \mu}{3\sigma}, \dfrac{\mu - LSL}{3\sigma}\right]$	$P_{pk} = \min\left[\dfrac{USL - \bar{X}}{3s}, \dfrac{\bar{X} - LSL}{3s}\right]$
$C_{pm} = \dfrac{USL - LSL}{6\sqrt{\sigma^2 + (\mu - T)^2}}$	$P_{pm} = \dfrac{USL - LSL}{6\sqrt{s^2 + (\bar{X} - T)^2}}$

Source: Quality Planning and Analysis, Copyright 2007. Used by permission.

TABLE 20.21 Process Capability and Process Performance Indexes

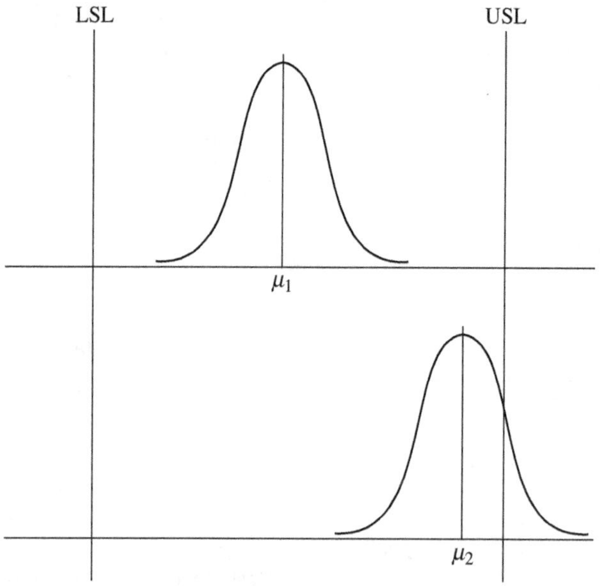

Figure 20.31 Process with equal process capability but different aim. (*Quality Planning and Analysis. Copyright 2007. Used by permission.*)

Thus, the C_p index measures potential capability, assuming that the process average is equal to the midpoint of the specification limits and the process is operating in statistical control; because the average is often not at the midpoint, it is useful to have a capability index that reflects both variation and the location of the process average. Such an index is C_{pk}.

C_{pk} reflects the current process mean's proximity to either the USL or LSL. C_{pk} is estimated by:

$$\hat{C}_{pk} = \min\left[\frac{\overline{X} - \text{LSL}}{3s}, \frac{\text{USL} - \overline{X}}{3s}\right]$$

In an example from Kane (1986),

$$\text{USL} = 20\overline{X} = 16$$

$$\text{LSL} = 8 \quad s = 2$$

The standard capability ratio is estimated as:

$$\frac{\text{USL} - \text{LSL}}{6\sigma} = \frac{20 - 8}{12} = 1.0$$

which implies that if the process were centered between the specification limits (at 14), then only a small proportion (about 0.27 percent) of product would be defective.
However, when we calculate C_{pk}, we obtain:

$$\hat{C}_{pk} = \min\left[\frac{16 - 8}{6}, \frac{20 - 16}{12}\right] = 0.67$$

which indicates that the process mean is currently nearer the USL. (Note that if the process were centered at 14, the value of C_{pk} would be 1.0.) An acceptable process will require reducing the standard deviation and/or centering the mean. Also note that if the actual average is equal to the midpoint of the specification range, then $C_{pk} = C_p$.

The higher the value of C_p, the lower the amount of product outside specification limits. In certifying suppliers, some organizations use C_{pk} as one element of certification criteria. In these applications, the value of C_{pk} desired from suppliers can be a function of the type of commodity purchased.

A capability index can also be calculated around a target value rather than the actual average. This index, called C_{pm} or the Taguchi index, focuses on reduction of variation from a target value rather than reduction of variability to meet specifications.

Most capability indexes assume that the quality characteristic is normally distributed. Krishnamoorthi and Khatwani (2000) propose a capability index for handling normal and nonnormal characteristics by first fitting the data to a Weibull distribution.

Two types of process capability studies are as follows:

1. *Study of process potential.* In this study, an estimate is obtained of what the process can do under certain conditions (i.e., variability under short-run defined conditions for a process in a state of statistical control). The C_p index estimates the potential process capability.
2. *Study of process performance.* In this study, an estimate of capability provides a picture of what the process is doing over an extended period. A state of statistical control is also assumed. The C_{pk} index estimates the performance capability.

Estimating Inherent or Potential Capability from Control Chart Analysis

In a process potential study, data are collected from a process operating without changes in material batches, workers, tools, or process settings. This short-term evaluation uses consecutive production over one time period. Such an analysis should be preceded by a control chart analysis in which any assignable causes have been detected and eliminated from the process.

Because specification limits usually apply to individual values, control limits for sample averages cannot be compared to specification limits. To make a comparison, we must first convert R to the standard deviation for individual values, calculate the 3s limits, and compare them to the specification limits. This process is explained below.

If a process is in statistical control, it is operating with the minimum amount of variation possible (the variation due to chance causes). If, and only if, a process is in statistical control, the following relationship holds for using s as an estimate of σ:

$$s = \frac{\overline{R}}{d_2}$$

Tables 20.22 and 20.23 provide values of d_2. If the standard deviation is known, process capability limits can be set at 3s, and this value used as an estimate of 3.

For the data shown in Fig. 20.32 (machine N-5),

$$s = \frac{\overline{R}}{d_2} = \frac{6.0}{2.534} = 2.37$$

and

$$\pm 3s = \pm 3(2.37) = 7.11$$

Factors for $\bar{X}$ and R Control Charts;* Factors for Estimating s from R†				
Number of Observations in Sample	A_2	D_3	D_4	Factor for Estimate from $\bar{R}$: $d_2 = \bar{R}/s$
2	1.880	0	3.268	1.128
3	1.023	0	2.574	1.693
4	0.729	0	2.282	2.059
5	0.577	0	2.114	2.326
6	0.483	0	2.004	2.534
7	0.419	0.076	1.924	2.704
8	0.373	0.136	1.864	2.847
9	0.337	0.184	1.816	2.970
10	0.308	0.223	1.777	3.078
11	0.285	0.256	1.744	3.173
12	0.266	0.284	1.717	3.258
13	0.249	0.308	1.692	3.336
14	0.235	0.329	1.671	3.407
15	0.223	0.348	1.652	3.472

* Upper control limited for $\bar{X} = UCL_{\bar{X}} = \bar{\bar{X}} + A_2\bar{R}$

Lower control limited for $\bar{X} = UCL_{\bar{X}} = \bar{\bar{X}} + A_2\bar{R}$

† Upper control limited for $R = UCL_R = D_4\bar{R}$

Lower control limited for $R = LCR_R = D_3\bar{R}$

$s = \bar{R}/d_2$

From *1950 ASTM Manual on Quality Control of Materials* and *ASTM Manual on Presentation of Data, 1945.* American Society for Testing and Materials. Copyright ASTM International. Reprinted with permission.

Source: Quality Planning and Analysis, Copyright 1997. Used by permission.

TABLE 20.22 Factors for $\bar{X}$ and R Control Charts

n	A_2	D_3	D_4	d_2
2	1.880	0	3.268	1.128
3	1.023	0	2.574	1.693
4	0.729	0	2.282	2.059
5	0.577	0	2.114	2.326
6	0.483	0	2.004	2.534
7	0.419	0.076	1.924	2.704
8	0.373	0.136	1.864	2.847
9	0.337	0.184	1.816	2.970
10	0.308	0.223	1.777	3.079

Source: Quality Planning and Analysis, Copyright 2007. Used by permission.

TABLE 20.23 Constants for $\bar{X}$ and R Chart

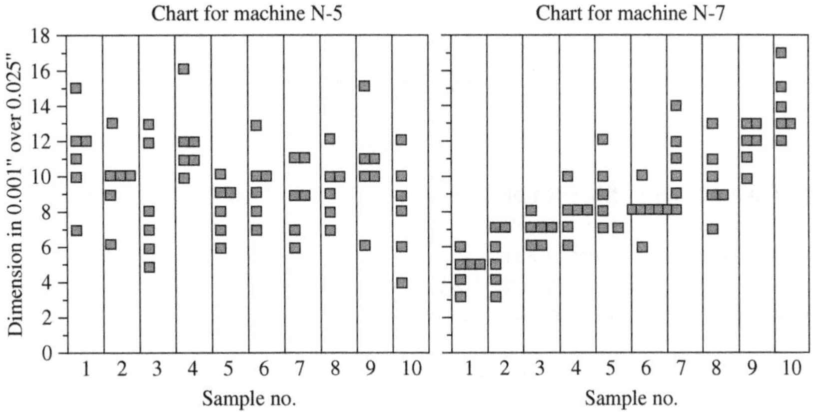

For machine N-5:

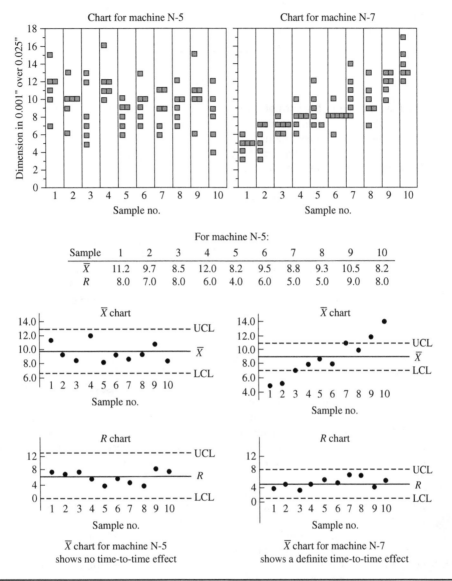

Sample	1	2	3	4	5	6	7	8	9	10
$\overline{X}$	11.2	9.7	8.5	12.0	8.2	9.5	8.8	9.3	10.5	8.2
R	8.0	7.0	8.0	6.0	4.0	6.0	5.0	5.0	9.0	8.0

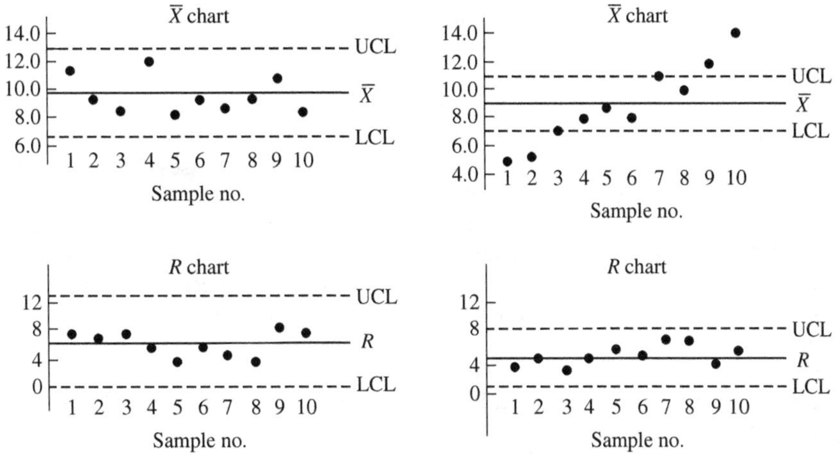

$\overline{X}$ chart for machine N-5 shows no time-to-time effect

$\overline{X}$ chart for machine N-7 shows a definite time-to-time effect

FIGURE 20.32 $\overline{X}$ and R charts confirm. (*Quality Planning and Analysis. Copyright 2007. Used by permission.*)

or

$$6s = 14.22 \text{ (or 0.0124 in the original data units)}$$

The specification limit was 0.258 ± 0.005.
Thus,

$$\text{USL} = 0.263$$

$$\text{LSL} = 0.253$$

Then,

$$C_p = \frac{\text{USL} - \text{LSL}}{6s} = \frac{0.263 - 0.253}{0.0142} = 0.72$$

Even if the process is perfectly centered at 0.258 (and it was not), it is not capable.

The Assumption of Statistical Control and Its Effect on Process Capability

All statistical predictions assume a stable population. In a statistical sense, a stable population is one that is repeatable (i.e., a population that is in a state of statistical control). The statistician rightfully insists that this be the case before predictions can be made. The manufacturing engineer also insists that the process conditions (feeds, speeds, etc.) be fully defined.

In practice, the original control chart analysis will often show that the process is out of statistical control. (It may or may not be meeting product specifications.) However, an investigation may show that the causes cannot be economically eliminated from the process. In theory, process capability should not be predicted until the process is in statistical control. However, in practice, some kind of comparison of capability to specifications is needed. The danger in delaying the comparison is that the assignable causes may never be eliminated from the process. The resulting indecision will thereby prolong interdepartmental bickering on whether "the specification is too tight" or "manufacturing is too careless."

A good way to start is by plotting individual measurements against specification limits. This step may show that the process can meet the product specifications even with assignable causes present. If a process has assignable causes of variation but is able to meet the specifications, usually no economic problem exists. The statistician can properly point out that a process with assignable variation is unpredictable. This point is well taken, but in establishing priorities of quality improvement efforts, processes that are meeting specifications are seldom given high priority.

If a process is out of control and the causes cannot be economically eliminated, the standard deviation and process capability limits can nevertheless be computed (with the out-of-control points included). These limits will be inflated because the process will not be operating at its best. In addition, the instability of the process means that the prediction is approximate.

It is important to distinguish between a process that is in a state of statistical control and a process that is meeting specifications. A state of statistical control does not necessarily mean that the product from the process conforms to specifications. Statistical control limits on sample averages cannot be compared to specification limits because specification limits refer to individual units. For some processes that are not in control, the specifications are being met and no action is required; other processes are in control, but the specifications are not being met, and action is needed.

In summary, we need processes that are both stable (in statistical control) and capable (meeting product specifications).

The increasing use of capability indexes has also led to the failure to understand and verify some important assumptions that are essential for statistical validity of the results. Five key assumptions are:

1. *Process stability*. Statistical validity requires a state of statistical control with no drift or oscillation.
2. *Normality of the characteristic being measured*. Unless nonparametric methods or alternative distributions are used, normality is needed to draw statistical inferences about the population.

3. *Sufficient data.* Sufficient data are necessary to minimize the sampling error for the capability indexes.
4. *Representativeness of samples.* Random samples must be included.
5. *Independent measurements.* Consecutive measurements cannot be correlated.

These assumptions are not theoretical refinements—they are important conditions for properly applying capability indexes. Before applying capability indexes, readers are urged to read the paper by Pignatiello and Ramberg (1993). It is always best to compare the indexes with the full data versus specifications depicted in a histogram.

Measuring Process Performance

A process performance study collects data from a process that is operating under typical conditions but includes normal changes in material batches, workers, tools, or process settings. This study, which spans a longer term than the process potential study, also requires that the process be in statistical control.

The capability index for a process performance study is:

$$C_{pk} = \min\left[\frac{\overline{X} - \text{LSL}}{3s}, \frac{\text{USL} - \overline{X}}{3s}\right]$$

Problem Consider a pump cassette used to deliver intravenous solutions (Baxter Travenol Laboratories, 1986). A key quality characteristic is the volume of solution delivered in a predefined time. The specification limits are

$$\text{USL} = 103.5 \quad \text{LSL} = 94.5$$

A control chart was run for one month, and no out-of-control points were encountered. From the control chart data, we know that

$$\overline{X} = 98.2 \text{ and } s = 0.98$$

Figure 20.33 shows the process data and the specification limits.

Solution The capability index is

$$C_{pk} = \min\left[\frac{98.2 - 94.5}{3(0.98)}, \frac{103.5 - 98.2}{3(0.98)}\right]$$

$$C_{pk} = 1.26$$

For many applications, 1.26 is an acceptable value of C_{pk}.

Interpretation of C_{pk}

In using C_{pk} to evaluate a process, we must recognize that C_{pk} is an abbreviation of two parameters—the average and the standard deviation. Such an abbreviation can inadvertently mask important detail in these parameters. For example, Fig. 20.34 shows that three extremely different processes can all have the same C_{pk} (in this case $C_{pk} = 1$).

Increasing the value of C_{pk} may require a change in the process average, the process standard deviation, or both. For some processes, increasing the value of C_{pk} by changing the average value (perhaps by a simple adjustment of the process aim) may be easier than

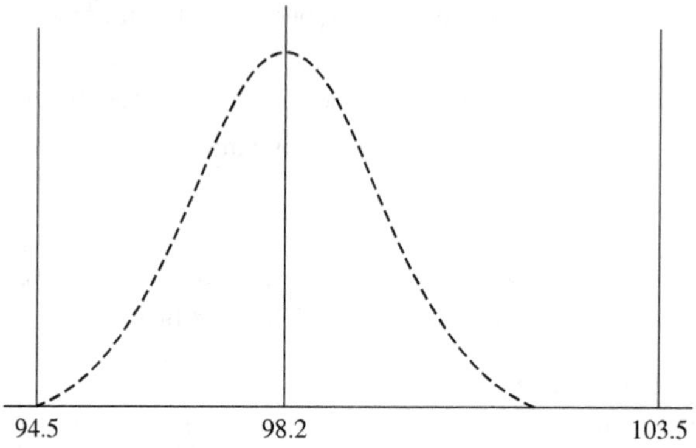

Figure 20.33 Delivered volume of solution.

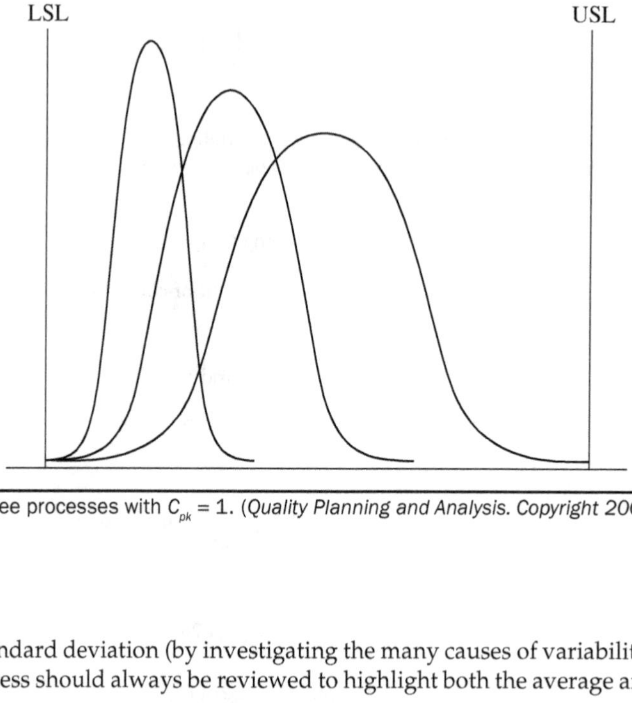

Figure 20.34 Three processes with $C_{pk} = 1$. (*Quality Planning and Analysis. Copyright 2007. Used by permission.*)

reducing the standard deviation (by investigating the many causes of variability). The histogram of the process should always be reviewed to highlight both the average and the spread of the process.

Note that Table 20.21 also includes the capability index C_{pm}. This index measures the capability around a target value T rather than the mean value. When the target value equals the mean value, the C_{pm} index is identical to the C_{pk} index.

Attribute (or Categorical) Data Analysis

The methods discussed earlier assume that numerical measurements are available from the process. Sometimes, however, the only data available are in attribute or categorical form (i.e., the number of nonconforming units and the number acceptable).

	Policy Writer						
Error Type	A	B	C	D	E	F	Total
1	0	0	1	0	2	1	4
2	1	0	0	0	1	0	2
3	0	(16)	1	0	2	0	(19)
4	0	0	0	0	1	0	1
5	2	1	3	1	4	2	(13)
6	0	0	0	0	3	0	3
.							
.							
.							
27							
28							
29							
Total	6	(20)	8	3	(36)	7	80

Source: Quality Planning and Analysis, Copyright 2007. Used by permission.

TABLE 20.24 Matrix of Errors by Insurance Policy Writers

The data in Table 20.24 on errors in preparing insurance policies also can be used to illustrate process capability for attribute data. The data reported 80 errors from six policy writers, or 13.3 errors per writer—the current performance. The process capability can be calculated by excluding the abnormal performance identified in the study—type 3 errors by worker B, type 5 errors, and errors of worker E. The error data for the remaining five writers becomes 4, 3, 5, 2, and 5, with an average of 3.8 errors per writer. The process capability estimate of 3.8 compares with the original performance estimate of 13.3.

This example calculates process capability in terms of errors or mistakes rather than the variability of a process parameter. Hinckley and Barkan (1995) point out that in many processes, nonconforming product can be caused by excessive variability or by mistakes (e.g., missing parts, wrong parts, wrong information, or other processing errors). For some processes, mistakes can be a major cause of failing to meet customer quality goals. The actions required to reduce mistakes are different from those required to reduce variability of a parameter.

Readers are directed to DeVor et al. (1992) for a good background in process control charting.

Software

While many of the tools mentioned in this chapter can be applied using programs such as Microsoft Excel, numerous software packages are available that provide more specialized assistance. Some of these packages and vendors are listed here, according to their primary emphasis. Most vendors have multiple software options.

Basic statistics:

- QI Macros
- SigmaXL
- StatPlus

Advanced statistics:

- JMP
- Minitab
- Systat

Design of experiments:

- StatSoft STATISTICA
- Stat-Ease
- STRATEGY
- Statgraphics

Monte Carlo, discrete event simulation:

- @Risk
- Crystal Ball
- iGrafx

Reliability, availability:

- Isograph
- Relex 2009
- ReliaSoft

Control charting:

- CHARTRunner
- Statit

References

Anand, K. N. (1996). The Role of Statistics in Determining Product and Part Specifications: A Few Indian Experiences, *Quality Engineering*, vol. 9, no. 2, pp. 187–193.

Automotive Industry Action Group (2003). *Measurement Systems Analysis* (3rd ed.). Chrysler, Ford, General Motors Suppler Quality Requirements Task Force, Southfield, MI.

Barnett, V. and Lewis, T. (1994). *Outliers in Statistical Data* (3rd ed.). John Wiley & Sons, New York.

Bender, A. (1975). Statistical Tolerancing as It Relates to Quality Control and the Designer, *Automotive Division Newsletter of ASQC*, April, p. 12.

Bothe, D. R. (1997). *Measuring Process Capability*. McGraw-Hill, New York.

Box, G. E. P. and Draper, N. R. (1969). *Evolutionary Operation: A Statistical Method for Process Improvement*. John Wiley & Sons, New York.

Box, G. E. P., Hunter, J. S., and Hunter, W. G. (2005). *Statistics for Experimenters: Design, Innovation and Discovery* (2nd ed.). Wiley-Interscience, Hoboken, NJ.

Box, G. E. P. and Luceño, A. (1997). *Statistical Control by Monitoring and Adjustment*. Wiley, New York.

Burdick, R. K. and Larsen, G. A. (1997). Confidence Intervals on Measures of Variability in R&R Studies, *Journal of Quality Technology*, vol. 29, no. 3, pp. 261–273.

Carter, C. W. (1996). Sequenced Levels Experimental Designs, *Quality Engineering*, vol. 8, no. 1, (pp. 181–188), no. 2 (pp. 361–366), no. 3 (pp. 499–504), no. 4 (pp. 695–698).

Case, K. E., Bennett, G. K., and Schmidt, J. W. (1975). The Effect of Inspector Error on Average Outgoing Quality, *Journal of Quality Technology*, vol. 7, no. 1, pp. 1–12.

Coleman, S., Greenfield, T., Stewardson, D., and Montgomery, D. C. (2008). *Statistical Practice in Business and Industry*. John Wiley & Sons, Hoboken, NJ. (See Chapter 13).

Davison, A. C. and Hinkley, D. (2006). *Bootstrap Methods and Their Applications* (8th ed.). Cambridge: Cambridge Series in Statistical and Probabilistic Mathematics, Davison Hinkley, Cambridge University Press, Cambridge.

del Castillo, E. (2007). *Process Optimization: A Statistical Approach*. Springer Science and Business Media, New York.

DeVor, R. E., Chang, T., and Sutherland, J. W. (1992). *Statistical Quality Design and Control: Contemporary Concepts and Methods*. Prentice Hall, Upper Saddle River, NJ.

Dodson, B. (1999). Reliability Modeling with Spreadsheets, *Proceedings of the Annual Quality Congress*, ASQ, Milwaukee, pp. 575–585.

Eagle, A. R. (1954). A Method for Handling Errors in Testing and Measurement, *Industrial Quality Control*, March, pp. 10–14.

Early, J. F. (1989). Quality Improvement Tools, *The Power of Quality*, The Health Care Forum, June.

Emanuel, J. T., Palanisamy, M. (2000). *Sequential Experimentation Using Two-Level Factorials*. Department of Industrial and Manufacturing Engineering and Technology, Bradley University, Peoria, IL.

Emory, W. C. and Cooper, D. R. (1991). *Business Research Methods* (4th ed.). Irwin/McGraw-Hill, Boston, MA.

Engel, J. and DeVries, B. (1997). Evaluating a Well-Known Criterion for Measurement Precision, *Journal of Quality Technology*, vol. 29, no. 4, pp. 469–476.

Gomer, P. (1998). Design for Tolerancing of Dynamic Mechanical Assemblies, *Annual Quality Congress Proceedings*, ASQ, Milwaukee, pp. 490–500.

Graves, S. B. (1997). How to Reduce Costs Using a Tolerance Analysis Formula Tailored to your Organization, Report no. 157, Center for Quality and Productivity Improvement, University of Wisconsin, Madison.

Griffith, G. K. (1996). *Statistical Process Control Methods for Long and Short Runs* (2nd ed.). ASQ Quality Press, Milwaukee, WI.

Gryna, F. M., Chua, R. C., and De Feo, J. A. (2007). *Juran's Quality Planning and Analysis* (5th ed.). McGraw Hill, New York.

Hinckley, C. M. and Barkan, P. (1995). The Role of Variation, Mistakes, and Complexity in Producing Nonconformities, *Journal of Quality Technology*, vol. 27, no. 3, pp. 242–249.

Hoag, L. L., Foote, B. L., and Mount-Cambell, C. (1975). The Effect of Inspector Accuracy on Type I and II Errors of Common Sampling Techniques, *Journal of Quality Technology*, vol. 7, no. 4, pp. 157–164.

Hough, L. D. and Pond, A. D. (1995). Adjustable Individual Control Charts for Short Runs. *Proceedings of the 40th Annual Quality Congress*, ASQ, Milwaukee, pp. 1117–1125.

Ireson, W. G., Coombs, C. F., Jr., and Moss, R. Y. (1996). *Handbook of Reliability Engineering and Management* (2nd ed.). McGraw-Hill, New York.

Jones, J. and Hayes, J. (1999). A Comparison of Electronic Reliability Prediction Models, *IEEE Transactions of Reliability*, vol. 48, no. 2, pp. 127–134.

Kane, V. E. (1986). Process Capability Indices, *Journal of Quality Technology*, vol. 18, no. 1, pp. 41–52.

Krishnamoorthi, I. S. and Khatwani, S. (2000). *Statistical Process Control for Health Care*. Duxbury, Paciric Grove, CA.

Kutner, M., Nachtsheim, C., Neter, J., and Li, W. (2004). *Applied Linear Statistical Models* (2nd ed.). Irwin/McGraw-Hill, New York.

Kvam, P. H. and Vidakovic, B. (2007). *Nonparametric Statistics with Applications to Science and Engineering*. John Wiley & Sons, Hoboken, NJ.

Law, A. M. and Kelton, W. D. (2000). *Simulation Modeling and Analysis* (3rd ed.). McGraw-Hill, Boston, MA.

Ledolter, J. and Swersey, A. (1997). An Evaluation of Pre-Control, *Journal of Quality Technology*, vol. 29, no. 1, pp. 163–171.

Ledolter, J. and Swersey, A. J. (2007). *Testing 1-2-3: Experimental Design with Applications in Marketing and Service Operations*. Stanford University Press, Palo Alto, CA.

Meeker, W. Q. and Escobar, L. A. (1998). *Statistical Methods for Reliability Data*. John Wiley & Sons, New York.

Montgomery, D. C. (2000). *Introduction to Statistical Quality Control* (4th ed.). Wiley, New York.

Myers, R. H., Montgomery, D. C., and Anderson-Cook, C. M. (2009). *Response Surface Methodology: Process and Product Optimization Using Designed Experiments*. John Wiley & Sons, Hoboken, NJ.

Nelson, L. S. (1989). Standardization of Shewhart Control Charts, *Journal of Quality Technology*, vol. 21, pp. 287–289.

O'Connor, P. D. T. (1995). *Practical Reliability Engineering* (3rd ed.) rev. John Wiley and Sons, New York.

Pignatiello, J. H., Jr. and Ramberg, J. S. (1993). Process Capability Indices: Just Say No, *ASQC Quality Congress Transactions 1993*, American Society for Quality, Milwaukee.

Pyzdek, T. (1993). Process Control for Short and Small Runs, *Quality Progress*, April, pp. 51–60.

Ross, P. J. (1996). *Taguchi Techniques for Quality Engineering*. McGraw-Hill, New York.

Sprent, P. and Smeeton, N. C. (2001). *Applied Nonparametric Statistical Methods* (3rd ed.). Chapman and Hall/CRC Press, Boca Raton, FL.

Steiner, S. H. (1997). Pre-Control and some Simple Alternatives, *Quality Engineering*, vol. 10, no. 1, pp. 65–74.

Tsai, P. (1988). Variable Gauge Repeatability and Reproducibility Study Using the Analysis of Variance Method, *Quality Engineering*, vol. 1, no. 1, pp. 107–115.

Wheeler, D. J. (1991). *Short Run SPC*. SPC Press, Inc, Knoxville, TN.

Wong, D. and Baker, C. (1988). Pain in Children: Comparison of Assessment Scales, *Pediatric Nursing*, vol. 14, no. 1, pp. 9–17.

Reference Charts for Table 20.3

Ninety-five percent confidence belts for population proportion

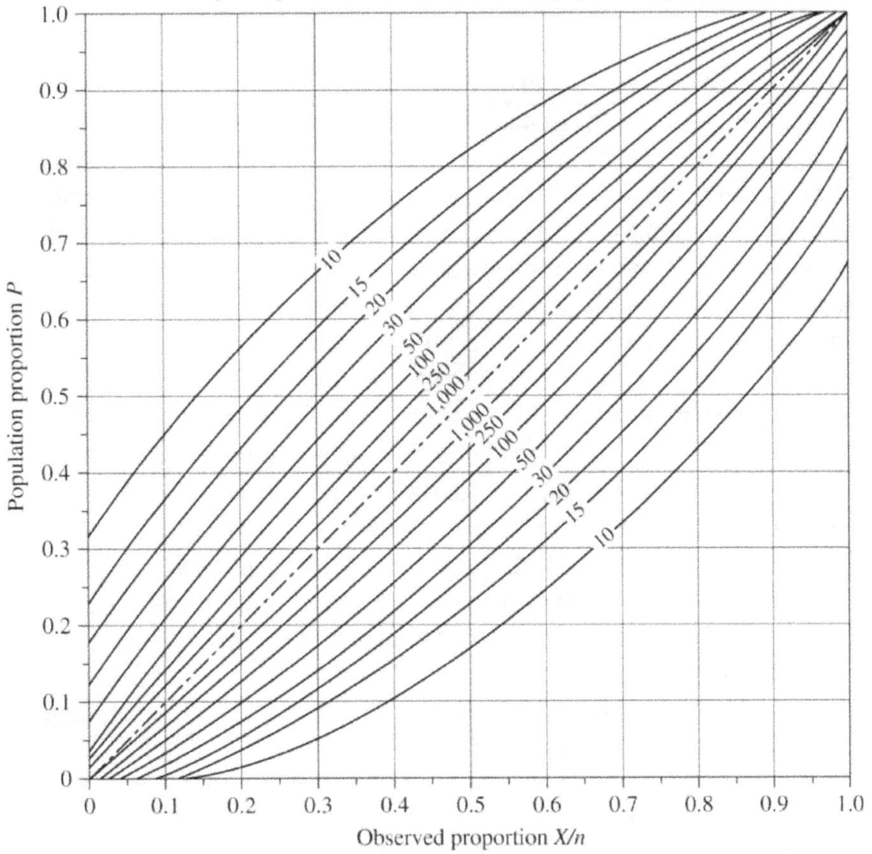

Example in a sample of 10 items, 8 were defective (X/n – 8/10). The 95% confidence limits on the population proportion defective are read from the two curves (for n – 10) as 0.43 and 0.98.

Binomial Distribution*

**Probability of r or fewer occurrences of an event in n trials, where p is the probability of occurrence on each trial.*

		P									
n	r	0.05	0.10	0.15	0.20	0.25	0.30	0.35	0.40	0.45	0.50
2	0	0.9025	0.8100	0.7225	0.6400	0.5625	0.4900	0.4225	0.3600	0.3025	0.2500
	1	0.9975	0.9900	0.9775	0.9600	0.9375	0.9100	0.8775	0.8400	0.7975	0.7500
3	0	0.8574	0.7290	0.6141	0.5120	0.4219	0.3430	0.2746	0.2160	0.1664	0.1250
	1	0.9928	0.9720	0.9392	0.8960	0.8438	0.7840	0.7182	0.6480	0.5748	0.5000
	2	0.9999	0.9990	0.9966	0.9920	0.9844	0.9730	0.9571	0.9360	0.9089	0.8750
4	0	0.8145	0.6561	0.5220	0.4096	0.3164	0.2401	0.1785	0.1296	0.0915	0.0625
	1	0.9860	0.9477	0.8905	0.8192	0.7383	0.6517	0.5630	0.4752	0.3910	0.3125
	2	0.9995	0.9963	0.9880	0.9728	0.9492	0.9163	0.8735	0.8208	0.7585	0.6875
	3	1.0000	0.9999	0.9995	0.9984	0.9961	0.9919	0.9850	0.9744	0.9590	0.9375
5	0	0.7738	0.5905	0.4437	0.3277	0.2373	0.1681	0.1160	0.0778	0.0503	0.0312
	1	0.9774	0.9185	0.8352	0.7373	0.6328	0.5282	0.4284	0.3370	0.2562	0.1875
	2	0.9988	0.9914	0.9734	0.9421	0.8965	0.8369	0.7648	0.6826	0.5931	0.5000
	3	1.0000	0.9995	0.9978	0.9933	0.9844	0.9692	0.9460	0.9130	0.8688	0.8125
	4	1.0000	1.0000	0.9999	0.9997	0.9990	0.9976	0.9947	0.9898	0.9815	0.9688
6	0	0.7351	0.5314	0.3771	0.2621	0.1780	0.1176	0.0754	0.0467	0.0277	0.0156
	1	0.9672	0.8857	0.7765	0.6554	0.5339	0.4202	0.3191	0.2333	0.1636	0.1094
	2	0.9978	0.9842	0.9527	0.9011	0.8306	0.7443	0.6471	0.5443	0.4415	0.3438
	3	0.9999	0.9987	0.9941	0.9830	0.9624	0.9295	0.8826	0.8208	0.7447	0.6562
	4	1.0000	0.9999	0.9996	0.9984	0.9954	0.9891	0.9777	0.9590	0.9308	0.8906
	5	1.0000	1.0000	1.0000	0.9999	0.9998	0.9993	0.9982	0.9959	0.9917	0.9844
7	0	0.6983	0.4783	0.3206	0.2097	0.1335	0.0824	0.0490	0.0280	0.0152	0.0078
	1	0.9556	0.8503	0.7166	0.5767	0.4449	0.3294	0.2338	0.1586	0.1024	0.0625
	2	0.9962	0.9743	0.9262	0.8520	0.7564	0.6471	0.5323	0.4199	0.3164	0.2266
	3	0.9998	0.9973	0.9879	0.9667	0.9294	0.8740	0.8002	0.7102	0.6083	0.5000
	4	1.0000	0.9998	0.9988	0.9953	0.9871	0.9712	0.9444	0.9037	0.8471	0.7734
	5	1.0000	1.0000	0.9999	0.9996	0.9987	0.9962	0.9910	0.9812	0.9643	0.9375
	6	1.0000	1.0000	1.0000	1.0000	0.9999	0.9998	0.0994	0.9984	0.9963	0.9922
8	0	0.6634	0.4305	0.2725	0.1678	0.1001	0.0576	0.0319	0.0168	0.0084	0.0039
	1	0.9428	0.8131	0.6572	0.5033	0.3671	0.2553	0.1691	0.1064	0.0632	0.0352
	2	0.9942	0.9619	0.8948	0.7969	0.6785	0.5518	0.4278	0.3154	0.2201	0.1445
	3	0.9996	0.9950	0.9786	0.9437	0.8862	0.8059	0.7064	0.5941	0.4770	0.3633
	4	1.0000	0.9996	0.9971	0.9896	0.9727	0.9420	0.8939	0.8263	0.7396	0.6367
	5	1.0000	1.0000	0.9998	0.9988	0.9958	0.9887	0.9747	0.9502	0.9115	0.8555
	6	1.0000	1.0000	1.0000	0.9999	0.9996	0.9987	0.9964	0.9915	0.9819	0.9648
	7	1.0000	1.0000	1.0000	1.0000	1.0000	0.9999	0.9998	0.9993	0.9983	0.9961

n	r	P									
		0.05	0.10	0.15	0.20	0.25	0.30	0.35	0.40	0.45	0.50
9	0	0.6302	0.3874	0.2316	0.1342	0.0751	0.0404	0.0207	0.0101	0.0046	0.0020
	1	0.9288	0.7748	0.5995	0.4362	0.3003	0.1960	0.1211	0.0705	0.0385	0.0195
	2	0.9916	0.9470	0.8591	0.7382	0.6007	0.4628	0.3373	0.2318	0.1495	0.0898
	3	0.9994	0.9917	0.9661	0.9144	0.8343	0.7297	0.6089	0.4826	0.3614	0.2539
	4	1.0000	0.9991	0.9944	0.9804	0.9511	0.9012	0.8283	0.7334	0.6214	0.5000
	5	1.0000	0.9999	0.9994	0.9969	0.9900	0.9747	0.9464	0.9006	0.8342	0.7461
	6	1.0000	1.0000	1.0000	0.9997	0.9987	0.9957	0.9888	0.9750	0.9502	0.9102
	7	1.0000	1.0000	1.0000	1.0000	0.9999	0.9996	0.9986	0.9962	0.9909	0.9805
	8	1.0000	1.0000	1.0000	1.0000	1.0000	1.0000	0.9999	0.9997	0.9992	0.9980
10	0	0.5987	0.3487	0.1969	0.1074	0.0563	0.0282	0.0135	0.0060	0.0025	0.0010
	1	0.9139	0.7361	0.5443	0.3758	0.2440	0.1493	0.0860	0.0464	0.0232	0.0107
	2	0.9885	0.9298	0.8202	0.6778	0.5256	0.3828	0.2616	0.1673	0.0996	0.0547
	3	0.9990	0.9872	0.9500	0.8791	0.7759	0.6496	0.5138	0.3823	0.2660	0.1719
	4	0.9999	0.9984	0.9901	0.9672	0.9219	0.8497	0.7515	0.6331	0.5044	0.3770
	5	1.0000	0.9999	0.9986	0.9936	0.9803	0.9527	0.9051	0.8338	0.7384	0.6230
	6	1.0000	1.0000	0.9999	0.9991	0.9965	0.9894	0.9740	0.9452	0.8980	0.8281
	7	1.0000	1.0000	1.0000	0.9999	0.9996	0.9984	0.9952	0.9877	0.9726	0.9453
	8	1.0000	1.0000	1.0000	1.0000	1.0000	0.9999	0.9995	0.9983	0.9955	0.9893
	9	1.0000	1.0000	1.0000	1.0000	1.0000	1.0000	1.0000	0.9999	0.9997	0.9990

CHAPTER 21

Managing Quality in Operations: Service

Joseph A. De Feo

High Points of This Chapter

1. Activities to integrate quality in service planning have two objectives: to incorporate product features and to prevent defects (minimize variability).
2. For self-control, we must provide personnel with the knowledge of what they are supposed to do, knowledge of what they are actually doing, and a process that can meet specifications and can be regulated.
3. The basic steps for controlling quality can be applied to service operations, and process quality audits apply to any activity that can affect final service quality.
4. Three factors are important in frontline customer contact: selection, training, and empowerment of personnel.
5. Various types of quality teams play a key role in service processes.
6. Quality measures must be designed for each type of service organization.

The Service Sector

Service based organizations refer to businesses serving customers through some means other than manufacturing or the production of goods. The service producing segment is made up of utilities; wholesale trade; retail trade; transportation and warehousing; information; finance, insurance, real estate, rental, and leasing; professional and business services; educational services, health care, and social assistance; arts, entertainment, recreation, accommodation, and food services; and other services, except government.

Another set of services is the new wave of services based on non-human interactions between service providers and service consumers has started to flood our contemporary life with incredible speed and range. These are the self-services and they are virtually everywhere-streets, shops, banks, restaurants, airports, and obviously on the Internet. Their astonishing growth has been propelled by the recent revolution in telecommunication and information technologies and most notably by the expansion of the Internet. Self-services are transforming the way we live and do business. Their full impact on our social, cultural and business environments remains yet to be seen.

In many service organizations, the cost of poor quality ranges from 25 to 40 percent of operating expenses. For example, in one large bank, the cost of poor quality is 37 percent for automatic teller machines (ATMs), 26 percent for customer inquiry centers, and 50 percent for commercial loan operations (Juran's Quality Handbook, Fifth Edition [JQH5], p. 33.11). Such numbers prove the need for addressing quality in the service sector.

As defined in Juran's Quality Handbook, Sixth Edition, p. 74, a service is work that is performed for someone else. A carpenter builds a home for a homeowner, the user; an automotive technician repairs cars for their owners; a nurse cares for patients; and a web browser provides information to meet the needs of its users. The overriding characteristic is that a service is intangible: it cannot be touched, gripped, handled, looked at, smelled, or tasted. Other characteristics are that the service is provided when the customer requests it, the service output is created as it is delivered, the service usually cannot be stored in an inventory, and completion time is critical. Services are often, but not necessarily, involved with the delivery of a tangible, physical good (a restaurant meal or plastic pellets).

Services account for more than two-thirds of the US economy. The service sector consists of a wide spectrum of industries. The major categories of service industries are listed in Table 21.1. Additionally, almost all companies providing tangible goods also provide services to their customers.

- Transportation (railroads, airlines, bus lines, subways, common carrier trucking, pipelines)
- Public utilities (telephone communication, energy services, sanitation services)
- Marketing (retail, food, apparel, automotive, wholesale trade, department stores)
- Finance (banks, insurance, sales finance, investment)
- Real estate
- Restaurants, hotels, and motels
- News media
- Business services (advertising, credit services, computer services)
- Health services (nursing, hospitals, medical laboratories)
- Personal services (amusements, laundry and cleaning, barber and beauty shops)
- Professional services (lawyers, doctors)
- Repair services (garages, television and home repairs)
- Government (defense, health, education, welfare, municipal services)

TABLE 21.1 Categories of Service Industries

As in the case of a manufactured product, quality means customer satisfaction and loyalty. Satisfaction and loyalty are achieved through two components: product features and freedom from deficiencies.

Research by Cronin and Taylor (1992) indicates that service quality is only one component in a consumer's buying preference—the total value of the offering (service quality, price, convenience, and availability) must be considered. Additionally, the research confirms that service quality is perceived and should be measured differently among service industries, specifically between high-involvement service providers (health care) and low-involvement providers (fast food). Finally, we note that because customer expectations of service continue to change rapidly, service design that incorporates the product features and a process design free of deficiencies must evolve rapidly.

Initial Planning for Quality

Chapter 2, Developing an Excellence Culture, presents a road map for planning the quality of a new product. The steps are: establish the project, identify the customers, discover customer needs, develop the product, develop the process, and establish process controls/ transfer to operations.

In service industries, the service design defines the features of the output provided to customers to meet their needs. Thus, the concepts presented in Chap. 4, Quality Planning and Design of New Goods and Services apply to both the manufacturing and service sectors. The service design is turned into a reality by the service process, i.e., the process features such as the work activities, people, equipment, and physical environment to meet customer needs.

The service process can be reviewed for quality by several means: analyze the process flow diagram, reduce the cycle time, error-proof the process, provide for supplier quality, qualify the process by validating the process and measurement capability, and plan for personnel self-control. These topics are discussed here and in Chap. 23, Managing Quality in the Supply Chain.

Analysis of the Process Flow Diagram

Process flow diagrams (alias process maps or process blueprints) are also discussed in Chap. 19, Graphical Tools to Improve Process Performance. Figure 21.1 shows a flow diagram for handling a request for adjustments to customer bills (AT&T). The "line of interaction" is the boundary of activities where the customer and frontline employees ("online group") have discussions. The "line of invisibility" separates activities that are seen or not seen by customers. The "organization boundary" shows which activities occur in the three departments involved in the process. Note how this example illustrates both frontline direct customer contact and backroom or back-office operations. Also note that the time required is shown for some activities.

The symbol P denotes process points at which problems could occur that would cause customer dissatisfaction. To prevent problems, we must identify potential problems, usually based on past data or an analysis of the flow diagram. In the service sector, problems arise for recurring reasons (AT&T):

- Promises are not kept.
- Customers must contact several people to achieve problem resolution.
- Only partial service is provided, the service is performed incorrectly, the wrong service is performed, and the wrong information is provided.
- Customers do not understand the service provided.
- Service is not provided when needed or takes too long.
- The customer is inconvenienced with paperwork or other matters.

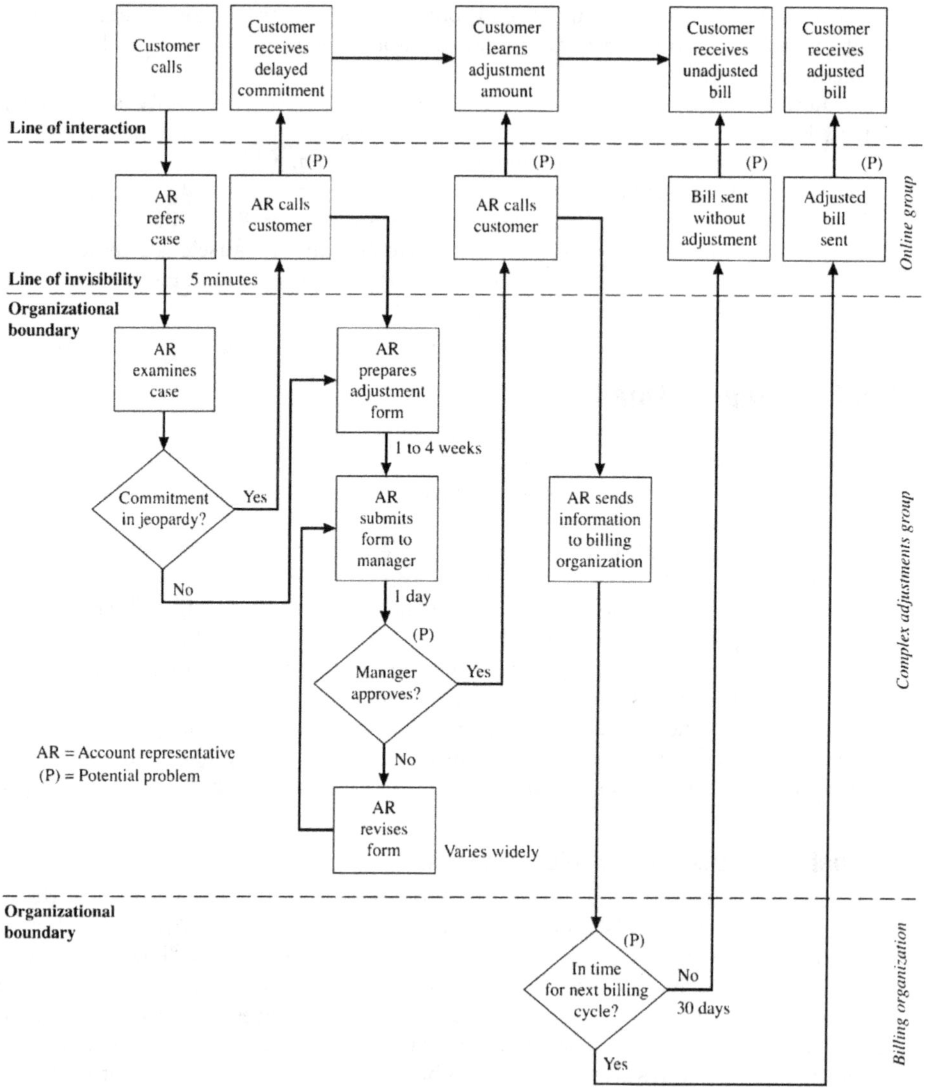

FIGURE 21.1 Service Blueprint Diagram. (Reproduced with permission of AT&T.)

These known or potential problems should be identified on the flow diagram and preventive actions put in place. If necessary, the problems can be regarded as process failures and analysis tracked using the approach of a failure mode, effect, and criticality analysis.

Reduction of Process Cycle Time

One of the most widely used metrics to measure service delivery is time. Customers continually demand shorter delivery times for services provided, and creating and analyzing a process flow diagram is an excellent tool to reduce cycle time in the context of a quality improvement project. See Chap. 16, Root Cause Analysis, for the universal, structured approach for diagnosing the causes and taking corrective action. This universal approach underpins the Six Sigma project approach. Lean efforts, focusing on the elimination of waste, are often used in used in service organizations to reduce cycle time. Among the common actions typically taken are:

- Eliminate rework loops to correct process errors.
- Eliminate or simplify steps of marginal value to the customer.
- Eliminate redundant steps such as inspections or reviews (but not until the causes of errors have been determined and eliminated).
- Combine steps and have them done by one worker, by several workers.
- Transfer approval steps to lower levels.
- Change the sequence of activities from consecutive steps to simultaneous steps.
- Perform a step after serving a customer rather than before.
- Use technology to perform both routine and complex steps.

Marriott Hotels analyzed the elapsed time of the guest registration process. The analysis led to the integration of two sub-processes—the front-desk function and the bell-person function (to escort new guests to their room and carry their luggage). Under a revised process the front desk now pre-assigns a suitable guest room and assembles registration information and keys for the bell-person who greets and escorts the guests directly to their rooms (Hadley, 1995).

Error-Proofing the Process

An important element of prevention of errors is error-proofing the process. Five important principles of error-proofing a process are:

- Eliminate the error-prone activity.
- Substitute a more reliable process.
- Make the work easier for the worker.
- Detect errors earlier.
- Minimize the effect of errors.

In the service sector, examples of error-proofing include eliminating non-value- added activities, automating routine or unpleasant human tasks, using software to detect missing or grossly erroneous information, using bar codes at the supermarket checkout, using countdowns and checklists during surgical operations, entering medical prescriptions into a computer to replace handwritten prescriptions, and performing automatic built-in inspections. Obviously, technology is an important source of error- proofing aids.

Input from Information Technology as an Internal Supplier

Service industries need to plan, control, and improve the quality of inputs from external suppliers. The basic approach is provided in Chap. 23, Managing Quality in the Supply Chain. The operations function within a service industry also has internal suppliers, and one of them, information technology (IT), deserves particular note.

Many of the significant advances in providing new service product features with amazing speed are due to the unending contributions of computers and related IT software and equipment.

These contributions have, unfortunately, been accompanied by significant problems in the services provided to operations people by the IT activity. Operations people report that these problems are sporadic and chronic quality-related problems that result in customer dissatisfaction and higher costs.

Research provides some understanding of the nature of these IT-related problems (Kittner et al., 1999). Operations managers from the financial services sector (but not from the IT or quality functions) were brought together to determine the perceived impact of the quality of IT services on the quality of the output of the operations function. The "operations" activities were process customer transactions, including both direct contact with external customers and backroom activities that did not involve direct contact.

First, the participants identified the outputs the IT department provided to operations as summarized in 11 categories in Table 21.2.

Next, participants were asked to identify problems involved in these services. The managers identified 115 issues—some were general, such as "downtime," and others were quite specific, such as "can't hire C++ programmers fast enough to keep up with user demand." These issues were further analyzed to identify 57 issues related to customer satisfaction. Participants were then asked to score each item on a scale of 1 to 10, where 10 was most important and 1 was least important. The mean scores of the top 10 issues are shown in Table 21.3.

In summary, Table 21.3 shows the perceptions of operations managers of problems with input received from IT and the relative importance of these problems to output of the operations function. Conclusion: Significant sporadic and chronic problems exist that cannot be solved by the traditional IT help desk. Chronic problems require the structured approach to quality improvement described in Chap. 16; sporadic problems require the troubleshooting approach.

Response Category	Examples
Provide reports	Summary of sales activities, list of "at risk" accounts
Provide hardware and software support	Personal computer support services, guidance on hardware and purchasing
Record information	Posting to accounts, current rate information
Provide data communications support	Local area network connections and software access, data security
Provide for system availability and maintenance	Online availability of all systems, ensure e-mail systems are up and running
Provide online information	Automatic teller machine online, customer account information
Provide support for data processing	Download all accounts for customer information; download all workplace banking customers and their profiles
Develop new systems	Application development for new systems; systems development, testing, and design
Implement new systems	Assist in the development of testing plans and rollout, implementation of new systems
Provide programming support	Provide programming support
Provide training	Software training

TABLE 21.2 IT Services Provided to Operations

Issue	Mean
Accuracy of information	9.33
System downtime	8.92
System response time	8.58
Network reliability	8.42
System performance	8.33
Level of expertise	8.17
Hardware performance	8.09
Thorough testing—both IT and users need to be involved	8.08
Hardware problems	7.92
Lack of training	7.75

TABLE 21.3 Top 10 Quality Issues

The demands continually made on the IT function for information services using current software and the development of new software make it almost impossible to address chronic problems effectively. But these problems often have a serious impact on external customer satisfaction. A quality department can help by taking a strong initiative to collect data to prove that action is necessary. Such proof can help to convince upper management to provide the IT and operations functions with the resources and a frame-work for improvement based on the project-by-project approach described in Chap. 5, Quality Improvement and Breakthrough Performance.

Input from Customers as an External Supplier

In service organizations, the customer is often a supplier. Customers may provide the detailed data for a service transaction, e.g., buy 100 shares of American Express stock at a price not to exceed $Y, order a dress of size 16 in the rose color, reserve a hotel room for March 12 to March 18. The customer input can be in error, resulting in delays to the customer and extra time and costs to the organization that must rectify the customer's error. JQH5, p. 33.17, describes three approaches that financial service providers use to prevent customer errors: customer education, error-proofing, and monitoring and measuring customer input.

Measuring Process Capability

We need to ensure that the process can meet customer needs and goals under normal operating conditions. When customer needs can be quantified in particular parameters, then process capability indexes can be used to evaluate the process. In the Six Sigma approach, the process capability can be described in units of sigma, e.g., a process might be at a level of 4.8 Sigma out of the ideal of 6 Sigma. These matters are discussed in Chap. 12, The Role of Statistics and Probability. An important quantitative parameter in the service sector is the time to complete a service transaction.

A preliminary measure of capability can be obtained by simply collecting a sample of data and comparing it to the specifications for a process. For example, consider the process for repairing ATM machines. An analysis concluded that the service organization should be able to respond in 10 minutes, and customers indicated that a machine should not be out of service for more than 90 minutes. A team looked at historical data measuring the repair time each day for a 6-month period. The results are shown in the histogram of Fig. 21.2. Note that

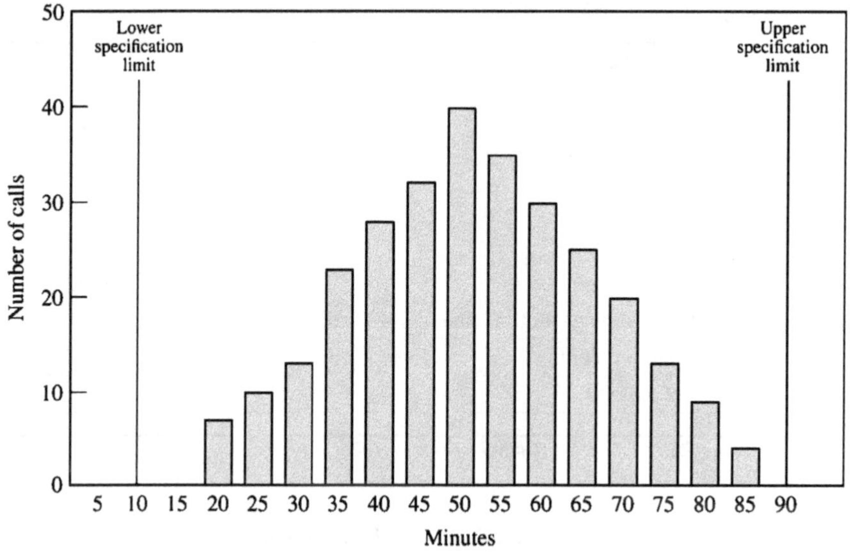

FIGURE 21.2 Time to repair machines. (From Juran Institute, Inc.)

some observations exceeded the maximum time of 90 minutes. Thus the process is not capable. The process must be changed and tested again to verify that it is capable. This preliminary capability study is based on available data. A full study under controlled conditions should be conducted following the methods described in Chap. 20, Accurate and Reliable Data and Measurement Systems.

A broader approach to evaluating a service process measures four parameters: effectiveness (of output), efficiency, adaptability, and cycle time.

IBM defines five levels of process maturity. The highest level, level 1, designates a business process that operates at maximum effectiveness and efficiency and is a benchmark or leader; the lowest level, level 5, suggests a process that is ineffective and may have major deficiencies. Melan (1993) defines the specific criteria for each level. Criteria include organizational matters (e.g., a process owner) and technical matters (e.g., measurements of effectiveness and efficiency).

Planning for Self-Control

This section applies the concept to service industries and provides checklists to evaluate planning for quality in services.

For self-control, we must provide people with:

1. Knowledge of what they are supposed to do
   a. Clear and complete work procedures
   b. Clear and complete performance standards
   c. Adequate selection and training of personnel
2. Knowledge of what they are actually doing (performance)
   a. Adequate review of work
   b. Feedback of review results

3. Ability and desire to regulate the process for minimum variation
   a. A process and job design capable of meeting quality objectives
   b. Process adjustments that will minimize variation
   c. Adequate worker training in adjusting the process
   d. Process maintenance to maintain the inherent process capability
   e. A strong quality culture and environment

In designing for self-control, technology is of increasing help—for "supposed to do," information can be placed on line and kept up-to-date instantly; for "knowledge on how we are doing," we can now often provide instant feedback; for "regulating a process," mechanisms can be integrated into a process to provide adjustments when necessary.

Most service organizations do not adhere to the three elements and 10 sub-elements of self-control. Collins and Collins (1993) provide six examples from the manufacturing and service sectors illustrating problems that were originally blamed on people but were management controllable. Berry et al. (1994) identifies 10 lessons learned in improving service quality. Three of these are service design (the real culprit is poor service system design), employee research (ask employees why service problems occur and what they need to do their jobs), and servant leadership (managers must serve by coaching, teaching, and listening to employees). Note that these three lessons are directly related to the concept of self-control.

The concept of self-control applies to the service sector—both to frontline direct customer contact operations and to backroom operations. Based on research with personnel in the financial services industry, Shirley and Gryna (1998) developed checklists for self-control that are presented here.

Criterion 1: Knowledge of "Supposed to Do"

Providing personnel in the service sector with the knowledge of what they are supposed to do is essential for self-control. The following checklist can help to evaluate this criterion.

Work Procedures

1. Are job descriptions published, available, and up-to-date?
2. Do personnel know who their customers are? Have personnel and customers ever met?
3. Do personnel who perform the job have any impact on the formulation of the job procedure?
4. Are job techniques and terminology consistent with the background and training of personnel?
5. Do guides and aids (e.g., computer prompts) lead personnel to the next step in a job?
6. Do provisions exist to audit procedures periodically and make changes? Are changes communicated to all affected personnel?
7. Do provisions exist for deviations from corporate directives to meet local conditions?
8. Are procedures "reader friendly"?
9. Does supervision have a thorough knowledge of operations and provide assistance when problems arise?
10. Do procedures given to personnel fully apply to the job they do in practice?
11. Have personnel responsibilities been clearly defined in terms of decisions and actions?

12. Do personnel know what happens to their output in the next stage of operations and understand the consequences of not doing the job correctly?
13. If appropriate, is job rotation used?

Performance Standards

1. Are formal job standards for quality and quantity needed? If yes, do they exist? Are they in written form?
2. Have personnel been told about the relative priority of quality versus quantity of output? Do personnel really understand the explanation?
3. Are job standards reviewed and changed when more tasks are added to a job?
4. Do personnel feel accountable for their output, or do they believe that shortcomings are not under their control?
5. Does information from a supervisor about doing a job always agree with information received from a higher level manager?

Training

1. Are personnel given an overview of the entire organization?
2. Is there regularly scheduled training to provide personnel with current information on customer needs and new technology?
3. Do personnel and their managers provide input to their training needs?
4. Does training include the why, not just the what?
5. Does the design of the training program consider the background of those to be trained?
6. Do the people doing the training provide enough detail? Do they know how to do the job?
7. Where appropriate, are personnel who are new to a job provided with mentors?

Criterion 2: Knowledge of Performance

For self-control, people must have the means of knowing whether their performance conforms to standard for product and process characteristics. The following checklist can help to evaluate this criterion.

Review of Work

1. Are personnel provided with the time and instructions for performing self-review of their work?
2. Can errors be detected easily?
3. Are independent checks on quality needed? Are they performed? Are these checks performed by peer personnel or others?
4. Is a review of work performed at various checkpoints in process, not just when work is complete? Is the sample size sufficient?
5. Does an independent audit of an entire process ensure that individual work assignments are integrated to achieve process objectives?
6. Where appropriate, are detailed logs of customer contacts kept?

Feedback

1. Do upper management and supervision provide the same message and actions on the importance of quality versus quantity?
2. If needed, do standards exist for making corrections to output?
3. Where appropriate, is feedback provided to both individuals and a group of personnel? Is time provided for discussion with the supervisor, and does the discussion occur?
4. Is feedback provided to those who need it? Is it timely? Is it personnel-specific?
5. Does feedback provide the level of detail needed particularly to correct problem areas? Have personnel been asked what detail is needed in the feedback?
6. Is feedback provided from customers (external or internal) to show the importance of the output and its quality?
7. Does feedback include information on both quality and quantity?
8. Are both positive and negative (corrective) feedback provided?
9. Is negative feedback given in private?
10. Do personnel receive a detailed report of errors by specific type of error?
11. Where appropriate, are reports prepared to describe trends in quality (in terms of specific errors)? Are such reports prepared for individual personnel and for an entire process performed by a group of people?
12. Are certain types of errors tracked with feedback from external customers? Could some of these be tracked with an internal early indicator?

A credit card provider has identified 18 key processes such as credit screening and payment processing. For the total of 18 processes, more than 100 internal and supplier process measures were identified. Daily and monthly performance results are available through video monitors and are also posted. Each morning, the head of operations meets with senior managers to discuss the latest results, identify problems, and propose solutions. Employees can access a summary of this meeting via telephone or email. The measurement system is linked to compensation by a daily bonus system that provides up to 12 percent of base salary for non-managers and 8 to 12 percent for managers (Davis et al., 1995).

Criterion 3: Ability to Regulate

The process given to personnel must be capable of meeting requirements, and the job design must include the necessary steps and authority for personnel to regulate the process. Here is a checklist to evaluate the ability to regulate.

Job Design

1. Can the process (including procedures, equipment, software, etc.) given to personnel meet standards for quality and quantity of output? Has this capability been verified by trial under normal operating conditions?
2. Has the design of the job used the principles of error-proofing?
3. Does the job design minimize monotonous or unpleasant tasks?
4. Does the job design anticipate and minimize errors due to normal interruptions in the work cycle?

5. Can special checks be created (e.g., balancing of accounts) to detect errors?
6. Can steps be incorporated in data entry processes to reject incorrect entries?
7. Does the job design include provisions for action when wrong information is submitted or information is missing as an input to a job?
8. Is paperwork periodically examined, and are obsolete records destroyed to simplify working conditions?
9. When the volume of work changes significantly, do provisions exist to adjust individual responsibilities or add resources?
10. Do external factors (e.g., no account number on a check, cash received instead of a check) hinder the ability to perform a task?
11. Are enough personnel cross trained to provide an adequate supply of experienced personnel to fill in when needed?
12. If appropriate, is a "productive hour" scheduled each day in which phone calls and other interruptions are not allowed, thus providing time to be away from the work location to attend to other tasks?
13. Has equipment, including any software, been designed to be compatible with the abilities and limitations of personnel?
14. Does an adequate preventive maintenance program exist for computers and other equipment used by personnel?
15. Do some personnel possess a hidden knack that needs to be discovered and explained to all personnel?
16. For a job requiring special skills, have personnel been selected to ensure the best match of their skills and job requirements?

Changes in Job Design

1. Are proposed changes limited by technology (e.g., address fields on forms)?
2. Can personnel institute changes in a job when they show that the change will provide benefits? Are personnel encouraged to suggest changes?
3. What levels of management approval are required for instituting proposed changes? Could certain types of changes be identified that do not need any level of management approval?
4. Do management actions confirm that they are open to recommendations from all personnel?

Handling Problems

1. Have personnel been provided with the time and training to identify problems, analyze problems, and develop solutions? Does this training include diagnostic training to look for patterns of errors and determine sources and causes?
2. Are personnel permitted to exceed process limits (e.g., maximum time on a customer phone call) if they believe it is necessary?
3. When personnel encounter an obstacle on a job, do they know where to seek assistance? Is the assistance conveniently available?

These checklists can help operations in the design (and redesign) of jobs to prevent errors, the diagnosis of quality problems on individual jobs, and the identification of common weaknesses in many jobs. The checklists can also help supervisors coach personnel, prepare for process audits, and develop training classes on quality. We proceed now from planning for quality in service operations to the control of quality in operations.

Control of Quality in Service Operations

The manufacturing sector has a long history of detailed procedures for controlling quality during operations. The formalization of such procedures in the service sector is still evolving. The basic steps for controlling quality are described in Chap. 6, Quality Control to Assure Compliance to Customer Requirements. These steps are: choose control subjects, establish measurement, establish standards of performance, measure actual performance, compare performance to standards, and take action on the difference. Applying these steps to service operations is discussed next.

Choose Control Subjects

To choose control subjects, we must first identify a major work process, identify the process objective, describe the process, identify customers, and discover customer needs. These steps, discussed in Chap. 6, lead to the selection of control subjects. Examples of control subjects in several service industries are given in Table 21.4.

Establish Measurement

In this step, we develop a unit of measure and the means of measurement (the sensor)—see Chap. 20. The unit of measure must be understandable to all, specific enough for decision making, and customer focused.

Control subjects can be a mixture of features of the product, features of the process, and side effects of the process. Quantification of control subjects involves two kinds of indicators that must be explicit for those running the process:

1. Performance indicators. These indicators measure the output of the process and its conformance to customer needs, as defined by the unit of measure for the control subject.

Major Work Product	Major Work Process	Control Subjects
Medical insurance	Claims processing	Accuracy of claim form
		Completeness of supporting documentation
Catering services	Food preparation	Freshness of ingredients
		Oven temperature
24-hour banking services	Maintenance of ATM machines	Availability of cash
		Number of service people available
Photo developing	Film processing	Maintenance of chemicals
		Accuracy of placement of film on spool

Source: Juran Institute. Inc.

Table 21.4 Control Subjects

Indicator	Weight
Abandoned calls	1
Complaints reopened	5
Damaged packages	10
International	1
Invoice adjustments requested	1
Lost packages	10
Missed pickups	10
Missing proofs of delivery	1
Overgoods (lost and found)	5
Right-day late deliveries	1
Traces	1
Wrong-day late deliveries	5

Source: American Management Association (1992).

TABLE 21.5 Federal Express Service Quality Indicators

2. Process indicators. These indicators measure activities or variation within the process that affect the performance indicators.

Table 21.5 shows 12 quality indicators at Federal Express along with the importance weights assigned to each indicator.

These measures are tracked every day, both individually and in total. Figure 21.3 shows how a division of AT&T related internal metrics to business processes and customer needs.

Specific measurements can be related to the underlying factors that dominate a process: setup, time, component, worker, and information.

Deyong and Case (1998) provides a methodology for linking customer satisfaction attributes with process metrics in service industries.

Establish Standards of Performance

This step involves setting target values for each performance indicator and each process indicator and also maximum and minimum limits.

Measure Actual Performance

Sometimes the measurement process can be automated. For example, the number of calls waiting to be answered in an insurance service center is clearly displayed; the elapsed time in filling a customer order at a fast-food franchise is clearly visible (the goal is 45 seconds).

In other cases, the measuring process is more complex. For example, to measure the quality of customer contact activities, banks and other organizations use "mystery shoppers"—researchers posing as customers to assess key dimensions of quality delivery.

Compare Performance to Standards

Sometimes the comparison must go beyond a simple analysis of average values. For example, in Fig. 21.4, two banks are compared on the time to process a loan. Both banks average

Overall quality	Business process	Customer need	Internal metric
	Product (30%)	Reliability (40%)	% repair call
		Easy to use (20%)	% calls for help
		Features/functions (40%)	Function performance test
	Sales (30%)	Knowledge (30%)	Supervisor observations
		Response (25%)	% proposals made on time
		Follow-up (10%)	% follow-ups made
	Installation (10%)	Delivery interval (30%)	Average order interval
		Does not break (25%)	% repair reports
		Installed when promised (10%)	% installed on due date
	Repair (15%)	No repeat trouble (30%)	% repeat reports
		Fixed fast (25%)	Average speed of repair
		Kept informed (10%)	% customers informed
	Billing (15%)	Accuracy, no surprises (45%)	% billing inquiries
		Resolve on first call (35%)	% resolved first calls
		Easy to understand (10%)	% billing inquiries

FIGURE 21.3 Relating internal metrics to business processes and customer needs. (From Kordupleski, R. E., Rust, R. T., and Zahorik, A. J. (1993). "Why Improving Quality Doesn't Improve Quality (Or Whatever Happened to Marketing?)," *California Management Review*, Spring, vol. 35, no. 3, pp. 82–95. [Copyright © 1993, by The Regents of the University of California. Reprinted from the California Management Review, vol. 35, no. 3. By permission of The Regents.])

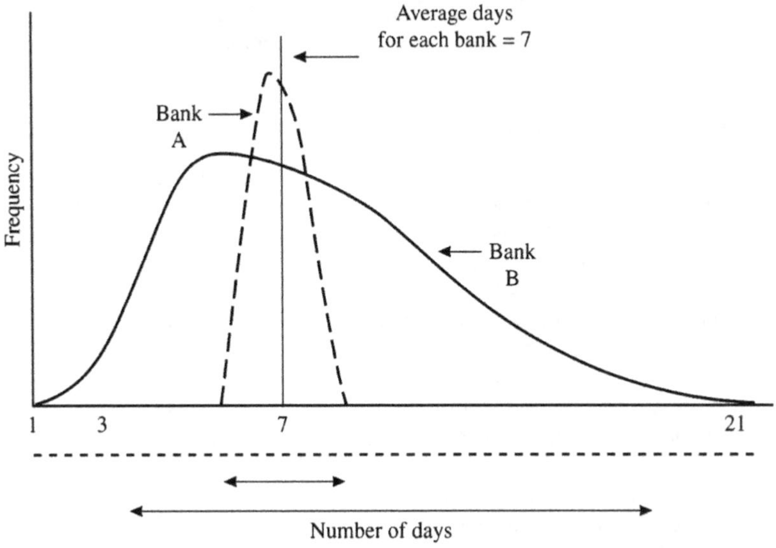

FIGURE 21.4 Time to process a loan. (From Juran Institute, Inc.)

about 7 days. For bank A, almost all loans are processed in a 5- to 8-day window; bank B takes an average of 7 days, but some loans take about 21 days to process.

The comparison of actual to a target may use statistical process control techniques to distinguish between common and special causes. This approach is explained in Chap. 19, Graphical Tools to Improve Process Performance.

Take Action on the Difference

Failure to meet standards may require one of three actions: troubleshooting, quality improvement, or quality planning. The steps for troubleshooting are similar to those for diagnosis in quality improvement. This process includes developing theories on possible causes and testing the theories (using data) to find the cause of the problem. The process for troubleshooting and diagnosing sporadic problems is usually simpler than the process for achieving improvement on chronic problems.

Process Quality Audits

A quality audit is an independent review conducted to compare some aspect of quality performance with a standard for that performance. A discussion of various types of quality audits is presented in Chap. 9, Quality Assurance and Audits.

A process quality audit includes any activity that can affect final product quality. This onsite audit is usually made on a specific process and uses process operating procedures. Adherence to existing procedures is emphasized, but audits often uncover situations of inadequate or nonexisting procedures. The self-control checklists presented earlier in this chapter can suggest useful specific subjects for process audits.

A major airline employs process audits to evaluate service in three areas: airport arrivals and departures, aircraft interior and exterior, and airport facilities. Forty-seven specific activities are audited periodically, and then performance measurements are made and compared with numerical goals. Two examples on the aircraft are the condition (appearance) of carpets inside the planes and the adhesion of paint on the planes.

McDonald's Corporation audits restaurants through a series of announced and unannounced visits. The audit includes quality, service, cleanliness, and sanitation. Highly detailed audit items include numerical standards on food-processing variables.

Frontline Customer Contact

A basic activity in the service industries is the service encounter, i.e., the contact made with the client when meeting a customer's need. Typically, examples involve a bank teller processing deposits or withdrawals of money, a flight attendant providing services on an airplane, or a hotel clerk registering a guest. In all cases, the quality of the transaction involves both the technical adequacy of the result and the social skills of the "frontline" person who conducts the transaction. Three factors emerge as important: selection of the frontline employee, training of that employee, and "empowerment" of the employee to act to meet customer needs.

In person-to-person type transactions, employee selection often has an immediate, direct, and lasting impact on customer perception. Some people have the necessary personal characteristics for frontline personnel; some people do not have these characteristics, even with training. Proper selection requires that we identify the personal characteristics required for a position (how about asking experienced employees?), use multiple interviews, train managers in interview procedures, identify nominees among present employees, and ask for recommendations from present employees. As one example, Federal Express selects people using scientifically prepared profiles of successful performers.

Training, of course, is essential. Superior-quality service organizations devote from 1 to 5 percent of employee working hours to training. The content of the training depends on the job requirements, but often it stresses product knowledge. In addition, training involves activities such as role playing to handle situations when a transaction goes wrong and handling an irate customer who has a complaint. Such training takes time and special effort, e.g., at Lands' End, each mail-order telephone representative spends time in a warehouse viewing the products. Isn't it impressive when such a person is able to describe the color "medium gray" to a customer? Basically, however, the training must enable the employee to provide customers with dependable and regular service. Customers are impressed when they know that they can depend on uniformly good service from an organization. When an extraordinary situation arises and it is handled well, we have a delighted customer.

Empowerment, a key step beyond training, involves giving a new degree of authority to frontline employees. The term usually means encouraging employees to handle unusual situations that standard procedures don't cover. In the past, the employee would check with a superior—while the customer waited—and waited. The concept minimizes the use of the rule book and maximizes the use of the frontline employee's knowledge, initiative, and judgment to take the action necessary to meet the need of a customer standing at a service counter. For example, the policy manual at Nordstrom's department store states: "Use your own best judgment at all times." Risky? Some people would think so, but the opportunities for customer satisfaction and the employee attitude toward "ownership" of his or her job are strongly convincing.

Bank One Corp. uses an extensive shopper survey to measure the performance of bank tellers. The survey queries customers on specific contact items such as a friendly greeting, employee identification, eye contact, a smile, use of the customer's name during the transaction, undivided employee attention, accurate processing of the transaction, ability to give clear explanations, a professional appearance, and a neatly organized work area.

The investment in selection, training, and empowerment of employees can lead to the fulfillment of a company's dream—delighted customers. Table 21.6 provides some categories of action and examples of remarkable exploits to generate delighted customers.

Action	Example
Providing a service far beyond the scope of the company service	An airline flight attendant accompanied a sick passenger and her daughter to a hospital.
Providing a service beyond the call of a normal effort	To eliminate the sound of tinkling glass in a chandelier due to air conditioning, hotel employees removed every second piece of glass 30 minutes before the start of the meeting.
Providing extraordinary recognition of customer inconvenience	An automobile manufacturer paid a customer $985 for "lost time" incurred by the customer on excessive and improperly made repairs.
Providing a recognition of personal customer loss	A customer reported that she lost a pen, having sentimental value, in a grocery store. A clerk searched for the pen but with no success. The clerk presented the lady with three $20 gift certificates.

TABLE 21.6 Some Action Taken to Achieve Delighted Customers

The service encounter goes beyond "speed, content, and attitude." A research study of five service organizations and nine manufacturers identified 15 competencies that exemplify superior frontline service (Jeffrey, 1995). The top six competencies were considered the most important by the organizations and their customers:

1. Build customer loyalty and confidence. Meet customer needs and do what is sensible to maintain customer goodwill.
2. Empathize with customers. Be sensitive to customer feelings and show genuine concern and respect.
3. Communicate effectively. Be articulate and diplomatic.
4. Handle stress. Stay organized and calm and show patience.
5. Listen actively. Interpret the meaning of the customer's words.
6. Demonstrate mental alertness. Process information quickly.

These competencies are not easy to achieve. Organizations must invest in careful selection of employees, training, and empowerment and then develop the employees into professionals to ensure retention of these key people.

Frontline personnel can serve as "listening posts" for an organization. When the listening posts are well designed to probe and ask specific questions, the information generated can help to develop new products and to sell existing products of which the customer is not aware. Some frontline personnel are uncomfortable asking certain questions that may lead to answers that the customer and the bank teller often prefer not to discuss, e.g., Why are you transferring your account to another bank? In some organizations, sales personnel and customer transaction personnel use every interaction as an opportunity to probe and ask specific questions. See Table 21.7. When the listening posts are well designed and the information they generate is well coordinated, their contribution can help to develop new products and to sell existing product that is already developed of which the customer is not aware.

USAA and American Express are companies that collect and analyze information generated in all types of customer contact. All written and telephone complaints, compliments, and inquiries are collected, classified, and analyzed. The resulting information provides a clear indication of what is going well, what needs to be improved, and what new needs are emerging from customers.

Another way to gather customer information for future product development is to use customer management software to track every encounter with a customer.

Call centers and help desks are now a main mechanism for a service encounter. When the call concerns a customer problem, typically, resolution of the problem is not achieved with the first phone call reporting the problem. At one health maintenance organization, 36 percent of the calls were related to a lack of complete, accurate, and timely information between suppliers and customers. Cross (2000) discusses some of the issues involved.

- Is there anything else I can do for you?
- Are there any problems or needs today you have in managing your financial affairs?
- Are there any products or services you wish we offered?
- Are you aware of any products on the market that we don't offer?

Table 21.7 Questions to Discover Customer Needs

Organizing for Quality in Service Operations

As in all industries, top-down management understanding, appreciation, and support of any organization-wide quality program is mandatory. It all starts at the top. If only nodding approval or a "let's wait and see" attitude is perceived by the employees, a quality improvement program will fail. The employees watch what management does much more than what management says. A quality council made up of senior executives should be established to oversee the quality improvement program as soon as possible. The council should meet regularly—quarterly would be a desirable interval. Their interest and guidance to the program will speak loudly to the organization.

Teams are a common means of driving improvement in the service industries. Teams may be ad hoc to address a specific problem or may be permanent to be responsible for a specific activity. For example, the American Express Consumer Card Group uses semiautonomous work teams. A team consists of 10 to 12 employees in the natural work group. Team members do customer service work, manage quality, inventory, and attendance; prepare work schedules; and prepare production reports and forecasts. Individual roles are defined to handle these team responsibilities. The team leader focuses on coaching, feedback, and special human resource issues.

One fast-food firm creates teams of crew members (workers at one location) who are trained to manage the site without a full-time manager (Harvard Business School, 1994). This approach means installing online sensor technology such as the time to prepare an order and providing crew members with the same operating and financial information provided to a restaurant general manager to run the site. Crew members make operating decisions such as ordering food materials. Thus knowledge that long separated "brain workers" from "hand workers" now resides in a computer on the operations floor.

The Ritz-Carlton Hotel company uses self-directed teams. These process teams are aligned with the way customers come in contact with the hotel: (1) pre-arrival team; (2) arrival, stayover, and departure team; (3) dining services team; (4) banquet services team; and (5) engineering and security team. In a self-directed work team, members may have specific individual roles, but the team shares accountability for meeting performance objectives.

Kaiser Permanente uses quality-in-daily-work teams. These frontline work teams focus on both work process improvements and the definition of a service guarantee to communicate service performance levels. The projects span across departments and include both clinical and support services. Table 21.8 shows a few examples.

Department	Customer	Essential Service	Service Guarantee
Nursing unit 7-West	Ambulance emergency department	Pick up patients within specified time frame	The critical care unit staff will pick up the patient within 15 minutes of emergency room's call
Admitting	Elective surgery patients	Admission process including explanation of forms	Provide service within five minutes of entering department
Education	Diabetic patients	Outpatient diabetic education program	95% of all diabetic patients will be scheduled for diabetic education
Pediatrics	Member	Telephone access	98% of all calls will be handled within two minutes

TABLE 21.8 Examples of Quality-in-Daily-Work Projects

Reduce cycle time of credit issuance
Reduce trading-call length
Reduce card issuance cycle time
Reduce encoding errors
Improve business travel contract-to-billing process
Eliminate nonreceived renewals
Eliminate incorrect fee adjustments
Improve XXX payment accuracy
Reduce YYY writeoffs

Table 21.9 Examples of Six Sigma Projects at a Financial Services Organization

For a discussion of research conducted on teams, see Katzenbach and Smith (1993). Mann (1994) explains how managers need to develop skills as coaches, developers, and managers of activities that reside in different departments ("boundary managers").

Six Sigma Projects in Service Industries

The Six Sigma approach to improvement includes the phases of define, measure, analyze, improve, and control. These phases are explained in Chap. 15. Six Sigma is increasingly being applied in the service sector. Table 21.9 shows examples of Six Sigma projects at American Express. Note the wide variety of projects in Table 21.9. Hahn et al. (2000) describes how GE Capital applied Six Sigma (including a modified full-factorial design of experiment) to reduce losses due to delinquent credit card customers. Bott et al. (2000) explains Six Sigma steps for two projects at American Express. Bottome and Chua (2005) applies Six Sigma and lean to reduce documentation errors at Genentech. Alonso et al. (2003) describes the implementation of Six Sigma at Telefonica de España (Telefonica), Spain's leading telecommunications company. Six Sigma is deployed as part of a far-reaching corporate culture change program, focused on customer satisfaction and quality as key differentiators.

Quality Measurement in Service Operations

The management of key work processes must include provisions for measuring process control and also for monitoring overall operations. Readers should review the basics of quality measurement in Chap. 20.

Measuring quality in service-based industries is often perceived as more difficult than in manufacturing. Modern manufacturing is old, with many scholars tracing its roots to the eighteenth century. By its nature, manufacturing has always relied on measurements, quantities, weights, and so on. Measurements seem to be quantifiable and intrinsic to manufacturing. Service-based industries, however, have often relied on qualitative measurements such as good, better, best and economy, deluxe, and luxury.

One approach to measuring service quality is the SERVQUAL model (Zeithaml et al., 1990). This model identifies five dimensions of quality:

- Tangibles: appearance of facilities, equipment, personnel, materials.
- Reliability: ability to perform dependably and accurately.
- Responsiveness: provide timely service.

Measure Number	Quality Measure
148	Number of abandoned calls/total calls
454	Average time to call pickup, seconds
458	Number of statement insertion errors
460	Number of applications not processed within standard
466	Number of hours credit card system down
467	Number of payments not posted

Source: JQH5, page 33.21.

TABLE 21.10 Quality Measures in a Credit Card Company

- Assurance: trust and confidence in employees.
- Empathy: individualized attention to customers.

Research by Cronin and Taylor (1992) confirms that service quality should be measured differently among service industries, specifically between high-involvement service providers (health care) and low-involvement providers (fast food).

Table 21.10 shows a few examples of quality measurements of overall operations in a credit card company. This organization can easily calculate the measures from basic raw data that are routinely collected in daily operations.

King and Dickinson (1996) present a framework (Fig. 21.5) of three measures of process effectiveness that are particularly suited for service processes.

Results measures are primarily customer perceptions of outcomes. These measures drive priorities for improvement and monitor performance improvement. Overview in- process measures predict results and are lead indicators of process outcomes. Changes in an overview measure will thus lead to changes in the related results measure(s). Overview measures help to initiate the search for root causes of poor performance.

Detailed in-process measures predict overview measures and are lead indicators of sub-process outcomes. Detailed in-process measures control the day-to-day operation of the process and provide early warnings and diagnostic information for improvement. Thus the approach forms a hierarchy of process control measures—exterior and interior to the organization and at different levels of a process.

As an example, consider the process of having a new phone installed at an arranged time. The results measure is the proportion of customers who report that the technician

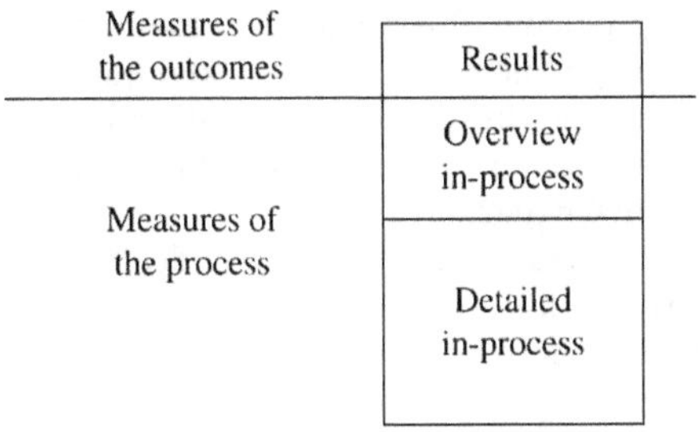

FIGURE 21.5 Framework of process effectiveness measures. (From King, M. N. and Dickinson, T. (1996). "A Framework for Process Effectiveness Measures," *Quality Engineering*, vol. 9, no. 1, pp. 45–50. [Reprinted with permission from Quality Engineering, © 1996 American Society for Quality.])

arrived on time. The overview in-process measure is also the proportion of appointments met, but this measure is logged by the technicians. Thus the results and the overview measures in this case are the same measure but collected both within and outside the process. The detailed in-process measures are many and might include proportion of customers who could not be given appointments at the time they requested or incidences and reasons for staff shortages.

Quality measurements are candidates for data analysis using statistical techniques such as the control charts discussed in Chap. 19. But the more basic point is that the reporting of data demonstrates to operating personnel that management regards quality as a high priority.

Figure 21.6 presents a flowchart for defining, collecting, and analyzing metrics in a service organization.

The Problem of "No Data!"

A frequent first lament of someone starting a quality program in a service industry is, "There are no data!" Although it is true in some instances that there are little useful or accurate data when starting out on a quality program, there often are more data than we think. Approach the information technology (IT) department of the organization early in a quality improvement program. It often has vast quantities of data that are routinely collected but not disseminated. A quick query of existing databases may yield a surprising quantity of data. The veracity of all data should be checked before using or relying on analysis of the data.

If there truly are little or no data available or if the data are corrupt and unreliable, then start gathering new data that will be useful. As with all change, suddenly starting to gather data where none had been gathered before can be alarming to employees or customers or both. There is a natural fear of change of being measured when one has not been measured in the past. Appropriate introduction of change methodologies should be used prior to gathering new data to allay such fears. If this step is skipped and misunderstanding of the real purpose of gathering the data exists, it can doom a quality program from the start. Too many individuals taking time out to just gather data or explain the purpose of gathering the data or the value of a quality system may seem a waste of time. The eagerness of the starting moment may be hard to resist. Keep in mind the old saying that applies here well. "Go slow to go fast." This saying is applicable to quality improvement efforts. This does not mean to imply that you should drag your feet or that implementation of a quality program is slow. It simply means to portray the obvious notion that to do things poorly fast is not improvement.

Maintaining a Focus on Continuous Improvement

Operations personnel in the service sector (and the manufacturing sector) are involved in addressing sporadic problems (fire drills) and also chronic problems. The actions required include troubleshooting, quality improvement, and quality planning. Maintaining the focus on improvement clearly requires a positive quality culture in an organization.

The broad approach to the three actions and the key elements of a positive quality culture are summarized in Chap. 2.

Finally, the operations function must be provided with the support to maintain the focus on improvement. The quality department should regard operations as a key internal customer and provide training and technical quality expertise to support operations. In addition, operations managers must be guided in reviewing the mountain of reports they receive to identify and prioritize quality problems and set up the teams and other mechanisms to address those problems. Also, the quality department can urge upper management to set up cross-functional teams to address operations problems that may be caused by other functional departments such as IT, marketing, and purchasing.

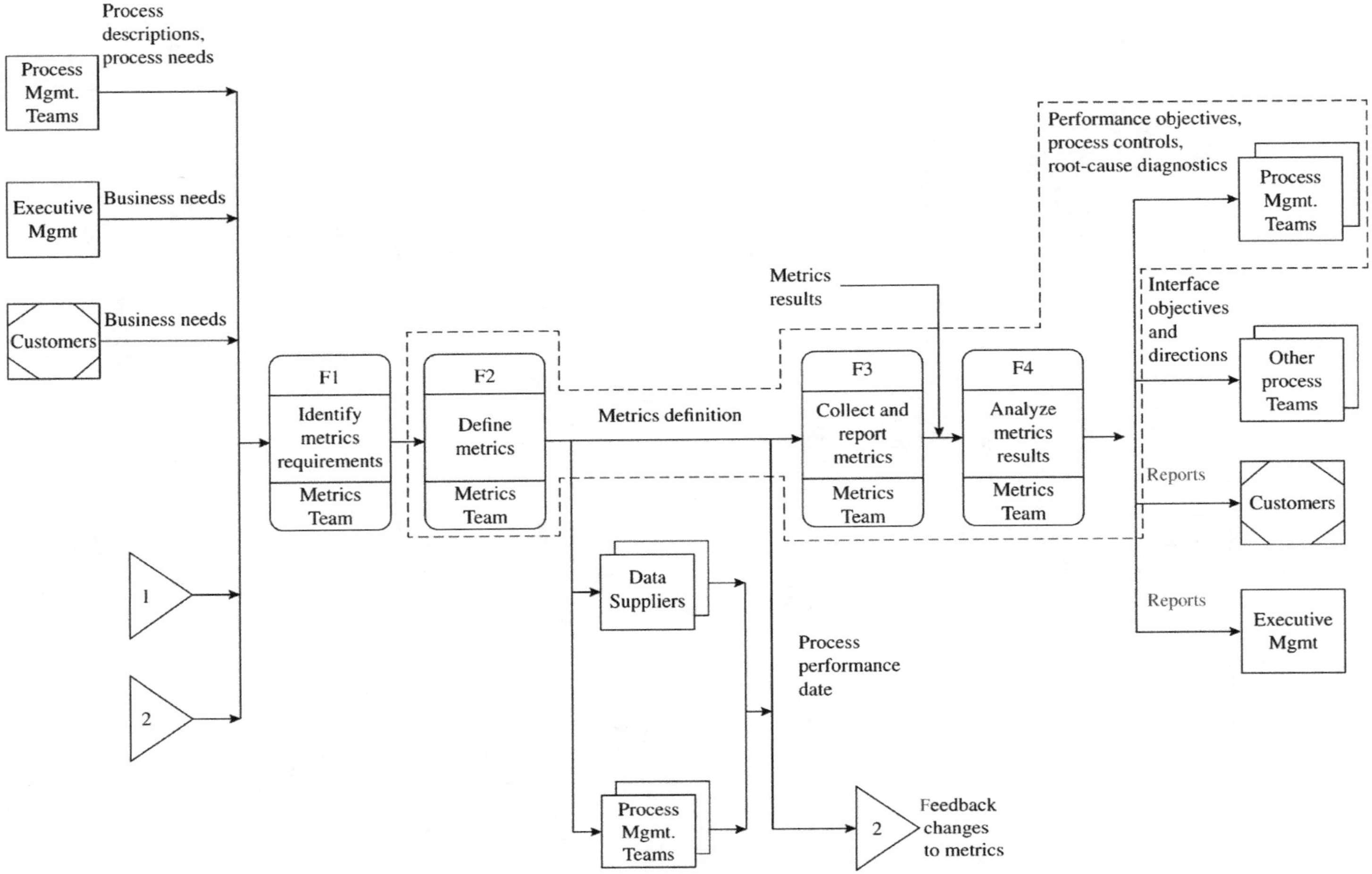

FIGURE 21.6 Flowchart for Defining, Collecting, and Analyzing Metrics in a Service Organization.

References

Alonso, F., del Rey, G., and de Ansorena, R. (2003). "How Telefonica Makes Its Management Connections," *European Quality*, vol. 8, no. 6, pp. 4–10.

American Management Association (1992). *Blueprints for Service Quality*, American Management Association, New York, pp. 51–64.

AT&T Quality Steering Committee (1990). *Achieving Customer Satisfaction*, AT&T Customer Information Center, Indianapolis, IN.

Berry, L. L., Parasuraman, A., and Zeithmal, V. A. (1994). "Improving Service Quality in America: Lessons Learned," *Academy of Management Executive*, vol. 8, no. 2, pp. 32–52.

Bott, C., Keim, E., Kim, S., and Palser, L. (2000). "Service Quality Six Sigma Case Studies," *Annual Quality Congress Proceedings*, ASQ, Milwaukee, pp. 225–231.

Bottome, R. and Chua, R. C. H. (2005). "Genentech Error-Proofs Its Batch Records," *Quality Progress*, July 2005, pp. 25–34.

Collins, W. H. and Collins, C. B. (1993). "Differentiating System and Execution Problems," *Quality Progress*, February, pp. 59–62.

Cronin, J. Joseph, Jr. and Taylor, Steven A. (1992). "Measuring Service Quality: A Reexamination and Extension," *Journal of Marketing*, July, pp. 55–68.

Cross, K. F. (2000). "Call Resolution: The Wrong Focus for Service Quality," *Quality Progress*, February, pp. 64–67.

Davis, R., Rosegrant, S., and Watkins, M. (1995). "Managing the Link between Measurement and Compensation," *Quality Progress*, February, pp. 101–106.

Deyong, C. F. and Case, K. E. (1998). "Linking Customer Satisfaction Attributes with Progress Metrics in Service Industries," *Quality Management Journal*, vol. 5, no. 2, pp. 76–90.

Hadley, H. (1995). Private Communication to Patrick Mene, Marriott Hotels and Resorts, Washington, DC.

Hahn, G. J., Doganaksoy, N., and Hoerl, R. (2000). "The Evolution of Six Sigma," *Quality Engineering*, vol. 12, no. 3, pp. 317–326.

Harvard Business School (1994). Case 9-694-076, Taco Bell, Boston.

Jeffrey, J. R. (1995). "Preparing the Front Line," *Quality Progress*, February, pp. 79–82.

Juran, J. M. and Godfrey, A. B. (1998). *Juran's Quality Handbook, Fifth Ed.* McGraw-Hill, New York, NY. Juran, J. M. and De Feo, J. A. (2010). *Juran's Quality Handbook, Sixth, Ed.* McGraw-Hill, New York, NY.

Katzenbach, J. R. and Smith, D. K. (1993). *Wisdom of Teams: Creating the High Performance Organization*, Harvard Business School Press, Boston, MA.

King, M. N. and Dickinson, T. (1996). "A Framework for Process Effectiveness Measures," *Quality Engineering*, vol. 9, no. 1, pp. 45–50.

Kittner, M., Jeffries, M., and Gryna, F. M. (1999). "Operational Quality Issues in the Financial Sector: An Exploratory Study on Perception and Prescription for Information Technology," *Journal of Information Technology Management*, vol. X, nos. 1–2, pp. 29–39.

Kordupleski, R. E., Rust, R. T., and Zahorik, A. J. (1993). "Why Improving Quality Doesn't Improve Quality (Or Whatever Happened to Marketing?)," *California Management Review*, Spring, vol. 35, no. 3, pp. 82–95.

Mann, D. W. (1994). "Re-engineering the Manager's Role," *ASQC Quality Congress Transactions*, Milwaukee, pp. 155–159.

Melan, E. H. (1993). *Process Management*, McGraw-Hill, New York.

Shirley, B. M. and Gryna F. M. (1998). "Work Design for Self-Control in Financial Services," *Quality Progress*, May, pp. 67–71.

Zeithaml, V. A., Parasuraman A., and Berry, L. L. (1990). *Delivering Service Quality*, Free Press, New York.

CHAPTER 22

Managing Quality in Operations: Manufacturing

Joseph A. De Feo

High Points of This Chapter

1. Activities to integrate quality in manufacturing planning have two objectives: to prevent defects and to minimize variability. To meet these objectives, we must discover the relationships between process features and product results.
2. By creating a flow diagram, we can dissect a manufacturing process and plan for quality at each workstation.
3. Error-proofing a process is an important element of prevention.

4. For human beings to be in a state of self-control, they must have knowledge of what they are supposed to do, knowledge of what they are actually doing, and a process that is capable of meeting specifications and can be regulated. Failure to meet all three of these criteria means that the quality problem is management controllable, and about 85 percent of quality problems are management controllable.
5. The key factors of dominance in manufacturing are setup, time, component, worker, and information.
6. A process quality audit is an independent evaluation of any activity that can affect final product quality.

Quality in Manufacturing in the Twenty-First Century

Manufacturing operations is the nerve tactical center of an organization—where the action is. This chapter covers operations in the manufacturing sector. Chapter 21 covers operations in the service sector.

In manufacturing entities, operations are activities, typically carried out in a factory, that transform material into goods. Before we consider the planning, control, and improvement of manufacturing activities, we must recognize four important issues that will transform traditional manufacturing of the twentieth century to a different manufacturing in the twenty-first century. As Dr. Juran clearly articulated "the twentieth century was the century of productivity; the twenty-first century will be the century of quality." These issues are discussed next.

Customer Demands for Higher Quality, Reduced Inventories, and Faster Response Time

As products and processes become more complex, "world class" quality levels are now common and expected. To remain competitive, manufacturers must continually and intently focus on quality in the big Q framework. World-class enterprises can no longer view quality as being solely product focused. They now need to view it as business- and enterprise-focused (big Q). For many products, quality levels of 1 to 3 percent defective are being replaced by 1 to 10 defects per million parts (or 3.4 defects per million as in the Six Sigma approach). Customers demand reduced inventory levels based on the "just-in-time" (JIT) production system. Under JIT, the concept of large lot sizes is challenged by pull systems, reducing changeover times, redesigning processes, and standardizing jobs. The results are smaller lot sizes and lower inventory.

But JIT works only if product quality of supplied materials is high because little or no inventory exists to replace defective product. Finally, customers want faster response time from suppliers—to develop and manufacture new products (conception to commercialization). That faster response time puts pressure on the product development process and can result in inadequate review of new designs or enhancements to existing designs for product performance and for manufacturability. In some organizations, this results in a concept known as "tailoring," which is streamlining the integrated product development schedule often at the expense of quality. Collectively, these three parameters (quality, inventories, and response time) place a heavy burden on operations.

Agile Competition

An agile organization is able to respond to constantly changing customer demands. This characteristic means agile supply chains, flexible manufacturing, mixed-model scheduling, single or pitch flow, changing over from one product to another quickly, manufacturing goods to customer order in small lot sizes, customizing goods for individual customers, and using the expertise of people and facilities within the company and also among groups of cooperating companies (partners). Goldman et al. (1995) describes the concept and includes examples.

This concept includes the "virtual" organization or complete value stream—a group of companies linked by an electronic network to enable the partners to satisfy a common customer objective. The virtual organization may be partially created by transferring complete functions to a supplier—outsourcing.

Impact of Technology

Technology (including computer information systems) is clearly improving quality by providing a wider variety of outputs and also more consistent output. The infusion of technology makes some jobs more complex, thereby requiring extensive job skills and quality planning; technology also makes other jobs less complex but may contribute to job monotony.

These issues suggest that quality during operations can no longer focus on inspection and checking (reactive) but must respond to ever-increasing customer demands and changing competitive conditions (proactive).

We proceed now to examine specific methods for planning, controlling, and improving quality during manufacturing operations.

Lean Manufacturing and Value Stream Management

Lean manufacturing is the process of optimizing operating systems by eliminating, or at least reducing, the waste within them. Anything that does not provide value to the customer is considered waste. The emphasis is on eliminating non-value-added activities such as producing defective product, excess inventory charges due to work-in-process and finished goods inventory, excess internal and external transportation of product, excessive inspection, and idle time of equipment or workers due to poor balance of work steps in a sequential process. The goal of lean manufacturing has long been one of the goals of industrial engineering. Shuker (2000) provides a useful introduction to lean manufacturing based on the Toyota production system.

Lean Manufacturing

The history of manufacturing is summarized in Fig. 22.1.

The lean mission is to have the following throughout the entire supply chain so as to win the marketplace:

- Shortest possible lead time
- Optimum level of strategic inventory
- Highest practical customer order service
- Highest possible quality (low defect rate)
- Lowest possible waste (low COPQ)

Lean is based on creating a "pull system." This is accomplished by synchronizing the flow of work (both internal and external to the company) to the "drumbeat" of the customer's requirements. All kinds of waste are driven out (time, material, labor, space, and motion).

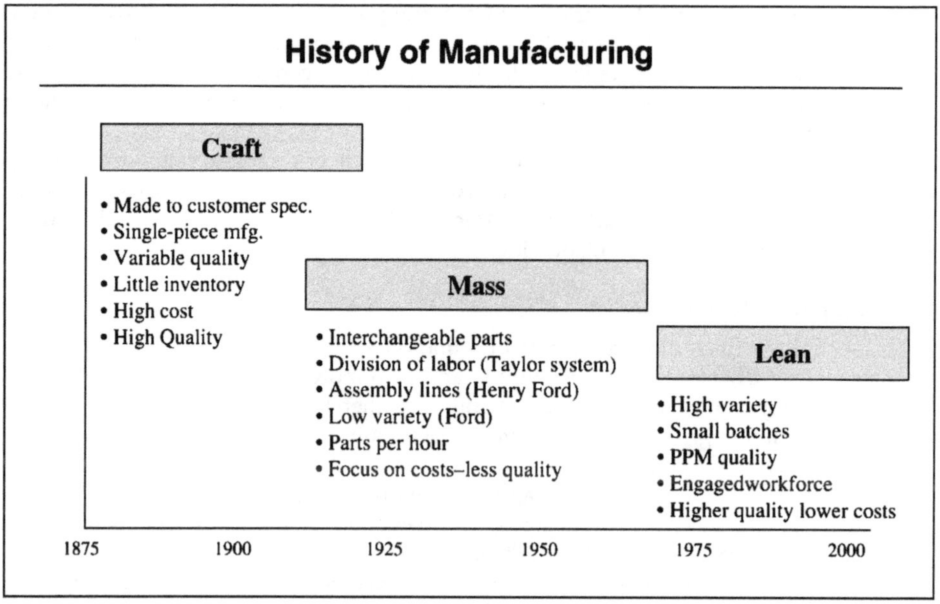

FIGURE 22.1 History of manufacturing.

The overall intent is to reduce variation and drive out waste by letting customers pull value through the entire value stream (or supply chain).

The key principles of lean are:

- Define value in the eyes of the customer.
- Measure the value stream for each product.
- Analyze value flow without interruptions.
- Improve process pull by reducing defects in products and deficiencies in processes.
- Control processes by driving out variation (short and long term) and implementing sound controls to maintain improvements.

Value is created by the customer. Lean starts by attempting to define value in terms of products and capabilities provided to the customer at the right time and appropriate price (see Fig. 22.2).

The following are summaries of key concepts used in lean (in the context of both lean manufacturing and lean service).

The Eight Wastes

Taiischi Ohno (1988) identifies various types of waste. The following is an adapted list:

1. Overproduction—making or doing more than is required or earlier than needed
2. Waiting—for information, materials, people, maintenance, etc.
3. Transport—moving people or goods around or between sites
4. Poor process design—too many/too few steps, non-standardization, inspection rather than prevention, etc.

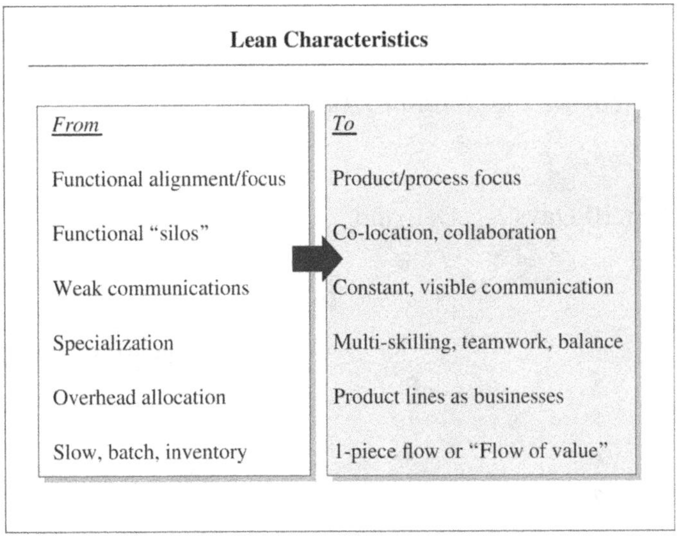

FIGURE 22.2 The characteristics of lean versus traditional manufacturing practices.

5. Inventory—raw materials, work in progress, finished goods, papers, electronic files, etc.
6. Motion—inefficient layouts at workstations, in offices, poor ergonomics
7. Defects—errors, scrap, rework, nonconformance
8. Underutilized personnel resources and creativity—ideas that are not listened to, skills that are not used

Flow and Takt Time

Traditional operations have worked within a "push" system. A push system computes start times and then pushes product into operations based on artificial demand or based on overhead absorption. The concept of "flow" requires the rearrangement of mental thoughts regarding "typical" production processes. One must not think of just "functions" and "departments." We need to redefine how functions, departments, and organizations work to make a positive contribution to the value stream. Flow production requires that we produce at the customer's purchase rate and if necessary, make every product every day to meet customer's orders, i.e., to meet the pace or "drum-beat." The pace or drumbeat is determined by takt time. Takt time comes from the German word for meter, as in music, which establishes the pace, or beat, of the music. It is the time which reflects the rate at which customers buy 1 unit.

$$\text{Takt time} = \frac{\text{Available time (in a day)}}{\text{Average daily demand}}$$

For example, in Fig. 22.3, the pace or takt time is calculated for the demand shown during a 10-day period.

Determine Pace

Takt Time Calculation Example:

Over 10 Days	Demand
1	30
2	40
3	50
4	60
5	10
6	30
7	40
8	20
9	60
10	40
10	**380**

Per Day:

$$\frac{\text{Time available in period (840 min.)}}{\text{Average demand (38)}} = 22.1 \text{ minutes}$$

Based on 2 shifts of 7 hours

FIGURE 22.3 Takt time calculation example.

Demand cannot be understated. A key component to satisfying the customer is understanding their demand (including seasonal or other particular demand). Demand variability can be mix driven, quantity driven, or both.

To be practical, takt time may need to be modified depending on the variability of the process. When modifying takt time beyond the simple equation, another name should be used, such as cell takt, practical takt, or machine takt. Although modifiers may be planned, they are still waste or planned waste. Manpower staffing requirements can then be determined:

$$\text{Minimum staffing required} = \frac{\text{Total labor time in process}}{\text{Takt time}}$$

Once takt time has been calculated, constraints (such as long setup times) should be identified and managed (or eliminated) to enable smaller batches or ideally, single piece flow, to eliminate overproduction and excess inventory. Pull production scheduling techniques are used so that customer demand pulls demand through the value stream (from supplier to production to the customer). In pull production, materials are staged at the point of consumption. As they are consumed, a signal is sent back to previous steps in the production process to pull forward sufficient materials to replenish only what has been consumed.

The steps for improvement teams (or kaizen teams) to lean out an operation are:

1. Determine pace (takt time and manpower).
2. Establish sequence and replenishment (product family turnover and setup/changeover required).
3. Design the process (proximity, sequence, interdependence).
4. Feed the process (strategic inventory, standard WIP [SWIP], Murphy buffer).
5. Balance the process (load, standard work).
6. Stabilize and refine (6S [which includes safety], continuous improvement).

Value Stream Management

The value stream consists of all activities required to bring a product from conception to commercialization. It includes detailed design, sales, marketing, order taking, scheduling, production, and delivery. Understanding the value stream allows one to see value-added steps, non-value-added but needed steps, and non-value-added steps. Value-added activities transform or shape material or information to meet customer requirements. Non-valued-added activities take time or resources, but do not add value to the customer's requirement (but may meet company requirements).

The value stream improvement journey typically starts with training the team on key concepts in Lean Value-Stream Mapping (LVSM), mapping the current state using value stream maps, which document materials and information flow as well as any pertinent information on the process (such as wait times, processing times, and inventory levels). Improvements are identified. The desired future state is then documented as a future state value stream map, and the improvements and control features are implemented to drive toward the future state goal.

An example of a value stream map for a paint line for the current state and future state are shown in Figs. 22.4a and 22.4b.

Initial Planning for Quality

Planning starts with a review of product designs (see Chap. 4). Then we review the process designs to identify key product and process characteristics, determine the importance of product characteristics (typically can be classified as critical, major, minor, and incidental), analyze the process flow diagram, determine the vital process characteristics, determine process control and capability, error-proof the process, plan for a neat and clean workplace, validate the measurement systems, and plan for operator self-control. These elements are discussed next.

Review of Product Designs

There is a clear advantage to having a new product design reviewed by operations personnel before the design is finalized for the marketplace. In practice, the extent of such review varies greatly—from essentially nothing ("tossing it over the wall" to the operations people) to a structured review using formal criteria and follow-up on open issues (typically evidenced by sound-integrated product development schedules with provisions for robust design for manufacturability during various stages). Although a product design review often occurs during the design and development process (see Chap. 4), the emphasis is on the adequacy of field performance.

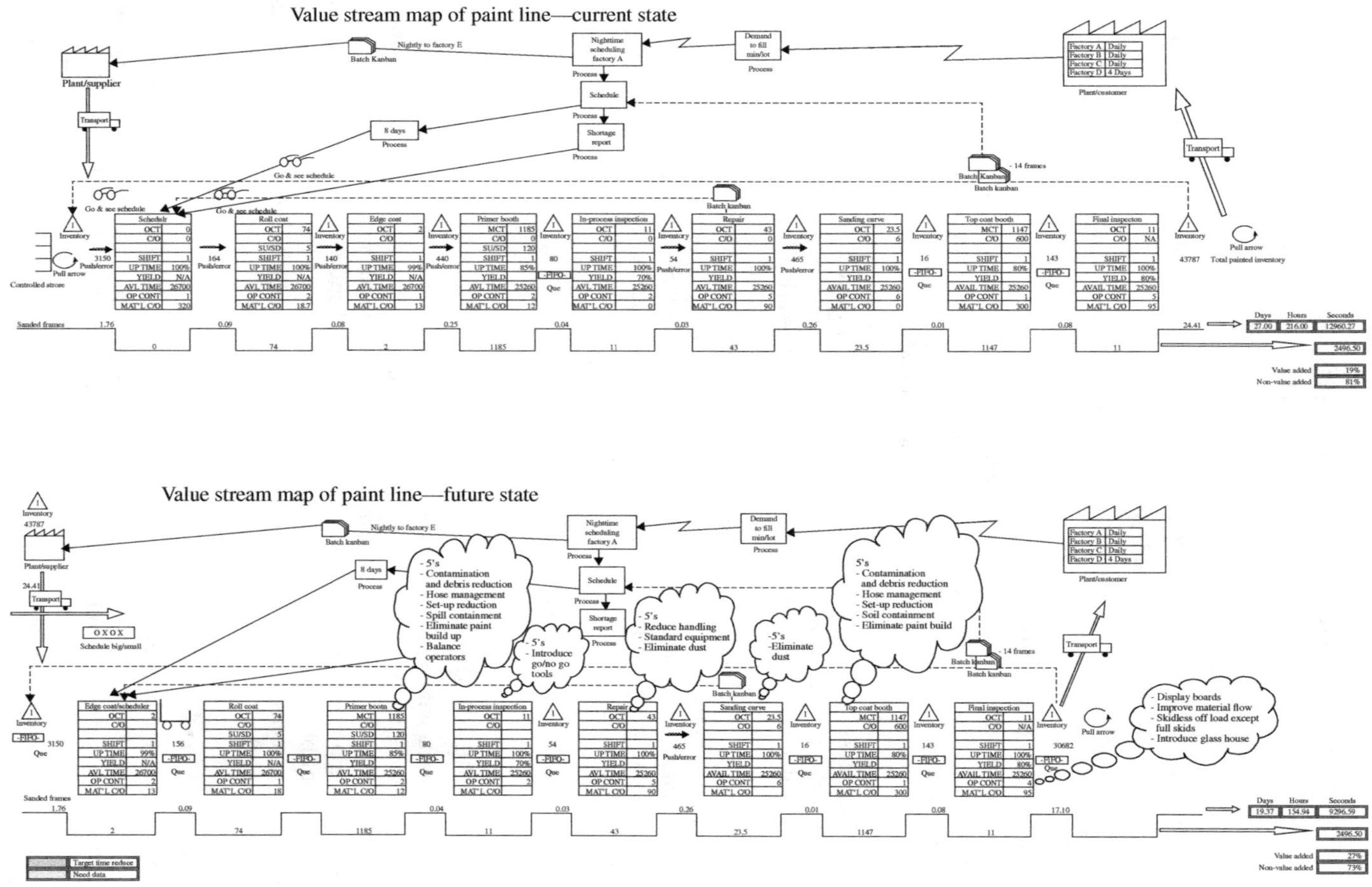

FIGURE 22.4 (*a*) Value Stream Map of Paint Line—Current State (*b*) Value Stream Map of Paint Line—Future State.

Review of product designs prior to release to operations must include an evaluation of manufacturability. This can only occur after there is a clear understanding of critical customer requirements (CCRs). This evaluation includes the following issues:

1. Identification of key product characteristics and key control characteristics (process features).
2. Relative importance of key product characteristics and key control characteristics.
3. Design for manufacturability.
4. *Process robustness.* A process is robust if it is flexible, easy to operate, and error proof and its performance will tolerate uncontrollable variations in factors internal and external to the process. Such an ideal is approached by careful planning of all process elements, e.g., cross-functional training of personnel to cover vacations. For a discussion of robustness, see Snee (1993).
5. Availability of capable manufacturing processes to meet product requirements, i.e., processes that not only meet specifications but do so with minimum variation.
6. Availability of capable measurement processes.
7. Identification of special needs for the product, e.g., handling, transportation, and storage during manufacture.
8. Material control, e.g., identification, traceability, segregation, and contamination control.
9. Special skills required of operations personnel.

This review of the product design must be supplemented by a review of the process design. The process review includes manufacturability issues initially raised in the product design review.

Identification of Key Product and Process Characteristics

Key product characteristics are outputs from a process that are measurable on, within, or about the product itself. They are the outputs perceived by the customer. Key process characteristics (KPCs) are inputs (process factors) that affect the outputs. They are unseen by the customer and are measurable only when they occur. Product and process characteristics can be identified by using inputs from market research, quality function deployment, design review, and failure mode and effect analysis. Somerton and Mlinar (1996) describes the use of these and other tools to identify key characteristics.

Relative Importance of Product Characteristics

Planners are better able to allocate resources when they know the relative importance of the many product characteristics.

One technique for establishing the relative importance is the identification of critical items. Critical items are the product characteristics that require a high level of attention to ensure that all requirements are met. One company identifies "quality sensitive parts" by using criteria such as part complexity and high-failure-rate parts. For such parts, special planning includes supplier involvement before and during the contract, process capability studies, reliability verification, and other activities.

Another technique is the classification of characteristics as described earlier in the section. Under this system, the relative importance of characteristics is determined and indicated on drawings and other documents. The classification can be simply "functional" (or "critical to quality") or "nonfunctional." Another system uses several degrees of importance such as critical, major, minor, and incidental. The classification uses criteria that reflect safety, operating failure, performance, service, and manufacture.

Analysis of the Process Flow Diagram

A process design can be reviewed by laying out the overall process in a flow diagram. Several types are useful. One type shows the paths followed by materials through their progression into a finished product. An example for a coating process at the James River Graphics Company is shown in Fig. 22.5. Planners use such a diagram to divide the flow into logical sections called workstations. For each workstation, they prepare a formal document listing such items as operations to be performed, sequence of operations, facilities and instruments to be employed, and process conditions to be maintained. This formal document becomes the plan to be carried out by the production supervisors and workforce. The document is the basis for control activities by the inspectors. It also becomes the standard against which the process audits are conducted.

Correlation of process variables with product results

A critical aspect of planning during manufacture is to discover, by data analysis, the relationships between process features or variables and product features or results.

Such knowledge enables a planner to create process control features, including limits and regulating mechanisms on the variables, to keep the process in a steady state and achieve the specified product results. In Fig. 22.5, each process feature is shown in a rectangle

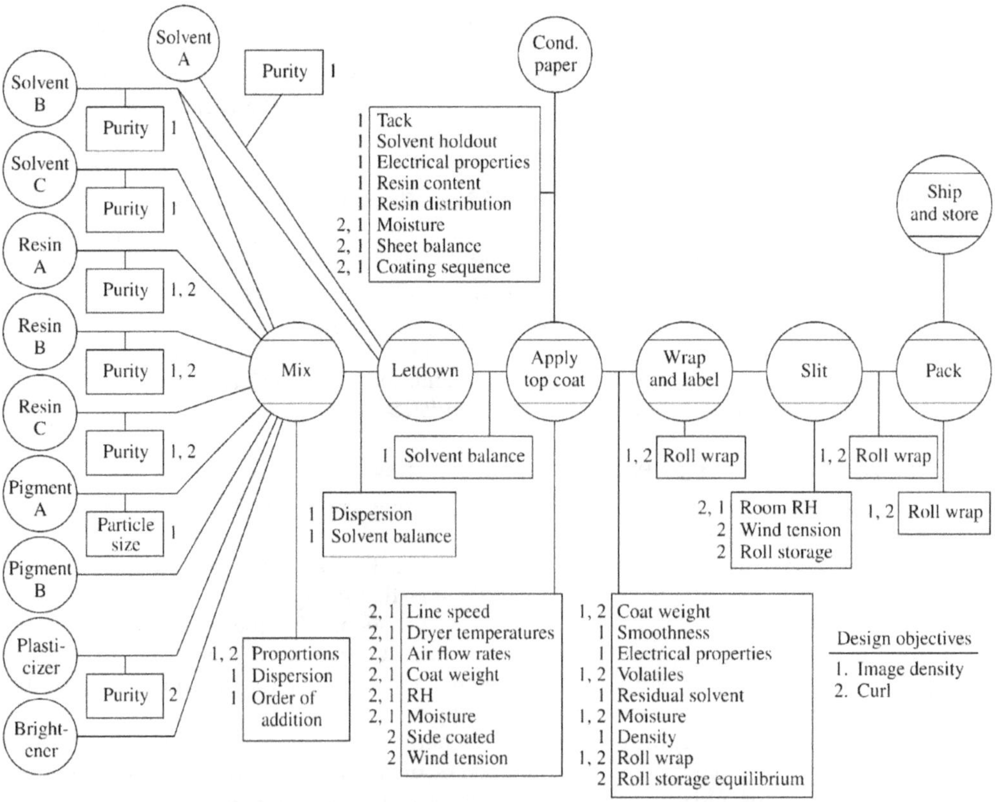

FIGURE 22.5 Product and process analysis chart. (From Siff, W. C. [1984]. "The Strategic Plan of Control—A Tool for Participative Management," ASQC Quality Congress Transactions, Milwaukee, pp. 384–390.)

attached to the circle representing an operation; product results are listed in rectangles between operations at the point where conformance can be verified. Some characteristics (e.g., coat weight) are both process features and product results.

For each control station in a process, designers identify the numerous control subjects over which control is to be exercised. Each control subject requires a feedback loop made up of multiple process control features. A process control spreadsheet or control plan helps to summarize the detail. For elaboration, see Chap. 6, Quality Control to Assure Compliance to Customer Requirements.

Determining the optimal settings and tolerances for process features sometimes requires much data collection and analysis. Eibl et al. (1992) discusses such planning and analysis for a paint-coating process for which little information was available about the relationship between process features and product results.

Many companies have not studied the relationship between process features and product results. The consequences of this lack of knowledge can be severe. In the electronic component manufacturing industry, some yields are shockingly low and will likely remain that way until the process features are studied in depth. In all industries, the imposition of new quality demands such as Six Sigma requires much deeper understanding of product results and process features than in the past.

To understand fully the relationship between process features and product results, we often need to apply the concept of statistical design of experiments. Under the Six Sigma approach, factorial experiments are becoming necessary to understand the interactions among several variables and product results. But upper management must supply the missing elements, i.e., the resources for full-time personnel to design and analyze the experiments and the training of process engineers to integrate these concepts in process planning.

Error-Proofing the Process

An important element of prevention is the concept of designing the process to be error free through "error-proofing" (the Japanese call it poka-yoke).

A widely used form of error-proofing is the design (or redesign of the machines and tools, the "hardware") to make human error improbable or even impossible. For example, components and tools may be designed with lugs and notches to achieve a lock-and-key effect, which makes it impossible to misassemble them. Tools may be designed to sense the presence and correctness of prior operations automatically or to stop the process on sensing depletion of the material supply. For example, in the textile industry a break in a thread releases a spring-loaded device that stops the machine. Protective systems, e.g., fire detection, can be designed to be "fail-safe" and to sound alarms as well as all-clear signals.

In a classic study, Nakajo and Kume (1985) discuss five fundamental principles of error-proofing developed from an analysis of about 1000 examples collected mainly from assembly lines. These principles are elimination, replacement, facilitation, detection, and mitigation (see Table 22.1). See JQH6, pp. 11.351–11.353, for further examples of error-proofing.

An adjunct to error-proofing is the use of poka-yoke systems. These are control systems built into the process that stop the equipment when some type of irregularity occurs and signal the operator to address the problem.

Plan for Neat and Clean Workplaces

How obvious, but the reality is that many workplaces are dirty and disorganized. The benefits of a good workplace include the prevention of defects; prevention of accidents; and the elimination of time wasted searching for tools, documentation, and other ingredients of manufacture. A simple body of knowledge now provides us with a framework to create a

Principle	Objective	Example
Elimination	Eliminating the possibility of error	Redesigning the process or product so that the task is no longer necessary
Replacement	Substituting a more reliable process for the worker	Using robotics (e.g., in welding or painting)
Facilitation	Making the work easier to perform	Color-coding parts
Detection	Detecting the error before further processing	Developing computer software that notifies the worker when a wrong type of keyboard entry is made (e.g., alpha versus numeric)
Mitigation	Minimizing the effect of the error	Using fuses for overloaded circuits

TABLE 22.1 Summary of Error-Proofing Principles

neat and clean workplace. The approach is called 6S for sort, set in order, sweep and shine, standardize, self-discipline, and safety (Fig. 22.6). The steps are as follows:

1. *Sort*. Remove all items from the workplace that are not needed for current operations.
2. *Set in order*. Arrange workplace items so that they are easy to find, to use, and to put away.
3. *Sweep and shine*. Sweep, wipe, and keep the workplace clean.
4. *Standardize*. Make "shine" become a habit.
5. *Self-discipline*. Create the conditions (e.g., time, resources, rewards) to maintain a commitment to the 6S approach.
6. *Safety*. Implement behavioral-based safety processes and procedures that drive zero recordable injuries and zero lost time accidents.

The Japanese developed the 6S technique to remove clutter from the work area, organize the workplace to help prevent defects from happening, and improve work flow. It is one of

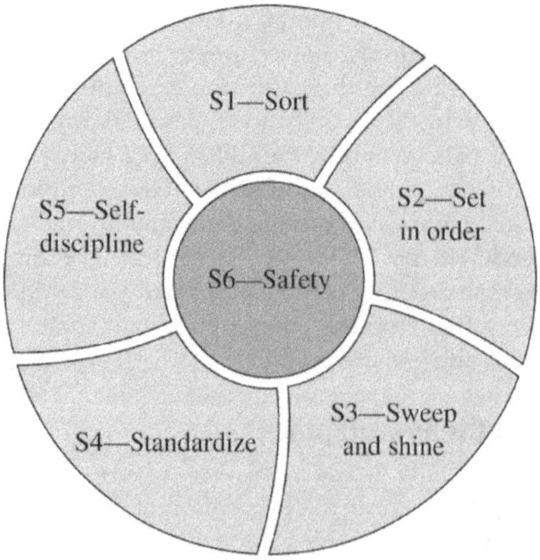

FIGURE 22.6 The 6S techniques.

the basic elements of lean that an organization can implement. It is a great improvement tool that is very easy to install and brings "order out of chaos."

This technique should be put in place prior to the project solution being implemented because it can yield additional results on its own merits. Also, the final solution implementation will be easier to put in place if this technique is done first.

Decades ago, industries producing critical items (health care, aerospace) learned that clean and neat workplaces are essential in achieving extremely low levels of defects. The quality levels demanded by the Six Sigma approach now provide the same impetus.

Perhaps the significance of the 6S approach is its simplicity. The benefits are obvious; the tools are the simplest work-simplification tools; the tools are easy to understand and apply. Simple tools sometimes get dramatic results, and that's what has happened with 6S.

Validate the Measurement System

Particularly with the low defect levels demanded under the Six Sigma approach, it is important to understand the capability of the manufacturing process and also the capability of the measurement process. Thus planning and control of the measurement process become part of the Six Sigma approach. Previous studies assumed that variation of the measurement process was small compared to variation caused by the manufacturing process and thus could be ignored (in practice, the assumption was rarely tested in most industries). When variation due to the measurement process alone is even moderately large, the result will be mistakes in determining whether a product meets specifications—some "good" product will be incorrectly classified as defective, and some "bad" product will be incorrectly classified as good. Thus the time has come to evaluate measurement capability and to determine whether measuring equipment is accurately measuring process output.

Figure 22.7 provides perspective on the measurement issue. Note that the observed process variation (i.e., the variation of recorded measurements) is from two sources: variation in the process manufacturing the product and variation in the measurement process. Further, these two sources contain various components. The components shown in Fig. 22.7 can be

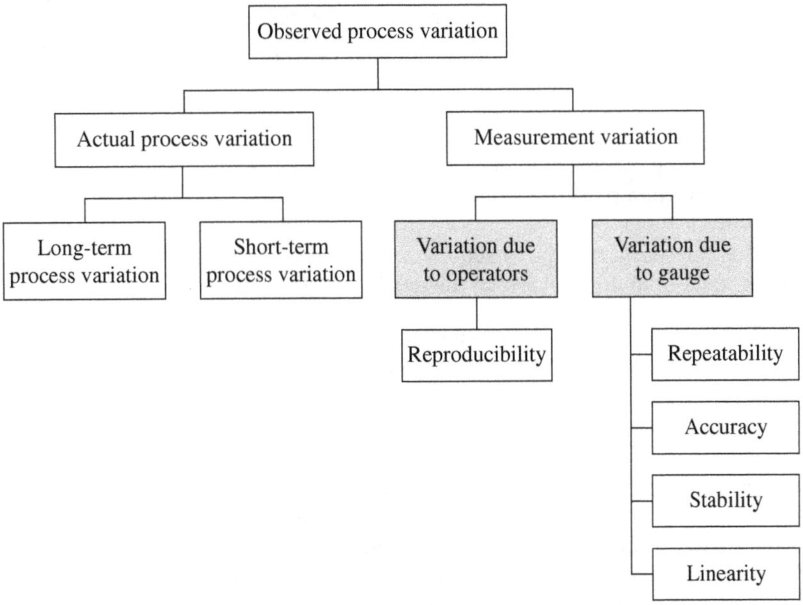

FIGURE 22.7 Possible sources of variation. (*Source:* Juran Institute, Inc.)

quantified and analyzed to determine measurement capability. The variation due to the process manufacturing the product can also be quantified.

Even in consumer product industries such as the manufacture of razor blades, tolerances can be on the order of the wavelength of visible light. Such tolerances are a long way from the days of tolerances in thousands or 10 thousands and using measuring instruments like micrometers and even supermicrometers. The measurement process must be capable of handling these conditions.

Concept of Controllability: Self-Control

An ideal objective for manufacturing planning is to place human beings in a state of self-control, i.e., to provide them with all they need to meet quality objectives. To do so, we must provide people with the following:

1. Knowledge of what they are supposed to do
   a. Clear and complete work procedures
   b. Clear and complete performance standards
   c. Adequate selection and training of personnel
2. Knowledge of what they are actually doing (performance)
   a. Adequate review of work
   b. Feedback of review results
3. Ability and desire to regulate the process for minimum variation
   a. A process and job design capable of meeting quality objectives
   b. Process adjustments that will minimize variation
   c. Adequate worker training in adjusting the process
   d. Process maintenance to maintain the inherent process capability
   e. A strong quality culture and environment

As we will see, most organizations do not adhere to the three elements and 10 subelements of self-control.

The concept of self-control has objectives that are similar to those of the Toyota production system. For a perceptive dissection of the Toyota system into four basic rules, see Spear and Bowen (1999).

The three basic criteria for self-control make possible a separation of defects into categories of "controllability," of which the most important are

1. *Worker controllable.* A defect or nonconformity is worker controllable if all three criteria for self-control have been met.
2. *Management controllable.* A defect or nonconformity is management controllable if one or more of the criteria for self-control have not been met.

Only management can provide the means for meeting the criteria for self-control. Hence, any failure to meet these criteria is a failure of management, and the resulting defects are therefore beyond the control of the workers. This theory is not 100 percent sound. Workers commonly have a duty to call management's attention to deficiencies in the system of control, and sometimes they do not do so. (Sometimes they do, and management fails to act.) However, the theory is much more right than wrong.

Whether the defects or nonconformities in a plant are mainly management controllable or worker controllable is of the highest order of importance. Reducing the former requires a program in which the main contributions must come from managers, supervisors, and technical specialists. Reducing the latter requires a different kind of program in which much of the contribution comes from workers. The great difference between these two kinds of programs suggests that managers should quantify their knowledge of the state of controllability before embarking on major programs.

An example of a controllability study is given in Table 22.2. A diagnostic team was set up to study scrap and rework reports in six machine shop departments for 17 working days. The defect cause was entered on each report by a quality engineer who was assigned to collect the data. When the cause was not apparent, the team reviewed the defect and, when necessary, contacted other specialists (who had been alerted by management about the priority of the project) to identify the cause. The purpose of the study was to resolve a lack of agreement on the causes of chronically high scrap and rework. It did the job. The study was decisive in obtaining agreement on the focus of the improvement program. In less than one year, more than $2 million was saved, and important strides were made in reducing production backlogs.

Controllability can also be evaluated by posing specific questions for each of the three criteria of self-control. (Typical questions that can be posed are presented in this chapter.) Although this approach does not yield a quantitative evaluation of management and worker-controllable defects, it does show whether the defects are primarily management or worker controllable.

In the experience of the author, defects are about 85 percent management controllable. This figure does not vary much from industry to industry, but it does vary greatly among

Category	Percent
Management controllable	
Inadequate training	15
Machine inadequate	8
Machine maintenance inadequate	8
Other process problems	8
Material handling inadequate	7
Tool, gage, fixture (TGF) maintenance inadequate	6
TGF inadequate	5
Wrong material	3
Operation run out of sequence	3
Miscellaneous	5
Total	68
Worker controllable	
Failure to check work	11
Improper operation of machine	11
Other (e.g., piece mislocated)	10
Total	32

TABLE 22.2 Controllability Study in a Machine Shop

processes. Other investigators in Japan, Sweden, the Netherlands, and Czechoslovakia have reached similar conclusions.

Although the available quantitative studies make clear that defects are mainly management controllable, many industrial managers do not know this or are unable to accept the data. Their long-standing beliefs are that most defects are the result of worker carelessness, indifference, and even sabotage. Such managers are easily persuaded to embark on worker-motivation schemes that, under the usual state of affairs, aim at a small minority of the problems and hence are doomed to achieve minor results at best. The issue is not whether quality problems in an industry are management controllable.

The need is to determine the answer in a given enterprise. This answer requires solid facts, preferably through a controllability study of actual defects, as shown in Table 22.2.

We now discuss the three main criteria for self-control.

Criterion 1: Knowledge of "Supposed to Do"

This knowledge commonly consists of the following:

1. The product standard, which may be a written specification, a product sample, or other definition of the end result to be attained.
2. The process standard, which may be a written process specification, written process instructions, an oral instruction, or other definition of "means to an end."
3. A definition of responsibility, i.e., what decisions to make and what actions to take. In developing product specifications, some essential precautions must be observed.

Unequivocal Information Must Be Provided

Specifications should be quantitative. If such specifications are not available, physical or photographic standards should be provided. But beyond the need for clear product specifications, there is also a need for consistent and credible specifications. In some organizations, production supervisors or even workers have a secret "black book" that contains the "real" specification limits used by inspectors for accepting product. The author has found this activity to be prevalent in many types of industries (electrical enclosures, automotive, construction equipment, flat glass, and even aircraft manufacturing to name a few). A further problem is communicating changes in specifications, especially when there is a constant parade of changes.

Information on Specifications Must Be Provided

All specifications contain multiple characteristics, and these are not equally important. Production personnel must be guided and trained to meet all specification limits. But they must also be given information on the relative importance of each characteristic to focus on priorities.

Reasons Must Be Explained

Explanation of the purposes served by both the product and the specification helps workers to understand why both the nominal specification value and the limits must be met.

Process Specifications Must Be Provided

Work methods and process conditions (e.g., temperature, pressure, and time cycles) must be unequivocally clear.

LTV, a steel manufacturer, uses a highly structured system of identifying key process variables, defining process control standards, communicating the information to the workforce, monitoring performance, and accomplishing diagnosis when problems arise. The process specification is a collection of process control standard procedures (Fig. 22.8). A procedure is

LTV Steel

INTEGRATED PROCESS CONTROL
Standard Procedures
Process Control

Plant: Indiana Harbor
Dept: No. 3 Sheet Mill

File No.: 716-2.2.2
Date Orig Issue:
Revision No.: 1
Date Revised:

Control Area	Control Point	Control Element	No.
Tandem Mill	Rolling	Rolling Solutions	2.2.2

Control Task
To maintain rolling solution characteristics at the proper levels.

Responsible for Control
Solution Attendant

Process Standard
- Oil concentration must be 2.5% to 3.5%
- Solution temperature must be 110°F–120°F
- SAP value must be above 120
- Iron fines must be less than 600 ppm

Reason for Control
- To provide the correct lubricity between work rolls and strip for reduced roll wear and control of strip temperature. This helps control strip flatness and avoids friction scratches.

Measurement
Tools/Equipment – Standard chem. test set-up
Frequency – Twice/turn
By – Solution Attendant

Routine Reporting of Data
Form No. Solution Attendant's Report
By – Solution Attendant

Control Chart
Type – X & Moving Range
By – Solution Att.

Corrective Action
- Solution concentration approaching limits - add rolling oil or water.
- Solution temp. approaching limits - adjust temperature control.
- SAP reading between 100 and 120, skim solution tank and add new oil. SAP reading below 100, retest immediately and contact Operating Supervisor. If retest below 100, shut down mill and switch to alternative solution tank.
- Iron fines approaching limit, skim tank for 2 hours and add 100 gallons oil. Retest after 30 minutes second time; repeat procedure if still near or above limit.

Operating Procedure
See attached sheet

Disposition of Non-compliant Product
Identify coil(s) for special surface evaluation. Notify Metallurgical Supervisor.

Review Procedure
Once per turn the Operating Supervisor will
- Check Solution Attendant's Report
- Visually check temperature of solution

Developed By: Robert Gordkin (IPC Coordinator), Richard H. Burg

Approved:
Department Superintendent/Manager — Manager–Quality Control — General Superintendent/Print Manager

Figure 22.8 Process control standard procedure. (From LTV Steel.)

developed for controlling the key process (variables that must be controlled to meet specification limits for the product). The procedure answers the following issues:

- What the process standards are.
- Why control is needed.
- Who is responsible for control.
- What and how to measure.
- When to measure.
- How to report routine data.
- Who is responsible for data reporting.
- How to audit.
- Who is responsible for audit.
- What to do with product that is out of compliance.
- Who developed the standard.

Often, detailed process instructions are not known until workers have become experienced with the process. Updating of process instructions based on job experience can be conveniently accomplished by posting a cause-and-effect diagram in the production department and attaching index cards to the diagram. Each card states additional process instructions based on recent experience. A checklist must be created.

The preceding discussion covers the first criterion of self-control: People must have the means of knowing what they are supposed to do. To evaluate adherence to this criterion, a checklist of questions can be created, including the following.

Adequate and Complete Work Procedures

1. Are there written product specifications, process specifications, standard work, and work instructions? If written down in more than one place, do they all agree? Are they legible? Are they conveniently accessible to the worker? Are they under formal document control?
2. Does the worker receive specification changes automatically and promptly?
3. Does the worker know what to do with defective raw or work-in-progress material?
4. Have responsibilities for decisions and actions been clearly defined?

Adequate and Complete Performance Standards

1. Do workers consider the standards attainable?
2. Does the specification define the relative importance of different quality characteristics? If control charts or other control techniques are to be used, is their relationship to product specifications clear, and do we also review for process control as well as process capability?
3. Are standards for visual defects displayed in the work area?
4. Are the written, physical, or pictorial specifications given to the worker the same as the criteria used by inspectors? Are deviations from the specification often allowed?

5. Does the worker know how the product is used?
6. Does the worker know the effect on future operations and product performance if the specification is not met?

Adequate Selection and Training

1. Does the personnel selection process adequately match worker skills with job requirements?
2. Has the worker been adequately trained to understand the specification and perform the steps needed to meet the specification?
3. Has the worker been evaluated by testing or other means to see whether he or she is qualified?

Criterion 2: Knowledge of Performance

For self-control, people must have the means of knowing whether their performance conforms to a standard. This conformance applies to:

- The product in the form of specifications for product characteristics.
- The process in the form of specifications for process variables.

This knowledge is secured from three primary sources: measurements inherent in the process, measurements by production workers, and measurements by inspectors.

Criteria for Good Feedback to Workers

The needs of production workers (as distinguished from supervisors or technical specialists) require that the data feedback can be read at a glance, deal only with the few important defects, deal only with worker-controllable defects, provide prompt information about symptom and cause, and provide enough information to guide corrective actions. Good feedback should:

- Be recognizable at a glance. The pace of events on the factory floor is swift. Workers should be able to review the feedback while in motion. Where a worker needs information about process performance over time, charts can provide an excellent form of feedback, provided they are designed to be consistent with the assigned responsibility of the worker (Table 22.3). It is useful to use visual displays to highlight recurrent problems. A problem described as "outer hopper switch installed backwards" displayed on a wall chart in large block letters has much more impact than the same message buried away as a marginal note in a work folder.
- Deal only with the few important defects. Overwhelming workers with data on all defects will result in diverting attention from the vital few.
- Deal only with worker-controllable defects. Any other course provides a basis for argument that will be unfruitful.
- Provide prompt information about symptoms and causes. Timeliness is a basic test of good feedback; the closer the system is to "real time" signaling, the better.
- Provide enough information to guide corrective action. The signal should be in terms that make it easy to decide on remedial action.

Responsibility of the Worker is to	Chart Should be Designed to Show
Make individual units of product meet a product specification	Measurements of individual units of product compared to product specification limits
Hold process conditions to the requirements of a process specification	Measurements of the process conditions compared to the process specification limits
Hold averages and ranges to specified statistical control limits	Averages and ranges compared to the stati stical control limits
Hold percentage nonconforming below some prescribed level	Actual percentage nonconforming compared to the limiting level

Table 22.3 Worker Responsibility Versus Chart Design

Feedback Related to Workers' Actions

The workers need to know what kind of process change to make to respond to a product deviation. Sources of this knowledge are:

- The process specifications (see Fig. 22.4)
- Cut-and-try experience by the workers
- The fact that the units of measure for both product and process are identical

If they lack all of these, workers can only cut and try further or stop the process and sound the alarm.

Sometimes the data feedback can be converted into a form that makes the workers' decisions easier about what action to take on the process.

For example, a copper cap had six critical dimensions. It was easy to measure the dimensions and to discover the nature of product deviation. However, it was difficult to translate the product data into process changes. To simplify this translation, use was made of a position–dimensions (P–D) diagram. The six measurements were first "corrected" (i.e., coded) by subtracting the thinnest from all the others. These corrected data were then plotted on a P–D diagram as shown in Fig. 22.9. Such diagrams provided a way of analyzing the tool setup.

Feedback to Supervisors

Beyond the need for feedback at workstations, there is a need to provide supervisors with short-term summaries. These take several forms.

Matrix summary. A common form of matrix is workers versus defects; the vertical columns are headed by worker names and the horizontal rows by the names of defect types. The matrix makes clear which defect types predominate, which workers have the most defects, and what the interaction is. Other matrices include machine number versus defect type and defect type versus calendar week. When the summary is published, it is usual to circle matrix cells to highlight the vital few situations that call for attention.

An elaboration of the matrix is to split the cell diagonally, thus permitting the entry of two numbers, e.g., number defective and number produced.

Pareto analysis. Some companies prefer to minimize detail and provide information on the total defects for each day plus a list of the top three (or so) defects encountered and

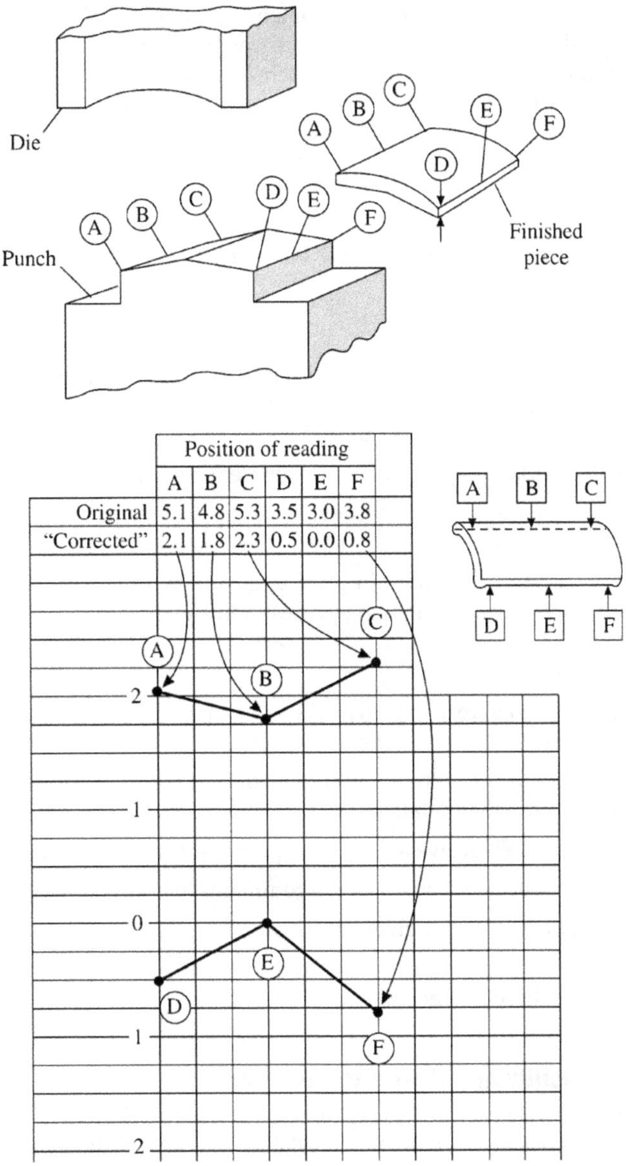

Figure 22.9 Method of drawing P–D diagram.

how many of each there were. In some industries, a "chart room" displays performance against goals by product and by department.

Automated quality information. Some situations justify the mechanization of both the recording and analysis of data. Entry of data into computer terminals on production floors is common now. Many varieties of software are available for analyzing, processing, and presenting quality information collected on the production floor.

The term quality information equipment (QIE) designates the physical apparatus that measures products and processes, summarizes the information, and feeds the information back for decision making. Sometimes such equipment has its own product development cycle to meet various product effectiveness parameters for the QIE. Remember QIE is also subject to sound measurement system analysis and gauge repeatability and reproducibility evaluation.

Checklist

A checklist for evaluating the second criterion of self-control includes questions such as these:

Adequate Review of Work

1. Are gauges provided to the worker? Do they provide numerical measurements rather than simply sorting good from bad? Are they precise enough? Are they regularly checked for accuracy?
2. Is the worker told how often to sample the work? Is sufficient time allowed?
3. Is the worker told how to evaluate measurements to decide when to adjust the process and when to leave it alone?
4. Is a checking procedure in place to ensure that the worker follows instructions on sampling work and making process adjustments?

Adequate Feedback

1. Are inspection results provided to the worker, and are these results reviewed by the supervisor with the worker?
2. Is the feedback timely and in enough detail to correct problem areas? Have personnel been asked what detail is needed in the feedback?
3. Do personnel receive a detailed report of errors by specific type of error?
4. Does feedback include positive comments in addition to negative?
5. Is negative feedback given in private?
6. Are certain types of errors tracked with feedback from external customers? Could some of these be tracked with an internal early indicator?

Criterion 3: Ability and Desire to Regulate

Ability and desire to regulate is the third criterion for self-control. Regulating the process depends on various management-controllable factors, including those explained below.

The process must be capable of meeting the tolerances. This factor is of paramount importance. In some organizations the credibility of specifications is a serious problem. Typically, a manufacturing process is created after release of the product design, a few trials are run, and full production commences. In cases where quality problems arise during full production, diagnosis sometimes reveals that the process is not capable of consistently meeting the design specifications. Costly delays in production then occur while the problem is resolved by changing the process or changing the specification. The capability of the manufacturing process should be verified during the product development cycle before the product design is released for full production.

The process must be responsive to regulatory action in a predictable cause-and-effect relationship to minimize variation around a target value.

Example 22.1 In a process for making polyethylene film, workers were required to meet multiple product parameters. The equipment had various regulatory devices, each of which could vary performance with respect to one or more parameters. However, the workers could not "dial in" a predetermined list of settings that would meet all parameters. Instead, it was necessary to cut and try to meet all parameters simultaneously. During the period of cut and try, the machine produced nonconforming product to an extent that interfered with meeting standards for productivity and delivery. The workers were unable to predict how long the cut-and-try process would go on before full conformance was achieved. Consequently, it became the practice to stop cut and try after a reasonable amount of time and to let the process run, whether in conformance or not.

The worker must be trained on how to use the regulating mechanisms and procedures. This training should cover the entire spectrum of action—under what conditions to act, what kind and extent of changes to make, how to use the regulating devices, and why these actions need to be taken.

Example 22.2 Of three qualified workers on a food process, only one operated the process every week and became proficient. The other two workers were used only when the primary worker was on vacation or was ill, and thus they never became proficient. Continuous training of the relief people was considered uneconomical, and agreements with the union prohibited their use except under the situations cited earlier. This problem is management controllable, i.e., additional training or a change in union agreements is necessary.

The act of adjustment should not be personally distasteful to the worker, e.g., should not require undue physical exertion.

Example 22.3 In a plant making glass bottles, one adjustment mechanism was located next to a furnace area. During the summer months, this area was so hot that workers tended to keep out of it as much as possible. When the regulation consists of varying the human component of the operation, the question of process capability arises in a new form: Does the worker have the capability to regulate?

The process must be maintained sufficiently to retain its inherent capability. Without adequate maintenance, equipment breaks down and requires frequent adjustments—often with an increase in both defects and variability around a nominal value. Clearly, such maintenance must be both preventive and corrective. The importance of maintenance has given rise to the concept of total productive maintenance (TPM) and reliability-centered maintenance (RCM). Under this approach, teams are formed to identify, analyze, and solve maintenance problems to maximize the uptime of process equipment. These teams consist of production line workers, maintenance personnel, process engineers, and others as needed. Problems are kept narrow in scope to encourage a steady stream of small improvements. Examples of improvements include a reduction in the number of tools lost, simplification of process adjustments, increased uptime, and gains in asset effectiveness as well as overall equipment effectiveness (OEE), particularly through improvements in unexpected losses.

Control Systems and the Concept of Dominance

Specific systems for controlling characteristics can be related to the underlying factors that dominate a process. The main categories of dominance include the following:

- *Setup dominant*. Such processes have high reproducibility and stability for the entire length of the batch to be made. Hence, the control system emphasizes verification of the setup before production proceeds. Examples of such processes are drilling, labeling, heat sealing, printing, and presswork.

- *Time dominant*. Such a process is subject to progressive change with time (wearing of tools, depletion of a reagent, heating up of a machine). The associated control system will feature a schedule of process checks with feedback to enable the worker to make compensatory changes. Screw machining, volume filling, wool carding, and papermaking are examples of time-dominant processes.
- *Component dominant*. Here the quality of the input materials and components is the most influential. The control system is strongly oriented toward supplier relations along with incoming inspection and sorting of inferior lots. Many assembly operations and food formulation processes are component dominant.
- *Worker dominant*. For such processes, quality depends mainly on the skill and knack possessed by the production worker. The control system emphasizes such features as training courses and certification for workers, error-proofing, and rating of workers and quality. Workers are dominant in processes such as welding, painting, and order filling.
- *Information dominant*. In these processes, the job information usually undergoes frequent change. Hence, the control system emphasizes the accuracy and up-to-dateness of the information provided to the worker (and everyone else). Examples include order editing and "travelers" used in job shops.

The various types of dominance differ also in the tools used for process control. Table 22.4 lists the forms of process dominance along with the usual tools used for process control.

Checklist

A checklist for evaluating the third criterion of self-control typically includes questions such as these:

- Process capable
  1. Has the quality capability of the process been measured to include both inherent variability and variability due to time? Is the capability checked periodically?
  2. Has the design of the job used the principles of error-proofing?
  3. Has equipment, including any software, been designed to be compatible with the abilities and limitations of workers?
- Process adjustments
  4. Has the worker been told how often to reset the process or how to evaluate measurements to decide when the process should be reset?
  5. Can the worker make a process adjustment to eliminate defects? Under what conditions should the worker adjust the process? When should the worker shut down the machine and seek more help? Whose help?
  6. Have the workers' actions that cause defects and the necessary preventive action been communicated to the worker, preferably in written form?
  7. Can workers institute job changes that they show will provide benefits? Are workers encouraged to suggest changes?
- Worker training in adjustments
  8. Do some workers possess a hidden knack that needs to be discovered and transmitted to all workers?

Setup Dominant	Time Dominant	Component Dominant	Worker Dominant	Information Dominant
Inspection of process conditions	Periodic inspection	Supplier rating	Acceptance inspection	Computer-generated information
First-piece inspection	$\bar{X}$ chart	Incoming inspection	*p* chart	"Active" checking of documentation
Lot plot	Median chart	Prior operation control	*c* chart	Bar codes and electronic entry
Precontrol	$\bar{X}$ and *R* chart	Acceptance inspection	Operator scoring	Process audits
Narrow-limit gaging	Precontrol	Mockup evaluation	Recertification of workers	
Attribute visual inspection	Narrow-limit gaging		Process audits	
	p chart			
	Process variables check			
	Automatic recording			
	Process audits			

TABLE 22.4 Control Tools for Forms of Process Dominance

9. Have workers been provided with the time and training to identify problems, analyze problems, and develop solutions? Does the training include diagnostic training to look for patterns of errors and determine sources and causes?

- Process maintenance

10. Is there an adequate preventive maintenance program for the process?

- Strong quality culture/environment

11. Is there sufficient effort to create and maintain awareness of quality?
12. Is there evidence of management leadership?
13. Have provisions been made for self-development and empowerment of personnel?
14. Have provisions been made for participation of personnel as a means of inspiring action?
15. Have provisions been made for recognition and rewards for personnel?

Use of Checklists on Self-Control

These checklists can help operations in the design (and redesign) of jobs to prevent errors; diagnose quality problems on individual jobs; identify common weaknesses in many jobs; and assist supervisors to function as coaches with personnel, prepare for process audits, and conduct training classes in quality.

Automated Manufacturing

The march to automation proceeds unabated. Several terms have become important:

- Computer-integrated manufacturing (CIM). CIM is the process of applying a computer in a planned fashion from design through manufacturing and shipping of the product.
- Computer-aided manufacturing (CAM). CAM is the process in which a computer is used to plan and control the work of specific equipment.
- Computer-aided design (CAD). CAD is the process by which a computer assists in the creation or modification of a design.

This trio of concepts is producing huge increases in factory productivity and efficiency. But automation, with proper planning, can also benefit product quality in several other ways:

- Automation can eliminate some of the monotony or fatiguing tasks that result in human errors. For example, when a manual seam-welding operation was turned over to a robot, the scrap rate plunged from 15 percent to zero, and manual grinding (a very dirty task) was virtually eliminated.
- Process variation can be reduced by the automatic monitoring and continuous adjustment of process variables.
- An important source of process troubles, i.e., the number of machine setups, can be reduced.
- Machines can automatically measure product and also record, summarize, and display the data for production line operators and staff personnel. Feedback to the worker can be immediate, thus providing early warning of impending troubles.

- With CAD, the quality engineer can provide inputs early in the design stage. When a design is placed into the computer, the quality engineer can review that design over and over again and thus keep abreast of design changes.

Collectively, these and other computer-based technologies for manufacturing are sometimes called "advanced manufacturing systems" (AMS).

With the continued enhancements of electronic information network provided by the Internet, a group of companies can operate as one virtual factory. This environment enables companies to exchange and act on information concerning inventory levels, delivery schedules, supplier lists, product specifications, and test data. It also means that CAD/CAM information and other manufacturing process information can be exchanged, data can be transferred to machines in a supplier's plant, and supplier software can be used to analyze producibility and to begin actual manufacturing.

Achieving the benefits of automated manufacturing requires a spectrum of concepts and techniques. Three of these are discussed here: the key functions of CIM, group technology, and flexible manufacturing systems.

Key Functions of Computer-Integrated Manufacturing

CIM integrates engineering and production with suppliers and customers (even globally) to interactively design, plan, and conduct the manufacturing activities. CIM activities include providing design tools to support innovation in remote sites, manufacturing planning, computer simulation models to evaluate and predict process performance, sensing and control tools to monitor processes, intelligence tools to gather and organize process data to share with other sites, information processing and transferring among geographically dispersed participants, and automated translation of text between different languages. For elaboration, see Lee (1995). This is tremendously important today with the continued emergence of manufacturing in developing countries around the world.

Group Technology

Group technology (or accelerated design replication) is the process of examining all items manufactured by a company to identify those with sufficient similarity that common design or manufacturing plans can be used. The aim is to reduce the number of new designs or new manufacturing plans. In addition to the savings in resources, group technology can improve both the quality of design and the quality of conformance by using proven designs and manufacturing plans. In many companies, only 20 percent of the parts initially thought to require a new design actually need it; of the remaining new parts, 40 percent could be built from an existing design, and the other 40 percent could be created by modifying an existing design. Accelerated replication of designs can reduce product design cycles from months to only weeks or even days.

Location of production machines can also benefit from the group technology concept. Machines are grouped according to value streams and can be sorted into cells of machines, each cell producing one or several part families (thus "cellular manufacturing"). For a discussion of shifting from traditional manufacturing organization to cellular manufacture and the impact on first-level supervisors, see JQH5, pp. 11.332–11.349.

Flexible Manufacturing System

A flexible manufacturing system (FMS) is a group of several computer-controlled machine tools, linked by a materials handling system and a computer, to accommodate varying production requirements. The system can be reprogrammed to accommodate design changes or new parts. In contrast, in a fixed automation system, machinery, materials handling equipment, and controllers are organized and programmed for production of a single part or limited range of parts.

Typically, the individual machines are robots or other types of numerically controlled machine tools, each of which is run by a microcomputer. Several such tools are linked by a minicomputer, and then several of these minicomputers are tied into the mainframe computer.

From the one extreme of the (typically) mass production in automated industries, we can shift to the other extreme of the (typically) low-volume production in job shop industries.

Even traditional craft industries now benefit from technology. For example, the manufacture of fine pianos uses programmable logic controllers to make process adjustments; in wine making, aerial images of vineyards, taken by using digital sensors to collect data on the chlorophyll content of the vines, provide a strong indicator of taste (the aircraft will soon be replaced by satellites).

The potential benefits of the automated factory require significant time and resources for planning. Automation, however, will not be total—there will never be robot plumbers in the factory.

Overall Review of Manufacturing Planning

We incur great risk in going directly from a proposed manufacturing process plan into regular production. The delays and extra costs involved in quality failures require a review of the proposed process, including software used with the process. Such a review is most effectively accomplished through preproduction trials and runs.

Ideally, product lots should be put through the entire system, with deficiencies found and corrected before full-scale production begins. This should include a feasibility stage, prototype stage, preproduction verification and validation stage, and finally production verification and validation. In practice, companies usually make some compromises with this ideal approach. "Preproduction" may be merely the first runs of regular production but with special provision for prompt feedback and correction of errors as found. Alternatively, the preproduction run may be limited to features of product and process design that are so new that prior experience cannot reliably provide a basis for good risk taking. Although some companies do adhere to a strict rule of proving the product and process through preproduction lots, the more usual approach is one of flexibility, in which the use of preproduction lots depends on various factors:

- The extent to which the product embodies new or untested quality features.
- The extent to which the design of the manufacturing process embodies new or untried machines, tools, etc.
- The amount and value of product that will be out in the field before process, product, and use difficulties are fully known.

These trials sometimes include "production validation tests" to ensure that the full-scale process can meet the design intent.

The terms process qualification and process certification are also used to describe the review of manufacturing processes. Black (1993) describes how Caterpillar developed an internal certification program similar to its supplier certification program. The certification program is built around a 12-step manufacturing procedure to achieve process control and includes the identification of critical product characteristics and critical process parameters, the determination of process capability indexes, and actions for continuous quality improvement.

Preproduction trials and runs provide the ultimate evaluation—by manufacturing real product. Other techniques provide an even earlier warning before any product is made. For example, the failure mode, effect, and criticality analysis is useful in analyzing a proposed product design. The same technique can dissect potential failure modes and their effects on a proposed process design. Another technique uses highly detailed checklists to review proposed processes.

Organizing for Quality in Manufacturing Operations

The organization of today is influenced by the interaction of two systems: the technical system (design, equipment, procedures) and the social system (people, roles)—thus the name "sociotechnical systems" (STSs).

New ways of organizing work, particularly at the workforce level, have emerged. For example, supervisors are becoming "coaches"; they teach and empower rather than assign and direct. Operators are becoming "technicians"; they perform multi-skilled jobs with broad decision making rather than narrow jobs with limited decision making. Team concepts play an important role in these new approaches.

In some organizations more than 40 percent of the people participate on teams; some organizations have a goal of 80 percent or greater. Permanent teams (e.g., process teams, self-managing teams) are responsible for all output parameters, including quality; ad hoc teams (e.g., quality project teams) are typically responsible for improvement in quality.

Although these various types of quality teams are showing significant results, the reality is that, for most organizations, daily work in a department is managed by a supervisor who has a complement of workers performing various tasks. This configuration is the "natural work team" in operations. But team concepts can certainly be applied to daily work. One framework for a team in daily operations work is the control process from the trilogy of quality processes. As applied to daily work, the steps are choose control subjects, establish measurement, establish standards of performance, measure actual performance, compare to standards, and take action on the difference. When the natural work team of the department (self-directed work team) is trained in these concepts, the work team gains greater control over the key work processes so that it can meet customer needs while increasing employee involvement and empowerment.

Planning for Evaluation of Product

The planning must recognize the need for formal evaluation of product to determine its suitability for the marketplace. Three activities are involved:

1. Measuring the product for conformance to specifications.
2. Taking action on nonconforming product.
3. Communicating information on the disposition of nonconforming product.

What has evolved is the concept of self-inspection combined with a product audit. Under this concept, all inspection and all conformance decisions, both on the process and on the product, are made by the production worker. (Decisions on the action to be taken on a nonconforming product are not, however, delegated to the worker.) However, an independent audit of these decisions is made. The quality department inspects a random sample periodically to ensure that the decision-making process used by workers to accept or reject a product is still valid. The audit verifies the decision process. Note that, under a pure audit concept, inspectors are not transferred to do inspection work in the production department. Except for those necessary to do audits, inspection positions are eliminated.

If an audit reveals that wrong decisions have been made by the workers, the product evaluated since the last audit is re-inspected—often by the workers themselves.

Self-inspection has decided advantages over the traditional delegation of inspection to a separate department:

- Production workers are made to feel more responsible for the quality of their work.
- Feedback on performance is immediate, thereby facilitating process adjustments. Traditional inspection also has the psychological disadvantage of an "outsider" reporting the defects to a worker.
- The costs of a separate inspection department can be reduced.
- The job enlargement that takes place by adding inspection to the production activity of the worker helps to reduce the monotony and boredom inherent in many jobs.
- Elimination of a specific station for inspecting all products reduces the total manufacturing cycle time.

Example 22.4 In a coning operation of textile yarn, the traditional method of inspection often resulted in finished cones sitting in the inspection department for several days, thereby delaying any feedback to production. Under self-inspection, workers received immediate feedback and could get machines repaired and setups improved more promptly. Overall, the program reduced nonconformities from 8 to 3 percent. An audit inspection of the products that were classified by the workers as "good" showed that virtually all of them were correctly classified. In this company, workers can also classify product as "doubtful." In one analysis, worker inspections classified 3 percent of the product as doubtful, after which an independent inspector reviewed the doubtful product and classified 2 percent as acceptable and 1 percent as nonconforming.

Example 22.5 A pharmaceutical manufacturer employed a variety of tests and inspections before a capsule product was released for sale. These checks included chemical tests, weight checks, and visual inspections of the capsules. A 100 percent visual inspection had traditionally been conducted by an inspection department. Defects ranged from "critical" (e.g., an empty capsule) to "minor" (e.g., faulty print). This inspection was time-consuming and frequently caused delays in production flow. A trial experiment of self-inspection by machine operators was instituted. Operators visually inspected a sample of 500 capsules. If the sample was acceptable, the operator shipped the full container to the warehouse; if the sample was not acceptable, the full container was sent to the inspection department for 100 percent inspection. During the experiment, both the samples and the full containers were sent to the inspection department for 100 percent inspection with re-inspection of the sample recorded separately. The experiment reached two conclusions: (1) the sample inspection by the operators gave results consistent with the sample inspection by the inspectors, and (2) the sample of 500 gave results consistent with the results of 100 percent inspection.

The experiment convinced all parties to switch to sample inspection by operators. Under the new system, good product was released to the warehouse sooner, and marginal product received a highly focused 100 percent inspection. In addition, the level of defects decreased. The improved quality level was attributed to the stronger sense of responsibility of operators (they themselves decided if product was ready for sale) and the immediate feedback received by operators from self-inspection. But there was another benefit—the inspection force was reduced by 50 people. These 50 people were shifted to other types of work, including experimentation and analysis on the various types of defects.

Criteria for Self-Inspection

For self-inspection, the following criteria must be met:

- Quality must be the number one priority within an organization. If this requirement is not clear, a worker may succumb to schedule and cost pressures and classify products as acceptable that should be rejected.

- Mutual confidence is necessary. Managers must have sufficient confidence in the workforce to be willing to give workers the responsibility for deciding whether the product conforms to specification. In turn, workers must have enough confidence in management to be willing to accept this responsibility.
- The criteria for self-control must be met. Failure to eliminate the management-controllable causes of defects suggests that management does not view quality as a high priority, and this environment may bias the workers during inspections. Workers must be trained to understand the specifications and perform the inspection.
- Specifications must be unequivocally clear. Workers should understand the use that will be made of their products (internally and externally) to grasp the importance of a conformance decision.
- The process must permit assignment of clear responsibility for decision making. An easy case for application is a worker running one machine because there is clear responsibility for making both the product and the product-conformance decision. In contrast, a long assembly line or the numerous steps taken in a chemical process make it difficult to assign clear responsibility. Application of self-inspection to such multi-step processes is best deferred until experience is gained with some simple processes.

Self-inspection should apply only to products and processes that are stabilized and meet specifications and only to personnel who have demonstrated their competence. It is important to note again that the organization must demonstrate the process is not only capable, but in a state of statistical control.

Worker response to such delegation of authority is generally favorable; the concept of job enlargement is a significant factor. However, workers who do qualify for self-inspection commonly demand some form of compensation for this achievement, e.g., a higher grade, more pay. Companies invariably make a constructive response to these demands because the economics of delegating are favorable. In addition, the resulting differential tends to act as a stimulus to nonqualified workers to qualify themselves.

An adjunct to self-inspection is the use of poka-yoke devices as part of inspection. These devices are installed in a machine to inspect the process conditions and product results and provide immediate feedback to the operator. Devices such as limit switches and interference pins, used to ensure proper positioning of materials on machines, are poka-yoke devices for inspecting process conditions. Go/no go gauges are examples of poka-yoke devices for inspecting product. For elaboration, see The Productivity Press Development Team (1997). Grout and Downs (1998) compares the use of poka-yoke devices for controlling processes with statistical process control charts.

Another concept related to self-inspection is jidoka. Jidoka is one of the two main pillars of the Toyota Production System. It refers to the ability to stop production lines, by man or machine, in the event of problems such as equipment malfunction, quality issues, or late work. Jidoka helps prevent the passing of defects, helps identify and correct problem areas using localization and isolation, and makes it possible to "build" quality at the production process. The three responsibilities of every operator are to:

- Check incoming work to ensure that it is defect free.
- Verify his or her work is free of defects.
- Never knowingly pass a defective product.
- Start with your existing quality system and enhance by asking the following:
  - What are the critical factors to the customer?

- How are the factors measured and recorded?
- Is the measurement system effective in identifying and preventing defects from reaching the customer?

For successful implementation of jidoka, an organization should:

1. Map and analyze its process.
2. Understand the customers' needs and translate them to its process.
3. Develop "in-station quality."
4. Mistake-proof the process and install automatic notification devices where possible.
5. Construct Andons signaling at points that are not 100 percent mistake-proofed.
6. Create a support structure—specific by issue.
7. Utilize root cause correction action when problems arise.
8. Implement visual control and Standard work throughout the process.
9. Develop a "stop-the-line" culture.

Conducting Process Quality Audits

A quality audit is an independent review to compare some aspect of quality performance with a standard for that performance. Application to manufacturing has been extensive and includes both audits of activities (process audits) and audits of product (product audits). A full discussion of quality audits is given in Chap. 9, "Quality Assurance and Audits."

A process quality audit includes any activity that can affect final product quality. This onsite audit is usually done on a specific process by one or more persons and uses the process operating procedures. Adherence to existing procedures is emphasized, but audits often uncover situations of inadequate or nonexistent procedures. The checklists presented earlier in this chapter on the three criteria for self-control can suggest useful specific subjects for process audits. Audits must be based on a foundation of hard facts that are presented in the audit report in a way that will help those responsible to determine and execute the required corrective action.

Peña (1990) explains an audit approach for processes. Two types of audits are employed: engineering and monitor. The engineering process audit is conducted by a quality assurance engineer and entails an intense review of all process steps including process parameters, handling techniques, and statistical process control. Table 22.5 shows the audit checklist.

The monitor process audit covers a broad range of issues, e.g., whether specifications are correct and whether logs are filled in and maintained. Discrepancies (critical, major, or minor) are documented, and corrective action is required in writing. Critical defects must be corrected immediately; majors and minors must be resolved within five working days.

A product audit involves the re-inspection of product to verify the adequacy of acceptance and rejection decisions. In theory, such product audits should not be needed. In practice, they can often be justified by field complaints. Such audits can take place at each inspection station for the product or after final assembly and packing. Sometimes an audit is required before a product may be moved to the next operation.

1. Is the specification accessible to production staff?
2. Is the current revision on file?
3. Is the copy on file in good condition and all pages accounted for?
4. If referenced documents are posted on equipment, do they match the specification?
5. If the log sheet is referenced in specifications, is a sample included in the specification?
6. Is the operator completing the log sheet according to specifications?
7. Are lots with out-of-specification readings authorized and taken care of in writing by the engineering department or the proper supervisor?
8. Are corrections to paperwork made according to specification?
9. Are equipment time settings according to specification?
10. Are equipment temperature settings according to specification?
11. Is the calibration sticker on equipment current?
12. Do chemicals or gases listed in the specification match actual usage?
13. Do quantities listed in the specification match the line setup?
14. Are changes of chemicals or gases made according to specification?
15. Is the production operator certified? If not, is this person authorized by the supervisor?
16. Is the production operating procedure according to specification?
17. Is the operator performing the written cleaning procedure according to specification?
18. If safety requirements are listed in the specification, are they being followed?
19. If process control procedures are written in the specification, are the actions performed by process control verifiable?
20. If equipment maintenance procedures are written in the specification, are the actions performed verifiable? according to specification?

From "Motorola's Secret to Total Quality Control" by E. Peña *Quality Progress*, October. 1990. Reprinted with permission form *Quality Progress*, © 1990 American Society for Quality.

TABLE 22.5 Audit Checklist

Quality Measurement in Manufacturing Operations

The management of key work processes must include provision for measurement. Table 22.6 shows examples for manufacturing activities. Note that many of the control subjects are forms of work output. In reviewing current units in use, a fruitful starting point is the measure of productivity. Productivity is usually defined as the amount of output related to input resources. Surprisingly, some organizations still mistakenly calculate only one measure of output, i.e., the total (acceptable and nonacceptable). Clearly, the pertinent output measure is that which is usable by customers (i.e., acceptable output).

The units in Table 22.6 become candidates for data analysis using statistical techniques such as control charts.. But there is a more basic point—the selection of the unit of measure and the periodic collection and reporting of data demonstrate to operating personnel that management regards the subject as having priority importance. This atmosphere sets the stage for improvement!

Subject	Unit of Measure
Quality of manufacturing output	Percentage of output meeting specifications at inspection ("first-time yield")
	Percentage of output meeting specifications at intermediate and final inspection
	Amount of scrap (quantity, cost, percentage, etc.), amount of rework (quantity, cost, percentage, etc.)
	Percentage of output shipped under waiver of specifications
	Number of defects found in product audit (after inspection)
	Warranty costs due to manufacturing defects
	Overall measure of product quality (defects in parts per million, weighted defects per unit, variability for critical characteristics, etc.)
	Amount of downgraded output
Quality of input to manufacturing	Percentage of critical operations with certified workers
	Amount of downtime of manufacturing equipment
	Percentage of product input meeting specifications
	Percentage of instruments meeting calibration schedules
	Percentage of specifications requiring changes after release

TABLE 22.6 Examples of Quality Measurement in Manufacturing

Maintaining a Focus on Continuous Improvement

Historically, the operations function has always been involved in troubleshooting sporadic problems. As chronic problems were identified, these were addressed using various approaches, such as quality improvement teams Often the remedies for improvement involve quality planning or replanning. These three types of action are summarized in Table 22.7.

Kannan et al. (1999) presents the results of a survey of the application of 38 quality management practices (e.g., use of benchmark data to improve quality practices), 39 tools and techniques (e.g., statistical process control), 29 areas of documentation (e.g., quality assurance manual), and 12 quality measurements (e.g., customer complaints) at the operations level in manufacturing industries. The analysis includes the degree of usage and the impact on five organization performance measures.

Maintaining the focus on improvement clearly requires a positive quality culture in the organization. Therefore, we must first determine the present quality culture and then take the steps to change the culture to one that will foster continuous improvement (see Chap. 2, "Developing an Excellence Culture"). In addition, the operations function must be provided with the support to maintain the focus on improvement. A key source of that support should be the quality department. Thus the quality department should view operations as its key internal customer and provide the training, technical quality expertise, and other forms of support to enable operations to maintain the focus on improvement. Also, a quality department can urge upper management to set up cross-functional teams to address operations problems that may be caused by other functional departments such as engineering, purchasing, and information technology.

Type of Action to Take	When to Take Action	Basic Steps
Troubleshooting (part of quality control)	Performance indicator outside control limits Performance indicator in clear trend toward control limits	Identify problem Diagnose problem Take remedial action
Quality improvement	The control limits are so wide that it is possible for the process to be in control and still miss the targets Performance indicator frequently misses its target	Identify project Establish project Diagnose cause Remedy the cause Hold the gains
Quality planning	Many performance indicators for this process miss their targets frequently Customers have significant needs that the product does not meet	Establish project Identify customers Discover customer needs Develop product Develop process Design controls

Source: Adapted from Juran Institute, Inc. (1995, pp. 5–7).

TABLE 22.7 Three Types of Action

Case Study on Error-Proofing

Case Study on Error-Proofing Documentation in a Biotech Environment Using Six Sigma and Lean Projects

Background

In FDA-regulated manufacturing environments such as those in the life sciences, biotech, and pharmaceutical industries, complete and accurate documentation of processes and the completion of process steps are critical. However, such documentation is prone to a variety of human errors: slips, lapses, mistakes, and violations. This is a common problem across these industries, but the majority of document errors do not impact product quality.

A decision by Genentech's senior management to error-proof such documentation resulted in the achievement of sustainable breakthrough improvements through the focused deployment of Six Sigma and lean project teams.

Type of Error	Description
Attentional lapse	• Forgot to record storage location and time and transfer after moving product from cold room • Forgot to make required entry on one of three equipment use logs tied to transfer lines and tanks • Forgot to enter information just entered on a ticket on a use log
Slip	• Transposed a lot number • Forgot to initial and date a page reviewed for completeness
Rule-based mistake	• Incorrectly exclude entry as "not applicable"

TABLE 22.8 Types of Document Errors Caught During Batch Record Review

The Error-Proofing Project was chartered in September 2003 by senior management (GMP Core Team) at Genentech. The objective of the project was to reduce the inspection risk associated with document errors and lost tickets. This objective was expressed as the following dual goal:

Goal 1. Reduce the rate of document errors recorded in the discrepancy management system by 50 percent relative to the Q1/Q2 2003 baseline by September 2004. Maintain the document error rate within this new "zone of control": less than five document errors per 100 tickets (main batch record documents) for three months or more.

Goal 2. Reduce the overall volume of master ticket errors by 50 percent relative to the Q1/Q2 2003 baseline.

Six error-proofing teams were chartered in October 2003 to investigate the drivers for this pattern of error and variability. Consultants worked with each team. The teams were chartered such that their implementations would begin in March 2004 and deliver measurable document error rate reductions by June 2004.

The six projects were

1. Timely Feedback (a Six Sigma DMAIC project)
2. Change Volume and Timing (a Six Sigma DMAIC project)
3. Value-Stream Mapping in Growth Hormone and Recovery Operations (a Lean/Value-Stream Mapping project)
4. Document Rule Clarification (a Six Sigma DMAIC project)
5. Document Development (a Design for Six Sigma (DMADV) project)
6. Document Complexity (a Design for Six Sigma (DMADV) project)

Key Findings

1. *Timely feedback.* Based on interviews across all shifts, the project team concluded that the timing and utility of feedback provided varied across production teams: in particular, the project team found that a mechanism for timely feedback was present only in certain instances. They set out to correlate the presence or absence of timely feedback to document error rates. As shown in the following hypothesis test (chi-square tables test of independence), the document error rate is dependent on the presence or absence of timely feedback.

Chi-Square Test: <1wk FB S, 0 FB S

Expected counts are printed below observed counts
Chi-square contributions are printed below expected counts

	<1wk FB S	**0 FB S**	**Total**
1	25	19	44
	20.20	23.80	
	1.138	0.967	
2	20	34	54
	24.80	29.20	
	0.928	0.788	
Total	45	53	98

Chi-sq = 3.820, DF = 1, p-Value = .051

>There is a statistical relationship of dependence between the lack of feedback and the groups we identified as high and low performers (on document errors).

Teams with timely feedback performed 50 percent better than those with incomplete or delayed feedback. As predicted by the team, the document error rate took its first significant shift toward a new zone of control in April 2004 immediately after the timely feedback mechanisms were introduced.

2. *Change timing*. The Change Volume and Timing project team (a DMAIC project) set out to investigate the relationship between document effective and release dates with regard to document errors per document. By correlating the interval between released and effective dates to the document errors recorded per execution of the ticket, the team was able to show that the documents made effective within three days of their release dates carried a disproportionate amount of document error discrepancies (Fig. 22.10). Because 45.7 percent of tickets had less than 2 days between release and effective dates, these documents represent a significant driver of document errors on the floor.

   As the team investigated further, they were able to confirm that 1–3 day intervals do not permit meaningful document training across shifts.

3. *Change volume*. Another driver of document errors confirmed by the change volume and timing team is the number of times a document gets changed in a year. More than 1440 ticket changes were initiated by SSF manufacturing in 2003. By comparing the discrepancy rates recorded for documents that were changed more than three times in a year to those that were changed less often, the team was able to confirm that frequent revisions drive document errors.

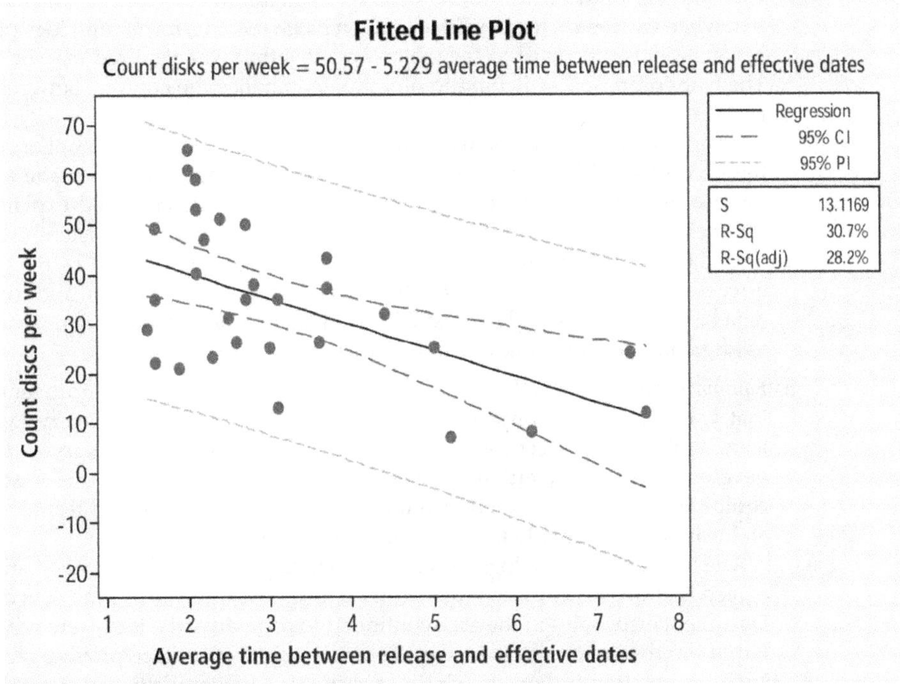

Figure 22.10 Relationship between the average time between release and effective dates for documents and discrepancies.

Change Frequency	N	Median	Ave Rank	Z
0.08	215	0.1268	188.4	23.06
0.17	127	0.1429	208.0	0.29
0.25	45	0.1875	243.8	2.30
0.33	20	0.5227	269.0	2.46
0.42	2	0.6483	317.0	1.33
0.58	1	0.6800	344.0	1.17
Overall	410		205.5	
H = 18.09 DF = 5 p = 0.003				
H = 18.28 DF = 5 p = 0.003 (adjusted for ties)				

TABLE 22.9 Kruskall–Wallis Test Shows Significant Differences in Median Document Errors Across Different Document Change Frequencies

Enforcement of a mechanism to pull documents with more than three changes per year, coupled with a system to make revisions more visible were proposed. The team remains convinced that these measures would yield further document error reduction benefits.

4. *Ink Color*: The Lean/Value Stream Mapping team in Growth Hormone Recovery operations generated a list of error-proofing implementation ideas during their rapid improvement phase. One of the ideas with evident universal utility grew out of an analysis of the factors that contribute to omission noted errors. The team noted that when entries are made in black ink on batch records, forms, and logs printed in black ink, it can be hard to see where an entry that was required may be missing. The team proposed switching to blue ink to enhance the contrast between entries and the printed text.

   After negotiating the details with the quality assurance and compliance group, the necessary SOP (standard operating procedure) changes were made to permit the use of blue ink without requiring it. A memo was circulated explaining the rationale for the change and asking managers to enforce the consistent use of blue ink. The blue ink initiative was implemented in March 2004 just as the omission error rate dropped below 5 for the first time. Manufacturing online auditor (MOA) data collected in July and August 2004 suggest that omissions were detected twice as often now that blue ink was in routine use.

5. *Equipment log location and other rapid improvements*. Another rapid improvement idea that was successfully implemented by the growth hormone recovery lean/VSM team looked at the centralization of equipment use logs. A survey tool was developed and spaghetti maps were used to test the hypothesis that having equipment use logs scattered throughout the work area increased technician wear and tear and incidentally the risk of document errors.

   After centralizing the logs in one location, 95 percent of the survey respondents agreed that the system was much easier to use (less travel, less time, less difficult). Document error rates in the area continued to drop after the logs were centralized. Rapid improvement efforts to identify single points of contact, provide radios, and structure communications for clarity, directness, and simplicity have contributed to a reduction in the impact of uncontrolled communications with the floor.

6. *Documentation rule confusion.* Interviews with operators who execute tickets confirmed that a high degree of confusion existed around the rules for cross outs, ruling out sections as not applicable, entering time correctly, and other fine points of batch record creation. This confusion was identified as a potentially significant driver of documentation errors.

   Even though many of the errors that result from confusion are crossed out (Fig. 22.11) and then corrected during online review, cross-outs, as such, can be viewed as a concern by a FDA auditor because they reflect the operator's struggle to create the batch record correctly the first time. The team systematically catalogued the points of confusion (sub-Y's) and then completed a cause and effect analysis for each one to identify the drivers [f (x's)]. Examples of these validated sources of confusion (proven X's) include

   - Documents use ambiguous language (e.g., "approximately"), which causes confusion around doc rules and leads to doc errors.
   - Some documents include information only around alert limits; other documents include information around alert limits and action limits. This inconsistency causes confusion around doc rules, which leads to doc errors.
   - Documentation rules keep changing, which causes confusion around doc rules and leads to doc errors.
   - Operators are trained by different trainers and content is not consistent, which causes confusion around doc rules and leads to doc errors.
   - Justification behind the rules (criteria for the rules) is not communicated, which causes confusion around doc rules and leads to doc errors.

   These validated causes of confusion were converted to customer needs and used as the basis for designing a clear, central SOP for documentation and associated training. These necessary tools were made effective and provided to operators in May 2004—just as the document error rate dropped below 5 per 100 tickets for the first time.

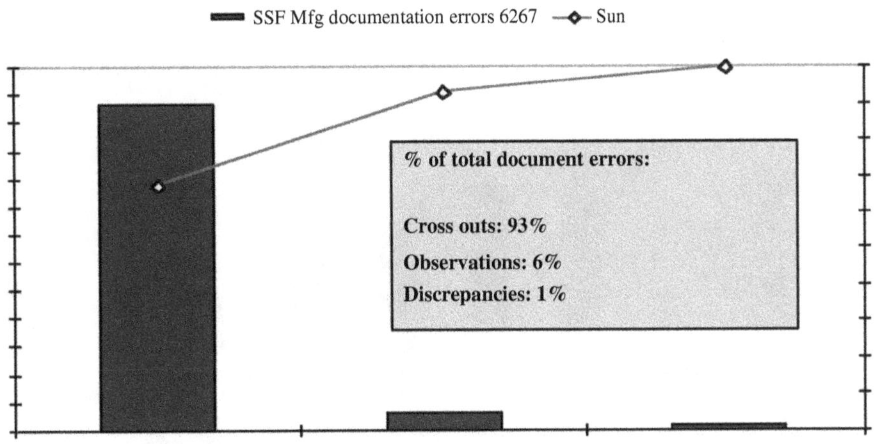

FIGURE 22.11 Baseline document error performance based on a review of documents completed before February 2004.

7. *Low accountability for word processing errors and no central coordination for document changes.* The document development and change process team (a DMADV team) defined its problem statement around the inability of the current process to reliably produce accurate/error-free documents within customer-defined timelines.

 Data from Q1/Q2 of 2003 were analyzed by the team to identify master ticket errors and map them to process steps where the error could have originated or caught (individual errors were often mapped to multiple locations). This analysis identified the "Document Development" stage of the process (including word processing and SME redline steps) as the most significant generator of ticket errors (Fig. 22.12).

 A value-stream map of the process was created and used to show that only 4 of the 54 steps were value-add, 2 were non-value-add, and the remaining 48 were all classified as "non-value-add needed."

 An FMEA (Failure Modes Effects Analysis) of the process ranked the various identified failure modes, and root causes were assigned to the highest severity failure modes. The top 10 RPN (Risk Priority Number) scores could be assigned to three root causes: communication gaps, business process inadequacies, and resource limitations.

 An improved process was designed using DMADV methodology to address the identified issues; the three root causes were converted to customer needs. The improved process (which features a centralized coordinator, automated step references, resource management tools, and development checklists) was piloted in the fall of 2004. Compared to the baseline data, the campaign preparation effort delivered accurate documents in accordance with defined timelines.

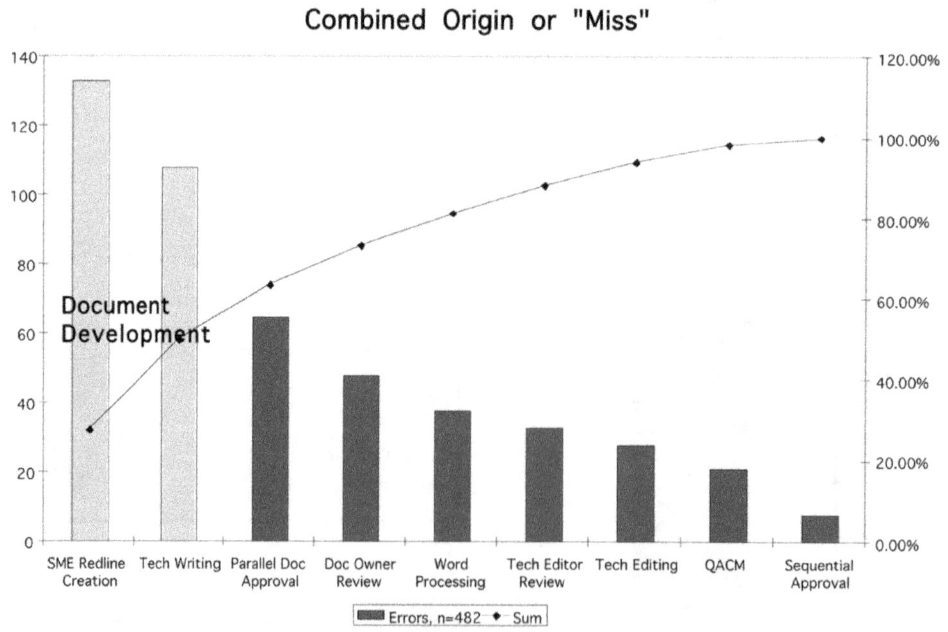

FIGURE 22.12 Pareto of performance gaps identified in Q1/Q2 2003 data, showing combined origin and "miss" steps where error could have been caught.

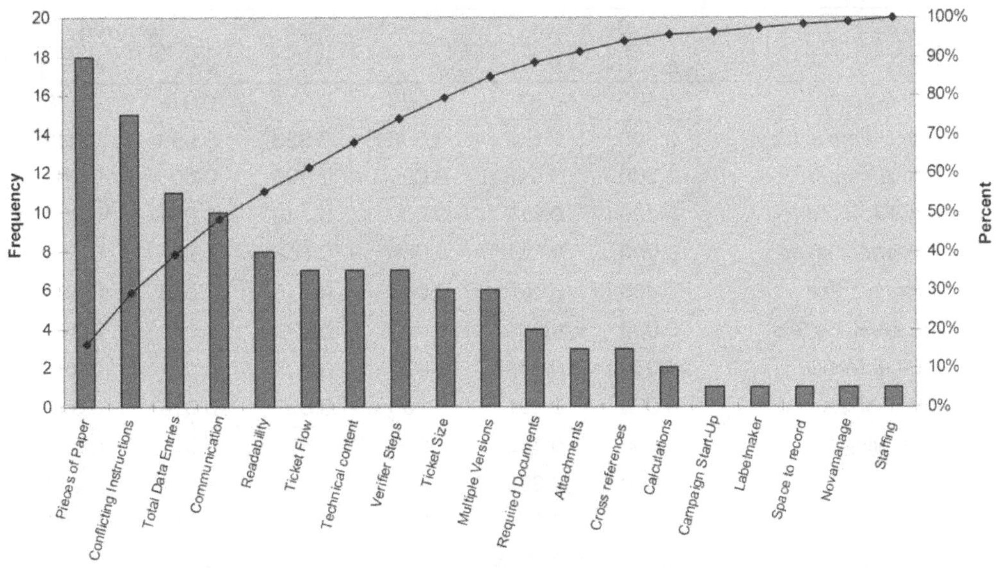

FIGURE 22.13 Pareto of customer needs for clear instructions.

8. *Document complexity.* The problem statement devised by the document complexity team (a DMADV project) emphasized the link between document errors and instructions to manufacturing and QA staff that are often overly complicated or inconsistent.

   The team interviewed customers (document users) across functions and prioritized their identified needs. A Pareto chart of these findings is provided in Fig. 22.13.

   The top three needs (identified most often) were fewer pieces of paper, no conflicting instructions, and fewer data entries.

   Using these data, the team was able to devise a system for scoring instruction complexity and applied this to a set of standard instructions. These scores were then correlated to discrepancy rates per document. As shown in Table 22.10, for the "Combined" and "Ferm Only" data, the error rate correlated strongly with complexity score and 11 of 12 complexity elements (recovery data correlate strongly with only 9 of 12 elements).

Regression analyses were used to confirm that complexity scores could predict error rates and significant complexity elements.

The results of the analysis were used to identify two quick fixes which were successfully implemented: preprinted POMS labels and other attachments in a binder and a departmental controlled document to list use logs. Both of these fixes were implemented by June 2004.

Breakthrough Results and New Zone of Control

As illustrated in Fig. 22.14, the first goal was achieved in May 2004, and the document error has been held within the new zone of control ever since.

The second goal was achieved almost immediately, and the incidence of master ticket errors has not returned to Q1/Q2 2003 levels since the project began (Fig. 22.15).

	Combined		Fermentation		Recovery	
	p	*R*	*p*	*R*	*p*	*R*
Complexity score	0.000	0.673	0.000	0.859	0.018	0.338
Total data entries	0.000	0.632	0.000	0.829	0.093	0.242
Total pages	0.000	0.618	0.000	0.925	0.006	0.385
SOP/SR refs	0.000	0.617	0.000	0.748	0.001	0.457
Operator steps	0.000	0.611	0.000	0.868	0.010	0.366
FNs to refer	0.000	0.600	0.000	0.773	0.018	0.336
Total SOPs/SRs	0.000	0.584	0.000	0.699	0.005	0.399
Other G-code	0.000	0.575	0.000	0.713	0.007	0.380
Verifier steps	0.000	0.491	0.000	0.877	0.016	0.343
Labeling steps	0.000	0.442	0.000	0.838	0.021	0.328
Calculations	0.000	0.395	0.019	0.466	0.134	0.217
Attachments	0.032	0.249	0.000	0.738	0.210	0.182
Ver/Op	0.470	20.085	0.748	20.068	0.832	0.031

Table 22.10 Correlation Results, Complexity Score, and Error Rate

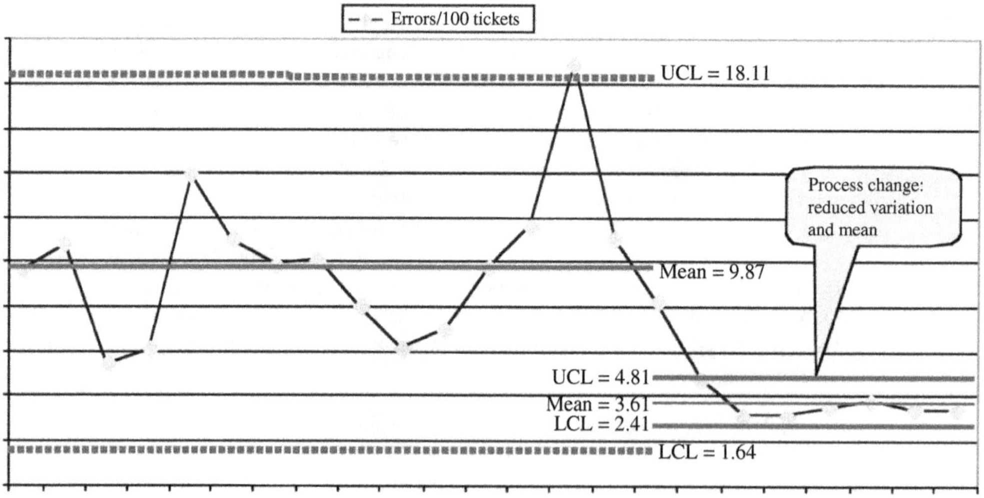

Figure 22.14 Document error rate: Jan 2003 through Nov 2004.

Lessons Learned

Achieving and sustaining these breakthrough results demonstrated the power of applying project-by-project improvement using the right tools (in this case, Six Sigma and lean/value-stream management) coupled with the right management priority and commitment of the right resources and support. Lessons learned include

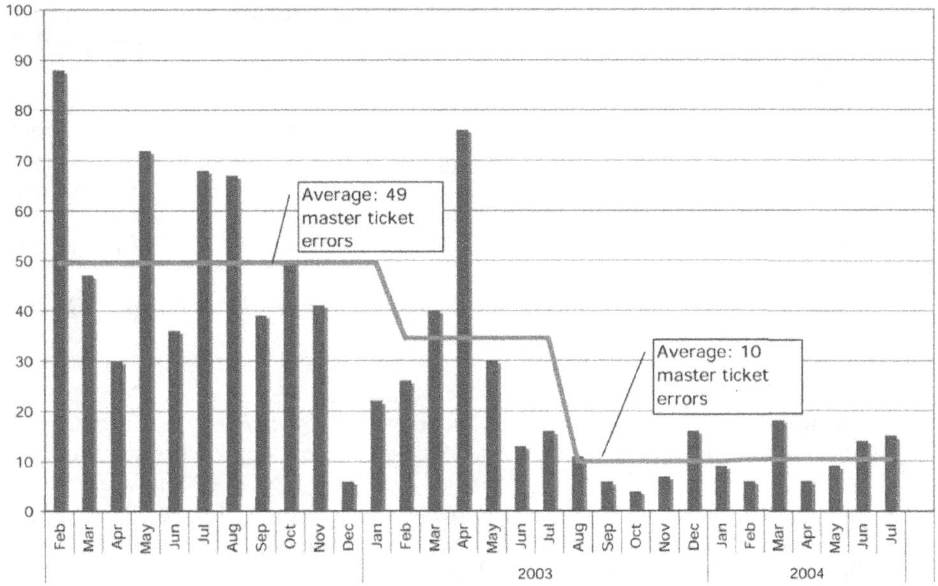

Figure 22.15 Master ticket errors per month.

- Dedicated resources perform better than part-time teams; effective sponsors ensure that team members are not over booked.
- Active and engaged sponsorship is key, especially during the initial scope of work definition and chartering.
- Management must resist the impulse to direct teams to a solution prematurely while providing clear boundaries to the scope of the inquiry and the timing of the project.
- Take care to prepare for training events such that logistical issues, absent sponsors, and confusion about project goals do not distract participants.

In addition to these success factors, this effort taught management one truly fundamental lesson and that is, put your best people on the problem and make the investment to provide them with world class tools and support and they will deliver lasting results.

References

Black, S. P. (1993). "Internal Certification: The Key to Continuous Quality Success," *Quality Progress*, January, vol. 26, no. 1, pp. 67–68.

Eibl, S., Kess, U., and Pukelsheim, F. (1992). "Achieving a Target Value for a Manufacturing Process," *Journal of Quality Technology*, January, vol. 24, no. 1, pp. 22–26.

Goldman, S. L., Nagel, R. N., and Preiss, K. (1995). *Agile Competitors and Virtual Organizations*, Van Nostrand Reinhold, New York.

Grout, J. R. and Downs, B. T. (1998). "Mistake-Proofing and Measurement Control Charts," *Quality Management Journal*, vol. 5, no. 2, pp. 67–75.

Juran, J. M., (1992). *Juran on Quality by Design*, Free Press, New York.

Juran, J. M., (2010). *Juran's Quality Handbook: The Complete Guide to Performance Excellence*, McGraw-Hill, New York, p. 209.

Lee, J. (1995). "Perspective and Overview of Manufacturing Initiatives in the United States," *International Journal of Reliability, Quality and Safety Engineering*, vol. 2, no. 3, pp. 227–233.

Nakajo, T. and Kume, H. (1985). "The Principles of Foolproofing and Their Application in Manufacturing," *Reports of Statistical Application Research, Union of Japanese Scientists and Engineers*, Tokyo, vol. 32, no. 2, June, pp. 10–29.

Ohno, T. (1988). The Toyota Production System—Beyond Large Scale Production, Productivity Inc., Portland, Oregon.

Peña, E. (1990). "Motorola's Secret to Total Quality Control," *Quality Progress*, October, vol. 23, no. 10, pp. 43–45.

Shuker, T. J. (2000). "The Leap to Lean," *Annual Quality Congress Proceedings*, ASQ, Milwaukee, pp. 105–112.

Siff, W. C. (1984). "The Strategic Plan of Control—A Tool for Participative Management," *ASQC Quality Congress Transactions*, Milwaukee, pp. 384–390.

Snee, R. D. (1993). "Creating Robust Work Processes," *Quality Progress*, February, pp. 37–41.

Somerton, D. G. and Mlinar, S. E. (1996). "What's Key? Tool Approaches for Determining Key Characteristics," *Proceedings of the Annual Quality Congress*, ASQ, Milwaukee, pp. 364–369.

Spear, S. and Bowen, H. K. (1999). "Decoding the DNA of the Toyota Production System," *Harvard Business Review*, September–October, pp. 96–106.

The Productivity Press Development Team, (1997). *Mistake Proofing for Operators: The ZQC System*, Productivity Press, Portland, OR.

CHAPTER 23

Managing Quality in the Supply Chain

Joseph A. De Feo

High Points of This Chapter

1. A revolution in the relationship between buyers and suppliers has emerged in the form of supplier partnerships and the supply chain.
2. Quality specifications often define requirements for both the product and the quality system.
3. Evaluating supplier quality capability involves qualifying the supplier's design and the manufacturing process.
4. Supplier partnerships require joint economic planning, joint technological planning, and cooperation during contract execution.
5. A certified supplier is one who, after extensive investigation, supplies material of such quality that routine testing on each lot received is not necessary.

6. Measurements for supplier relations should be based on input from customers; provide for both evaluation and feedback; and include early, concurrent, and lagging indicators of performance.
7. Supplier quality improvement needs upper management involvement at all links in the supply chain.

Supplier Relations—a Revolution

This step on the spiral of quality concerns the purchase of goods or services from suppliers, or vendors.

For many companies, purchases account for more than 60 percent of the sales dollar and are the source of over half of the quality problems. Poor quality of supplier items results in extra costs for the purchaser, for example, for one appliance manufacturer, 75 percent of all warranty claims were traced to components purchased for the appliances.

Current emphasis on inventory reduction provides a further focus on quality. Under the just-in-time inventory concept, goods are received from suppliers only in the quantity and at the time that they are needed for production. The buyer stocks no inventories. If a portion of the purchased product is defective, production at the buyer's plant is disrupted because of the lack of backup inventory. With conventional purchasing, supplier quality problems can be hidden by excess inventory; with the just-in-time concept, purchased product must meet quality requirements.

The interdependence of buyers and suppliers has increased dramatically. Sometimes the interdependence takes the form of integrated facilities, for example, a can manufacturer locates next door to a brewery. Sometimes technological skills are involved, for example, an automobile manufacturer asks a supplier to propose a design for a purchased item. The supplier becomes an extension of the buyer's organization—a virtual department.

These circumstances have led to a revolution in the relationship between buyers and suppliers. In the past, the parties were often adversaries; some purchasers viewed suppliers as potential criminals who might try to sneak some defective product past the purchaser's incoming inspection. Today, the key phrase is partnership alliance, working closely together for the mutual benefit of both parties.

This new view of supplier relations requires changing the purchasing process from a traditional view to a strategic view. An overview of some changes is shown in Table 23.1.

Part of the revolution in supplier relations is the expansion of the traditional supplier concept to the broader supply chain concept (Fig. 23.1). Donovan and Maresca (in JQH5, Section 21) define the supply chain as the tasks, activities, events, processes, and interactions undertaken by all suppliers and all end users in the development, procurement, production, delivery, and consumption of a specific good or service. Note that this definition includes end users, prime suppliers or distributors, and multiple tiers of suppliers to prime manufacturing or service organizations. Supply chain management is reserved for items that are of strategic importance to an organization.

The purchasing function has the primary role of managing the supply chain to achieve high quality and value throughout the supply chain. Admittedly, this ideal is lofty, but it highlights a new focus—from managing purchasing transactions and troubleshooting to managing processes and supplier relationships. Under supply chain management, mechanisms must be put in place to ensure adequate linkages among parties in the supply chain. Such mechanisms include clear contractual requirements and continuous feedback and communication. For further discussion of the supply chain concept, see JQH5, pages 21.4–21.9. Of course, managing supply chains is difficult. Fisher (1997) describes some of these difficulties and suggests a framework for supply chains based on the nature of the product demand and whether the

Aspect in the Purchasing Process	Traditional View	Strategic View
Supplier relationship	Adversarial, competitive, distrusting	Cooperative, partnership, based on trust
Length of relationship	Short term	Long term; indefinite
Quality assurance	Inspection upon receipt	No incoming inspection necessary
Supplier base	Many suppliers, managed in aggregate	Few suppliers, carefully selected and managed
Purchasing business plans	Independent of end-user organization business plans	Integrated with end-user organization business plans
Focus of purchasing decisions	Price	Total cost of ownership

Source: Adapted from *JQH5*. p. 21.6.

TABLE 23.1 Traditional Versus Strategic View of the Purchasing Process

products are primarily functional or primarily innovative. Supply chains also apply to the service sector. For an example involving physicians in a hospital and health care system, see Zimmerli (1996). This example also illustrates the use of internal customer surveys for the major processes of distribution, purchasing, and sterile processing.

Much has been written about the importance of developing trust to replace the adversarial relationships of the past. Significant progress has been made, but in practice, suppliers (both large and small) still report pockets of arrogance exhibited by some purchasers.

Scope of Activities for Supplier Quality

A purchasing system includes three key activities: specification of requirements, selection of a supplier, and supply chain management. The overall quality objective is to meet the needs of the purchaser (and the ultimate user) with a minimum of incoming inspection or later corrective action; this objective in turn leads to minimizing overall cost.

To achieve this quality objective, certain primary activities must be identified and responsibilities assigned. Table 23.2 shows a typical list of responsibilities assigned in one company. These activities are discussed in this chapter.

The responsibility matrix in Table 23.2 shows that the quality department has the principal responsibility for many supplier quality activities. Under an alternative policy, the pur-

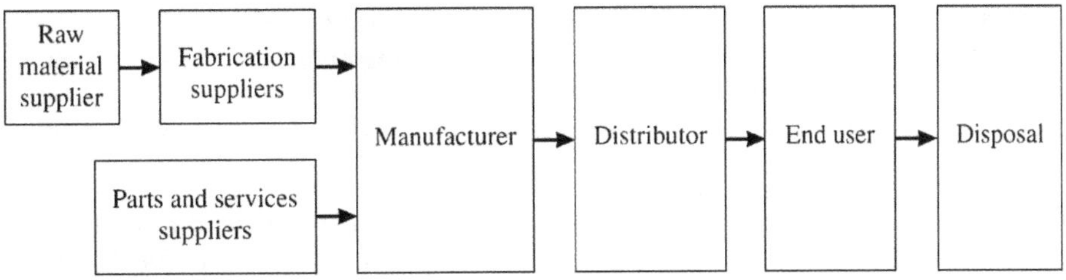

FIGURE 23.1 Elements of a supply chain. (JQH5, p. 21.4.)

Activity	Product Development	Purchasing	Quality
Defining product and program quality requirements	XX		X
Evaluating alternative suppliers	X	X	X X
Selecting suppliers		X X	
Conducting joint quality planning	X		X X
Cooperating with the supplier during the execution of the contract	X	X	XX
Obtaining proof of conformance to requirements	X		XX
Certifying qualified suppliers	X	X	XX
Conducting quality improvement programs as required	X	X	XX
Creating and utilizing supplier quality ratings		XX	X

Note: X X, principal responsibility; X, collateral responsibility.

TABLE 23.2 Responsibility Matrix-Supplier Relations

chasing department has the principal responsibility for quality, whereas others (e.g., product development and quality) have collateral responsibility. Such a shift in responsibility places a stronger focus on quality in setting priorities for delivery schedules, price, and quality. To meet this responsibility, most purchasing departments would need to supplement their technical capabilities. Some organizations have met that need by transferring technical specialists into the purchasing department.

To reflect a broad view of suppliers and the supply chain concept, some organizations are shifting from a function-based organization for purchasing transactions to a process-based organization for managing the supply chain. The process-based organization uses a cross-functional team and process owner to focus on the total cost of ownership (rather than on the initial price of a purchased item), to identify opportunities for increased value, and to achieve a competitive advantage.

Next we examine how quality relates to the three key activities of supply management: specification of requirements, selection of suppliers, and management of the supply chain.

Specification of Quality Requirements for Suppliers

Goals and requirements for suppliers must be aligned with those for each link in the supply chain, particularly the end user and the purchasing organization. These goals and requirements include quality parameters and general business issues.

For modern products, quality planning starts before a contract is signed. Such planning must recognize two issues:

1. The buyer must transmit to the supplier a full understanding of the use to be made of the product. Communicating usage requirements can be difficult even for a simple product.

2. The buyer must obtain information to be sure that the supplier can provide a product that meets all fitness-for-use requirements.

The complexity of many modern products makes it difficult to communicate usage needs to a supplier in a specification. Not only are the field usage conditions of a complex product sometimes poorly known, but the internal environments surrounding a particular component may not be known until the complete product is designed and tested. For example, specifying accurate temperature and vibration requirements to a supplier of an electric component may not be feasible until the complete system is developed. Such cases require, at the least, continuous cooperation between supplier and buyer. In special cases, it may be necessary to award separate development and production contracts to discover how to finalize requirements.

Circumstances may require two kinds of specifications:

1. Specifications defining the product requirements.
2. Specifications defining the quality-related activities expected of the supplier, that is, the supplier's quality system.

Definition of Numerical Quality and Reliability Requirements for Lots

Beyond the quality and reliability requirements imposed on individual units or products, there is usually a need for added numerical criteria to judge conformance of lots of products.

These criteria are typically needed in acceptance sampling procedures, which makes it possible to accept or reject an entire lot of product based on the inspection and test result of a random sample from the lot. The application of sampling procedures is facilitated if lot quality requirements are defined in numerical terms. Examples of numerical indexes are shown in Table 23.3.

The selection of numerical values for these criteria depends on several factors and also on probability considerations. These criteria are also a means of indexing sampling plans developed from statistical concepts. Unfortunately, many suppliers do not understand the statistical concepts and make incorrect interpretations of the quality-level requirement and also the results of sampling inspection (see Table 23.3). These criteria can also be a source of confusion in product liability discussions. Suppliers must understand that all product submitted is expected to meet specifications.

For complex and/or time-oriented products, numerical reliability requirements can be defined in supplier purchasing documents. Sometimes such requirements are stated in terms of mean time between failures. Numerical reliability requirements can help to clarify what a customer means by "high reliability."

Example 23.1 A capacitor manufacturer requested bids on a unit of manufacturing equipment that was to perform several manufacturing operations. Reliability of the equipment was important to maintaining production schedules, so a numerical requirement on "mean time between jams" (MTBJ) was specified to prospective bidders. (Previously, reliability had not been treated quantitatively. Equipment manufacturers had always promised high reliability, but results had been disappointing.) After several rounds of discussion with bidders, the manufacturer concluded that the desired level of reliability was unrealistic if the machine were to perform several operations. The capacitor manufacturer finally decided to revise the requirement for several operations and thereby reduce its complexity. The effort to specify a numerical requirement in the procurement document forced a clear understanding of reliability. Suppliers can also be required to demonstrate, by test, specified levels of reliability.

Quality Index	Meaning	Typical Values, %	Common Misinterpretation
Parts per million (ppm)	Number of defects per million items	5–1000	—
Acceptable quality level (AQL)*	Percentage defective that has a high probability (say $2 \geq .90$) of being accepted by the sampling plan	0.01–10.0	All accepted lots are at least as good as the AQL; all rejected lots are worse than the AQL
Lot tolerance percentage defective (LTPD)	Percentage defective that has a low probability (say $\leq .10$) of being accepted by the sampling plan	0.5–10.0	All lots better than the LTPD will be accepted; all lots worse than the LTPD will be rejected
Average outgoing quality limit (AOQL)	Worse average percentage defective over many lots after sampling inspection has been performed and rejected lots 100% inspected	0.1–10.0	All accepted lots are at least as good as the AOQL; all rejected lots are worse than the AOQL

*Some sampling tables and other sources define AQL as the maximum percentage defective considered satisfactory as a process average.

Table 23.3 Forms of Numerical Sampling Criteria

Definition of the Supplier Quality System

The second type of specification is a departure from the traditional practice of not telling a supplier how to run his or her plant. Defining required activities within a supplier's plant is sometimes necessary to ensure that a supplier has the expertise to conduct the full program needed for a satisfactory product. For some products, government regulations require that a buyer impose certain processing requirements (e.g., sanitary conditions for manufacturing pharmaceutical products) on suppliers. For other products, such as a complex mechanical or electronic subsystem, the overall system requirements may result in a need for a supplier to meet a numerical reliability or maintainability requirement and to conduct certain activities to ensure that such requirements are met. For still other products, suppliers are required to use statistical process control techniques on selected product characteristics or process parameters. Documents such as the ISO 9000 series and TS-16949 (formerly QS 9000), which define the elements of quality programs, can be cited as requirements in a contract with a supplier.

Example 23.2 Several suppliers were asked to submit bids on a battery needed in a space program. They were given a numerical reliability goal and asked to include in their bid proposal a description of the reliability activities that would be conducted to help meet the goal. Most of the prospective suppliers included a reliability program consisting of appropriate reliability activities for a battery. However, one supplier apparently had no expertise in formal reliability methodology and submitted a surprising write-up.

That supplier made a word-for-word copy of a reliability program write-up previously published for a missile system (the word battery was substituted for missile). This led to a suspicion, later confirmed, that the supplier knew little about reliability programs.

For complex products for which a supplier is asked to design and manufacture a product, the supplier can be required to include in the proposal a preliminary reliability prediction; a failure mode, effect, and criticality analysis; a reliability test plan; or other reliability analyses. The supplier's response provides some assurance on the design concept and also shows that the supplier has the reliability expertise to conduct the program and has included the funds and schedule time in the proposal.

Supplier Selection: Outsourcing

Should we make or buy? This decision requires an analysis of factors such as the skills and facilities needed, available internal capacity, ability to meet delivery schedules, expected costs of making or buying, and other matters. This question brings us to the issue of outsourcing.

Outsourcing

Outsourcing is the process of subcontracting to a supplier external to the organization an activity that is currently conducted in house. Outsourcing is undertaken to reduce costs (the primary impetus), reduce cycle time, or improve quality. Estimates suggest that at least 85 percent of major corporations now outsource at least some activities. A trade association, the Outsourcing Institute, now exists.

But what activities should be outsourced? One principle holds that outsourcing should be confined to activities that are required but do not provide a competitive advantage, for example, security, facility maintenance, administration of health benefits. Activities that are strategic and involve core competencies should not be outsourced. In practice, some organizations outsource significant (core) functional activities such as customer service, marketing, product design, and information technology.

Many important business issues enter into decisions about outsourcing. Bettis et al. (1992) offer important cautions about outsourcing core activities such as design and manufacturing. The more that outsourcing results in a supplier obtaining technical knowledge and market knowledge, the higher the risks to the company doing the outsourcing.

Outsourcing reduces internal costs by reducing personnel because the outsourcer (supplier) companies have the technology and knowledge to perform certain tasks more efficiently than some companies can internally. But there can be a serious impact on product quality if the supplier does not assign a high priority to quality. Outsourcing can also undermine employee morale and loyalty by creating fear that other activities will also be outsourced, resulting in a further loss of jobs. Some forward-looking companies like Eastman Chemical follow a policy that, if an activity is outsourced, no one will lose a job—people are retrained to take the positions of those who are normally retiring. Outsourcing also assumes that a capable supplier can be found and that adequate monitoring of the contract will ensure high quality. Sometimes these issues are glossed over in the zeal to reduce costs, and the result can be significant quality problems in the purchased items. Peterson (1998) analyzes some potential shortcomings when dealing with contract manufacturers. Bossert (1994) provides a checklist of 11 elements (e.g., inspection instructions, sufficient manufacturing controls) to compare contract manufacturing services.

Sharman (2002) discusses how the Internet is a driver of outsourcing and is accelerating supply chain trends. Ericson (2003) identifies six factors to help decide what and when to outsource. Meseck (2004) provides risk management strategies to help identify appropriate outsourcing opportunities and calculate the financial impact of global outsourcing. Weidenbaum (2004) discusses the high-level pros and cons of outsourcing, including when to and why to outsource and how it affects the United States. Hussey and Jenster (2003) provide an in-depth discussion of the types of outsourcing and the management issues associated with each from the supplier viewpoint. Ramachandran and Voleti (2004) discuss outsourcing from the perspective of suppliers in India, together with success factors and the need for suppliers to manage both growth and consolidation. Both Soliman (2003) and Weerakkody et al. (2003) discuss global outsourcing of Application Service Providers (ASPs). Karmaker (2004) argues that the offshoring issue should be one of competitiveness, not job loss. He examines current changes in the service industry and recommends strategies to realign, redesign, and restructure. Clott (2004) provides background facts and figures on outsourcing, its impact on work with emphasis on IT. He also discusses proponent and dissenter views and briefly discusses labor and ethics factors. Lee (2004) argues that the best supply chains are not just fast and cost-effective; they are agile and adaptable, and they ensure that all their companies' interests stay aligned.

Once skills are lost through outsourcing, it is difficult to reverse the process if later events require that the activity be brought back into an organization. This situation could be devastating for activities such as product design and selected operations. Clearly, outsourcing can be a sensible, viable business decision—after all, the subcontracting of selected manufacturing activities has been a part of manufacturing history. But the desire for cost reduction may be taking outsourcing too far. Perhaps we should first study the entire activity to be outsourced as a quality improvement project using the road map provided, "Improving Quality While Decreasing Cost." Thus, suppose customer service is a candidate for outsourcing. The process of customer service would be studied for both effectiveness and efficiency, and the necessary internal changes would be made. The result (quality and costs) could then be compared to those of outside suppliers. Some organizations even set up the activity under question as a separate profit center (to compete against outside suppliers) to spur internal improvement—the threat of job loss is a powerful spur.

To sum up on a sensitive issue: Outsourcing can provide superior quality and lower costs for an activity that a company cannot easily develop and maintain on its own, for example, information technology. Outsourcing can also enable a company to focus resources on the core competencies that are important for competitive advantage, for example, product design, operations, marketing. But these core activities vary by organization. The authors believe that the core competencies must be carefully identified within each organization, and once identified they should be performed internally and not be outsourced.

Multiple Suppliers versus Single Source

Multiple sources of supply have advantages. Competition can result in better quality, lower costs, better service, and minimum disruption of supply from strikes or other catastrophes.

A single source of supply also has advantages. The size of the contract given to a single source will be larger than that with multiple sources, and the supplier will attach more significance to the contract. With a single source, communications are simplified and more time is available for working closely with the supplier. The most dramatic examples of single sources are multidivisional companies in which some divisions are suppliers to others.

A clear trend has emerged. Organizations are significantly reducing the number of multiple suppliers. Since about 1980, reductions of 50 to 70 percent in the supplier base have become common. This trend does not necessarily mean that businesses are going to have single sources for all purchases; it does mean a single source for some purchases and fewer multiple suppliers for other purchases. Working with a smaller number of suppliers helps to achieve useful partnerships by providing the time and skills necessary to facilitate in-depth cooperation. The forms of cooperation are discussed later in this chapter.

Whether a single source or multiple suppliers, selection must be based on the reputation of the supplier, qualification tests of the supplier's design, survey of the supplier's manufacturing facility, and information from data banks and other sources on supplier quality.

Assessment of Supplier Capability

Evaluating supplier quality capability involves one or both of the following actions:

1. Qualifying the supplier's design through the evaluation of product samples.
2. Qualifying the supplier's capability to meet quality requirements on production lots, that is, the supplier's quality system.

Qualifying the Supplier's Design

In some cases, the supplier is asked to create a new design to meet the functions desired by the purchaser. In these cases, the supplier makes samples based on the proposed design. (Such samples are often made in an engineering model shop because a manufacturing process for the new design has not yet been created.) The samples are tested (the "qualification test") either by the purchaser or by the supplier, who then submits the results to the purchaser. Qualification test results are often rejected. Two reasons are common: (1) The test results show that the design does not provide the product functions desired, or (2) the test procedure is not adequate to evaluate the performance of the product. Such rejections (and ensuing delays in shipments) can be prevented by starting with a rigorous definition of product requirements and by requiring an approval of the test procedure before the tests commence.

Qualification test results do show whether the supplier has created a design that meets the performance requirements; such test results do not show whether the supplier is capable of manufacturing the item under production conditions.

A supplier may be required to submit a failure mode, effects, and criticality analysis as evidence of analyses to prevent product or process failures. Increasingly, this requirement is part of the Six Sigma approach to quality during design.

Qualifying the Supplier's Manufacturing Process

Evaluation of the supplier's manufacturing capability can be done by reviewing past data on similar products, performing process capability analysis, or evaluating the supplier's quality system through a quality survey.

Data showing the supplier's past performance on the same or similar products may be available within the local buyer's organization, other divisions of the same corporation, government data banks, or industry data banks.

With the process capability analysis approach, data on key product characteristics are collected from the process and evaluated by using statistical indexes for process capability. All evaluation occurs before the supplier is authorized to proceed with full production. Typically, process capability analysis of a supplier's process is reserved for significant product characteristics, safety-related items, or products requiring compliance with government regulations. The third approach, a quality survey, is explained next.

When all three approaches can be used, the information collected can provide a sound prediction of supplier capability.

Supplier Quality Survey (Supplier Quality Evaluation)

A supplier quality survey is an evaluation of the ability of a supplier's quality system to meet quality requirements on production lots, that is, to prevent, identify, and remove any product that does not meet requirements. The results of the survey are used in the supplier selection process, or, if the supplier has already been chosen, the survey alerts the purchaser to areas where the supplier may need help in meeting requirements. The survey can vary from a simple questionnaire mailed to the supplier to a visit to the supplier's facility.

The questionnaire poses explicit questions such as these submitted to suppliers of a manufacturer of medical devices:

- Has your company received the quality requirements on the product and agreed that they can be fully met?
- Are your final inspection results documented?
- Do you agree to provide the purchaser with advance notice of any changes in your product design?
- What protective garments do your employees wear to reduce product contamination?
- Describe the air-filtration system in your manufacturing areas.

The more formal quality survey consists of a visit to the supplier's facility by a team of observers from departments such as quality, engineering, manufacturing, and purchasing. Such a visit may be part of a broader survey of the supplier covering financial, managerial, and technological competence. Depending on the product involved, the activities included in the quality portion of the survey can be chosen from the following list:

- Management: philosophy, quality policies, organization structure, indoctrination, commitment to quality.
- Design: organization, systems in use, caliber of specifications, orientation to modern techniques, attention to reliability, engineering change control, development laboratories.
- Manufacture: physical facilities, maintenance, special processes, process capability, production capacity, caliber of planning, lot identification and traceability.
- Purchasing: specifications, supplier relations, procedures.
- Quality: organizational structure, availability of quality and reliability engineers, quality planning (materials, in-process, finished goods, packing, storage, shipping, usage, field service), audit of adherence to plan.
- Inspection and test: laboratories, special tests, instruments, measurement control.
- Quality coordination: organization for coordination, order analysis, control over subcontractors, quality cost analysis, corrective action loop, disposition of nonconforming product.
- Data systems: facilities, procedures, effective use reports.
- Personnel: indoctrination, training motivation.
- Quality results: performance attained, self-use of product, prestigious customers, prestigious subcontractors.

Following the survey, the team reports its findings. These consist of (1) some objective findings as to the supplier's facilities (or lack of facilities), (2) subjective judgments on the effectiveness of the supplier's operations, (3) a further judgment on the extent of assistance needed by the supplier, and (4) a highly subjective prediction whether the supplier will deliver a good product if awarded a contract.

The quality survey is a technique for evaluating the supplier's ability to meet quality requirements on production lots. The evaluation of various quality activities can be quantified by a scoring system.

A scoring system that includes importance weights for activities is illustrated in Table 23.4. This system is used by a manufacturer of electronic assemblies. In this case, the importance weights (*W*) vary from 1 to 4 and must total 25 for each of the three areas surveyed. The weights show the relative importance of the various activities in the overall index. The actual ratings (R) of the activities observed are assigned as follows:

- The specific activity is satisfactory in every respect (or does not apply).
- The activity meets minimum requirements but improvements could be made.
- The activity is unsatisfactory.

Supplier quality surveys have both merits and limitations. On the positive side, such surveys can identify important weaknesses such as a lack of special test equipment or an absence of essential training programs. Further, the survey opens up lines of communication and can stimulate action on quality by the supplier's upper management. On the negative side, surveys that emphasize the supplier's organization, procedures, and documentation have had only limited success in predicting future performance of the product.

	Receiving Inspection			Manufacturing			Final Inspection		
Activity	*R*	*W*	*R* × *W*	*R*	*W*	*R* × *W*	*R*	*W*	*R* × *W*
Quality management	8	3	24	8	3	24	8	3	24
Quality planning	8	4	32	8	4	32	10	4	40
Inspection equipment	10	3	30	10	3	30	10	3	30
Calibration	0	3	0	10	3	30	0	3	0
Drawing control	0	3	0	10	2	20	10	2	20
Corrective action	10	3	30	8	3	24	8	3	24
Handling rejects	10	2	20	8	2	16	10	3	30
Storage and shipping	10	1	10	10	1	10	10	1	10
Environment	8	1	8	8	1	8	8	1	8
Personnel experience	10	2	20	10	3	30	10	2	20
Area total			174			224			206

Note: R, rating; *W*, weight.

Interpretation of area totals:

Fully approved: Each of the three area totals is 250.

Approved: None of the three area totals is less than 200.

Conditionally approved: No single total is less than 180.

Unapproved: One or more of the area totals is less than 180.

TABLE 23.4 Scoring of a Supplier Quality Survey

Suppliers in some industries have been burdened with quality surveys from many purchasers. These repeat surveys (called "multiple assessment") are time consuming for suppliers. In another approach, a standard specification of the elements of a quality system (e.g., the ISO 9000 series) is created and assessors are trained to use the specification to evaluate supplier capability. A list of suppliers that have passed the assessment is published, and other purchasers are encouraged to use these results instead of making their own assessment of a supplier. The assessors are independent of the supplier or purchasing organization—thus the term third-party assessment. In some countries, a national standards organization acts in this role.

Bossert (1998) describes how supplier evaluation can start with an ISO 9000 assessment and then be supplemented by a quality survey including a supplier visit. The supplier visit covers contract and specification review, process audit, process risk analysis (using a failure mode and effects analysis), and statistical techniques (including the measurement process).

On to the third phase of supplier relations—management of the supply chain through quality planning, quality control, and quality improvement.

Supply Chain Quality Planning

Donovan and Maresca (in Section 21 of JQH5) suggest the following steps for a purchasing process that involves the purchasing organization, suppliers, and end users. This approach is sometimes called a "sourcing process."

1. Document the organization's historic, current, and future procurement activity.
2. Identify a commodity from the procurement activity that represents both high expenditure and high criticality to the business.
3. For this commodity, assemble a cross-functional team.
4. Determine the sourcing needs of the customer through data collection, survey, and other activities.
5. Analyze the supply industry's structure, capabilities, and trends.
6. Analyze the cost components of the commodity's total cost of ownership.
7. Translate the customer needs into a sourcing process that will satisfy the customer and provide the opportunity to manage and optimize the total cost of ownership.
8. Obtain management endorsement to transfer the sourcing strategy into operation. Implement it.

In doing the detailed quality planning with suppliers, three approaches emerge:

1. *Inspection*. The focus is on various forms of product inspection.
2. *Prevention*. The premise is that quality must be built in by the supplier with the purchaser's help. But there is still an arm's-length relationship between purchaser and supplier.
3. *Partnership*. Suppliers are offered the financial security of a long-term relationship in exchange for a supplier's commitment to quality that includes a strong teamwork relationship with the buyer.

Partnership—involving not just quality but also other business issues—is clearly the wave of the future. Teamwork actions vary greatly, for example, training a supplier's staff in quality techniques, including suppliers in a design review meeting to gain ideas on how supplier parts can best be used, sharing confidential sales projections with suppliers to assist in supplier production scheduling. Such partnerships often lead to formation of supplier quality councils, which help provide new approaches for the benefits of both the buyer and suppliers. Various opportunities for teamwork are discussed later. But such teamwork depends on truly open communication between buyers and suppliers.

Such cooperation can best be achieved by setting up multiple channels of communication: Designers must communicate directly with designers, quality specialists with quality specialists, etc. These multiple channels are a drastic departure from the single channel, which is the method in common use for purchase of traditional products. In the single-channel approach, a specialist in the buyer's organization must work through the purchasing agent, who in turn speaks with the salesperson in the supplier's organization, to obtain information. Of course, the concept of multiple channels seems sensible, but wouldn't it be useful to determine whether multiple channels yield better results in quality? Carter and Miller (1989) did just that.

Example 23.3 In an innovative research study, they compared quality levels for two communication structures: serial (single channel) and parallel (multiple channel). At a manufacturer of mechanical seals, one section of a plant followed the serial communication concept while a second area used parallel communication. Over a 19-month period, the section using parallel communication improved the average percentage of items rejected from 30.3 to 15.0 percent, a statistically significant difference; the section with serial communication had no such improvement—in fact, its rejection percentage increased slightly.

Next we address how partnership can be achieved through joint economic planning, joint technological planning, and cooperation during contract execution.

Joint Economic Planning

The economic aspects of joint quality planning concentrate on two major approaches:

1. *Value rather than conformance to specification.* The technique used is to analyze the value of what is being bought and to try to effect an improvement. The organized approach is known as value engineering. Applied to supplier quality relations, value engineering looks for excessive costs due to (1) over specification for the use to which the product will be put, for example, a special product ordered when a standard product would do; (2) emphasis on original price rather than on cost of use over the life of the product; and (3) emphasis on conformance to specification, not fitness for use. Suppliers are encouraged to make recommendations on design or other requirements that will improve or maintain quality at a lower cost.
2. *Total cost of ownership.* The buyer must add a whole array of quality-related costs to the purchase price: incoming inspection, materials review, production delays, downtime, extra inventories, etc. However, the supplier also has a set of costs it is trying to optimize. The buyer should put together the data needed to understand the life-cycle costs or the cost of use and then press for a result that will optimize them.

Example 23.4 A heavy-equipment manufacturer bought 11,000 castings per year from several suppliers. It was decided to calculate the total cost of the purchased casting as the original purchase price plus incoming inspection costs plus the costs of rejections detected later in assembly. The unit purchase price on a contract given to the lowest bidder was $19.

The inspection and rejection costs amounted to an additional $2.11. The variation among bid prices was $2. Thus the lowest bid does not always result in the lowest total cost.

Joint Technological Planning

The standard elements of such planning include:

1. Agreement on the meaning of performance requirements in the specifications
2. Quantification of quality, reliability, and maintainability requirements

Example 23.5 A supplier was given a contract to provide an air-conditioning system with a mean time between failures of at least 2000 hours. As part of joint planning, the supplier was required to submit a detailed reliability program early in the design phase. The program write-up was submitted and included a provision to impose the same 2000-hour requirement on each supplier of parts for the system. This revealed a complete lack of understanding by the supplier of the multiplication rule. "The Relationship between Part and System Reliability".

3. Definition of reliability and maintainability tasks to be conducted by the supplier.
4. Preparation of a process control plan for the manufacturing process. The supplier can be asked to submit a plan summarizing the specific activities which will be conducted during the manufacture of the product. Typically, the plan must include statistical process control techniques to prevent defects by detecting problems early.
5. Definition of special tasks required of the supplier. These may include activities to ensure that good manufacturing practices are met, special analyses are prepared for critical items, etc.
6. Seriousness classification of defects to help the supplier understand where to concentrate efforts.
7. Establishment of sensory standards for qualities that require use of the human being as an instrument.

Example 23.6 The federal government was faced with the problem of defining the limits of color on a military uniform. It was finally decided to prepare physical samples of the lightest and darkest acceptable colors. Such standards were then sent out with the provision that the standards would be replaced periodically because of color fading.

8. Standardization of test methods and test conditions between supplier and buyer to ensure their compatibility.

Example 23.7 A carpet manufacturer repeatedly complained to a yarn supplier about yarn weight. The supplier visited the customer to verify the test methods. Their test methods were alike. Next an impartial testing lab verified the tests at the carpet plant. Finally, the mystery was solved. The supplier was spinning (and measuring) the yarn in bone-dry conditions, but the carpet manufacturer measured at standard conditions. During this period, $62000 more was spent for yarn than if it had been purchased at standard weight.

9. Establishment of sampling plans and other criteria relative to inspection and test activity. From the supplier's viewpoint, the plan should accept lots having the usual process average. For the buyer, the critical factor is the amount of damage caused by one defect getting through the sampling screen. Balancing the cost of sorting versus sampling can be a useful input in designing a sampling plan. In addition to sampling criteria, error of measurement can also be a problem.
10. Establishment of quality levels. In the past, suppliers were often given "acceptable quality levels" (AQL). The AQL value was just one point on the "operating characteristic" curve that described the risks of sampling plans. A typical AQL

value might be 2.0 percent. Many suppliers interpreted this to mean that product which included 2 percent defective was acceptable. It is best to make clear to the supplier through the contract that all product submitted is expected to meet specifications and that any nonconforming product may be returned for replacement. In many industries, the unit of measurement is defects per million (DPM).

11. Establishment of a system of lot identification and traceability. This concept has always been present in some degree, for example, heat numbers of steel, lot numbers of pharmaceutical products. More recently, with intensified attention to product reliability, this procedure is more acutely needed to simplify the localization of trouble, to reduce the volume of product recall, and to fix responsibility. These traceability systems, though demanding some extra effort to preserve the order of manufacture and identify the product, make greater precision in sampling possible.
12. Establishment of a system of timely response to alarm signals resulting from defects. Under many contracts, the buyer and supplier are yoked to a common timetable for completion of the final product. Usually, a separate department (e.g., materials management) presides over major aspects of scheduling. However, upper management properly looks to the people associated with the quality function to set up alarm signals to detect quality failures and to act positively on these signals to avoid deterioration, whether in quality, cost, or delivery.

Such depth of joint technological planning bears no resemblance to the old approach of sending a supplier a blueprint with a fixed design and a schedule.

Supply Chain Quality Control

Donovan and Maresca (JQH5) suggest these steps for successful supplier control:

1. Create a cross-functional team.
2. Determine critical performance metrics.
3. Determine minimum standards of performance.
4. Reduce the supplier base to those able to meet minimum performance requirements.
5. Assess supplier performance:
   a. Supplier quality systems assessment.
   b. Supplier business management.
   c. Supplier product fitness for use.

The detailed quality control activities focus on cooperation during contract execution, supplier certification, supplier rating, and quality measurement for supplier relations. These activities emphasize continuous feedback to suppliers.

Cooperation during Contract Execution

This cooperation usually concentrates on the following activities.

Evaluation of Initial Samples of Product

Under many circumstances, the supplier must submit test results of a small initial sample produced from production tooling and a sample from the first production shipment before the full shipment is made. The latter evaluation can be accomplished by having a buyer's representative visit the supplier's plant and observe the inspection of a random sample selected from the first production lot. A review can also be made of process capability of process control type data from that lot.

Design Information and Changes

Design changes may take place at the initiative of either the buyer or the supplier. Either way, the supplier should be treated like an in-house department when developing procedures for processing design changes. This need is especially acute for modern products, for which design changes can affect products, processes, tools, instruments, stored materials, procedures, etc. Some of these effects are obvious, but others are subtle, requiring a complete analysis to identify the effects. Failure to provide adequate design change information to suppliers has been a distinct obstacle to good supplier relations.

Surveillance of Supplier Quality

Quality surveillance is the continuing monitoring and verification of the status of procedures, methods, conditions, processes, products, services, and analysis of records in relation to stated references to ensure that specified requirements for quality are being met (ISO 8402). Surveillance by the buyer can take several forms: inspecting the product, meeting with suppliers to review quality status, auditing elements of the supplier quality program, monitoring of the manufacturing practices of the supplier, reviewing statistical process control data, and witnessing specific operations or tests. Major or critical contracts require on-site presence or repeat visits.

Evaluating Delivered Product

Evaluation of supplier product can be achieved by using one of the methods listed in Table 23.5.

In previous decades, incoming inspection often consumed a large amount of time and effort. With the advent of modern complex products, many companies have found that they do not

Method	Approach	Application
100% inspection	Every item in a lot is evaluated for all or some of the characteristics in the specification	Critical items where the cost of inspection is justified by the cost of risk of defectives; also used to establish the quality level of new suppliers
Sampling inspection	A sample of each lot is evaluated by a predefined sampling plan and a decision is made to accept or reject the lot	Important items where the supplier has established an adequate quality record by the prior history of lots submitted
Identifying inspection	The product is examined to ensure that the supplier sent the correct product; no inspection of characteristics is made	Items of less importance where the reliability of the supplier laboratory has been established in addition to the quality level of the product
No inspection	The lot is sent directly to a storeroom or processing department	For purchase of standard materials or goods not used in the product, e.g., office supplies
Using supplier data (supplier certification)	Data of the supplier inspection is used in place of incoming inspection	Items for which a supplier has established a strong quality record

Table 23.5 Methods of Evaluating Supplier Product

have the necessary inspection skills or equipment. This situation has forced them to rely more on the supplier's quality system or inspection and test data, as discussed later in this chapter.

The choice of evaluation method depends on a variety of factors:

- Prior quality history on the part and supplier.
- Criticality of the part on overall system performance.
- Criticality on later manufacturing operations.
- Warranty or use history.
- Supplier process capability information.
- The nature of the manufacturing process. For example, a press operation depends primarily on the adequacy of the setup. Information on the first few pieces and last few pieces in a production run is usually sufficient to draw conclusions about the entire run.
- Product homogeneity. For example, fluid products are homogeneous, and the need for large sample sizes is thus less.
- Availability of required inspection skills and equipment.

A useful tool for learning about a supplier's process and comparing several suppliers' manufacturing product to the same specification is the histogram. A random sample is selected from a lot, and measurements are made on selected quality characteristics. The data are charted as frequency histograms. The analysis consists of comparing the histograms to the specification limits.

An application of histograms to evaluating the hardenability of a particular grade of steel from three suppliers is shown in Fig. 23.2. The specification was a maximum Rockwell C reading of 43 measured at Jominy position J8.

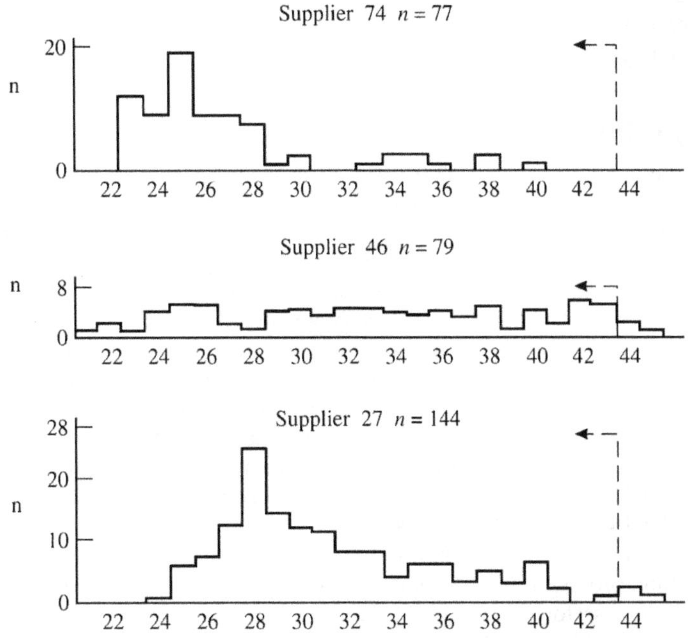

FIGURE 23.2 Histograms on hardenability.

Histograms were also prepared for carbon, manganese, nickel, and chromium content. Analysis revealed:

- Supplier 46 had a process without any strong central tendency. The histogram on nickel for this supplier was also rectangular in shape, indicating a lack of control of the nickel content and resulting in several heats of steel with excessively high Rockwell values.
- Supplier 27 had several heats above the maximum, although the process had a central value of about 28. The histograms for manganese, nickel, and chromium showed several values above and apart from the main histogram.
- Supplier 74 showed much less variability than the others. Analysis of other histograms for this supplier suggested that about half of the original heats of steel had been screened out and used for other applications.

Note how these analyses can be made without visiting the supplier plants, that is, "the product tells on the process." Histograms have limitations but they are an effective tool for incoming inspection.

Action on Nonconforming Product

During the performance of the contract, there will arise instances of nonconformance. These may be on the product itself or on process requirements or procedural requirements. Priority effort should go to cases where a product is unfit for use.

Communications to the supplier on nonconformance must include a precise description of the symptoms of the defects. The best description is in the form of samples, but if this is not possible, the supplier should have the opportunity to visit the site of the trouble. There are numerous related questions: What disposition is to be made of the defective items? Who will sort or repair them? Who will pay the costs? What were the causes? What steps are needed to avoid a recurrence? These questions are outside the scope of pure defect detection; they require discussion among departments within each company and further discussions between buyer and supplier.

Supplier Certification

A "certified" supplier is one whose quality data record establishes that it is not necessary to perform routine inspection and test on each lot or batch received. A "preferred" supplier produces quality better than the minimum. An "approved" supplier meets minimum requirements. Some organizations use different terms and even different rankings, but usually certified suppliers are the ideal. Unfortunately, they are in the minority. Spooner and Collins (1995) describe how criteria were developed at Walker Manufacturing to define these categories, for example, a certified supplier has performed "at an overall 90 percent compliance level and met the individual rating component requirements for four consecutive quarters within 2 years."

ASQ recommends eight criteria for certification. These are summarized in Table 23.6.

Supplier certification provides a model for the low DPM levels necessary for just-in-time manufacture, drastically reduces buyer inspection costs, and identifies suppliers for partnerships. Certified suppliers receive preference in competitive bidding and achieve industry recognition by their certified status.

Schneider et al. (1995) explain how process capability indexes are used as part of the certification process at Dow Chemical. The concept of supplier certification applies equally to the service sector. Brown (1998) describes the approach used by a telecommunications company for suppliers of building leasing, maintenance, security, and food service.

Criteria	Examples
No product-related lot rejections for at least 1 year	An alternative is volume related, e.g., no rejects in 20 consecutive lots
No non-product-related rejections for at least six months	The marking on a container or the timeliness of an analysis document
No production-related negative incidents for at least six months	Ease with which the supplier's product can be used in the buyer's process or product
Passed a recent on-site quality system evaluation	A supplier survey on defined criteria
Has a totally agreed on specification	No ambiguous phrases like "characteristic odor" or "clear of contamination"
Fully documented process and quality system	The system must include plans for continuous improvement
Timely copies of inspection and test data	Real-time availability of data
Process is stable and in control	Statistical control and process capability studies

Source: Adapted from Maass et al. (1990).

Table 23.6 Criteria for Supplier Certification

Supplier Quality Rating

Supplier quality rating provides a quantitative summary of supplier quality over a period of time. This type of rating is useful in deciding how to allocate purchases among suppliers. Rating furnishes both buyer and supplier with common factual information that becomes a key input for identification and tracking of improvement efforts and for allocating future purchases among suppliers.

To create a single numerical quality score is difficult because there are several units of measure, such as:

- The quality of multiple lots expressed as lots rejected versus lots inspected.
- The quality of multiple parts expressed as percentage nonconforming.
- The quality of specific characteristics expressed in numerous natural units, for example, ohmic resistance, percentage of active ingredient, mean time between failures.
- The economic consequences of bad quality, expressed in dollars.

Because these units of measure vary in importance among companies, published rating schemes differ markedly in emphasis.

Measures in Use

Supplier quality rating plans are based on one or more of the following measures.

Product Percentage Nonconforming

This measure is a ratio of the amount of defective items received to the total number of items received. On a lot-by-lot basis, the formula is the number of lots rejected divided by the number of lots received; on an individual piece basis, the formula is the number of individual pieces rejected divided by the number of individual pieces received.

Overall Product Quality

This plan summarizes supplier performance at incoming inspection and later phases of product application. Points are assigned for each phase; the maximum number of points is given when no problems are encountered. Table 23.7 shows an example from AT&T. Note that the phases are incoming inspection, production failures, vendor response to problems, and AT&T customer complaints. Each rating element has further detailed criteria that are used to assign points for the element, for example, if 3 percent of the lots are rejected for "visual/mechanical" reasons in a rating period, then one point is deducted from the maximum of five for that element. Note that the overall rating evaluates the supplier response to problems, whereas detailed criteria include both timeliness and adequacy of the response.

Economic Analysis

This type of plan compares suppliers on the total dollar cost for specific purchases. The total dollar cost includes the quoted price plus quality costs for defect prevention, detection, and correction.

Composite Plan

Supplier performance is not limited to quality. It includes delivery against schedule, price, and other performance categories. These multiple needs suggest that the supplier rating should include overall supplier performance rather than just supplied quality performance. The purchasing department is a strong advocate of this principle and has valid grounds for this advocacy. Table 23.8 illustrates this approach with an example from Tecumseh Products Company. The overall rating of 92.46 is calculated by combining the four ratings using weights of 40 percent for quality, 30 percent for delivery, 20 percent for cost, and 10 percent for responsiveness to problems. Walker Manufacturing uses categories and weights of 35 percent for quality, 35 percent for delivery, 20 percent for price, and 10 percent for supplier support (Spooner and Collins, 1995).

Rating Element	Maximum Points
Incoming inspection	
Visual mechanical PPM	10
Visual/mechanical-percentage of lot rejections	5
Testing PPM	10
Testing-lot rejections	5
Ship-to-stock credit	—
Production failures	
Shop complaints	20
Quality appraisal	10
Vendor response	
Response to problems	10
Failure analysis response	20
AT&T customer complaints	10
Total	100

Source: Nocera et al. (1989).

TABLE 23.7 AT&T Quality Performance Rating

Overall combined rating	92.46
Total quality rating	99.05
Total delivery rating	95.58
Total cost rating	79.22
Total response rating	83.30
Total lots received	18
Total parts received	398,351
Total parts rejected	3804

Source: Wind (1991).

TABLE 23.8 Supplier Rating Report

Some organizations use a periodic supplier rating to determine the share of future purchases given to each supplier. The rating system and its effect on market share are fully explained to all suppliers. The approach has been used successfully by both automotive and appliance manufacturers to highlight the importance of quality to their suppliers.

Quality Measurement in Supplier Relations

The management of quality-related activities in supplier relations must include provision for measurement. Table 23.9 shows units of measure for various subject areas of supplier relations. Klenz (2000) discusses the use of a data warehouse for supplier quality analysis.

Supply Chain Quality Improvement

Donovan and Maresca (JQH5) propose a sequence of five tiers of progression for improvement:

1. Create a joint team of the end user and supplier to align goals, analyze the supply chain business process, and work on chronic problems.
2. Focus on cost reduction, including the cost of poor quality.
3. Evaluate the value added by each link in the supply chain.
4. Exchange information and ideas routinely throughout the chain.
5. Have the supply chain work as a single process with all parties routinely collaborating on improvement opportunities to generate value for customers as well as suppliers.

The general approach to handling chronic supplier problems follows the step-by-step approach to improvement. This process includes the early steps of establishing the proof of the need for the supplier to take action and the application of Pareto analysis to identify the vital few problems. The following section on Pareto analysis of suppliers explains the form of such analyses of suppliers' problems.

Cooperation often requires providing technical assistance to suppliers. Miller and Kegaris (1986) describe how businesses may need to share proprietary information on a "need to know" basis. This often represents a major breakthrough in communications.

Sometimes upper management must provide the leadership in obtaining action from suppliers. Amazing results can be achieved when the initial step in an improvement program

Subject	Units of Measure
Quality of submitted lots	Percentage of lots rejected
	Cost of poor quality
	Percentage of lots accepted on waiver
	Number of rejected lots classified "use as is"
	Percentage of suppliers certified
Supplier relations program	Percentage of suppliers classified acceptable as a result of a supplier survey
	Percentage of qualification test *procedures* approved on first submission
	Percentage of qualification test *results* approved on first submission
	Percentage of initial product samples approved on first submission
	Percentage of first production shipments approved on first submission
	Percentage of suppliers submitting data
	Average time to resolve problems
Business relationships	Average number of multiple suppliers per item
	Percentages of purchases as single source
	Percentage of purchases to lowest bidder
	Average time to secure bids
	Average time to secure answers to technical inquiries
Adequacy of inventory	Percentage of stockouts
Service to suppliers	Average number of days to pay supplier invoice
	Number of accounts payable beyond X days

TABLE 23.9 Examples of Quality Measurement In Supplier Relations

is a meeting of both the buyer's and supplier's upper management teams, who plan the action steps for improvement together. Such discussions have much more impact than a meeting of the two quality managers.

Example 23.8 For an appliance manufacturer, 75 percent of the warranty costs were due to suppliers' items. The president and his staff met individually with the counterpart team from each of 10 key suppliers. Warranty data were presented to establish the "proof of the need." A goal was set for a 50 percent reduction in warranty costs over a 5-year period. Each supplier was asked to develop a quality improvement program. The purchaser provided an eight-hour training session for the president and staff members of the key suppliers. Follow-up meetings were held. A system of supplier recognition awards was set up and purchasing practices were changed to transfer business to the best suppliers. The result: a decline in service calls from 41 to 13 calls per 100 products and a saving of $16 per unit in warranty costs.

Pareto Analysis of Suppliers

Supplier improvement programs can fail because the vital few problems are not identified and attacked. Instead, the programs consist of broad attempts to tighten up all procedures. The Pareto analysis can be used to identify the problem in a number of forms:

1. *Analysis of losses (defects, lot rejections, etc.)* by material number or part number. Such analysis serves a useful purpose applied to catalog numbers involving substantial or frequent purchases.

2. *Analysis of losses by product family*. This process identifies the vital few product families present in small but numerous purchases of common product families, for example, fasteners, paints.
3. *Analysis of losses by process*, that is, classification of the defects or lot rejections in terms of the processes to which they relate, for example, plating, swaging, coil winding.
4. *Analysis by supplier across the entire spectrum of purchases*. This process can help to identify weaknesses in the supplier's managerial approach, as contrasted with the technological, which is usually correlated with products and processes. One company had 222 suppliers on the active list. Of these, 38 (or 17 percent) accounted for 53 percent of the lot rejections and 45 percent of the bad parts.
5. *Analysis by total cost of the parts*. In one company, 37 percent of the part numbers purchased accounted for only 5 percent of the total dollar volume of purchases but for a much higher percentage of the total incoming inspection cost. The conclusion was that these "useful many" parts should be purchased from the best suppliers, even at top prices. The alternative of relying on incoming inspection would be even more costly.
6. *Analysis by failure mode*. This technique is used to discover major defects in the management system. For example, suppose that studies disclose multiple instances of working to the wrong issue of the specification. In such cases, the system used for specification revision should be reexamined. If value analysis discovers multiple instances of over specification, the design procedures for choosing components should be reexamined. These analyses by failure mode can reveal how the buyer is contributing to his or her own problems.

The cross-functional team approach for quality improvement also applies to supplier quality. This means that there should be joint customer-supplier teams and also that suppliers must be encouraged to set up an infrastructure (quality council, formation of teams, identification of projects, execution of projects) internally to address quality. Chen and Batson (1996) describe how Johnson & Johnson Consumer Products use 17 steps in this approach to supplier quality improvement. JQH5, presents quality improvement in the automotive industry. In the service sector, Sun Health Alliance employs an innovative approach to stimulate improvement (Nussman, 1993). Sun provides grant funding to partner hospitals and corporate partners to support quality improvement demonstration projects. The projects include patient care topics (e.g., establishing clinical pathways for specific diagnoses), nonclinical topics (e.g., reducing turnaround time for lab results), and employee-specific topics (e.g., reducing turnover or employee "needle sticks").

Handfield et al. (2000) discuss the results of research to identify "pitfalls" in supplier development. The research involved 84 companies in the fields of telecommunications, automobiles, electronics, computers, services, chemicals, consumer nondurable goods, and aerospace. The pitfalls were mainly concerned with identifying key projects, defining the details of the agreement between the buyer and supplier organizations, and monitoring the status and modifying strategies when necessary.

Some of the pitfalls were supplier specific; some were buyer specific; some were specific to the supplier–buyer relationship.

The supplier-specific pitfalls stemmed chiefly from the suppliers' lack of commitment and lack of technical or human resources. To avoid these pitfalls, companies took these actions:

1. Show suppliers where they stand.
2. Tie business relationships to performance improvement.
3. Illustrate supplier benefits clearly.

4. Ensure follow-up through a supplier champion (a supplier employee).
5. Keep initial improvements simple.
6. Draw on buyer's resources.
7. Offer personnel support.
8. Build training centers.

The buyer-specific pitfalls occur when buyers see no obvious potential benefits from working on supplier development. To avoid this situation, companies found these tactics helpful:

1. Consolidate to fewer suppliers.
2. Keep a long-term focus.
3. Determine the total cost of ownership.
4. Set small goals.
5. Make executive commitment in the buyer organization a priority.

The supplier–buyer relationship pitfalls involved lack of trust between the organizations, poor alignment of cultures, and insufficient inducements to suppliers. Constructive solutions were to:

1. Delegate an ombudsman from the buyer organization.
2. Make provisions for handling confidential information.
3. Spell out clearly a cooperative purchasing relationship with well-defined objectives beyond the purchase price.
4. Minimize legal involvement.
5. Adapt to local cultures.
6. Create a road map that defines responsibilities and expectations for both organizations.
7. Offer financial incentives.
8. Show suppliers how they can become "designed in" to buyer products and thus have greater potential for future business.
9. Offer repeat business as an incentive.

Supplier quality improvement needs upper management at all links in the supply chain to provide a structured approach to improvement. Cheerleading and flag waving will not work.

References

Bettis, R. A., S. P. Bradley, and G. Hamel (1992). "Outsourcing and Industrial Decline," *Academy of Management Executive*, vol. 6, no. 1, pp. 7–22.

Bossert, J. L., ed. (1994). *Supplier Management Handbook*, ASQ Quality Press, Milwaukee, p. 212.

Bossert, J. L. (1998). "Considerations for Global Supplier Quality," *Quality Progress*, January, pp. 29–32.

Brown, J. O. (1998). "A Practical Approach to Service-Supplier Certification," *Quality Progress*, January, pp. 35–39.

Carter, J. R. and J. G. Miller (1989). "The Impact of Alternative Vendor/Buyer Communication Structures on the Quality of Purchased Materials," *Decision Sciences*, Fall, pp. 759–776.

Chen, B. A. and R. G. Batson (1996). "A Team Based Supplier Quality Improvement Process," *Annual Quality Congress Proceedings*, ASQ, Milwaukee, pp. 537–544.
Clott, C. B. (2004). "Perspectives on Global Outsourcing and the Changing Nature of Work," *Business and Society Review*, vol. 109, no. 2, pp. 153–170.
Ericson, C. (2003). "A Global Look at e-Sourcing," *Supply Chain Management Review*, vol. 7, no. 6, p. 13.
Fisher, M. L. (1997). "What Is the Right Supply Chain for Your Product?" *Harvard Business Review*, March–April, pp. 105–116.
Handfield, R. B., D. R. Krause, T. V. Scannell, and R. M. Monczka (2000). "Avoid the Pitfalls in Supplier Development," *Sloan Management Review*, vol. 41, no. 2, pp. 37–49. By permission of the publisher. All rights reserved.
Hussey, D. and P. Jenster (2003). "Outsourcing: The Supplier Viewpoint," *Strategic Change*, vol. 12, no. 1, pp. 7–20.
Karmaker, U. (2004). "Will You Survive the Services Revolution?" *Harvard Business Review*, June.
Klenz, B. W. (2000). "Leveraging the Data Warehouse for Supplier Quality Analysis," *Annual Quality Congress Proceedings*, ASQ, Milwaukee, pp. 519–528.
Lee, H. L. (2004). "The Triple-A Supply Chain," *Harvard Business Review*, October, pp. 102–112.
Maass, R. A., J. O. Brown, and J. L. Bossert (1990). *Supplier Certification: A Continuous Improvement Strategy*, ASQ Quality Press, Milwaukee.
Meseck, G. (2004). "Risky Business: How to Calculate the Financial Impact of Global Outsourcing," *Logistics Today*, vol. 45, no. 8, p. 34.
Miller, G. D. and R. J. Kegaris (1986). "An Alcoa-Kodak Joint Team," Juran Report Number Six, Juran Institute, Inc., Wilton, CT, pp. 29–34.
Nocera, C. D., M. K. Foliano, and R. E. Blalock (1989). "Vendor Rating and Certification," *Impro Conference Proceedings*, Juran Institute, Inc., Wilton, CT, pp. 9A-29 to 9A-38.
Nussman, H. B. (1993). "The Sun Health Alliance for Quality—A Unique Customer-Supplier Partnership," *Impro Conference Proceedings*, Juran Institute, Inc., Wilton, CT, pp. 3A.1-1 to 3A.1-4.
Peterson, Y. S. (1998). "Outsourcing: Opportunity or Burden," *Quality Progress*, June, pp. 63–64.
Ramachandran, K. and S. Voleti (2004). "Business Process Outsourcing (BPO): Emerging Scenario and Strategic Options for IT-Enabled Services," *Vikalpa*, vol. 29, no. 1, pp. 49–62.
Schneider, H., J. Pruett, and C. Lagrange (1995). "Uses of Process Capability Indices in the Supplier Certification Process," *Quality Engineering*, vol. 8, no. 1, pp. 225–235.
Sharman, G. (2002). "How the Internet Is Accelerating Supply Chain Trends," *Supply Chain Management Review*, vol. 6, no. 2, p. 18.
Soliman, K. S. (2003). "A Framework for Global IS Outsourcing by Application Service Providers," *Business Process Management Journal*, vol. 9, no. 60, p. 375.
Spooner, G. R. and D. W. Collins (1995). "A Cross Functional Approach to Supplier Evaluation," *Proceedings of the Annual Quality Congress*, ASQ, Milwaukee, pp. 825–832.
Weerakkody, V., W. L. Currie, and Y. Ekanayake (2003). "Re-Engineering Business Processes Through Application Service Providers: Challenges, Issues and Complexities," *Business Process Management Journal*, vol. 9, no. 6, p. 776.
Weidenbaum, M. (2004). "Outsourcing; Pros and Cons," *Executive Speeches*, vol. 19, no. 1, 31–35.
Wind, J. F. (1991). "Revolutionize Supplier Rating by Computerization," *Quality Congress Transactions*, ASQ, Milwaukee, pp. 556–564.
Zimmerli, B. (1996). "Re-Engineering the Supply Chain," *Impro Conference Proceedings*, Juran Institute, Inc., Wilton, CT, pp. 4F-1 to 4F-18.

CHAPTER 24

Inspection, Test, and Measurement

Joseph A. De Feo

High Points of This Chapter

1. Product acceptance involves three decisions: conformance, fitness for use, and communication.
2. In deciding whether or not nonconforming product is fit for use, input must be secured from several sources.
3. The communication decision involves both outsiders (customers) and insiders.
4. Inspection planning includes the designation of inspection stations and spelling out the instructions and facilities required.
5. A classification of characteristics is a list of quality characteristics derived from the specifications; a classification of defects is a list of symptoms of nonconformance during manufacture and field use.
6. The amount of inspection necessary depends mainly on the amount of prior knowledge about product quality, homogeneity of the lot, and the allowable risk.

The Terminology of Inspection

Inspection and test typically include measurement of an output and comparison to specified requirements to determine conformity. Inspection and test activities ensure that manufactured products, individual components, and multicomponent systems are adequate for their intended purpose. Inspection and testing are the operational parts of quality control, which is the most important factor to the survival of any manufacturing company. Quality control directly supports the other factors of cost, productivity, on-time delivery, and market share. Therefore, all quality standards needed to produce the components of a product and perform its assembly must be specified in a manner such that customers' expectations are met. Inspection is performed for a wide variety of purposes, for example, distinguishing between good and bad product, determining whether a process is changing, measuring process capability, rating product quality, securing product design information, rating the inspectors' accuracy, and determining the precision of measuring instruments. Each of these purposes has its special influence on the nature of the inspection and on the manner of doing it.

The distinction between "inspection" and "test" has become blurred. Whereas inspection is the activity of examining the product or its components to determine if they meet the design standards, testing is a procedure in which the item is observed during operation in order to determine whether it functions properly for a reasonable period of time. Inspection, typically performed under static conditions on items such as components, can vary from simple visual or destructive examination to a series of complex measurements. The emphasis in inspection is to determine conformance to a standard. Test, on the other hand, is performed under either static or dynamic conditions and is typically performed on more complex items such as subassemblies or systems. Test results determine conformance and can also be used as input for other analyses such as evaluating a new design, diagnosing problems, or making physical adjustments on products. Some industries have their own terms for inspection or test, for example, assay is used in the mining and pharmaceutical industries.

Although the terms inspection and test usually refer to manufacturing industries, the concepts also apply to other industries. In service industries, different terms are used, for example, review, checking, reconciliation, examination. The evaluation of the correctness of an income tax return, the cleanliness of a hotel room, or the accuracy of a bank teller's closing balance are forms of inspection—a measurement, a comparison to a standard, and a decision.

Conformance to Specification and Fitness for Use

Of all the purposes of inspection, the most ancient and the most extensively used is product acceptance, that is, determining whether a product conforms to standard and therefore should be accepted. Product can mean a discrete unit, a collection of discrete units (a "lot"), a bulk product (a tank car of chemicals), or a complex system.

Product can also mean a service, such as a transaction at a bank; an inquiry to an agency about tax regulations; or the performance of personnel before, during, and after an airline flight. In these examples, inspection characteristics can be identified, standards set, and conformance judged.

Product acceptance involves the disposition of product based on its quality. This disposition involves several important decisions:

1. *Conformance.* Judging whether the product conforms to specification.
2. *Fitness for use.* Deciding whether nonconforming product is fit for use.
3. *Communication.* Deciding what to communicate to insiders and outsiders.

The Conformance Decision

Except in small companies, the number of conformance decisions made each year is huge. There is no possibility that the supervisory body can become involved in the details of so many decisions. Hence the work is organized so that inspectors or production workers can make these decisions themselves. To this end, they are trained to understand the products, the standards, and the instruments. Once trained, they are given the jobs of inspecting and judging conformance. (In many cases, the delegation is to automated instruments.)

Associated with the conformance decision is the disposition of conforming product. The inspector is authorized to identify the product ("stamp it up") as acceptable product. This identification then serves to inform packers, shippers, etc., that the product should proceed to its next destination (further processing, storeroom, customer). Strictly speaking, this decision to "ship" is made not by inspectors but by management. With some exceptions, product that conforms to specification is also fit for use. Hence company procedures (which are established by the managers) provide that conforming products should be shipped as a regular practice.

The Fitness-for-Use Decision

In the case of nonconforming products, a new question arises: Is this nonconforming product fit or unfit for use? Product features are said to possess "fitness-for-use" if they are able to meet customer demands, protect human safety, and protect the environment. Unfit product is disposed of in various ways: scrap, sort, rework, return to supplier, sell at discount, etc. In some cases, the answer is obvious—the nonconformance is so severe that the product is clearly unfit. Hence it is scrapped or, if economically repairable, brought to conformance. However, in many cases, the answer as to fitness for use is not obvious. In such cases, if enough is at stake, a study is made to determine fitness for use. This study involves securing inputs such as the following:

- *Who will the user be?* A technologically sophisticated user may be able to deal successfully with the nonconformance; a consumer may not. A nearby user may have easy access to field service; a distant or foreign user may lack such easy access.
- *How will the product be used?* For many materials and standard products, the specifications are broad enough to cover a variety of possible uses, and the actual

use to which the product will be put is not known at the time of manufacture. For example, sheet steel may be cut up to serve as decorative plates or as structural members; a television receiver may be stationed at a comfortable range or at an extreme range; chemical intermediates may be employed in numerous formulas.

- *Are there risks to human safety or to structural integrity?* Where such risks are significant, all else is academic.
- *What is the urgency?* For some applications, the client cannot wait because the product in question is critical to putting some broader system into operation. Hence the product may demand delivery now and cause repairs in the field.
- *What are the company's and the users' economics?* For some nonconformances, the economics of repair are so forbidding that the product must be used as is, although at a price discount. In some industries, for example, textiles, the price structure formalizes this concept by use of a separate grade—"seconds."
- *What are the users' measures of fitness for use?* These may differ significantly from those available to the manufacturer. For example, a manufacturer of abrasive cloth used a laboratory test to judge the ability of the cloth to polish metal; a major client evaluated the cost per 1000 pieces polished.

These and other inputs may be needed at several levels of fitness for use, that is, the effects on the economics of subsequent processors, the marketability requirements of the merchants, the qualities that determine fitness for the ultimate user, and the qualities that influence field maintenance. The internal costs can be estimated to arrive at an economic optimum. However, the effects can go well beyond money: Schedules are disrupted, people are blamed, etc.

The job of securing such inputs is often assigned to a staff specialist, for example, a quality engineer who "makes the rounds," contacting the various departments that are able to provide pertinent information. There may be a need to contact the customer and even to conduct an actual tryout. A typical list of sources is shown in Table 24.1. Once all the information has been collected and analyzed, the fitness-for-use decision can be made. If the amount of money at stake is small, this decision will be delegated to a staff specialist, to the quality manager, or to some continuing decision-making committee such as a material review board. If the amount at stake is large, the decision will usually be made by a team of upper managers.

Deliberations on the fitness-for-use decision are often a dramatic blend of voices—some balanced and judicious, others bowing to the pressures of delivery deadlines, even if it means tossing up gems of earnest nonsense.

Input	Usual Sources
Who will the user be?	Marketing
How will this product be used?	Marketing; client
Are there risks to human safety or to structural integrity?	Product research and design
What is the urgency?	Marketing; client
What are the company's and users' economics?	All departments; client
What are the users' measures of fitness for use?	Market research; marketing; client

Table 24.1 Sources of Information

The Communication Decision

The conformance and fitness-for-use decisions are a source of essential information, although some of the data is not well communicated.

Data on nonconforming products are usually communicated to the producing departments to aid them in preventing a recurrence. More elaborate data collection systems may require periodic summaries to identify "repeaters," which then become the subject of special studies.

When nonconforming products are sent out as fit for use, the need for two additional categories of communication arises:

1. *Communication to "outsiders".* (Usually customers) who have a right and a need to know. All too often, manufacturing companies neglect or avoid informing their customers when shipping nonconforming products. Such avoidance can be the result of bad experience, that is, some customers will seize on such nonconformances to secure a price discount despite the fact that use of the product will not add to their own costs. Neglect is more usually a failure even to face the question of what to communicate. A major factor here is the design of the forms used to record the decisions (nonconforming material control as part of the quality management system). With rare exceptions, these forms lack provisions that force those involved to make recommendations and decisions on (a) whether to inform the outsiders and (b) what to communicate to them.
2. *Communication to insiders.* When nonconforming goods are shipped as fit for use, the reasons for doing so are not always communicated to inspectors and especially not to production workers. The resulting vacuum of knowledge has been known to breed some bad practices. When the same type of nonconformance has been shipped several times, an inspector may conclude (in the absence of knowing why) that it is a waste of time to report such nonconformances in the first place. Yet in some future case, the special reasons that were the basis for the decision to ship the nonconforming goods may not be present. In like manner, a production worker may conclude that it is a waste of time to exert all that effort to avoid some nonconformance that will be shipped anyway. Such reactions by well-meaning employees can be minimized if the company squarely faces the question, what shall we communicate to insiders?

Disposition of Nonconforming Product

Once an inspector finds that a lot of product is nonconforming, he or she prepares a report to that effect. Copies of this report are sent to the implicated departments. This action sets a planned sequence of events into motion. The lot is marked "hold" and is often sent to a special holding area to reduce the risk of mix-ups. The product is put into quarantine. Schedulers look into the possibility of shortages and the need for replacement. An investigator is assigned to collect the type of information needed as inputs for the fitness-for-use decision, as discussed earlier.

Decision Not to Ship

The investigation may conclude that the lot should not be shipped as is. In that event, the economics are studied to find the best disposition: sorting, repairing, downgrading, scrapping, etc. Supplemental accounting efforts may charge the costs to the responsible source, especially if supplier responsibility is involved. Some degree of action to prevent a recurrence also occurs.

Decision to Ship

This decision may come about in several ways:

- *Waiver by the designer.* Such a waiver is a change in specification as to the lot in question that thereby puts the lot into a state of conformance. These decisions should be managed through systems such as engineering change notices or engineering change requests.
- *Waiver by the customer* or by the marketing department on behalf of the customer. Such a waiver in effect supersedes the specification. (The waiver may have been "bought" by a change in warranty or by a discount in price.)
- *Waiver by the quality department* under its delegation to make fitness-for-use decisions on noncritical matters. The criteria for "noncritical" may be based on prior seriousness classification of characteristics, on the low cost of the product involved, or on still other bases. For minor categories of seriousness, the delegation may even be made by the quality engineers or by inspection supervisors. However, for major and critical defects, the delegation is typically by the technical manager, the quality manager, or some team of managers.
- *Waiver by a formal material review board.* This board concept was originally evolved by military buyers of defense products to expedite decisions on nonconforming lots. Membership on the board includes the military representative plus the designer and the quality specialist. A unanimous decision is required to ship nonconforming product. The board procedures provide for formal documentation of the facts and conclusions, thereby creating a data source of great potential value.
- *Waiver by upper managers.* This part of the procedure is restricted to cases of a critical nature involving risks to human safety, marketability of the product, or risk of loss of large sums of money. For such cases, the stakes are too high to warrant decision making by a single department. Hence, the managerial team takes over. Waivers, however, have an insidious way of becoming part of a culture. It is valuable to continuously track the amount of product shipped under the waiver of specifications, for example, the percentage of lots shipped each month under waiver.

The service sector also has a variety of actions for disposition of a nonconforming service. Peach (1997, p. 435) provides some examples:

- For a cable television distribution failure, the disposition is to repair by reinitializing the connection.
- For an incorrect crediting in an account by a bank, the disposition is to rework by crediting the correct account and debiting the incorrect amount.
- For tainted food in a restaurant, the disposition is to scrap by returning the food to the supplier, disposing in the trash, and notifying local health authorities.

Corrective Action

Aside from a need to dispose of the nonconforming lot, there is a need to prevent a recurrence. This prevention process is of two types, depending on the origin of the nonconformance.

1. Some nonconformances originate in some isolated, sporadic change that took place in an otherwise well-behaved process. Examples are a mix-up in the materials

used, an instrument that is out of calibration, or a human mistake in turning a valve too soon. For such cases, local supervision is often able to identify what went wrong and to restore the process to its normal good behavior. Sometimes, this troubleshooting may require the assistance of a staff specialist. In any case, no changes of a fundamental nature are involved because manufacturing planning has already established an adequate process. Documentation of these decisions should at best follow simple investigation and resolutions such as through an 8D process.

2. Other nonconformances are "repeaters." They arise over and over again, as evidenced from their recurring need for disposition by the material review board or other such agency. Such recurrences point to a chronic condition that must be diagnosed and remedied if the problem is to be solved. Local supervision is seldom able to find the cause of these chronic nonconformances, mainly because the responsibility for diagnosis is vague. Lacking agreement on the cause, the problem goes on and on amid earnest debates about who or what is to blame—unrealistic design, incapable process, poor motivation, etc. The need is not for troubleshooting to restore the normal good behavior because the normal behavior is bad. Instead, the need is to organize for an improvement project, as discussed in Chap. 5, Quality Improvement and Breakthrough Performance. In these instances, one should use more sophisticated techniques such as root cause analysis or Six Sigma methodologies.

Inspection Planning

Inspection planning is the activity of (1) designating the "stations" at which inspection should take place and (2) providing those stations with the means for knowing what to do plus the facilities for doing it. For simple, routine quality characteristics, the planning is often done by the inspector. For complex products made in large multidepartmental companies, the planning is usually done by specialists such as quality engineers.

Locating the Inspection Stations

The basic tool for choosing the location of inspection stations is the flowchart (see Fig. 6.8). The most usual locations are:

- At receipt of goods from suppliers, usually called "incoming inspection" or "supplier inspection."
- Following the setup of a production process to provide added assurance against producing a defective batch. In some cases, this "setup approval" also becomes approval of the batch.
- During the running of critical or costly operations, usually called "process inspection."
- Prior to delivery of goods from one processing department to another, usually called "lot approval" or "tollgate inspection."
- Prior to shipping completed products to storage or to customers, usually called "finished-goods inspection."

- Before performing a costly, irreversible operation, for example, pouring a melt of steel.
- At natural "peepholes" in the process.

The concept of inspection stations also applies to the service sector. For example, in arranging travel for customers, "receiving inspection" often includes verifying information on a customer credit card; "process inspection" includes verifying that the customer has a passport, visa, and driver's license; "final inspection" includes confirming that tickets match agreements (Peach, 1997, p. 432).

The inspection station is not necessarily a fixed zone where the work comes to the inspector. In some cases, the inspector goes to the work by patrolling a large area and performing inspections at numerous locations, including the shipping area, at the supplier's plant, or on the customer's premises (typically known as roving inspection). Process inspection is often within the responsibility of production operators. An adjunct to this approach is the use of "poka-yoke" devices as part of the inspection.

These devices are installed in the machine to inspect process conditions and product results and provide immediate feedback to the operator (see Chap. 22).

Choosing and Interpreting Quality Characteristics

The planner prepares a list of which quality characteristics are to be checked at which inspection stations. The planner should use classification of characteristics from engineering (critical, major, minor, or incidental) to help determine those vital few product or process characteristics for inspection. For some of these characteristics, the planner may find it necessary to provide information that supplements the specifications. Product specifications are prepared by comparatively few people, each generally aware of the needs of fitness for use. In contrast, these specifications must be used by numerous inspectors and operators, most of whom lack such awareness. The planner can help bridge this gap in a number of ways:

- By providing inspection and test environments that simulate the conditions of use. This principle is widely used, for example, in testing electrical appliances. It is also extended to applications such as the type of lighting used for inspecting textiles.
- By providing supplementary information that goes beyond the specifications prepared by product designers and process engineers. Some of this information is available in published standards—company, industry, and national. Other information is specifically prepared to meet the specific needs of the product under consideration. For example, in an optical goods factory, the generic term beauty *defects* was used to describe several conditions that differed widely as to their effect on fitness for use. A scratch on a lens surface in the focal plane of a microscope made the lens unfit for use. A scratch on the large lens of a pair of binoculars, although not functionally serious, was visible to the user and hence was not acceptable. Two other species of scratches were neither adverse to fitness for use nor visible to the user and hence were unimportant. These distinctions were clarified through planning analysis and woven into the procedures.
- By helping to train inspectors and supervisors to understand the conditions of use and the "why" of the specification requirements.
- By providing seriousness classification (see the section "Seriousness Classification").

Detailed Inspection Planning

For each quality characteristic, the planner determines the detailed work to be done. This determination covers matters such as:

- The type of inspection or test to be done. This area may require a detailed description of the testing environment, testing equipment, testing procedure, and tolerances for accuracy.
- The number of units to be inspected or tested (sample size).
- The method of selecting the samples to be inspected or tested.
- The type of measurement (attributes, variables, other).
- Conformance criteria for the units, usually the specified product tolerance limits.

Beyond this detailed planning for the characteristics and units is further detailed planning applicable to the product, the process, and the data system:

- Conformance criteria for the lot, usually consisting of the allowable number of nonconforming units in the sample.
- The physical disposition to be made of the product—the conforming lots, the nonconforming lots, and the units tested.
- Criteria for decisions on the process—should it run or stop?
- Data to be recorded, forms to be used, reports to be prepared.

This planning is usually included in a formal document that must be approved by the planner and the inspection supervisor. For an example from Baxter Travenol, see Fig. 24.1.

Sensory Characteristics

Sensory characteristics are those for which we lack measuring instruments and for which the senses of human beings must be used as measuring instruments. Sensory qualities may involve technological performance of a product (e.g., adhesion of a protective coating),

Part Number: XXXX — Part Name: YYYY

Process	Characteristics	C_p[1] Index	C_{pk}[1] Index	Frequency[2]	Sample size[2]	Analysis methods	Out-of-control conditions are encountered[4]
Incoming inspection	Stock thickness	1.6	1.0	Every shipment	—	Review control charts provided with each lot	Impound lot—contact supplier for resolution
In-process inspection	Thickness	1.9	1.1	Every 1000 parts	2 pieces	Micrometers/$\bar{X}$ and s chart	Correct process
	Width	1.5	1.4	Every 10,000 parts	5 pieces	Micrometer/median chart	Correct process
	Length	1.6	1.2	Every 4 hours	75 pieces	Tapered ring gage/p chart	Correct process
Assembly area	Thickness	2.0	1.8	Hourly	30 pieces	Special gage/p chart	Correct process
	Width	2.2	1.9	Chart hourly	100%	Automatic tester/u chart	Repair by responsible operator
Outgoing[3]	Complete assembly	2.8	1.9	Hourly	20 pieces	Automatic tester/$\bar{X}$ and s chart	Correct process
	Complete assembly	NA	1500 DPM	Each lot	50 pieces	Complete visual inspection plus gage and test stand/c chart	Reject lot and sort for identified nonconformance

[1]Explanations and formulas are contained in the SPC Guideline.
[2]The frequencies and sample size are determined from the performance study of the stability of each process. They are periodically reviewed and updated as required.
[3]After 6 months production experience, the process control and inspection records will be reviewed to determine if outgoing inspection can be reduced.
[4]If any nonconforming products are found in the process samples, then there will be performed 100% inspection of all products produced since the last in control point.

FIGURE 24.1 Control plan. (From Baxter Travenol Laboratories, 1986.)

aesthetic characteristics (e.g., odor of a perfume), taste (e.g., food), or human services characteristics (e.g., the spectrum of hotel services).

An important category of sensory characteristics is the visual quality characteristic. Typically, written specifications are not clear because of their inability to quantify the characteristics. Among the approaches employed to describe the limits for characteristics are the following:

1. Providing photographs to define the limits of acceptability of the product.

EXAMPLE 24.1. A fast-food enterprise has the problem of defining quality standards for suppliers of hamburger buns. The solution is photographs showing the ideal and the maximum and minimum acceptable limits for "golden brown" color, symmetry of bun, and distribution of sesame seeds.

2. Providing physical standards to define the limits of acceptability.

EXAMPLE 24.2. A government agency needed to define the lightest and darkest acceptable shades of khaki for suppliers of uniforms. Color swatches of cloth were prepared for the limiting shades and issued to inspectors. Imagine the follow-up required to periodically replace the swatches when fading was imminent!

3. Specifying the conditions of inspection instead of trying to explicitly define the limits of acceptability.

EXAMPLE 24.3. Riley (1979) describes a special inspection procedure for cosmetic (appearance) defects of electronic calculators. Part drawings indicate the relative importance of different surfaces, using a system of category numbers and class letters. Three categories identify the surface being inspected:

1. Plastic window (critical areas only)
2. External
3. Internal

Three classes indicate the frequency with which the surface will be viewed by the user:

A. Usually seen by the user.

B. Seldom seen by the user.

C. Never seen by user (except during maintenance).

For example, a sheet-metal part that will seldom be seen carries a grade of Coating IIB.

The conditions of inspection are stated in terms of viewing distance, viewing time, and lighting conditions. The distance and time are specified for each combination of surface being inspected and the frequency of viewing by the user. Lighting conditions must be between 75 and 150 ft-candles from a nondirectional source.

The guidelines help to establish cosmetic gradings on parts drawings. However, a judgment must still be made by the inspector whether or not the end user would consider the flaw(s) objectionable, using the specified time and distance.

Seriousness Classification

Quality characteristics are decidedly unequal in their effect on fitness for use. A relative few are "serious," that is, of critical importance; many are of minor importance. Clearly, the more important the characteristic, the greater the attention it should receive in matters such as extent of quality planning; precision of processes, tooling, and instruments; sizes of samples; strictness of criteria for conformance; etc. However, making such discrimination requires that the relative importance of the characteristics be made known to the various decision makers involved: process engineers, quality planners, inspection supervisors, etc. To this end, many companies use formal systems of seriousness classification. The resulting classification is used in inspection and quality planning and also in specification writing, supplier relations, product audits, executive reports on quality, etc. This multiple use of seriousness classification dictates that the system be prepared by an interdepartmental committee which then:

1. Decides how many classes or strata of seriousness to create (usually three or four).
2. Defines each class.
3. Classifies each characteristic into its proper class of seriousness.

Characteristics and Defects

There are actually two lists that need to be classified. One is the list of quality characteristics derived from the specifications. The other is the list of "defects," that is, symptoms of nonconformance during manufacture and of field failure during use. There is a good deal of commonality between these two lists, but there are differences as well. (For example, the list of defects found on glass bottles has little resemblance to the list of characteristics.) In addition, the two lists do not behave alike. The design characteristic "diameter," for example, gives rise to two defects—oversize and undersize. The amount by which the diameter is oversize may be decisive as to seriousness classification.

Normally, it is feasible to make one system of classification applicable to both lists. However, the uses for the resulting classifications are sufficiently varied to make it convenient to publish the lists separately.

Definitions for the Classes

Most sets of definitions show the influence of the pioneering work of the Bell System in the 1920s. Study of numerous such systems reveals an inner pattern that is a useful guide to any committee faced with applying the concept to its own company. Table 24.2 shows the nature of this inner pattern applied to a company in the food industry.

Classification

Classification is a long and tedious but essential task. However, it yields some welcome by-products by pointing out misconceptions and confusion among departments and thereby opening the way to clear up vagueness and misunderstandings. Then, when the final seriousness classification is applied to several different purposes, it is subjected to several new challenges that provide still further clarification of vagueness. A problem often encountered is the reluctance of the designers to become involved in seriousness classification of characteristics. They may offer plausible reasons: All characteristics are critical; the tightness of the tolerance is an index of seriousness, etc. Yet the real reasons may be unawareness of the benefits, a feeling that other matters have higher departmental priority, etc. In such cases, it may be worthwhile to demonstrate the benefits of classification by working out a small-scale example. In one company, the classification-of-characteristics program reduced the number of dimensions that had to be checked from 682 to 279 and reduced inspection time from 215 to 120 minutes.

Automated Inspection

Automated inspection and testing are widely used to reduce inspection costs, reduce error rates, alleviate personnel shortages, shorten inspection time, avoid inspector monotony, and provide still other advantages. Applications of automation have successfully been made to mechanical gauging, electronic testing (for high volumes of components as well as

Defect	Effect on Consumer Safety	Effect on Usage	Consumer Relations	Loss to Company	Effect on Conformance to Government Regulations
Critical	Will surely cause personal injury or illness	Will render the product totally unfit for use	Will offend consumers' sensibilities due to odor, appearance, etc.	Will lose customers and will result in losses greater than value of product	Fails to conform to regulations for purity, toxicity, identification
Major A	Very unlikely to cause personal injury or illness	May render the product unfit for use and may cause rejection by the user	Will likely be noticed by consumers and will likely reduce product salability	May lose customers and may result in losses greater than the value of the product; will substantially reduce production yields	Fails to conform to regulations on weight, volume, or batch control
Major B	Will not cause injury or illness	Will make the product more difficult to use, e.g., removal from package, or will require improvisation by the user; affects appearance, neatness	May be noticed by some consumers and may be an annoyance if noticed	Unlikely to lose customers; may require product replacement; may result in loss equal to product value	Minor nonconformance to regulations on weight, volume, or batch control, e.g., completeness of documentation
Minor	Will not cause injury or illness	Will not affect usability of the product; may affect appearance, neatness	Unlikely to be noticed by consumers and of little concern if noticed	Unlikely to result in loss	Conforms fully to regulations

Table 24.2 Composite Definitions for Seriousness Classification in the Food Industry

system circuitry), nondestructive tests of many kinds, chemical analyses, color discrimination, visual inspection (e.g., of large-scale integrated circuits), etc. In addition, automated testing is extensively used as a part of scheduled maintenance programs for equipment in the field.

Examples in nonmanufacturing activities range from the spelling check provided within word processors to the checking of bank transactions for errors.

A company contemplating the use of automated inspection first identifies the few tests that dominate the inspection budgets and use of personnel. The economics of automation are computed, and trials are made on some likely candidates for a good return on investment. As experience is gained, the concept is extended further and further.

With the emphasis on defect levels in the parts-per-million range, many industries are increasingly accepting on-machine automated 100 percent inspection and testing. Orkin and Olivier (1999) provide an extensive table that identifies seven categories of potential applications of automated inspection ranging from dimensional gauging to nondestructive testing.

A dramatic example of automated inspection is the concept of "machine vision," where electronic eyes inspect and guide an array of industrial processes. The applications include steering robots to place doors on cars, finding blemishes on vegetables in a frozen food processing line, examining wood for knots in veneer panels, and checking that the right color drug capsule goes into correctly labeled packages before being shipped to pharmacies (*Fortune*, February 16, 1998, p. 104B). High-speed visual inspection devices are either integrated or are slightly off-line from manufacturing operations. Products can automatically undergo multicharacteristic checks. At the end of the inspection cycle, the computer monitor tells the operator whether the product is acceptable and then updates quality process statistics. Human inspection methods usually detect only 80 to 90 percent of the defects (see later discussion); with machine visual inspection, essentially all defects are detected. For elaboration, see JQH5, page 23.15. A critical requirement for all automated test equipment is precision measurement, that is, repeated measurements on the same unit of product should yield the "same" test results within some acceptable range of variation. This repeatability is inherent in the design of the equipment and can be quantified by the methods discussed in the section "Errors of Measurement." In addition, means must be provided to keep the equipment "accurate," that is, in calibration with respect to standards for the units of measure involved.

Still another aspect of automated test equipment is the problem of processing the data that are generated by the tests. Modern systems of electronic data processing allow entering these test data directly from the test equipment into the computer without the need for intermediate documents. Such direct entry supports the prompt preparation of data summaries, conformance calculations, comparisons with prior lots, etc. In turn, it is feasible to program the computer to issue instructions to the test equipment with respect to frequency of test, disposition of units tested, alarm signals relative to improbable results, etc.

Cooper (1997) discusses how to design, test, implement, and maintain a paperless inspection system.

Inspection Accuracy

Inspection accuracy depends on (1) the completeness of the inspection planning (see earlier discussion), (2) the bias and precision of the instruments (see later in this section), and (3) the level of human error.

High error rates are particularly prevalent in inspection tasks having a high degree of monotony, for example, viewing jars of a food product for foreign particles, screening luggage at an airport security gate. Surprisingly, monotony that causes an inspector to miss defects can build up in a short time. With monotonous inspection, inspectors detect about 80 to 90 percent of the defects and miss the remainder. Thus 100 percent inspection that is monotonous is not 100 percent effective in detecting defects. One of the advantages of automated inspection is the elimination of human error.

Human errors in inspection arise from multiple causes, of which four are most important: technique errors, inadvertent errors, conscious errors, and communication errors. The nature of these errors is similar to the same categories for personnel in other activities.

Measure of Inspector Accuracy

Some companies carry out regular evaluations of inspector accuracy as part of the overall evaluation of inspection performance. The plans employ a check inspector who periodically reviews random samples of work previously inspected by the various inspectors. The check inspection findings are then summarized, weighted, and converted into some index of inspector performance. JQH5, pages 23.51–23.53, explains this procedure.

Harris and Chaney (1969) provided some early research on inspector accuracy. Among their findings were that inspection accuracy decreases with reductions in defect rates, inspection accuracy increases with repeated inspections (up to a total of six), inspection accuracy decreases with additional product complexity, and the effect cannot be overcome by increasing the allowable inspection time. These conclusions are sobering.

Errors of Measurement

Variation in a process has two sources: variation of the process making the product and variation of the measurement process. Particularly with the low defect levels demanded under the Six Sigma approach, we must understand the capability of the manufacturing process and also the capability of the measurement process.

Even when correctly used, a measuring instrument may not give a true reading of a characteristic. The difference between the true value and the measured value can be due to one or more of five sources of variation (Fig. 24.2).

There is much confusion as to terminology. Bias is sometimes referred to as "accuracy." Because accuracy has several meanings in the literature (especially in measuring instrument catalogs), its use as an alternative for "bias" is not recommended.

The distinction between repeatability and bias is illustrated in Fig. 24.3.

Repeatability is often referred to as "precision." Any statement of bias and repeatability (precision) must be preceded by three conditions:

1. Definition of the test method. This definition includes the step-by-step procedure, equipment to be used, preparation of test specimens, test conditions, etc.
2. Definition of the system of causes of variability, such as material, analysts, apparatus, laboratories, days, etc. ASTM recommends that modifiers of the word precision be used to clarify the scope of the precision measure. Examples of such modifiers are single operator, single analyst, single-laboratory-operator-material-day, and multilaboratory.
3. Existence of a statistically controlled measurement process. The measurement process must have stability for the statements on bias and precision to be valid. This stability can be verified by a control chart.

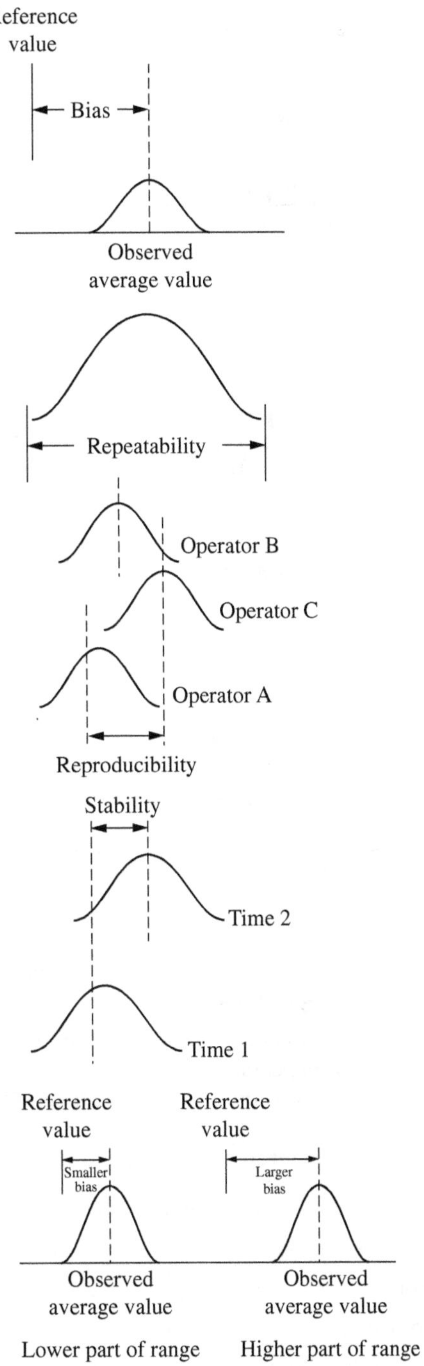

FIGURE 24.2 Five sources of measurement variation. (Reprinted with permission from the MSA Manual. DaimlerChrysler, Ford, General Motors Supplier Quality Requirements Task Force.)

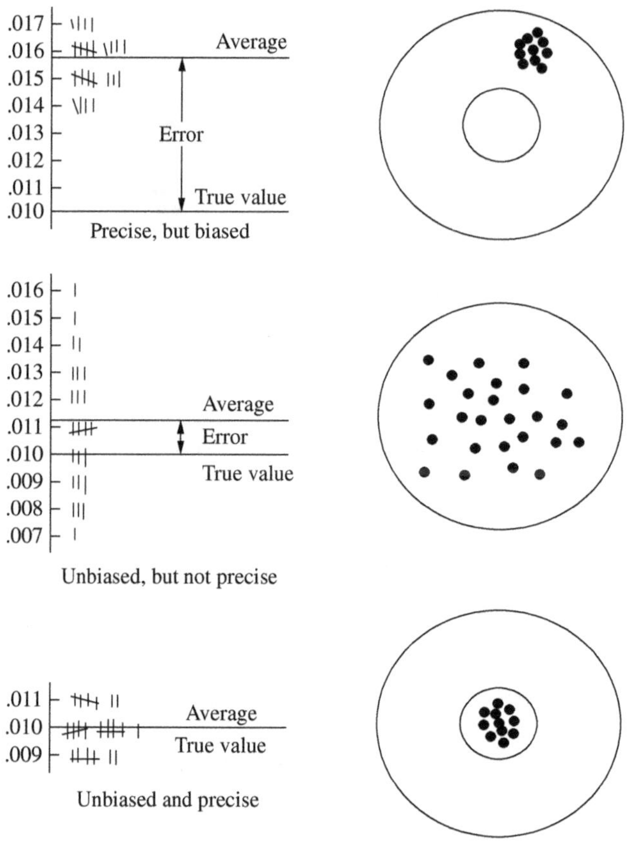

Figure 24.3 Distinction between bias and repeatability (precision).

How Much Inspection Is Necessary?

The amount of inspection to decide the acceptability of a lot can vary from no inspection to a sample to 100 percent inspection. The decision is governed mainly by the amount of prior knowledge available as to quality, the homogeneity of the lot, and the allowable degree of risk.

Prior knowledge that is helpful in deciding on the amount of inspection includes:

- Previous quality history on the product item and the supplier (internal or external).
- Criticality of the item on overall system performance.
- Criticality on later manufacturing or service operations.
- Warranty or use history.
- Process capability information. A process that is in statistical control with good uniformity around a target value will require minimum inspection.
- Measurement capability information, for example, the availability of accurate and precise instruments.

- The nature of the manufacturing process. For example, some operations depend primarily on the adequacy of the setup.
- Inspection of the first few and the last few items in a production run. This is usually sufficient.
- Product homogeneity. For example, a fluid product is homogeneous and reduces the need for large sample sizes.
- Data on process variables and process conditions, for example, as provided by automatic recording charts.
- Degree of adherence to the three elements of self-control for the personnel operating the process.

Competition to reduce costs has resulted in pressures to reduce the amount of inspection. The concept of inspection by the producers (self-inspection) has added to the focus of reducing inspection. Opportunities do exist for cost reduction in inspection activities. First, however, the causes of the high failure costs must be diagnosed and removed, and the prerequisites for self-inspection must be met.

EXAMPLE 24.4. The Datapoint Corporation manufactures office and computer products (Adams, 1987). Part of the operation was 100 percent in-line inspection of visual characteristics by the quality staff. A dramatic shift was planned—production personnel would do their own visual inspection; the quality staff would perform an audit inspection and do diagnostic work on the causes of nonconformities. But a number of steps were required: a quality education process for first-line management, supervisors, and line personnel; special training in workmanship standards to help people recognize nonconformances; an 18-month implementation plan to phase in the new approach; use of data from functional acceptance tests for process yield reports; analysis of data; and a process audit system to review documentation, tools, materials, and people.

The results were dramatic: The staff of 35 in-line inspectors was reduced to five process auditors, and scrap and rework plunged from 15 to 2 percent.

Economics of Inspection

We have several alternatives for evaluating lots:

1. *No inspection*. This approach is appropriate if the same lot has already been inspected by qualified laboratories, for example, in other divisions of the same company or in supplier companies. Prior inspections by qualified production workers have the same effect.
2. *Small samples*. Small samples can be adequate if the process is inherently uniform and the order of production can be preserved. For example, in some punch press operations, the stamping dies are made with a high degree of stability. As a result, the successive pieces stamped out by such dies exhibit a high degree of uniformity for certain dimensional characteristics. For such characteristics, if the first and last pieces are correct, the remaining pieces are also correct, even for lot sizes running to many thousands of pieces. In its generalized form, the press example is one of a high degree of process capability combined with "stratified" sampling—sampling based on knowledge of the order of production.

   Small samples can also be used when the product is homogeneous due to its fluidity (gases, liquids) or to prior mixing operations. This homogeneity need not be assumed—it can be verified by sampling. Even solid materials may be homogeneous due to prior fluidity. Once homogeneity has been established, the sampling needed is minimal.

For a product with a continuing history of good quality, sampling can be periodic, for example, "skip-lot" or "chain sampling"

3. *Large samples*. In the absence of prior knowledge, the information about lot quality must be derived solely from sampling, which means random sampling and hence relatively large samples. The actual sample sizes depend on two main variables: (1) the tolerable percentage of defects and (2) the risks that can be accepted. Once values have been assigned to these variables, the sample sizes can be determined scientifically in accordance with the laws of probability. However, the choices of defect levels and risks are based largely on empirical judgments.

   Random sampling is clearly needed in cases where there is no ready access to prior knowledge, for example, purchases from certain suppliers. However, there remain many cases in which random sampling is used despite the availability of inputs such as process capability, order of manufacture, fluidity, etc. A major obstacle is a lack of publications which show how to design sampling plans in ways which use these inputs. In the absence of such publications, quality planners are faced with creating their own designs. This means added work amid the absence of protection derived from the use of recognized, authoritative published materials.

4. *One hundred percent inspection*. This technique is used when the results of sampling show that the level of defects present is too high for the product to go on to the users. In critical cases, added provisions may be needed to guard against inspector fallibility, for example, automated inspection or redundant 200 percent inspection.

An economic evaluation of these alternatives requires a comparison of total costs under each one.

- Let N = number of items in lot.
- n = number of items in sample.
- p = proportion defective in lot.
- A = damage cost incurred if a defective slips through inspection.
- I = inspection cost per item.
- P_a = probability that lot will be accepted by sampling plan.
- A and I are sometimes denoted as $k1$, and $k2$, respectively.

Consider the comparison of sampling inspection versus 100 percent inspection. Suppose it is assumed that no inspection errors occur and the cost to replace a defective found in inspection is borne by the producer or is small compared to the damage or inconvenience caused by a defective. The total costs are summarized in Table 24.3. These costs reflect both inspection costs and damage costs and recognize the probability of accepting or rejecting a lot under sampling inspection. The expressions can be equated to determine

Alternative	Total Cost
No inspection	NpA
Sampling	$nI + (N - n)pAP_a + (N - n)(I - P_a)I$
100% inspection	NI

Table 24.3 Economic Comparison of Inspection Alternatives

a breakeven point. If it is assumed that the sample size is small compared to the lot size, the breakeven point, p_b, is

- $p_b = I/A$

If the lot quality (p) is less than p_b, the total cost will be lowest with sampling inspection or no inspection. If p is greater than p_b, 100 percent inspection is best. This principle is often called the Deming *kp* rule.

For example, a microcomputer device costs $.50 per unit to inspect. A damage cost of $10.00 is incurred if a defective device is installed in the larger system. Therefore:

- Formula: $p_b = 0.50/10.00 = 0.05 = 5.0\%$

If the percentage defective is expected to be greater than 5 percent, then 100 percent inspection should be used. Otherwise, use sampling or no inspection.

The variability in quality from lot to lot is important. If past history shows that the quality level is much better than the breakeven point and is stable from lot to lot, little if any inspection may be needed. If the level is much worse than the breakeven point and consistently so, it will usually be cheaper to use 100 percent inspection rather than sampling. If the quality is at neither of these extremes, a detailed economic comparison of no inspection, sampling, and 100 percent inspection should be made. Sampling is usually best when the product is a mixture of high-quality lots and low-quality lots or when the producer's process is not in a state of statistical control.

The high costs of component failures in complex electronic equipment coupled with the development of automatic testing equipment for components has resulted in the economic justification of 100 percent inspection for some electronic components. The cost of finding and correcting a defective can increase by a ratio of 10 for each major stage that the product moves to from production to the customer, that is, if it costs $1 at incoming inspection, the cost increases to $10 at the printed circuit board stage, $100 at the system level, and $1000 in the field.

The Concept of Acceptance Sampling

Acceptance sampling is the process of evaluating a portion of the product in a lot for the purpose of accepting or rejecting the entire lot. It involves the application of specific sampling plans to a designated lot or sequence of lots. Any acceptance sampling application must distinguish whether the purpose is to accumulate information on the immediate product being sampled or on the process which produced an immediate product at hand.

The main advantage of sampling is economy. Despite some added costs for designing and administering sampling plans, the lower costs of inspecting only part of the lot result in an overall cost reduction.

In addition to this main advantage, there are others:

- The smaller inspection staff is less complex and less costly to administer.
- There is less damage to the product, that is, handling incidental to inspection is itself a source of defects.
- The lot is disposed of in shorter (calendar) time so that scheduling and delivery are improved.
- The problem of monotony and inspector error induced by 100 percent inspection is minimized.

- Rejection (rather than sorting) of nonconforming lots tends to dramatize the quality deficiencies and to urge the organization to look for preventive measures.
- Proper design of the sampling plan commonly requires study of the actual level of quality required by the user. The resulting knowledge is a useful input to the overall quality planning.

The disadvantages are sampling risks, greater administrative costs, and less information about the product than provided by 100 percent inspection.

Acceptance sampling is used when (1) the cost of inspection is high in relation to the damage cost resulting from passing a defective product, (2) 100 percent inspection is monotonous and causes inspection errors, or (3) the inspection is destructive. Acceptance sampling is most effective when it is preceded by a prevention program that achieves an acceptable level of quality of conformance.

We must also emphasize what acceptance sampling does not do. It does not provide refined estimates of lot quality. (It does determine, with specified risks, an acceptance or rejection decision on each lot.) Also, acceptance sampling does not provide judgments whether or not rejected product is fit for use. (It does give a decision on a lot with respect to the defined quality specification.)

In recent years, the emphasis on statistical process control has led some practitioners to conclude that acceptance sampling is no longer a valid concept. Their belief, stated here in oversimplified terms, is that only two levels of inspection are valid—no inspection or 100 percent inspection. This text takes the viewpoint that the concept of prevention (using statistical process control and other statistical and managerial techniques) is the foundation for meeting product requirements. Acceptance sampling procedures are, however, important in a program of acceptance control. Under this latter approach, described at the end of this chapter, sampling procedures are continually matched to process history and quality results. This step ultimately leads to phasing out acceptance sampling in favor of supplier certification and process control.

This chapter presents examples of specific acceptance sampling plans.

For some perceptive discussions of the modern role of acceptance sampling, see Schilling (1994) and Taylor (1994).

Sampling Risks: The Operating Characteristic Curve

Neither sampling nor 100 percent inspection can guarantee that every defective item in a lot will be found. Sampling involves a risk that the sample will not adequately reflect the conditions in the lot; 100 percent inspection has the risk that monotony and other factors will result in inspectors missing some of the defectives (see section "Inspection Accuracy"). Both of these risks can be quantified. Sampling risks are of two kinds:

1. Good lots can be rejected (the producer's risk). This risk corresponds to the α risk.
2. Bad lots can be accepted (the consumer's risk). This risk corresponds to the β risk.

The operating characteristic (OC) curve for a sampling plan quantifies these risks. The OC curve for an attribute plan is a graph of the percentage defective in a lot versus the probability that the sampling plan will accept a lot. Because p is unknown, the probability must be stated for all possible values of p. It is assumed that an infinite number of lots are produced. Figure 24.4 shows an "ideal" OC curve where it is desired to accept all lots 1.5 percent defective or less and reject all lots having a quality level greater than 1.5 percent defective. All lots less than 1.5 percent defective have a probability of acceptance of 1.0 (certainty); all lots

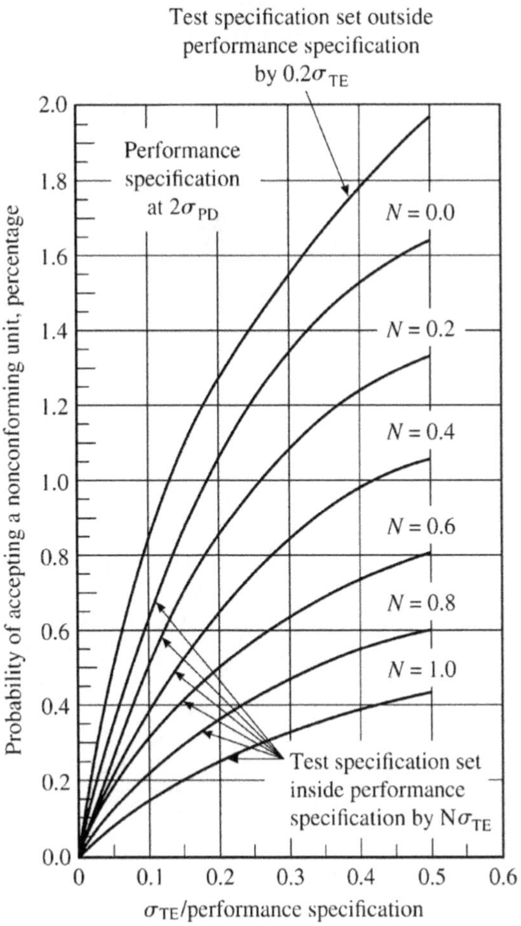

FIGURE 24.4 Ideal OC curve.

greater than 1.5 percent defective have a probability of acceptance of zero. Actually, however, no sampling plan exists that can discriminate perfectly; there always remains some risk that a "good" lot will be rejected or that a "bad" lot will be accepted. The best that can be achieved is to make the acceptance of good lots more likely than the acceptance of bad lots.

An acceptance sampling plan basically consists of a sample size (n) and an acceptance criterion (c). For example, a sample of 125 units is to be randomly selected from the lot. If five or fewer defectives are found, the lot is accepted. If six or more defectives are found, the lot is rejected.

The sample of 125 could, by the laws of chance, contain 0, 1, 2, 3, even up to 125 defectives. It is this sampling variation that causes some good lots to be rejected and some bad lots to be accepted. The OC curve for $n = 125$ and $c = 5$ is curve A (Fig. 24.5). (The other curves will be discussed later.) A 1.5 percent defective lot has about a 98 percent chance of being accepted. A much worse lot, say 6 percent defective, has a 23 percent chance of being accepted. With the risks stated quantitatively, the adequacy of the sampling plan can be judged.

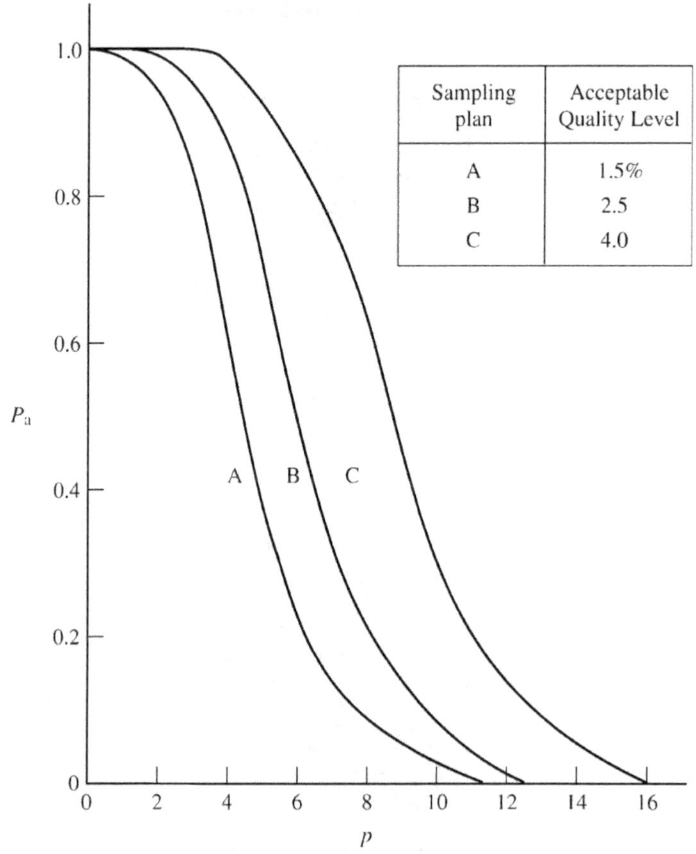

FIGURE 24.5 OC curves.

The OC curve for a specific plan states only the chance that a lot having p percent defective will be accepted by the sampling plan. The OC curve does not:

- Predict the quality of lots submitted for inspection. For example (see Fig. 24.5), it is incorrect to say that there is a 36 percent chance that the lot quality is 5 percent defective.
- State a "confidence level" in connection with a specific percentage defective.
- Predict the final quality achieved after all inspections are completed.

These and other myths about the OC curve require careful explanation of the concept to those using it. (Acceptable quality level is explained in the section "Quality Indexes for Acceptance Sampling Plans.")

Constructing the Operating Characteristic Curve

An OC curve can be developed by determining the probability of acceptance for several values of incoming quality, p. The probability of acceptance is the probability that the number of defectives in the sample is equal to or less than the acceptance number for the sampling plan.

Three distributions can be used to find the probability of acceptance: the hypergeometric, binomial, and Poisson distributions. When its assumptions can be met, the Poisson distribution is preferable because of the ease of calculation.

Grant and Leavenworth (1996, pp. 183–193) describe the use of hypergeometric and binomial distributions.

The Poisson distribution yields a good approximation for acceptance sampling when the sample size is at least 16, the lot size is at least 10 times the sample size, and p is less than 0.1. The Poisson distribution function as applied to acceptance sampling is

The equation can be solved by using a calculator or by using Table C in App. II. This table gives the probability of r or fewer defectives in a sample of n from a lot having a fraction defective of p. To illustrate Table C, consider the sampling plan previously cited: $n = 125$ and $c = 5$. To find the probability of accepting a 4 percent defective lot, calculate np as 125(0.04) = 5.00. Table C then gives the probability of five or fewer defectives as 0.616. Figure 24.5 (curve A) shows this probability as the value of P_a for 4 percent defective lot quality.

The preceding discussion of sampling risks assumes that the proportion defective of incoming lots is reasonably constant. This assumption is often made in practice. Chun and Rinks (1998) derive modified producer's and consumer's risks when incoming quality is not constant.

Quality Indexes for Acceptance Sampling Plans

Many of the published plans can be categorized in terms of one of several quality indexes:

1. *Acceptable quality level (AQL)*. The units of quality level can be selected to meet the particular needs of a product. Thus ANSI/ASQC Z1.4 (1993) defines AQL as "the maximum percent nonconforming (or the maximum number of nonconformities per hundred units) that, for purposes of sampling inspection, can be considered satisfactory as a process average." If a unit of product can have a number of different defects of varying seriousness, then demerits can be assigned to each type and product quality measured in terms of demerits. Because an AQL is an acceptable level, the probability of acceptance for an AQL lot should be high (Fig. 24.6).
2. *Limiting quality level (LQL)*. LQL defines unsatisfactory quality. Different titles are sometimes used to denote an LQL; for example, Dodge–Romig plans use the term lot tolerance percentage defective (LTPD). Because an LQL is an unacceptable level, the probability of acceptance for an LQL lot should be low (see Fig. 24.6). In some tables, this probability, known as the consumer's risk, is designated Pc, and has been standardized at 0.1. The consumer's risk is not the probability that the consumer will actually receive product at the LQL. The consumer will, in fact, not receive 1 lot in 10 at LQL fraction defective. What the consumer actually gets depends on the actual quality in the lots before inspection and on the probability of acceptance.
3. *Indifference quality level (IQL)*. IQL is a quality level somewhere between the AQL and LQL. It is frequently defined as the quality level that has a probability of acceptance of 0.5 for a given sampling plan (see Fig. 24. 6).

   It should be emphasized to both internal and external suppliers that all product submitted for inspection is expected to meet specifications. An acceptable quality level does not mean that submitting a certain amount of nonconforming product is approved. The AQL simply recognizes that, under sampling, some nonconforming product will pass through the sampling scheme.

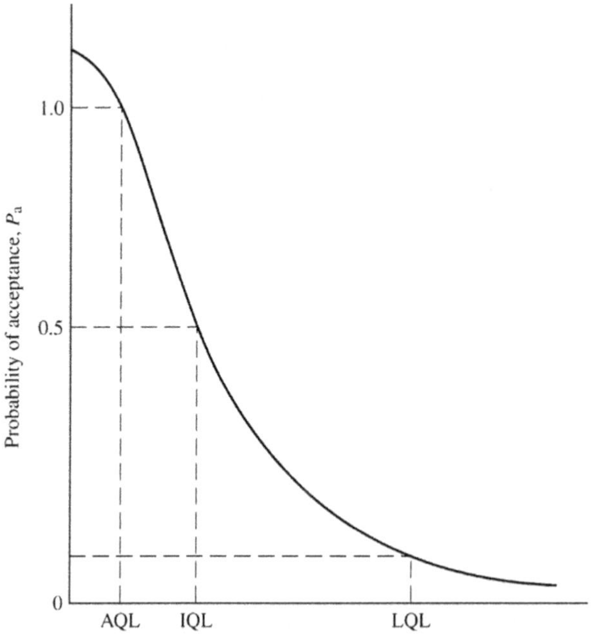

FIGURE 24.6 Quality indexes for sampling plans.

4. *Average outgoing quality limit (AOQL)*. An approximate relationship exists between the fraction defective in the material before inspection (incoming quality p) and the fraction defective remaining after inspection (outgoing quality AOQ):
   - AOQ = pP_a. When incoming quality is perfect, outgoing quality must also be perfect. However, when incoming quality is bad, outgoing quality will also be perfect (assuming no inspection errors) because the sampling plan will cause all lots to be rejected and inspected in detail. Thus at either extreme—incoming quality excellent or terrible—the outgoing quality will tend to be good. Between these extremes is the point at which the percentage of defectives in the outgoing material will reach its maximum. This point is the average outgoing quality limit (AOQL).
   - These indexes apply primarily when the production occurs in a continuing series of lots. The LQL concept is recommended for isolated lots. The indexes were originally developed by statisticians to help describe the characteristics of sampling plans.
   - Misinterpretations (particularly of the AQL) are common and are similar to those mentioned in the section "Disposition of Nonconforming Product." For example, a sampling plan based on AQL will accept some lots that have a quality level worse than the AQL.

Types of Sampling Plans

Sampling plans are of two types:

1. *Attributes plans*. A random sample is taken from the lot, and each unit is classified as acceptable or defective. The number defective is then compared with the allowable number stated in the plan, and a decision is made to accept or reject the lot. This chapter illustrates attributes plans based on AQL.

2. *Variables plans.* A sample is taken, and a measurement of a specified quality characteristic is made on each unit. These measurements are then summarized into a sample statistic (e.g., sample average), and the observed value is compared with an allowable value defined in the plan. A decision is then made to accept or reject the lot.

The key advantage of a variables sampling plan is the additional information provided in each sample that, in turn, results in smaller sample sizes compared with an attributes plan that has the same risks. However, if a product has several important quality characteristics, each must be evaluated against a separate variables acceptance criterion (e.g., numerical values must be obtained and the average and standard deviation for each characteristic calculated). In a corresponding attributes plan, the sample size required may be higher, but the several characteristics can be treated as a group and evaluated against one set of acceptance criteria.

Single, Double, and Multiple Sampling

Many published sampling tables give a choice among single, double, and multiple sampling. In single-sampling plans, a random sample of n items is drawn from the lot. If the number of defectives is less than or equal to the acceptance number (c), the lot is accepted. Otherwise, the lot is rejected. In double-sampling plans (Fig. 24.7), a smaller initial sample is usually drawn, and a decision to accept or reject is reached on the basis of this smaller first

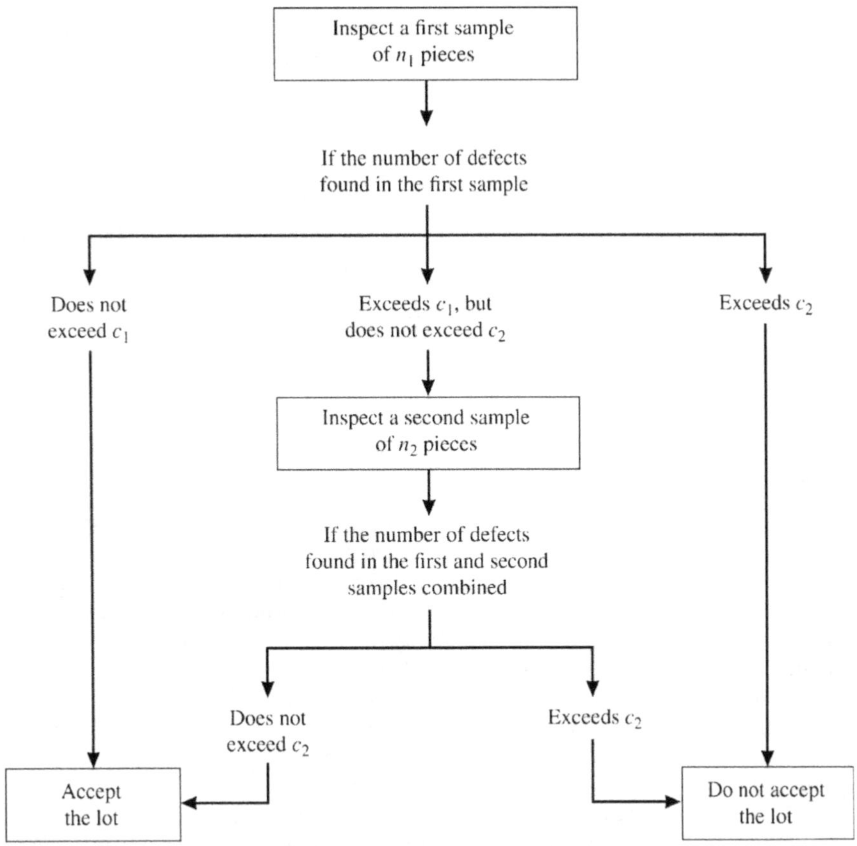

FIGURE 24.7 Schematic operation of double sampling.

sample if the number of defectives is either quite large or quite small. A second sample is taken if the results of the first are not decisive. Because it is necessary to draw and inspect the second sample only in borderline cases, the average number of pieces inspected per lot is generally smaller in double sampling. In multiple-sampling plans, one, two, or several still smaller samples are taken, usually continuing as needed until a decision to accept or reject is obtained. Thus, double- and multiple-sampling plans may mean less inspection but are more complicated to administer.

In general, it is possible to derive single-, double-, or multiple-sampling schemes with essentially identical OC curves.

Characteristics of a Good Acceptance Plan

An acceptance sampling plan should have these characteristics:

- The index (AQL, AOQL, etc.) used to define "quality" should reflect the needs of the consumer and the producer and not be chosen primarily for statistical convenience.
- The sampling risks should be known in quantitative terms (the OC curve). The producer should have adequate protection against the rejection of good lots; the consumer should be protected against the acceptance of bad lots.
- The plan should minimize the total cost of inspection of all products. This requires careful evaluation of the pros and cons of attributes and variables plans, as well as single, double, and multiple sampling. It should also reflect product priorities, particularly from the fitness-for-use viewpoint.
- The plan should use other knowledge, such as process capability, supplier data, and other information.
- The plan should have built-in flexibility to reflect changes in lot sizes, quality of product submitted, and any other pertinent factors.
- The measurements required by the plan should provide information useful in estimating individual lot quality and long-run quality.
- The plan should be simple to explain and administer.

Fortunately, published tables are available that meet many of these characteristics. We now proceed to a discussion of AQL plans.

ANSI/ASQC Z1.4

ANSI/ASQC Z1.4 (1993) is an attributes sampling system. Its quality index is the acceptable quality level (AQL). The AQL is the maximum percentage nonconforming (or the maximum number of nonconformities per 100 units) that, for purposes of sampling inspection, can be considered satisfactory as a process average. (The standard uses the term nonconformity rather than defective unit.) The probability of accepting material of AQL quality is always high but not exactly the same for all plans. For lot quality just equal to the AQL, the "percentage of lots expected to be accepted" ranges from about 89 to 99. The choice may be made from 26 available AQL values ranging from 0.010 to 1000.0. (AQL values of 10.0 or less may be interpreted as percentage nonconforming or nonconformities per 100 units; values above 10.0 are interpreted as nonconformities per 100 units.)

The tables specify the relative amount of inspection to be used as "inspection level" I, II, or III; level II is regarded as normal. The inspection-level concept permits the user to balance

Lot or Batch Size	Special Inspection Levels				General Inspection Levels		
	S-1	S-2	S-3	S-4	I	II	III
2–8	A	A	A	A	A	A	B
9–15	A	A	A	A	A	B	C
16–25	A	A	B	B	B	C	D
26–50	A	B	B	C	C	D	E
51–90	B	B	C	C	C	E	F
91–150	B	B	C	D	D	F	G
151–280	B	C	D	E	E	G	H
281–500	B	C	D	E	F	H	J
501–1200	C	C	E	F	G	J	K
1201–3200	C	D	E	G	H	K	L
3201–10,000	C	D	F	G	J	L	M
10,001–35,000	C	D	F	H	K	M	N
35,001–150,000	D	E	G	J	L	N	P
150,001–500,000	D	E	G	J	M	P	Q
500,001 and above	D	E	H	K	N	Q	R

Table 24.4 Sample-Size Code Letters

the cost of inspection against the amount of protection required. The three levels involve inspection in amounts roughly in the ratio 0.4:1.0:1.6. (Four additional inspection levels are provided for situations requiring "small-sample inspection.")

A plan is chosen from the tables as follows:

1. The following information must be known:
   a. AQL
   b. Lot size
   c. Type of sampling (single, double, or multiple)
   d. Inspection level (usually level II)
2. Knowing the lot size and inspection level, a code letter is obtained from Table 24.4.
3. Knowing the code letter, AQL, and type of sampling, the sampling plan is read from Table 24.5. (Table 24.5 is for single sampling; the standard also provides tables for double and multiple sampling.)

For example, suppose that a purchasing agency has contracted for a 1.5 percent AQL. Sup- pose also that the parts are bought in lots of 1500 pieces. The table of sample-size code letters (Table 24.4) shows that K plans are required for inspection level II. Table 24.5 states that the sample size is 125. For AQL = 1.5, the acceptance number is given as five and the rejection number as six. Therefore, the entire lot of 1500 articles may be accepted if five or fewer nonconforming articles are found but must not be accepted (rejected) if six or more are found.

Sampling risks are defined by the OC curve published in the standard. The curve for this plan is shown as curve A in Fig. 24.5.

The standard provides single, double, and multiple plans for each code letter (i.e., lot-size category). The plans for code letter K are shown in Table 24.6. Thus, the three plans can be found under the AQL column of 1.5. For example, the double-sampling plan calls for a

Sample Size Code Letter	Sample Size	Acceptable Quality Levels (Normal Inspection)																							
		0.010		0.015		0.025		0.040		0.065		0.10		0.15		0.25		0.40		0.65		1.0		1.5	
		Ac	Re	Ac	Re	Ac	Re	Ac	Re	Ac	Re	Ac	Re	Ac	Re	Ac	Re	Ac	Re	Ac	Re	Ac	Re	Ac	Re
A	2																								
B	3																								
C	5																							↓	
D	8																					↓		0	1
E	13																			↓		0	1	↑	
F	20																	↓		0	1	↑		↓	
G	32															↓		0	1	↑		↓		1	2
H	50													↓		0	1	↑		↓		1	2	2	3
J	80											↓		0	1	↑		↓		1	2	2	3	3	4
K	125									↓		0	1	↑		↓		1	2	2	3	3	4	5	6
L	200							↓		0	1	↑		↓		1	2	2	3	3	4	5	6	7	8
M	315					↓		0	1	↑		↓		1	2	2	3	3	4	5	6	7	8	10	11
N	500			↓		0	1	↑		↓		1	2	2	3	3	4	5	6	7	8	10	11	14	15
P	800	↓		0	1	↑		↓		1	2	2	3	3	4	5	6	7	8	10	11	14	15	21	22
Q	1250	0	1	↑		↓		1	2	2	3	3	4	5	6	7	8	10	11	14	15	21	22	↑	
R	2000	↑				1	2	2	3	3	4	5	6	7	8	10	11	14	15	21	22	↑			

TABLE 24.5 Master Table for Normal Inspection (Single Sampling)

Acceptable Quality Levels (Normal Inspection)																											
2.5		4.0		6.5		10		15		25		40		65		100		150		250		400		650		1000	
Ac	Re	Ac	Re	Ac	Re	Ac	Re	Ac	Re	Ac	Re	Ac	Re	Ac	Re	Ac	Re	Ac	Re	Ac	Re	Ac	Re	Ac	Re	Ac	Re
		↓		0	1	↓		↓		1	2	2	3	3	4	5	6	7	8	10	11	14	15	21	22	30	31
↓		0	1	↑		↓		1	2	2	3	3	4	5	6	7	8	10	11	14	15	21	22	30	31	44	45
0	1	↑		↓		1	2	2	3	3	4	5	6	7	8	10	11	14	15	21	22	30	31	44	45	↑	
↑		↓		1	2	2	3	3	4	5	6	7	8	10	11	14	15	21	22	30	31	44	45	↑			
↓		1	2	2	3	3	4	5	6	7	8	10	11	14	15	21	22	30	31	44	45	↑					
1	2	2	3	3	4	5	6	7	8	10	11	14	15	21	22	↑		↑		↑							
2	3	3	4	5	6	7	8	10	11	14	15	21	22	↑													
3	4	5	6	7	8	10	11	14	15	21	22	↑															
5	6	7	8	10	11	14	15	21	22	↑																	
7	8	10	11	14	15	21	22	↑																			
10	11	14	15	21	22	↑																					
14	15	21	22	↑																							
21	22	↑																									
↑																											

Notes: ↓, use first sampling plan below arrow. If sample size equals, or exceeds, lot of batch size, do 100 percent inspection.
↑, use first sampling plan above arrow. Ac, acceptance number.
Re, rejection number.

Table 24.5 Master Table for Normal Inspection (Single Sampling) (*Continued*)

Type of Sampling Plan	Cumulative Sampling Size	Acceptable Quality Levels (Normal Inspection)											
		Less than 0.10		0.10		0.15		× …		1.0		1.5 …	
		Ac	Re	Ac	Re	Ac	Re	Ac	Re	Ac	Re	Ac	Re
Single	125	∇		0	1	0				3	4	5	6
Double	80	∇		*		Use letter J		Use letter M		1	4	2	5
	160									4	5	6	7
Multiple	32	∇		*						#	3	#	4
	64									0	3	1	5
	96									1	4	2	6
	128									2	5	3	7
	160									3	6	5	8
	192									4	6	7	9
	224									6	7	9	10
		Less than 0.15		0.15		×		0.25		1.5		2.5	

X		4.0		×		… ×		1.0		Higher than 10		Cumulative Sample Size
Ac	Re	Ac	Re	Ac	Re	Ac	Re	Ac	Re	Ac	Re	
8	9	10	11	12	13	18	19	21	22	Δ	10	125
3	7	5	9	6	10	9	14	11	16	Δ		80
11	12	12	13	15	16	23	24	26	27			160
0	4	0	5	0	6	1	8	2	9	Δ		32
2	7	3	8	3	9	6	12	7	14			64
4	9	6	10	7	12	11	17	13	19			96
6	11	8	13	10	15	16	22	19	25			128
9	12	11	15	14	17	22	25	25	29			160
12	14	14	17	18	20	27	29	31	33			192
14	15	18	19	21	22	32	33	37	38			224
4.0		×		6.5		10		×		Higher than 10		

Notes: Δ, use next preceding sample-size code letter for which acceptance and rejection numbers are available.
∇, use next subsequent sample-size code letter for which acceptance and rejection numbers are available.
Ac, acceptance number.
Re, rejection number.
*, use single-sampling plan above (or alternatively use letter N).
#, acceptance not permitted at this sample size.

Table 24.6 Sampling Plan for Sample-Size Code Letter K

first sample of 80 units. If two or fewer nonconforming are found, the lot is accepted. If five or more nonconforming are found, the lot is not accepted. If three or four nonconforming are found in the sample of 80, then a second sample of 80 is taken, giving a cumulative sample size of 160. If the total number of nonconforming in both samples is six or less, the lot is accepted; seven or more nonconforming means lot rejection.

Switching Procedures in ANSI/ASQC Z1.4

ANSI/ASQC Z1.4 includes provision for tightened inspection if quality deteriorates. If two out of five consecutive lots are not acceptable (rejected) on original inspection, a tightened inspection plan is imposed. The sample size is the same as usual, but the acceptance number is reduced. (The tightened plans do require larger sample sizes if the probability of acceptance for an AQL lot is less than 0.75.) For the example, previously cited, the tightened plan can be read from Table 24.6 as a sample size of 125 and an acceptance number of three.

ANSI/ASQC Z1.4 also provides for reduced inspection when the supplier's record has been good. The preceding 10 lots must have had a normal inspection with all lots accepted. A table of lower limits for the process average is provided to help decide whether the supplier's record has been good enough to switch to reduced inspection. The plan does, however, provide an option of switching to reduced inspection without using the table of lower limits. Under reduced sampling, the sample size is usually 40 percent of the normal sample size.

These switching rules apply when production is submitted at a steady rate. The plan provides other rules for using normal, tightened, and reduced inspection.

Other Provisions of ANSI/ASQC Z1.4

The standard provides OC curves for most of the individual plans along with "limiting quality" values for a probability of acceptance of 10 and 5 percent. Average sample-size curves for double and multiple sampling are also included. The latter curves show the average sample sizes expected as a function of the product quality submitted. Although the OC curves are roughly the same for single, double, and multiple sampling, the average sample-size curves vary considerably because of the inherent differences among the three types of sampling. The standard also states the AOQL that would result if all rejected lots were screened for nonconforming units.

In ANSI/ASQC Z1.4, a sampling scheme is defined as "a combination of sampling plans with switching rules and possibly a provision for discontinuance of inspection." For the sampling schemes associated with the individual plans, the standard provides OC curves and information on AOQL, limiting quality, and average sample sizes—all for single sampling.

Dodge–Romig Sampling Tables

Dodge and Romig (1959) provide four sets of attributes plans emphasizing either lot-by-lot quality (LTPD) or long-run quality (AOQL):

- Lot tolerance percentage defective (LTPD): single sampling, ouble sampling
- Average outgoing quality limit (AOQL): single sampling, double sampling

These plans differ from those in ANSI/ASQC Z1.4 in that the Dodge–Romig plans assume that all rejected lots are 100 percent inspected and the defectives are replaced with acceptable items. Plans with this feature are called rectifying inspection plans. The tables provide protection against poor quality on either a lot-by-lot basis or average long-run quality. The LTPD plans assure that a lot of poor quality will have a low probability of acceptance, that is, the probability of acceptance (or consumer's risk) is 0.1 for a lot with LTPD quality. LTPD values range from 0.5 to 10.0 percent defective. The AOQL plans assure that, after all sampling and 100 percent inspection of rejected lots, the average quality over many lots will not exceed the AOQL. The AOQL values range from 0.1 to 10.0 percent. Each LTPD plan lists the corresponding AOQL, and each AOQL plan lists the LTPD.

Selection of a Numerical Value of the Quality Index

The problem of selecting a value of the quality index (e.g., AQL, AOQL, or lot tolerance percentage defective) is one of balancing the cost of finding and correcting a defective against the loss incurred if a defective slips through an inspection procedure.

Enell (1954), in a classic paper, has suggested using the breakeven point (see section "How Much Inspection Is Necessary") in the selection of an AQL. The breakeven point for inspection is defined as the cost to inspect one piece divided by the damage done by one defective. For the example cited, the breakeven point was 5 percent defective.

Because a 5 percent defective quality level is the breakeven point between sorting and sampling, the appropriate sampling plan should provide for a lot to have a 50 percent probability of being sorted or sampled, that is, the probability of acceptance for the plan should be 0.50 at a 5 percent defective quality level. The OC curves in a set of sampling tables such as ANSI/ASQC Z1.4 can now be examined to determine an AQL. For example, suppose that the device is inspected in lots of 3000 pieces. The OC curves for this case (code letter K) are shown in ANSI/ASQC Z1.4 and Fig. 24.5. The plan closest to having a Pa of 0.50 for a 5 percent level is the plan for an AQL of 1.5 percent. Therefore, this is the plan to adopt.

Some plans include a classification of defects to help determine the numerical value of the AQL. Defects are first classified as critical, major, or minor according to definitions provided in the standard. Different AQLs may be designated for groups of defects considered collectively or for individual defects. Critical defects may have a 0 percent AQL, whereas major defects may be assigned a low AQL, say 1 percent, and minor defects a higher AQL, say 4 percent. Some manufacturers of complex products specify quality in terms of the number of defects per million parts.

In practice, quantification of the quality index is a matter of judgment based on the following factors: past performance in quality, effect of nonconforming product on later production steps, effect of nonconforming product on fitness for use, urgency of delivery requirements, and cost of achieving the specified quality level.

A thorough discussion of this difficult issue is provided by Schilling (1982, pp. 571–586) in the classic, comprehensive book on acceptance sampling.

How to Select the Proper Sampling Procedures

The methods of acceptance sampling are many and varied. It is essential to select a sampling procedure appropriate to the acceptance sampling situation to which it is applied. Sampling procedures can serve different purposes. As itemized by Schilling (1982), they include:

- Guaranteeing quality levels at stated risks
- Maintaining quality at AQL level or better
- Guaranteeing an AOQL
- Reducing inspection after good history
- Checking inspection
- Ensuring compliance with mandatory standards
- Reliability sampling
- Checking inspection accuracy

For each purpose, Schilling recommends specific attributes or variables sampling plans. Selection of a plan depends on the purpose, the quality history, and the extent of knowledge of the process.

The steps involved in the selection and application of a sampling procedure are shown in Fig. 24.8. Emphasis is on the feedback of information necessary for the proper application,

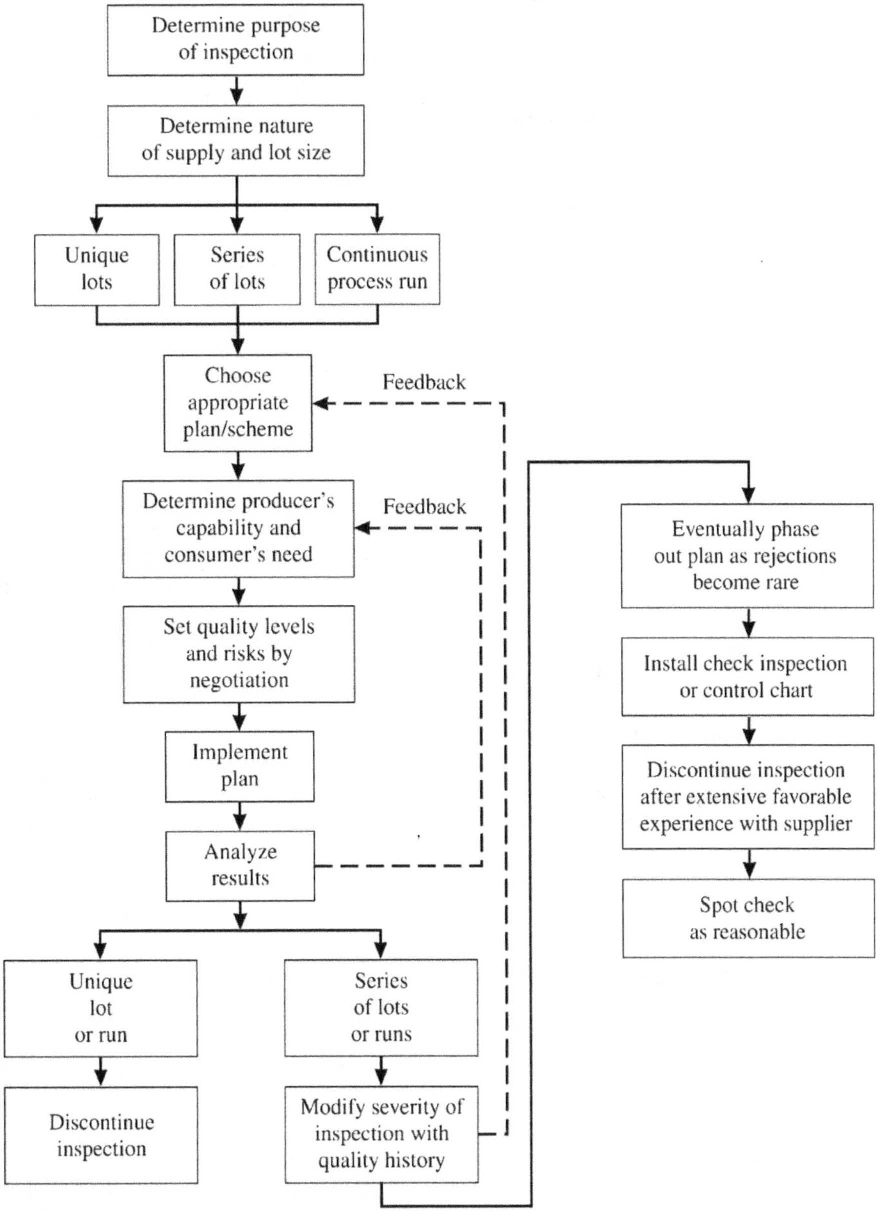

FIGURE 24.8 Check sequence for implementation of sampling procedure. From Acceptance Sampling in Quality Control by Edward G. Schilling. Copyright 1982 by Taylor & Francis Group LLC. (Reproduced with permission of Taylor & Francis Group LLC in the format Textbook via Copyright Clearance Center.)

modification, and evolution of sampling to encourage continuous improvement and reduced inspection costs. This can be achieved by moving from a system of acceptance sampling to acceptance control.

Moving from Acceptance Sampling to Acceptance Control

Acceptance sampling is the process of evaluating a portion of the product in a lot for the purpose of accepting or rejecting the entire lot as either conforming or not conforming to a quality specification. Acceptance control is a "continuing strategy of selection, application, and modification of acceptance sampling procedures to a changing inspection environment" (Schilling, 1982, p. 564). This evaluation of a sampling plan application is shown in the life cycle of acceptance control (Table 24.7). The cycle is applied over the lifetime of a product to achieve (1) quality improvement (using process control and process capability concepts) and (2) reduction and elimination of inspection (using acceptance sampling).

Life cycle of acceptance control application. From acceptance sampling in quality control by Edward G. Schilling. (Copyright 1982 by Taylor & Francis Group LLC. Reproduced with permission of Taylor & Francis Group LLC in the format Textbook via Copyright Clearance Center.)

Stage	Step	Method
Preparatory	Choose plan appropriate to purpose Determine producer capability Determine consumer needs Set quality levels and risks Determine plans	Analysis of quality system to define the exact need for the procedure Process performance evaluation using control charts Process capability study using control charts Economic analysis and negotiation Standard procedures if possible
Initiation	Train inspector Apply plan properly Analyze results	Include plan, procedure, records, and action Ensure random sampling Keep record and control charts
Operational	Assess protection Adjust plan Decrease sample size if warranted	Periodically check quality history and OC curves When possible, change severity to reflect quality history and cost Modify to use appropriate sampling plans taking advantage of credibility of supplier with cumulative results
Phase out	Eliminate inspection effort where possible	Use demerit rating or check inspection procedures when quality is consistently good Keep control charts
Elimination	Spot check only	Remove all inspection when warranted by extensive favorable history

TABLE 24.7 Life Cycle of Acceptance Control Application

References

Adams, R. (1987). "Moving from Inspection to Audit," *Quality Progress*, January, pp. 30–31.

ANSI/ASQC Z1.4 (1993). Sampling Procedures and Tables for Inspection by Attributes, American Society for Quality Control, Milwaukee.

Baxter Travenol Laboratories (1986). Statistical Process Control Guideline, Baxter Travenol Laboratories, Deerfield, IL, p. 23.

Chun, Y. H. and D. B. Rinks (1998). "Three Types of Producer's and Consumer's Risks in the Single Sampling Plan," *Journal of Quality Technology*, vol. 30, no. 3, pp. 254–268.

Cooper, J. (1997). "Implementing a Paperless Inspection System," *Annual Quality Congress Proceedings*, ASQ, Milwaukee, pp. 231–235.

Dodge, H. F. and H. G. Romig (1959). *Sampling Inspection Tables*, 2nd ed., John Wiley & Sons, New York.

Enell, J. W. (1954). "What Sampling Plan Shall I Choose?" *Industrial Quality Control*, vol. 10, no. 6, pp. 96–100.

Grant, E. L. and R. S. Leavenworth (1996). *Statistical Quality Control*, 7th ed., McGraw-Hill, New York.

Harris, D. H. and F. B. Chaney (1969). *Human Factors in Quality Assurance*, John Wiley and Sons, New York, pp. 77–85.

Orkin, F. I. and D. P. Olivier (1999). In JQH5, Table 10.2.

Peach, R. W., ed. (1997). *The ISO 9000 Handbook*, 3d ed., McGraw-Hill, New York.

Riley, F. D. (1979). "Visual Inspection—Time and Distance Method," ASQC Annual Technical Conference Transactions, Milwaukee, p. 483.

Schilling, E. G. (1982). Acceptance Sampling in Quality Control, Marcel Dekker, New York.

Schilling, E. G. (1994). "The Importance of Sampling in Inspection," *Annual Quality Congress Proceedings*, ASQ, Milwaukee, pp. 809–812.

Taylor, W. A. (1994). "Acceptance Sampling in the 90s," *Annual Quality Congress Proceedings*, ASQ, Milwaukee, pp. 591–598.

CHAPTER 25

Organization-Wide Assessment of Quality

Joseph A. De Feo and Er Ralston

High Points of This Chapter

1. All organizations should conduct an annual company-wide assessment of business excellence and quality.
2. Common assessments comprise four elements: cost of poorly performing processes or COPQ^3, standing in the marketplace, employee culture, and overall health of the operating and quality systems.
3. The Malcolm Baldrige National Award for Excellence and similar national awards provide an additional means of assessing an organization. These awards recognize superior achievement.
4. The ISO 9000 standards provide minimum criteria for a quality system. These documents provide some assurance to potential customers that an organization certified as meeting the standard has an adequate quality system.

Why Assess Performance?

An early step to transform with a strategy of quality is to understand the baseline of your culture, system, processes, people, and costs. A key role of the quality office is to conduct these assessments to educate your leaders so they understand what needs to be changed or improved. System assessments provide a comprehensive, cost-effective review that provides an objective baseline to assess the performance of organizations. There are many organizational assessments that can be used. We will examine six areas are important to being competitive in quality:

1. Assessing quality risk
2. Estimating the costs of poorly performing processes and cost of poor quality
3. Assessing performance and standing in the marketplace
4. Assessing using national performance standards and awards
5. Assessing to the international system standards
6. Competitive benchmarking best practices

Quality Risk Assessment

From 1970 to 1986, the Juran Institute was one of the first to document and conduct comprehensive assessments for quality. These assessments were based on Dr. Juran's early writings and the need for improved competitiveness in many industries in the United States. The Institute developed a set of guidelines that, if met, were and still are critical for business success and leadership in quality.

The Juran comprehensive assessment of quality existed prior to the ISO 9000 standards and the national quality awards that came later in the 1980s. Today, the Juran Quality Risk Assessment has been used by many organizations to better under- stand and create a plan to improve performance. For some organizations, the ISO Standards are used; others use regulatory requirements such as FDA standards or Sarbanes-Oxley; and still others use the U.S. Malcolm Baldrige Criteria for Performance Excellence, the EFQM Excellence Model, or their own country equivalent.

Quantifying the costs related to poorly performing quality (COP^3; DeFeo 2003) is important to understand the total loss due to poor quality. Estimating the COP^3 is one of the best means to get management's attention. As you will see, the COP^3 for most organizations can exceed the value of their profits!

Assessing market standing and the acceptance of your quality from your customers is important. Knowing what the paying customers like and dislike can be beneficial in driving the organization to a new level of performance.

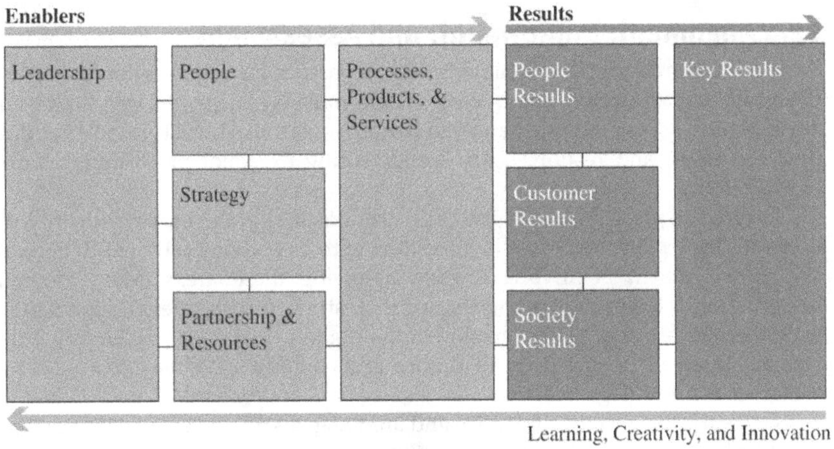

FIGURE 25.1 EFQM excellence model.

Each of these assessments brings with it a business model for integrating quality into the organization. One example is the EFQM model shown in Fig. 25.1.

The Baldrige Model was named after a former Secretary of Commerce, Malcolm Baldrige. This model incorporates best practices from many companies since its inception in 1980s. It is used to measure and evaluate the total business system against a set of competencies and best practices (Fig. 25.2).

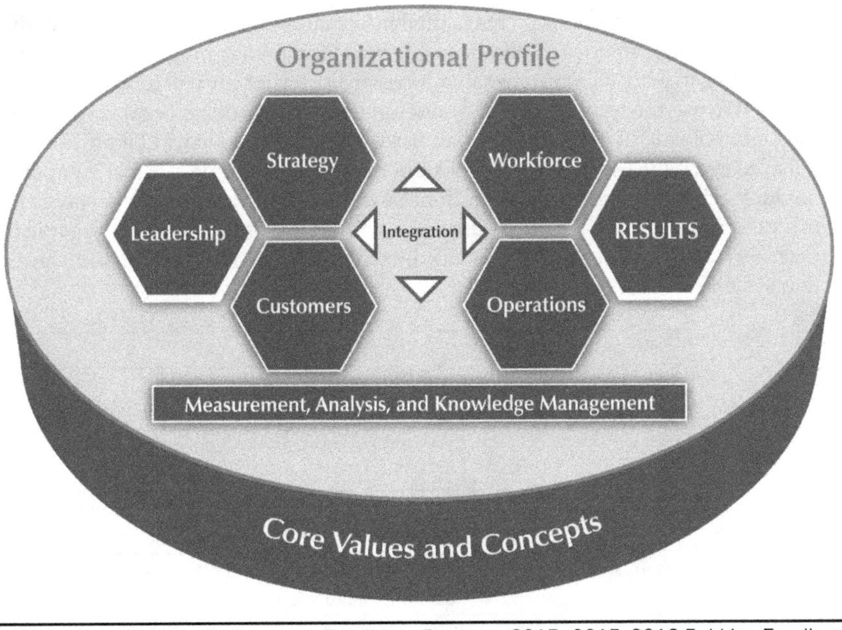

FIGURE 25.2 From Baldrige Performance Excellence Program. 2015. 2015–2016 Baldrige Excellence Framework: A Systems Approach to Improving Your Organization's Performance. Gaithersburg, MD: U.S. Department of Commerce, National Institute of Standards and Technology. http://www.nist.gov/baldrige.

Strategic Alignment, Deployment, and Assessment

We use the term business system to assess all parts of the organization rather than the quality system infrastructure. The quality system must be integrated into the business system. An effective assessment must analyze the organizational structure, roles and responsibilities, processes, and resources for implementing excellent performance throughout the organization.

In order to properly assess any organization, start at the top and identify the organization's strategic plan and how it is associated with day-to-day activities. It is crucial that top-level management be involved when assessing these areas. Many organizations are negatively affected by not having aligned their strategic objectives and goals to specific daily activities and projects. The leadership of the organization needs to identify if their mission, vision, values, and policies are all aligned and communicated effectively, keeping the end-customer in mind (Fig. 25.3).

From these objectives, strategies and annual goals are developed. Strategies are a means of achieving the vision of the company. Strategies are typically long term in nature and associated with key success factors such as increase revenue, reduce costs, or develop new products. In order to achieve these long-term goals, the leadership teams create annual goals that are short term in nature. As reassurance, the leaders develop policies that help manage the goals on day-to-day activities.

Figure 25.4 shows how the vision breaks down into key strategies, strategic goals, annual goals, and initiatives and projects. Typically, there many different levels of management from the executive group; when deploying the strategic plan, a lot of information and direction gets lost in translation. This leads to the strategic plan not being aligned with individual projects, which could lead to projects of low importance and very important resources being wasted. The organizational assessment will identify those areas that are disjointed to realign them.

The assessment approach described in this chapter provides a few options that can be used to stimulate upper management to take action on quality (that's why assessment is presented early in this book). This factual information not only inspires immediate action on quality, but also provides a solid basis for developing long-range strategies for quality.

We begin with a comprehensive assessment based on the Juran Organization Health Check. We use the term quality risk assessment to describe an organization-wide review of the status of quality. The quality risk assessment has proven to be efficient and effective. Other assessments, such as the ISO 9000 standards, are at times seen as a paper chase and do not look at business results. The National Award Criteria are comprehensive and do include business results, but their main use is to win the award. Therefore, organizations should choose what is most important when developing their own system and the assessment of that system.

Selected definitions	
Mission	What business we are in
Vision	Desired future state of organization
Values	Principles to be observed to meet vision or Principle to be served by meeting vision
Policy	How we will operate and our commitment to customers and society

Figure 25.3 Organizational vision and mission. (Juran Institute Inc., Southbury, CT.)

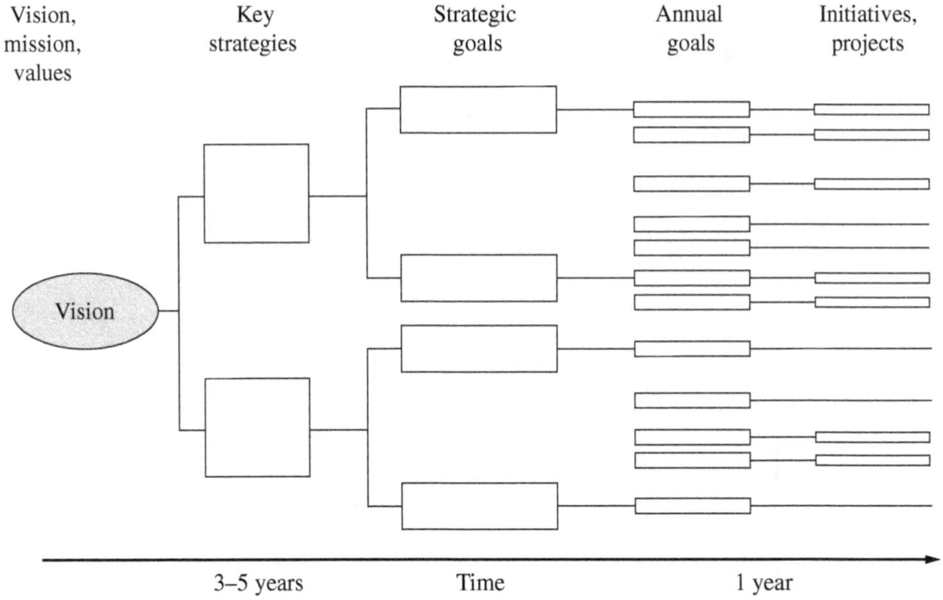

FIGURE 25.4 Deploying the vision. (Juran Institute Inc., Southbury, CT.)

An effective quality risk assessment should include three areas or three criteria for success:

1. *Approach.* What are the organizational approach, intent, and design, used within the organization?
2. *Deployment.* How broadly has the approach been deployed or executed in the organization? In other words, how many employees are knowledgeable and have been trained, understand, and can implement the approach?
3. *Results.* What are the measurable outcomes that demonstrate the approach and deployment are valid?

An effective assessment is conducted by observation, data collection, and interviews of key functions within the organization. Specific strengths and weaknesses are documented at the detailed dimension level. Alignment and themes are identified across each category and are then scored. The Juran Organization Health Check categories and points are shown in Table 25.1.

The results of the quality risk assessment must provide an organization with a review of the organization's performance, in terms of strengths, weaknesses, opportunities, and threats (SWOT) and become a formal part of the SWOT analysis for overall operations of an organization. Other elements may be added as circumstances require. An annual or biannual assessment is usually warranted. This assessment can be conducted for an entire organization or for a division, a plant, a department, or a process.

We begin to discuss how crucial an assessment is to ensure that the strategic plan is aligned and properly deployed. Later we discuss the different assessments that are used to drill down deeper into specific categories such as cost of poor quality, organization culture, and market standing.

Categories and Items (Point Values)
1. Strategic alignment and deployment (100)
1.1 Strategic planning (30)
1.2 Strategic deployment (45)
1.3 Stakeholder (executive point of view) (15)
1.4 Key support systems (10)
2. Quality management system (100)
2.1 Quality system (40)
2.2 Quality control (10)
2.3 Continuous improvement (50)
3. Measurement and analysis capabilities (75)
3.1 Measurement systems (25)
3.2 Customer relationship and requirements (25)
3.3 Competitors (5)
3.4 Product and service quality (5)
3.5 Support processes (5)
4. Effectiveness and efficiency of business processes (100)
4.1 Core business processes (30)
4.2 Voice of the customer (20)
4.3 Product and service creation (20)
4.4 Service and order fulfillment (10)
4.5 Key support processes (10)
4.6 Process efficiency and cost (10)
5. Employee engagement and culture (100)
5.1 Human resource structure (30)
5.2 Talent development (40)
5.3 Culture (30)
6. Supplier management (75)
6.1 Supply chain policy (25)
6.2 Supplier procedures (50)
7. Scorecard and results (450)
7.1 Customer (80)
7.2 Product and service (80)
7.3 Culture (80)
7.4 Supplier performance (80)
7.5 Financial results (130)
Total points = 1,000

TABLE 25.1 Juran's Organization Health Check Scoring Guide

Plan the Assessment

The assessment is an evaluation of current element activities in the organization. Such an evaluation could cover a wide range of both scope (global to a few activities) and examination (cursory to detailed). Our discussion concentrates on an organization-wide examination in sufficient depth to meet the needs of the overall quality assessment.

Often, an organization starts by having an outsider conduct a three- to five-day assessment of the current baseline of performance. The assessment is structured by initiating the assessment, evaluating the organization's elements, and preparing a detailed report. Proper planning will ensure that the assessment is efficient and completed quickly to allow leadership to make timely decisions. Planning requires that the organization:

1. Define the team.
2. Define objectives.
3. Define scope and criteria.
4. Establish a communication plan between the team and leadership.
5. Complete the pre-assessment (generalized version of OHC).

Once the planning is complete, the assessment can begin. To collect the right data, there may be one-on-one discussions with selected employees typically from all levels of the organization within the scope of the assessment. These employees will validate and demonstrate the existence and effectiveness of the total business system and subsequent critical business processes.

Examples of items targeted during task observations, focus groups, ad hoc interviews, and reviews of operational results are as follows:

- Is your organization's quality management system responsible for creating a focus on the customer and a quality foundation?
- Is the quality management system aligned to the organization's strategies, goals, and actions for creating a quality culture?
- Are improvement methodologies used to effectively identify projects and diagnose root causes of problems within the organization?
- Are there measurement systems in place to improve the overall performance of your organization?
- Does your organization have an effective process for measuring customer satisfaction and loyalty?
- Does your organization perform cost of poor quality assessments? Has its standing in the marketplace and the quality culture been assessed?
- Are your key business processes and support processes identified, evaluated, measured, and continuously improved?
- Does your organization incorporate customer requirements into the key business processes?
- Are the organization's workforce policies known and implemented?
- Is there an effective evaluation process in place to measure employees' capabilities?
- Does the workforce have control over the work processes?
- Are your suppliers certified and continuously verified to maintain performance?

- Are the trends in key measures for your product and service quality performance trending positively over the past 5 years?
- Is the financial impact of the results communicated, and does the impact trigger improvements where necessary?
- Are your results from improvement activities sustained over time?

As the assessment is completed, the observations, data collected, and information received from interviews are documented and measured against the quality risk assessment scoring criteria described in Table 25.1 and the level of approach, deployment, and results.

For example, if an organization scores 20 percent of the total points, there may be no systematic approach and an undocumented deployment, and there may be not data reported or poor results are reported. Additionally, if the organization scores 95 percent of the total points, the organization will probably have a strong, integrated prevention and fact-based system in place in all areas. There are continuous evaluation and improvement cycles with breakthrough refinements. The approach is fully deployed with no gaps, and the results are being sustained. Juran considers performance at this level "best in class" or "world-class quality." Typically, no organization can ever score a 100 percent because it will always be continuously improving the seven elements.

Cost of Poor Quality

During the 1950s, the concept of "quality costs" emerged. Different people assigned different meanings to the term. Some people equated quality costs with the costs of attaining quality; some people equated the term with the extra costs incurred because of poor quality. This book focuses on the cost of poor quality because this component of assessment is important in reducing costs and customer dissatisfaction.

The cost of poor quality is the annual monetary loss of products and processes that are not achieving their quality objectives. In De Feo and Barnard (2003), the cost of poor quality (COPQ) is appropriately renamed the cost of poorly performing processes (or COP^3, pronounced, C.O.P. cubed). This is to emphasize the fact that the cost of poor quality is not limited to quality, but is essentially the cost of waste associated with poor performance of processes.

Companies estimate the cost of poor quality for several reasons:

1. Quantifying the size of the quality problem in the language of money improves communication between middle managers and upper managers. In some companies, the need to improve communication on quality-related matters has become a major objective for embarking on a study of the costs of poor quality. Some managers say, "We don't need to spend time to translate the defects into dollars. We realize that quality is important, and we already know what the major problems are." But typically, when a study is done, these managers are surprised by two results. First, the quality costs turn out to be much higher than the managers had thought—in many industries in excess of 20 percent of sales. Second, though the distribution of the quality costs confirms some of the known problem areas, it also reveals other problem areas that were not previously recognized.
2. Major opportunities for cost reduction can be identified. Costs of poor quality do not exist as a homogeneous mass. Instead, they are the total of specific segments, each traceable to some specific cause. These segments are unequal in size, and relatively few of the segments account for the bulk of the costs. A major by-product of this evaluation is the identification of these vital few segments.

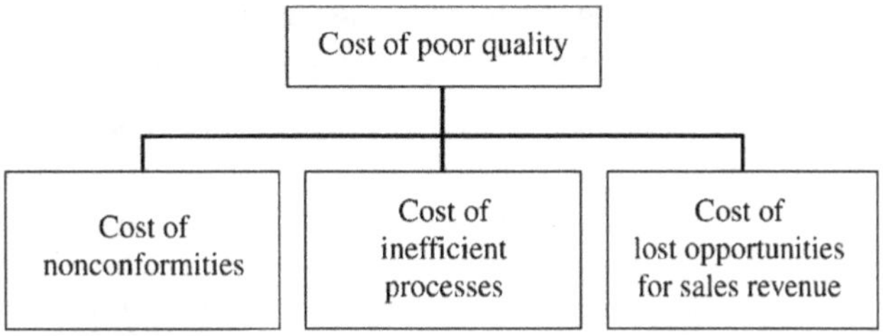

Figure 25.5 Cost of poor quality.

3. Opportunities for reducing customer dissatisfaction and associated threats to product salability can be identified. Some costs of poor quality are the result of product failures after a sale. In part, these costs are paid by the manufacturer in the form of warranty charges, claims, etc. But whether or not the costs are paid by the manufacturer, the failures add to customers' costs because of downtime and other forms of disturbance. Analysis of the manufacturer's costs, supplemented by market research into customers' costs of poor quality, can identify the vital few areas of high costs. These areas then lead to problem identification.
4. Measuring this cost provides a means of evaluating the progress of quality improvement activities and spotlighting obstacles to improvements.
5. Knowing the cost of poor quality (and the three other assessment elements) leads to the development of a strategic quality plan that is consistent with overall organization of goals.

The main components of the cost of poor quality (which apply to both manufacturing and service organizations) are shown in Fig. 25.5. Note that this framework reflects not only the cost of nonconformities (sometimes called "quality costs") but also process inefficiencies and the impact of quality on sales revenue. Each organization must decide which cost elements to include in its cost of poor quality.

Categories of Quality Costs

Many organizations summarize the costs associated with quality in four categories: internal failures, external failures, appraisal, and prevention. These categories are discussed later. A useful reference on definitions, categories, and many other aspects is Campanella (1999). For an exhaustive listing of elements within the four categories, see Atkinson et al. (1994).

Collectively, the four categories are often called the "cost of quality." The cost of poor quality includes the internal and external failure categories, whereas the appraisal and prevention categories are viewed as investments to achieve quality objectives.

Internal Failure Costs

Internal failure costs are the cost of deficiencies discovered before delivery that are associated with the failure to meet explicit requirements or implicit needs of customers. Also included are avoidable process losses and inefficiencies that occur even when requirements and needs are met. These costs would disappear if no deficiencies existed. Internal failure costs consist of (1) the cost of failure to meet customer requirements and needs and (2) the cost of inefficient processes.

Failure to Meet Customer Requirements and Needs

Examples of subcategories are costs associated with the following:

Scrap. The labor, material, and (usually) overhead on defective product that cannot be repaired economically. The titles are numerous—scrap, spoilage, defectives, etc.

Rework. Correcting defects in physical products or errors in service products. Lost or missing information. Retrieving information that should have been supplied. Failure analysis. Analyzing nonconforming goods or services to determine causes. Scrap and rework—supplier. Scrap and rework because of nonconforming product received from suppliers. This area also includes the costs to the buyer of resolving supplier quality problems.

One hundred percent sorting inspection. Finding defective units in product lots that contain unacceptably high levels of defectives.

Reinspection, retest. Reinspection and retest of products that have undergone rework or other revision.

Changing processes. Modifying manufacturing or service processes to correct deficiencies.

Redesign of hardware. Changing designs of hardware to correct deficiencies.

Redesign of software. Changing designs of software to correct deficiencies.

Scrapping of obsolete product. Disposing of products that have been superseded.

Scrap in support operations. Defective items in indirect operations.

Rework in internal support operations. Correcting defective items in indirect operations.

Downgrading. The difference between the normal selling price and the reduced price because of poor quality.

Cost of Inefficient Processes

Examples of subcategories include the following:

Variability of product characteristics. Losses that occur even with conforming product (e.g., overfill of packages due to variability of filling and measuring equipment).

Unplanned downtime of equipment. Loss of capacity of equipment due to failures.

Inventory shrinkage. Loss due to the difference between actual and recorded inventory amounts.

Variation of process characteristics from "best practice." Losses due to cycle time and costs of process compared to best practices in providing the same output. The best-practice process may be internal or external to the organization.

Non-value-added activities. Redundant operations, sorting inspections, and other non-value-added activities. A value-added activity increases the usefulness of a product to the customer; a non-value-added activity does not. The concept is similar to the 1950s idea of value engineering and value analysis.

External Failure Costs

External failure costs are associated with deficiencies that are found after the customer receives the product. Also included are lost opportunities for sales revenue. These costs also disappear if there were no deficiencies.

Failure to Meet Customer Requirements and Needs

Examples of subcategories include the following:

Warranty charges. The costs involved in replacing or making repairs to products that are still within the warranty period.

Complaint adjustments. The costs of investigation and adjustment of justified complaints attributable to defective product or installation.

Returned material. The costs associated with receipt and replacement of defective product received from the field.

Allowances. The costs of concessions made to customers due to substandard products accepted by the customer as is or to conforming product that does not meet customer needs.

Penalties due to poor quality. This category applies to goods or services delivered or to internal processes such as late payment of an invoice resulting in a lost discount for paying on time.

Rework on support operations. Correcting errors on billing and other external processes.

Revenue losses in support operations. An example is the failure to collect receivables from some customers.

Lost Opportunities for Sales Revenue

Examples include the following:

Customer defections. Profit on potential customers lost because of poor quality.

New customers lost because of lack of capability to meet customer needs. Profit on potential revenue lost because of inadequate processes to meet customer needs.

Appraisal Costs

Appraisal costs are incurred to determine the degree of conformance to quality requirements. Examples include the following:

Incoming inspection and test. Determining the quality of purchased product, whether by inspection on receipt, by inspection at the source, or by surveillance.

In-process inspection and test. In-process evaluation of conformance to requirements.

Final inspection and test. Evaluation of conformance to requirements for product acceptance.

Document review. Examination of paperwork to be sent to customer.

Balancing. Examination of various accounts to assure internal consistency.

Product quality audits. Performing quality audits on in-process or finished products.

Maintaining accuracy of test equipment. Keeping measuring instruments and equipment calibrated.

Inspection and test materials and services. Materials and supplies in inspection and test work (e.g., x-ray film) and services (e.g., electric power) where significant.

Evaluation of stocks. Testing products in field storage or in stock to evaluate degradation.

In collecting appraisal costs, the decisive factor is the kind of work done and not the department name (the work may be done by chemists in the laboratory, by sorters in the operations department, by testers in the inspection department, or by an external firm engaged for the purpose of testing). Also note that industries use various terms for "appraisal," for example, checking, balancing, reconciliation, review.

Prevention Costs

Prevention costs are incurred to keep failure and appraisal costs to a minimum. Examples include the following:

Quality planning. This category includes the broad array of activities that collectively create the overall quality plan and the numerous specialized plans.

It includes also the preparation of procedures needed to communicate these plans to all concerned.

New-products review. Reliability engineering and other quality-related activities associated with the launching of a new design.

Process planning. Process capability studies, inspection planning, and other activities associated with the manufacturing and service processes.

Process control. In-process inspection and test to determine the status of the process (rather than for product acceptance).

Quality audits. Evaluating the execution of activities in the overall quality plan.

Supplier quality evaluation. Evaluating supplier quality activities prior to supplier selection, auditing the activities during the contract, and performing associated effort with suppliers.

Training. Preparing and conducting quality-related training programs. As in the case of appraisal costs, some of this work may be done by personnel who are not on the payroll of the quality department. The decisive criterion is again the type of work, not the name of the department performing the work.

Note that prevention costs are costs of special planning, review, and analysis activities for quality. Prevention costs do not include basic activities such as product design, process design, process maintenance, and customer service.

The compilation of prevention costs is initially important because it highlights the small investment currently made (usually) in prevention activities and suggests the potential for an increase in prevention costs to reduce failure costs. Upper management immediately grasps this point. Continuing measurement of prevention costs can usually be excluded, however, to focus on the major opportunity, that is, failure costs.

One of the issues in calculating the cost of poor quality is how to handle overhead costs. Three approaches are common: include total overhead using direct labor or some other base, include variable overhead only (the usual approach), or do not include overhead at all. The allocation of overhead has an impact on the total cost of poor quality and on the distribution over various departments. Activity-based costing (ABC) can help to provide a realistic allocation of overhead costs. ABC is an accounting method that allocates overhead based on the activities that cause overhead cost elements to be incurred. Cokins (1999) discusses the impact of traditional accounting and ABC on quality.

An example for one plant of a tire manufacturer is shown in Table 25.2. This example resulted in some conclusions that are typical of these studies.

- The total of almost $900,000 per year is large.
- Most (79.1 percent) of the total is concentrated in failure costs, specifically in waste scrap and consumer adjustments.
- Failure costs are about five times the appraisal costs.
- A small amount (4.3 percent) is spent on prevention.
- Some consequences of poor quality cannot be conveniently quantified, for example, customer ill will and customer policy adjustment. However, these factors are listed as a reminder of their existence.

As a result of this study, management decided to increase the budget for prevention activities. Three engineers were assigned to identify and pursue specific quality improvement projects.

Cost of quality failures—losses		
Defective stock	$3,276	0.37%
Repairs to product	73,229	8.31
Collect scrap	2,288	0.26
Waste—scrap	187,428	21.26
Consumer adjustments	408,200	46.31
Down grading products	22,838	2.59
Customer ill will	Not counted	
Customer policy adjustment	Not counted	
Total	$697,259	79.10%
Cost of appraisal		
Incoming inspection	$32,655	2.68
Inspection 1	32,582	3.70
Inspection 2	25,200	2.86
Spot-check inspection	65,910	7.37
Total	$147,347	16.61%
Cost of prevention		
Local plant quality		
Control engineering	7,848	0.89
Corporate quality		
Control engineering	30,000	3.40
Total	$37,848	4.29%
Grand Total	$882,454	100.00%

TABLE 25.2 Annual Quality Cost—Tire Manufacturer

Strictly defined, the cost of poor quality is the sum of the internal and external failure cost categories. This definition assumes that the elements of appraisal costs—for example, 100 percent sorting inspection—necessitated by inadequate processes are classified under internal failures. Some practitioners use the term cost of quality for the four broad categories.

Although many organizations use the categories of internal failure, external failure, appraisal, and prevention, the structure may not apply in all cases. A different approach employed by a bank is shown in Table 25.3.

Here the cost of poor service is calculated based on customer satisfaction data and customer loyalty/retention data. In this case, the cost of poor quality is the cost of lost opportunities for sales revenue (see Fig. 25.5).

In still another approach, some organizations define the cost of poor quality with a focus on the cost of key activities or processes, that is, the difference between actual costs and the cost in an organization that has the best practice for that activity. This approach addresses the cost of inefficient processes (see Fig. 25.5). For a discussion of the traditional quality cost approach (four categories), the process cost approach, and the "quality loss approach," see Schottmiller (1996).

$10,000,000	Annual customer service revenue
1,000	Number of customers
× 25%	Percent dissatisfied
250	Number of dissatisfied
× 75%	Percent of switchers (60–90% of dissatisfied)
188	Number of switchers
× $10,000	Average revenue per customer
$1,880,000	Revenue lost through poor quality

Source: The University of Tampa.

TABLE 25.3 Revenue Lost Through Poor Quality

Additional examples of cost-of-poor-quality studies in both manufacturing and service industries are provided by Campanella (1999) and Atkinson et al. (1994).

Hidden Quality Costs

The cost of poor quality may be understated because of costs that are difficult to estimate. "Hidden" costs occur in both manufacturing and service industries and include the following:

1. Potential lost sales.
2. Costs of redesign of products due to poor quality.
3. Costs of changing processes due to inability to meet quality requirements for products.
4. Costs of software changes due to quality reasons.
5. Costs of downtime of equipment and systems including computer information systems.
6. Costs included in standards because history shows that a certain level of defects is inevitable and allowances should be included in standards.
   a. *Extra material purchased.* The purchasing buyer orders 6 percent more than the production quantity needed.
   b. *Allowances for scrap and rework during production.* History shows that 3 percent is "normal," and accountants have built this allowance into the cost standards. One accountant said, "Our scrap cost is zero. The production departments are able to stay within the 3% that has been added in the standard cost and therefore the scrap cost is zero." Ah, for the make-believe numbers game.
   c. *Allowances in time standards for scrap and rework.* One manufacturer allows 9.6 percent in the time standard for certain operations to cover scrap and rework.
   d. *Extra process equipment capacity.* One manufacturer plans for 5 percent unscheduled downtime of equipment and provides extra equipment to cover the downtime. In such cases, the alarm signals ring only when the standard value is exceeded. Even when operating within those standards, however, the costs should be a part of the cost of poor quality. They represent opportunities for improvement.

7. Extra indirect costs due to defects and errors. Examples are space charges and inventory charges.
8. Scrap and errors not reported. Scrap may never be reported because employees fear reprisals, or scrap may be charged to a general ledger account without being identified as scrap.
9. Extra process costs due to excessive product variability (even though within specification limits). For example, a process for filling packages with a dry soap mix meets requirements for label weight on the contents. The process aim, however, is set above label weight to account for variability in the filling process. See "Cost of Inefficient Processes" under "Internal Failure Costs."
10. Cost of errors in support operations, for example, order filling, shipping, customer service, billing.
11. Cost of poor quality within a supplier's company. Such costs are included in the purchase price.

Hidden costs can accumulate to a large amount, three or four times the reported failure cost. Where agreement can be reached to include some of these costs and where credible data or estimates are available, then they should be included in the study. Otherwise, they should be left for future exploration.

Progress has been made in quantifying certain hidden costs, and therefore some of them have been included in the four categories discussed. Obvious costs of poor quality are the tip of the iceberg (Fig. 25.6).

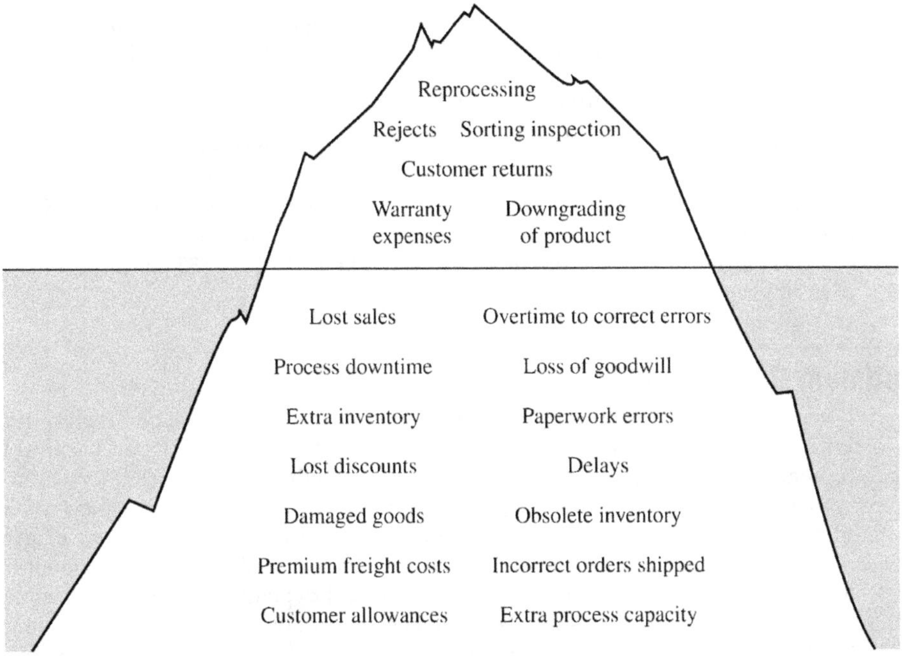

FIGURE 25.6 Hidden costs of poor quality.

Money (annual cost of poor quality)
24% of sales revenue
15% of manufacturing cost
13 cents per share of common stock
$7.5 million per year for scrap and rework compared to a profit of $1.5 million per year
$176 million per year
40% of the operating cost of a department
Other languages
The equivalent of one plant in the company making 100% defective work all year
32% of engineering resources spent in fincling and correcting design weaknesses
25% of manufacturing capacity devoted to correcting quality problems
13% of sales orders canceled
70% of inventory carried attributed to poor quality levels
25% of manufacturing personnel assigned to correcting quality problems

Table 25.4 Languages of Management

Relating the Cost of Poor Quality to Business Measures

Interpretation of the cost of poor quality is aided by relating this cost to other figures with which managers are familiar. Table 25.4 shows actual examples of the annual cost of poor quality related to various business measures.

Reducing the cost of poor quality has a dramatic impact on company financial performance, as illustrated by the Dupont Financial Model (Werner and Stoner 1995). This model states,

$$\text{Return on assets} = \text{Profit margin} \times \text{Asset turnover}$$

Suppose that the cost of poor quality (COPQ) was 10 percent of sales revenue, the profit margin was 7 percent, and the asset turnover was 3.0. The return on assets is then 7.0 × 3.0 or 21 percent. Further suppose that a quality improvement effort reduced the COPQ from 10 to 6 percent and the asset turnover remained at 3.0. The profit margin would then be 7.0 + 4.0 or 11 percent, and the return on assets would be 11.0 × 3.0 or 33 percent. Note the impact of asset turnover.

Optimum Cost of Quality

When cost summaries on quality are first presented to managers, one typical question is, What are the right costs? The managers are looking for a standard ("par") against which to compare their actual costs so that they can judge whether there is a need for action.

Unfortunately, few credible data are available because (1) companies almost never publish such data and (2) the definition of cost of poor quality varies by company. But we can cite some numbers. For manufacturing organizations, the annual cost of poor quality is about 15 percent of sales income, varying from about 5 to 35 percent depending on product complexity. For service organizations the average is about 30 percent of operating expenses, varying from 25 to 40 percent depending on service complexity. In one study of healthcare costs, Mortimer et al. (2003) estimate that as much as 30 percent of all direct healthcare outlays today are the result of poor quality care, consisting primarily of overuse, misuse, and waste. With U.S. national

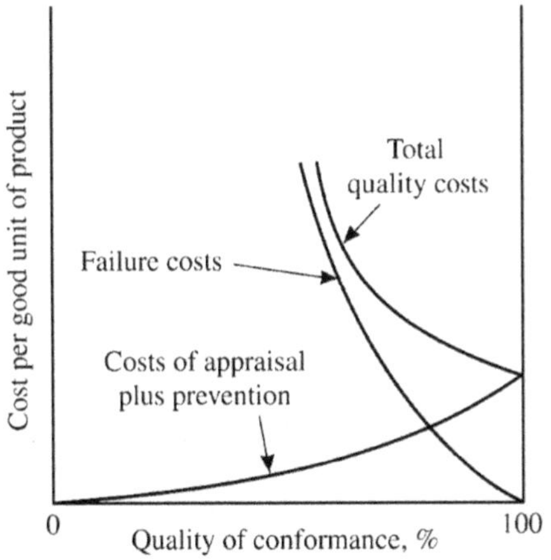

FIGURE 25.7 Model for optimum costs. (JQH4, p. 4.19.)

health expenditures of roughly $1.4 trillion in 2001, the 30 percent figure translates into $420 billion spent each year as a direct result of poor quality. In addition, the indirect costs of poor quality (e.g., reduced productivity due to absenteeism) add an estimated 25 to 50 percent—$105 to $210 billion—to the national bill. Private purchasers absorb about one-third of these costs. In that study, it is estimated that poor quality healthcare costs the typical employer between $1,900 and $2,250 per covered employee each year. But three conclusions on cost data do stand out: Total costs are highest for complex industries, failure costs are the largest percentage of the total, and prevention costs are a small percentage of the total.

The study of the distribution of quality costs over the major categories can be further explored by using the model shown in Fig. 25.7. The model shows three curves:

1. *The failure costs.* These costs equal zero when the product is 100 percent good and rise to infinity when the product is 100 percent defective. (Note that the vertical scale is cost per good unit of product. At 100 percent defective, the number of good units is zero, and hence the cost per good unit is infinity.)
2. *The costs of appraisal plus prevention.* These costs are zero at 100 percent defective and rise as perfection is approached.
3. *The sum of curves 1 and 2.* The third curve, marked "total quality costs," represents the total cost of quality per good unit of product.

Figure 25.7 suggests that the minimum level of total quality costs occurs when the quality of conformance is 100 percent, that is, perfection. This result has not always been the case. During most of the 20th century, the predominant role of (fallible) human beings limited the efforts to attain perfection to finite costs. Also, the inability to quantify the impact of quality failures on sales revenue resulted in underestimating the failure costs. The result was to view the optimum value of quality of conformance as less than 100 percent.

Although perfection is obviously the goal for the long run, perfection is not necessarily the most economic goal for the short run or for every situation. Industries, however, are facing increasing pressure to reach perfection. The prospect is that the trend to 100 percent conformance will extend to more and more goods and services of greater and greater complexity.

To evaluate whether quality improvement has reached the economic limit, we need to compare the benefits possible from specific projects with the costs involved in achieving these benefits. When no justifiable projects can be found, the optimum has been reached.

Let's identify a few key points about the cost of poor quality. This cost is usually high (sometimes larger than the annual profit), quantifying this cost can be the key to gaining approval from management to assign resources to quality improvement, and the main uses of the cost-of-poor-quality study are to identify opportunities for improvement projects and to provide supporting data to assist in improvement.

Next, we move to another element of assessment—standing in the marketplace.

Assessment and Standing in the Marketplace

Estimating the cost of poor quality is an essential part of the assessment. But it is not enough. We also need to understand where the company stands on quality in the marketplace, relative to the competition. This component of assessment will prove important in increasing sales income.

Similar to the assessment of the cost of poor quality, the market study (1) gives a snapshot of standing relative to competition and (2) identifies opportunities and threats. The approach must be based on a market research study. Such studies should be planned not by any one department but by a team involving members from marketing, product development, quality, operations, and other areas as needed. This team must agree beforehand on what questions need to be answered by the field study. Three types of questions should be considered: (1) What is the relative importance of various product qualities as seen by the user? (2) For each of the key qualities, how does our product compare with competitors' products, as seen by users? (3) How likely is the customer to purchase from us again or recommend us to others?

Chapter 4 includes other questions to help in developing new or modified products and services. Answers to such questions must be based on input from current customers, lost customers, and noncustomers. Opinions of company personnel, no matter how extensive the experience base is, cannot and should not substitute for the voice of the customer.

Examples of Field Studies

The first example comes from a manufacturer of health products. In a multiattribute study, customers were asked to consider several product attributes and indicate both their relative importance and a competitive rating. The results for one product are shown in Table 25.5. Note that an overall score is obtained for each manufacturer by multiplying the relative importance by the score for that attribute and then adding these results.

Attribute	Relative importance, %	Company X		Competitor A		Competitor B	
		Rating	Weighted Rating	Rating	Weighted Rating	Rating	Weighted Rating
Safety	28	6	168	5	140	4.5	126
Performance	20	6	120	7	140	6.5	130
Quality	20	6	120	7	140	4	80
Field service	12	4	48	8	96	5	60
Ease of use	8	4	32	6	48	5	40
Company image	8	8	64	4	32	4	32
Plant service	4	7.5	30	7.5	30	5	20
Total			582		626		488

TABLE 25.5 Multiattribute Study

Attribute	Comparison to Competition, %		
	Superior	Competitive	Inferior
Analysis of customer needs			
Preparation of quality requirements and purchase order			
Preparation of specifications and technical documentation			
Quality of equipment			
Quality and availability of spare parts			
Quality of field repair service			

TABLE 25.6 Heavy Equipment Case

In another example, a manufacturer of equipment was experiencing a decline in market share. Complaint's about quality led to a proposal to "beef up the inspection." Discussions within the company revealed uncertainty about the nature of the complaints, so management decided to conduct a field study to learn more about customer views. A team was formed to plan and conduct the study. About 50 customers were visited.

In one part of the study, six attributes were identified, and customers were asked to rate the company as superior, competitive, or inferior to the competition on each attribute (see Table 25.6). The results were a surprise. The equipment problems were confirmed, but the study revealed the presence of both design and manufacturing causes. Also, documentation and field repair service were identified as weak areas; these were surprises to the company involved. The company took dramatic action. This manufacturer created a broad approach to quality, starting with the initial design and continuing throughout the spiral of all activities affecting fitness for use. This outcome was in stark contrast to the original proposal of adding inspectors. The study required about seven worker-months of effort, including planning, customer visits, analysis of results, and preparation of a report—a small price to pay to develop a proper strategy.

Many organizations in service industries have extensive experience in market research. For example, one bank periodically conducts market research as part of a quality system. This research probes 20 attributes of banking service by asking consumers about the relative importance of attributes and consumer degree of satisfaction. Table 25.7 shows the format of the summarized results.

	Very Satisfied, %	Sample Size	Important and Low in Satisfaction, %	Important and High in Satisfaction, %
1. Greeting you with a smile				
8. Processing transactions without error				
14. Easy-to-read and understand bank statements				
20. Prompt follow-up to questions and problems				

TABLE 25.7 Market Research at a Bank

In an example from the public sector, the United States Postal Service has used a Customer Satisfaction Index survey that asks customers for the relative importance and satisfaction rating on 10 attributes (e.g., responsiveness, carrier services, complaint handling). A separate question asks about "customer willingness to switch to another mail service."

Graphing the market research results can be helpful. Rust et al. (1994) show an example of how the mapping of satisfaction and importance ratings can relate customer views and potential action (Fig. 25.8). In this approach, attributes in which importance is high and satisfaction is poor represent the greatest potential for gain.

In Fig. 25.8, the four quadrants are roughly defined by the averages on the two axes. Interpretation of the quadrants is typically as follows:

Upper left (satisfaction strong, importance low). Maintain the status quo.

Upper right (satisfaction strong, importance high). Leverage this competitive strength through advertising and personal selling.

Lower left (satisfaction weak, importance low). Assign little or no priority to action.

Lower right (satisfaction weak, importance high). Add resources to achieve an improvement.

Collecting data on customer loyalty and retention goes beyond customer satisfaction research. Lauter (1997) reports on customer retention results at a bank. The actual customer retention percentage was calculated for each quarter in 1996—the percentage retained was typically more than 95 percent. Further analysis of the customers lost revealed a surprising result (Fig. 25.9). For example, in the first quarter, 120,000 households (HHs) were lost, and this loss translated into $26 million of profit. Although the information has been modified because of its proprietary nature, a small percentage of customers lost from a high volume of customers can result in a significant absolute dollar amount of revenue.

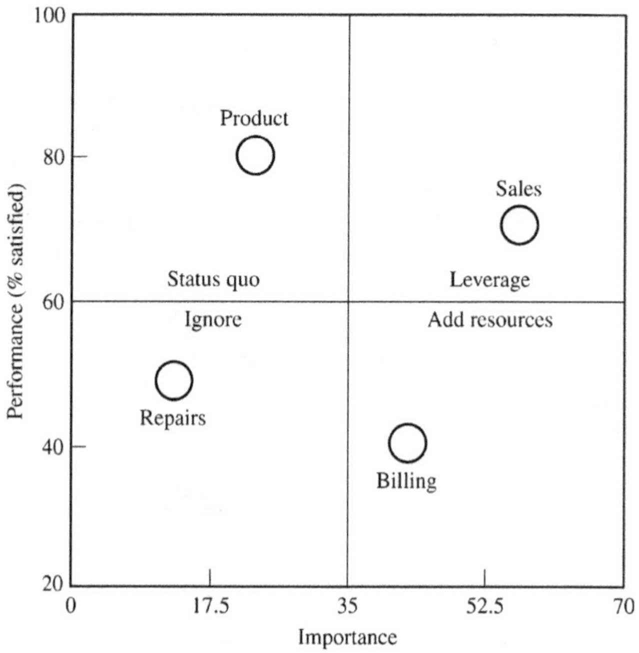

FIGURE 25.8 Performance versus importance in driving satisfaction: quadrant map. (Rust et al., 1994.)

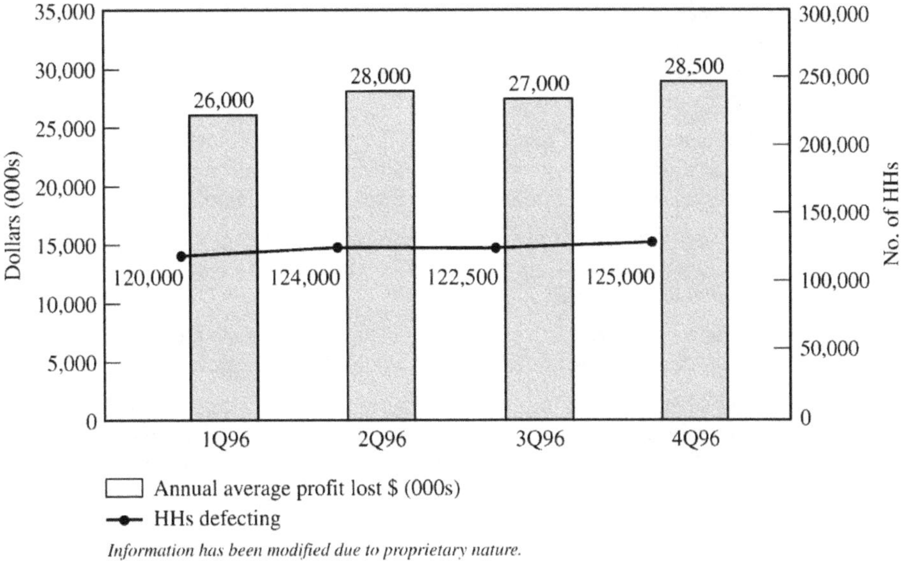

Figure 25.9 Average "opportunity cost of attrition." (Lauter, 1997.)

Assessing Using National Performance Standards and Awards

The national quality and performance excellence criteria and awards have become a major influence on attaining superior organization results. In fact, it could be argued that had it not been for the recognition Motorola received as one of the first recipients of the Malcolm Baldrige National Quality Award, it is possible that the practice of Six Sigma methodology may have never gained the level of exposure and practice that it has today. National quality awards in particular have greatly influenced the way many organizations manage their systems for quality and performance excellence. The prestige of winning a national quality award has provided increased recognition to the award recipients' role-model approaches and created additional incentives for many organizations to apply for the award and stick to their approaches. However, the real benefit stated by many is not the award itself, but the organizational improvement that can occur from adopting the award models and criteria as a means for pursuing excellence.

Malcolm Baldrige National Quality Award Criteria

The Baldrige performance excellence criteria are a framework that any organization can use to improve overall performance.

The seven categories that make up the award criteria are connected and integrated as depicted earlier in the framework of Fig. 25.2. The following is a list of the Baldrige Award categories:

1. *Leadership*. Examines how senior executives guide the organization and how the organization addresses its responsibilities to the public and practices good governance and citizenship.
2. *Strategic planning*. Examines how the organization sets strategic directions and how it develops key action plans.

3. *Customer and market focus*. Examines how the organization engages its customers for long-term marketplace success, builds a customer-focused culture, and listens to the voice of its customers and uses this information to improve and identify opportunities for innovation.
4. *Measurement, analysis, and knowledge management*. Examines the management, effective use, analysis, and improvement of data and information to support key organization processes and the organization's performance management system.
5. *Workforce focus*. Examines how the organization engages, manages, and develops its workforce to develop its full potential and how the workforce is aligned with the organization's objectives.
6. *Process management*. Examines aspects of how work systems are designed and how key work processes are designed, managed, and improved.
7. *Results*. Examines the organization's performance and improvement in its key business areas: product and service, customer-focused, financial and marketplace, workforce-focused, process effectiveness, and leadership. The category also examines how the organization performs relative to competitors and other organizations with similar product offerings.

Within each category of the Baldrige Criteria are the requirements and questions that are used as part of an assessment or an awards application. There are 18 criteria (Fig. 25.10). In addition, there are 36 areas to address contained within the 18 criteria. The numerous requirements are expressed as individual criteria questions.

Malcolm Baldrige National Quality Award Core Values

There are 11 core values and concepts embodied in the award criteria (Fig. 25.11):

1. *Visionary leadership*. An organization's senior leaders should set directions and create a customer focus, clear and visible values, and high expectations.
2. *Customer-driven excellence*. Performance and quality are judged by an organization's customers. Thus, an organization must take into account all product features and characteristics and all modes of customer access and support that contribute value to its customers.
3. *Organizational and personal learning*. Achieving the highest levels of organizational performance requires a well-executed approach to organizational and personal learning that includes sharing knowledge via systematic processes.
4. *Valuing workforce members and partners*. An organization's success depends increasingly on an engaged workforce that benefits from meaningful work, clear organizational direction, and performance accountability and that has a safe, trusting, and cooperative environment. Additionally, the successful organization capitalizes on the diverse backgrounds, knowledge, skills, creativity, and motivation of its workforce and partners.
5. *Agility*. Success in today's ever-changing, globally competitive environment demands agility—a capacity for rapid change and flexibility. Organizations face shorter cycles for the introduction of new/improved products, and nonprofit and government organizations are increasingly being asked to respond rapidly to new or emerging social issues. Major improvements in response times often require new work systems, simplification of work units and processes, or the ability for rapid changeover from one process to another.

P	**Preface: Organiztional Profile**		
	P.1 Organizational description		
	P.2 Organizational situation		
Categories and Items			**Point values**
1	**Leadership**		**120**
	1.1 Senior leadership	70	
	1.2 Governance and societal responsibilities	50	
2	**Strategic Planning**		**85**
	2.1 Strategy development	40	
	2.2 Strategy deployment	45	
3	**Customer Focus**		**85**
	3.1 Customer engagement	40	
	3.2 Voice of the customer	45	
4	**Measurement, Analysis, and Knowledge Management**		**90**
	4.1 Measurement, analysis, and improvement of organizational performance	45	
	4.2 Management of information, knowledge, and information technology	45	
5	**Workforce Focus**		**85**
	5.1 Workforce engagement	45	
	5.2 Workforce environment	40	
6	**Process Management**		**85**
	6.1 Work systems	35	
	6.2 Work processes	50	
7	**Results**		**450**
	7.1 Product outcomes	100	
	7.2 Customer-focused outcomes	70	
	7.3 Financial and market outcomes	70	
	7.4 Workforce-focused outcomes	70	
	7.5 Process effectiveness outcomes	70	
	7.6 Leadership outcomes	70	
	TOTAL POINTS		**1,000**

FIGURE 25.10 Criteria for performance excellence—item listing. (Baldrige National Quality Program.)

6. *Focus on the future*. Creating a sustainable organization requires understanding the short- and long-term factors that affect an organization and marketplace.
7. *Managing for innovation*. Innovation means making meaningful change to improve an organization's products, services, programs, processes, operations, and business model to create new value for the organization's stakeholders. Innovation should lead an organization to new dimensions of performance.
8. *Management by fact*. Organizations depend on the measurement and analysis of performance. Such measurements should derive from business needs and strategy, and they should provide critical data and information about key processes, outputs, and results. Many types of data and information are needed for performance management.

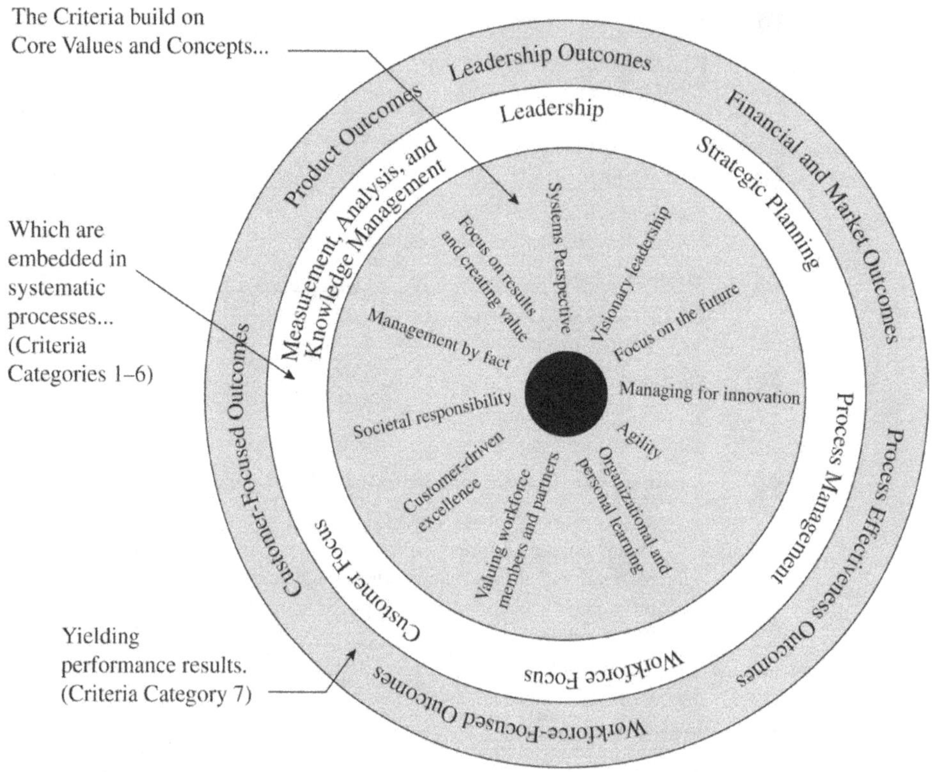

Figure 25.11 The role of core values and concepts. (Baldrige National Quality Program.)

9. *Societal responsibility*. An organization's leaders should stress responsibilities to the public, ethical behavior, and the need to consider societal well-being and benefit. Leaders should be role models for the organization in focusing on ethics and the protection of public health, safety, and the environment.

10. *Focus on results and creating value*. An organization's performance measurements need to focus on key results. Results should be used to create and balance value for key stakeholders—customers, workforce, stockholders, suppliers, and partners; the public; and the community. By creating value for key stakeholders, an organization builds loyalty, contributes to growing the economy, and contributes to society.

11. *Systems perspective*. The Baldrige Criteria provide a systems perspective for managing an organization and its key processes to achieve results—and to strive for performance excellence. The seven Baldrige Criteria Categories, the Core Values, and the Scoring Guidelines form the building blocks and the integrating mechanism for the system. However, successful management of overall performance requires organization-specific synthesis, alignment, and integration. Synthesis means looking at the organization as a whole and building on key business attributes, including core competencies, strategic objectives, action plans, and work systems. Alignment means using the key linkages among requirements given in the Baldrige Criteria Categories to ensure consistency of plans, processes, measures, and actions.

Baldrige Scoring System

The Baldrige scoring system is based on a 0–1,000 point scale. The available points are distributed among the categories and items as shown in Fig. 25.10. There is a heavy focus on results with Category 7, Business Results, making up 450 of the total 1,000 points available. During an assessment or award application review, examiners assign scores based upon scoring guidelines. There are two evaluation dimensions with scoring guidelines for each: (1) process and (2) results.

Process refers to the work methods an organization uses and improves to address the item requirements in categories 1 through 6. Four factors used to evaluate process are approach, deployment, learning, and integration (ADLI):

1. *Approach.* Refers to the methods used to accomplish the process; the appropriateness of the methods to the item requirements and the organization's operating environment; the effectiveness of the use of the methods; the degree to which the approach is repeatable and based on reliable data and information (i.e., systematic).
2. *Deployment.* Refers to the extent to which the approach is applied in addressing item requirements relevant and important to the organization; the approach is consistently applied; the approach is used (executed) by all appropriate work units.
3. *Learning.* Refers to refining the approach through cycles of evaluation and improvement; encouraging breakthrough change to the approach through innovation; sharing refinements and innovations with other relevant work units and processes in the organization.
4. *Integration.* Refers to the extent to which the approach is aligned with the organizational needs identified in the organizational profile and other process items; the measures, information, and improvement systems are complementary across processes and work units; the plans, processes, results, analyses, learning, and actions are harmonized across processes and work units to support organization-wide goals.

Results refers to an organization's outputs and outcomes in achieving the requirements in items 7.1–7.6 (Category 7). The four factors used to evaluate results are levels, trends, comparisons, and integration (LeTCI):

1. *Levels.* Refers to the current level of performance.
2. *Trends.* Refers to the rate of performance improvements or the sustainability of good performance (i.e., the slope of trend data); the breadth (i.e., the extent of deployment) of the performance results.
3. *Comparisons.* Refers to performance relative to appropriate comparisons, such as similar competitors or organizations; the performance relative to benchmarks or industry leaders.
4. *Integration.* Refers to the extent to which the results measure (often through segmentation) and address important customer, product, market, process, and action plan performance requirements identified in the organizational profile and in process items; results include valid indicators of future performance; results are harmonized across processes and work units to support organization-wide goals.

Using Quality Awards as a System Assessments Tool

Many organizations use the criteria from national quality awards for the benefit of getting a thorough organizational assessment, often with no intent of even applying for award recognition. There are various methods that can be used to complete an assessment based upon many factors, including the size and geography of the organization, the number of facilities, and the availability of internal expertise to conduct an assessment.

Written Responses

Without completing a formal application, organizations can use a question-and-answer approach to respond to the criteria questions from an awards program. There could be multiple responses for each question if more than one input is desired for the response. Some more-advanced formats for written responses could seek more probing information for each question. For example, a Baldrige-based written response questionnaire may be formatted to seek a specific response for approach, deployment, learning, and integration for each process-related question. This provides richer information for the purposes of scoring the application using the awards process scoring guidelines.

Survey

This approach can be used to gather assessment input from a large number of people. Questions are designed to gather the collective input on the performance of the organization as it relates to the awards criteria. The Baldrige National Quality Program provides a free survey for this purpose called, "Are We Making Progress" (2008). This is a 40-question survey that can be completed in about 10 minutes. A survey can be used as the primary information gathering method, or it can be used as supplemental or additional information as part of more comprehensive assessment.

Application

A formal application can be prepared that simulates an actual awards process application. This requires more effort than simple written responses but can also be more revealing because of the thought process of the writer in attempting to use consistent language, identify linkages, and identify the most important strengths of the organization. Organizations that are just starting out may decide to answer only the higher-level questions rather than the more specific multiple requirements in awards criteria. For example, if assessing the Category 3, Customer Focus, section of the Baldrige Criteria, the self-assessment team may only respond to the item-level questions:

1. *Customer engagement*. How do you engage customers to serve their needs and build relationships?
2. *Voice of the customer*. How do you obtain and use information from your customers?

This is a much simpler process, whereas the multiple requirements from Category 3, Customer Focus, include 35 separate questions compared to only 2 questions at the item level.

Interview

In this approach, the assessment team would schedule interviews with key individuals to respond to criteria questions. During the interviews, the assessment team will record notes about the approaches used and results attained for further evaluation after all of the information has been collected. This approach requires less preparation time than written responses. The disadvantage is that the respondents may not provide complete information during the interview setting, depending on their understanding of the question and their recall of all the relevant information in a single interview setting. There is also less learning on the part of the respondents when compared to written responses because the level of engagement is less.

Focus Group

This is similar to the interview approach except multiple people are involved in the sessions for different sections of the criteria review. For example, an assessment for item 1.1, Senior Leadership, of the Baldrige Criteria may involve several members of the senior leadership team in a focus group setting. Like the interview approach, the assessment team will record notes about the approaches used and results attained for further evaluation after all of the information has been collected. This approach has an advantage over individual interviews because of the shared knowledge of the group and the ability to build on the responses of others to provide complete assessment information.

Collaborative Assessment

This is a variation on a focus group approach. Instead of gathering specific answers to the approach questions, the respondents provide their opinions about strengths and opportunities for improvement for each section of the assessment criteria. Responses are captured in real time for review by the focus group. At the end of each focus group session, each individual can provide a rating of the importance of each opportunity for improvement. This collaborative approach to assessment has the advantage of involving many people in gathering the information and building consensus on both the strengths and opportunities for improvement within the organization (Hoyt and Ralston 2001).

Assessing to the International Standards

Standards exist principally to facilitate international trade and to avoid harming customers and society. In the prestandardization era (pre-1980), there were various national and multinational standards. Standards for electrical, mechanical, and chemical process compatibility have been around for decades. Other standards, such as military standards, were developed for military and others for the nuclear power industry, and, to a lesser extent, for commercial and industrial use. These standards have commonalities and historical linkages. However, they often were not consistent in terminology or content for widespread use in international trade. As a result, organizations were left to re-create their own standards or adapt the existing ones. This only led to even less commonality. In the 1980s, as most of the first world industrial organizations began to improve quality and safety at record paces, there became a need to fill a void. That void was a common quality management system that would be a nonbinding "contract" between the customer and the supplier. This void was filled by the ISO 176 Technical Committee in the form of the ISO 9000 set of standards. This was later followed by filling a similar void for environmental standards with ISO 14000. Many organizations globally began using these standards as a "certified" standard for performance. While their intent was important, the standards became more of an opportunity to get a certificate of compliance that could be used to "wow" customers, rather than a set of requirements that, when met, ensure customer needs are met.

Certain industry/economic sectors then began developing industrywide quality system standards, based on the verbatim adoption of ISO 9000, together with industrywide supplemental requirements. The automotive industry (QS 9000), the pharmaceutical and medical devices industry (cGMPs), government regulatory agencies, and military procurement agencies (AS 9100 and MAP) are adopting this approach in many places worldwide. Even software development uses the CMMi standard of software quality created in the early 1990s at Carnegie Mellon University to ensure a common approach to managing software quality. The standards play an important—but not always understood—role in the management of quality.

ISO 9000 Quality Management System Standard

The ISO 9000 standards have had great impact on international trade and quality systems implementation by organizations worldwide. These international standards have been adopted as national standards by more than 100 countries. They have been applied in a wide range of industry/economic sectors and government regulatory areas. The ISO 9000 standards deal with the management systems used by organizations to ensure quality in design, production, delivery, and support products. The standards apply to all generic product categories: hardware, software, processed materials, and services. The complete family of ISO 9000 standards provide quality management guidance, quality assurance requirements, and supporting technology for an organization's quality management system. The standards provide guidelines or requirements on what features are to be present in the management system of an organization but do not prescribe how the features are to be implemented. This nonprescriptive character gives the standards their wide applicability for various products and situations. Upon implementing ISO 9000, an organization can be registered as a Certified Quality Management System. See Chap. 11, ISO 9000 Quality Management System, for more information on implementing and applying the standards.

Quality System Certification/Registration

The earliest users of quality assurance requirements standards were large customer organizations such as electric power providers and military organizations. These customers often purchase complex products to specific functional design. In such situations, the quality assurance requirements are called up in a two-party contract, where the providing organization (i.e., the supplier) is referred to as the "first party" and the customer organization is referred to as the "second party." Such quality assurance requirements typically include provisions for the providing organization to have internal audits sponsored by its management to verify that its quality system meets the contract requirements. These are first-party audits. Such contracts typically also include provisions to have external audits sponsored by the management of the customer organization to verify that the supplier organization's quality system meets the contract requirements. These are second-party audits. Within a contractual arrangement between two such parties, it is possible to tailor the requirements, as appropriate, and to maintain an ongoing dialogue between customer and supplier.

When such assurance arrangements become a widespread practice throughout the economy, the two-party, individual-contract approach becomes burdensome. There develops a situation where each organization in the supply chain is subject to periodic management system audits by many customers and is itself subjecting many of its subsuppliers to such audits. There is a lot of redundant effort throughout the supply chain because each organization is audited multiple times for essentially the same requirements. The conduct of audits becomes a significant cost element for both the auditor organizations and audited organizations.

Certification/Registration-Level Activities

The development of quality system certification/registration is a means to reduce the redundant, non-value-adding effort of these multiple audits. A third-party organization, which is called a *certification body* in some countries, or a *registrar* in other countries (including the United States), conducts a formal audit of a supplier organization to assess conformance to the appropriate quality system standard, say, ISO 9001 or ISO 9002. When the supplier organization is judged to be in complete conformance, the third party issues a certificate to the supplying organization and registers the organization's quality system in a publicly available register. Thus, the terms certification and registration carry the same marketplace meaning because they are two successive steps signifying successful completion of the same process.

To maintain its registered status, the supplier organization must pass periodic surveillance audits by the registrar. Surveillance audits are often conducted semiannually. They may be less comprehensive than the full audit. If so, a full audit is performed every few years.

In the world today, there are hundreds of certification bodies/registrars. Most of them are private, for-profit organizations. Their services are valued by the supplier organizations they register—and by the customer organizations of the supplier organizations—because the registration service adds value in the supply chain. It is critical that the registrars do their work competently and objectively and that all registrars meet standard requirements for their business activities. They are, in fact, supplier organizations that provide a needed service product in the economy.

Industry-Specific Adoptions and Extensions of ISO 9000 Standards

In some sectors of the global economy, there are industry-specific adoptions and extensions of the ISO 9000 standards. These situations are a classic example of a problem opportunity. As problems, such adaptations and extensions strain the goal of nonproliferation. As opportunities, they have been found effective in a very few industries where there are special circumstances and where appropriate ground rules can be developed and implemented consistently. These special circumstances have been characterized by:

1. Industries where the product impact on the health, safety, or environmental aspects is potentially severe; as a consequence most nations have regulatory requirements regarding the quality management system of a supplier.
2. Industries that have had well-established, internationally deployed industry-specific or supplier specific quality system requirements documents prior to publication of the ISO 9000 standards.

Fortunately, in the very few instances so far, the operational nonproliferation criteria of the ISO/IEC directives have been followed.

Medical Device Industry

Circumstance 1 relates to the medical device manufacturing industry. For example, in the United States, the Food and Drug Administration (FDA) developed and promulgated the Good Manufacturing Practice (GMP) regulations. The GMPs operate under the legal imprimatur of the FDA regulations, which predate the ISO 9000 standards. The FDA regularly inspects medical device manufacturers for compliance with the GMP requirements. Many of these requirements are quality management system requirements that parallel the subsequently published ISO 9002:1987 requirements. Other GMP regulatory requirements relate more specifically to health, safety, or environmental aspects. Many other nations have similar regulatory requirements for such products.

In the United States, the FDA has created revised GMPs that parallel closely the ISO 9000 standard, plus specific regulatory requirements related to health, safety, or environment. The expansion of scope to include quality system requirements related to product design reflects the recognition of the importance of product design and the greater maturity of quality management practices in the medical device industry worldwide. Similar trends are taking place in other nations, many of which are adopting ISO 9001 verbatim for their equivalent of the GMP regulations.

Current Good Manufacturing Practices (cGMPs) for human pharmaceuticals affect every American. Consumers expect that each batch of medicines they take will meet quality standards so that they will be safe and effective. Most people, however, are not aware of cGMPs, or how the FDA ensures that drug manufacturing processes meet these basic objectives. Recently, the FDA announced a number of regulatory actions taken against drug manufacturers based on the lack of cGMPs.

What Are cGMPs?

Current Good Manufacturing Practice regulations—or cGMPs—are enforced by the U.S. Food and Drug Administration (FDA). cGMPs provide for systems that ensure proper design, monitoring, and control of manufacturing processes and facilities. Adherence to the cGMP regulations ensures the identity, strength, quality, and purity of drug products by requiring that manufacturers of medications adequately control manufacturing operations. This includes establishing strong quality management systems, obtaining appropriate quality raw materials, establishing robust operating procedures, detecting and investigating product quality deviations, and maintaining reliable testing laboratories. This formal system of controls at a pharmaceutical organization, if adequately put into practice, helps to prevent instances of contamination, mix-ups, deviations, failures, and errors and ensures that drug products meet their quality standards.

The cGMP requirements were established to be flexible in order to allow each manufacturer to decide individually how to best implement the necessary controls by using scientifically sound design, processing methods, and testing procedures. The flexibility in these regulations allows companies to use modern technologies and innovative approaches to achieve higher quality through continual improvement.

Accordingly, the "c" in cGMP stands for "current," requiring companies to use technologies and systems that are up-to-date in order to comply with the regulations. Systems and equipment that may have been "top-of-the-line" to prevent contamination, mix-ups, and errors 10 or 20 years ago may be less than adequate by today's standards. It is important to note that cGMPs are minimum requirements. Many pharmaceutical manufacturers are already implementing comprehensive, modern quality systems and risk management approaches that exceed these minimum standards.

Why Are cGMPs Important to Software Development?

A consumer usually cannot detect (through smell, touch, or sight) that a drug product is safe or if it will work. While cGMPs require testing, testing alone is not adequate to ensure quality. In most instances, testing is done on a small sample of a batch (e.g., a drug manufacturer may test 100 tablets from a batch that contains 2 million tablets), so that most of the batch can be used for patients rather than destroyed by testing. Therefore, it is important that drugs are manufactured under conditions and practices required by the cGMP regulations to ensure that quality is built into the design and manufacturing process at every step. Facilities that are in good condition, equipment that is properly maintained and calibrated, employees who are qualified and fully trained, and processes that are reliable and reproducible are a few examples of how cGMP requirements help to ensure the safety and efficacy of drug products.

How Does the FDA Determine If an Organization Is Complying with cGMP Regulations?

The FDA inspects pharmaceutical manufacturing facilities worldwide using scientifically and cGMP-trained individuals whose job it is to evaluate whether the organization is following the cGMP regulations. The FDA also relies on reports of potentially defective drug products from the public and the industry. The FDA will often use these reports to identify sites for which an inspection or investigation is needed. Most companies that are inspected are found to be fully compliant with the cGMP regulations.

In August 2002, the FDA announced the Pharmaceutical CGMPs for the 21st Century Initiative. In that announcement, the FDA explained the agency's intent to integrate quality systems and risk management approaches into its existing programs with the goal of encouraging industry to adopt modern and innovative manufacturing technologies. The CGMP initiative was spurred by the fact that since 1978, when the last major revision of the CGMP regulations was published, there have been many advances in manufacturing science and in our understanding of quality systems. In addition, many pharmaceutical manufacturers are already implementing comprehensive, modern quality systems and risk management approaches. This guidance is intended to help manufacturers implementing modern quality systems and risk management approaches to meet the requirements of the agency's CGMP regulations. The agency also saw a need to harmonize the CGMPs with other non-U.S. pharmaceutical regulatory systems and with the FDA's own medical device quality systems regulations. This guidance supports these goals. It also supports the objectives of the Critical Path Initiative, which intends to make the development of innovative medical products more efficient so that safe and effective therapies can reach patients sooner.

The CGMPs for the 21st Century Initiative steering committee created a Quality System Guidance Development working group (QS working group) to compare the current CGMP regulations, which call for some specific quality management elements, to other existing quality management systems. The QS working group mapped the relationship between CGMP regulations (parts 210 and 211 and the 1978 Preamble to the CGMP regulations) and various quality system models, such as the Drug Manufacturing Inspections Program (i.e., systems-based inspectional program), the Environmental Protection Agency's Guidance for Developing Quality Systems for Environmental Programs, ISO Quality Standards, other quality publications, and experience from regulatory cases. The QS working group determined that, although the CGMP regulations do provide great flexibility, they do not incorporate explicitly all of the elements that today constitute most quality management systems.

The CGMP regulations and other quality management systems differ somewhat in organization and in certain constituent elements; however, they are very similar and share underlying principles. For example, the CGMP regulations stress quality control. More recently developed quality systems stress quality management, quality assurance, and the use of risk management tools in addition to quality control. The QS working group decided that it would be very useful to examine exactly how the CGMP regulations and the elements of a modern, comprehensive quality system fit together in today's manufacturing world. This guidance is the result of that examination.

In ISO, a new technical committee, ISO/TC210, has been formed specifically for medical device systems. TC210 has developed standards that provide supplements to ISO 9001 clauses. These supplements primarily reflect the health, safety, and environment aspects of medical devices and tend to parallel the regulatory requirements in various nations. These standards are in late stages of development and international approval at this time.

Software Quality

The global economy has become permeated with electronic information technology (IT). The IT industry now plays a major role in shaping and driving the global economy. As in past major technological advances, the world seems fundamentally very different and, paradoxically, fundamentally the same. Computer software development occupies a central position in this paradox.

First, it should be noted that computer software development is not so much an industry as it is a discipline.

Second, many IT practitioners emphasize that computer software issues are complicated by the multiplicity of ways that computer software quality may be critical in a supplier organization's business. For example,

- The supplier's product may be complex software whose functional design requirements are specified by the customer.
- The supplier may actually write most of its software product or may integrate off-the-shelf packaged software from subsuppliers.
- The supplier may incorporate computer software/firmware into its product, which may be primarily hardware and/or services.
- The supplier may develop and/or purchase from subsuppliers software that will be used in the supplier's own design and/or production processes of its product.

However, it is important to acknowledge that hardware, processed materials, and services are often involved in a supplier organization's business in these same multiple ways, too.

What, then, are the issues in applying ISO 9001 to computer software development? There is general consensus worldwide that

- The generic quality management system activities and associated requirements in ISO 9001 are relevant to computer software, just as they are relevant in other generic product categories (hardware, other forms of software, processed materials, and services).
- There are some things that are different in applying ISO 9001 to computer software.
- There is, at this time, no worldwide consensus as to which things, if any, are different enough to make a difference and what to do about any things that are different enough to make a difference.

ISO/TC176 developed and published ISO 9000-3:1991 as a means of dealing with this important, paradoxical issue. ISO 9000-3 provides guidelines for applying ISO 9001 to the development, supply, and maintenance of (computer) software. ISO 9000-3 has been useful and is widely used. ISO 9000-3 offers guidance that goes beyond the requirements of ISO 9001, and it makes some assumptions about the life-cycle model for software development, supply, and maintenance. In the United Kingdom, a separate certification scheme (TickIT) for software development has operated for several years, using a combination of ISO 9001 and ISO 9003. The scheme has received both praise and criticism from various constituencies worldwide. Those who praise the scheme claim that it:

- Addresses an important need in the economy to provide assurance for customer organizations that the requirements for quality in software they purchase (as a separate product, or incorporated in a hardware product) will be satisfied.

- Includes explicit provisions beyond those for conventional certification to ISO 9001 to ensure competency of software auditors, their training, and audit program administration by the certification body.
- Provides a separate certification scheme and logo to exhibit this status publicly.

Those who criticize the scheme claim that it:

- Is inflexible and attempts to prescribe a particular life-cycle approach to computer software development that is out of tune with current best practices for developing many types of computer software.
- Includes unrealistically stringent auditor qualifications in the technology aspects of software development and qualifications whose technical depth is not necessary for effective auditing of management systems for software development.
- Is almost totally redundant with conventional third-party certification to ISO 9001, under which the certification body/registrar already is responsible for competency of auditors and accreditation bodies verify the competency as part of accreditation procedures.
- Adds substantial cost beyond conventional certification to ISO 9001 and provides little added value to the supply chain.

In the United States a proposal to adopt a TickIT-like software scheme was presented to the ANSI/RAB accreditation program. The proposal was rejected, primarily on the basis that there was not consensus and support in the IT industry and the IT-user community.

CMMi: Software and Systems Development

Another standard that gained popularity is the Capability Maturity Model (CMM), which is a service mark owned by Carnegie-Mellon University (CMU) and refers to a development model elicited from actual data. The data was collected from organizations that contracted with the U.S. Department of Defense, which funded the research and became the foundation from which CMU created the Software Engineering Institute (SEI). Like any model, it is an abstraction of an existing system. Unlike many that are derived in academia, this model is based on observation rather than on theory. When applied to an existing organization's software development processes, CMM allowed an effective approach toward improving them. Eventually, it became clear that the model could be applied to other processes, which gave rise to a more general concept that is applied to business processes and to developing people.

The Capability Maturity Model was originally developed as a tool for objectively assessing the ability of government contractors' processes to perform a contracted software project. CMM is based on the process maturity framework first described in the 1989 book *Managing the Software Process* by Watts Humphrey. It was later published in a report in 1993 (Technical Report CMU/SEI-93-TR-024 ESC-TR-93-177 February 1993, Capability Maturity Model SM for Software, Version 1.1) and as a book by the same authors in 1995.

Though the CMM comes from the field of software development, it is used as a general model to aid in improving organizational business processes in diverse areas; for example, in software engineering, system engineering, project management, software maintenance, risk management, system acquisition, information technology (IT), services, business processes generally, and human capital management. The CMM has been used extensively worldwide in government, commerce, industry, and software development organizations.

An organization may be assessed by an SEI-authorized lead appraiser and will then be able to claim that it has been assessed as CMM level X, where X is from 1 to 5 (maturity levels).

Maturity Level 1 = Initial; Maturity Level 2 = Managed; Maturity Level 3 = Defined; Maturity Level 4 = Quantitatively Managed; Maturity Level 5 = Optimizing (read further for more explanation of the levels). Although sometimes called CMM certification, the SEI doesn't use this term due to certain legal implications.

In the 1970s, the use of computers became more widespread, flexible, and less expensive. Organizations began to adopt computerized information systems, and the demand for software development grew significantly. The processes for software development were in their infancy, with few standard or "best practice" approaches defined.

As a result, the growth was accompanied by growing pains: Project failure was common, the field of computer science was still in its infancy, and the ambitions for project scale and complexity exceeded the market capability to deliver.

In the 1980s, several U.S. military projects involving software subcontractors ran over budget and were completed much later than planned, if they were completed at all. In an effort to determine why this was occurring, the U.S. Air Force funded a study at the SEI.

The Standard CMMI Appraisal Method for Process Improvement (SCAMPI) is the official SEI method to provide benchmark-quality ratings relative to CMMI models. The CMMI model is used as a "ruler" to measure an organization's process definition as the model is a collection of process best practices assimilated into process areas. The SCAMPI appraisal methodology is utilized to measure how well an organization has institutionalized the process definition into its everyday way of doing business. SCAMPI appraisals are used to identify strengths and weaknesses of current processes, reveal development/acquisition risks, and determine capability and maturity level ratings. They are mostly used either as part of a process improvement program or for rating prospective suppliers. The method defines the appraisal process as consisting of preparation; on-site activities; preliminary observations, findings, and ratings; final reporting; and follow-on activities.

Active development of the model by the U.S. Department of Defense Software Engineering Institute (SEI) began in 1986 when Humphrey joined the Software Engineering Institute located at Carnegie Mellon University in Pittsburgh, Pennsylvania, after retiring from IBM. At the request of the U.S. Air Force, he began formalizing his Process Maturity Framework to aid the U.S. Department of Defense in evaluating the capability of software contractors as part of awarding contracts.

Benchmarking Best Practices: Moving Toward Sustainability

There are several organizations that ask, "We have been implementing improvement activities and have received amazing results. We continue to reduce costs, improve efficiency, and add Belts to the workforce. If we have done all of this, what is next to continually manage quality and ensure that results are sustained?" The answer to this question is to perform internal benchmarks of the organization or perform external benchmarks. This may seem like a methodology that has been done before, but with the technology capabilities many organizations have, benchmarking can be used to facilitate communication among plants, regions, and corporate headquarters. The type of communication that may occur is how to solve a quality problem faster and before it happens. For example, there are two plants that manufacture the same car, one in the United States and one in South Korea. The plant in South Korea may encounter a problem; through benchmarking, it could reach out to the plant in the United States to see if it had a similar problem. Most likely, the plants will have similar issues. The U.S. plant may have found a solution and can provide that solution to the South Korean plant and quickly fix the issue, rather than going through the process of diagnosing the root cause. This type of thinking and communication is how organizations can continually evolve, adapt, and create more efficiencies by itself. A person may consider this as an organization self-actualizing its maximum potential.

Our experience and research has led us to the development of the 7-Step Benchmarking Process©. The key to the 7-Step Benchmarking Process is the importance of taking benchmarking beyond its conventional level of analysis to facilitate the understanding of best practices and their implementation. In this connection, the performance measurement part of our benchmarking process, positioning analysis, is augmented with the proprietary performance measurement methodology that is expressed in the form of a single index, the Juran Quality Index©.

Furthermore, the learning phase of the benchmarking process puts in place mechanisms that encourage the best practitioners to share their knowledge and experience with the rest of the organization. We have developed a unique and effective method for data normalization based upon the technical complexity of assets and processes. The Juran Complexity Factor© enables the direct comparison of differing assets and processes across an entire organization. Figure 25.12 is the model of the seven-step process.

Step 1: Preparation and Planning

Thorough advance planning is critical to success of benchmarking. As a result of the preparation and planning, an organization will:

- Determine the areas to be included.
- Review metric models, key performance indicators (KPIs), and KPI definitions.
- Create scorecards and structure to report findings.
- Determine logistics and project roll-out of the plan.
- Communicate key issues.
- Agree on roles and responsibilities.

It is very important to clearly and unambiguously define metrics and KPIs in step 1 to ensure consistency in data collection.

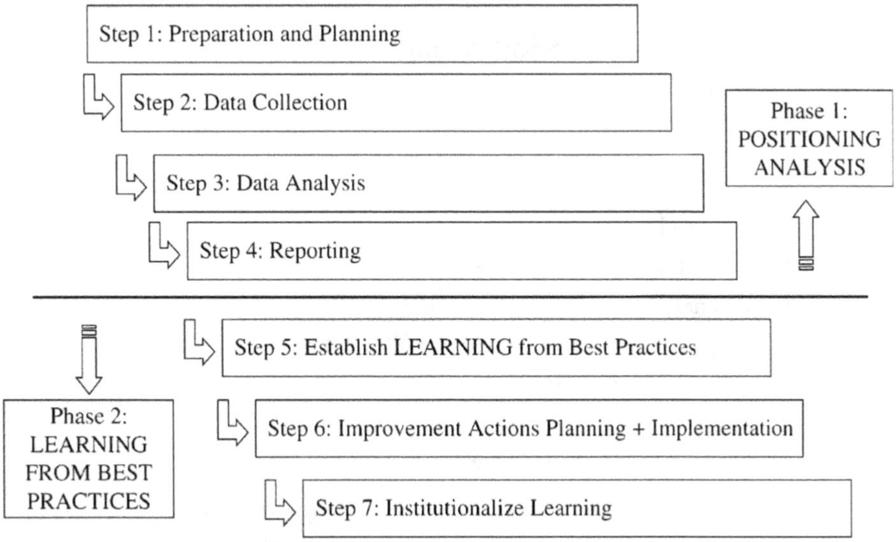

Figure 25.12 The Juran 7-Step Benchmarking Process. (Juran Institute, Inc. Copyright 1994. Used by permission.)

Step 2: Data Collection and Validation

Valid data are the key to success of any benchmarking program. Incorrect or inaccurate data can easily result in misguided conclusions and inappropriate actions and can lead to the failure of any improvement program.

Step 3: Data Analysis

Effective data normalization is essential to meaningful benchmarking because it enables "apples to apples" comparisons in greatly differing operational situations. In this program, the Juran Complexity Factor is used to normalize data, and the Juran Quality Index is used for the comparative analysis between sites.

The Juran Complexity Factor (CF) is an overall measurement of the complexity of the routine operation and maintenance of any type of site. The CF is calculated by breaking down the processes and assets at each site into modules of standard equipment. It is based upon an empirical system based on 25 years of worldwide experience in operation maintenance. The system has subsequently been refined and modified by Juran. The CF relies on a weighting factor that is calculated depending on the types of process and products and the inputs and outputs to that process. The identification of items is systematic and consistent in order to avoid omissions and duplications. In order to maintain an objective and unbiased assessment of each of the processes or sites, the calculation of the CF will be carried out by a Juran consultant. The Juran Quality Index is a cumulative indicator of overall performance based on both efficiency and effectiveness. It is a numerical value indicating the percentage deviation of a site's performance, relative to the corresponding group of sites. Thus, it provides a balanced viewpoint, accounting for all aspects of sites' performance and comparing these to the average. It has the advantage of providing a single index figure that immediately gives a high-level indication of relative performance, but it is also possible to further break down the index into its main constituent elements. The following example of a "Management Dashboard" shows how the index may be broken down into its constituent elements and represented graphically. Juran will supply a JQI for each site.

Step 4: Reporting—Benchmark Report

All data will be analyzed and conclusions drawn from all sites. The report will include all normalized data sets/charts, all KPIs, key findings per system (gap analysis with prioritization, strengths and weaknesses, and improvement suggestions). The benchmarking report will be made up of seven components:

- High-level overview
- Key findings and site comparisons
- Gap analysis
- Complexity factor assessments for all benchmarked processes and assets
- KPI/metric definitions
- Process and asset descriptions
- Next steps and sustainability

This step is similar to the one identified in the Organization Health Check but continually expands and more information is extracted and shared amongst the organization.

Step 5: Establish Learning from Best Practices

After the benchmark report has been delivered to the leadership, the organization should conduct workshops and have representatives from each area included in the benchmark. The objective of this workshop is the identification of a master class of opportunities and the transfer of knowledge from the best sites to other sites. The site that has the best practices will present to the other sites the "why's and the how's" of their best practices. The intention is that the audience learns from these presentations to help them subsequently formulate their own improvement programs. The best practice workshops are recommended to be conducted on an annual basis to ensure sustainability of desired goals.

Step 6: Improvement Action Planning and Implementation

This road map shows how the organization will continue to sustain the best practices and continually improve and meet the organization's strategic goals in a timely manner.

Step 7: Institutionalized Learning

As part of the action planning, there may be requirements to teach or reteach to achieve the best practices identified. This may seem like a repetitious step, but it is essential for an organization to evolve and adapt on its own.

Data Normalization

The single biggest problem in any benchmarking exercise is how to compare benchmarked subjects on a like-for-like basis (i.e., how to compare apples with pears). In some circumstances, the benchmarkers will be similar enough to enable direct comparisons of performance between them. However, more typically, the subjects being benchmarked will all be different from each other—be they organizations as a whole, business units, different sites, or different functional groups, business processes, or products. No two subjects will be identical, although the extent of difference between them will vary considerably depending on what and who is being benchmarked. Thus, in order to be able to compare differences in performance levels, some intervention must be exercised. Some form of data normalization is usually required to enable like comparisons to be made between what may be very different subjects. Without it, direct comparisons of performance are normally impossible and may lead to misinformed conclusions. Normalization can be made on the basis of a wide range of factors including scope, scale, contractual arrangements, regulatory requirements, and geographical and political differences.

One solution is to organize benchmarkers into categories or peer groups with other benchmarkers or datasets with similar characteristics. The key is to be able to identify the factors that are driving the performance and then develop a method by which these drivers can be considered when comparing performance metrics. In its simplest form, this may involve stratifying the data according to underlying criteria. For example, if a health authority wishes to compare death rates of people in different regions, it may stratify these according to gender or age. Another example comes from the chemical industry. A series of chemical organizations may decide to benchmark their performance in the field of managing for the environment, and in so doing, they may wish to compare emission levels for a variety of polluting gases (e.g., oxides of nitrogen and sulfur, carbon dioxide, and methane). These data could be stratified according to their harmful impact to the environment through the use of a standardized measurement such as the environmental impact unit (EIU). A further example might be those organizations comparing the efficiency of their R&D activities using a KPI that measures the percentage of sales attributable to new products or services (e.g., products that have been on the market for less than two years).

But, of course, even within these groups there may be differences between the benchmarking subjects. In order to be able to conduct a valid comparison of performance, these differences in characteristics need to be taken into consideration in the analysis. The most effective way of doing this is through normalization of the performance data.

Normalization is essentially the process of converting metrics into a form that enables their comparison on a like-for-like basis, accounting for all (or as many as possible) of the variation between the benchmarking subjects. What is essential is that the normalizing factor used is truly a driver for the performance being benchmarked. For example, if benchmarking the operating costs of the invoicing function of an organization, perhaps a suitable normalizing factor is the number of invoices raised. For example, the costs could be compared on a per-invoice-raised basis. However, it may be that some invoices are more complicated to produce than others (e.g., they may contain more line items or be for a higher total value, which requires more checks before the invoice is raised), so this way of normalizing may not after all be appropriate. The most common way of normalizing is by looking at performance per unit or per hour. For example, if we are measuring the cost of manufacturing a motor car, we might compare the cost per vehicle produced, or if we are looking at the time taken to treat a hospital patient with a given ailment, we might consider the number of patients examined per hour.

In some cases, a simple measurement per unit is not sufficient to accommodate for the variation observed between benchmarking subjects, and a more sophisticated approach has to be developed. In such cases, the use of weighting factors that represent the variation of the different benchmarking subject is often a very effective way of normalization. Weighting factors may be developed in relation to cost, time, and efficacy. An example of a highly effective weighting factor is the Juran Complexity Factor (JCF). The JCF was developed to enable like-for-like comparisons to be made between oil and gas production facilities of very different size and design. The normalizing factor takes into consideration the equipment present in the facility and the time it takes to operate and maintain this equipment under normal conditions. The JCF is then used to normalize all cost performance between facilities in the benchmarking. This enables organizations to directly benchmark their facilities with those of other organizations, even though they may be very different is design and size.

The efficacy of any normalization method should be fully tested before it is implemented. As mentioned, for a normalizing factor to be effective it must be representative of the driving force for the performance subjects being benchmarked. This means that there must be a good relationship between the performance metric and the normalizing factor. A good way of testing this is to examine the correlation between the normalizing factor and the performance metric being normalized. There should be a strong direct relationship between the two. For example, an increase in the normalizing factor should lead to an increase in the metric being benchmarked (and vice versa), although this relationship may or may not be linear.

Analysis and Identification of Best Practices

The aim of the analysis is to determine the findings from the data collected in the benchmarking in conjunction, where appropriate, with other pertinent data and information from a number of different sources, including the public domain, the participants themselves, and from any previous editions of the benchmarking study. The level of analysis will depend on the scope and objectives agreed at the commencement of the benchmarking.

It is essential that the analysis be impartial and totally objective. It must also be aligned to the benchmarking objectives, and to be of value, it must indicate the benchmarker's strengths and weaknesses and determine and (where possible) quantify the gaps to the best performers, and identify (as far as is possible) the reasons for these gaps. It is important that the metrics are considered collectively and not in isolation because the results from one metric may help to explain those of another. The strategies and working practices of each of the participants should also be explored and used to determine how they may influence their performance.

The performance data and any normalization data streams are analyzed to compare participant performance and determine performance gaps. It is also important to consider a level of statistical testing of the data to ensure that the comparisons being made are statistically significant and the conclusions drawn thereafter are valid.

Quantitative analyses are typically made in relation to the top quartile (i.e., the boundary to the 25th percentile), the best in class (i.e., the single best performer), or the average (mean) of the benchmarking population. There are pros and cons to comparisons to each of these criteria. Analysis of the gap to the best in class is probably the most common and, on the face of it, seems the most obvious—after all the objective is to close the gap to the best performer. However, making comparisons to the data of a single benchmarker at a single point in time will always carry a risk that there is error in the data value (although the validation process should minimize this error) or that the performance level reported is not sustainable in the longer term and therefore not realistic. In contrast, comparison to the top quartile and, in particular, to the average is more stable and reliable because these comprise data from more than one participant.

Reasons for apparent differences in performance should be considered during the analyses. With multinational or globally benchmarking studies, it is important to consider the impact that may be attributed to differences in geographical location. For example, when analyzing costs, it is clear that cost levels (e.g., salaries) in the West cannot be easily compared to Asia, the Middle East, Russia, Africa, or Latin America. In addition, fluctuations in exchange rates between currencies can have a dramatic effect. Likewise, different tax regimes, regulatory requirements, political policies, and cultural difference can all significantly influence performance.

Report Development

Once the analysis is complete, it must be reported to the benchmarking participants. The content of the report and the medium used for reporting will have been agreed upon at the outset of the benchmarking exercise and will, in part, be determined by the type of benchmarking being undertaken.

Reports may be delivered online or electronically or in paper, hard-copy format. Whatever the medium selected, the report must present the benchmarking findings in a clear, concise, and easily understood form. Optimum use of color, diagrams, pictures, and charts should be made in order to optimize communication of the findings. Charts and tables should be annotated to provide guidance to the reader. The analysis should be reported in full, together with recommendations for the focus of performance improvement efforts required in order to close gaps.

One very important point that must be addressed is the level of data anonymization that will be employed in the report. There is always a tradeoff between confidentiality and learning opportunity. The higher the level of confidentiality, the lower the potential for learning. If the identity of the superior performers cannot be revealed, then the opportunity to learn from them is minimized. However, compliance with antitrust legislation is always a primary requirement. Thus, the report must always be in line with any confidentiality agreement

Attribute	Company Score	Competitor Scores
Overall quality	38.6	36.6–46.3
Local dial quality	40.6	36.7–47.9
Billing quality	34.5	28.7–37.2
Installation quality	41.2	43.2–53.3
Long distance quality	47.5	40.9–55.3
Operator quality	41.5	35.0–47.1

Table 25.8 Benchmark Data for Local Telephone Service

Clause Titles	
1	Scope
2	Normative reference
3	Definitions
4	Quality system requirements
4.1	Management responsibility
4.2	Quality system
4.3	Contract review
4.4	Design control
4.5	Document and data control
4.6	Purchasing
4.7	Control of customer-supplied product
4.8	Product identification and traceability
4.9	Process control
4.10	Inspection and testing
4.11	Control of inspection, measuring, and test equipment
4.12	Inspection and test status
4.13	Control of nonconforming product
4.14	Corrective and preventive action
4.15	Handling, storage, packaging, preservation, and delivery
4.16	Control of quality records
4.17	Internal quality audits
4.18	Training
4.19	Servicing
4.20	Statistical techniques

Table 25.9 The Clauses of ISO 9001 and Their Typical Structure

made between the benchmarkers and must meet legal requirements. But, in order to maximize the learning potential, the degree of openness should also be maximized.

Unfortunately, many benchmarking exercises will stop at this point. But to maximize the value gained from benchmarking, organizations must go further to try to understand the practices that enable the leaders to attain their superior performance levels. This is the purpose of phase 2 of Juran's 7-Step Benchmarking Process.

References

Atkinson, H., J. Hamburg, and C. Ittner (1994). Linking Quality to Profits, ASQ Quality Press, Milwaukee, and Institute of Management Accountants, Montvale, NJ.

Campanella, J. ed. (1999). *Principles of Quality Costs*, 3rd ed., ASQ, Milwaukee.

Cokins, G. (1999). "Why Is Traditional Accounting Failing Quality Managers? Activity Based Costing Is the Solution," *Annual Quality Congress Proceedings*, ASQ, Milwaukee.

De Feo, J. A. and W. Barnard (2003). *Juran's Six Sigma Breakthrough and Beyond*, McGraw-Hill, New York.

Lauter, B. E. (1997). "Determining the State of Your Customers," Sterling Quality Conference, Orlando, FL.

Mortimer, J., J. A. DeFeo, and L. Stepnick (2003). "Reducing the Costs of Poor Quality Health Care" a report published by the Midwest Business Group on Health in collaboration with the Juran Institute, Inc. and The Severyn Group, Inc.

Rust, R. T., A. J. Zahorik, and T. L. Keiningham (1994). Return on Quality, Probus, Chicago.

Schottmiller, J. C. (1996). "ISO 9000 and Quality Costs," *Annual Quality Congress Proceedings*, ASQ, Milwaukee, pp. 194–199.

Werner, F. M. and J. A. F. Stoner (1995). *Modern Financial Managing*, Harper Collins College Publishers, New York, pp. 143–144.

APPENDIX I

Sample Competency Matrices and Job Profiles

Title	Global Division Vice President of Quality
Primary Purpose	The Division Vice President of Quality will contribute to company profitability by delivering industry leading quality performance to customers. This executive will provide visionary leadership on long-range objectives and quality programs that will systemically drive organizational success in both current and future business environments. This executive will instill the right competencies and processes to achieve industry-leading teams at all levels of the organization to deliver customer focused quality solutions. He/she will interact frequently with sales, marketing, engineering, procurement and manufacturing on a global basis.
Duties	Duties and responsibilities will include 1. Realigning current resources to support a proactive approach to predict, prevent, and protect the company and its customers from quality performance issues. 2. Responsibility for company quality processes and development of strategic quality, vision, mission, goals and objectives in conjunction with company-wide direction. 3. Creating overall direction for consistent quality methodologies, processes and procedures across all operations of the division. 4. Working in collaboration with the global organization to ensure the management of performance, qualification process, and use of metrics. 5. Acting as quality advocate inside and outside of the company. 6. Drives the implementation and maintenance of Total Quality Management and Lean Six Sigma methods in the division globally. 7. Responsible to ensure all locations have knowledgeable and effective management representatives to oversee maintenance and continual improvement of quality system standards. 8. Extensive travel is required as is accessibility via phone to participate in and lead global conference calls as required.

(Continued)

Title	Global Division Vice President of Quality
	Initiatives include 1. Drive Lean and Six Sigma initiatives globally within the division. 2. Ensure quality system compliance to standards and customer requirements across the division. 3. Complete process and system compliance auditing and improvements at regular intervals. 4. Conduct customer satisfaction monitoring and report results quarterly to division leadership. 5. Drive quality metrics reporting through all of the global division operations. 6. Serve as customer advocate to business teams overseeing the division's effectiveness at meeting customer quality standards as well as other requirements. 7. Establish and implement standardized policies, standards, processes, metrics, and controls surrounding customer quality. 8. Ensure the implementation of employee/supplier training and education. Programs to develop a consistent understanding of the company quality process. 9. Facilitate the use and development of division resources for addressing internal and external quality issues early in the product life cycle and to evolve the team to higher value proactive activities. 10. Work with global quality and IT to develop standardized data collection and reporting systems. 11. Facilitate the linkage of quality performance to sourcing decisions. 12. Meet annual quality goals. 13. Collaborate with company quality leadership to leverage best practices across the company. Current year goals include 1. Quality direction—form and chair quality council. Lead the development of quality strategy, establish quality scorecard, define the "right" quality tools and metrics, and drive effectiveness of the quality system. 2. Customer advocate—provide for customer issue escalation and timely resolution. 3. Attract and develop industry-leading talent. 4. Report quality metrics. 5. Organize and conduct regular formal quality reviews. 6. Develop and manage execution of a global quality improvement plan for the division. Link priorities to business goals.
Education Required	An undergraduate degree is required. A Master of Business Administration degree is desirable.
Education Preferred	Certification as a Six Sigma Black Belt or Master Black Belt. Certified Quality Engineer.

(Continued)

Title	Global Division Vice President of Quality
Work Experience Required	Candidate must have 7 to 10 years experience in component-level quality environments. Must have extensive experience with Six Sigma methods and a record of successful application of Six Sigma to drive improvements in engineering, manufacturing, cost improvement, and sourcing. Must be able to drive Six Sigma into internal and external processes. Must have a track record of effectively identifying root causes and organizational levers to successfully address barriers to industry-leading business performance. Must have enthusiasm for new industry initiatives and be able to influence others to address resistance to change. Must have proven ability to lead diverse global teams to achieve cost, quality, and time to market commitments. Additionally, must have 1. Strong negotiation and relationship building skills. 2. Skills to create the path by which the division can achieve a quality driven culture focused on the customer. 3. High level of energy to create support for quality initiatives. 4. Drive key functional leaders to embrace quality as a critical element of the Division's success. 5. Provide objective, accurate, and truthful data to drive improvement in quality. 6. Track record of developing, attracting, and retaining top talent and developing effective teams. 7. Success in collaborating with key work partners. 8. Success in managing change through all levels of the organization. 9. Strong presentation and writing skills. 10. Executive presence.
Skills Required	Required knowledge includes 1. Fundamentals of global industry quality standards including ISO-TS-16949, ISO-9000-2000 and major customer/industry requirements. 2. Analytical nature and discipline. 3. How to achieve the metrics. 4. Ability to push issues with key stakeholders. 5. Technical knowledge and ability to hire the right people. 6. Demonstrated ability to build a strong team. 7. Global authority and cultural sensitivity. 8. Must demonstrate extensive knowledge of automotive quality standards and requirements. 9. Detailed understanding of business performance and quality metrics. 10. Must be able to leverage a personal command of business statistics and performance metrics to build a compelling case for specific decisions and recommendations.

(Continued)

Title	Global Division Vice President of Quality
Other	A broad based, visible senior business leader who has successfully utilized quality tools and philosophy in leading an organization to higher levels of performance; a leader who can generate instant credibility both within the company and with external customers and suppliers creating a path by which the division can achieve a customer-focused, quality driven culture. Must be able to lead key functional leaders to embrace quality as a critical success factor in the business. Must have the presence to be able to interact effectively at the highest levels of the company's, customer and supplier leadership. 1. An individual with a vision for the future, a view of how things might be different. 2. A leader who questions assumptions and conventional thinking. Challenges tradition, existing business models, and is an effective change agent. Able to communicate enthusiasm for new industry initiatives and influence others to address resistance to change. 3. Able to build a strong, cohesive quality management team, working with diverse cultures, personalities, and ambitions. 4. Able to translate strategy into rigorous operating plans complete with goals, accountabilities, timetables, and measures. 5. A high-energy leader with a positive attitude in the face of difficult challenges or adversity, who delivers on commitments to customers. 6. Professional and decisive. Reaches closure in a timely fashion on difficult or complex problems and demonstrates courage and strength of conviction. 7. An articulate, effective communicator who sets clear standards and goals and holds individuals accountable. An exemplary behavior that is consistent with ethical principles such as avoiding conflicts of interest, avoiding compromising business situations and handling confidential information appropriately, understanding that professional integrity is critical to building the trust and respect required to influence and lead others.

Title	Director, Global Customer Quality
Primary Purpose	Provides internal and external direction and communication necessary to develop and sustain customer confidence in the organization's quality performance. Activities include both proactive and reactive components of customer quality relationship management on behalf of the organization.
Duties	1. Coordinate internal communication and integration of customer quality requirements. 2. Coordinate global responses to complaints and quality performance problems. 3. Primary contact for GAM customer quality issues. 4. Maintain global actions list and/or project plans for achieving/sustaining preferred supplier status to major customers. 5. Maintain current customer scorecards on intranet site for access by all company entities globally. 6. Ensure customer requirements for change notification are available, updated and accessible by all entities. 7. Drive global improvement in QN reductions based on improved QN cause data, effective problem solving, and systemic analysis and improvement. 8. Participate in the Company Customer Review Process when it is conducted for major accounts. 9. Communicate best practices in achieving customer requirements.
Education Required	Bachelor's degree
Education Preferred	MBA, Certified Black Belt, ASQ certifications. Microsoft Office proficiency (Excel, Word, PowerPoint, etc.)
Work Experience Required	A minimum of 10 to 15 years experience in quality roles or related manufacturing areas where a substantial understanding of company's quality processes in design and manufacturing has been gained. Global experience is preferred.
Skills Required	The person in this job must possess good leadership, and excellent planning and project management skills. He/she will be able to champion for the needed elements of change and harness the necessary resources to bring about change. He/she must be the voice of the customer while working to actively effect improvements necessary to exceed customer expectations. The jobholder must be professional, credible, and able to influence key regional owners and stakeholders around the world. As a result, cross-cultural sensitivity and acceptability is essential. Analytical and creative thinking, independent judgment and the ability to present information and ideas clearly and concisely are also required. An independent thinker, the jobholder must be able to organize their own work, anticipating, planning, and monitoring their workload. The person in this job must possess strong written and oral communication skills. A solid experienced-based understanding of quality standards, systems and tools is a requirement.
Other	The jobholder must be prepared to travel both domestically and internationally as required. He/she must be a team player and a self-directed, self-starter who sees what needs to be done and can get objectives accomplished with minimal supervision. It would be expected that the person in this role would be qualified to fit into a leadership role within the quality organization. This must be a part of the selection criteria to be sure the best person is selected.

Title	Senior Manager, Global Supplier Quality
Primary Purpose	This position is responsible for defining and driving global SQA quality and improvement programs for the company in both lead as well as individual contributor roles. As the SQA lead, this position is responsible for collaborating with all divisional SQA personnel to create and deliver quality-focused initiatives around material and product qualification, testing plans, schedules, quality issues and risks in a fast-paced development and manufacturing environment. The job requires experience in successfully building, leading and executing quality programs in a manufacturing environment, has demonstrated quality-management/Six Sigma experience, and has demonstrated technical and practical quality experience in proactive and reactive environments.
Duties	1. Defining, setting and delivering SQA quality improvement initiatives with the SQA team leads that involve all elements of planning/execution and release-based quality/risk assessment. 2. Drive specific process, delivery and tool improvements in development, manufacturing, and release quality measurement and assessment with division SQA personnel. 3. Contribute to and support release-specific SQA programs led by division SQA personnel that support quality improvement initiatives. 4. Collaborate and partner with Program Management, Development, Support leads to build an end-end SQA focus on process improvement that helps drive "quality upstream" across the supply base globally. 5. Responsible to ensure development and monitoring of annual supplier quality improvement plans globally. 6. Defines, develops and directs compliance to a global SQA standard by working with all divisions for one set of common standards applied equally on a global basis. 7. Other duties and responsibilities as may be determined by management.
Education Required	Four-year BA/BS degree or related field. Dedicated to concepts and principles of supplier quality assurance and management.
Education Preferred	As above
Work Experience Required	Three years of demonstrated experience in a quality role in a multi-site manufacturing environment. Previous experience managing a successful quality improvement process with multi-functional involvement (engineering or manufacturing). A minimum of 5 years working experience in a quality management role including responsibility for interfacing with customers on SQA requirements.
Skills Required	• Supplier quality management. • Manufacturing quality management. • Advance quality planning. • Advanced quality tools including DOE. • Excellent communication and presentation skills. • Effective coordination and project management skills. • Action and results oriented.
Other	The jobholder must be willing to travel globally, 50 percent minimum. Chinese language skills a significant plus.

Title	Corporate Quality Engineer
Primary Purpose	This position is to be the internal global resource and driver of the tools and methods necessary for supporting the implementation of global TQM, Six Sigma and Lean techniques.
Duties	1. Primary quality training and development coach for Lead Teams, Sponsors, Six Sigma Black Belts, and Green Belts. 2. Facilitate knowledge transfer from related external consultants to company employees. 3. Leader and facilitator for select project teams. 4. Problem solving facilitator and resource for select problems requiring immediate resolution. 5. Coordinate internal communication and integration of customer quality requirements as required. 6. Cultivate a global network of internal company expertise in TQM and Six Sigma. 7. Identify internal and external (supplier) opportunities for improvement to assist entities in meeting quality and cost reduction targets. Assist in the management of consultant schedules and expenses. Provide input on selecting future Black Belts and Master Black Belts. 8. Supports company TQM program by following approved policies and procedures. 9. Performs other related duties as assigned by management.
Education Required	Bachelor's degree and Certified Six Sigma Master Black Belt.
Education Preferred	MBA
Work Experience Required	A minimum of 5 to 10 years of plant manufacturing experience in a quality or manufacturing role with demonstrated successful, measurable performance in the application of Six Sigma. Must have achieved Master Black Belt certification as a result of successful projects implemented in a manufacturing environment. Trained and experienced in team facilitation skills as well as the technical tools associated with TQM, Lean, and Six Sigma. Knowledge and successful experience in a leading new product development process (APQP) is essential. Must be willing to travel extensively to work with company locations and suppliers worldwide.
Work Experience Preferred	Management experience overseeing the implementation of a successful Six Sigma initiative in a leading manufacturing company. Global experience is preferred. Language skills beyond English are highly desirable.
Skills Required	The person in this job must possess good leadership, and excellent planning and project management skills. Demonstrated successful facilitation and teamwork skills are also required. The jobholder must be professional, credible, and able to influence key regional owners and stakeholders around the world. As a result, cross-cultural sensitivity and acceptability is essential. Analytical and creative thinking, independent judgment and the ability to present and teach the quality tools and methods clearly and concisely are also required. The person in this job must possess strong written and oral communication skills with the ability to work effectively and efficiently with management as well as hourly employees.
Other	It would be expected that the person in this role would be qualified to fit into a management role within the quality organization in the future. This must be a part of the selection criteria to be sure the best person is selected.

Position	Director of Auditing, Quality Systems and Environmental Compliance
Overview	The Global Director of Auditing for quality systems and environmental compliance conducts, coordinates and schedules audits for management to assess effectiveness of management systems and controls for compliance to quality and environmental standards/regulations. This includes examining records, interviewing workers to ensure recording of transactions and compliance with applicable standards and/or laws and regulations. This person coordinates the process by which company locations will assess management systems to determine their efficiency and protective value. As part of this process the position enforces uniform methods to review records pertaining to operations, emissions and waste management. This requires the analysis of data obtained for evidence of deficiencies in controls, duplication of effort or lack of compliance with laws, government regulations, and management policies or procedures. This person oversees a standard approach to preparing reports of findings and recommendations for local, divisional and corporate management. This person may conduct special studies for management. Regularly works to verify that adequate internal controls are in place to minimize risk and exposure to the company, under the general guidance of the VP of Global Quality.
Job Description	The Director of Audits will work with each of the global division's at the entity level to plan, coordinate and sometimes participate in quality and environmental audits and internal control assessments of all operations. In so doing, the Audit Manager will perform assessments and establish appropriate staffing resource and corrective action recommendations for the entities and divisions. This will include cross-entity and cross-regional audits as appropriate to leverage best practices and share successes. Additionally the audit manager will be responsible to work with all divisions to ensure company optimizes regional resources to serve all divisions.
Primary Responsibilities	Review prior audit reports and initiate discussions with local quality and site management to assess quality and/or environmental system compliance. The result will be a documented report of the risk of operations being examined due to noncompliance. This person ensures coordination with appropriate local quality and/or environmental resources to meet standards and regulatory requirements. • Plan and execute audits in a professional manner to ensure timely completion of summary reports. It will be the site management's responsibility to develop and implement appropriate improvement plans to address audit findings. • Communicate results of all audit findings and recommendations for improvement to management through concise, high quality audit reports. • Participate in special projects when necessary including but not limited to occasional assistance in preparation for customer quality and/or environmental audits. • Play an active role in the continuous improvement of the company quality and environmental audit function. • Combined domestic and international travel is estimated to be no more than 40 percent.

(*Continued*)

Position	Director of Auditing, Quality Systems and Environmental Compliance
Requirements	• Bachelor's degree in a business or relevant technical discipline. Extensive lead audit and audit management experience in quality systems and environmental compliance will be considered in place of a formal educational degree. • Professional designation/certification as a Lead Auditor is required. • 7 + years work experience in auditing preferably in a manufacturing industry environment. • ISO-9000-2000, ISO-TS-16949 and ISO-14001 assessment experience required. Familiarity with these and any new or revised requirements related to quality and environmental systems compliance. • Strong work ethic, systemic and process thinking, and organization skills required. • Strong interpersonal and communication (written and verbal) skills necessary to deal with all levels of personnel. • Microsoft Office proficiency (Excel, Word, PowerPoint, etc.)
Broad Content of the Role	The Audit Manager is responsible for optimizing and coordinating with divisions and entities for managing the audit program within the company. This includes coordinating the planning, scheduling, performing, tracking closure, and reporting of audits.
Key Tasks	• Owns the audit process and ensures this is well documented, understood, effective and up to date with the most current standards and regulatory requirements. • Creates and maintains the annual global audit schedule and status reports. • Trains auditors to function as internal auditors. • Participate frequently as part of audit teams. • Track audit actions to closure. • Reports on audit program key performance indicators to process owners and company management: e.g., average age of overdue actions, recurrent findings, concentration of findings per department/location. • Ensures company management is knowledgeable of risks of noncompliance to quality standards and environmental regulations. • Acts as a consultant to the business. Develops and maintains expertise and acts as the Subject Matter Expertise (SME) for quality systems and environmental compliance.

(*Continued*)

Position	Director of Auditing, Quality Systems and Environmental Compliance
Essential Experience	• Documenting process definition and improvement. • Implementation of recognized industry process/quality/environmental standards and/or regulations. • Design and delivery of training for standards, regulations, and auditing. • Interfacing with external registrars/regulatory bodies for resolution of major findings.
Essential Skills	Excellent communicator and negotiator at all levels. Good analytical and problem solving skills.
Essential Disposition	• Committed and enthusiastic about quality, environment, and process improvement. • Attention to detail. • Flexible and adaptable, good team player, practical, and pragmatic. • Proactive, able to work on own initiative.
Job Challenges	Remote and dotted line supervision of audit resources. The ability to assimilate information quickly and to deal with senior management often in uncooperative situations. Will need to be pragmatic without compromising the integrity of the communication. Multitasking of several assignments and initiatives will be expected.

Master Black Belt Profile

Master Black Belts are company-wide Six Sigma or quality experts. The Master Black Belt is qualified to teach other Six Sigma Green and Black Belts the methodologies, tools, and applications in all functions and levels of the company. In addition, the Master Black Belt is able to provide leadership integrating the Six Sigma approach into the business strategy of the company, and contributes to creating and carrying out the organization's strategic business and operational plans. As a Black Belt, the Master Black Belt candidate has personally led successful project teams.

Key Roles

- Provides technical support and mentoring.
- Facilitates multiple projects.
- Provides advice to Champions and executive management.
- Trains others on the Lean and Six Sigma tools and techniques.
- Provides leadership to management groups in the integration of the Lean and Six Sigma approaches with the organization's business strategy.
- Contributes to creating and carrying out the organization's strategic business and operational plans.
- Is trained on advanced tools, strategic deployment, and train-the-trainer workshops.

Black Belt Profile

Black Belts are technical specialists' assigned full responsibility to implement Six Sigma projects through a business unit, function, or process. They will become viewed as "initiators" of improvement activity, and are full-time on-site project leaders.

Key Roles

- Keeps Champion informed of project progress.
- Develops, coaches, and leads multifunctional improvement teams.
- Mentors and advises management on prioritizing, charting, and launching projects.
- Uses and teaches tools and methods to Green Belts, Yellow Belts, and Subject Matter Experts.
- Actively seeks to use the Six Sigma breakthrough steps to solve chronic problems, remove waste, and plan new services or products.
- Learns to align projects to local business objectives.
- Provides project management, facilitates, and leads teams.
- Is trained and certified in the appropriate toolset.

Green Belt Profile

Green Belts are employees with sufficient knowledge to support and participate on Lean and Lean Six Sigma projects. They can be a team leader or a team member.

Key Roles

- May lead projects.
- May be a core project team member.
- Actively participates and contributes expertise to larger Black Belt projects.
- Uses Lean and Six Sigma steps to solve problems.
- Uses Lean to remove waste.
- Completes multiple projects over time, one at a time.
- Is trained and certified in the Green Belt toolset.

Lean Master Profile

Lean Masters are company-wide Lean or Quality experts. The Lean Master is qualified to teach other Lean Experts and team members the methodologies, tools, and applications in all functions and levels of the company. In addition, the Lean Master is able to provide leadership integrating the Lean approach into the business strategy of the company, and contributes to creating and carrying out the organization's strategic business and operational plans. As a Lean Expert, the Lean Master candidate has personally led successful lean project teams.

Key Roles

- Provides technical support and mentoring.
- Facilitates multiple projects.
- Provides advice to Champions and executive management.
- Mentors and advises management on prioritizing, charting, and launching projects.
- Trains others on the Lean tools and techniques.
- Provides leadership to management groups in the integration of Lean with the organization's business strategy.
- Contributes to creating and carrying out the organization's strategic business and operational goals.
- Is trained on advanced tools, strategic deployment, and train-the-trainer workshops.

Lean Expert Profile

Lean Experts are technical specialists assigned full responsibility to implement Lean projects through a business unit, function, or process. They will become viewed as "initiators" of improvement activity, and are full-time on-site project leaders.

Key Roles

- Keeps Champion informed of project progress.
- Develops, coaches, and leads multifunctional improvement teams.
- Mentors and advises management on prioritizing, charting, and launching projects.
- Uses and teaches tools and methods to Lean managers and subject matter experts.
- Actively seeks to use the Lean steps to solve chronic problems, remove waste, and plan new services or products.
- Learns to align projects to local business objectives.
- Provides project management, facilitates, and leads teams.
- Is trained and certified in the appropriate toolset.

APPENDIX II

Tables and Charts

Observations in Sample, n	Chart for Averages: Factors for Control Limits			Chart for Standard Deviations: Factors for Central Line		Factors for Control Limits				Chart for Ranges: Factors for Central Line		Factors for Control Limits				
	A	A_2	A_3	c_4	$1/c_4$	B_3	B_4	B_5	B_6	d_2	$1/d_2$	d_3	D_1	D_2	D_3	D_4
2	2.121	1.880	2.659	0.7979	1.2533	0	3.267	0	2.606	1.128	0.8865	0.853	0	3.686	0	3.267
3	1.732	1.023	1.954	0.8862	1.1284	0	2.568	0	2.276	1.693	0.5907	0.888	0	4.358	0	2.574
4	1.500	0.729	1.628	0.9213	1.0854	0	2.266	0	2.088	2.059	0.4857	0.880	0	4.698	0	2.282
5	1.342	0.577	1.427	0.9400	1.0638	0	2.089	0	1.964	2.326	0.4299	0.864	0	4.918	0	2.114
6	1.225	0.483	1.287	0.9515	1.0510	0.030	1.970	0.029	1.874	2.534	0.3946	0.848	0	5.078	0	2.004
7	1.134	0.419	1.182	0.9594	1.0423	0.118	1.882	0.113	1.806	2.704	0.3698	0.833	0.204	5.204	0.076	1.924
8	1.061	0.373	1.099	0.9650	1.0363	0.185	1.815	0.179	1.751	2.847	0.3512	0.820	0.388	5.306	0.136	1.864
9	1.000	0.337	1.032	0.9693	1.0317	0.239	1.761	0.232	1.707	2.970	0.3367	0.808	0.547	5.393	0.184	1.816
10	0.949	0.308	0.975	0.9727	1.0281	0.284	1.716	0.276	1.669	3.078	0.3249	0.797	0.687	5.469	0.223	1.777
11	0.905	0.285	0.927	0.9754	1.0252	0.321	1.679	0.313	1.637	3.173	0.3152	0.787	0.811	5.535	0.256	1.744
12	0.866	0.266	0.886	0.9776	1.0229	0.354	1.646	0.346	1.610	3.258	0.3069	0.778	0.922	5.594	0.283	1.717
13	0.832	0.249	0.850	0.9794	1.0210	0.382	1.618	0.374	1.585	3.336	0.2998	0.770	1.025	5.647	0.307	1.693
14	0.802	0.235	0.817	0.9810	1.0194	0.406	1.594	0.399	1.563	3.407	0.2935	0.763	1.118	5.696	0.328	1.672
15	0.775	0.223	0.789	0.9823	1.0180	0.428	1.572	0.421	1.544	3.472	0.2880	0.756	1.203	5.741	0.347	1.653
16	0.750	0.212	0.763	0.9835	1.0168	0.448	1.552	0.440	1.526	3.532	0.2831	0.750	1.282	5.782	0.363	1.637
17	0.728	0.203	0.739	0.9845	1.0157	0.466	1.534	0.458	1.511	3.588	0.2787	0.744	1.356	5.820	0.378	1.622
18	0.707	0.194	0.718	0.9854	1.0148	0.482	1.518	0.475	1.496	3.640	0.2747	0.739	1.424	5.856	0.391	1.608
19	0.688	0.187	0.698	0.9862	1.0140	0.497	1.503	0.490	1.483	3.689	0.2711	0.734	1.487	5.891	0.403	1.597
20	0.671	0.180	0.680	0.9869	1.0133	0.510	1.490	0.504	1.470	3.735	0.2677	0.729	1.549	5.921	0.415	1.585
21	0.655	0.173	0.663	0.9876	1.0126	0.523	1.477	0.516	1.459	3.778	0.2647	0.724	1.605	5.951	0.425	1.575
22	0.640	0.167	0.647	0.9882	1.0119	0.534	1.466	0.528	1.448	3.819	0.2618	0.720	1.659	5.979	0.434	1.566
23	0.626	0.162	0.633	0.9887	1.0114	0.545	1.455	0.539	1.438	3.858	0.2592	0.716	1.710	6.006	0.443	1.557
24	0.612	0.157	0.619	0.9892	1.0109	0.555	1.445	0.549	1.429	3.895	0.2567	0.712	1.759	6.031	0.451	1.548
25	0.600	0.153	0.606	0.9896	1.0105	0.565	1.435	0.559	1.420	3.931	0.2544	0.708	1.806	6.056	0.459	1.541

*The above table is a copy of Table 27 in *ASTM Manual on Presentation of Data and Control Chart Analysis*. (1976). ASTM Publication STP15D, American Society for Testing and Materials, Philadelphia, pp. 134–135. Used with permission.

Notes: For $n > 25$, $A = 3/\sqrt{n}$, $A_3 = 3/c_4\sqrt{n}$, $c_4 \approx 4(n-1)/(4n-3)$; $B_3 = 1 - 3/c_4\sqrt{2(n-1)}$, $B_4 = 1 + 3/c_4\sqrt{2(n-1)}$, $B_5 = c_4 - 3/\sqrt{2(n-1)}$, $B_6 = c_4 + 3/\sqrt{2(n-1)}$

TABLE A Factors for Computing Control Chart Lines*

Purpose of Chart	Chart for	Central Line	3-Sigma Control Limits
For analyzing past inspection data for control ($\overline{\overline{X}}$, $\overline{s}$, $\overline{R}$ are average values for the data being analyzed)	Averages Standard deviations Ranges	$\overline{\overline{X}}$ $\overline{s}$ $\overline{R}$	$\overline{\overline{X}} \pm A_3\overline{s}$, or $\overline{\overline{X}} \pm A_2\overline{R}$ $B_3\overline{s}$ and $B_4\overline{s}$ $D_3\overline{R}$ and $D_4\overline{R}$
For controlling quality during production ($\overline{X}_0$, σ_0, R_0, are selected standard values; $R_0 = d_2\sigma_0$ for samples of size n)	Averages Standard deviations Ranges	$\overline{X}_0$ s_0 or $c_4\sigma_0$ R_0 or $d_2\sigma_0$	$\overline{X}_0 \pm A\sigma_0$ or $\overline{X}_0 \pm A_2R_0$ $B_5\sigma_0$ and $B_6\sigma_0$ $D_1\sigma_0$ and $D_2\sigma_0$

FORMULAS

Proportion of total area under the curve from $-\infty$ to $K = \frac{X-\mu}{\sigma}$. To illustrate: when $K = +2.0$, the probability is 0.9773 of obtaining a value equal to or less than X.

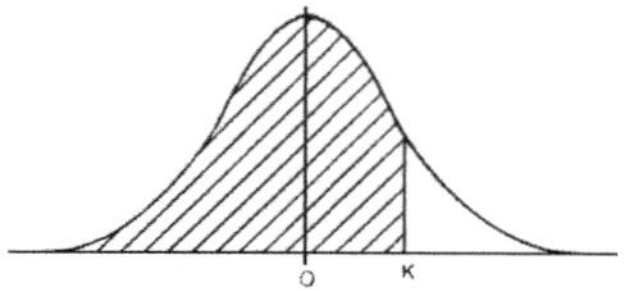

K	0.09	0.08	0.07	0.06	0.05	0.04	0.03	0.02	0.01	0.00
–3.5	0.00017	0.00017	0.00018	0.00019	0.00019	0.00020	0.00021	0.00022	0.00022	0.00023
–3.4	0.00024	0.00025	0.00026	0.00027	0.00028	0.00029	0.00030	0.00031	0.00033	0.00034
–3.3	0.00035	0.00036	0.00038	0.00039	0.00040	0.00042	0.00043	0.00045	0.00047	0.00048
–3.2	0.00050	0.00052	0.00054	0.00056	0.00058	0.00060	0.00062	0.00064	0.00066	0.00069
–3.1	0.00071	0.00074	0.00076	0.00079	0.00082	0.00085	0.00087	0.00090	0.00094	0.00097
–3.0	0.00100	0.00104	0.00107	0.00111	0.00114	0.00118	0.00122	0.00126	0.00131	0.00135
–2.9	0.0014	0.0014	0.0015	0.0015	0.0016	0.0016	0.0017	0.0017	0.0018	0.0019
–2.8	0.0019	0.0020	0.0021	0.0021	0.0022	0.0023	0.0023	0.0024	0.0025	0.0026
–2.7	0.0026	0.0027	0.0028	0.0029	0.0030	0.0031	0.0032	0.0033	0.0034	0.0035
–2.6	0.0036	0.0037	0.0038	0.0039	0.0040	0.0041	0.0043	0.0044	0.0045	0.0047
–2.5	0.0048	0.0049	0.0051	0.0052	0.0054	0.0055	0.0057	0.0059	0.0060	0.0062
–2.4	0.0064	0.0066	0.0068	0.0069	0.0071	0.0073	0.0075	0.0078	0.0080	0.0082
–2.3	0.0084	0.0087	0.0089	0.0091	0.0094	0.0096	0.0099	0.0102	0.0104	0.0107
–2.2	0.0110	0.0113	0.0116	0.0119	0.0122	0.0125	0.0129	0.0132	0.0136	0.0139
–2.1	0.0143	0.0146	0.0150	0.0154	0.0158	0.0162	0.0166	0.0170	0.0174	0.0179
–2.0	0.0183	0.0188	0.0192	0.0197	0.0202	0.0207	0.0212	0.0217	0.0222	0.0228
–1.9	0.0233	0.0239	0.0244	0.0250	0.0256	0.0262	0.0268	0.0274	0.0281	0.0287
–1.8	0.0294	0.0301	0.0307	0.0314	0.0322	0.0329	0.0336	0.0344	0.0351	0.0359
–1.7	0.0367	0.0375	0.0384	0.0392	0.0401	0.0409	0.0418	0.0427	0.0436	0.0446
–1.6	0.0455	0.0465	0.0475	0.0485	0.0495	0.0505	0.0516	0.0526	0.0537	0.0548
–1.5	0.0559	0.0571	0.0582	0.0594	0.0606	0.0618	0.0630	0.0643	0.0655	0.0668
–1.4	0.0681	0.0694	0.0708	0.0721	0.0735	0.0749	0.0764	0.0778	0.0793	0.0808
–1.3	0.0823	0.0838	0.0853	0.0869	0.0885	0.0901	0.0918	0.0934	0.0951	0.0968
–1.2	0.0985	0.1003	0.1020	0.1038	0.1057	0.1075	0.1093	0.1112	0.1131	0.1151
–1.1	0.1170	0.1190	0.1210	0.1230	0.1251	0.1271	0.1292	0.1314	0.1335	0.1357

–1.0	0.1379	0.1401	0.1423	0.1446	0.1469	0.1492	0.1515	0.1539	0.1562	0.1587
–0.9	0.1611	0.1635	0.1660	0.1685	0.1711	0.1736	0.1762	0.1788	0.1814	0.1841
–0.8	0.1867	0.1894	0.1922	0.1949	0.1977	0.2005	0.2033	0.2061	0.2090	0.2119
–0.7	0.2148	0.2177	0.2207	0.2236	0.2266	0.2297	0.2327	0.2358	0.2389	0.2420
–0.6	0.2451	0.2483	0.2514	0.2546	0.2578	0.2611	0.2643	0.2676	0.2709	0.2743
–0.5	0.2776	0.2810	0.2843	0.2877	0.2912	0.2946	0.2981	0.3015	0.3050	0.3085
–0.4	0.3121	0.3156	0.3192	0.3228	0.3264	0.3300	0.3336	0.3372	0.3409	0.3446
–0.3	0.3483	0.3520	0.3557	0.3594	0.3632	0.3669	0.3707	0.3745	0.3783	0.3821
–0.2	0.3859	0.3897	0.3936	0.3974	0.4013	0.4052	0.4090	0.4129	0.4168	0.4207
–0.1	0.4247	0.4286	0.4325	0.4364	0.4404	0.4443	0.4483	0.4562	0.4562	0.4602
–0.0	0.4641	0.4681	0.4721	0.4761	0.4801	0.4840	0.4880	0.4920	0.4960	0.5000
K	**0.00**	**0.01**	**0.02**	**0.03**	**0.04**	**0.05**	**0.06**	**0.07**	**0.08**	**0.09**
+0.0	0.5000	0.5040	0.5080	0.5120	0.5160	0.5199	0.5239	0.5279	0.5319	0.5359
+0.1	0.5398	0.5438	0.5478	0.5517	0.5557	0.5596	0.5636	0.5675	0.5714	0.5753
+0.2	0.5793	0.5832	0.5871	0.5910	0.5948	0.5987	0.6026	0.6064	0.6103	0.6141
+0.3	0.6179	0.6217	0.6255	0.6293	0.6331	0.6368	0.6406	0.6443	0.6480	0.6517
+0.4	0.6554	0.6591	0.6628	0.6664	0.6700	0.6736	0.6772	0.6808	0.6844	0.6879
+0.5	0.6915	0.6950	0.6985	0.7019	0.7054	0.7088	0.7123	0.7157	0.7190	0.7224
+0.6	0.7257	0.7291	0.7324	0.7357	0.7389	0.7422	0.7454	0.7486	0.7517	0.7549
+0.7	0.7580	0.7611	0.7642	0.7673	0.7704	0.7734	0.7764	0.7794	0.7823	0.7852
+0.8	0.7881	0.7910	0.7939	0.7967	0.7995	0.8023	0.8051	0.8079	0.8106	0.8133
+0.9	0.8159	0.8186	0.8212	0.8238	0.8264	0.8289	0.8315	0.8340	0.8365	0.8389
+1.0	0.8413	0.8438	0.8461	0.8485	0.8508	0.8531	0.8554	0.8577	0.8599	0.8621
+1.1	0.8643	0.8665	0.8686	0.8708	0.8729	0.8749	0.8770	0.8790	0.8810	0.8830
+1.2	0.8849	0.8869	0.8888	0.8907	0.8925	0.8944	0.8962	0.8980	0.8997	0.9015
+1.3	0.9032	0.9049	0.9066	0.9082	0.9099	0.9115	0.9131	0.9147	0.9162	0.9177
+1.4	0.9192	0.9207	0.9222	0.9236	0.9251	0.9265	0.9279	0.9292	0.9306	0.9319
+1.5	0.9332	0.9345	0.9357	0.9370	0.9382	0.9394	0.9406	0.9418	0.9429	0.9441

TABLE B Normal Distribution

K	0.00	0.01	0.02	0.03	0.04	0.05	0.06	0.07	0.08	0.09
+1.6	0.9452	0.9463	0.9474	0.9484	0.9495	0.9505	0.9515	0.9525	0.9535	0.9545
+1.7	0.9554	0.9564	0.9573	0.9582	0.9591	0.9599	0.9608	0.9616	0.9625	0.9633
+1.8	0.9641	0.9649	0.9656	0.9664	0.9671	0.9678	0.9686	0.9693	0.9699	0.9706
+1.9	0.9713	0.9719	0.9726	0.9732	0.9738	0.9744	0.9750	0.9756	0.9761	0.9767
+2.0	0.9773	0.9778	0.9783	0.9788	0.9793	0.9798	0.9803	0.9808	0.9812	0.9817
+2.1	0.9821	0.9826	0.9830	0.9834	0.9838	0.9842	0.9846	0.9850	0.9854	0.9857
+2.2	0.9861	0.9864	0.9868	0.9871	0.9875	0.9878	0.9881	0.9884	0.9887	0.9890
+2.3	0.9893	0.9896	0.9898	0.9901	0.9904	0.9906	0.9909	0.9911	0.9913	0.9916
+2.4	0.9918	0.9920	0.9922	0.9925	0.9927	0.9929	0.9931	0.9932	0.9934	0.9936
+2.5	0.9938	0.9940	0.9941	0.9943	0.9945	0.9946	0.9948	0.9949	0.9951	0.9952
+2.6	0.9953	0.9955	0.9956	0.9957	0.9959	0.9960	0.9961	0.9962	0.9963	0.9964
+2.7	0.9965	0.9966	0.9967	0.9968	0.9969	0.9970	0.9971	0.9972	0.9973	0.9974
+2.8	0.9974	0.9975	0.9976	0.9977	0.9977	0.9978	0.9979	0.9979	0.9980	0.9981
+2.9	0.9981	0.9982	0.9983	0.9983	0.9984	0.9984	0.9985	0.9985	0.9986	0.9986
+3.0	0.99865	0.99869	0.99874	0.99878	0.99882	0.99886	0.99889	0.99893	0.99896	0.99900
+3.1	0.99903	0.99906	0.99910	0.99913	0.99915	0.99918	0.99921	0.99924	0.99926	0.99929
+3.2	0.99931	0.99934	0.99936	0.99938	0.99940	0.99942	0.99944	0.99946	0.99948	0.99950
+3.3	0.99952	0.99953	0.99955	0.99957	0.99958	0.99960	0.99961	0.99962	0.99964	0.99965
+3.4	0.99966	0.99967	0.99969	0.99970	0.99971	0.99972	0.99973	0.99974	0.99S75	0.99976
+3.5	0.99977	0.99978	0.99978	0.99979	0.99980	0.99981	0.99981	0.99982	0.99983	0.99983

*Adapted with permission from Grant, Eugene L. and Leavenworth, Richard S. (1972). *Statistical Quality Control*, 4th ed. McGraw-Hill, New York, pp. 642–643.

Table B Normal Distribution* *(Continued)*

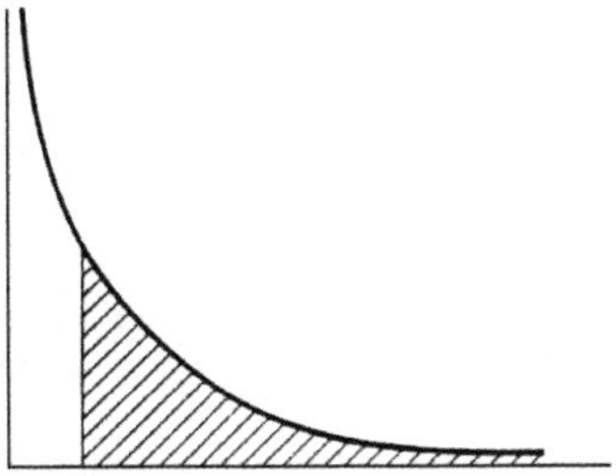

$\frac{X}{\mu}$	**0.00**	**0.01**	**0.02**	**0.03**	**0.04**	**0.05**	**0.06**	**0.07**	**0.08**	**0.09**
0.0	1.000	0.9900	0.9802	0.9704	0.9608	0.9512	0.9418	0.9324	0.9231	0.9139
0.1	0.9048	0.8958	0.8860	0.8781	0.8694	0.8607	0.8521	0.8437	0.8353	0.8270
0.2	0.8187	0.8106	0.8025	0.7945	0.7866	0.7788	0.7711	0.7634	0.7758	0.7483
0.3	0.7408	0.7334	0.7261	0.7189	0.7118	0.7047	0.6977	0.6907	0.6839	0.6771
0.4	0.6703	0.6637	0.6570	0.6505	0.6440	0.6376	0.6313	0.6250	0.6188	0.6126
0.5	0.6065	0.6005	0.5945	0.5886	0.5827	0.5769	0.5712	0.5655	0.5599	0.5543
0.6	0.5488	0.5434	0.5379	0.5326	0.5273	0.5220	0.5169	0.5117	0.5066	0.5016
0.7	0.4966	0.4916	0.4868	0.4819	0.4771	0.4724	0.4677	0.4630	0.4584	0.4538
0.8	0.4493	0.4449	0.4404	0.4360	0.4317	0.4274	0.4232	0.4190	0.4148	0.4107
0.9	0.4066	0.4025	0.3985	0.3946	0.3906	0.3867	0.3829	0.3791	0.3753	0.3716
	0.0	**0.1**	**0.2**	**0.3**	**0.4**	**0.5**	**0.6**	**0.7**	**0.8**	**0.9**
1.0	0.3679	0.3329	0.3012	0.2725	0.2466	0.2231	0.2019	0.1827	0.1653	0.1496
2.0	0.1353	0.1225	0.1108	0.1003	0.0907	0.0821	0.0743	0.0672	0.0608	0.0550
3.0	0.0498	0.0450	0.0408	0.0369	0.0334	0.0302	0.0273	0.0247	0.0224	0.0202
4.0	0.0183	0.0166	0.0150	0.0130	0.0123	0.0111	0.0101	0.0091	0.0082	0.0074
5.0	0.0067	0.0061	0.0055	0.0050	0.0045	0.0041	0.0037	0.0033	0.0030	0.0027
6.0	0.0025	0.0022	0.0020	0.0018	0.0017	0.0015	0.0014	0.0012	0.0011	0.0010

*Adapted with permission from Selby, S. M. (ed.) (1969). *CRC Standard Mathematical Tables*, 17th ed. The Chemical Rubber Co., pp. 201–207.

TABLE C Exponential Distribution*

	Sample Size = *n*																			
	1	**2**	**3**	**4**	**5**	**6**	**7**	**8**	**9**	**10**	**11**	**12**	**13**	**14**	**15**	**16**	**17**	**18**	**19**	**20**
1	.5000	.2929	.2063	.1591	.1294	.1091	.0943	.0830	.0741	.0670	.0611	.0561	.0519	.0483	.0452	.0424	.0400	.0378	.0358	.0341
2		.7071	.5000	.3864	.3147	.2655	.2295	.2021	.1806	.1632	.1489	.1368	.1266	.1178	.1101	.1034	.0975	.0922	.0874	.0831
3			.7937	.6136	.5000	.4218	.3648	.3213	.2871	.2594	.2366	.2175	.2013	.1873	.1751	.1644	.1550	.1465	.1390	.1322
4				.8409	.6853	.5782	.5000	.4404	.3935	.3557	.3244	.2982	.2760	.2568	.2401	.2254	.2125	.2009	.1905	.1812
5					.8706	.7345	.6352	.5596	.5000	.4519	.4122	.2789	.3506	.3263	.3051	.2865	.2700	.2553	.2421	.2302
6						.8909	.7705	.6787	.6065	.5481	.5000	.4596	.4253	.3958	.3700	.3475	.3275	.3097	.2937	.2793
7							.9057	.7979	.7129	.6443	.5878	.5404	.5000	.4653	.4350	.4085	.3850	.3641	.3453	.3283
8								.9170	.8194	.7406	.6756	.6211	.5747	.5347	.5000	.4695	.4425	.4184	.3968	.3774
9									.9259	.8368	.7634	.7018	.6494	.6042	.5650	.5305	.5000	.4728	.4484	.4264
10										.9330	.8511	.7825	.7240	.6737	.6300	.5915	.5575	.5272	.5000	.4755
11											.9389	.8632	.7987	.7432	.6949	.6525	.6150	.5816	.5516	.5245
12												.9439	.8734	.8127	.7599	.7135	.6725	.6359	.6032	.5736
13													.9481	.8822	.8249	.7746	.7300	.6903	.6547	.6226
14														.9517	.8899	.8356	.7875	.7447	.7063	.6717
15															.9548	.8966	.8450	.7991	.7579	.7207
16																.9576	.9025	.8535	.8095	.7698
17																	.9600	.9078	.8610	.8188
18																		.9622	.9126	.8678
19																			.9642	.9169
20																				.9659

*Adapted with permission from "The Table of Median Ranks of Sample Values on Their Population with an Application to Certain Fatigue Studies." (1951). *Industrial Mathematics*, no. 2, p. 7.

TABLE D Median Ranks*

1000 × probability of r or fewer occurrences of event that has average number of occurrences equal to np.

np \ r	0	1	2	3	4	5	6	7	8	9
0.02	980	1,000								
0.04	961	999	1,000							
0.06	942	998	1,000							
0.08	923	997	1,000							
0.10	905	995	1,000							
0.15	861	990	999	1,000						
0.20	819	982	999	1,000						
0.25	779	974	998	1,000						
0.30	741	963	996	1,000						
0.35	705	951	994	1,000						
0.40	670	938	992	999	1,000					
0.45	638	925	989	999	1,000					
0.50	607	910	986	998	1,000					
0.55	577	894	982	998	1,000					
0.60	549	878	977	997	1,000					
0.65	522	861	972	996	999	1,000				
0.70	497	844	966	994	999	1,000				
0.75	472	827	959	993	999	1,000				
0.80	449	809	953	991	999	1,000				
0.85	427	791	945	989	998	1,000				
0.90	407	772	937	987	998	1,000				
0.95	387	754	929	984	997	1,000				
1.00	368	736	920	981	996	999	1,000			
1.1	333	699	900	974	995	999	1,000			
1.2	301	663	879	966	992	998	1,000			
1.3	273	627	857	957	989	998	1,000			
1.4	247	592	833	946	986	997	999	1,000		
1.5	223	558	809	934	981	996	999	1,000		
1.6	202	525	783	921	976	994	999	1,000		
1.7	183	493	757	907	970	992	998	1,000		
1.8	165	463	731	891	964	990	997	999	1,000	
1.9	150	434	704	875	956	987	997	999	1,000	
2.0	135	406	677	857	947	983	995	999	1,000	

Table E Poisson Distribution

np \ r	0	1	2	3	4	5	6	7	8	9
2.2	111	355	623	819	928	975	993	998	1,000	
2.4	091	308	570	779	904	964	988	997	999	1,000
2.6	074	267	518	736	877	951	983	995	999	1,000
2.8	061	231	469	692	848	935	976	992	998	999
3.0	050	199	423	647	815	916	966	988	996	999
3.2	041	171	380	603	781	895	955	983	994	998
3.4	033	147	340	558	744	871	942	977	992	997
3.6	027	126	303	515	706	844	927	969	988	996
3.8	022	107	269	473	668	816	909	960	984	994
4.0	018	092	238	433	629	785	889	949	979	992
4.2	015	078	210	395	590	753	867	936	972	989
4.4	012	066	185	359	551	720	844	921	964	985
4.6	010	056	163	326	513	686	818	905	955	980
4.8	008	048	143	294	476	651	791	887	944	975
5.0	007	040	125	265	440	616	762	867	932	968
5.2	006	034	109	238	406	581	732	845	918	960
5.4	005	029	095	213	373	546	702	822	903	951
5.6	004	024	082	191	342	512	670	797	886	941
5.8	003	021	072	170	313	478	638	771	867	929
6.0	002	017	062	151	285	446	606	744	847	916
	10	**11**	**12**	**13**	**14**	**15**	**16**			
2.8	1,000									
3.0	1,000									
3.2	1,000									
3.4	999	1,000								
3.6	999	1,000								
3.8	998	999	1,000							
4.0	997	999	1,000							
4.2	996	999	1,000							
4.4	994	998	999	1,000						
4.6	992	997	999	1,000						
4.8	990	996	999	1,000						
5.0	986	995	998	999	1,000					
5.2	982	993	997	999	1,000					
5.4	977	990	996	999	1,000					
5.6	972	988	995	998	999	1,000				
5.8	965	984	993	997	999	1,000				
6.0	957	980	991	996	999	999	1,000			

Table E Poisson Distribution (*Continued*)

np \ r	0	1	2	3	4	5	6	7	8	9
6.2	002	015	054	134	259	414	574	716	826	902
6.4	002	012	046	119	235	384	542	687	803	886
6.6	001	010	040	105	213	355	511	658	780	869
6.8	001	009	034	093	192	327	480	628	755	850
7.0	001	007	030	082	173	301	450	599	729	830
7.2	001	006	025	072	156	276	420	569	703	810
7.4	001	005	022	063	140	253	392	539	676	788
7.6	001	004	019	055	125	231	365	510	648	765
7.8	000	004	016	048	112	210	338	481	620	741
8.0	000	003	014	042	100	191	313	453	593	717
8.5	000	002	009	030	074	150	256	386	523	653
9.0	000	001	006	021	055	116	207	324	456	587
9.5	000	001	004	015	040	089	165	269	392	522
10.0	000	000	003	010	029	067	130	220	333	458
	10	**11**	**12**	**13**	**14**	**15**	**16**	**17**	**18**	**19**
6.2	949	975	989	995	998	999	1,000			
6.4	939	969	986	994	997	999	1,000			
6.6	927	963	982	992	997	999	999	1,000		
6.8	915	955	978	990	996	998	999	1,000		
7.0	901	947	973	987	994	998	999	1,000		
7.2	887	937	967	984	993	997	999	999	1,000	
7.4	871	926	961	980	991	996	998	999	1,000	
7.6	854	915	954	976	989	995	998	999	1,000	
7.8	835	902	945	971	986	993	997	999	1,000	
8.0	816	888	936	966	983	992	996	998	999	1,000
8.5	763	849	909	949	973	986	993	997	999	999
9.0	706	803	876	926	959	978	989	995	998	999
9.5	645	752	836	898	940	967	982	991	996	998
10.0	583	697	792	864	917	951	973	986	993	997
	20	**21**	**22**							
8.5	1,000									
9.0	1,000									
9.5	999	1,000								
10.0	998	999	1,000							

Table E Poisson Distribution (*Continued*)

np \ r	0	1	2	3	4	5	6	7	8	9
10.5	000	000	002	007	021	050	102	179	279	397
11.0	000	000	001	005	015	038	079	143	232	341
11.5	000	000	001	003	011	028	060	114	191	289
12.0	000	000	001	002	008	020	046	090	155	242
12.5	000	000	000	002	005	015	035	070	125	201
13.0	000	000	000	001	004	011	026	054	100	166
13.5	000	000	000	001	003	008	019	041	079	135
14.0	000	000	000	000	002	006	014	032	062	109
14.5	000	000	000	000	001	004	010	024	048	088
15.0	000	000	000	000	001	003	008	018	037	070
	10	**11**	**12**	**13**	**14**	**15**	**16**	**17**	**18**	**19**
10.5	521	639	742	825	888	932	960	978	988	994
11.0	460	579	689	781	854	907	944	968	982	991
11.5	402	520	633	733	815	878	924	954	974	986
12.0	347	462	576	682	772	844	899	937	963	979
12.5	297	406	519	628	725	806	869	916	948	969
13.0	252	353	463	573	675	764	835	890	930	957
13.5	211	304	409	518	623	718	798	861	908	942
14.0	176	260	358	464	570	669	756	827	883	923
14.5	145	220	311	413	518	619	711	790	853	901
15.0	118	185	268	363	466	568	664	749	819	875
	20	**21**	**22**	**23**	**24**	**25**	**26**	**27**	**28**	**29**
10.5	997	999	999	1,000						
11.0	995	998	999	1,000						
11.5	992	996	998	999	1,000					
12.0	988	994	997	999	999	1,000				
12.5	983	991	995	998	999	999	1,000			
13.0	975	986	992	996	998	999	1,000			
13.5	965	980	989	994	997	998	999	1,000		
14.0	952	971	983	991	995	997	999	999	1,000	
14.5	936	960	976	986	992	996	998	999	999	1,000
15.0	917	947	967	981	989	994	997	999	999	1,000

*Adapted with permission from Grant, E. L. and Leavenworth, Richard S. (1972). *Statistical Quality Control*, 4th ed. McGraw-Hill, New York.

Table E Poisson Distribution* (*Continued*)

Probability of r or fewer occurrences of an event in n trials, where p is the probability of occurrence on each trial.

		p									
n	***r***	**0.05**	**0.10**	**0.15**	**0.20**	**0.25**	**0.30**	**0.35**	**0.40**	**0.45**	**0.50**
2	0	0.9025	0.8100	0.7225	0.6400	0.5625	0.4900	0.4225	0.3600	0.3025	0.2500
	1	0.9975	0.9900	0.9775	0.9600	0.9375	0.9100	0.8775	0.8400	0.7975	0.7500
3	0	0.8574	0.7290	0.6141	0.5120	0.4219	0.3430	0.2746	0.2160	0.1664	0.1250
	1	0.9928	0.9720	0.9392	0.8960	0.8438	0.7840	0.7182	0.6480	0.5748	0.5000
	2	0.9999	0.9990	0.9966	0.9920	0.9844	0.9730	0.9571	0.9360	0.9089	0.8750
4	0	0.8145	0.6561	0.5220	0.4096	0.3164	0.2401	0.1785	0.1296	0.0915	0.0625
	1	0.9860	0.9477	0.8905	0.8192	0.7383	0.6517	0.5630	0.4752	0.3910	0.3125
	2	0.9995	0.9963	0.9880	0.9728	0.9492	0.9163	0.8735	0.8208	0.7585	0.6875
	3	1.0000	0.9999	0.9995	0.9984	0.9961	0.9919	0.9850	0.9744	0.9590	0.9375
5	0	0.7738	0.5905	0.4437	0.3277	0.2373	0.1681	0.1160	0.0778	0.0503	0.0312
	1	0.9774	0.9185	0.8352	0.7373	0.6328	0.5282	0.4284	0.3370	0.2562	0.1875
	2	0.9988	0.9914	0.9734	0.9421	0.8965	0.8369	0.7648	0.6826	0.5931	0.5000
	3	1.0000	0.9995	0.9978	0.9933	0.9844	0.9692	0.9460	0.9130	0.8688	0.8125
	4	1.0000	1.0000	0.9999	0.9997	0.9990	0.9976	0.9947	0.9898	0.9815	0.9688
6	0	0.7351	0.5314	0.3771	0.2621	0.1780	0.1176	0.0754	0.0467	0.0277	0.0156
	1	0.9672	0.8857	0.7765	0.6554	0.5339	0.4202	0.3191	0.2333	0.1636	0.1094
	2	0.9978	0.9842	0.9527	0.9011	0.8306	0.7443	0.6471	0.5443	0.4415	0.3438
	3	0.9999	0.9987	0.9941	0.9830	0.9624	0.9295	0.8826	0.8208	0.7447	0.6562
	4	1.0000	0.9999	0.9996	0.9984	0.9954	0.9891	0.9777	0.9590	0.9308	0.8906
	5	1.0000	1.0000	1.0000	0.9999	0.9998	0.9993	0.9982	0.9959	0.9917	0.9844

TABLE F Binomial Distribution

		p									
n	***r***	**0.05**	**0.10**	**0.15**	**0.20**	**0.25**	**0.30**	**0.35**	**0.40**	**0.45**	**0.50**
7	0	0.6983	0.4783	0.3206	0.2097	0.1335	0.0824	0.0490	0.0280	0.0152	0.0078
	1	0.9556	0.8503	0.7166	0.5767	0.4449	0.3294	0.2338	0.1586	0.1024	0.0625
	2	0.9962	0.9743	0.9262	0.8520	0.7564	0.6471	0.5323	0.4199	0.3164	0.2266
	3	0.9998	0.9973	0.9879	0.9667	0.9294	0.8740	0.8002	0.7102	0.6083	0.5000
	4	1.0000	0.9998	0.9988	0.9953	0.9871	0.9712	0.9444	0.9037	0.8471	0.7734
	5	1.0000	1.0000	0.9999	0.9996	0.9987	0.9962	0.9910	0.9812	0.9643	0.9375
	6	1.0000	1.0000	1.0000	1.0000	0.9999	0.9998	0.9994	0.9984	0.9963	0.9922
8	0	0.6634	0.4305	0.2725	0.1678	0.1001	0.0576	0.0319	0.0168	0.0084	0.0039
	1	0.9428	0.8131	0.6572	0.5033	0.3671	0.2553	0.1691	0.1064	0.0632	0.0352
	2	0.9942	0.9619	0.8948	0.7969	0.6785	0.5518	0.4278	0.3154	0.2201	0.1445
	3	0.9996	0.9950	0.9786	0.9437	0.8862	0.8059	0.7064	0.5941	0.4770	0.3633
	4	1.0000	0.9996	0.9971	0.9896	0.9727	0.9420	0.8939	0.8263	0.7396	0.6367
	5	1.0000	1.0000	0.9998	0.9988	0.9958	0.9887	0.9747	0.9502	0.9115	0.8555
	6	1.0000	1.0000	1.0000	0.9999	0.9996	0.9987	0.9964	0.9915	0.9819	0.9648
	7	1.0000	1.0000	1.0000	1.0000	1.0000	0.9999	0.9998	0.9993	0.9983	0.9961
9	0	0.6302	0.3874	0.2316	0.1342	0.0751	0.0404	0.0207	0.0101	0.0046	0.0020
	1	0.9288	0.7748	0.5995	0.4362	0.3003	0.1960	0.1211	0.0705	0.0385	0.0195
	2	0.9916	0.9470	0.8591	0.7382	0.6007	0.4628	0.3373	0.2318	0.1495	0.0898
	3	0.9994	0.9917	0.9661	0.9144	0.8343	0.7297	0.6089	0.4826	0.3614	0.2539
	4	1.0000	0.9991	0.9944	0.9804	0.9511	0.9012	0.8283	0.7334	0.6214	0.5000
	5	1.0000	0.9999	0.9994	0.9969	0.9900	0.9747	0.9464	0.9006	0.8342	0.7461
	6	1.0000	1.0000	1.0000	0.9997	0.9987	0.9957	0.9888	0.9750	0.9502	0.9102
	7	1.0000	1.0000	1.0000	1.0000	0.9999	0.9996	0.9986	0.9962	0.9909	0.9805
	8	1.0000	1.0000	1.0000	1.0000	1.0000	1.0000	0.9999	0.9997	0.9992	0.9980

10	0	0.5987	0.3487	0.1969	0.1074	0.0563	0.0282	0.0135	0.0060	0.0025	0.0010
	1	0.9139	0.7361	0.5443	0.3758	0.2440	0.1493	0.0860	0.0464	0.0232	0.0107
	2	0.9885	0.9298	0.8202	0.6778	0.5256	0.3828	0.2616	0.1673	0.0996	0.0547
	3	0.9990	0.9872	0.9500	0.8791	0.7759	0.6496	0.5138	0.3823	0.2660	0.1719
	4	0.9999	0.9984	0.9901	0.9672	0.9219	0.8497	0.7515	0.6331	0.5044	0.3770
	5	1.0000	0.9999	0.9986	0.9936	0.9803	0.9527	0.9051	0.8338	0.7384	0.6230
	6	1.0000	1.0000	0.9999	0.9991	0.9965	0.9894	0.9740	0.9452	0.8980	0.8281
	7	1.0000	1.0000	1.0000	0.9999	0.9996	0.9984	0.9952	0.9877	0.9726	0.9453
	8	1.0000	1.0000	1.0000	1.0000	1.0000	0.9999	0.9995	0.9983	0.9955	0.9893
	9	1.0000	1.0000	1.0000	1.0000	1.0000	1.0000	1.0000	0.9999	0.9997	0.9990

*Adapted with permission from Miller, Irwin and Freund, John E. (1965). Probability and Statistics for Engineers. Prentice-Hall, Englewood Cliffs, NJ, pp. 388–389.

For more extensive tables see The Staff of Harvard University Computation Laboratory (1955). *Tables of Cumulative Binomial Probability Distribution*. Harvard University Press, Cambridge, MA. See also Robertson, W. H. (1960). *Tables of the Binomial Distribution Function for Small Values of p*. Sandia Corp. Monograph, available from the Office of Technical Services, Department of Commerce, Washington, DC.

Table F Binomial Distribution* (*Continued*)

Value of t corresponding to certain selected probabilities (i.e., tail areas under the curve). To illustrate: The probability is 0.975 that a sample with 20 degrees of freedom would have t = +2.086 or smaller.

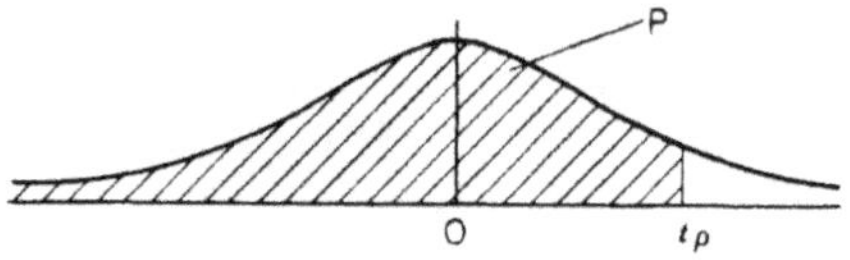

DF	$t_{.60}$	$t_{.70}$	$t_{.80}$	$t_{.90}$	$t_{.95}$	$t_{.975}$	$t_{.99}$	$t_{.995}$
1	0.325	0.727	1.376	3.078	6.314	12.706	31.821	63.657
2	0.289	0.617	1.061	1.886	2.920	4.303	6.965	9.925
3	0.277	0.584	0.978	1.638	2.353	3.182	4.541	5.841
4	0.271	0.569	0.941	1.533	2.132	2.776	3.747	4.604
5	0.267	0.559	0.920	1.476	2.015	2.571	3.365	4.032
6	0.265	0.553	0.906	1.440	1.943	2.447	3.143	3.707
7	0.263	0.549	0.896	1.415	1.895	2.365	2.998	3.499
8	0.262	0.546	0.889	1.397	1.860	2.306	2.896	3.355
9	0.261	0.543	0.883	1.383	1.833	2.262	2.821	3.250
10	0.260	0.542	0.879	1.372	1.812	2.228	2.764	3.169
11	0.260	0.540	0.876	1.363	1.796	2.201	2.718	3.106
12	0.259	0.539	0.873	1.356	1.782	2.179	2.681	3.055
13	0.259	0.538	0.870	1.350	1.771	2.160	2.650	3.012
14	0.258	0.537	0.868	1.345	1.761	2.145	2.624	2.977
15	0.258	0.536	0.866	1.341	1.753	2.131	2.602	2.947
16	0.258	0.535	0.865	1.337	1.746	2.120	2.583	2.921
17	0.257	0.534	0.863	1.333	1.740	2.110	2.567	2.898
18	0.257	0.534	0.862	1.330	1.734	2.101	2.552	2.878
19	0.257	0.533	0.861	1.328	1.729	2.093	2.539	2.861
20	0.257	0.533	0.860	1.325	1.725	2.086	2.528	2.845
21	0.257	0.532	0.859	1.323	1.721	2.080	2.518	2.831
22	0.256	0.532	0.858	1.321	1.717	2.074	2.508	2.819
23	0.256	0.532	0.858	1.319	1.714	2.069	2.500	2.807
24	0.256	0.531	0.857	1.318	1.711	2.064	2.492	2.797
25	0.256	0.531	0.856	1.316	1.708	2.060	2.485	2.787
26	0.256	0.531	0.856	1.315	1.706	2.056	2.479	2.779
27	0.256	0.531	0.855	1.314	1.703	2.052	2.473	2.771
28	0.256	0.530	0.855	1.313	1.701	2.048	2.467	2.763
29	0.256	0.530	0.854	1.311	1.699	2.045	2.462	2.756
30	0.256	0.530	0.854	1.310	1.697	2.042	2.457	2.750
40	0.255	0.529	0.851	1.303	1.684	2.021	2.423	2.704
60	0.254	0.527	0.848	1.296	1.671	2.000	2.390	2.660
120	0.254	0.526	0.845	1.289	1.658	1.980	2.358	2.617
∞	0.253	0.524	0.842	1.282	1.645	1.960	2.326	2.576

*Adapted with permission from Dixon, W. J. and Massey, F. J., Jr. (1969). *Introduction to Statistical Analysis*, 3rd ed. McGraw-Hill, New York. Entries originally from Fisher, R. A. and Yates, F. *Statistical Tables*. Oliver & Boyd, London, Table III.

TABLE G* Distribution of *t*

Sample Size	$\phi_{.95}$	$\phi_{.975}$	$\phi_{.99}$
2	3.175	6.353	15.910
3	0.885	1.304	2.111
4	0.529	0.717	1.023
5	0.388	0.507	0.685
6	0.312	0.399	0.523
7	0.263	0.333	0.429
8	0.230	0.288	0.366
9	0.205	0.255	0.322
10	0.186	0.230	0.288
11	0.170	0.210	0.262
12	0.158	0.194	0.241
13	0.147	0.181	0.224
14	0.138	0.170	0.209
15	0.131	0.160	0.197
16	0.124	0.151	0.186
17	0.118	0.144	0.177
18	0.113	0.137	0.168
19	0.108	0.131	0.161
20	0.104	0.126	0.154

*Adapted with permission from Lord, E. (1957). "The Use of the Range in Place of the Standard Deviation in the *t* Test." *Biometrika*, vol. 34.

Table H* Percentile for $\tau_d = \dfrac{\overline{X} - \mu_0}{R}$

Percentages are values for α for a two-tail test. (Two-tail percentage points are given for the binomial for p = 0.05.)

N	1%	5%	10%	25%
1				
2				
3				0
4				0
5			0	0
6		0	0	1
7		0	0	1
8	0	0	1	1
9	0	1	1	2
10	0	1	1	2
11	0	1	2	3
12	1	2	2	3
13	1	2	3	3
14	1	2	3	4
15	2	3	3	4
16	2	3	4	5
17	2	4	4	5
18	3	4	5	6
19	3	4	5	6
20	3	5	5	6
21	4	5	6	7
22	4	5	6	7
23	4	6	7	8
24	5	6	7	8
25	5	7	7	9
26	6	7	8	9
27	6	7	8	10
28	6	8	9	10
29	7	8	9	10
30	7	9	10	11
31	7	9	10	11
32	8	9	10	12
33	8	10	11	12
34	9	10	11	13
35	9	11	12	13
36	9	11	12	14
37	10	12	13	14
38	10	12	13	14
39	11	12	13	15
40	11	13	14	15

Table I Critical Values of *r* for the Sign Test

N	1%	5%	10%	25%
41	11	13	14	16
42	12	14	15	16
43	12	14	15	17
44	13	15	16	17
45	13	15	16	18
46	13	15	16	18
47	14	16	17	19
48	14	16	17	19
49	15	17	18	19
50	15	17	18	20

*Adapted with permission from Dixon, W. J. and Massey, F. J., Jr. (1969). *Introduction to Statistical Analysis*, 3rd ed. McGraw-Hill, NewYork.

TABLE I Critical Values of *r* for the Sign Test* (*Continued*)

$n = n_A = n_B$	$\phi'_{.95}$	$\phi'_{.975}$	$\phi'_{.99}$
2	2.322	3.427	5.553
3	0.974	1.272	1.715
4	0.644	0.813	1.047
5	0.493	0.613	0.772
6	0.405	0.499	0.621
7	0.347	0.426	0.525
8	0.306	0.373	0.459
9	0.275	0.334	0.409
10	0.250	0.304	0.371
11	0.233	0.280	0.340
12	0.214	0.260	0.315
13	0.201	0.243	0.294
14	0.189	0.189	0.276
15	0.179	0.216	0.261
16	0.170	0.205	0.247
17	0.162	0.195	0.236
18	0.155	0.187	0.225
19	0.149	0.179	0.216
20	0.143	0.172	0.207

*Adapted with permission from Lord, E. (1947). "The Use of the Range in Place of the Standard Deviation in the *t* Test." *Biometrika*, vol. 34.

TABLE J* Percentile for $\tau_d = \dfrac{\bar{X}_1 - \bar{X}_2}{\frac{1}{2}(R_1 + R_2)}$

Values of F corresponding to certain selected probabilities (i.e., tail areas under the curve). To illustrate: The probability is 0.05 that the ratio of two sample variances obtained with 20 and 10 degrees of freedom in numerator and denominator, respectively, would have F = 2.77 or larger. For a two-sided test, a lower limit is found by taking the reciprocal of the tabulated F value for the degrees of freedom in reverse. For the above example, with 10 and 20 degrees of freedom in numerator and denominator, respectively, F is 2.35 and 1/F is 1/2.35, or 0.43. The probability is 0.10 that F is 0.43 or smaller or 2.77 or larger.

n_2 \ n_1	1	2	3	4	5	6	7	8	9
					$F_{.95}(n_1, n_2)$				
1	161.4	199.5	215.7	224.6	230.2	234.0	236.8	238.9	240.5
2	18.51	19.00	19.16	19.25	19.30	19.33	19.35	19.37	19.38
3	10.13	9.55	9.28	9.12	9.01	8.94	8.89	8.85	8.81
4	7.71	6.94	6.59	6.39	6.26	6.16	6.09	6.04	6.00
5	6.61	5.79	5.41	5.19	5.05	4.95	4.88	4.82	4.77
6	5.99	5.14	4.76	4.53	4.39	4.28	4.21	4.15	4.10
7	5.59	4.74	4.35	4.12	3.97	3.87	3.79	3.73	3.68
8	5.32	4.46	4.07	3.84	3.69	3.58	3.50	3.44	3.39
9	5.12	4.26	3.86	3.63	3.48	3.37	3.29	3.23	3.18
10	4.96	4.10	3.71	3.48	3.33	3.22	3.14	3.07	3.02
11	4.84	3.98	3.59	3.36	3.20	3.09	3.01	2.95	2.90
12	4.75	3.89	3.49	3.26	3.11	3.00	2.91	2.85	2.80
13	4.67	3.81	3.41	3.18	3.03	2.92	2.83	2.77	2.71
14	4.60	3.74	3.34	3.11	2.96	2.85	2.76	2.70	2.65
15	4.54	3.68	3.29	3.06	2.90	2.79	2.71	2.64	2.59
16	4.49	3.63	3.24	3.01	2.85	2.74	2.66	2.59	2.54
17	4.45	3.59	3.20	2.96	2.81	2.70	2.61	2.55	2.49
18	4.41	3.55	3.16	2.93	2.77	2.66	2.58	2.51	2.46
19	4.38	3.52	3.13	2.90	2.74	2.63	2.54	2.48	2.42
20	4.35	3.49	3.10	2.87	2.71	2.60	2.51	2.45	2.39
21	4.32	3.47	3.07	2.84	2.68	2.57	2.49	2.42	2.37
22	4.30	3.44	3.05	2.82	2.66	2.55	2.46	2.40	2.34
23	4.28	3.42	3.03	2.80	2.64	2.53	2.44	2.37	2.32
24	4.26	3.40	3.01	2.78	2.62	2.51	2.42	2.36	2.30
25	4.24	3.39	2.99	2.76	2.60	2.49	2.40	2.34	2.28
26	4.23	3.37	2.98	2.74	2.59	2.47	2.39	2.32	2.27
27	4.21	3.35	2.96	2.73	2.57	2.46	2.37	2.31	2.25
28	4.20	3.34	2.95	2.71	2.56	2.45	2.36	2.29	2.24
29	4.18	3.33	2.93	2.70	2.55	2.43	2.35	2.28	2.22
30	4.17	3.32	2.92	2.69	2.53	2.42	2.33	2.27	2.21
40	4.08	3.23	2.84	2.61	2.45	2.34	2.25	2.18	2.12
60	4.00	3.15	2.76	2.53	2.37	2.25	2.17	2.10	2.04
120	3.92	3.07	2.68	2.45	2.29	2.17	2.09	2.02	1.96
∞	3.84	3.00	2.60	2.37	2.21	2.10	2.01	1.94	1.88

TABLE K* Distribution of *F*

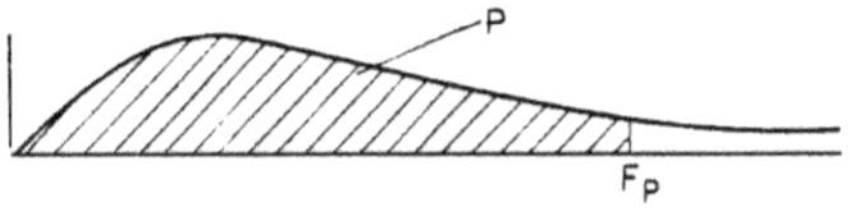

10	12	15	20	24	30	40	60	120	∞
$F_{.95}(n_1, n_2)$									
241.9	243.9	245.9	248.0	249.1	250.1	251.1	252.2	253.3	254.3
19.40	19.41	19.43	19.45	19.45	19.46	19.47	19.48	19.49	19.50
8.79	8.74	8.70	8.66	8.64	8.62	8.59	8.57	8.55	8.53
5.96	5.91	5.86	5.80	5.77	5.75	5.72	5.69	5.66	5.63
4.74	4.68	4.62	4.56	4.53	4.50	4.46	4.43	4.40	4.36
4.06	4.00	3.94	3.87	3.84	3.81	3.77	3.74	3.70	3.67
3.64	3.57	3.51	3.44	3.41	3.38	3.34	3.30	3.27	3.23
3.35	3.28	3.22	3.15	3.12	3.08	3.04	3.01	2.97	2.93
3.14	3.07	3.01	2.94	2.90	2.86	2.83	2.79	2.75	2.71
2.98	2.91	2.85	2.77	2.74	2.70	2.66	2.62	2.58	2.54
2.85	2.79	2.72	2.65	2.61	2.57	2.53	2.49	2.45	2.40
2.75	2.69	2.62	2.54	2.51	2.47	2.43	2.38	2.34	2.30
2.67	2.60	2.53	2.46	2.42	2.38	2.34	2.30	2.25	2.21
2.60	2.53	2.46	2.39	2.35	2.31	2.27	2.22	2.18	2.13
2.54	2.48	2.40	2.33	2.29	2.25	2.20	2.16	2.11	2.07
2.49	2.42	2.35	2.28	2.24	2.19	2.15	2.11	2.06	2.01
2.45	2.38	2.31	2.23	2.19	2.15	2.10	2.06	2.01	1.96
2.41	2.34	2.27	2.19	2.15	2.11	2.06	2.02	1.97	1.92
2.38	2.31	2.23	2.16	2.11	2.07	2.03	1.98	1.93	1.88
2.35	2.28	2.20	2.12	2.08	2.04	1.99	1.95	1.90	1.84
2.32	2.25	2.18	2.10	2.05	2.01	1.96	1.92	1.87	1.81
2.30	2.23	2.15	2.07	2.03	1.98	1.94	1.89	1.84	1.78
2.27	2.20	2.13	2.05	2.01	1.96	1.91	1.86	1.81	1.76
2.25	2.18	2.11	2.03	1.98	1.94	1.89	1.84	1.79	1.73
2.24	2.16	2.09	2.01	1.96	1.92	1.87	1.82	1.77	1.71
2.22	2.15	2.07	1.99	1.95	1.90	1.85	1.80	1.75	1.69
2.20	2.13	2.06	1.97	1.93	1.88	1.84	1.79	1.73	1.67
2.19	2.12	2.04	1.96	1.91	1.87	1.82	1.77	1.71	1.65
2.18	2.10	2.03	1.94	1.90	1.85	1.81	1.75	1.70	1.64
2.16	2.09	2.01	1.93	1.89	1.84	1.79	1.74	1.68	1.62
2.08	2.00	1.92	1.84	1.79	1.74	1.69	1.64	1.58	1.51
1.99	1.92	1.84	1.75	1.70	1.65	1.59	1.53	1.47	1.39
1.91	1.83	1.75	1.66	1.61	1.55	1.50	1.43	1.35	1.25
1.83	1.75	1.67	1.57	1.52	1.46	1.39	1.32	1.22	1.00

TABLE K* Distribution of *F* *(Continued)*

n_2 \ n_1	1	2	3	4	5	6	7	8	9
					$F_{.975}(n_1, n_2)$				
1	647.8	799.5	864.2	899.6	921.8	937.1	948.2	956.7	963.3
2	38.51	39.00	39.17	39.25	39.30	39.33	39.36	39.37	39.39
3	17.44	16.04	15.44	15.10	14.88	14.73	14.62	14.54	14.47
4	12.22	10.65	9.98	9.60	9.36	9.20	9.07	8.98	8.90
5	10.01	8.43	7.76	7.39	7.15	6.98	6.85	6.76	6.68
6	8.81	7.26	6.60	6.23	5.99	5.82	5.70	5.60	5.52
7	8.07	6.54	5.89	5.52	5.29	5.12	4.99	4.90	4.82
8	7.57	6.06	5.42	5.05	4.82	4.65	4.53	4.43	4.36
9	7.21	5.71	5.08	4.72	4.48	4.32	4.20	4.10	4.03
10	6.94	5.46	4.83	4.47	4.24	4.07	3.95	3.85	3.78
11	6.72	5.26	4.63	4.28	4.04	3.88	3.76	3.66	3.59
12	6.55	5.10	4.47	4.12	3.89	3.73	3.61	3.51	3.44
13	6.41	4.97	4.35	4.00	3.77	3.60	3.48	3.39	3.31
14	6.30	4.86	4.24	3.89	3.66	3.50	3.38	3.29	3.21
15	6.20	4.77	4.15	3.80	3.58	3.41	3.29	3.20	3.12
16	6.12	4.69	4.08	3.73	3.50	3.34	3.22	3.12	3.05
17	6.04	4.62	4.01	3.66	3.44	3.28	3.16	3.06	2.98
18	5.98	4.56	3.95	3.61	3.38	3.22	3.10	3.01	2.93
19	5.92	4.51	3.90	3.56	3.33	3.17	3.05	2.96	2.88
20	5.87	4.46	3.86	3.51	3.29	3.13	3.01	2.91	2.84
21	5.83	4.42	3.82	3.48	3.25	3.09	2.97	2.87	2.80
22	5.79	4.38	3.78	3.44	3.22	3.05	2.93	2.84	2.76
23	5.75	4.35	3.75	3.41	3.18	3.02	2.90	2.81	2.73
24	5.72	4.32	3.72	3.38	3.15	2.99	2.87	2.78	2.70
25	5.69	4.29	3.69	3.35	3.13	2.97	2.85	2.75	2.68
26	5.66	4.27	3.67	3.33	3.10	2.94	2.82	2.73	2.65
27	5.63	4.24	3.65	3.31	3.08	2.92	2.80	2.71	2.63
28	5.61	4.22	3.63	3.29	3.06	2.90	2.78	2.69	2.61
29	5.59	4.20	3.61	3.27	3.04	2.88	2.76	2.67	2.59
30	5.57	4.18	3.59	3.25	3.03	2.87	2.75	2.65	2.57
40	5.42	4.05	3.46	3.13	2.90	2.74	2.62	2.53	2.45
60	5.29	3.93	3.34	3.01	2.79	2.63	2.51	2.41	2.33
120	5.15	3.80	3.23	2.89	2.67	2.52	2.39	2.30	2.22
∞	5.02	3.69	3.12	2.79	2.57	2.41	2.29	2.19	2.11

TABLE K* Distribution of *F* (*Continued*)

10	12	15	20	24	30	40	60	120	∞
					$F_{.975}(n_1, n_2)$				
968.6	976.7	984.9	993.1	997.2	1,001	1,006	1,010	1,014	1,018
39.40	39.41	39.43	39.45	39.46	39.46	39.47	39.48	39.49	39.50
14.42	14.34	14.25	14.17	14.12	14.08	14.04	13.99	13.95	13.90
8.84	8.75	8.66	8.56	8.51	8.46	8.41	8.36	8.31	8.26
6.62	6.52	6.43	6.33	6.28	6.23	6.18	6.12	6.07	6.02
5.46	5.37	5.27	5.17	5.12	5.07	5.01	4.96	4.90	4.85
4.76	4.67	4.57	4.47	4.42	4.36	4.31	4.25	4.20	4.14
4.30	4.20	4.10	4.00	3.95	3.89	3.84	3.78	3.73	3.67
3.96	3.87	3.77	3.67	3.61	3.56	3.51	3.45	3.39	3.33
3.72	3.62	3.52	3.42	3.37	3.31	3.26	3.20	3.14	3.08
3.53	3.43	3.33	3.23	3.17	3.12	3.06	3.00	2.94	2.88
3.37	3.28	3.18	3.07	3.02	2.96	2.91	2.85	2.79	2.72
3.25	3.15	3.05	2.95	2.89	2.84	2.78	2.72	2.66	2.60
3.15	3.05	2.95	2.84	2.79	2.73	2.67	2.61	2.55	2.49
3.06	2.96	2.86	2.76	2.70	2.64	2.59	2.52	2.46	2.40
2.99	2.89	2.79	2.68	2.63	2.57	2.51	2.45	2.38	2.32
2.92	2.82	2.72	2.62	2.56	2.50	2.44	2.38	2.32	2.25
2.87	2.77	2.67	2.56	2.50	2.44	2.38	2.32	2.26	2.19
2.82	2.72	2.62	2.51	2.45	2.39	2.33	2.27	2.20	2.13
2.77	2.68	2.57	2.46	2.41	2.35	2.29	2.22	2.16	2.09
2.73	2.64	2.53	2.42	2.37	2.31	2.25	2.18	2.11	2.04
2.70	2.60	2.50	2.39	2.33	2.27	2.21	2.14	2.08	2.00
2.67	2.57	2.47	2.36	2.30	2.24	2.18	2.11	2.04	1.97
2.64	2.54	2.44	2.33	2.27	2.21	2.15	2.08	2.01	1.94
2.61	2.51	2.41	2.30	2.24	2.18	2.12	2.05	1.98	1.91
2.59	2.49	2.39	2.28	2.22	2.16	2.09	2.03	1.95	1.88
2.57	2.47	2.36	2.25	2.19	2.13	2.07	2.00	1.93	1.85
2.55	2.45	2.34	2.23	2.17	2.11	2.05	1.98	1.91	1.83
2.53	2.43	2.32	2.21	2.15	2.09	2.03	1.96	1.89	1.81
2.51	2.41	2.31	2.20	2.14	2.07	2.01	1.94	1.87	1.79
2.39	2.29	2.18	2.07	2.01	1.94	1.88	1.80	1.72	1.64
2.27	2.17	2.06	1.94	1.88	1.82	1.74	1.67	1.58	1.48
2.16	2.05	1.94	1.82	1.76	1.69	1.61	1.53	1.43	1.31
2.05	1.94	1.83	1.71	1.64	1.57	1.48	1.39	1.27	1.00

TABLE K* Distribution of *F* (*Continued*)

n_2 \ n_1	1	2	3	4	5	6	7	8	9
					$F_{.99}(n_1, n_2)$				
1	4,052	4,999.5	5,403	5,625	5,764	5,859	5,928	5,982	6,022
2	98.50	99.00	99.17	99.25	99.30	99.33	99.36	99.37	99.39
3	34.12	30.82	29.46	28.71	28.24	27.91	27.67	27.49	27.35
4	21.20	18.00	16.69	15.98	15.52	15.21	14.98	14.80	14.66
5	16.26	13.27	12.06	11.39	10.97	10.67	10.46	10.29	10.16
6	13.75	10.92	9.78	9.15	8.75	8.47	8.26	8.10	7.98
7	12.25	9.55	8.45	7.85	7.46	7.19	6.99	6.84	6.72
8	11.26	8.65	7.59	7.01	6.63	6.37	6.18	6.03	5.91
9	10.56	8.02	6.99	6.42	6.06	5.80	5.61	5.47	5.35
10	10.04	7.56	6.55	5.99	5.64	5.39	5.20	5.06	4.94
11	9.65	7.21	6.22	5.67	5.32	5.07	4.89	4.74	4.63
12	9.33	6.93	5.95	5.41	5.06	4.82	4.64	4.50	4.39
13	9.07	6.70	5.74	5.21	4.86	4.62	4.44	4.30	4.19
14	8.86	6.51	5.56	5.04	4.69	4.46	4.28	4.14	4.03
15	8.68	6.36	5.42	4.89	4.56	4.32	4.14	4.00	3.89
16	8.53	6.23	5.29	4.77	4.44	4.20	4.03	3.89	3.78
17	8.40	6.11	5.18	4.67	4.34	4.10	3.93	3.79	3.68
18	8.29	6.01	5.09	4.58	4.25	4.01	3.84	3.71	3.60
19	8.18	5.93	5.01	4.50	4.17	3.94	3.77	3.63	3.52
20	8.10	5.85	4.94	4.43	4.10	3.87	3.70	3.56	3.46
21	8.02	5.78	4.87	4.37	4.04	3.81	3.64	3.51	3.40
22	7.95	5.72	4.82	4.31	3.99	3.76	3.59	3.45	3.35
23	7.88	5.66	4.76	4.26	3.94	3.71	3.54	3.41	3.30
24	7.82	5.61	4.72	4.22	3.90	3.67	3.50	3.36	3.26
25	7.77	5.57	4.68	4.18	3.85	3.63	3.46	3.32	3.22
26	7.72	5.53	4.64	4.14	3.82	3.59	3.42	3.29	3.18
27	7.68	5.49	4.60	4.11	3.78	3.56	3.39	3.26	3.15
28	7.64	5.45	4.57	4.07	3.75	3.53	3.36	3.23	3.12
29	7.60	5.42	4.54	4.04	3.73	3.50	3.33	3.20	3.09
30	7.56	5.39	4.51	4.02	3.70	3.47	3.30	3.17	3.07
40	7.31	5.18	4.31	3.83	3.51	3.29	3.12	2.99	2.89
60	7.08	4.98	4.13	3.65	3.34	3.12	2.95	2.82	2.72
120	6.85	4.79	3.95	3.48	3.17	2.96	2.79	2.66	2.56
∞	6.63	4.61	3.78	3.32	3.02	2.80	2.64	2.51	2.41

TABLE K* Distribution of *F* (*Continued*)

10	12	15	20	24	30	40	60	120	∞
$F_{.99}(n_1, n_2)$									
6,056	6,106	6,157	6,209	6,235	6,261	6,287	6,313	6,339	6,366
99.40	99.42	99.43	99.45	99.46	99.47	99.47	99.48	99.49	99.50
27.23	27.05	26.87	26.69	26.60	26.50	26.41	26.32	26.22	26.13
14.55	14.37	14.20	14.02	13.93	13.84	13.75	13.65	13.56	13.46
10.05	9.89	9.72	9.55	9.47	9.38	9.29	9.20	9.11	9.02
7.87	7.72	7.56	7.40	7.31	7.23	7.14	7.06	6.97	6.88
6.62	6.47	6.31	6.16	6.07	5.99	5.91	5.82	5.74	5.65
5.81	5.67	5.52	5.36	5.28	5.20	5.12	5.03	4.95	4.86
5.26	5.11	4.96	4.81	4.73	4.65	4.57	4.48	4.40	4.31
4.85	4.71	4.56	4.41	4.33	4.25	4.17	4.08	4.00	3.91
4.54	4.40	4.25	4.10	4.02	3.94	3.86	3.78	3.69	3.60
4.30	4.16	4.01	3.86	3.78	3.70	3.62	3.54	3.45	3.36
4.10	3.96	3.82	3.66	3.59	3.51	3.43	3.34	3.25	3.17
3.94	3.80	3.66	3.51	3.43	3.35	3.27	3.18	3.09	3.00
3.80	3.67	3.52	3.37	3.29	3.21	3.13	3.05	2.96	2.87
3.69	3.55	3.41	3.26	3.18	3.10	3.02	2.93	2.84	2.75
3.59	3.46	3.31	3.16	3.08	3.00	2.92	2.83	2.75	2.65
3.51	3.37	3.23	3.08	3.00	2.92	2.84	2.75	2.66	2.57
3.43	3.30	3.15	3.00	2.92	2.84	2.76	2.67	2.58	2.49
3.37	3.23	3.09	2.94	2.86	2.78	2.69	2.61	2.52	2.42
3.31	3.17	3.03	2.88	2.80	2.72	2.64	2.55	2.46	2.36
3.26	3.12	2.98	2.83	2.75	2.67	2.58	2.50	2.40	2.31
3.21	3.07	2.93	2.78	2.70	2.62	2.54	2.45	2.35	2.26
3.17	3.03	2.89	2.74	2.66	2.58	2.49	2.40	2.31	2.21
3.13	2.99	2.85	2.70	2.62	2.54	2.45	2.36	2.27	2.17
3.09	2.96	2.81	2.66	2.58	2.50	2.42	2.33	2.23	2.13
3.06	2.93	2.78	2.63	2.55	2.47	2.38	2.29	2.20	2.10
3.03	2.90	2.75	2.60	2.52	2.44	2.35	2.26	2.17	2.06
3.00	2.87	2.73	2.57	2.49	2.41	2.33	2.23	2.14	2.03
2.98	2.84	2.70	2.55	2.47	2.39	2.30	2.21	2.11	2.01
2.80	2.66	2.52	2.37	2.29	2.20	2.11	2.02	1.92	1.80
2.63	2.50	2.35	2.20	2.12	2.03	1.94	1.84	1.73	1.60
2.47	2.34	2.19	2.03	1.95	1.86	1.76	1.66	1.53	1.38
2.32	2.18	2.04	1.88	1.79	1.70	1.59	1.47	1.32	1.00

*Adapted with permission from Pearson, E. S. and Hartley, H. O. (eds.) (1958). *Biometrika Tables for Statisticians*, 2nd ed. Cambridge University Press, New York, vol. I.

Note: n_1 = degrees of freedom for numerator. n_2 = degrees of freedom for denominator.

TABLE K* Distribution of *F* (*Continued*)

Values of χ^2 corresponding to certain selected probabilities (i.e., tail areas under the curve). To illustrate: The probability is 0.95 that a sample with 20 degrees of freedom, taken from a normal distribution, would have $\chi^2 = 31.41$ or smaller.

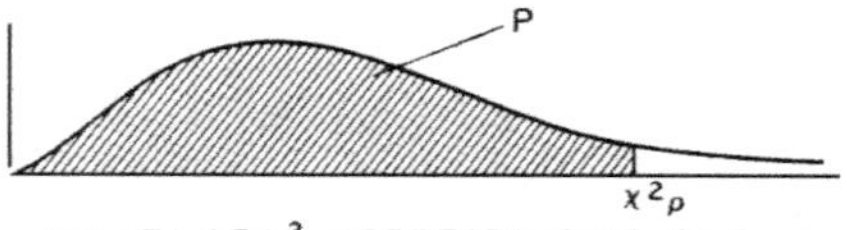

VALUES OF χ^2_P CORRESPONDING TO P

DF	$\chi^2_{.005}$	$\chi^2_{.01}$	$\chi^2_{.025}$	$\chi^2_{.05}$	$\chi^2_{.10}$	$\chi^2_{.90}$	$\chi^2_{.95}$	$\chi^2_{.975}$	$\chi^2_{.99}$	$\chi^2_{.995}$
1	0.000039	0.00016	0.00098	0.0039	0.0158	2.71	3.84	5.02	6.63	7.88
2	0.0100	0.0201	0.0506	0.1026	0.2107	4.61	5.99	7.38	9.21	10.60
3	0.0717	0.115	0.216	0.352	0.584	6.25	7.81	9.35	11.34	12.84
4	0.207	0.297	0.484	0.711	1.064	7.78	9.49	11.14	13.28	14.86
5	0.412	0.554	0.831	1.15	1.61	9.24	11.07	12.83	15.09	16.75
6	0.676	0.872	1.24	1.64	2.20	10.64	12.59	14.45	16.81	18.55
7	0.989	1.24	1.69	2.17	2.83	12.02	14.07	16.01	18.48	20.28
8	1.34	1.65	2.18	2.73	3.49	13.36	15.51	17.53	20.09	21.96
9	1.73	2.09	2.70	3.33	4.17	14.68	16.92	19.02	21.67	23.59
10	2.16	2.56	3.25	3.94	4.87	15.99	18.31	20.48	23.21	25.19
11	2.60	3.05	3.82	4.57	5.58	17.28	19.68	21.92	24.73	26.76
12	3.07	3.57	4.40	5.23	6.30	18.55	21.03	23.34	26.22	28.30
13	3.57	4.11	5.01	5.89	7.04	19.81	22.36	24.74	27.69	29.82
14	4.07	4.66	5.63	6.57	7.79	21.06	23.68	26.12	29.14	31.32
15	4.60	5.23	6.26	7.26	8.55	22.31	25.00	27.49	30.58	32.80
16	5.14	5.81	6.91	7.96	9.31	23.54	26.30	28.85	32.00	34.27
18	6.26	7.01	8.23	9.39	10.86	25.99	28.87	31.53	34.81	37.16
20	7.43	8.26	9.59	10.85	12.44	28.41	31.41	34.17	37.57	40.00
24	9.89	10.86	12.40	13.85	15.66	33.20	36.42	39.36	42.98	45.56
30	13.79	14.95	16.79	18.49	20.60	40.26	43.77	46.98	50.89	53.67
40	20.71	22.16	24.43	26.51	29.05	51.81	55.76	59.34	63.69	66.77
60	35.53	37.48	40.48	43.19	46.46	74.40	79.08	83.30	88.38	91.95
120	83.85	86.92	91.58	95.70	100.62	140.23	146.57	152.21	158.95	163.64

*Adapted with permission from Dixon, W. J. and Massey, F. J., Jr. (1969). *Introduction to Statistical Analysis*, 3rd ed. McGraw-Hill, New York.

TABLE L* Distribution of χ^2

Values of F′ corresponding to certain selected cumulative probabilities. To illustrate: The probability is 0.95 that the ratio of sample ranges R_1/R_2 is 2.6 or less when $n_1 = n_2 = 5$.

n_2	Cumulative Probability	n_1								
		2	3	4	5	6	7	8	9	10
2	0.025	0.039	0.217	0.37	0.50	0.60	0.68	0.74	0.79	0.83
	0.05	0.079	0.31	0.50	0.62	0.74	0.80	0.86	0.91	0.95
	0.95	12.7	19.1	23	26	29	30	32	34	35
	0.975	25.5	38.2	52	57	60	62	64	67	68
3	0.025	0.026	0.160	0.28	0.39	0.47	0.54	0.59	0.64	0.68
	0.05	0.052	0.23	0.37	0.49	0.57	0.64	0.70	0.75	0.80
	0.95	3.19	4.4	5.0	5.7	6.2	6.6	6.9	7.2	7.4
	0.975	4.61	6.3	7.3	8.0	8.7	9.3	9.8	10.2	10.5
4	0.025	0.019	0.137	0.25	0.34	0.42	0.48	0.53	0.57	0.61
	0.05	0.043	0.20	0.32	0.42	0.50	0.57	0.62	0.67	0.70
	0.95	2.02	2.7	3.1	3.4	3.6	3.8	4.0	4.2	4.4
	0.975	2.72	3.5	4.0	4.4	4.7	5.0	5.2	5.4	5.6
5	0.025	0.018	0.124	0.23	0.32	0.38	0.44	0.49	0.53	0.57
	0.05	0.038	0.18	0.29	0.40	0.46	0.52	0.57	0.61	0.65
	0.95	1.61	2.1	2.4	2.6	2.8	2.9	3.0	3.1	3.2
	0.975	2.01	2.6	2.9	3.2	3.4	3.6	3.7	3.8	3.9
6	0.025	0.017	0.115	0.21	0.30	0.36	0.42	0.46	0.50	0.54
	0.05	0.035	0.16	0.27	0.36	0.43	0.49	0.54	0.58	0.61
	0.95	1.36	1.8	2.0	2.2	2.3	2.4	2.5	2.6	2.7
	0.975	1.67	2.1	2.4	2.6	2.8	2.9	3.0	3.1	3.2
7	0.025	0.016	0.107	0.20	0.28	0.34	0.40	0.44	0.48	0.52
	0.05	0.032	0.15	0.26	0.35	0.41	0.47	0.51	0.55	0.59
	0.95	1.26	1.6	1.8	1.9	2.0	2.1	2.2	2.3	2.4
	0.975	1.48	1.9	2.1	2.3	2.4	2.5	2.6	2.7	2.8
8	0.025	0.016	0.102	0.19	0.27	0.33	0.38	0.43	0.47	0.50
	0.05	0.031	0.14	0.25	0.33	0.40	0.45	0.50	0.53	0.57
	0.95	1.17	1.4	1.6	1.8	1.9	1.9	2.0	2.1	2.1
	0.975	1.36	1.7	1.9	2.0	2.2	2.3	2.3	2.4	2.5
9	0.025	0.015	0.098	0.18	0.26	0.32	0.37	0.42	0.46	0.49
	0.05	0.030	0.14	0.24	0.32	0.38	0.44	0.48	0.52	0.55
	0.95	1.10	1.3	1.5	1.6	1.7	1.8	1.9	1.9	2.0
	0.975	1.27	1.6	1.8	1.9	2.0	2.1	2.1	2.2	2.3
10	0.025	0.015	0.095	0.18	0.25	0.31	0.36	0.41	0.44	0.48
	0.05	0.029	0.13	0.23	0.31	0.37	0.43	0.47	0.51	0.54
	0.95	1.05	1.3	1.4	1.5	1.6	1.7	1.8	1.8	1.9
	0.975	1.21	1.5	1.6	1.8	1.9	1.9	2.0	2.0	2.1

*Adapted with permission from Dixon, W. J. and Massey, F. J., Jr. (1969). *Introduction to Statistical Analysis*, 3rd ed. McGraw-Hill, New York.

TABLE M* Percentiles $F' = \frac{R_1}{R_2}$

Enter the horizontal scale with the sample fraction defective. Rise vertically to the upper and lower curves for the stated sample size. Read the corresponding upper and lower confidence limits on the vertical scale. To illustrate: If a sample of 50 is 20% defective, the 95% confidence limits on the population fraction defective are 10 and 35%.

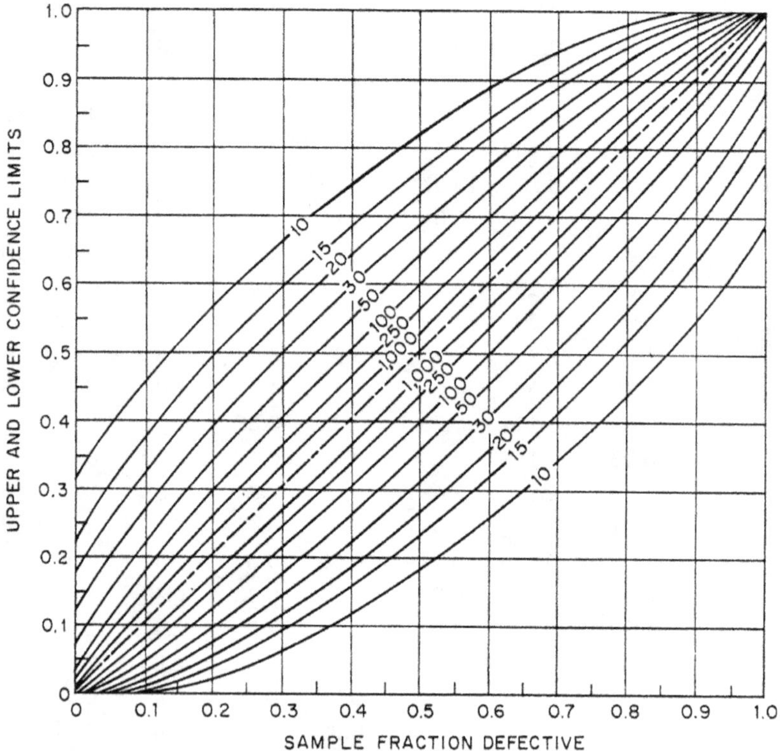

---

*By permission of Prof. E. S. Pearson from Clopper, C. J. and Pearson, E. S. (1934). "The Use of Confidence or Fiducial Limits Illustrated in the Case of the Binomial." *Biometrika*, vol. 26, p. 404.

CHART N Confidence Limits for Fraction Defective*

n_2	α for 2-Sided Test	α for 1-Sided Test	n_1 (Smaller Sample) 1	2	3	4	5	6	7	8	9	10	11	12
3	0.20	0.10		3	7									
	0.10	0.05			6									
	0.05	0.025												
	0.01	0.005												
4	0.20	0.10		3	7	13								
	0.10	0.05			6	11								
	0.05	0.025				10								
	0.01	0.005												
5	0.20	0.10		4	8	14	20							
	0.10	0.05		3	7	12	19							
	0.05	0.025			6	11	17							
	0.01	0.005					15							
6	0.20	0.10		4	9	15	22	30						
	0.10	0.05		3	8	13	20	28						
	0.05	0.025			7	12	18	26						
	0.01	0.005				10	16	23						
7	0.20	0.10		4	10	16	23	32	41					
	0.10	0.05		3	8	14	21	29	39					
	0.05	0.025			7	13	20	27	36					
	0.01	0.005				10	16	24	32					
8	0.20	0.10		5	11	17	25	34	44	55				
	0.10	0.05		4	9	15	23	31	41	51				
	0.05	0.025		3	8	14	21	29	38	49				
	0.01	0.005				11	17	25	34	43				
9	0.20	0.10	1	5	11	19	27	36	46	58	70			
	0.10	0.05		4	9	16	24	33	43	54	66			
	0.05	0.025		3	8	14	22	31	40	51	62			
	0.01	0.005			6	11	18	26	35	45	56			
10	0.20	0.10	1	6	12	20	28	38	49	60	73	87		
	0.10	0.05		4	10	17	26	35	45	56	69	82		
	0.05	0.025		3	9	15	23	32	42	53	65	78		
	0.01	0.005			6	12	19	27	37	47	58	71		
11	0.20	0.10	1	6	13	21	30	40	51	63	76	91	106	
	0.10	0.05		4	11	18	27	37	47	59	72	86	100	
	0.05	0.025		3	9	16	24	34	44	55	68	81	96	
	0.01	0.005			6	12	20	28	38	49	61	73	87	
12	0.20	0.10	1	7	14	22	32	42	54	66	80	94	110	127
	0.10	0.05		5	11	19	28	38	49	62	75	89	104	120
	0.05	0.025		4	10	17	26	35	46	58	71	84	99	115
	0.01	0.005			7	13	21	30	40	51	63	76	90	105

*Reproduced with permission from Tate, M. W. and Clelland, R. C. (1957). *Non-parametric and Shortcut Statistics*. The Interstate Printers & Publishers, Danville, IL.

TABLE O Critical Values of Smaller Rank Sum for the Wilcoxon-Mann-Whitney Test*

n_1 = number of values above the median and n_2 = number of values below the median.

	Probability of an Equal or Smaller Number of Runs		Probability of an Equalor Larger Number of Runs	
$m = n_1 = n_2$	$\alpha = 0.05$	$\alpha = 0.01$	$\alpha = 0.05$	$\alpha = 0.01$
5	3	2	9	10
6	3	2	11	12
7	4	3	12	13
8	5	4	13	14
9	6	4	14	16
10	6	5	16	17
11	7	6	17	18
12	8	7	18	19
13	9	7	19	21
14	10	8	20	22
15	11	9	21	23
16	11	10	23	24
17	12	10	24	26
18	13	11	25	27
19	14	12	26	28
20	15	13	27	29
21	16	14	28	30
22	17	14	29	32
23	17	15	31	33
24	18	16	32	34
25	19	17	33	35
26	20	18	34	36
27	21	19	35	37
28	22	19	36	39
29	23	20	37	40
30	24	21	38	41

*Reproduced with permission from Swed, Freda S. and Eisenhart, C. (1943). "Tables for Testing Randomness of Grouping in a Sequence of Alternatives." *Annals of Mathematical Statistics*, vol. XIV, pp. 66 and 87, Tables II and III.

TABLE P Limiting Values for Number of Runs above and below the Median of a Set of Values*

Statistic	No. of Observations	$P_{.90}$	$P_{.95}$	$P_{.98}$	$P_{.99}$
$r_{10} = \frac{X_2 - X_1}{X_n - X_1}$	3	0.886	0.941	0.976	0.988
	4	0.679	0.765	0.846	0.889
	5	0.557	0.642	0.729	0.780
	6	0.482	0.560	0.644	0.698
	7	0.434	0.507	0.586	0.637
$r_{11} = \frac{X_2 - X_1}{X_{n-1} - X_1}$	8	0.479	0.554	0.631	0.683
	9	0.441	0.512	0.587	0.635
	10	0.409	0.477	0.551	0.597
$r_{21} = \frac{X_3 - X_1}{X_{n-1} - X_1}$	11	0.517	0.576	0.638	0.679
	12	0.490	0.546	0.605	0.642
	13	0.467	0.521	0.578	0.615
$r_{22} = \frac{X_3 - X_1}{X_{n-2} - X_1}$	14	0.492	0.546	0.602	0.641
	15	0.472	0.525	0.579	0.616
	16	0.454	0.507	0.559	0.595
	17	0.438	0.490	0.542	0.577
	18	0.424	0.475	0.527	0.561
	19	0.412	0.462	0.514	0.547
	20	0.401	0.450	0.502	0.535
	21	0.391	0.440	0.491	0.524
	22	0.382	0.430	0.481	0.514
	23	0.374	0.421	0.472	0.505
	24	0.367	0.413	0.464	0.497
	25	0.360	0.406	0.457	0.489

*Adapted with persmission from Dixon, W. J. and Massey, F. J., Jr. (1969). *Introduction to Statistical Analysis*, 3rd ed. McGraw-Hill, New York.

Table Q Criteria for Testing for Extreme Mean*

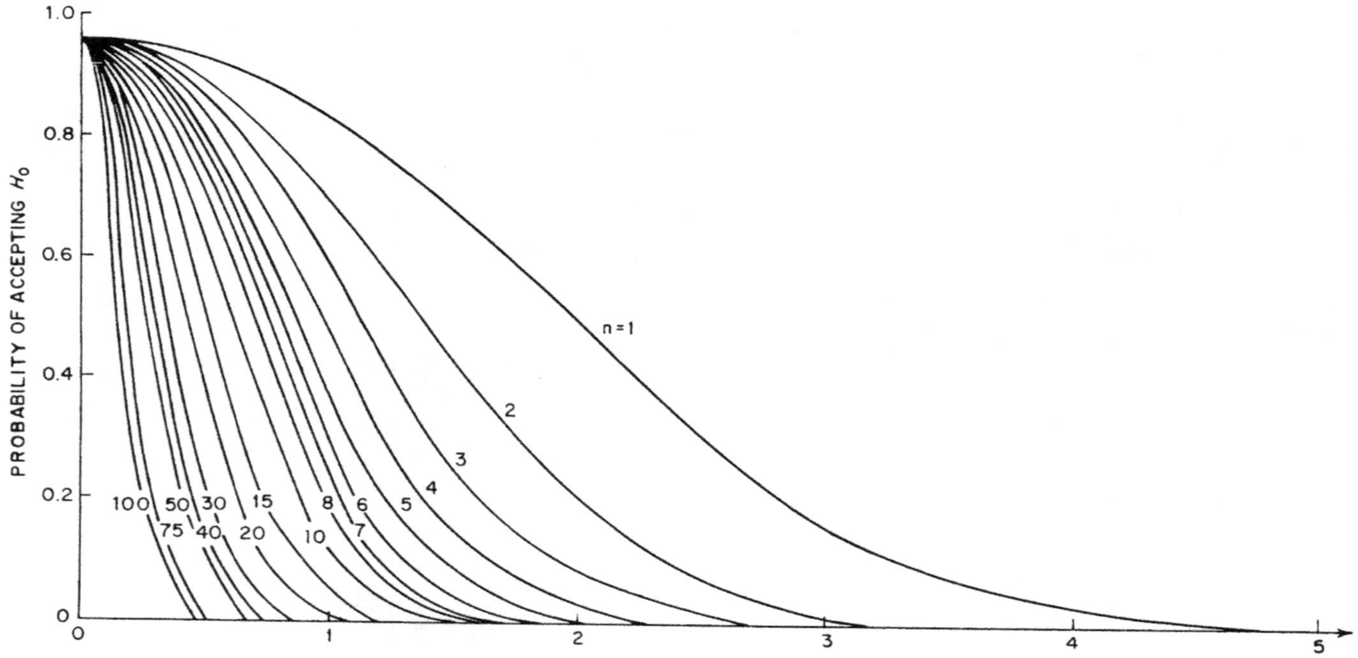

*Adapted with permission from Ferris, Charles D., Grubbs, Frank E., and Weaver, Chalmers L. (1946). "Operating Characteristics for the Common Statistical Tests of Significance." *Annals of Mathematical Statistics*, June.

CHART R Operating Characteristics of the Two-Sided Normal Test for a Level of Significance Equal to 0.05*

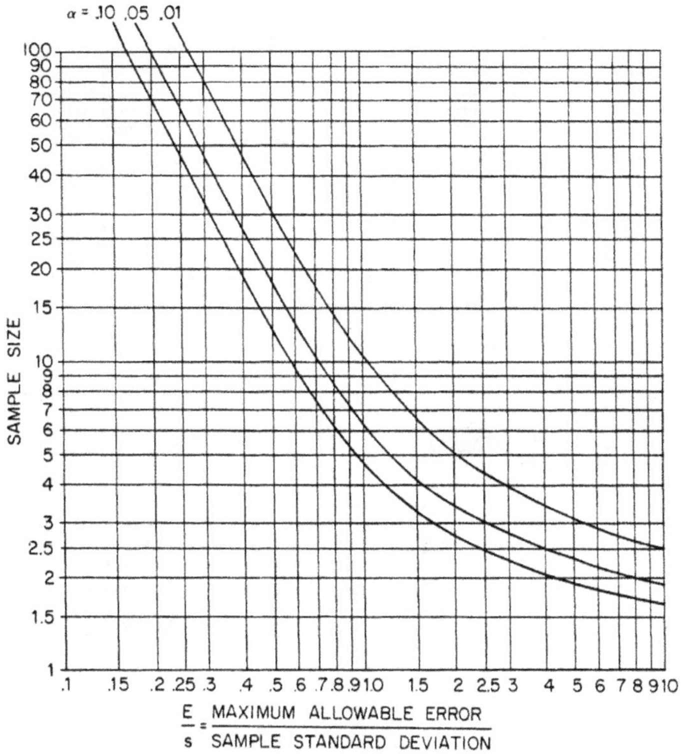

*Reproduced with permission from Weida, Frank M. and Lum, Mary D. (1953). *Statistical Inference, Reliability, and Significance.* WADC Technical Report 53-149, U.S. Air Force.

CHART S Size of Sample for Arithmetic Mean When σ Is Unknown*

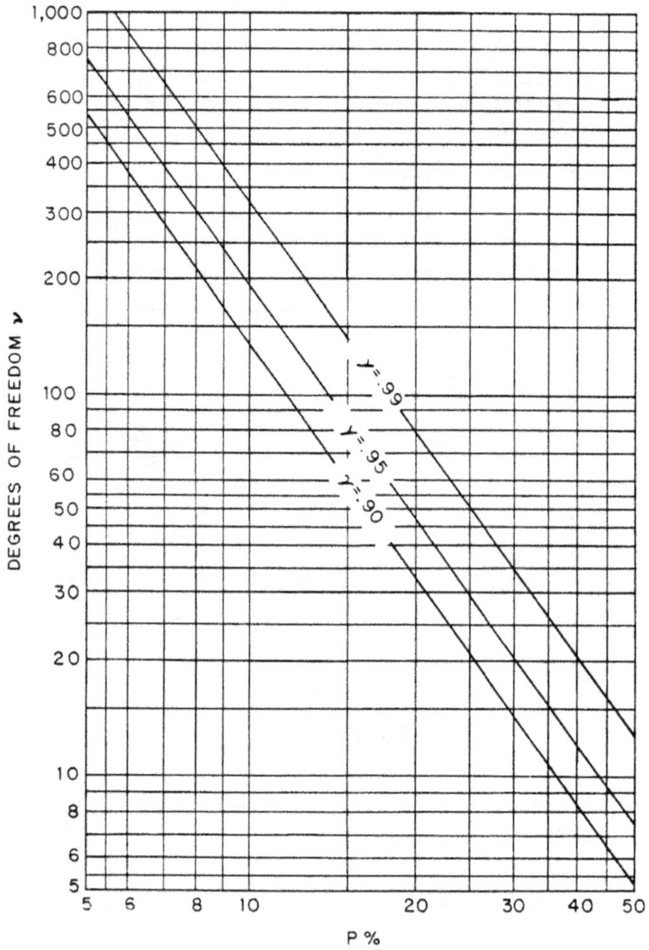

*Adapted with permission from Greenwood, J. A. and Sandomire, M. M. (1950). "Statistics Manual, Sample Size Required for Estimating the Standard Deviation as a Percent of Its True Value." *Journal of the American Statistical Association*. vol. 45, p. 258. The manner of graphing is adapted with permission from Crow, E. L., Davis, F. A., and Maxfield, M. W. (1955). *NAVORD Report 3369*. NOTS 948, U.S. Naval Ordnance Test Station, China Lake, CA. (Reprinted by Dover Publications, New York, 1960.)

CHART T Number of Degrees of Freedom Required to Estimate the Standard Deviation within *P*% of Its True Value with Confidence Coefficient γ

Factors K_1 such that the probability is γ that at least a proportion P of the distribution will be included between $\overline{X} \pm K_1 R$ where $\overline{X}$ is the mean and R is the range in a sample of size n.

P	$\gamma = 0.90$				$\gamma = 0.95$				$\gamma = 0.99$			
n	0.90	0.95	0.99	0.999	0.90	0.95	0.99	0.999	0.90	0.95	0.99	0.999
2	11.298	13.294	17.090	21.374	22.635	26.634	34.238	42.821	113.429	133.469	171.576	214.588
3	3.069	3.631	4.711	5.936	4.399	5.206	6.752	8.509	9.951	11.776	15.275	19.249
4	1.877	2.227	2.902	3.672	2.422	2.873	3.744	4.737	4.233	5.021	6.543	8.279
5	1.428	1.697	2.216	2.812	1.749	2.078	2.715	3.444	2.709	3.219	4.205	5.335
6	1.194	1.420	1.857	2.360	1.418	1.686	2.206	2.803	2.042	2.429	3.178	4.038
7	1.050	1.248	1.635	2.080	1.222	1.453	1.903	2.420	1.678	1.996	2.615	3.325
8	0.951	1.131	1.483	1.888	1.090	1.297	1.700	2.165	1.449	1.724	2.261	2.878
9	0.879	1.046	1.372	1.747	0.997	1.187	1.556	1.981	1.290	1.536	2.014	2.565
10	0.824	0.981	1.286	1.639	0.926	1.103	1.446	1.843	1.176	1.400	1.836	2.340
11	0.780	0.929	1.219	1.554	0.871	1.037	1.361	1.735	1.088	1.296	1.701	2.168
12	0.745	0.887	1.164	1.484	0.827	0.985	1.292	1.648	1.020	1.215	1.594	2.033
13	0.715	0.852	1.118	1.426	0.790	0.940	1.235	1.575	0.964	1.148	1.507	1.922
14	0.690	0.822	1.079	1.377	0.759	0.904	1.187	1.514	0.917	1.093	1.435	1.830
15	0.669	0.797	1.046	1.334	0.733	0.873	1.146	1.462	0.878	1.046	1.373	1.753
16	0.650	0.774	1.016	1.297	0.710	0.845	1.110	1.417	0.845	1.007	1.322	1.687
17	0.633	0.755	0.991	1.265	0.690	0.822	1.109	1.377	0.816	0.972	1.277	1.630
18	0.619	0.737	0.968	1.235	0.672	0.801	1.051	1.342	0.790	0.941	1.236	1.578
19	0.605	0.721	0.947	1.209	0.656	0.782	1.027	1.311	0.768	0.916	1.203	1.535
20	0.594	0.707	0.929	1.186	0.642	0.765	1.005	1.282	0.748	0.892	1.171	1.495

*Adapted with permission from Mitra, S. K. (1957). "Tables for Tolerance Limits for a Normal Population Based on Sample Mean and Range on Mean Range." *Journal of the American Statistical Association*, vol. 52, no. 277, March, p. 92.

TABLE U Tolerance Factors for Normal Distribution*

Factors k such that the probability is γ that at least a proportion P of the distribution will be less than $\overline{X} + ks$ (or greater than $\overline{X} - ks$) where $\overline{X}$ and s are estimates of the mean and standard deviation computed from a sample size of n. Two-sided factors cover $\overline{X} \pm ks$.

P / n	$\gamma = 0.90$				$\gamma = 0.95$				$\gamma = 0.99$			
	0.90	**0.95**	**0.99**	**0.999**	**0.90**	**0.95**	**0.99**	**0.999**	**0.90**	**0.95**	**0.99**	**0.999**
	One-Sided Factors*											
3	4.258	5.310	7.340	9.651	6.158	7.655	10.552	13.857				
4	3.187	3.957	5.437	7.128	4.163	5.145	7.042	9.215				
5	2.742	3.400	4.666	6.112	3.407	4.202	5.741	7.501				
6	2.494	3.091	4.242	5.556	3.006	3.707	5.062	6.612	4.408	5.409	7.334	9.540
7	2.333	2.894	3.972	5.201	2.755	3.399	4.641	6.061	3.856	4.730	6.411	8.348
8	2.219	2.755	3.783	4.955	2.582	3.188	4.353	5.686	3.496	4.287	5.811	7.566
9	2.133	2.649	3.641	4.772	2.454	3.031	4.143	5.414	3.242	3.971	5.389	7.014
10	2.065	2.568	3.532	4.629	2.355	2.911	3.981	5.203	3.048	3.739	5.075	6.603
11	2.012	2.503	3.444	4.515	2.275	2.815	3.852	5.036	2.897	3.557	4.828	6.284
12	1.966	2.448	3.371	4.420	2.210	2.736	3.747	4.900	2.773	3.410	4.633	6.032
13	1.928	2.403	3.310	4.341	2.155	2.670	3.659	4.787	2.677	3.290	4.472	5.826
14	1.895	2.363	3.257	4.274	2.108	2.614	3.585	4.690	2.592	3.189	4.336	5.651
15	1.866	2.329	3.212	4.215	2.068	2.566	3.520	4.607	2.521	3.102	4.224	5.507
16	1.842	2.299	3.172	4.164	2.032	2.523	3.463	4.534	2.458	3.028	4.124	5.374
17	1.820	2.272	3.136	4.118	2.001	2.486	3.415	4.471	2.405	2.962	4.038	5.268
18	1.800	2.249	3.106	4.078	1.974	2.453	3.370	4.415	2.357	2.906	3.961	5.167
19	1.781	2.228	3.078	4.041	1.949	2.423	3.331	4.364	2.315	2.855	3.893	5.078
20	1.765	2.208	3.052	4.009	1.926	2.396	3.295	4.319	2.275	2.807	3.832	5.003
21	1.750	2.190	3.028	3.979	1.905	2.371	3.262	4.276	2.241	2.768	3.776	4.932
22	1.736	2.174	3.007	3.952	1.887	2.350	3.233	4.238	2.208	2.729	3.727	4.866
23	1.724	2.159	2.987	3.927	1.869	2.329	3.206	4.204	2.179	2.693	3.680	4.806
24	1.712	2.145	2.969	3.904	1.853	2.309	3.181	4.171	2.154	2.663	3.638	4.755
25	1.702	2.132	2.952	3.882	1.838	2.292	3.158	4.143	2.129	2.632	3.601	4.706
30	1.657	2.080	2.884	3.794	1.778	2.220	3.064	4.022	2.029	2.516	3.446	4.508
35	1.623	2.041	2.833	3.730	1.732	2.166	2.994	3.934	1.957	2.431	3.334	4.364
40	1.598	2.010	2.793	3.679	1.697	2.126	2.941	3.866	1.902	2.365	3.250	4.255
45	1.577	1.986	2.762	3.638	1.669	2.092	2.897	3.811	1.857	2.313	3.181	4.168
50	1.560	1.965	2.735	3.604	1.646	2.065	2.863	3.766	1.821	2.296	3.124	4.096

Two-Sided Factors†												
2	15.978	18.800	24.167	30.227	32.019	37.674	48.430	60.573	160.193	188.491	242.300	303.054
3	5.847	6.919	8.974	11.309	8.380	9.916	12.861	16.208	18.930	22.401	29.055	36.616
4	4.166	4.943	6.440	8.149	5.369	6.370	8.299	10.502	9.398	11.150	14.527	18.383
5	3.494	4.152	5.423	6.879	4.275	5.079	6.634	8.415	6.612	7.855	10.260	13.015
6	3.131	3.723	4.870	6.188	3.712	4.414	5.775	7.337	5.337	6.345	8.301	10.548
7	2.902	3.452	4.521	5.750	3.369	4.007	5.248	6.676	4.613	5.488	7.187	9.142
8	2.743	3.264	4.278	5.446	3.136	3.732	4.891	6.226	4.147	4.936	6.468	8.234
9	2.626	3.125	4.098	5.220	2.967	3.532	4.631	5.899	3.822	4.550	5.966	7.600
10	2.535	3.018	3.959	5.046	2.839	3.379	4.433	5.649	3.582	4.265	5.594	7.129
11	2.463	2.933	3.849	4.906	2.737	3.259	4.277	5.452	3.397	4.045	5.308	6.766
12	2.404	2.863	3.758	4.792	2.655	3.162	4.150	5.291	3.250	3.870	5.079	6.477
13	2.355	2.805	3.682	4.697	2.587	3.081	4.044	5.158	3.130	3.727	4.893	6.240
14	2.314	2.756	3.618	4.615	2.529	3.012	3.955	5.045	3.029	3.608	4.737	6.043
15	2.278	2.713	3.562	4.545	2.480	2.954	3.878	4.949	2.945	3.507	4.605	5.876
16	2.246	2.676	3.514	4.484	2.437	2.903	3.812	4.865	2.872	3.421	4.492	5.732
17	2.219	2.643	3.471	4.430	2.400	2.858	3.754	4.791	2.808	3.345	4.393	5.607
18	2.194	2.614	3.433	4.382	2.366	2.819	3.702	4.]25	2.753	3.279	4.307	5.497
19	2.172	2.588	3.399	4.339	2.337	2.784	3.656	4.667	2.703	3.221	4.230	5.399
20	2.152	2.564	3.368	4.300	2.310	2.752	3.615	4.614	2.659	3.168	4.161	5.312

*Adapted from Lieberman, Gerald J. (1958). "Tables for One-Sided Tolerance Limits." *Industrial Quality Control*, vol. XIV, no. 10, April, p. 8. Adapted with permission of the American Society for Quality Control.

†Adapted with permission from Eisenhart, C., Hastay, M. W., and Wallis, W. A. (1947). *Techniques of Statistical Analysis*. McGraw-Hill, New York.

TABLE V One-Sided and Two-sided Statistical Tolerance Limit Factors k for a Normal Distribution

n \ P	$\gamma = 0.90$				$\gamma = 0.95$				$\gamma = 0.99$			
	0.90	0.95	0.99	0.999	0.90	0.95	0.99	0.999	0.90	0.95	0.99	0.999
Two-Sided Factors†												
21	2.135	2.543	3.340	4.264	2.286	2.723	3.577	4.567	2.620	3.121	4.100	5.234
22	2.118	2.524	3.315	4.232	2.264	2.697	3.543	4.523	2.584	3.078	4.044	5.163
23	2.103	2.506	3.292	4.203	2.244	2.673	3.512	4.484	2.551	3.040	3.993	5.098
24	2.089	2.480	3.270	4.176	2.225	2.651	3.483	4.447	2.522	3.004	3.947	5.039
25	2.077	2.474	3.251	4.151	2.208	2.631	3.457	4.413	2.494	2.972	3.904	4.985
26	2.065	2.460	3.232	4.127	2.193	2.612	3.432	4.382	2.460	2.941	3.865	4.935
27	2.054	2.447	3.215	4.106	2.178	2.595	3.409	4.353	2.446	2.914	3.828	4.888
30	2.025	2.413	3.170	4.049	2.140	2.549	3.350	4.278	2.385	2.841	3.733	4.768
35	1.988	2.368	3.112	3.974	2.090	2.490	3.272	4.179	2.306	2.748	3.611	4.611
40	1.959	2.334	3.066	3.917	2.052	2.445	3.213	4.104	2.247	2.677	3.518	4.493
45	1.935	2.306	3.030	3.871	2.021	2.408	3.165	4.042	2.200	2.621	3.444	4.399
50	1.916	2.284	3.001	3.833-	1.996	2.379	3.126	3.993	2.162	2.576	3.385	4.323

TABLE V One-Sided and Two-sided Statistical Tolerance Limit Factors k for a Normal Distribution* (*Continued*)

γ is the probability that an interval will cover a proportion P of the population with a random sample of size N.

N \ γ	0.5	0.7	0.9	0.95	0.99	0.995
2	0.293	0.164	0.052	0.026	0.006	0.003
4	0.615	0.492	0.321	0.249	0.141	0.111
6	0.736	0.640	0.490	0.419	0.295	0.254
10	0.838	0.774	0.664	0.606	0.496	0.456
20	0.918	0.883	0.820	0.784	0.712	0.683
40	0.959	0.941	0.907	0.887	0.846	0.829
60	0.973	0.960	0.937	0.924	0.895	0.883
80	0.980	0.970	0.953	0.943	0.920	0.911
100	0.984	0.976	0.962	0.954	0.936	0.929
150	0.990	0.984	0.975	0.969	0.957	0.952
200	0.992	0.988	0.981	0.977	0.968	0.961
500	0.997	0.996	0.993	0.991	0.987	0.986
1,000	0.999	0.998	0.997	0.996	0.994	0.993

*Adapted with permission from Dixon, W. J. and Massey, F. J., Jr. (1969). Introduction to Statistical Analysis, 3rd ed. McGraw-Hill, NewYork.

TABLE W *P* for Interval between Sample Extremes*

N \ γ	0.50	0.70	0.90	0.95	0.99	0.995
0.995	336	488	777	947	1,325	1,483
0.99	168	244	388	473	662	740
0.95	34	49	77	93	130	146
0.90	17	24	38	46	64	72
0.85	11	16	25	30	42	47
0.80	9	12	18	22	31	34
0.75	7	10	15	18	24	27
0.70	6	8	12	14	20	22
0.60	4	6	9	10	14	16
0.50	3	5	7	8	11	12

*Adapted with permission from Dixon, W. J. and Massey, F. J., Jr. (1969). *Introduction to Statistical Analysis*, 3rd ed. McGraw-Hill, New York.

TABLE X *N* for Interval between Sample Extremes*

Number of Observations in Subgroup	E_2
2	2.660
3	1.772
4	1.457
5	1.290
6	1.184
7	1.109
8	1.054
9	1.010
10	0.975
11	0.946
12	0.921
13	0.899
14	0.881
15	0.864

TABLE Y E_2 Factors for Control Charts

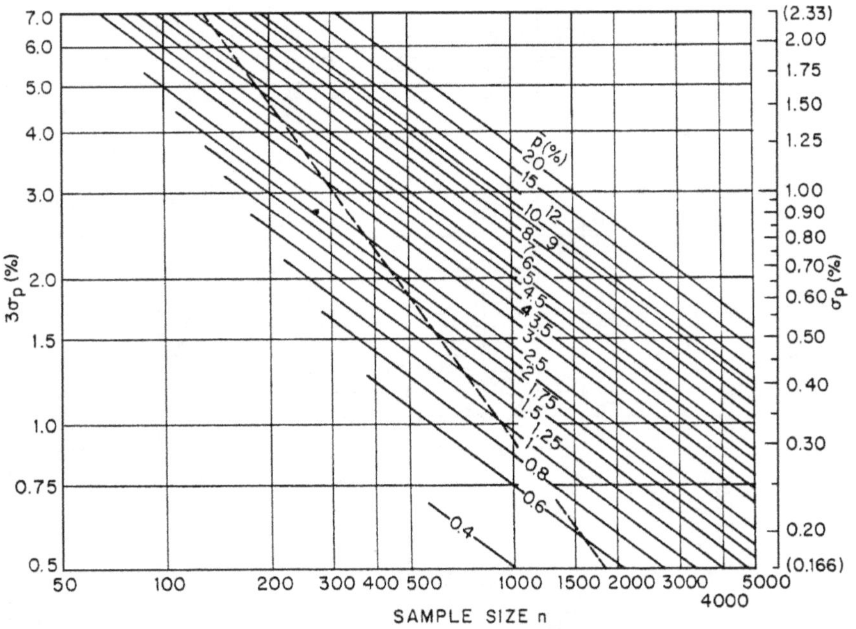

CHART Z Control Limits of *p* Charts

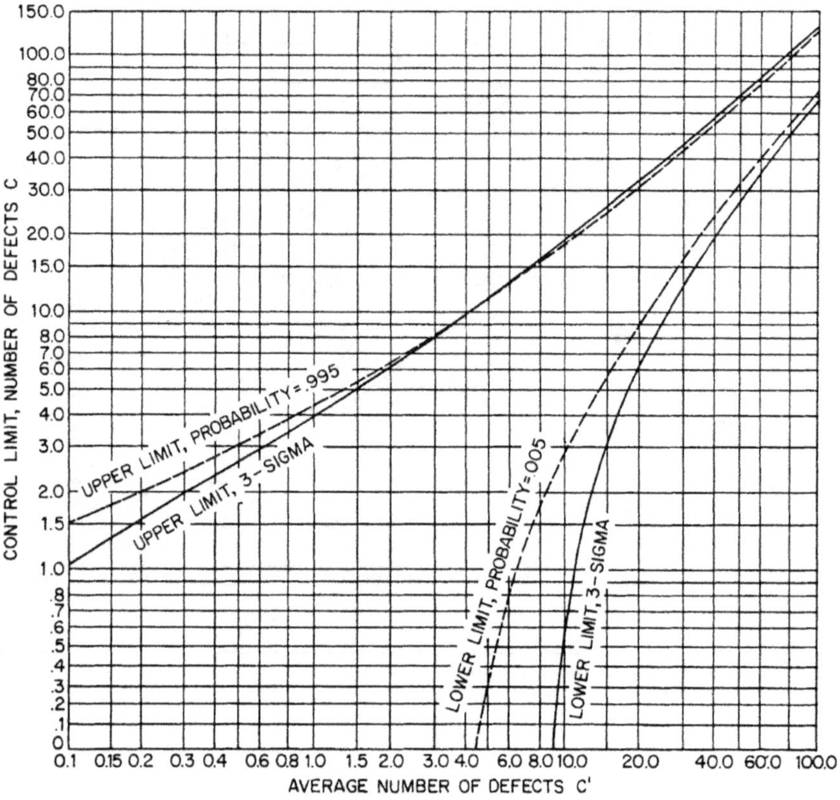

*Reproduced by permission from American War Standard Z1.3-1942. American Standards Association, New York.

TABLE AA Control Limits for *c*, Number of Defects per Sample*

$2\,\alpha_0 = 0.0027$
$\alpha_0 = 0.00135$[†]

δ	θ	d
0.2	5°43′	330.4
0.4	11°19′	82.6
0.5	14°00′	52.9
0.6	16°42′	36.7
0.8	21°48′	20.6
1.0	26°34′	13.2
1.2	30°58′	9.2
1.3	32°59′	7.8
1.4	35°00′	6.7
1.6	38°40′	5.2
1.8	41°59′	4.1
2.0	45°00′	3.3
2.2	47°44′	2.7
2.4	50°12′	2.3
2.6	52°26′	2.0
2.8	54°28′	1.7
3.0	56°19′	1.5

*Adapted with permission from Johnson, Norman L. and Leone, Fred C. (1964). *Statistics and Experimental Design in Engineering and Physical Sciences*. John Wiley & Sons, New York, vol. I, p. 322.

†For limits comparable with the 3-sigma limits used in the Shewhart control chart.

Table BB Factors for Cumulative Sum Control Chart*

1306	1189	5731	3968	5606	5084	8947	3897	1636	7810
0422	2431	0649	8085	5053	4722	6598	5044	9040	5121
6597	2022	6168	5060	8656	6733	6364	7649	1871	4328
7965	6541	5645	6243	7658	6903	9911	5740	7824	8520
7695	6937	0406	8894	0441	8135	9797	7285	5905	9539
5160	7851	8464	6789	3938	4197	6511	0407	9239	2232
2961	0551	0539	8288	7478	7565	5581	5771	5442	8761
1428	4183	4312	5445	4854	9157	9158	5218	1464	3634
3666	5642	4539	1561	7849	7520	2547	0756	1206	2033
6543	6799	7454	9052	6689	1946	2574	9386	0304	7945
9975	6080	7423	3175	9377	6951	6519	8287	8994	5532
4866	0956	7545	7723	8085	4948	2228	9583	4415	7065
8239	7068	6694	5168	3117	1586	0237	6160	9585	1133
8722	9191	3386	3443	0434	4586	4150	1224	6204	0937
1330	9120	8785	8382	2929	7089	3109	6742	2468	7025
2296	2952	4764	9070	6356	9192	4012	0618	2219	1109
3582	7052	3132	4519	9250	2486	0830	8472	2160	7046
5872	9207	7222	6494	8973	3545	6967	8490	5264	9821
1134	6324	6201	3792	5651	0538	4676	2064	0584	7996
1403	4497	7390	8503	8239	4236	8022	2914	4368	4529
3393	7025	3381	3553	2128	1021	8353	6413	5161	8583
1137	7896	3602	0060	7850	7626	0854	6565	4260	6220
7437	5198	8772	6927	8527	6851	2709	5992	7383	1071
8414	8820	3917	7238	9821	6073	6658	1280	9643	7761
8398	5224	2749	7311	5740	9771	7826	9533	3800	4553
0995	8935	2939	3092	2496	0359	0318	4697	7181	4035
6657	0755	9685	4017	6581	7292	5643	5064	1142	1297
8875	8369	7868	0190	9278	1709	4253	9346	4335	3769
8399	6702	0586	6428	7985	2979	4513	1970	1989	3105
6703	1024	2064	0393	6815	8502	1375	4171	6970	1201
4730	1653	9032	9855	0957	7366	0325	5178	7959	5371
8400	6834	3187	8688	1079	1480	6776	9888	7585	9998
3647	8002	6726	0877	4552	3238	7542	7804	3933	9475
6789	5197	8037	2354	9262	5497	0005	3986	1767	7981
2630	2721	2810	2185	6323	5679	4931	8336	6662	3566

Table CC Random Numbers

1374	8625	1644	3342	1587	0762	6057	8011	2666	3759
1572	7625	9110	4409	0239	7059	3415	5537	2250	7292
9678	2877	7579	4935	0449	8119	6969	5383	1717	6719
0882	6781	3538	4090	3092	2365	6001	3446	9985	6007
0006	4205	2389	4365	1981	8158	7784	6256	3842	5603
4611	9861	7916	93CS	2074	9462	0254	4827	9198	3974
1093	3784	4190	6332	1175	8599	9735	8584	6581	7194
3374	3545	6865	8819	3342	1676	2264	6014	5012	2458
3650	9676	1436	4374	4716	5548	8276	6235	6742	2154
7292	5749	7977	7602	9205	3599	3880	9537	4423	2330
2353	8319	2850	4026	3027	1708	3518	7034	7132	6903
1094	2009	8919	5676	7283	4982	9642	7235	8167	3366
0568	4002	0587	7165	1094	2006	7471	0940	4366	9554
5606	4070	5233	4339	6543	6695	5799	5821	3953	9458
8285	7537	1181	2300	5294	6892	1627	3372	1952	3028

*Adapted with permission from Owen, Donald B. (1962). *Handbook of Statistical Tables*. Addison-Wesley, Reading, MA. Courtesy U.S. Atomic Energy Commission.

TABLE CC Random Numbers* (*Continued*)

$\alpha = 0.01; \beta = 0.2$

DF_2 \ DF_1	1	2	3	4	5	6	7	8	9
2	80.37	106.63	119.75	127.62	132.87	136.63	139.45	141.63	143.38
4	17.28	18.58	18.95	19.11	19.18	19.21	19.23	19.24	19.24
6	11.36	11.12	10.77	10.49	10.27	10.11	9.97	9.86	9.77
8	9.41	8.76	8.21	7.83	7.54	7.32	7.15	7.01	6.89
10	8.47	7.63	7.02	6.58	6.26	6.03	5.84	5.68	5.56
12	7.91	6.98	6.33	5.87	5.54	5.29	5.09	4.93	4.80
14	7.55	6.56	5.88	5.41	5.07	4.81	4.61	4.45	4.31
16	7.30	6.26	5.56	5.09	4.75	4.49	4.28	4.11	3.98
18	7.11	6.05	5.35	4.86	4.51	4.24	4.04	3.87	3.73
20	6.96	5.89	5.17	4.68	4.33	4.06	3.85	3.68	3.54
24	6.76	5.66	4.93	4.41	4.08	3.80	3.57	3.42	3.28
30	6.55	5.42	4.68	4.19	3.82	3.55	3.33	3.16	3.02
40	6.35	5.20	4.45	3.96	3.57	3.31	3.10	2.92	2.79
60	6.18	5.00	4.25	3.74	3.37	3.10	2.88	2.70	2.55
80	6.10	4.88	4.16	3.65	3.28	2.99	2.76	2.59	2.43
120	6.00	4.80	4.04	3.53	3.17	2.89	2.66	2.50	2.34
240	5.90	4.71	3.96	3.46	3.06	2.79	2.56	2.40	2.25
∞	5.84	4.62	3.87	3.35	2.98	2.70	2.47	2.29	2.14

Table DD Values of ϕ^2 for Determining Sample Size in Analysis of Variance

10	12	15	20	24	30	40	60	120	∞
144.82	147.02	149.30	151.63	152.84	154.06	155.30	156.55	157.83	159.09
19.24	19.24	19.24	19.22	19.21	19.21	19.19	19.18	19.18	19.17
9.69	9.57	9.44	9.30	9.22	9.14	9.07	8.99	8.90	8.81
6.80	6.64	6.48	6.31	6.21	6.12	6.02	5.91	5.81	5.70
5.45	5.29	5.11	4.92	4.82	4.71	4.61	4.49	4.38	4.26
4.69	4.52	4.33	4.13	4.02	3.91	3.80	3.68	3.56	3.43
4.20	4.02	3.83	3.63	3.52	3.40	3.28	3.16	3.03	2.89
3.86	3.68	3.48	3.27	3.16	3.04	2.92	2.80	2.66	2.52
3.61	3.43	3.23	3.01	2.90	2.78	2.66	2.53	2.39	2.24
3.42	3.23	3.03	2.82	2.70	2.58	2.46	2.32	2.18	2.03
3.13	2.96	2.76	2.53	2.43	2.31	2.16	2.02	1.88	1.72
2.90	2.70	2.50	2.27	2.16	2.02	1.88	1.74	1.59	1.42
2.66	2.46	2.25	2.02	1.90	1.77	1.61	1.46	1.30	1.13
2.43	2.23	2.02	1.78	1.66	1.52	1.37	1.21	1.04	0.841
2.31	2.13	1.90	1.66	1.54	1.39	1.25	1.08	0.902	0.689
2.22	2.02	1.80	1.56	1.44	1.28	1.12	0.960	0.766	0.528
2.13	1.90	1.69	1.44	1.32	1.17	1.00	0.828	0.624	0.345
2.02	1.81	1.58	1.34	1.21	1.05	0.884	0.704	0.472	0.000

*These tables are computed from Lehmer, Emma (1944). "Inverse Tables of Probabtilities of Errors of Second Kind." *Annals of Mathematical Statistics*, vol. 15, p. 390. Reproduced from Dixon, W. J. and Massey, F. J., Jr. *Introduction to Statistical Analysis*, 1st ed. McGraw-Hill, NewYork, p. 330.

Table DD Values of ϕ^2 for Determining Sample Size in Analysis of Variance* (*Continued*)

Glossary of Acronyms

ACE—Award for Corporate Excellence

ACH—Automated clearing house

ADLI—Approach, deployment, learning, and integration

AHT—Average handle time

ALOS—Average length of stay

AMI—Acute myocardial infarction

ANOVA—Analysis of variance

ANSI—American National Standards Institute

ANZSIC—Australian and New Zealand Standard Industrial Classification

AOQ—Average outgoing quality

APQC—American Productivity and Quality Center

AQAP—Allied quality assurance publications

AQIP—Annual quality improvement plan

ARDEC—Armament Research, Development, and Engineering Center

ARL—Army Research Laboratory

ARPAnet—Advanced Research Projects Agency Network

ASME—American Society of Mechanical Engineers

ASQ—American Society for Quality

ASTM—American Society for Testing and Materials

ATM—Automatic teller machine

BAM—Business activity monitoring

BBB—Better Business Bureau

BBE—Behavior-based expectation

BPM—Business process management

BPO—Business process outsourcing

BPQM—Business process quality management

BPR—Business process reengineering

CAFÉ—Connecticut Award for Excellence

CCL—Critical-components list

CDP—Carbon disclosure project

CE—European Conformity

CEN—European Committee for Standardization

CEO—Chief executive officer

CFO—Chief financial officer

CFR—Code of Federal Regulations

cGMP—Current good manufacturing practice

CHF—Congestive heart failure

CI—Continuous improvement

CIn—Continuous innovation

CIO—Chief information officer

CMM—Capability maturity model

CMMi—Capability maturity model integration

CMS—Centers for Medicare and Medicaid Services

CMU—Carnegie-Mellon University

COBIT—Control objectives for information and related technology

COO—Chief operating officer

COP^3—Cost of poorly performing processes

COPQ—Cost of poor quality

COTS—Commercial off-the-shelf

C_p—Process capability

C_{pk}—Process capability index

CPM—Critical path method

C_{pm}—Taguchi capability index

CQE—Certified quality engineer

CTO—Chief technology officer

CTQ—Critical to quality

CUMSUM/CUSUM—Cumulative sum

DA—Decision analysis

DC—Discharge

DCE—Defect containment effectiveness

DES—Discrete event simulation

DF—Degree of freedom

DFA—Design for assembly

DFM—Design for manufacturing

DFMA—Design for manufacture and assembly

DFMEA—Design failure mode and effects analysis

DFSS—Design for Six Sigma

DMADV—Define, measure, analyze, design, verify

DMAIC—Define, measure, analyze, improve, control

DOE—Design of experiments

DpKLOC—Defects per thousand lines of code

DPMO—Defects per million opportunities

DRM—Digital Rights Management

DVT—Design verification test

EDI—Electronic Data Interchange

EFQM—European Foundation for Quality Management

EFT—Electronic funds transfer

EHR—Electronic health record

EIU—Environmental impact unit

EMMA—Electronic Municipal Market Access

EMR—Electronic medical record

EMS—Environmental management systems

ENPV—Expected net present value

EP—Environmental protection

ERA—Emergency response action

ERP—Enterprise resource planning

ESD—Emergency services department

ESS—Employee self-service

EU—European Union

EVOP—Evolutionary operations

EWMA—Exponentially weighted moving average

FC—Family Council

FDA—Food and Drug Administration

FDM—Functional deployment matrix

FLC—Federal Laboratory Consortium

FMEA—Failure mode and effects analysis

FMECA—Failure mode effects criticality analysis

FRACAS—Failure reporting and corrective action systems

FTA—Fault tree analysis

FUNDIBEQ—The Ibero-American Foundation for Quality Management

GATT—General Agreement on Tariffs and Trade

GDP—Good distribution practice

GDP—Gross domestic product

GEM—Global excellence model

GHG—Greenhouse gas

GIDEP—Government-Industry Data Exchange Program

GMO—Genetically modified organism

GMP—Good manufacturing practice

H_a—Alternative hypothesis

HACCP—Hazard analysis and critical control points

H_o—Null hypothesis

HR—Human resources

IAQ—International Academy for Quality

IAQG—International Aerospace Quality Group

ICB—Industry classification benchmark

ICU—Intensive care unit

IEC—International Electrotechnical Commission

IEEE—Institute of Electrical and Electronics Engineers

IHI—Institute for Healthcare Improvement

IOD—Institute of Directors

IOM—Institute of Medicine

IPEC—International Pharmaceutical Excipients Council

IRPS—International Reliability Physics Symposium

ISIC—International Standard Industrial Classification of All Economic Activities

ISO—International Organization for Standardization

ISO/DIS—International Organization for Standardization/Draft International Standard

ISO/TS—International Organization for Standardization/Technical Specification

IT—Information technology

ITIL—Information Technology Infrastructure Library

IVR—Interactive voice response

JCAHO—Joint Commission on the Accreditation of Healthcare Organizations

JCF—Juran complexity factor

JDI—Just Do It

JIT—Just in time

JUSE—Union of Japanese Scientists and Engineers

KPA—Key process area

KPC—Key product characteristic(s)

KPI—Key performance indicator

KPIV—Key process input variable

KPOV—Key process output variable

KRI—Key result indicator

LCA—Life cycle assessment

LDMADV—Lean Design for Six Sigma (Define, Measure, Analyze, Design, Verify)

LeTCI—Levels, trends, comparisons, and integration

LSL—Lower specification limit

LSS—Lean Six Sigma

LTC—Long-term care

MANOVA, MANCOVA—Multivariate analysis of variance, multivariate analysis of covariance

MAP—Missile (Defense Agency) assurance provision

MBNQA—Malcolm Baldrige National Quality Award

MDA—Missile Defense Agency

MDT—Mean downtime

MRP—Material requirements planning

MSA—Measurement system analysis

MTBF—Mean time between failure

MTTF—Mean time to failure

MTTR—Mean time to repair

MVT—Manufacturing verification test

NACD—National Association for Corporate Directors

NACE—Statistical classification of economic activities in the European Community

NAFTA—North American Free Trade Association

NAICS—North American Industry Classification System

NASA—North American Space Administration

NATO—North Atlantic Treaty Organization

NHTSA—National Highway Traffic Safety Administration

NICE—Nonpersonal interactivity-Infrastructure availability-Controllability-Effort inevitability

NIST—National Institute of Standards and Technology

NMMC—North Mississippi Medical Center

NPI—New product introduction

NPR—NASA procedural requirements

NPV—Net present value

NQA—National quality award

NVCASE—National Voluntary Conformity Assessment System Evaluation

OA—Operational availability

OECD—Organisation for Economic Co-operation and Development

OEMs—Original equipment manufacturers

OFAT—One factor at a time

OKVED—Russian Economic Activities Classification System

OLA—Operating level agreement

ONE—Organizations of Noteworthy Excellence

OpEx—Operational excellence

OSHA—Occupational Safety and Health Administration

OSI—Owner satisfaction index

OST—On-stream time

OTT—Office of Technology Transfer

OVT—Operations verification test

P4P—Pay for performance

PC—Personal computer

PCA—Principal-component analysis

PCE—Phase containment effectiveness

P-D—Position-dimension

PDA—Personal digital assistant

PDCA—Plan, do, check, act

PDSA—Plan, do, study, act

PERT—Program evaluation and review technique

PFMEA—Process failure mode and effects analysis

PMBoK—Project management body of knowledge

PMI—Project Management Institute

POS—Point of sale

PT—Personal transporter

QA—Quality assurance

QA/RA—Quality assurance/regulatory assurance

QbD—Quality by design

QC—Quality control

QFD—Quality function deployment

QiCG—Quality in corporate governance

QIE—Quality information equipment

QIP—Quality improvement plan

QS—Quality system

QSAR—Quality System Assessment Recognition

R&D—Research and development

R&R—Repeatability and reproducibility

RADAR—Results, approach, deployment, assessment, and review

RAM—Reliability, availability, and maintainability

RCA—Root cause analysis

RCCA—Root cause corrective action

RCM—Reliability-centered maintenance

RIAC—Reliability information analysis center

RIE—Rapid improvement event

RoHS—Restriction of hazardous substances

ROI—Return on investment

RPN—Risk priority number

SCA—Special cause analysis

SCAMPI—Standard CMMI appraisal method for process improvement

SCIP—Surgical care improvement program

SDLC—Systems development life cycle

SEI—Software Engineering Institute

SERVQUAL—Service quality

SIPOC—Supplier-input-process-output-customer

SITA—Société Internationale de Telecommunications Aeronautiques

SL—Service line

SLA—Service level agreement

SME—Subject matter expert

SMED—Single-minute exchange of die

SOA—Service-oriented architecture

SOP—Standard operating procedure

SP—Strategic planning

SPC—Statistical process control

SQC—Statistical quality control

SQE—Supplier quality engineer

SQL—Simple Query Language

SRE—Society of Reliability Engineers

SST—Self-service technology

STS—Sociotechnical system

TCE—Total containment effectiveness

TCI—Test capability index

TPM—Total productive maintenance

TPS—Toyota Production System

TQC—Total quality control

TQM—Total quality management

TRIZ—Russian for "the theory of solving inventor's problems" or "the theory of inventor's problem solving"

TSR—Tele-service representative

UKSIC—United Kingdom Standard Industrial Classification of Economic Activities

UNIVAC—Universal Automatic Computer, first commercial computer

USDA—U.S. Department of Agriculture

USL—Upper specification limit

VOC—Voice of the customer

VoIP—Voice over Internet Protocol

VSM—Value stream map

WAIS—Wide-area information server

WCQ—World-class quality

WHO—World Health Organization

WWW—World Wide Web

XML—Extensible Markup Language

Glossary of Terms

Acceptance test: A highly structured form of testing a completed, integrated system to assess compliance with specifications; commonly applied to complex systems, such as computer systems.

Accuracy of a sensor: The degree to which a sensor tells the truth—the extent to which its evaluation of some phenomenon agrees with the "true" value as judged by an established standard.

Activities: The steps in a process or subprocess.

Administrator: An overseer or entity that is created and given powers to establish standards and to see that they are enforced.

Advertising: The process of publicizing a product or service to generate sales; requires technological and legal review of copy; activities to propagandize product safety through education and warnings.

Affinity diagram: A diagram that clusters together items of a similar type; a prelude to a cause-effect diagram used in quality improvement, and used in quality design to group together similar needs or features.

Annual goals: What an organization seeks to achieve over a one- (to several-) year period; the aim or end to which work effort is directed.

Arbitration: Adversarial process in which parties agree to be bound by the decision of a third party. It is an attractive form of resolving differences because it avoids the high cost and long delays inherent in most lawsuits.

Assembly tree: A process that incorporates the outputs of several subprocesses.

Autonomous process: A group of related activities that usually performed by one department or a single group of individuals.

Availability: In the context of product design, the probability that a product, when used under given conditions, will perform satisfactorily when called upon.

Avoidance of unnecessary constraints: Not overspecifying the product for the team.

Basis for establishing quality goals: In addition to the scope of the project, a goal statement must include the goal(s) of the project. An important consideration in establishing quality goals is the choice of the basis for which the goals are set.

Benchmarking: A recent label for the concept of setting goals based on knowing what has been achieved by others. It identifies the best in class and the methods behind it that make it best.

Bias: The presence or influence of any factor that causes the population or process being sampled to appear different from what it actually is.

Black belts: On-site implementation experts with the ability to develop, coach, and lead cross-functional process improvement teams.

Breakthrough: The organized creation of beneficial change and the attainment of unprecedented levels of performance.

Business process adaptability: The ability of a process to readily accommodate changes in both the requirements and the environment while maintaining its effectiveness and efficiency over time.

Business process management (BPM): A process to sustain the changes made from a portfolio of improvement projects.

Business process outsourcing (BPO): The growing practice of one organization outsourcing some number of its processes to a third party to execute the selected processes.

Capability maturity model integration: A process improvement methodology that enables organizations to better manage their processes across business units and projects, resulting in improved organization performance.

Carbon Disclosure Project (CDP): A nonprofit organization with the mission to provide information to investors and stakeholders regarding the opportunities and risks to commercial operations presented by climate change.

Carryover analyses: Typically, a matrix-based assessment of a design that depicts the degree of carryover of design elements from a prior version, with particular regard to failure proneness.

Cause: A proven reason for the existence of a defect. Often there are multiple causes, in which case they typically follow the Pareto principle—the vital few causes will dominate all the rest.

Change agent: The individual or group with responsibility for leading and implementing an organizational change; anyone within the organization, at whatever level, who has the desire and accepts the responsibility for initiating or leading the change effort.

Company-financed: Paid for by a company. In the context of testing, in this form, industrial organizations buy test services from independent test laboratories to secure the mark (certificate, seal, label) of the laboratory for their product(s).

Comparative performance: How the final product will perform vis-à-vis the competition.

Competitive analysis: Feature-by-feature comparison with competitors' products; usually a matrix depicting a feature-by-feature comparison to the competition, with particular regard to best-in-class targets.

Conceptual learning: The process of acquiring a better understanding of the cause-effect relationship, leading to "know-why."

Consortium: An association of business organizations. This form involves creating an association of organizations from various countries. The consortium usually is dedicated to a specific project.

Consumer-financed: Paid for by consumers. In the context of testing, in this form the test laboratory derives its income by publishing its test results, usually in a monthly journal plus an annual summary.

Consumerism: A popular name for the movement to help consumers solve their problems through collective action; an aspect of quality beyond mere technical specifications in which the expectations of the public are included.

Control: A universal managerial process to ensure that all key operational processes are stable over time, and to prevent adverse change to ensure planned performance targets are met.

Control chart: A graphical tool used to determine if a process is in a state of (usually statistical) control over time. Most popular are Shewhart statistical process control charts.

Control station: An area in which quality control takes place. In lower levels of organization, it is usually confined to a limited physical area.

Correlation: Statistically, any departure of two or more random variables from independence. For example, data on frequency of symptoms are plotted against data on the suspected cause to show a relationship.

Cost of poor quality (COPQ): The costs that would disappear in the organization if all failures were removed from a product, service, or process; typically measures of a percent of sales or total costs.

Costs: The total amount of money spent by an organization to meet customer needs. With respect to quality, costs include the expenditure to design and ensure delivery of high-quality goods and services, plus the costs or losses resulting from poor quality.

Council: An executive group formed to oversee and coordinate all strategic activities aimed at achieving the strategic plan, which is responsible for executing the strategic business plan and monitoring the key performance indicators.

Critical factors: Those aspects that present serious danger to human life, health, and the environment, or risk the loss of very large sums of money.

Criticality analysis: Means of identifying the "vital few" features that are vulnerable in the design so that they can receive priority for attention and resources; usually a matrix that depicts the degree of failure of a feature or component against the ranking of customer needs, along with responsibilities detailed for correction.

Cultural needs: The portion of customer needs, especially of internal customers, beyond products and processes that are instead related to preservation of status, job security, self-respect, respect of others, continuity of habit patterns, and still other elements of what is broadly called the cultural pattern. These seldom are stated openly.

Customer: Organization or person that receives a product. A customer can be internal or external.

Customer disloyalty: The negative state of a customer who no longer wants the producer's products or services. They find better-performing products and services and then become unfaithful to the producer to whom they had been previously loyal.

Customer dissatisfaction: Customer's negative perception of the degree to which the customer's requirements have been fulfilled.

Customer loyalty: The delighted state of a customer when the features of the good or service meet his or her needs and are delivered free from failure.

Customer needs spreadsheet: A spreadsheet tool depicting the relationship between customer communities and the statements of need. Needs strongly relating to a wide customer base subsequently rise in priority when features are considered. Advanced forms of this spreadsheet and others appear as the "house of quality," or quality function deployment (QFD).

Customer reaction: How customers will rate the product compared with others available.

Customer satisfaction: Customer's positive perception of the degree to which the customer's requirements have been fulfilled.

Customer service: Activities related to enhancing the customer experience, including such pursuits as observation of use of the product to discover the hazards inherent during use (and misuse); feeding the information back to all concerned; providing training and warnings to users.

Customs or traditions: Elements of culture that provide the precedents and premises that are guides to decisions and actions.

Cycle time: The time required to carry out processes, especially those that involve many steps performed sequentially in various departments.

Defect: Any state of unfitness for use or nonconformance to specification.

Deployment: In the context of strategy, the means of subdividing the goals and allocating the subgoals to lower levels.

Design for maintainability: Evaluation of particular designs for the ease and cost of maintaining them during their useful life.

Design for manufacture and assembly: Evaluation of the complexity and potential for problems during manufacture with a view to make assembly as simple and error-free as possible.

Design for quality: A structured process for developing products (both goods and services) that ensures that customer needs are met by the final output.

Design for Six Sigma (DFSS): A methodology to create both a design for a product and the process to produce it in such a way that defects in the product and the process are not only extremely rare, but also predictable.

Design network: A tree diagram depicting the events that occur either in parallel or sequentially when designing. Usually shown with the total time needed to complete the event, along with earliest start and subsequent stop dates, a design network is used to manage a particularly complex design effort.

Diagnosis: The process of studying symptoms, theorizing as to causes, testing theories (hypotheses), and discovering causes.

Diagnostic journey: From symptoms to theories about what may cause the symptom(s); from theories to testing of the theories; from tests to establishing root cause(s).

Documentation: Recording of information, especially to meet regulatory requirements. For example, the growth of safety legislation and of product liability has enormously increased the need for documentation.

Dominant cause: A major contributor to the existence of defects, and one that must be remedied before there can be an adequate performance breakthrough.

Dry run: A walk-through of the new process, with the planning team playing a dominant operating role in the process; a test of a process under operating conditions, in which effects of failure are mitigated (e.g., product is not delivered to a customer).

Ecoquality: The concept and associated activities intended to enable clients across industries to respond to demands from customers, regulatory agencies, and shareholders for accountability in producing products and services fit for ecological use, focusing on understanding carbon profiles and reducing them to appropriate levels.

Employee engagement: The levels of connection employees feel with their employer, as demonstrated by their willingness and ability to help their organization succeed, largely by providing discretionary effort on a sustained basis.

Empowerment: The process of enhancing the capacity of individuals or groups to make choices and to transform those choices into desired actions and outcomes.

Entropy: The tendency of all living things and all organizations to head toward their own extinction.

Equipment and supplies: Physical devices and other hard goods needed to perform the process.

Estimation: The process of analyzing a sample result to approximate the corresponding value of the population parameter.

External customers: People external to the company, organization, system, or agency who are affected by the use of the product or service. They receive value from the product of the organization. This is in contrast to internal customers, who are users within the organization.

Failure: Any fault, defect, or error that impairs a service or product from meeting the customer needs.

Failure mode and effects analysis (FMEA): A methodical approach to risk analysis that calculates the combined impact of the probability of a particular failure, the effects of that failure, and the probability that the failure can be detected and corrected, thereby establishing a priority ranking for designing in failure prevention countermeasures.

Fault tree analysis: An aid in the design of preventive countermeasures that traces all possible combinations of causes that could lead to a particular failure.

Feature: A property or characteristic possessed by a good or service that responds to customer needs.

Financial control: Process consists of evaluating actual financial performance, comparing this with the financial goals, and taking action on the difference—the accountant's "variance."

Financial improvement: This process aims to improve financial results. It takes many forms: cost reduction projects, new facilities, and new product development to increase sales, mergers and acquisitions, joint ventures, and so on.

Financial planning: Process that prepares the annual financial and operational budgets. It defines the deeds to be done in the year ahead; it translates those deeds into money—revenue, costs, and profits; and it determines the financial benefits of doing all those deeds.

Flow diagram: A popular depiction of a process, using standard symbols for activities and flow direction. It originated in software design during the 1950s and evolved into the process mapping widely used today.

Focus group: The popular technique of placing customers in a setting led by a trained facilitator to probe for the understanding of customer needs.

Glossary: The chief weapon used to remove the ambiguity of words and terms between parties, especially customers and providers. A working dictionary of in-context usage, such as, What does "comfortable" mean for an office chair?

Goal statement: In the context of a project, the written charter for the team that describes the intent and purpose of the project. It should incorporate the specific goal(s) of the project.

Handoff: A transfer of material or information from one person or entity to another, especially across departmental boundaries.

Hidden customers: An assortment of different customers who are easily overlooked because they may not come to mind readily. They can exert great influence over the product design.

Homogeneity: Uniformity that implies defects are spread throughout a production unit. Unlike an assembled product, a defective part cannot simply be removed and replaced.

Human resources (HR) function (or subfunction): In the context of quality, within an organization, bears the responsibility for implementing quality and performance excellence training and development strategy.

Inherent performance: How the final product will perform on one or more dimensions.

Innovation: A new way of doing something; incremental, radical, and revolutionary change of producing new products and services, or improving processes and systems.

Internal customers: Customers inside the producing organization. Everyone inside the organization plays three roles: supplier, processor, and customer.

Inventory: Raw materials, work in progress, finished goods, papers, electronic files, etc.

Key control characteristic: A process parameter for which variation must be controlled around some target value during manufacturing and assembly; inputs that affect outputs.

Language: Verbal means of communication. Many countries harbor multiple languages and numerous dialects that can be a serious barrier to communication.

Lean: The process of optimizing systems to reduce costs and improve efficiency by eliminating product and process waste; also the state of a system after such optimization.

Linearity: In the context of measurement system analysis, the difference in bias values at different points along the expected operating range of a measurement instrument.

Managing for quality: A set of universal methods that an enterprise, a business, an agency, a university, a hospital, or any organization can use to attain superior results by ensuring that all goods, services, and processes meet stakeholder needs.

Mandated government certification: Under this concept, products are required by law to be independently approved for adequacy before they may be sold to the public.

Market as a basis: Meeting or exceeding market quality as a means to establish quality goals that affect product salability.

Market experiments: Introducing and testing ideas for features in the market that allow one to analyze and evaluate concepts.

Market leadership: The result of entering a new market first and gaining superiority that marketers call a *franchise.*

Market research: Any of a variety of techniques aimed at answering the three fundamental questions: (1) What is important to the users? (2) What is the order of the importance? (3) How well do we do in meeting them in that order as compared to the competition?

Marketing: The process of promotion, including activities to provide product labeling for warnings, dangers, antidotes; training of the field force in the contract provisions; supplying of safety information to distributors and dealers; setup of exhibits on safety procedures; conducting of tests after installation and training of users in safety; publication of a list of dos and don'ts relative to safety; establishment of a customer relations climate that minimizes animosity and claims.

Materials: Tangible elements, data, facts, figures, or information (these, along with equipment and supplies, also may make up inputs required as well as what is to be done to them).

Mean: The average value of a list of numbers.

Mean time between failures (MTBF): The mean (or average) time between successive failures of a product.

Median: The middle value in a sequential list of numbers.

Mediation: Adversarial process in which a third party—the mediator—helps contestants work out a settlement.

Merchants: People who purchase products for resale, wholesalers, distributors, travel agents and brokers, and anyone who handles the product.

Method: The orderly arrangement of a series of tasks, activities, or procedures.

Mission statement: A short, memorable description of an organization's reason for existence; definition of the company's business, its objectives, and its approach to reach those objectives.

Mistake proofing: A proactive approach to reducing defects by eliminating the opportunity to create a defect by designing and implementing creative devices and procedures.

Mode: The value that occurs most often in a list of numbers.

Modular test: A test of individual segments of the process.

Needs analysis spreadsheet: Tool used to record the breakdown of primary needs into precise and measurable terms.

Ombudsman: A Swedish word used to designate an official whose job it is to receive citizens' complaints and to help them secure action from the government bureaucracy.

On-stream time (OST): The actual run time divided by available time for process equipment associated with the process; may best be set in the high 80 to 90 percentile for many processes.

Operational learning: The process of obtaining validation of action-outcome links, leading to "know-how."

Organization: Group of people and facilities with an arrangement of responsibilities, authorities, and relationships.

Perceived needs: Apparent, supposed, and potentially superficial needs expressed by customers based on their perceptions. These may differ entirely from the supplier's perceptions of what constitutes product quality.

Performance: Measure of whether the product does what it is supposed to in terms of the principal operating characteristics. This dimension is based on measurable attributes and superiority.

Performance excellence: The state achieved by an organization that is pursuing superior results with a set of universal methods aimed at improving the quality of its goods, services, processes, people, and financial performance.

Performance failure: How the product will perform with respect to product failure.

Performance management: A systematic, data-oriented approach to managing people at work that relies on positive reinforcement as the primary means to maximize performance.

Plan-Do-Study-Act (PDSA): A rapid-cycle change and control tool used to solve sporadic, day-to-day problems.

Planning network: It is used to manage a particularly complex planning effort.

Policies: A guide to managerial action. There may be policies in a number of areas such as quality, environment, safety, and human resources.

Potential customers: Those not currently using the product or service but capable of becoming customers.

PRE-Control: A statistical technique for detecting process conditions and changes that may cause defects (rather than changes that are statistically significant).

Precision of a sensor: A measure of the ability of a sensor to reproduce its results over and over on repeated tests.

Procedure: Specified way to carry out an activity or a process.

Process: Set of interrelated resources and activities that transform inputs into outputs.

Process analysis: A process flowchart technique that also shows the time necessary to do each task, the dependencies the task requires (such as access to a computer network), and the time "wasted" in between tasks. Usually it is interview-driven and requires a skilled process expert.

Process anatomy: A coherent structure that binds or holds the process together. This structure supports the creation of the goods or the delivery of the service.

Process capability: A method used to discover whether a process is consistently capable of meeting desired goals.

Process control: An ongoing managerial process in which the actual performance of the operating process is evaluated by measurements taken at the control points, comparing the measurements to the quality targets, and taking action on the difference.

Process feature: Any property, attribute, and so on needed to create the goods or deliver the service and achieve the product feature goals that will satisfy a customer need.

Process goal: The numeric target for a process.

Processor: Employees, departments, functions, business units, agencies that produce or carry out a process within the organization; also organizations and people who use the product or output as an input for producing their own product.

Product: Result of a process.

Product design: A creative process based largely on technological or functional expertise.

Product design spreadsheet: A method used to record and analyze product features and goals needed to meet customers' needs.

Product improvement: A common form of competition in quality through improving products so that they have greater appeal to the users and therefore can be sold successfully in the face of competition from existing products.

Product remediation: Situation in which final product made suffers from variation in quality among batches and the situation needs to be managed.

Productivity: An output performance index, such as units produced per person/hour.

Project goals: Specific objectives of a project; these should be measurable, attainable, realistic, and time-bound.

Psychological needs: For many products or services, customer needs that extend beyond the technological features of the good or service. The needs also include matters of a psychological nature.

Purchaser: Someone who buys the product for himself or herself or for someone else.

Quality: Degree to which an inherent characteristic fulfills requirements.

Quality control: A universal managerial process for conducting operations so as to provide stability over time, and to prevent adverse change and to maintain the status quo. Quality control takes place by use of the feedback loop. Quality control entails the maintenance or restoration of the operating status quo as measured by (meeting) the acceptable level of defects and provision of customer needs.

Quality function deployment: A valuable tool for collecting and organizing the required information needed to complete the operational quality planning process.

Quality management: All activities of the overall management function that determine the quality policy, objectives, and responsibilities and implement them by such means as quality planning, quality control, and quality improvement within the quality system.

Quality superiority: Exceptionally high quality, defined only in terms of the organization's internal standards. It must be clearly based on the customer needs and the benefits the customer is seeking.

Quality system: Organizational structure, procedures, processes, and resources needed to implement quality management.

Quality warranties: Assurances that stimulate producers to give priority to quality and stimulate sellers to seek out reliable sources of supply.

Range: The difference between the maximum and minimum values in a list of numbers.

Ranking: For defects, position in the order of frequency.

Rapid improvement events (RIEs): Focused efforts that are facilitated and conducted by Lean experts or Black Belts to enable Lean teams to analyze the value streams and quickly develop and implement solutions in a short time.

Recognition: Ceremonial actions taken to publicize meritorious performance, typically nonfinancial in nature.

Redundancy: The existence of more than one element for accomplishing a given task, where all elements must fail before there is an overall failure of the system.

Reliability: The ability of a product to perform a required function under stated conditions for a stated time; or more simply, the chance that a product will work for the required time. In the context of Lean, the ability to supply a product or service on or before the date promised.

Remedial journey: From root causes to remedial changes in the process to remove or go around the cause(s); from remedies to testing and proving the remedies under operating conditions; from workable remedies to dealing with resistance to change; from dealing with resistance to establishing new controls to hold the gains.

Remedial proposals: Plans to eliminate the causes of consumer problems at their source.

Remedy: A change that can eliminate or neutralize a cause of defects.

Repeatability: The variation in measurements obtained with one measurement instrument when used several times by an appraiser while measuring the identical characteristic on the same part.

Replenishment time: The time from placement of an order with the supplier until the order is received and can be used by the producer.

Reproducibility: The variation in the average of the measurements made by different appraisers using the same measuring instrument when measuring the identical characteristic on the same part.

Return on investment (ROI): The ratio of the estimated gain to the estimated resources needed.

Revenue: Gross receipts, whether from sales, budget appropriations, tuition, or government agency grants.

Review process: An examination of gaps between what has been achieved and the target, and between measurement of the current state and the target it is seeking. This increases the probability of reaching goals.

Rewards: Salaries, salary increases, bonuses, promotions, and so on often resulting from the annual review of employee performance. In the past this review has focused on meeting goals from traditional parameters: costs, productivity, schedule, and now breakthrough.

Root cause analysis (RCA): Compared to Plan-Do-Study-Act, a more in-depth analysis that identifies true root causes of events (a special cause may itself be a root cause, but is more readily pinpointed).

Salability analysis: A matrix tool used to depict the price willing to be borne, or the cost needed to deliver, a given feature of a product. It evaluates which features stimulate customers to be willing to buy the product and the price they are willing to pay.

Scatter diagram: The graphical technique of plotting one continuous variable against another, to determine correlation. This is a prelude to regression analyses to determine prediction equations.

Scorecards and key performance indicators: Measurements that are visible throughout the organization and are used to evaluate the degree to which a strategic plan is being achieved.

Selection matrix: A matrix tool showing the choices to be made, ranked according to agreed upon criteria. It is used in both improvement and design settings.

Self-inspection: A state in which decisions on the product are delegated to the workforce.

Sensor: A specialized detecting device or measurement tool designed to recognize the presence and intensity of certain phenomena and to convert this sense knowledge into information.

Simulation: A design and analysis technique that manipulates and observes a mathematical or physical model representing a real-world process, for which direct experiments may not be possible.

Six Sigma: A quality program that ultimately improves customers' experiences, lowers producers' costs, and builds better leaders.

Six Sigma DMAIC: A process that defines, measures, analyzes, improves, and controls existing processes that fall below the Six Sigma specification of only 3.4 defects per million opportunities over the long term.

Sporadic spike: A sudden, unplanned increase in waste arising from one or more unexpected sources.

Stability: In the context of measurement system analysis, the total variation in the measurements obtained with a measurement system on the same master or parts when measuring a single characteristic over an extended time.

Standard deviation: The square root of the variance.

Standard of performance: An established, aimed-at target toward which work is expended.

Statistical inference: The process of estimating, through sampling and application of statistical methods, certain characteristics of a population. In the world of quality, these estimates and statistical conclusions are used to draw practical conclusions, typically giving the practitioner confidence in taking subsequent action (or inaction) to improve a process.

Statistical quality control: Statistics-based methodologies including acceptance sampling and control charting, frequently employed to yield freedom from biases.

Statute: The enabling act that defines the purpose of a regulation and especially the subject matter to be regulated. It establishes the "rules of the game" and creates an agency to administer the act.

Steering team: Also called quality council or quality committee. This team plays the central role in directing and coordinating the organization's efforts to manage for quality.

Strategic deployment process: Procedures to carry out a strategy. It requires that the organization incorporate customer focus into the organization's vision, mission, values, policies, strategies, and long- and short-term goals and projects.

Strategic planning: The systematic approach to defining long-term business goals and planning the means to achieve them.

Strategy: A defined plan, idea, or course of action regarding how an organization can outperform competitors or achieve similar objectives.

Stratification: The separation of data into categories, usually as part of diagnosing a quality problem to identify causes of defects.

Subprocesses: Smaller units obtained by the decomposition of larger processes for both the development and operation of the process.

Supplier: A person or organization that provides a product to the customer. A supplier can be internal or external.

Support processes: Secretarial support, outsources of printing services, copying services, temporary help, and so on.

Survey: The passive technique of eliciting answers to preset questions about satisfaction or needs.

Suspicions: Prior history of hostilities resulting from ancient wars, religious differences, membership in different clans, and so on.

Symptom: The outward evidence of a defect, or that something is wrong. A defect may have multiple symptoms.

Technology as a basis: A traditional approach in many organizations to establish the quality goals on a technological basis.

Technology transfer: Conveyance of know-how (such as a method or invention). This is carried out in numerous ways: international professional societies and their committees; conferences; exchange visits; training courses and seminars; and university technology transfer offices.

Theory or hypothesis: In the context of quality, unproved assertions as to reasons for the existence of defects and symptoms. Usually, multiple theories are advanced to explain the presence of defects.

Total productive maintenance (TPM): An approach to maintenance in which equipment operators perform much of the routine maintenance, often on a continuous basis. TPM identifies the sources of losses and drives toward elimination of all of them; it focuses on zero losses.

Training: Transfer of skills and knowledge required to complete a process.

Transport: Moving of people or goods around or between sites.

Tree diagram: Any of a variety of diagrams depicting events that are completed in parallel or simultaneously as branches of a tree. It is less refined than the design network, but useful to understand the activities from a "big picture" perspective.

Understanding gap: The lack of understanding of customer needs.

Unit of measurement: A defined amount of some quality feature, permitting evaluation of that feature in numbers, e.g., hours of time to provide service, kilowatts of electric power, or concentration of a medication.

Upper managers: The highest leadership posts of an enterprise. Applied to a corporation, upper management includes the president (chief executive officer) plus the corporate vice presidents; applied to an autonomous division, upper management includes the general manager and the directly subordinate managers.

"User-friendly" needs: Needs that, when gratified, enable amateurs to use technological and other complex product or services with confidence and ease.

Value analysis: Calculation of both the incremental cost of specific features of the product and the cost of meeting specific customer needs and subsequent comparison of the costs of alternative designs.

Variance: The average squared deviation of each data point from the mean.

Vision (statement): A desired future state of the organization or enterprise. It should define the benefits that a customer, an employee, a shareholder, or society at large can expect from the organization.

Voice of market: Who are or will be the customers or target audience for a product, and what share of the market or market niche it will capture.

Name Index

D

E

F

G

H

S

T

V

W

Y

Z

Subject Index

B

C

D

E

F

L

M

N

O

P

Q

T